HUMAN NEUROANATOMY

HUMAN NEUROANATOMY

School of Medicine
University of South Carolina
Columbia, South Carolina

AMSTERDAM · BOSTON · HEIDELBERG · LONDON · NEW YORK · OXFORD
PARIS · SAN DIEGO · SAN FRANCISCO · SINGAPORE · SYDNEY · TOKYO
Academic Press is an imprint of Elsevier

Academic Press is an imprint of Elsevier
84 Theobald's Road, London WC1X 8RR, UK
30 Corporate Drive, Suite 400, Burlington, MA 01803, USA
525 B Street, Suite 1900, San Diego, CA 92101-4495, USA

First edition 2008

Notice
No responsibility is assumed by the publisher for any injury and/or damage to persons
or property as a matter of products liability, negligence or otherwise, or from any use
or operation of any methods, products, instructions or ideas contained in the material
herein. Because of rapid advances in the medical sciences, in particular, independent
verification of diagnoses and drug dosages should be made

Library of Congress Cataloguing in Publication Data
A catalogue record for this book is available from the Library of Congress

British Library Cataloguing in Publication Data
A catalogue record for this book is available from the British Library

ISBN: 978-0-12-068251-5

For information on all Academic Press publications
visit our web site at books.elsevier.com

Typeset by Charon Tec Ltd (A Macmillan Company), Chennai, India
www.charontec.com

Printed and bound by CPI Group (UK) Ltd, Croydon, CR0 4YY

Transferred to Digital Print 2011

Working together to grow
libraries in developing countries

www.elsevier.com | www.bookaid.org | www.sabre.org

ELSEVIER   BOOK AID
            International   Sabre Foundation

Contents

Chapter 9: The Reticular Formation   149

Chapter 10: The Auditory System   163

Chapter 11: The Vestibular System   179

Preface

It is a great privilege to write a book on the human brain. I have studied and taught human neuroanatomy for more than 30 years to medical students and graduate students from an assortment of disciplines (biomedical science, exercise science, neuroscience, physical therapy, psychology) as well as residents and practicing physicians.

My students have asked me thousands of questions over these many years. Their questions have encouraged me in my own personal study and have helped clarify my thinking about the structure and function of the human brain. These students, through their questions, have taught me many things and I am most grateful for that. This book is dedicated to my students as a way of thanking them for what they have taught me. It gives me an occasion to share our collective understanding of the human brain with others.

I am also grateful to Dr. Paul A. Young, Professor and Chairman Emeritus, Department of Anatomy and Neurobiology, Saint Louis University School of Medicine, who gave me the opportunity to begin my graduate studies in anatomy and served as a role model to me. Dr. Young is the epitome of a dedicated and excellent teacher and the author of an exceptional textbook on basic clinical neuroanatomy.

I am also grateful for the privilege of studying for some six years with the late Dr. Elizabeth C. Crosby. She was my teacher, fellow researcher and friend, whose ability I greatly admired and whose friendship I valued highly. Dr. Crosby had a profound understanding of the human nervous system based on her many years of study of the comparative anatomy of the nervous system of vertebrates including man. She had a long and distinguished career teaching medical students, residents, neurologists and neurosurgeons and she had many years of experience correlating neuroanatomy with neurology and neurosurgery in clinical conferences and on rounds. Because of that experience, one could gradually see the clinicians become more anatomically minded and the anatomists more clinically conscious. Dr. Crosby imparted to me her clinically conscious, anatomical-mindedness, that hopefully is reflected in my teaching and in this book.

The preparation of this book has come at a time when there has been an enormous explosion in our knowledge about the nervous system. One only has to Google the term 'brain' to discover some 155,000,000 references or search the same term in PubMed and find over a million references. To keep up with the literature in one's research field is an ever-present challenge. To keep up with current studies on the whole human brain and spinal cord is an impossible task. At the end of each chapter is a set of 'Further Readings' that the interested reader might want to consider should there be a desire to learn more about the topics covered in that chapter or gain a different perspective on a particular topic.

At the end of the entire text is a bibliography of some 2164 references. The information in this text was gleaned from those books and papers. The careful reader of that list will note that the papers are almost exclusively focused on work done in nonhuman primates and humans. The selection of the body of work on which this book is built was influenced by my particular interest in the nervous system of nonhuman primates and humans. Such an approach by necessity leaves out a great deal of information derived from studies in nonprimates, but that is far beyond the scope of this text.

The author is grateful for the tremendous assistance provided during the past five years by Erica Peake, the former Interlibrary Loan/Circulation Specialist at the USC School of Medicine. Erica retrieved hundreds of books and papers for me through our interlibrary loan system that were not available online. Very special thanks are also due to Glenda Fedricci of the Department of Pharmacology, Physiology and Neuroscience. Her careful, meticulous, and ever cheerful manner in looking over the references and the text as a whole has prevented many errors from making their way into print.

Glenda has been such a big help to me in so many ways that there is no way I can adequately express my thanks to her. Any errors that are present in the text, whatever their form, are the sole responsibility of the author.

Lastly, this book would not have been completed without the persistence of Johannes Menzel, senior publishing editor for neuroscience with Elsevier, who has worked with me on an almost daily basis for more than two years. With a time differential of five hours (earlier in London than here in Columbia), my email inbox would begin filling each day starting about 5 am EST with multiple emails containing helpful answers to my questions, guidance, gentle suggestions and corrections! Johannes has been unfailing in his support for this book, always encouraging, and ever optimistic. He has been patient with my tortoise-like pace and accepting of my failures and faults. While the imprint of

Academic Press may be on the outside of the book, Johannes' imprint is found on almost every page throughout the inside of this book.

It is my sincere hope that you the reader will enjoy reading this book and that in the process you will begin to grasp something of what little we do know about the structure and function of the human brain and spinal cord. It is my hope that by reading this book you will begin a lifelong study of the nervous system. It is also my hope that that study will lead you to do more than just write a book but rather make a discovery, find a cure, or engage in some worthwhile endeavor that will relieve the suffering of those with neurological disease and give them hope for a better life.

Soli Deo Gloria.

James R. Augustine

Neurology is the greatest and, I think, the most important, unexplored field in the whole of science. Certainly our ignorance and the amount that is to be learned is just as vast as that of outer space. And certainly too, what we learn in this field of neurology is more important to man. The secrets of the brain and the mind are hidden still. The interrelationship of brain and mind are perhaps something we shall never be quite sure of, but something toward which scientists and doctors will always struggle.

Wilder Penfield (1891–1976) (From the Penfield papers, Montreal
Neurological Institute, with permission of literary executors,
Theodore Rasmussen and William Feindel)

1

Introduction to the Nervous System

The human **nervous system** is a specialized complex of excitable cells, called **neurons**. There are many functions associated with neurons including: (1) **reception** of stimuli; (2) **transformation** of these stimuli into **nerve impulses**; (3) **conduction** of nerve impulses; (4) **neuron to neuron communication** at points of functional contact between neurons called **synapses**; and (5) the **integration, association, correlation**, and **interpretation** of impulses such that the nervous system may act on, or respond to, these impulses. The nervous system resembles a well-organized and extremely complex communicational system designed to receive information from the external and internal environment, assimilate, record, and use such information as a basis for immediate and intended behavior. The ability of neurons to communicate with one another is one way in which neurons differ from other cells in the body. Such communication between neurons often involves chemical messengers called **neurotransmitters**.

The human nervous system consists of the **central nervous system** (CNS) and the **peripheral nervous system** (PNS). The CNS, surrounded and protected by bones of the skull and vertebral column, consists of the brain and spinal cord. The term 'brain' refers to the following structures: **brain stem, cerebellum, diencephalon**, and the **cerebral hemispheres**. The PNS includes all **cranial, spinal**, and **autonomic nerves** as well as their **ganglia**, and **associated sensory and motor endings**.

1.1. NEURONS

The structural unit of the nervous system is the **neuron** with its **cell body** (or soma) and numerous elaborate **processes**. There are many contacts between neurons through these processes. The volume of cytoplasm in the processes of a neuron greatly exceeds that found in its cell body. A collection of neuronal cell bodies in the PNS is a **ganglion**; a population of neuronal cell bodies in the CNS is a **nucleus**. An example of the former is a **spinal ganglion** and of the latter is the **dorsal vagal nucleus** – a collection of neuronal cell bodies in the brain stem whose processes contribute to the formation of the vagal nerve [X].

1.1.1. Neuronal Cell Body (Soma)

The **neuronal cell body** (Fig. 1.1), with complex machinery for continuous protein synthesis, has a prominent, central **nucleus** (with a large **nucleolus**), various **organelles,** and **inclusions** such as the **chromatophil** (Nissl) **substance, neurofibrils** (aggregates of neurofilaments), **microtubules**, and **actin filaments** (microfilaments). The neuronal cell body also has a region devoid of chromatophil substance that corresponds to the point of origin of the axon called the **axon hillock** (Fig. 1.1). With proper staining and then examined microscopically, the chromatophil substance appears as intensely basophil aggregates of rough endoplasmic reticulum. There is an age-related increase of the endogenous pigment **lipofuscin** in lysosomes of postmitotic neurons and in some glial cells of the human brain. Thus lipofuscin, as a marker of cellular aging, is termed 'age pigment'. Lipofuscin consists of a pigment matrix in association with varying amounts of lipid droplets. Another age pigment, **neuromelanin** makes its appearance by 11–12 months of life in the human locus coeruleus and by about 3 years of life in the human substantia nigra. This brownish to black pigment undergoes age-related reduction in both these nuclear groups and is marker for catecholaminergic neurons.

Neuronal Cytoskeleton

Neurofibrils, microtubules, and actin filaments in the neuronal cell body make up the **neuronal cytoskeleton** that supports and organizes organelles and inclusions, determines cell shape, and generates mechanical forces in the cytoplasm. Injury to the neuronal cell body or its processes due to genetic causes, mechanical damage, or exposure to toxic substances will disrupt the neuronal

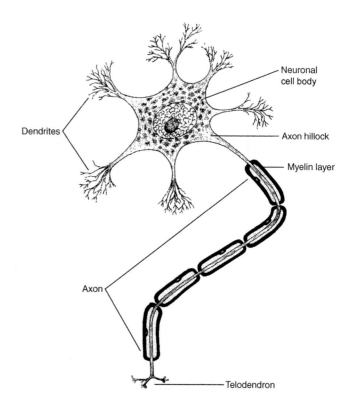

FIGURE 1.1.   Component parts of a typical neuron.

cytoskeleton. **Neurofibrils**, identifiable with the light microscope as linear fibrillary structures, are aggregates of neurofilaments when viewed with the electron microscope. **Neurofilaments**, a class of intermediate filaments, are slender tubular structures 8–14 nm in diameter occurring only in neurons. Neurofilaments help maintain the radius of larger axons. **Microtubules** are longer, have a hollow-core, and an outside diameter of about 22–25 nm. Their protein subunit is composed of α **and β tubulin**. They form paths or 'streets' through the center of the axoplasm that are traveled by substances transported from the neuronal cell body and destined for the axon terminal. In the terminal, such substances may participate in the renewal of axonal membranes and for making synaptic vesicles. **Actin filaments** (microfilaments, F-actin) found in the neuronal cell body and measuring about 7 nm in diameter are composed predominately of the protein **actin**.

Neurofibrillary Degenerations

Neurofilaments increase in number, thicken, or become tangled in normal aging and in certain diseases such as Alzheimer's disease (AD) and Down's syndrome. These diseases are termed **neurofibrillary degenerations** because of the involvement of neurofilaments. The eighth leading cause of death among adults in the

United States (53,852 deaths in 2002 or 2.2 deaths per 100,000 population), is Alzheimer's disease, an irreversible degenerative disease with an insidious onset, inexorable progression, and fatal outcome. Alzheimer's disease involves loss of memory and independent living skills, confusion, disorientation, language disturbances, and a generalized intellectual deficit involving personality changes that ultimately result in the loss of identity ('Mr. Jones is no longer the same person'). Progression of symptoms occurs over an average of 5 to 15 years. Eventually patients with Alzheimer's disease are disoriented, lose control of voluntary motor activity, become bedridden, incontinent, and unable to feed themselves. Approximately 4.5 million Americans suffer from Alzheimer's disease.

Neuritic Plaques, Neurofibrillary Tangles and Neuropil Threads

Small numbers of plaques and tangles characterize the brain of normal individuals 65 years of age and over. Neuritic plaques, neurofibrillary tangles, and neuropil threads, however, are structural changes characteristic of the brains of patients with Alzheimer's disease. These structural changes may occur in neuronal populations in various parts of the human brain. Other elements such as 10 nm and 15 nm straight neurofilaments, various-sized dense granules, and microtubule-associated proteins, especially the **tau protein**, also occur in this disease. **Neurofibrillary tangles** occur in the neuronal cytoplasm and have a paired helical structure that consists of pairs of 14–18 nm neurofilaments linked by thin cross-bridging filaments that coil around each other at regular 70–90 nm intervals. These **paired helical filaments**, unlike any neuronal organelle and unique to the human brain, are formed by one or more modified polypeptides that have unusual solubility properties but originate from neurofilament or other normal cytoskeletal proteins. Antibodies raised against the **microtubule-associated protein, tau**, are a useful marker that recognizes the presence of this protein in these neurofibrillary tangles. The tau protein helps organize and stabilize the neuronal cytoskeleton. Proponents of the 'Tau theory' of Alzheimer's disease suggest that the phosphorylated form of this protein is a central mediator of this disease as it loses its ability to maintain the neuronal cytoskeleton eventually aggregating into neurofibrillary tangles. **Neuropil threads** (curly fibers) are fine, extensively altered neurites in the cerebral cortex consisting of paired helical filaments or nonhelical straight filaments with no neurofilaments. They occur primarily in dendrites.

Degenerating neuronal processes along with an extracellular glycoprotein called **amyloid precursor protein** or β-amyloid protein (β-AP) form **neuritic plaques**. These plaques are of three types: primitive plaques composed of distorted neuronal processes with a few reactive cells, classical plaques of neuritic processes around an amyloid core, and end-stage plaques with a central amyloid core surrounded by few or no processes. Proponents of the 'amyloid hypothesis' of Alzheimer's disease regard the production and accumulation of β-amyloid protein in the brain and its consequent neuronal toxicity as a key event in this disease. In addition to the amyloid hypothesis and the 'Tau theory', other possible causes of Alzheimer's disease include inflammation or vascular factors.

1.1.2. Axon Hillock

The **axon hillock** (Fig. 1.1), a small prominence or elevation of the neuronal cell body, gives origin to the **initial segment** of an axon. Chromatophil substance is scattered throughout the neuronal cell body but reduced in the axon hillock appearing as a pale region on one side of the neuronal cell body.

1.1.3. Neuronal Processes – Axons and Dendrites

Since most stains do not mark them, **neuronal processes** or **neurites** often go unrecognized. Two types of neurites characteristic of neurons are **axons** and **dendrites** (Fig. 1.1). Axons transmit impulses away from the neuronal cell body while dendrites transmit impulses to it. The term **axon** applies to any long peripheral process extending from the spinal cord regardless of direction of impulse conduction.

Axons

The **axon hillock** (Fig. 1.1) arises from the neuronal cell body, tapers into an **axon initial segment** and then continues as an **axon** which remains near the cell body or extends for a considerable distance before ending as a **telodendron** [Greek, end tree] (Fig. 1.1). A 'considerable distance' might involve an axon leaving the spinal cord and passing to a limb to activate the fingers or toes. In a seven-foot-tall professional basketball player, the distance from the spinal cord to the tip of the fingers would certainly be 'a considerable distance'. Long axons usually give off **collateral branches** arising at right angles to the axon.

Beyond the initial segment, axonal cytoplasm lacks chromatophil substance but has various microtubule-associated proteins (MAPs), actin filaments, neurofilaments, and microtubules that provide support and assist in the transport of substances along the entire length of the axon. The structural component of axoplasm, the **axoplasmic matrix**, is distinguishable by the presence of abundant microtubules and neurofilaments that form distinct bundles in the center of the axon.

Myelin

Concentric layers of plasma membranes may insulate axons. These layers of lipoprotein wrapping material, called **myelin**, increase the efficiency and speed of saltatory conduction of impulses along the axon. **Oligodendrocyte**s, a type of neuroglial cell, are myelin-forming cells in the CNS whereas **neurilemmal (Schwann) cells** produce myelin in the PNS. Each **myelin layer** (Fig. 1.1) around an axon has periodic interruptions at **nerve fiber nodes** (of Ranvier). These nodes bound individual **internodal segments** of myelin layers.

A radiating process from a myelin-forming cell forms an internodal segment. The distal part of such a process forms a concentric spiral of lipid-rich surface membrane, the **myelin lamella**, around the axon. Multiple processes from a single oligodendrocyte form as many as 40 internodal segments in the CNS whereas in the PNS a single neurilemmal cell forms only one internodal segment. In certain demyelinating diseases, such as **multiple sclerosis** (MS), myelin layers, though normally formed, are disturbed or destroyed perhaps by anti-myelin antibodies. Impulses attempting to travel along disrupted or destroyed myelin layers are erratic, inefficient, or absent.

Dendrites

Although neurons have only one axon, they have many **dendrites** (Fig. 1.1). On leaving the neuronal cell body, dendrites taper, twist, and ramify in treelike manner. Dendritic trees grow continuously in adulthood. Dendrites are usually short and branching but rarely myelinated, with smooth proximal surfaces and branchlets covered by innumerable **dendritic spines** that give dendrites a surface area far greater than the neuronal cell body. With these innumerable spines, dendrites form a major receptive area of a neuron. Dendrites have few neurofilaments but many microtubules. Larger dendrites, but never axons, contain chromatophil substance. Dendrites in the PNS may have specialized **receptors** at their peripheral termination that respond selectively to stimuli and convert them into impulses, evoking sensations such as pain, touch, or temperature. Chapter 6 has additional information on these specialized endings.

1.2. CLASSIFICATION OF NEURONS

1.2.1. Neuronal Classification by Function

Based on function, there are three neuronal types: (1) **motor**; (2) **sensory**; and (3) **interneurons**. Motoneurons (motor neurons) carry impulses that influence the contraction of nonstriated and skeletal muscle or cause a gland to secrete. Ventral horn neurons of the spinal cord are examples of motoneurons. Sensory neurons such as dorsal horn neurons carry impulses that yield a variety of sensations such as pain, temperature, touch, pressure, vision, hearing, taste, or smell. Interneurons relate motor and sensory neurons by transmitting information from one neuronal type to another.

1.2.2. Neuronal Classification by Number of Processes

Based on the number of processes, there are four neuronal types: **unipolar**, **bipolar**, **pseudounipolar**, or **multipolar**. **Unipolar neurons** occur during development but are rare in the adult brain. **Bipolar neurons** (Fig. 1.2C) have two separate processes, one from each pole of the neuronal cell body. One process is an axon and the other a dendrite. Bipolar neurons are in the retina, olfactory epithelium, and ganglia of the vestibulocochlear nerve [VIII].

The term **pseudounipolar neuron** (Fig. 1.2A) refers to adult neurons that during development were bipolar

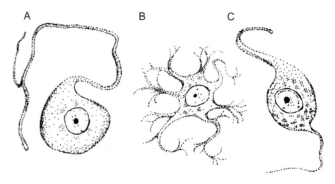

FIGURE 1.2. Neurons classified by the number of processes extending from the soma. **A**, Pseudounipolar neuron in the spinal ganglia. **B**, Multipolar neuron in the ventral horn of the spinal cord. **C**, Bipolar neuron typically in the retina, olfactory epithelium, and ganglia of the vestibulocochlear nerve [VIII].

but their two processes eventually came together and fused to form a single, short stem. Thus, they have a single T-shaped process that bifurcates, sending one branch to a peripheral tissue and the other branch into the spinal cord or brain stem. The peripheral branch functions as a dendrite – the central branch as an axon. **Pseudo-unipolar neurons** are sensory and in all spinal ganglia, the trigeminal ganglion, geniculate ganglion [VII], glossopharyngeal, and vagal ganglia. Both branches of a spinal ganglionic neuron have similar diameters and the same density of microtubules and neurofilaments. These organelles remain independent as they pass from the neuronal cell body and out into each branch. A special collection of pseudounipolar neurons in the CNS is the trigeminal mesencephalic nucleus.

Most neurons are **multipolar neurons** in that they have more than two processes – a single axon and numerous dendrites (Fig. 1.1). Examples include motoneurons and numerous small interneurons of the spinal cord, pyramidal neurons in the cerebral cortex, and Purkinje cells of the cerebellar cortex. **Multipolar neurons** are divisible into two groups according to the length of their axon. **Long-axon multipolar** (Golgi type I) **neurons** have axons that pass from their neuronal cell body and extend for a considerable distance before ending (Fig. 1.3A). These long axons form commissures, association, and projection fibers of the CNS.

Short-axon multipolar (Golgi type II) **neurons** have short axons that remain near their cell body of origin (Fig. 1.3B). Such neurons are numerous in the cerebral cortex, cerebellar cortex, and spinal cord.

1.3. THE SYNAPSE

Under normal conditions, the dendrites of a neuron receive impulses, carry them to its cell body, and then transmit those impulses away from the cell body via the neuronal axon to a muscle or gland causing movement or yielding a secretion. Because of this unidirectional flow of impulses (dendrite to cell body to axon), neurons are said to be **polarized**. Impulses also travel from one neuron to another through points of functional contact between neurons called **synapses** (Fig. 1.4). Such junctions are points of functional contact between two neurons for purposes of transmitting impulses. Simply put, the nervous system consists of chains of neurons linked together at synapses. Impulses travel from one neuron to the next through synapses. Since synapses occur between component parts of two adjacent neurons, the following terms describe most synapses: axodendritic, axosomatic, axoaxonic, somatodendritic, somatosomatic, and dendrodendritic. Axons may form symmetric or asymmetric synapses. **Asymmetric synapses** contain round or spherical vesicles and are distinguished by a thickened, postsynaptic density. They are presumably excitatory in function. **Symmetric synapses** contain flattened or elongated vesicles, pre- and postsynaptic membranes that are parallel to one another but lack a

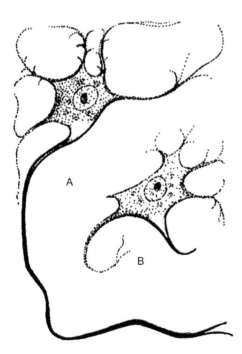

FIGURE 1.3. Multipolar neurons classified by the length of their axon. **A**, Long-axon multipolar (Golgi type I) neurons have extremely long axons; **B**, Short-axon (Golgi type II) multipolar neurons have short axons that end near their somal origin.

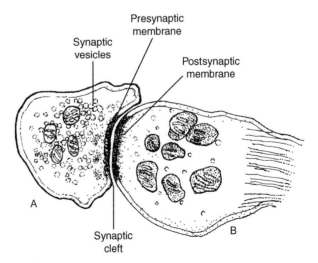

FIGURE 1.4. Ultrastructural appearance of an interneuronal synapse in the central nervous system with presynaptic (**A**) and postsynaptic (**B**) parts.

thickened postsynaptic density. Symmetric synapses are presumably inhibitory in function.

1.3.1. Components of a Synapse

Most synapses have a **presynaptic part** (Fig. 1.4A), an intervening measurable space or **synaptic cleft** of about 20 to 30 nm, and a **postsynaptic part** (Fig. 1.4B). The presynaptic part has a **presynaptic membrane** (Fig. 1.4) – the plasmalemma of a neuronal cell body or that of one of its processes, **associated cytoplasm** with mitochondria, neurofilaments, **synaptic vesicles** (Fig. 1.4), cisterns, vacuoles, and a **presynaptic vesicular grid** consisting of trigonally arranged dense projections that form a grid. Visualized at the ultrastructural level, presynaptic vesicles are either dense or clear in appearance, and occupy spaces in the grid. The grid with vesicles is a characteristic ultrastructural feature of central synapses.

Presynaptic vesicles store chemical substances or **neurotransmitters** synthesized in the neuronal cell body. Upon arrival of a nerve impulse at the presynaptic membrane, there is the release of small quantities (quantal emission) of a neurotransmitter through the presynaptic membrane by a process of exocytosis. Released neurotransmitter diffuses across the synaptic cleft to activate the **postsynaptic membrane** (Fig. 1.4) on the postsynaptic side of the synapse thus bringing about changes in postsynaptic activity. The postsynaptic part has a thickened postsynaptic membrane and some associated synaptic web material, collectively called the **postsynaptic density**, consisting of various proteins and other components plus certain polypeptides.

1.3.2. Neurotransmitters and Neuromodulators

Over 50 chemical substances are identifiable as **neurotransmitters**. Chemical substances that do not fit the classical definition of a neurotransmitter are termed **neuromodulators**. **Acetylcholine** (ACh), **histamine**, **serotonin** (5-HT), the **catecholamines** (dopamine, norepinephrine, and epinephrine), and certain **amino acids** (aspartate, glutamate, γ-aminobutyric acid, and glycine) are examples of neurotransmitters. **Neuropeptides** are derivatives of larger polypeptides that encompass more than three-dozen substances. Cholecystokinin (CCK), neuropeptide Y (NPY), somatostatin (SOM), substance P, and vasoactive intestinal polypeptide (VIP) are neurotransmitters. Classical neurotransmitters coexist in some neurons with a neuropeptide. Almost all of these neurotransmitters are in the human brain. On the one hand, neurological disease may alter certain neurotransmitters while on the other hand their alteration may lead to certain neurological disorders. Neurotransmitter deficiencies occur in Alzheimer's disease where there is a cholinergic and a noradrenergic deficit, perhaps a dopaminergic deficit, a loss of serotonergic activity, a possible deficit in glutamate, and a reduction in somatostatin and substance P.

1.3.3. Neuronal Plasticity

Because of enormous changes that occur during development, the nervous system is modifiable or said to be 'plastic'. The term **neuronal plasticity** refers to this phenomenon. Changes continue to occur in the mature nervous system at the synaptic level as we learn, create, store and recall memories, and as we age. Alterations in synaptic function, the development of new synapses, and the modification or elimination of those already existing are examples of **synaptic plasticity**. With experience and stimulation, the nervous system is able to organize and reorganize synaptic connections. Age-related synaptic loss occurs in the primary visual cortex, hippocampal formation, and cerebellar cortex in humans.

Another aspect of synaptic plasticity involves changes accompanying defective development and some neurological diseases. Defective development may result in spine loss and alterations in dendritic spine geometry in specific neuronal populations. A decrease in neuronal number, lower density of synapses, atrophy of the dendritic tree, abnormal dendritic spines, loss of dendritic spines, and the presence of long, thin spines occur in the brains of children with mental retardation. Deterioration of intellectual function seen in Alzheimer's disease may be due to neuronal loss and a distorted or reduced **dendritic plasticity** – the inability of dendrites of affected neurons to respond to, or compensate for, loss of inputs, loss of adjacent neurons, or other changes in the microenvironment.

Fetal Alcohol Syndrome

Prenatal exposure to alcohol, as would occur in an infant born to a chronic alcoholic mother, may result in **Fetal Alcohol Syndrome**. Decreased numbers of dendritic spines and a predominance of spines with long, thin pedicles characterize this condition. The significance of these dendritic alterations in mental retardation, Alzheimer's disease, Fetal Alcohol Syndrome, and other neurological diseases awaits further study.

1.3.4. The Neuropil

The precisely organized gray matter of the nervous system where most synaptic junctions and innumerable functional interconnections between neurons and their processes occur is termed the **neuropil**. The neuropil is the matrix or background of the nervous system.

1.4. NEUROGLIAL CELLS

Although the nervous system may include as many as 10^{12} neurons (estimates range between 10 billion and 1 trillion – the latter seems more likely), it has an even larger number of **neuroglial cells**. Neuroglial cells are in both the central and peripheral nervous system. **Ependymocytes, astrocytes, oligodendrocytes**, and **microglia** are examples of **central glia; neurilemmal cells** and **satellite cells** are examples of peripheral glia. Satellite cells surround the cell bodies of neurons.

Although astrocytes and oligodendrocytes arise from ectoderm, **microglial cells** arise from mesodermal elements (blood monocytes) that invade the brain in perinatal stages and after brain injury. In the developing human cerebral hemispheres, the appearance of microglial elements goes hand in hand with the appearance of vascularization.

1.4.1. Neuroglial Cells differ from Neurons

Neuroglial cells differ from neurons in at least the four following ways: (1) they have only one kind of process; (2) they are separated from neurons by an intercellular space of about 150–200 Å and from each other by gap junctions across which they communicate; (3) they cannot generate impulses but display uniform intracellular recordings and have a potassium-rich cytoplasm; (4) astrocytes and oligodendrocytes retain the ability to divide, especially after injury to the nervous system. Virchow, who coined the term 'neuroglia', thought these supporting cells represented the interstitial connective tissue of brain – a kind of 'nerve glue' ('Nervenkitt') in which neuronal elements were dispersed. An aqueous extracellular space separates neurons and neuroglial cells and accounts for about 20% of total brain volume. Neuroglial processes passing between the innumerable axons and dendrites in the neuropil, serve to compartmentalize the glycoprotein matrix of the extracellular space of the brain.

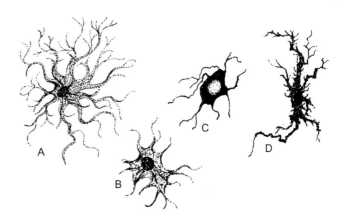

FIGURE 1.5. Types of neuroglial cells in humans. **A**, Protoplasmic astrocyte in the cerebral gray matter stained by Cajal's gold chloride sublimate method. **B**, Fibrous astrocyte in the cerebral white matter stained by Cajal's gold chloride sublimate method. This gliocyte usually has a vascular process extending to nearby blood vessels or to the cortical or ventricular surface. **C**, Oligodendrocyte revealed by a silver impregnation method. This small cell (10–20 μm in diameter) is in the deep layers of the cerebral cortex. **D**, Microglial cell revealed by the del Rio-Hortega silver carbonate method. These cells are evenly and abundantly distributed throughout the cerebral cortex.

1.4.2. Identification of Neuroglia

Identifying neuroglial cells in sections stained by routine methods such as hematoxylin and eosin is difficult. Their identification requires special methods such as metallic impregnation, histochemical, and immunocytochemical methods. Astrocytes are identifiable using Cajal's gold chloride sublimate technique, microglia by del Rio-Hortega's silver carbonate technique, and oligodendrocytes by silver impregnation methods. Immunocytochemical methods are available for the visualization of astrocytes using the intermediate filament cytoskeletal protein **glial fibrillary acidic protein** (GFAP). Various antibodies are available for the identification of oligodendrocytes and microglia. Microglial cells are identifiable in the normal human brain with a specific histochemical marker (lectin *Ricinus communis* agglutinin-1) or identified under various pathological conditions with a monoclonal antibody (AMC30).

Astrocytes

Two kinds of **astrocytes – protoplasmic** (Fig. 1.5A) and **fibrous** (Fig. 1.5B), are recognized. Astrocytes have a light homogeneous cytoplasm and nucleoplasm less dense than that in oligodendrocytes. Astrocytes are stellate with the usual cytoplasmic organelles and long, fine, **perikaryal filaments** and **particulate glycogen** as distinctive characteristics. These astroglial filaments are intermediate in size (7–11 nm) and composed of glial fibrillary acidic protein. Their radiating and tapering processes, with characteristic filaments and particles, often extend to the surface of blood vessels as **vascular processes** or underlie the pial covering on the surface of the brain as **pial processes**.

Protoplasmic astrocytes occur in areas of gray matter and have fewer fibrils than fibrous astrocytes. **Fibrous astrocytes** have numerous glial filaments and occur in white matter where their vascular processes expand in sheetlike manner to cover the entire surface of nearby blood vessels forming a **perivascular glial limiting membrane**. Processes of fibrous astrocytes completely cover and separate the cerebral cortex from the pia-arachnoid as a **superficial glial limiting membrane** whereas along the ventricular surfaces they form the **periventricular glial limiting membrane**. Astrocytic processes cover the surfaces of neuronal cell bodies and their dendrites, surround certain synapses, and separate bundles of axons in the central white matter. Fibrous astrocytes with abnormally thickened and beaded processes occur in epileptogenic foci removed during neurosurgical procedures.

Oligodendrocytes

The most numerous glial element in adults, called **oligodendrocytes** (Fig. 1.5C), are small **myelin-forming** cells ranging in diameter from 10–20 µm, with a dense nucleus and cytoplasm. This nuclear density results from a substantial amount of heterochromatin in the nuclear periphery. A thin rim of cytoplasm surrounds the nucleus and densely packed organelles balloon out on one side. Oligodendrocytes lack the perikaryal fibrils and particulate glycogen characteristic of astrocytes. Their cytoplasm is uniformly dark with abundant free ribosomes, ribosomal rosettes, and randomly arranged microtubules, 25 nm in diameter, that extend into the oligodendrocyte processes and become aligned parallel to each other. Accumulations of abnormal microtubules in the cytoplasm and processes of oligodendrocytes, called **oligodendroglial microtubular masses**, are present in brain tissue from patients with such neurodegenerative diseases as Alzheimer's disease and Pick's disease.

Oligodendrocytes are identifiable in various parts of the brain. **Interfascicular oligodendrocytes** accumulate in the deeper layers of the human cerebral cortex in rows parallel to bundles of myelinated and nonmyelinated fibers. **Perineuronal oligodendrocytes** form neuronal satellites in close association with neuronal cell bodies. The cell bodies of these perineuronal oligodendrocytes contact each other yet maintain their myelin-forming potential, especially during remyelination of the CNS. Perineuronal oligodendrocytes are the most metabolically active of the neuroglia. Associated with capillaries are the **perivascular oligodendrocytes**.

Microglial Cells

Microglial cells are rod-shaped with irregular processes arising at nearly right angles from the cell body (Fig.1.5D). They have elongated, dark nuclei and dense clumps of chromatophil substance around a nuclear envelope. Cytoplasmic density varies, with few mitochondria (often with dense granules), little endoplasmic reticulum, and occasional vacuoles. Microglia are often indented or impinged on by adjacent cellular processes and are evenly and abundantly distributed throughout the cerebral cortex. In certain diseases, microglial cells are transformable into different shapes, elongating and appearing as rod cells or collecting in clusters forming microglial nodules. Microglial cells are CNS-adapted macrophages derived from mesodermal elements (blood monocytes).

1.4.3. Neuroglial Function

Neuroglial cells are partners with neurons in the structure and function of the nervous system in that they **support, protect, insulate**, and **isolate neurons**. Neuroglial cells help maintain conditions favorable for neuronal excitability by maintaining ion homeostasis (external chloride, bicarbonate, and proton homeostasis and regulation of extracellular K^+ and Ca^{++}) while preventing the haphazard flow of impulses. Impairment of neuroglial control of neuronal excitability may be a cause of **epilepsy** (also called **focal seizures**) in humans. About 2.7 million people in the United States are afflicted with focal seizures consisting of sudden, excessive, rapid, and localized electrical discharge by small groups of neurons in the brain. Every year another 181,000 people develop this disorder.

Neuroglial cells control neuronal metabolism by regulating substances reaching neurons such as glucose and lipid precursors, and by serving as a dumping ground for waste products of metabolism. They are continually communicating with neurons serving as a metabolic interface between them and the extracellular fluid, releasing and transferring macromolecules, and altering the ionic composition of the microenvironment. They also supply necessary metabolites to axons. Neuroglial cells terminate synaptic transmission by removing chemical substances involved in synaptic transmission from synapses.

Astrocytes are involved in the response to injury involving the CNS. A glial scar (**astrocytic gliosis**) forms by proliferation of fibrous astrocytes. As neurons degenerate during the process of aging, astrocytes proliferate and occupy the vacant spaces. The brains of patients older than 70 may show increased numbers of fibrous astrocytes. The intimate relationship between neurons and astrocytes in the developing nervous system has led to the suggestion that this relationship is significant in normal development and that astrocytes are involved in neuronal migration and differentiation. Astrocytes in tissue culture are active in the metabolism and regulation of glutamate (an excitatory amino acid) and γ-aminobutyric acid (GABA) an inhibitory amino acid. Astrocytes remove potential synaptic transmitter substances such as adenosine and excess extracellular potassium.

Astrocytes may regulate local blood flow to and from neurons. A small number of substance P-immunoreactive astrocytes occur in relation to blood vessels of the human brain (especially in the deep white matter and deep gray matter in the cerebral hemispheres). Such astrocytes may cause an increase in blood flow in response to local metabolic changes. Astrocytes in tissue culture act as vehicles for the translocation of macromolecules from one cell to another.

Oligodendrocytes are the **myelin-forming cells** in the CNS and are equivalent to neurilemmal cells in the PNS. Each internodal segment of myelin originates from a single oligodendrocyte process, yet a single oligodendrocyte may contribute as many as 40 internodal segments as it gives off numerous sheetlike processes. A substantial number of oligodendrocytes in the white matter do not connect to myelin segments. Pathologic processes involving oligodendrocytes may result in demyelination. Oligodendrocytes related to capillaries likely **mediate iron mobilization and storage** in the human brain based on the immunocytochemical localization in human oligodendrocytes of transferrin (the major iron binding and transport protein), ferritin (an iron storage protein), and iron.

Microglia are evident after indirect neural trauma such as transection of a peripheral nerve in which case they interpose themselves between synaptic endings and the surface of injured neurons (a phenomenon called **synaptic stripping**). Microglial cells are also involved in pinocytosis perhaps to prevent the spread of exogenous proteins in the CNS extracellular space. They are dynamic elements in a variety of neurological conditions such as infections, autoimmune disease, and degeneration and regeneration. Microglial cells are likely antigen-presenting cells in the development of inflammatory lesions of the human brain such as **multiple sclerosis**.

Proliferation and accumulation of microglia occurs near degenerating neuronal processes and in close association with amyloid deposits in the cerebral and cerebellar cortices in **Alzheimer's disease**. Microglia may process neuronal amyloid precursor protein in these degenerating neurons leading to the formation and deposition of a polypeptide called β-amyloid in neuritic plaques. Thus, microglial cells are likely involved in the pathogenesis of amyloid deposition in Alzheimer's disease.

Based on their structure, distribution, macrophage-like behavior, and the observation that they can be induced to express major histocompatibility complex (MHC) antigens, microglia are thought to form a network of **immune competent cells** in the central nervous system. Microglial cells (and invading macrophages) are among the cellular targets for the **human immunodeficiency virus-1** (HIV-1) known to cause **acquired immunodeficiency syndrome** (AIDS). Infected microglia presumably function to release toxic substances capable of disrupting and perhaps destroying neurons leading to the neurological impairments associated with the AIDS. Another possibility is that destruction of the microglia causes an altered immune-mediated reaction to the AIDS virus and other pathogens in these patients.

1.4.4. Neuroglial Cells and Aging

Oligodendrocytes show few signs of aging but astrocytes and microglia may accumulate lipofuscin with age. There is a generalized, age-related increase in the number of microglia throughout the brain. Age-related astrocytic proliferation and hypertrophy are associated with neuronal loss. A demonstrated decrease in oligodendrocytes remains unexplained. Future studies of aging are sure to address the issue of neuroglial cell changes and their effect on neurons.

1.5. AXONAL TRANSPORT

Neuronal processes grow, regenerate, and replenish their complex machinery. They are able to do this because proteins synthesized in the neuronal cell body readily reach the neuronal processes. **Axonal transport** is the continuous flow (in axons and dendrites) of a range of membranous organelles, proteins, and enzymes at different rates and along the entire length of the neuronal process. A universal property of neurons, axonal transport, is ATP dependent, oxygen and temperature dependent, requires calcium, and probably involves calmodulin and the contractile proteins actin and myosin in association with microtubules. Axonal transport takes place from the periphery to the neuronal cell body (**retrograde transport**) and from the neuronal cell body to the terminal ending (**anterograde transport**).

Rapid or **fast axonal transport**, with a velocity of 50–400 mm/day, carries membranous organelles. **Slow axonal transport**, characterized by two subcomponents of different velocities, carries structural proteins, glycolytic enzymes, and proteins that regulate polymerization of structural proteins. The slower subcomponent (SCa) of slow axonal transport, with a velocity of 1–2 mm/day, carries assembled neurofilaments and microtubules. The faster subcomponent of slow axonal transport, with a velocity of 2–8 mm/day carries proteins that help maintain the cytoskeleton such as actin (the protein subunit of actin filaments), clathrin, fodrin, and calmodulin as well as tubulin (the protein subunit of microtubules), and glycolytic enzymes. The size of a neuronal process does not influence the pattern or rate of axonal transport.

1.5.1. Functions of Axonal Transport

Anterograde transport plays a vital role in the normal maintenance, nutrition, and growth of neuronal processes supplying the terminal endings with synaptic transmitters, certain synthetic and degradative enzymes, and with membrane constituents. One function of **retrograde transport** is to recirculate substances delivered by anterograde transport that are in excess of local needs. Structures in the neuronal cell body may degrade or resynthesize these excess substances as needed. Half the protein delivered to the distal process returns to the neuronal cell body. Retrograde transport, occurring at the rate of 150–200 mm/day, permits the transfer of worn-out organelles and membrane constituents to lysosomes in the neuronal cell body for digestion and disposal. **Survival** or **neurotrophic factors**, such as **nerve growth factor** (NGF) reach their neuronal target by this route. Tetanus toxin, the poliomyelitis virus, and herpes simplex virus gain access to neuronal cell bodies by retrograde transport. Retrograde axonal transport can thus convey both essential and harmful or noxious substances to the neuronal cell body.

1.5.2. Defective Axonal Transport

The phenomenon of **defective axonal transport** may cause disease in peripheral nerves, muscle, or neurons. Mechanical and vascular blockage of axonal transport in

the human optic nerve [II] causes swelling of the optic disk (papilledema). Senile muscular atrophy may result from age-related adverse effects on axoplasmic transport. Certain genetic disorders (Charcot–Marie–Tooth disease and Déjerine–Sottas disease), viral infections (herpes zoster, herpes simplex, poliomyelitis), and metabolic disorders (diabetes and uremia) manifest a reduction in the average velocity of axonal transport. Accumulation of transported materials in the axon terminal may lead to **terminal overloading** and **axonal breakdown** causing degeneration and denervation. Interference with axonal transport of neurofilaments may be a mechanism underlying the structural changes in Alzheimer's disease (neurofibrillary tangles and neuritic plaques) and other degenerative diseases of the CNS. In the future, retrograde transport may prove useful in the treatment of injured or diseased neurons by applying drugs to terminal processes for eventual transport back to the injured or diseased neuronal cell body.

Neurons are polarized transmitters of nerve impulses and active chemical processors with bi-directional communication through various small molecules, peptides, and proteins. Information exchange involving a chemical circuit is as essential as that exchanged by electrical conduction. These chemical and electrical circuits work in a complementary manner to achieve the extraordinary degree of complex functioning characteristic of the human nervous system.

1.6. DEGENERATION AND REGENERATION

The traditional view is that during development newly generated neurons and synapses form but that this process of neuronal generation, termed **neurogenesis**, does not occur in the adult primate brain. Recent observations in adult macaques indicate that neurogenesis does occur in several regions of the cerebral cortex and that these newly produced neurons extend axons. The areas of the brain where these new neurons appear are involved in learning and memory. With the exception of recently discovered human neuroblasts that migrate to the olfactory bulb via a lateral ventricular extension, once destroyed most mature neurons in the CNS die and are not replaced. After becoming committed to an adult class or population and synthesizing a neurotransmitter, a neuron loses the capacity for DNA synthesis and thus also for cell division. The implications of this are devastating for those who have suffered CNS injury. Some 222,000 to 285,000 people are living with spinal cord injuries with nearly 11,000 new cases every year in the United States. An additional 4860 individuals die each year before reaching the hospital. Another 2,000,000 patients have suffered brain trauma or other injury to the head with over 800,000 new cases each year. Thus, the inability of the adult nervous system to add neurons or replace damaged neurons as needed is a serious problem to those afflicted with CNS injury.

1.6.1. Axon or Retrograde Reaction

Degeneration of neurons is similar in the CNS and PNS. One exception is the difference in the myelin-forming **oligodendrocytes** in the CNS in contrast to the myelin-forming **neurilemmal cells** of the PNS. Only hours after injury to a neuronal process, perhaps because of a signal conveyed by retrograde axonal transport, a genetically programmed and predictable series of changes occur in a normal neuronal cell body (Fig. 1.6A). These collective changes in the neuronal cell body are termed the **axon** or **retrograde reaction**. One to three days after the initial injury the neuronal cell body swells and becomes rounded (Fig. 1.6B), the cell wall appears to thicken, and the nucleolus enlarges. These events are followed by displacement of the nucleus to an eccentric position (Fig. 1.6C), widening of the rough endoplasmic reticulum, and mitochondrial swelling. Chromatophil substance at this time undergoes conspicuous rearrangement – a process referred to as **chromatolysis** involving fragmentation and loss of concentration of chromatophil substance causing loss of basophil staining by injured

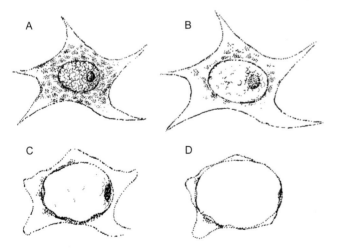

FIGURE 1.6. Changes in the neuronal cell body during the axon reaction. **A**, Normal cell. **B**, Swollen soma and nucleus with disruption of the chromatophil substance. **C**, and **D**, Additional swelling of the cell body and nucleus with eccentricity of the nucleus and loss of concentration of the chromatophil substance.

neurons (Fig. 1.6D). Chromatolysis is prominent about 15–20 days after injury.

Along with the axon reaction, there are alterations in protein and carbohydrate synthesis occurring in the chromatolytic neuron. DNA-dependent RNA synthesis seems to play a key role in this process. As the axon reaction continues, there is increased production of free polyribosomes, rough endoplasmic reticulum, neurofilaments, and an increase in the size and number of lysosomes. The axon reaction includes a dramatic proliferation in perineuronal microglia leading to a displacement of synaptic terminals on the neuronal cell body and stem dendrites causing electrophysiological disturbances.

The sequence of events characteristic of an axon reaction depends, in part, on the neuronal system and age as well as the severity and exact site of injury. If left unchecked the axon reaction leads to neuronal dissolution and death. If the initial injury is not severe, the neuronal nucleus returns to a central position, the chromatophil substance becomes concentrated, and the neuronal cell body returns to normal size.

Initial descriptions of chromatolysis suggested it was a degenerative process caused by neuronal injury. Recent work suggests chromatolysis represents neuronal reorganization leading to a regenerative process. As part of the axon reaction, the neuronal cell body shifts from production of neurotransmitters and high-energy ATP to the production of lipids and nucleotides needed for repair of cell membranes. Thus, chromatolysis may be the initial event in a series of metabolic changes involving the conservation of energy and leading to neuronal restoration.

1.6.2. Anterograde Degeneration

Transection of a peripheral nerve, such as traumatic section of the ulnar nerve at the elbow, yields proximal and distal segments of the transected nerve. Changes taking place throughout the entire length of the **distal segment** (Fig. 1.7) are termed **anterograde** (Wallerian) **degeneration**. Minutes after injury, swelling and retraction of neurilemmal cells occur at the nerve fiber nodal regions. Twenty-four hours after injury, the myelin layer loosens. During the next two to three days, the myelin layer swells, fragments, globules form, and then the myelin layer disrupts about day four. Disappearance of myelin layers by phagocytosis takes about six months. A significant aspect of this process is that the endoneurial tubes and basement membranes of the distal segment collapse and fold but maintain their continuity. About six weeks after injury

Proximal segment   Distal segment

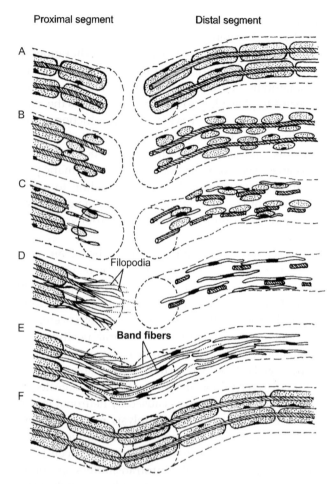

FIGURE 1.7. Sequential steps (A–F) in the degeneration and regeneration in the proximal and distal segments of a transected neuronal process. In the proximal segment, degeneration extends back to the first or second nerve fiber node. Anterograde degeneration exists throughout the entire distal segment. Proliferation of neurolemmocytes from the distal segment forms a bridge across the transection paving the way for an axonal sprout to find its way across the gap and eventually form a new process of normal diameter and length.

there is fragmentation and breakdown of the cytoplasm of the distal segment.

1.6.3. Retrograde Degeneration

Changes occurring in the **proximal segment** (Fig. 1.7) of a transected peripheral nerve are termed **retrograde degeneration**. One early event at the cut end of the proximal stump is the accumulation of proteins. As the stump seals the axon retracts, and a small knob or swelling develops. Firing stops as the injured neuron recovers its resting potential. Normal firing does not occur for several days. Other changes are similar to those taking place in the distal segment except that the process of retrograde degeneration in the proximal

segment extends back only to the first or second nerve fiber node and does not reach the neuronal cell body (unless the initial injury is near the soma).

1.6.4. Regeneration of Peripheral Nerves

Although the degenerative process is similar in the CNS and PNS, the process of regeneration is not comparable. In neither system is there regeneration of neuronal cell bodies or processes if the cell body is seriously injured. Severance of the neuronal process near the cell body will lead to death of the soma and no regeneration. For the neuronal process to regenerate, the neuronal cell body must survive the injury. Only about 25% of those patients with surgically approximated severed peripheral nerves will experience useful functional recovery.

Many events occur during regeneration of peripheral nerves. The timing and sequence of those events is unclear. Regenerating neurons shift their metabolic emphasis by decreasing production of transmitter-related enzymes while increasing production of substances necessary for the growth of a new cytoskeleton such as **actin** (the protein subunit of actin filaments) and **tubulin** (the protein subunit of microtubules). There is an increase in axonal transport of proteins and enzymes related to the hexose monophosphate shunt. Axonal sprouting from the proximal segment of a transected nerve during regeneration is a continuation of the process of cytoskeletal maintenance needed to sustain a neuronal process and its branches.

A tangible sign of regeneration, the proliferation of neurilemmal cells from the distal segment, takes place about day four and continues for three weeks. A 13-fold increase in these myelin-forming cells appears in the remains of the neurolemma, basal lamina, and the persisting endoneurial connective tissue. Mechanisms responsible for the induction of neurilemmal cell proliferation are unclear. Human neurilemmal cells maintained in cell culture will proliferate if they make contact with the exposed plasmalemma of demyelinated axons.

Band Fibers, Growth Cones and Filopodia

Proliferating neurilemmal cells send out cytoplasmic processes called **band fibers** (Fig. 1.7E) that bridge the gap between the proximal and distal segments of a severed nerve. As the band fibers become arranged in longitudinal rows, they serve as guidelines for the **growth cones**, bulbous and motile structures with a core of tubulin surrounded by actin that arise from the axonal sprouts of the proximal segment. Microtubules and neurofilaments, though rare in growth cones, occur behind them and extend into the base of the growth cone, following the growth cones as they advance. Cytoskeletal proteins from the neuronal cell body such as actin and tubulin enter the growth cones by slow axonal transport 24 hours after initial injury. The rate of construction of a new cytoskeleton behind the advancing growth cone limits the outgrowth of the regenerating process. Such construction depends on materials arriving by slow transport that are available at the time of axonal injury. The unstable surface of a parent growth cone yields two types of protrusions – many delicate, hairlike offspring called **filopodia** (microspikes) and thin, flat **lamellipodia** (lamella) both of which are filled with densely packed actin filaments forming the motile region of the growth cone. Neuronal **filopodia** (Fig. 1.7D) are 10–30 μm long and 0.2 μm in diameter and evident at the transection site extending from the proximal side and retracting as they try to find their way across the scaffold of neurilemmal cells. After they make contact with their targets, extension of the filopodia ceases. There is successive addition of actin monomers at the apex of the growth cone with an ensuing rearward translocation of the assembled actin filaments. Both guidance and elongation of neuronal processes are essential features underlying successful regeneration. Such guidance is probably due to the presence of signaling molecules in the extracellular environment. In addition to their role in regeneration, growth cones play a role in the development of the nervous system allowing neuronal processes to reach their appropriate targets.

At the transection site, growth cones progress at the rate of about 0.25 mm/day. If the distance between the proximal and distal stumps is not greater than 1.0–1.5 mm, the axonal sprouts from the proximal side eventually link up with the distal stump. As noted earlier, the endoneurial tubes and basement membranes of the distal segment collapse and fold but maintain their continuity. Growth cones invade the persisting endoneurial tubes and advance at a rate of about 1.0–1.5 mm/day. A general rule for growth of peripheral nerves in humans is one inch per month. After transection of the median nerve in the axilla, nine months may be required before motor function returns in the muscles innervated by that nerve and 15 months before sensory function returns in the hand. After injury to a major nerve to the lower limb, 9–18 months is required before motor function returns. If a motor nerve enters a sensory endoneurial tube or vice versa, the process of regeneration will cease. If one kind of sensory fiber (one that carries painful impulses) enters the endoneurial tube of another kind of sensory fiber (one that carries tactile impulses), then abnormal sensations called **paresthesias** (numbness, tingling, or prickling) may appear in the absence of specific stimulation.

After a regenerated process crosses the transection site and enters the appropriate endoneurial tube, regeneration is still incomplete. The new process must be of normal diameter and length, remyelination must occur, and the original site of termination be identified with eventual re-establishment of appropriate connections. If the regenerating nerve is a motor nerve, it must find the muscle it originally innervated. A regenerating sensory nerve must innervate an appropriate peripheral receptor. Reduced sensitivity and poor tactile discrimination with peripheral nerve injuries is a result of misguidance of regenerating fibers and poor reinnervation. Regrowing fibers may end in deeper tissues and in the palm rather than in the fingertips – the site of discriminative tactile receptors. Poor motor coordination for fine movements observed in muscles of the human hand after peripheral nerve section and repair may be the result of misdirection of regenerating motor axons.

Collateral Sprouting

Collateral sprouts may arise from the main axonal shaft of uninjured axons remaining in a denervated area. Such collateral sprouting, representing an attempt by uninjured axons to innervate an adjacent area that has lost its innervation, is often confused with axonal sprouts that originate from the proximal segment of injured or transected neuronal processes. Collateral sprouting from adjacent uninjured axons may lead to invasion of a denervated area and restoration of sensation in the absence of regeneration by injured axons thus leading to recovery of sensation.

Neuromas

If the distance between the severed ends of a transected process is too great to reestablish continuity, the growing fibers from the proximal side continue to proliferate forming a tangled mass of endings. The resulting swollen, overgrown mass of disorganized fibers and connective tissue is termed a **traumatic neuroma** or **nerve tumor**. A neuroma is usually firm, the size of a pea, and forms in about three weeks. When superficial, incorporated in a dense scar, and subject to compression and movement, a neuroma may be the source of considerable pain and paresthesias. Neuromas form in the brain stem, spinal cord, or on peripheral nerves. In most peripheral nerve injuries, the nerve is incompletely severed and function is only partially lost. Blunt or contusive lacerations, crushing injuries, fractures near nerves, stretching or traction on nerves, repeated concussion of a nerve and gunshot wounds may produce neuromas in continuity. Indeed, in about 60% of such cases, **neuromas in continuity** develop. A common example is Morton's metatarsalgia – an interdigital neuroma in continuity along the plantar digital nerves as they cross the transverse metatarsal ligament. Wearing ill-fitting high heeled shoes stretches these nerves bringing them into contact with the ligament. Other examples are intraoral neuromas that form on the branches of the inferior alveolar nerve (inferior dental branches and the mental nerve) or on branches of the maxillary nerve (superior dental plexus), amputation neuromas in those who have had limbs amputated, and bowler's thumb which results from repetitive trauma to a digital nerve.

1.6.5. Regeneration and Neurotrophic Factors

Regeneration of a peripheral nerve requires an **appropriate microenvironment** (a stable neuropil, sufficient capillaries, and neurilemmal cells), and the presence of certain **neurotrophic factors** such as nerve growth factor (NGF), brain-derived neurotrophic factor (BDNF), or neurotrophin-3. Absorption of these factors by the axonal tip and their retrograde transport will influence the metabolic state of the neuronal cell body and support neuronal survival and neurite growth. Other substances attract the tip of the growth cone or axonal sprout thus determining the direction of growth.

1.6.6. Regeneration in the Central Nervous System

Regeneration of axons occurs in certain nonmyelinated parts of the mammalian CNS such as the neurohypophysis (posterior lobe of the pituitary gland) in the dog, retinal ganglionic cell axons and olfactory nerves in mice, and in the corticospinal tract of neonatal hamsters. However, the process of CNS regeneration leading to restoration of function is invariably unsuccessful in humans. Several theories attempt to explain this situation. The **barrier hypothesis** suggests that mechanical obstruction and compression due to formation of a dense glial scar at the injury site impedes the process of axonal growth in the human CNS. Such dense scar formation or **astrocytic gliosis** is the result of the elaboration of astrocytes in response to injury. This glial scar forms an insurmountable barrier to effective regeneration in the CNS. Remyelination, accompanied by astrocytic gliosis, takes

place in the CNS if axonal continuity is preserved. Myelin, in the process of degeneration, releases active peptides such as **axonal growth inhibitory factor** (AGIF) and **fibroblast growth factor** (FGF). AGIFs may lead to abortive growth of most axons whereas the FGFs are apparently responsible for the deposition of a collagenous scar. The observation that the breakdown of myelin in the PNS is unaccompanied by elaboration of AGIFs seems to strengthen this hypothesis. The presence of these growth promoting and growth inhibiting molecules along with the formation of glial scars offers a great challenge to those seeking therapeutic methods to aid those with CNS injury.

Efforts are underway to determine if neurons of the CNS are missing the capability of activating necessary mechanisms to increase production of ribosomal RNA. Other attempts at restoring function in the injured spinal cord involve removing the injured cord region and then replacing it with tissue from the PNS.

Inherent neuronal abilities and the properties of the environment (neuropil, local capillaries, and the presence of repulsive substrates or inhibitors of neurite outgrowth) are responsible for the limited capacity for CNS regeneration. Neuroglial cells, by virtue of their ability to produce trophic and regulatory substances plus their ability to proliferate forming a physical barrier to regeneration, also play an essential role in regeneration. A minimum balance exists between the capacity of axons to regenerate and the ability of the environment to support regeneration. CNS regeneration in humans is an enigma awaiting innovative thinking and extensive research. Success in this endeavor will bring joy to millions of victims of CNS injury and their families.

1.7. NEURAL TRANSPLANTATION

In light of the absence of CNS regeneration leading to restoration of function in humans, there is a great deal of interest in the possibility of **neural transplantation** as a means of improving neurological impairment due to injury, aging, or disease. Sources of donor material for neural transplants are neural precursor cells from human embryonic stem cells, adult cells or umbilical cords, ganglia from the PNS (spinal and autonomic ganglia and adrenal medullary tissue), and cultured neurons. Other sources are genetically modified cell-lines capable of secreting neurotrophic factors or neurotransmitters.

Focal brain injuries, diseases of well-circumscribed chemically defined neuronal populations, identifiable high-density terminal fields, areas without highly specific point-to-point connections, or regions where simple one-way connections from the transplant would be functionally effective, are likely to profit from neural transplantation. Neurological diseases such as Alzheimer's and Parkinson's disease, are characterized by a complex set of signs and symptoms with damage to more than one region and more than one neurotransmitter involved, such that patients suffering from these diseases might not benefit from a single neural transplant but may require dissimilar transplants in different locations. Because these diseases are also progressive and degenerative, it is possible that the transplant itself will be subject to the same progressive and degenerative process. An equally disconcerting prospect is that with additional degeneration of the brain, the signs and symptoms ameliorated by the original transplant may disappear, replaced by a new set of signs and symptoms that might require a second transplant for their alleviation. Finally, because of the age of most patients with these diseases it is likely that they have other physical conditions that might necessitate selecting for treatment only those who do not have other underlying conditions or who have a very early stage of the disease.

Another approach to this problem that would circumvent the risks and ethical issues associated with neural transplantation would be to administer neurotrophic factors to support neuronal survival or promote growth of functional processes. An exciting development in this regard is the isolation of a protein called **glial cell line-derived neurotrophic factor** (GDNF), which promotes the survival of dopamine-producing neurons in experimental animals. In Parkinson's disease, there is restricted damage to a well-defined group of dopamine-producing neurons in the midbrain. Such a neurotrophic agent might prevent or reverse the signs and symptoms of this chronic, degenerative disease. An additional option would be to investigate the initial changes in the brain that lead to a particular neurological impairment and seek a means of preventing such changes. Much work remains before neural transplantation becomes a useful and practical form of therapy leading to complete functional recovery from neurological injuries, diseases, or age-related changes.

FURTHER READING

Abbott NJ, Ronnback L, Hansson E (2006) Astrocyte–endothelial interactions at the blood–brain barrier. Nat Rev Neurosci 7:41–53.

Allen NJ, Barres BA (2005) Signaling between glia and neurons: focus on synaptic plasticity. Curr Opin Neurobiol 15:542–548.

Ambrosi G, Virgintino D, Benagiano V, Maiorano E, Bertossi M, Roncali L (1995) Glial cells and blood–brain barrier in the human cerebral cortex. Ital J Anat Embryol 100 (Suppl 1): 177–184.

Antel J (2005) Oligodendrocyte/myelin injury and repair as a function of the central nervous system environment. Clin Neurol Neurosurg 108:245–249.

Baumann N, Pham-Dinh D (2001) Biology of oligodendrocyte and myelin in the mammalian central nervous system. Physiol Rev 81:871–927.

Farber K, Kettenmann H (2005) Physiology of microglial cells. Brain Res Rev 48:133–143.

Hering H, Sheng M (2001) Dendritic spines: structure, dynamics and regulation. Nat Rev Neurosci 2:880–888.

Hyman SE (2005) Neurotransmitters. Curr Biol 15: R154–R158.

Itzev DE, Ovtscharoff WA, Marani E, Usunoff KG (2002) Neuromelanin-containing, catecholaminergic neurons in the human brain: ontogenetic aspects, development and aging. Biomed Rev 13:39–47.

Koehler RC, Gebremedhin D, Harder DR (2006) Role of astrocytes in cerebrovascular regulation. J Appl Physiol 100: 307–317.

Masland RH (2004) Neuronal cell types. Curr Biol 14: R497–R500.

McLaurin JA, Yong VW (1995) Oligodendrocytes and myelin. Neurol Clin 13:23–49.

Newman EA (2003) New roles for astrocytes: regulation of synaptic transmission. Trends Neurosci 26:536–542.

Oberheim NA, Wang X, Goldman S, Nedergaard M (2006) Astrocytic complexity distinguishes the human brain. Trends Neurosci 29:547–553.

Pellerin L (2005) How astrocytes feed hungry neurons. Mol Neurobiol 32:59–72.

Riga D, Riga S, Halalau F, Schneider F (2006) Brain lipopigment accumulation in normal and pathological aging. Ann NY Acad Sci 1067:158–163.

Roy S, Zhang B, Lee VM, Trojanowski JQ (2005) Axonal transport defects: a common theme in neurodegenerative diseases. Acta Neuropathol (Berl) 109:5–13.

Sherman DL, Brophy PJ (2005) Mechanisms of axon ensheathment and myelin growth. Nat Rev Neurosci 6:683–90.

Stevens B (2003) Glia: much more than the neuron's sidekick. Curr Biol 13:R469–R472.

Torrealba F, Carrasco MA (2004) A review on electron microscopy and neurotransmitter systems. Brain Res Rev 47:5–17.

Tyler WJ, Murthy VN (2004) Synaptic vesicles. Curr Biol 14:R294–R297.

Volterra A, Meldolesi J (2005) Astrocytes, from brain glue to communication elements: the revolution continues. Nat Rev Neurosci 6:626–640.

Wigley C (2004) Nervous system. In: Standring S, Editor-in-Chief. *Gray's Anatomy*. 39th ed, pp 43–68. Elsevier/ Churchill Livingstone, London.

Although life is a continuous process, fertilization (which, incidentally, is not a 'moment') is a critical landmark because, under ordinary circumstances, a new, genetically distinct human organism is formed when the chromosomes of the male and female pronuclei blend in the oocyte.

Ronan O'Rahilly and Fabiola Müller, 1996

2

Development of the Nervous System

Human development is divisible into two primary periods: a prenatal period, or the nine months before birth, and a **postnatal period**, the time after birth. The postnatal period includes infancy, childhood, adolescence, and adulthood. Labor and delivery (childbirth) are dynamic events in the interim between these two periods. The prenatal period lasts from the time of fertilization until birth and can be divided into the **embryonic period proper** (the first two months or the first eight postovulatory weeks), and the **fetal period** (the third to the ninth month). Development of the nervous system begins in the embryonic period and extends into the postnatal period.

Our focus is on certain events in the embryonic period. In these eight weeks, the major brain regions and their subdivisions and future spinal cord develop from embryonic ectoderm, setting the stage for the adult nervous system.

Although development is a continuous process, this description focuses on weekly intervals. Table 2.1 summarizes the initial appearance of various features in the first five weeks of the embryonic period using the **Carnegie staging system** that correlates stage, age, and number of somites (only for a limited time). Each of 600 sectioned embryos is assigned to one of 23 stages (each two or three days in length) covering the first eight postovulatory weeks. *Superscripts in the text of this chapter refer to Carnegie embryonic stages.*

2.1. FIRST WEEK OF DEVELOPMENT (FERTILIZATION, FREE BLASTOCYST, ATTACHING BLASTOCYST)

2.1.1. Fertilization

Development begins at the time of **fertilization** – the union of two specialized cells, one from the male, a **spermatozoon** (Fig. 2.1A), and one from the female, an **oocyte** (Fig. 2.1B), to form a **zygote** (Fig. 2.1C). A zygote is the unicellular, fused product of these two cells with two sets of chromosomes (a maternal and a paternal set). This **unicellular embryo** is the **ultimate stem cell** in that it can develop into any type of embryonic tissue and can form an entire embryo. Fertilization, normally occurring in the lateral end of the uterine tube (ampulla), initiates a series of events leading to growth and differentiation of the organism.

2.1.2. From Two Cells to the Free Blastocyst

By 36 hours, the zygote divides into two cells (Fig. 2.1D) that then divide into four cells (Fig. 2.1E) at about 40 hours. Additional division of cells leads to the formation of a spherical, solid mass of a dozen or more cells (Fig. 2.1F). The term **morula** [Latin, mulberry] designates embryos[2] with a dozen or more cells present but no blastocystic cavity. In mammals the morula gives rise to both embryonic and nonembryonic (chorion, amnion) structures. Cells of the morula are **totipotent stem cells** that have total potential. They are capable of developing into

a complete organism or differentiating into any of its cells or tissues. By the third or fourth day of development as this cleaving embryo makes it way into the uterine cavity, fluid enters its center resulting in a spherical outer mass of cells called a **blastocyst** (Fig. 2.1G), surrounding a fluid-filled space, the **cavity of the blastocyst** (Fig. 2.1G). A blastocyst has two components: an **outer cell mass** or trophoblast (Fig. 2.1G) – a collection of ectodermal cells in a peripheral position, and an **inner cell mass** or embryoblast (Fig. 2.1G). The outer mass of trophoblastic cells nourishes the developing structure and forms protective membranes around it. The appearance of the cavity of the blastocyst indicates that the embryo has gone through a series of divisions and differentiations (a process known as determination) such that its cells lose their potential and gain differentiated function. Such cells are **pluripotent stem cells** that can give rise to most, but not all cells or tissues of an organism.

At this early time, the dorsoventral axis of the embryo becomes apparent. The surface of the inner cell mass facing the cavity of the blastocyst represents the ventral surface of the embryo while that surface adjoining the trophoblast represents its dorsal surface. A coronal plane is definable at this time. The embryo proper develops from the inner cell mass. The term **embryo** refers to the developing organism from the time of fertilization until the embryonic period ends. Once in the uterus, the blastocyst begins to implant in the endometrium. To achieve the best possible environment in which to develop, the blastocyst completely embeds in the endometrium.

2.2. SECOND WEEK OF DEVELOPMENT (IMPLANTATION, PRIMITIVE STREAK APPEARS, THREE LAYERS OF CELLS)

2.2.1. Implantation and the Appearance of Two Distinct Layers of Cells

During the second week of development, implantation of the blastocyst begins on the posterior uterine wall. Also during the second week, two distinct layers are distinguishable in the inner cell mass. An inner layer or **primary endoderm** (Fig. 2.2A) adjoins the cavity of the blastocyst and becomes the lining of the gastrointestinal tract, its glands, and the epithelial component of structures that arise as outgrowths of it. An outer layer or **epiblast** (Fig. 2.2A) is a pseudostratified columnar epithelium that becomes the **embryonic ectoderm** and forms the brain, spinal cord, all nerves, and sensory organs plus the skin, hair, and nails. At this time, the

TABLE 2.1. Early development of the human nervous system

Carnegie Stage	Age	Somites	Initial appearance of various features
First Week			
Stage 1	ca day 1		Unicellular embryo formed at fertilization
			New, genetically distinct human organism
Stage 2	ca 2–3 days		2 to ca 16 celled cleaving embryo proceeds along uterine tube
Stage 3	4–5 days		Embryo termed a blastocyst lies in uterine cavity
			Dorsal and ventral surfaces to embryonic disc
Stage 4	6 days		Attachment of blastocyst to uterine endometrium
Second Week			
Stage 5	ca 7–12 days		Continuation of implantation
Stage 6	ca 17 days		Primitive streak develops
			Embryo acquires right and left sides, rostral and caudal ends
Third Week			
Stage 7	ca 19 days		Notochordal process – an axial structure
Stage 8	ca 23 days		Epiblast transformed into neural plate
			The neural groove, the first morphological indication of the nervous system, becomes recognizable
			Neural groove bounded by faint neural folds
Stage 9	ca 26 days	1–3 pairs	Three major divisions of the brain; rhombencephalon, mesencephalon, prosencephalon
			Neural groove is long and deep
			Areas of neural crest
			Mesencephalic flexure
			Somites (occipital)
			Developing ear
			Neuromeres in open neural folds
Fourth Week			
Stage 10	ca 29 days	4–12 pairs	Neural folds begin to fuse
			Neural tube begins
			Telencephalon medium and diencephalon distinguishable
			Neural crest formation continues
			Optic primordia (developing eye)
Stage 11	ca 30 days	13–20 pairs	Rostral (or cranial) neuropore closes
			Spinal part of neural tube is lengthening rapidly
Stage 12	ca 31 days	21–29 pairs	Caudal neuropore closes
			Brain occupies 42% of neural tube
			Secondary neurulation begins
Stage 13	ca 32 days	30 or more pairs	Closed neural tube
			First appearance of the cerebellum
			Isthmus rhombencephali
			Cervical flexure
			Oculomotor and trochlear nuclei
			Three divisions of trigeminal ganglion
Fifth week			
Stage 14	ca 33 days		Future cerebral hemispheres become identifiable
			Pontine flexure appears
			Distinction between metencephalon and myelencephalon
			All 16 neuromeres are now identifiable
			Blood vessels now penetrate the wall of the brain
Stage 15	ca 35 days		Five major subdivisions: telencephalon, diencephalon, midbrain, pons, and medulla
			Most cranial nerves are present
Stage 16	ca 37–39 days		Presence of hippocampal thickening
			Evagination of the neurohypophysis is now evident

Modified from O'Rahilly and Müller, 2001, 2006.

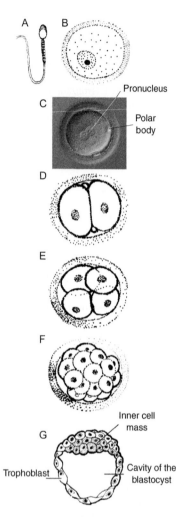

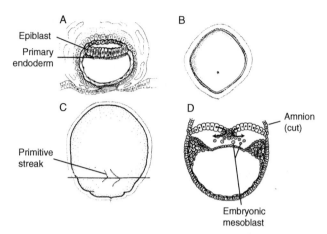

FIGURE 2.2. Second week of human development. **A**, Section of the middle of an implanted human embryo of 7–12 days with a bilaminar embryonic disc (after Hertig and Rock, 1941). **B**, Dorsal view of the bilaminar disc of a human embryo of 7–12 days (stage 5) (after Hertig and Rock, 1941). **C**, Dorsal view of a human embryo of 13 days (stage 6) depicting the initial appearance of the primitive streak (after O'Rahilly, 1973). **D**, Section of the embryonic disc in the region of the primitive streak showing the appearance of embryonic mesoblast cells from the primitive streak.

the embryonic disk is about 0.2 mm in length. The right and left sides, rostral and caudal ends, and dorsal and ventral surfaces of the embryo are distinguishable. Arising from the base of the primitive streak, between the endoderm and ectoderm, is a third layer of cells, the **embryonic mesoblast** (Fig. 2.2D). The embryonic mesoblast becomes the **mesoderm** that forms the skeleton, muscles, and many internal body organs. **Embryonic ectoderm, endoderm**, and **mesoderm** are collectively the **primary germ layers**. They develop in the first three weeks and form all tissues and organs of the body. The primitive streak diminishes in size, undergoes degenerative changes, and disappears. If it persists in the sacrococcygeal region, it may give rise to a tumor called a **teratoma**.

FIGURE 2.1. First week of human development. Formation of a free blastocyst from the fused product of fertilization. **A**, Spermatozoon (after Bloom and Fawcett, 1975). **B**, Oocyte (after Bloom and Fawcett, 1975). **C**, Zygote (after Shettles, 1955). **D**, Two cells (after Lewis and Hartman, 1933). **E**, Four cells (after Lewis and Hartman, 1933). **F**, 12 to 16 cells (after Lewis and Hartman, 1933). **G**, Free blastocyst (after Hertig et al., 1954).

epiblast and the primary endoderm collectively form a bilaminar, flat, circular plate of cells – the **embryonic disc** (Fig. 2.2B).

2.2.2. Primitive Streak and a Third Layer of Cells Appear

The circular embryonic disc becomes elongated and then pear-shaped by expansion mainly at its rostral end. A thickened band of pluripotential, epiblastic cells, the **primitive streak** (Fig. 2.2C), appears in the median plane in the caudal part of the embryonic disc. The longitudinal axis of the disc and future body, coinciding with the axis of the primitive streak, is established. At this stage,

2.3. THIRD WEEK OF DEVELOPMENT (NEURAL PLATE, GROOVE, AND FOLDS, THREE MAIN DIVISIONS OF THE BRAIN)

2.3.1. Primitive Node and Notochordal Process Appear

As the third week begins, the cranial **primitive streak** (Fig. 2.3) has a conspicuous knot of specialized ectodermal cells, the **primitive node** (Fig. 2.3). Embryonic mesoblast continues to spread from the primitive streak.

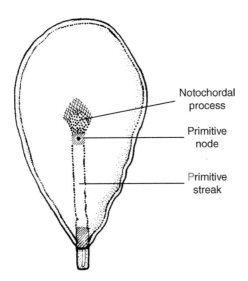

FIGURE 2.3. Third week of human development. Dorsal surface of the embryonic disc of a human embryo of 16 days with a visible primitive node and, extending from it, the notochordal process (after O'Rahilly, 1973).

A conspicuous rod of cells, the **notochordal process**, extends like a telescope from the primitive node and appears between the primary ectoderm and endoderm. At this stage, the embryonic disk is about 0.4mm in length. Later in development, the **notochordal process** (Fig. 2.3) is concerned with formation of the **notochord**. The notochord indicates the future bony vertebral column in humans but disappears as the developing vertebral bodies surround it. Notochordal remnants expand to form the **nucleus pulposus** of the adult intervertebral disc. The notochord ends rostrally near the adenohypophysis (anterior lobe of the pituitary gland)[14]. If remnants of notochord persist, they may develop into rare tumors called **chordomas**. These slimy, gelatinous tumors grow slowly, invade adjacent bone and soft tissue, and seldom metastasize. They account for less than 1% of all CNS tumors. Fifty percent of chordomas arise in the sacrum, 15% in other vertebrae, and the remaining 35% occur intracranially and frequently originate from the clivus.

2.3.2. Neural Plate, Groove, Folds and Neuromeres Appear

The third week of development, distinguished by rapid growth, coincides with the first missed menstrual period and the initial appearance of the brain. Before somites are visible, and when the embryonic disc is 0.5 to 2.0mm in length, a thickening of ectodermal cells in the median plane overlies the notochordal process that measures

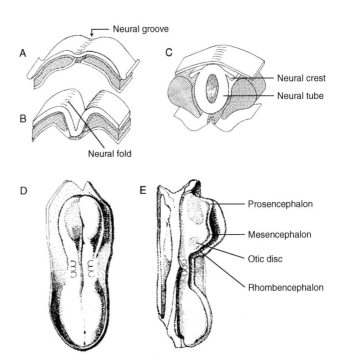

FIGURE 2.4. Third and fourth weeks of human development. **A, B, C.** Transverse sections of human embryos at stages 8, 9 and 10 showing the formation of the neural tube (after O'Rahilly and Müller, 1999). **D,** Dorsal view of a human embryo of 20 days (after Ingalls, 1920). **E,** Left lateral view of the embryo in part D, showing the three primary brain regions in the rostral half of the neural folds and the otic region (after Bartelmez and Evans, 1926).

some 0.4mm in length at this time. The ectodermal thickening or **general area of the neural plate** appears early in the third week of development[7–8]. Neural plate formation is induced by the prechordal plate, notochord, and surrounding mesoderm. The neural plate invaginates along the median plane forming the **neural groove** (Fig. 2.4A). Appearing during the third week of development[8], this shallow neural groove is the *first visible sign of the nervous system* before the heart or any other organs become visible. Raised margins on either side of the neural groove, also distinguishable at this time[8b], are the **neural folds** (Fig. 2.4B). The neural groove deepens and lengthens toward the end of the third week[9]. Appearing at three weeks[9] in the open neural folds are the six **primary neuromeres** (prosencephalon, mesencephalon, and rhombomeres A, B, C, and D). Neuromeres are not 'bulges' or 'segments' but rather transverse subdivisions perpendicular to the longitudinal axis of the developing brain and on both sides of the body (O'Rahilly and Müller, 2001). They appear at definite times and in a definite sequence. Smaller subdivisions, called **secondary neuromeres**, appear before[10–11] and after[12–14] closure of the neural tube. All 16 neuromeres are present at five weeks[14] in human embryos. The primitive streak and primitive node remain visible on the dorsal surface of the embryonic disc. The dorsal surface of the disc becomes the dorsal surface of the body.

2.3.3. Three Main Divisions of the Brain Identifiable

Characteristic of the end of the third week is the appearance of one to three pairs of somites. Neural folds elevate and become prominent as the neural groove deepens[9]. **Three major divisions** of the brain, (Fig. 2.4E) as well as the area of the future spinal cord are distinguishable in the completely open neural folds (before the neural tube begins to form). The future brain or **encephalon** develops from the rostral half of the neural folds. The three major divisions of the brain identifiable in the neural folds are the **prosencephalon (forebrain)**, **mesencephalon (midbrain)**, and **rhombencephalon (hindbrain)**. These rostral to caudal regions appear as enlargements separated by constrictions. The future **spinal cord** is caudal to the rhombencephalon. Recent evidence suggests that the first neurons of the future cerebral cortex in humans are in the forebrain of the neural folds at this stage. They likely originate from the subpallium by a particularly complex process termed tangential migration.

2.3.4. Mesencephalic Flexure Appears

An abrupt bend, the **mesencephalic flexure** (see Fig. 2.6), appears[9] in the neural folds at the mesencephalic level making it easier to identify the three major divisions of the brain – prosencephalon (forebrain), mesencephalon (midbrain), and rhombencephalon (hindbrain) (see Fig. 2.6). At the same time, the lips of the neural folds begin to close at the **cervical flexure** situated at the junction between the rhombencephalon (hindbrain) and the future spinal cord. The area of the future ear, the **otic disc** (Fig. 2.4), is recognizable and rostral to it areas of **neural crest** are beginning to develop.

2.4. FOURTH WEEK OF DEVELOPMENT (NEURAL TUBE FORMS AND CLOSES, NEURAL CREST FORMATION CONTINUES)

2.4.1. Formation of the Neural Tube

Circulation of blood, cardiac contraction, and fusion of the neural folds forming a **neural tube** (Fig. 2.5A–D, F)

characterize the onset of the fourth week[10] of development. Closure of the neural tube begins at rhombencephalic, upper cervical levels, or both but soon is identifiable in several sites. Consequently, this process does not occur in a zipperlike manner proceeding rostrally and caudally as has often been described. By the end of this stage, the neural tube extends from the rhombencephalon (hindbrain) in the otic disc (rostrally) to the latest-formed somite (caudally). At its rostral and caudal ends, the neural tube remains open (Fig. 2.5H). The neural groove now becomes the floor of the neural tube. The process of neural tube formation is termed **primary neurulation**. Three main divisions of the brain appear in the neural folds before the formation of the neural tube or any of its parts. During this stage, the rhombencephalon shows four rhombomeres (Rh.A, Rh.B, Rh.C, and Rh.D). Brain and spinal cord malformations occurring during the process of primary neurulation are **neural tube defects**.

2.4.2. Rostral and Caudal Neuropores Remain Temporarily Open

Large areas at the ends of the newly formed neural tube remain open. These slits diminish to become **rostral** and **caudal neuropores**, respectively (Fig. 2.5D). **Primary neurulation** coincides with embryonic elongation and elevation of the cranial part of the neural folds. The upper two-thirds of the embryonic neural tube is more advanced than the caudal third. Two subdivisions of the forebrain, the **telencephalon medium** (telencephalon impar) and the **diencephalon** become recognizable[10]. The diencephalon begins to show the **optic sulcus** and **optic primordium**, first signs of a developing eye[10]. The wall of the neural tube is about the same thickness throughout its extent. A mesencephalic flexure, having made its appearance earlier[9], is present in all embryos at this time.

2.4.3. Neural Crest Cells Emerge

During **primary neurulation**, a very rapid event, the neural tube separates from surface ectoderm. In embryos of about 3½ weeks[10], **neural crest cells** emerge from a wedge-shaped area in the dorsolateral edge of the neural tube, at mesencephalic and rhombencephalic levels, and migrate ventrolaterally reaching either side of the neural tube. The **neural crest cells** (Fig. 2.5E, F, G) in this location lie on the dorsolateral aspect of the neural tube as part of a column of neurectodermal cells that extends

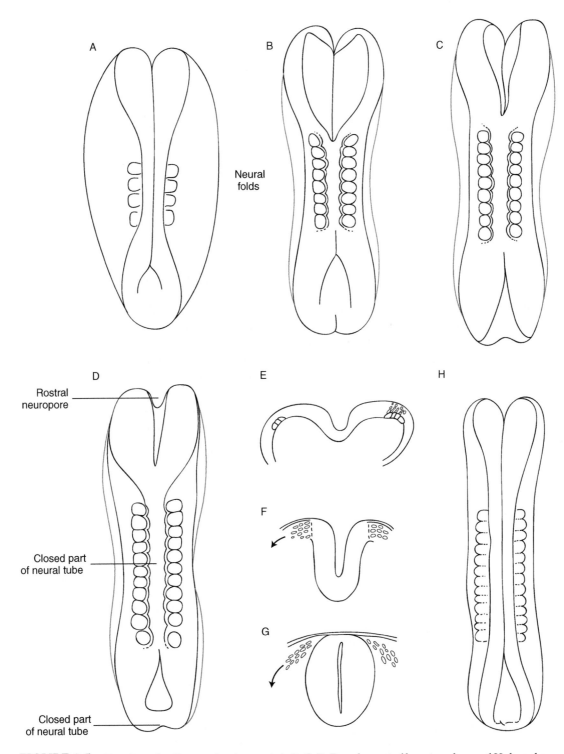

FIGURE 2.5. Fourth week of human development. **A, B, C, D,** Dorsal aspect of human embryos of 22 days showing successive steps in the fusion of the neural folds forming the neural tube (after Heuser and Corner, 1957). **E, F, G,** Transverse sections of a human embryo of 22 days depicting the emergence of cells of neural crest from the lateral edge of the neural folds. **H,** Dorsal surface of a human embryo of 24 days with a closed rostral neuropore (after Streeter, 1942).

from the mesencephalon (future midbrain) to the **caudal neuropore**. They form cranial, spinal, autonomic nerves, their ganglia, and supporting cells (the human vestibular and cochlear ganglia arise from neural crest cells given off from the wall of the otic vesicle). The **leptomeninges,** which form part of the coverings of the brain and spinal cord, especially the pia mater, originate from the neural crest. The neural crest contributes to the formation of the cartilaginous skull, the adjacent loose connective tissue of the face, forms melanocytes (in skin, hair, and irides) and neurosecretory cells. Evidence exists that the neural crest plays a major role in facial development. A common cellular origin from ectoderm of the neural folds provides a basis for this correlation.

Neural Crest Syndrome

An apparent abnormality thought to occur in the differentiation of neural crest in humans is a nearly complete and hypothetically predictable **neural crest syndrome**. In such instances, there is autonomic and sensory impairment, meningeal thickening and cavitation, and other clinical abnormalities. Other related conditions arise from the abnormal development of neural crest cells or from defects in the migration, growth, and differentiation of these cells.

2.4.4. Neural Canal – the Future Ventricular System

The neural tube determines the form of the embryo at this time. The central core of this hollow tube, the **neural canal**, will eventually become the fluid-filled **ventricular system** of the brain and the **central canal** of the spinal cord. The neural tube forms the entire central nervous system (CNS) or brain and spinal cord. The adult CNS, therefore, is a hollow tube with well-developed walls connected by peripheral nerves with various parts of the body.

2.4.5. Neuropores Close and the Closed Neural Tube is Filled with Fluid

The **rostral (or cephalic) neuropore** of the neural tube closes during the fourth week[11] when 13–20 pairs of somites are present. Consequently, there is no communication rostrally between the amniotic cavity and the ventricular system. The **caudal neuropore** gradually closes before 4½[12] weeks at the level of future somite 31 corresponding to the second sacral vertebral level. At

this level, the closed neural tube extends into neural tissue laid down as a **neural cord** and derived from pluripotent mesenchyme. This process is termed **secondary neurulation.**

Also during the fourth week of development, the closed neural tube separates from the surface ectoderm[13]. Secondary neuromeres appear before[10–11] and after[12–14] closure of the neural tube. The continuous cavity of the neural tube constitutes the future central canal of the spinal cord and early ventricular cavity of the brain. The closed neural tube fills with a protein-rich ependymal fluid, presumably produced by cells that line the ventricular system. Cells of the choroid plexus will not appear for another two weeks. (The choroid plexus is a specialized part of the ventricular system responsible for secreting cerebrospinal fluid [CSF] that eventually fills the ventricular system.) As the rostral part of the neural tube fills with fluid it expands, widens, and undergoes rostrocaudal enlargement before the choroid plexus appears. As the dorsal aspect of the embryo fills out, the embryo assumes a C-shape with the head bent toward the heart and the caudal part of the trunk bent toward the anterior surface of the body. The embryo maintains this C-shape throughout the remaining embryonic period. Neural crest cells continue to emerge from the rostral part of the neural tube. After formation of a completely closed neural tube and its separation from the ectoderm, the initial formation of a blood-vascular supply to the neural tube begins closely investing the surface of the neural tube.

Despite these events and the growth that has taken place during the first four weeks of development, brain volume is only about 0.003 ml and the embryo is 4–6 mm long. Arm and leg buds are visible at this time. The neural tube has an advanced and enlarged rostral end (the future brain) and an attenuated, immature, and elongated caudal end (the future spinal cord).

2.4.6. Cervical Flexure Present

The CNS is still the primary determiner of embryonic form with the outer contour of the neural tube the same as that of the embryo. Embryos of this stage have a more rostrally located **mesencephalic flexure** (Fig. 2.6, single arrow) and a second flexure, the **cervical flexure**, more caudally located at the rhombencephalic-spinal junction (Fig. 2.6, double arrows). The neural tube has areas of decreased cellular proliferation that appear as ridges or grooves. These bulges or cranial ganglia (Fig. 2.6), constitute reliable landmarks corresponding to future sites of the fifth, seventh, eighth, ninth, and tenth **cranial**

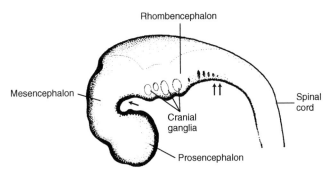

FIGURE 2.6. Fourth week of human development. A median section of a human embryo of 28 days showing the cervical flexure (double arrows) at the junction of the rhombencephalon with the spinal cord. An arrow indicates the region of the mesencephalic flexure. Future sites of the trigeminal, facial, glossopharyngeal and the accessory-vagal ganglia are present (after Streeter, 1945).

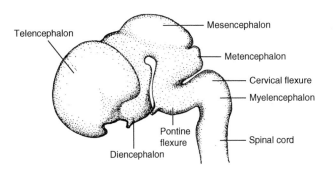

FIGURE 2.7. Additional development of the human brain. Lateral view of the brain at about 52 days (after O'Rahilly, Gardner, 1971).

nerves. The retina, its pigment layer, and a plate of ectoderm that will become the lens are evident at this time.

2.5. FIFTH WEEK OF DEVELOPMENT (FIVE SUBDIVISIONS OF THE BRAIN IDENTIFIABLE)

2.5.1. Simple Tube Transforms into Complex Organ System

In the fifth week, a simple tubular structure transforms into a complex organ system. Beginnings of basic functional areas are established, major regions are distinguishable, and new areas in these regions appear. All this developmental activity reflects early brain specialization. In the nine months before birth and continuing to the time that the adult brain develops, the forebrain grows at a faster rate than the remaining nervous system.

2.5.2. Five Subdivisions of the Brain Appear

In the fifth week of development, there is continued proliferation and enlargement of cells particularly in the rostral two-thirds of the neural tube. **Three main divisions of the brain** (Fig. 2.4E), evident in the neural folds during the third week of development[9], are transformed into **five subdivisions of the brain** (Fig. 2.7). The **prosencephalon (forebrain)** has differentiated into two subdivisions: the **telencephalon** and **diencephalon** are clearly distinguishable in the fourth week[10]. The **mesencephalon** undergoes slight change and remains as the **midbrain**, still distinguished by

the **mesencephalic flexure**. The **rhombencephalon (hindbrain)** transforms into the **metencephalon (pons and cerebellum**[13,14]**)** and **myelencephalon (medulla oblongata)**. Neurons of the trochlear nucleus which give rise to the trochlear nerve [IV] are derived from the relatively narrow region at the junction between the midbrain and the rhombencephalon known as the **isthmus rhombencephali** – an important subdivision of the embryonic brain.

Each of these five subdivisions of the brain (telencephalon, diencephalon, midbrain, pons, and medulla oblongata)[15] which can be discerned at five weeks has a fluid-filled cavity continuous with the adjacent region and with the neural canal of the future spinal cord. The telencephalon, recognizable as the future **cerebral hemispheres**[15], is associated with the **lateral ventricles** and rostral part of the **third ventricle**, the diencephalon with the middle and posterior parts of the **third ventricle**, and the midbrain with the **aqueduct of the midbrain**. Associated with the pons and medulla is the fourth ventricle and rostral part of the central canal. By growth and expansion of the walls of these five subdivisions identifiable during the fifth week of development[15], the adult brain develops. These five subdivisions have similar relations and comparable functions in birds, fish, mammals (humans, dogs, elephants and dolphins), amphibians and reptiles.

2.5.3. Brain Vesicles vs. Brain Regions

The concept of three brain vesicles arising from studies of the developing chick is not applicable to human development. Superficial to the mesencephalon in the chick are paired optic lobes. Comparable structures in mammals, the superior (optic) colliculi, are barely larger than the inferior (auditory) colliculi. In order to utilize the chick nervous system as a model for the mammalian

TABLE 2.2.   Divisions of the neural tube and their adult derivatives

Three major divisions of the brain	Five major subdivisions of the brain	Adult derivatives	Associated ventricular cavity
3rd week of development	5th week of development	Adult	
Prosencephalon (forebrain)	Telencephalon (endbrain)	Telencephalon medium Cerebral hemispheres Caudate and putamen Rhinencephalon	Lateral ventricles and rostral part of third ventricle
	Diencephalon (interbrain)	Epithalamus Hypothalamus Thalamus Subthalamus	Middle and posterior parts of third ventricle
Mesencephalon (midbrain)	Mesencephalon (midbrain)	Midbrain	Aqueduct of the midbrain
Rhombencephalon (hindbrain)	Metencephalon (afterbrain)	Pons Cerebellum	Fourth ventricle
	Myelencephalon (marrow brain)	Medulla oblongata	Fourth ventricle and central canal
Spinal cord Area of future spinal cord	Spinal cord	Spinal cord	Central canal

Modified from O'Rahilly and Müller, 2001, 2006.

nervous system one has to embrace the concept that the avian optic lobes represent a mesencephalic vesicle. Accepting this interpretation leads to the misinformed concept of three symmetrical vesicles. While easy to describe and diagram, such an interpretation is a misunderstanding of events. In humans, neural tube formation and closure takes place without ever having gone through a three-vesicle stage. Although appealing, the concept of three brain vesicles is incompatible with events occurring in early human development.

Three main divisions of the brain occur during the third week of development[9] as enlargements of the neural folds long before the neural tube closes (completed after the caudal neuropore closes during the fourth week[12]). Table 2.2 summarizes the divisions of the neural tube and their adult derivatives. It also indicates the associated ventricular cavity with each of the adult derivatives. The remaining chapters cover the structure and certain aspects of the function of these adult derivatives.

Brain volume at birth is 27% the adult value even though almost all neurons have formed. By the end of the first postnatal year, brain volume is about 75% of the adult value. Increase in brain volume results from an increase in (1) neuronal size; (2) extent of their dendritic tree; (3) axonal length and number of axonal collaterals; (4) formation of myelin layers about individual axons; and (5) the number of neuroglial cells and blood vessels.

2.6. VULNERABILITY OF THE DEVELOPING NERVOUS SYSTEM

Growth and differentiation of the nervous system takes place over a prolonged period and involves many different events leaving the brain susceptible to a multitude of injurious agents and conditions. Several **sensitive developmental periods** occur when the nervous system is vulnerable to injury. Before the time of fertilization, developmental disorders caused by genetic or chromosomal alterations may occur in the initial oocyte or spermatozoon. After fertilization and before implantation (during the second week of development), problems of maternal metabolism (iodine deficiency, uncontrolled insulin-dependent diabetes, and phenylketonuria) can affect development. After implantation and throughout the embryonic period, acute starvation or chronic dietary deficiency in the mother (micronutrients) can dramatically influence development.

In the fetal period, developmental vulnerability is associated with global nutrients consisting primarily of caloric intake in the diet. Poor maternal nutrition has a clear and substantial effect on birth weight as well as length and head circumference. Low birth weight often gives rise to neurological impairment and mental retardation. Exposure to toxins and other harmful agents may also injure the developing nervous system. **Alcohol related birth defects** (ARBD) may appear with 1–2 drinks per day. Children of mothers who drink in pregnancy are often born with physical and behavioral abnormalities because alcohol readily passes through the placenta to the developing fetus. **Fetal alcohol syndrome** (FAS) is an umbrella term for the resulting abnormalities including head and facial deformities, mental retardation, and growth deficiency with brain damage. The severity of damage depends on the quantity of alcohol consumed and the length of time of the alcohol ingestion. If the mother continues to drink after the baby is born, alcohol readily enters breast milk and can then pass to the nursing infant.

Maternal smoking habits also affect the fetus leading to growth retardation, learning disabilities, cognitive deficits and hyperactivity as well as perinatal mortality. Studies in nonhuman primates indicate that nicotine elicits neurodevelopmental damage that is highly selective for different brain regions, and that dietary supplements ordinarily thought to be neuroprotectant (choline or vitamin C) might worsen some of the adverse effects of nicotine on the fetal brain. At the cellular level nicotine is a neuroteratogen that leads to abnormalities of neuronal proliferation and differentiation, a loss of neurons, and compromises synaptic function.

Physical trauma or placental failure in labor and delivery can cause severe anoxia resulting in serious insult to the developing fetal brain leading to neurological damage. Newborns, no longer protected by the security of maternal membranes, are vulnerable to infections and metabolic disorders. With further growth, maturation, and myelination extending into the second year, the developing nervous system continues to remain vulnerable to injury.

2.7. CONGENITAL MALFORMATIONS OF THE NERVOUS SYSTEM

The term **congenital malformation** refers to structural abnormalities existing at birth. About 3% of newborns have congenital malformations. About 10% of all such malformations involve the CNS (occurring at an average rate of 80–100 per 10,000 births) causing considerable infant morbidity and mortality. Most CNS malformations arise in the third through the fifth week of development; about half are detectable at birth. Genetic transmission is the reason for such defects, though discrete environmental agents such as drugs and chemicals, infections, metabolic disturbances, and ionizing radiation also play a role. In about two-thirds of such malformations, the specific cause is unclear.

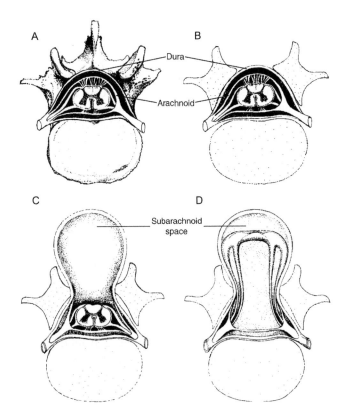

FIGURE 2.8. **A,** Horizontal section of the human spinal cord in the vertebral canal showing the meningeal coverings and the dorsal and ventral roots. **B,** Spina bifida occulta. **C,** Meningocele. **D,** Myelomeningocele.

2.7.1. Spinal Dysraphism

All forms of developmental malformations occurring in the median plane of the back are termed **spinal dysraphism**. Faulty closure in the median plane in the cerebral region is termed **cerebral dysraphism**. Dysraphia may involve skin, muscles, bones, blood vessels, and the brain or spinal cord. These conditions are clinically significant when they compress or stretch the spinal cord or spinal roots. **Spina bifida (rachischisis)**, the most common CNS malformation, refers to a limited cranial or spinal defect. Two forms of spina bifida, **spina bifida occulta and spina bifida cystica**, collectively make up about 15% of all developmental malformations and about 50% of all CNS malformations. The rate of spina bifida in the United States in 2002 was 20.13 per 100,000 live births.

Spina Bifida Occulta

Spina bifida occulta (Fig. 2.8B) is a form of spina bifida in which there is a bony spinal defect superficial to a normal spinal cord. The cord and its coverings remain in position but there is an abnormality of certain mesodermal elements surrounding the neural tube. Cleavage in one or more vertebral arches covered by soft tissue and intact skin makes the defect invisible on the surface but identifiable using radiography, ultrasound, or magnetic resonance imaging. Spinal cord and spinal roots may adhere to the outermost meningeal layer surrounding the spinal cord and attach to the skin by means of fibrous connections. These attachments bind the spinal cord and its roots in a fixed position, preventing them from moving in the vertebral canal during growth of the vertebral column and cord. Such a condition is termed a **tethered cord**. Owing to the differential growth rates of the vertebral column and the spinal cord, considerable tension and traction results on the slower growing spinal cord and spinal roots leading to neurological deficits that accompany spina bifida occulta.

Spina Bifida Cystica

In this serious form of spina bifida (Fig. 2.8), membranes surrounding the spinal cord herniate through a bony defect in one or more vertebral arches. In such instances, there is usually a translucent blue or bluish-white sac, protruding into or through the overlying skin. If only the meninges are involved in the cyst and if the cord remains in normal position, the condition is termed a **meningocele** (Fig. 2.8C). Normally no significant loss of neurological function occurs in infants with meningocele. If parts of the spinal cord and spinal roots are in the meningeal sac, then the term **myelomeningocele** is used (Fig. 2.8D). The frequency of this malformation is approximately one in 1200–1400 live births affecting 6000–11,000 newborns in the United States each year. Myelomeningocele appears to be second only to cerebral palsy as a cause of chronic locomotor disability in children. If no meningeal sac is evident but the spinal cord and spinal roots are evident on the surface, then the term **myelocele** is used.

About 80% of those with myelomeningocele also have **hydrocephaly** (an increase of cerebrospinal fluid in the brain ventricles). Since the total volume of cerebrospinal fluid in the ventricles is about 140 milliliters, and since the normal daily production is about 700 milliliters of this fluid, faulty absorption or the obstruction of cerebrospinal fluid may result in the accumulation of fluid that dilates the ventricles. Accompanying this expansion of the ventricles is often thinning of the brain and separation of the bones of the skull at the suture lines.

If the neural plate in the spinal region fails to form a closed neural tube (neurulation defect), neural tissue is exposed to the surface (open defect), a condition called

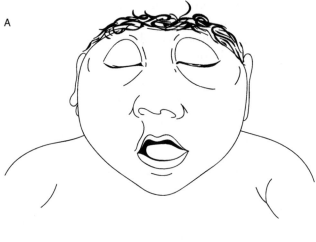

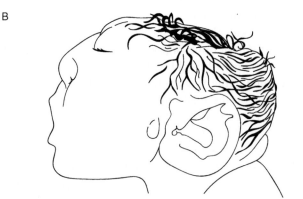

FIGURE 2.9. Anencephaly. **A**, Anterior and **B**, Lateral views. The vault of the skull and the brain are replaced by a spongy, vascular mass.

myeloschisis. Presence of an incompletely formed spinal cord in a membranous sac usually leads to severe signs and symptoms such as muscle weakness, bowel and bladder disturbances, and a variable degree of sensory loss.

2.7.2. Anencephaly

Another neural tube defect is **anencephaly** in which the **rostral neuropore** does not close resulting in partial or total brain absence and obvious defects of the skull (cranioschisis). The term **anencephalus** refers to a person afflicted with anencephaly. All bones of the calvaria are present but some are small fragments in abnormal positions. There may be an accompanying spina bifida. **Anencephaly** (Fig. 2.9) is a conspicuous malformation that is incompatible with life. The rate of anencephaly in the United States in 2002 was 9.55 per 100,000 live births. In this malformation, the eyes bulge and the vault of the skull and brain are no longer evident. Rather there remains a spongy, vascular mass called the **area**

cerebrovasculosa, into which proximal parts of the cranial nerves disappear. The amount of brain that remains is difficult to determine. The rostral part of the neural groove probably fails to close in anencephaly. Prolonged exposure of neural tissue to harmful excretory substances in amniotic fluid may result in degeneration of the brain. Although anencephaly may occur early in development, no instance of this malformation appears in the literature before approximately 32 postfertilizational days[13] (during the 5th week of development).

Spina bifida, anencephaly, and hydrocephaly, alone or in combination encompass over 90% of all congenital malformations of the nervous system, and nearly a third of all major congenital malformations recognizable at or shortly after birth. Most deaths attributable to CNS malformations occur in infants in the first year of life with a reported incidence in the United States of at least 94 deaths per 100,000 individuals under one year of age.

FURTHER READING

Bystron I, Molnar Z, Otellin V, Blakemore C (2005) Tangential networks of precocious neurons and early axonal outgrowth in the embryonic human forebrain. J Neurosci 25:2781–2792.

Bystron I, Rakic P, Molnar Z, Blakemore C (2006) The first neurons of the human cerebral cortex. Nat Neurosci 9:880–886.

Collins P (2004) Development of the nervous system. In: Standring S, Editor-in-Chief. *Gray's Anatomy*. 39th ed, pp 241–274. Elsevier/Churchill Livingstone, London.

Müller F, O'Rahilly R (1991) Development of anencephaly and its variants. Am J Anat 190:193–218.

Müller F, O'Rahilly R (1997) The timing and sequence of appearance of neuromeres and their derivatives in staged human embryos. Acta Anat (Basel) 158:83–99.

Müller F, O'Rahilly R (2003) Segmentation in staged human embryos: the occipitocervical region revisited. J Anat 203:297–315.

Müller F, O'Rahilly R (2003) The prechordal plate, the rostral end of the notochord and nearby median features in staged human embryos. Cells Tissues Organs 173:1–20.

Müller F, O'Rahilly R (2004) Embryonic development of the central nervous system. In: Paxinos G, Mai JK, eds. *The Human Nervous System*. 2nd ed. Chapter 2, pp 22–48. Elsevier/Academic Press, Amsterdam.

Müller F, O'Rahilly R (2004) Olfactory structures in staged human embryos. Cells Tissues Organs 178:93–116.

Müller F, O'Rahilly R (2004) The primitive streak, the caudal eminence and related structures in staged human embryos. Cells Tissues Organs 177:2–20.

Müller F, O'Rahilly R (2006) The amygdaloid complex and the medial and lateral ventricular eminences in staged human embryos. J Anat 208:547–564.

O'Rahilly R, Müller F (1994) Neurulation in the normal human embryo. Ciba Found Symp 181:70–82.

O'Rahilly R (1997) Making planes plain. Clin Anat 10:128–129.

O'Rahilly R, Müller F (1999) *The Embryonic Human Brain: An Atlas of Developmental Stages*. 2nd ed. Wiley-Liss, New York.

O'Rahilly R, Müller F (1999) Minireview: summary of the initial development of the human nervous system. Teratology 60(1):39–41.

O'Rahilly R, Müller F (2000) Prenatal ages and stages – measures and errors. Teratology 61:382–384.

O'Rahilly R, Müller F (2001) *Human Embryology & Teratology*. 3rd ed. Wiley-Liss, New York.

O'Rahilly R, Müller F (2001) Prenatal development of the brain. In: Timor-Tritsch I, Monteagudo A, Cohen H. *Ultrasonography of the Prenatal and Neonatal Brain*. 3rd ed, pp 1–12. McGraw-Hill, New York.

O'Rahilly R, Müller F (2002) The two sites of fusion of the neural folds and the two neuropores in the human embryo. Teratology 65:162–170.

O'Rahilly R, Müller F (2003) Somites, spinal ganglia, and centra. Enumeration and interrelationships in staged human embryos, and implications for neural tube defects. Cells Tissues Organs 173:75–92.

One should not assume that the spinal cord need not exist, nor that it would be preferable to have it located elsewhere than in the spine, nor, having been located in the spine that it was more protected from injuries than it actually is. For if the marrow did not wholly exist, one of two things would result: either all the parts of the animal located below the head would be completely deprived of movement, or it would be absolutely necessary that a nerve descend directly from the brain to each part.

Galen (ca 130–ca 200)

The Spinal Cord

3.1. EMBRYOLOGICAL CONSIDERATIONS

In the third week of human development, three main divisions of the brain and the future spinal cord are distinguishable in the neural folds. Shortly thereafter during the fourth week, the neural tube begins to form. The external surface of the future spinal cord at this time is smooth and shows no evidence of segmentation (Fig. 3.1). The broader rostral part of this closed tubular

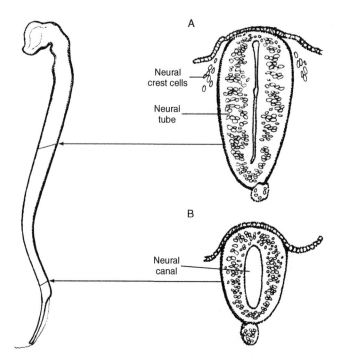

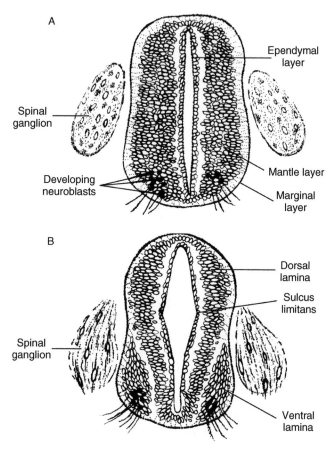

FIGURE 3.1. External surface of the neural tube at about 24 days. The spinal neural tube has smooth contours and shows no evidence of segmentation. **A,** Transverse section of the rostral end of the cord more advanced and definitely larger than **B,** a section of the lower end of the cord. In both sections, the spinal cord shows a uniform arrangement of cells (redrawn from Streeter, 1942). In **A,** the cells of neural crest have detached from the lateral edge of the neural tube.

FIGURE 3.2. **A,** Transverse section of the human cord at the end of the fourth week of development. Its wall has several cellular zones (after Sensenig, 1951). **B,** With continued growth, the central canal changes shape and a distinct sulcus limitans appears dividing the cord wall into alar and basal laminae (after Sensenig, 1957).

structure becomes the adult brain and brain stem while its long caudal part becomes the adult spinal cord.

If one examines the internal structure of the recently closed neural tube in the spinal cord region, a uniform arrangement of cells is apparent at all levels (Fig. 3.1A, B). Cells along the border of the **neural canal** (Fig. 3.1B) show numerous mitotic figures indicative of their intense proliferative activity. Cells formed in this layer migrate out to the surface of the neural tube increasing the bulk of the spinal cord wall and giving rise to additional layers of cells. The neural canal in the developing spinal cord becomes the **central canal** of the adult spinal cord. As such, the central canal is a caudal continuation of the ventricular system of the brain stem (fourth ventricle).

3.1.1. Layers of the Developing Spinal Cord

By the end of the fourth week, the spinal cord wall has several cellular layers (Fig. 3.2). A proliferating bed of cells along the edge of the neural canal, the **ependymal layer** (Fig. 3.2A) is the primary source of all spinal cord neurons and most neuroglial cells. Near the surface of the developing cord is a narrow **marginal layer** (Fig. 3.2A) that lacks neurons. Between the ependymal and

marginal layers is the ever-expanding **mantle layer** (Fig. 3.2A). At the end of this fourth week, new cells from the ependymal layer called **neuroblasts,** with processes emerging ventrolaterally from the cord, are identifiable. By continued growth and expansion of the walls of the spinal cord, the central canal elongates and a distinct **sulcus limitans** (Fig. 3.2B) appears along the ventricular surface on both sides of the central canal. At the beginning of the fifth week, the sulcus limitans is observable (Fig. 3.2B).

3.1.2. Formation of Ventral Gray Columns and Ventral Roots

Dividing the spinal cord wall into the **alar** (dorsal) and **basal** (ventral) **laminae** (Figs 3.2B, 3.3) is the **sulcus limitans.** Proliferation and migration of neuroblasts into the basal lamina causes a definitive **ventral gray column** that forms the **ventral horn** of the adult spinal cord (Fig. 3.3B).

A

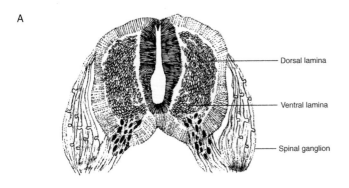

B

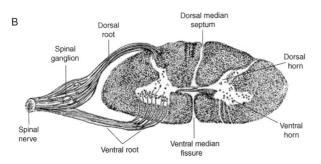

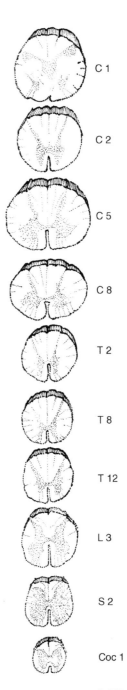

FIGURE 3.3. **A**, Transverse section of the human spinal cord at about 52 days (after Sensenig, 1951). Migration of neuroblasts into the alar and basal laminar areas foreshadows the adult condition shown in **B**. Definitive dorsal and ventral horns occur at the level of the cervical enlargement. Axons of these ventral horn neurons form the ventral root filaments that contribute to the formation of a spinal nerve. Dorsal horn neurons receive fibers from spinal ganglionic cells.

Because these neuronal populations have a longitudinal arrangement that parallels the long axis of the spinal cord, they are termed **gray columns**. Longitudinally oriented fiber bundles in the spinal cord are often termed 'white columns'. Thus the used of the term 'column' is often confusing. The term **funiculi** is a better term for the longitudinally oriented fiber bundles of the spinal cord with the term **column** best reserved for longitudinally oriented neuronal populations. Axons of neurons in the ventral horns emerge from the cord, form **ventral roots**, and contribute to the formation of **spinal nerves**. The spinal nerves enter skeletal muscles of the trunk and limbs and make contact with them at neuromuscular junctions at about 10 weeks of prenatal age.

3.1.3. Formation of Dorsal Gray Columns

Proliferation and aggregation of neuroblasts in the dorsal lamina leads to the formation of the sensory **dorsal gray column** that forms the **dorsal horn** of the adult cord (Fig. 3.3B). Spinal ganglionic cells in the intervertebral foramen have central processes that synapse with neurons of the dorsal horn.

FIGURE 3.4. Transverse sections of different spinal levels in humans showing the characteristic appearance of the gray and white matter (redrawn from Gray, 1973).

3.1.4. Dorsal and Ventral Horns vs. Dorsal and Ventral Gray Columns

The development of the dorsal and ventral laminae results in the formation of the dorsal and ventral gray columns of the adult cord and leads to an H-shaped configuration of the spinal gray matter characteristic of all levels of the spinal cord (Fig. 3.4). When viewed in transverse section, the dorsal and ventral gray columns are termed the **dorsal** and **ventral horns** (Fig. 3.3B). As the dorsal and ventral columns grow, the elongated central canal becomes a slitlike structure (Fig. 3.3B). Growth of the right and left alar and basal laminae toward the median plane indents the dorsal surface of the adult cord with a **dorsal median septum** (Fig. 3.3B) and the ventral surface of the cord with a **ventral median fissure** (Fig. 3.3B). Both are visible on the external surface of the adult cord.

3.1.5. Development of Neural Crest Cells

During neural tube formation, cells from the lateral edge of the neural folds emerge and migrate to a position on either side of the neural tube. These **neural crest cells** (Fig. 3.1A) form cranial, spinal, and autonomic nerves, their ganglia, and associated supportive cells. The human vestibular and cochlear ganglia, however, are from neural crest cells given off from the wall of the otic vesicle. The leptomeninges, especially the pia mater, which are part of the coverings of the brain and spinal cord, are from neural crest. Neural crest contributes to the formation of the cartilaginous skull, adjacent loose connective tissue, forms melanocytes in the skin, hair, and irides and neurosecretory cells. **Spinal ganglia** initially appear as aggregations of neurons dorsolateral to the spinal cord (Figs 3.2, 3.3). Central processes of the spinal ganglionic neurons collectively form **dorsal roots** (Fig. 3.3B) that pass to the alar lamina and synapse with dorsal horn neurons. Peripheral processes of the spinal ganglionic cells contribute to the formation of spinal nerves (Fig. 3.3B). Peripheral processes of neural crest-derived spinal ganglionic neurons travel in and contribute to the formation of spinal nerves. They transmit impulses from peripheral receptors to the spinal cord where such impulses may then enter the CNS. Each spinal nerve has a spinal ganglion with the possible exceptions of the first cervical nerve and the coccygeal nerve.

3.1.6. The Framework of the Adult Cord is Present at Birth

Neuronal cell bodies in the ventral horns send processes ventrolaterally from the cord to form the ventral roots. Spinal ganglionic neurons send processes into the cord where many synapse with neurons in the dorsal horns. Peripheral processes of these spinal ganglionic cells join ventral root filaments to form a typical spinal nerve (Fig. 3.5). Considerable **cellular and synaptic rearrangement** takes place in the alar laminae region in early

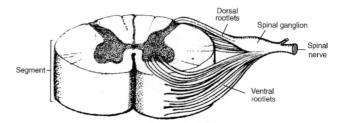

FIGURE 3.5. An individual human spinal segment showing the formation of a spinal nerve by the union of the dorsal and ventral roots. Note the presence of the spinal ganglion near the point at which the roots join.

development of nonhuman primates. Synaptic remodeling occurs through the death of certain neurons. Additional studies may identify other mechanisms responsible for the complex synaptic relationships in the spinal cord.

3.2. GROSS ANATOMY

3.2.1. Spinal Cord Weight and Length

A fresh spinal cord in adult humans weighs about 28.3 g (range = 23.5–35.8 g) and accounts for 2.06% of the weight of the CNS. A fresh brain weighing approximately 1400 g has a spinal cord weighing under 30 g. Spinal cord length measures 43.1 cm (range = 40.7–47.9 cm) with the adult male spinal cord (44.8 cm) slightly longer than the adult female cord (41.8 cm).

3.2.2. Spinal Segments, Regions, and Enlargements

The spinal cord is a thick, tubular structure with 2 enlargements, 5 regions, 31 segments, and 31 pairs of nerves. That part of the cord to which the dorsal and ventral rootlets of any one pair of spinal nerves attaches is a **spinal segment** (Fig. 3.5). Since the adult spinal cord has 31 pairs of spinal nerves (Fig. 3.6), there are also 31 spinal segments. Based on these segments, the human spinal cord consists of five distinct regions – **cervical, thoracic, lumbar, sacral,** and **coccygeal.** Each cervical region has eight segments and eight pairs of nerves, the thoracic region 12 segments and an equal number of pairs of nerves, the lumbar and sacral regions each have five segments and five pairs of nerves, and the coccygeal region has one to three vestigial segments and a similar number of pairs of nerves. A **cervical enlargement** (C4 to T1) is at the level of origin of those ventral roots contributing to the formation of the brachial plexus and innervating the

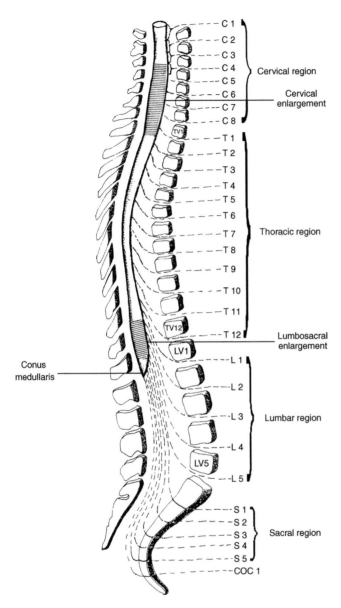

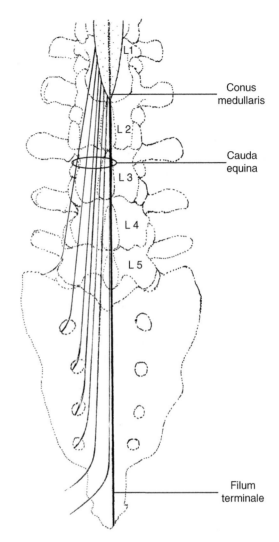

FIGURE 3.6. The human spinal cord in the vertebral canal. Note the 31 pairs of spinal nerves and the cervical (C4-T1) and lumbosacral (L2-S3) enlargement levels. The cord ends caudally in a cone-shaped region, the conus medullaris, at the level of the intervertebral disc between the first and second lumbar vertebrae.

FIGURE 3.7. Termination of the human spinal cord at the conus medullaris. A strand of non-neural tissue, the filum terminale, descends from the tail of the conus and anchors it to the coccygeal vertebrae. Some fibers of the cauda equina are visible. The view here is of these fibers projected over the lumbar and sacral vertebrae.

upper limb (Fig. 3.6). A **lumbosacral enlargement** (L2 to S3) is at the level of origin of those ventral roots contributing to the formation of the lumbosacral plexus and innervating the lower limb (Fig. 3.6).

3.2.3. Spinal Segments in Each Region are of Unequal Length

A cervical segment averages 12 mm in length (range = 7.2–14.3 mm), a thoracic segment 21 mm

(range = 12.8–25 mm), a lumbar segment 12.3 mm, and a sacral segment 7 mm in length (range = 4.7–8.4 mm). The cervical cord region accounts for 22% of the length of the cord, the thoracic region about 56%, and the lumbar region about 14% with 8% of the length of the cord composed of the sacral and coccygeal regions. Consequently, the thoracic region has more segments and is the longest, heaviest, and greatest region by volume.

3.2.4. Conus Medullaris, Filum Terminale, and Cauda Equina

The spinal cord gently tapers to end caudally in a cone-shaped region, the **conus medullaris** (Figs 3.6, 3.7).

A fibrocellular strand of non-neural tissue, the **filum terminale** (Fig. 3.7), descends from the conus and anchors it to the coccyx. The apex of the conus is the caudal end of the spinal cord. In the second month of development, the spinal cord is longer than the surrounding vertebral column. About the third prenatal month, the vertebral column begins to grow faster than the spinal cord causing the cord to end opposite the caudal margin of the second lumbar vertebra at birth. Such a differential rate of growth continues after birth and leads to an adult cord about a third shorter than the vertebral column. The resulting discrepancy between the spinal segments and the corresponding vertebrae is clinically significant. Because of the disparity in length of the spinal cord and the vertebral column, the lower-most roots extend beyond the caudal tip of the spinal cord as a bundle of fibers. Collectively this bundle of fibers is termed the **cauda equina** (Fig. 3.7). The **cauda equina** have an important relationship with the conus, filum, and appropriate vertebrae (Fig. 3.7). The inferior roots forming the adult sacral spinal nerves often achieve a length of more than 20 centimeters.

3.2.5. Termination of the Adult Spinal Cord

The spinal cord ends in adults opposite the intervertebral disk between the first and second lumbar vertebrae (Figs 3.6, 3.7) with a normal range of variation between the last thoracic and the third lumbar vertebra. The adult spinal cord reaches its level of termination about two months after birth, indicating that the vertebral column and the spinal cord grow at equal rates in childhood. Overall spinal cord length in humans increases 2.7 times before reaching its adult length of about 18 inches. In adults, the ratio between the length of the cord and the length of the vertebral column is 65:100.

3.2.6. Differential Rate of Growth: Vertebral Column vs. the Spinal Cord

One consequence of the differential rate of growth between the vertebral column and the spinal cord is that, as the dorsal and ventral rootlets elongate, they must pass obliquely from the cord in order to emerge from the intervertebral foramina. One way to appreciate the displacement of the spinal cord relative to the vertebral column and its influence on the spinal roots is by examining Figure 3.8. At two months of age, the rootlets of the first sacral nerve (Fig. 3.8), the S1 segment of the spinal cord, and the first sacral vertebra are all at the same level. The ganglion of the first sacral nerve remains fixed in

the intervertebral foramen between S1 and S2. As the vertebral column grows and lengthens, it takes the first sacral ganglion with it. The slower growing cord appears to ascend. Thus, the S1 spinal cord segment, initially at the level of the first sacral vertebra at two months of age (Fig. 3.8A) ends up opposite the second lumbar vertebra (Fig. 3.8D) at six months of age with elongated rootlets of the first sacral nerve. The ganglion of the first sacral nerve ends up in the intervertebral foramen between S1 and S2. Figure 3.8 also illustrates the change in level of the conus medullaris during this period of spinal cord and vertebral column growth.

3.2.7. Relationship between Spinal Segments and Vertebrae

To determine the injured spinal cord level after vertebral trauma, it is essential to understand the relation of the spinal segments to the bony vertebrae (Fig. 3.6). Cervical segments correspond closely to the appropriate cervical vertebrae even though there are eight cervical segments and a similar number of pairs of nerves but only seven cervical vertebrae. Thoracic segments and pairs of nerves appear out of position relative to the thoracic vertebrae, thus the spinous process of vertebra T8 is opposite spinal segment T10. All lumbar segments align with T11 and T12 vertebrae. Sacral and coccygeal segments are level with the first lumbar vertebral body.

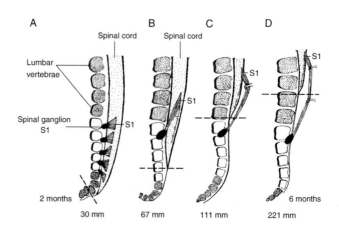

FIGURE 3.8. Topographical relations at the caudal end of the spinal cord in a series of human fetuses from two months to six months of age (A–D). Note that the spinal ganglion of S1 remains fixed in the intervertebral foramen between the first and second sacral vertebrae. As the vertebral column grows, it pulls the S1 spinal ganglion with it. Growing at a slower pace, the cord appears to ascend. Initially at the level of the first sacral vertebra, the S1 segment ends up opposite the second lumbar vertebra. The caudal tip of the spinal cord is indicated by the dashed line (after Streeter, 1919).

3.3. NUCLEAR GROUPS – GRAY MATTER

3.3.1. General Arrangement of Spinal Cord Gray Matter

The **gray matter** of the spinal cord is a well-organized population of neurons with a centrally located, H-shaped configuration (Fig. 3.9) surrounded by numerous bundles of fibers forming the **white matter of the spinal cord**. Smaller bundles of fibers called **funiculi** are distinguishable in the white matter. The adult spinal cord includes a **dorsal funiculus, ventral funiculus** and a **lateral funiculus** (Fig. 3.9). Connecting right and left halves of the gray matter at all spinal levels is a transverse **gray commissure** with a **central canal** (Fig. 3.9). Each half of the gray matter has dorsal and ventral gray columns arranged in a longitudinal manner paralleling the long axis of the spinal cord. When viewed in transverse section these dorsal and ventral gray columns are termed **dorsal and ventral horns** (Fig. 3.9). Their shape varies from one spinal segment to the next (Fig. 3.4) reflecting differences in the arrangement and size of the nuclear groups (neuronal populations) that form the gray columns. In addition to dorsal and ventral horns, transverse sections of spinal segments T1 to L1 or L2 and S2 to S4 have a **lateral horn** (Fig. 3.9) formed by a longitudinal column of neurons that occupies a position lateral and dorsolateral to the central canal on either side.

3.3.2. Gray Matter at Enlargement Levels

Because the limbs have more skeletal muscle than the thoracic wall, spinal segments that supply the limbs have a much larger ventral horn than do those segments that supply only trunk musculature (Fig. 3.4). The presence of a greatly enlarged ventral horn at limb levels causes grossly visible enlargements. The **cervical enlargement** (C4 to T1) is at the level of origin of ventral roots contributing to the formation of the brachial plexus and innervating the upper limb. The **lumbosacral enlargement** (L2 to S3) is at the level of origin of ventral roots contributing to the formation of the lumbosacral plexus and innervating the lower limb (Fig. 3.6). The area of skin innervated by sensory fibers is also greater at limb levels than at trunk levels. Consequently, more spinal ganglionic neurons are available to provide a larger number of sensory fibers to this larger area of skin that leads to a larger dorsal horn at limb levels.

Nuclear groups of the spinal cord, arranged in longitudinal columns, attain an adult configuration by prenatal week 14. The presence or absence of a particular neuronal group at a given cord level imparts a distinct appearance to each spinal segment. When viewing transverse sections of the cord, nuclear groups of the dorsal, ventral, and when present, lateral horn are used to identify a given segment. At certain levels, the dorsal horn has four identifiable nuclear groups, the lateral horn two, and the ventral horn as many as five definable nuclear groups. Such morphological variability is a reflection of functional differences. Table 3.1 lists certain nuclear groups in the human spinal cord and their approximate longitudinal extent. The following description presents nuclear groups in the human spinal cord as they appear in transverse section.

3.3.3. Spinal Laminae

A scheme of organization of the cat spinal gray matter, called **spinal laminae** (of Rexed), is applicable to the human spinal cord (Fig. 3.10). Using the method of Rexed, gray matter of the human spinal cord is partitioned into ten layers or laminae designated by Roman

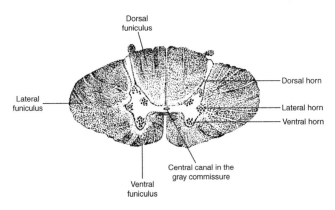

FIGURE 3.9. A transverse section of the human spinal cord at the sixth thoracic level. Note the lateral horn between the dorsal and ventral horns formed by preganglionic sympathetic neurons from T1-L1 or L2 spinal segments. Also, note the large divisions of the white matter.

TABLE 3.1. Certain nuclear groups in the human spinal cord and their approximate longitudinal extent

Nuclear group	Type of cells	Longitudinal extent
Intermediolateral column	Sympathetic neurons	from T1 to L1 or L2
Sacral parasympathetic nucleus	Parasympathetic nucleus	from S2 to S4
Onuf's nucleus	Ventral horn	S1, S2, S3

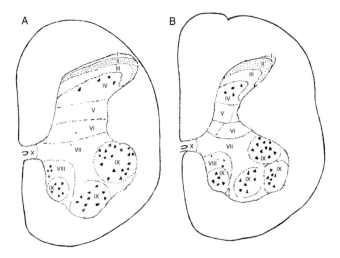

FIGURE 3.10. Two transverse sections of the human spinal cord at the level of the (**A**) first sacral and the (**B**) seventh cervical segments. Roman numerals indicate the spinal laminae of Rexed applied to the human spinal cord. Spinal lamina X includes the gray matter around the central canal (redrawn from Chambers and Liu, 1972).

numerals I–X (Fig. 3.10). The **marginal nucleus** at the apex of the dorsal horn is **spinal lamina I**, the head of the dorsal horn, corresponding to the **substantia gelatinosa**, is **spinal lamina II**, and the neck of the dorsal horn is designated **spinal laminae III, IV, and V**. The **nucleus proprius** corresponds to **spinal laminae III and IV**. The base of the **dorsal horn** corresponds to **spinal lamina VI** and the ventral horn is designated **spinal laminae VII to IX**. Gray matter around the central canal is **spinal lamina X**.

3.3.4. Dorsal Horn

The dorsal horn of the spinal cord corresponds to spinal lamina I–VI. At the apex of the dorsal horn is the **marginal nucleus** or spinal lamina I. At the head of the dorsal horn, extending the length of the spinal cord and being largest at limb levels is the **substantia gelatinosa** that corresponds to spinal lamina II. Beneath the substantia gelatinosa, and in the neck of the dorsal horn at all cord regions, is the **nucleus proprius** corresponding to spinal lamina III and IV. The **thoracic nucleus**, on the medial side of the base of the dorsal horn, extends from C7 to L2 spinal levels. Some neurons in this nucleus normally have a chromatolytic appearance. The **secondary visceral gray substance** is often difficult to identify at the base of the dorsal horn, to the lateral side of the thoracic nucleus. It extends from T1 to L2 and from S2 to S4 spinal levels. The secondary visceral substance resembles to some extent the substantia gelatinosa.

3.3.5. Lateral Horn

The lateral horn is formed by at least two neuronal groups; an **intermediolateral nucleus**, composed of preganglionic sympathetic neurons extending from T1 to L1 or L2, and a **sacral parasympathetic nucleus** extending from S2 to S4. The intermediolateral nucleus forms an extension of gray matter lateral to the central canal called the **lateral horn**. Some of these neurons retain a position near the central canal as an **intermediomedial nucleus**. These intermediolateral and the intermediomedial nuclei have neurons of origin for the preganglionic fibers of the sympathetic division of the autonomic nervous system. The **sacral parasympathetic nucleus** does not have medial and lateral subdivisions nor does it form a lateral horn. Fibers from it travel in the S2 to S4 ventral roots to synapse with postganglionic neurons in, on, or near pelvic structures.

3.3.6. Ventral Horn

The ventral horn of the spinal cord is divisible into **medial**, **lateral**, and **central** divisions. These divisions can be broken into subdivision formed by neuronal clusters that innervate small groups of muscles. These subdivisions also represent a functional pattern. All major nuclear groups of the ventral horn have large multipolar neurons with coarse chromatophil (Nissl) substance characteristic of efferent neurons. Axons of these ventral horn neurons enter the ventral root.

Medial Division of the Ventral Horn

The **medial division of the ventral horn** is evident at all cord levels. At regions other than the enlargements, it occupies almost the entire ventral horn. At enlargement levels, it is in the medial parts of the gray matter. Functionally this division is concerned with innervation of the neck, trunk, intercostal, and abdominal muscles.

Lateral Division of the Ventral Horn

The **lateral division of the ventral horn** provides innervation to the limbs, is at cervical and lumbosacral enlargement levels, appears as a lateral extension of the ventral horn, and has three nuclei. The **ventrolateral nucleus**, present from about C4 to C8 and L2 to S2, innervates muscles of the shoulder girdle, arm, and those of the hip and thigh. The **dorsolateral nucleus**, supplying muscles of the forearm and hand as well as the leg and foot, extends from C4 to T1 and L2 to S2. The **retrodorsolateral nucleus** consists of large neurons from

C8 to T1 and S1 to S3 that innervate intrinsic muscles of the hand and the corresponding small muscles of the foot that move the toes.

Central Division of the Ventral Horn

The **central division of the ventral horn** consists of at least two, but perhaps three, cell groups in the middle part of the ventral horn. These groups are the **phrenic nucleus** and the **accessory nucleus**. The phrenic nucleus extends from C3 to C5 though there is a difference of opinion about these levels. The phrenic nucleus innervates the diaphragm. Often, in the uppermost cervical region, there is a slight expansion of the ventral horn gray matter in a dorsolateral direction. In this area, neurons of the **accessory nucleus** are present. The **accessory nucleus** extends from C1 to about C5 or C6. The caudal part of the **accessory nucleus** supplies the trapezius and the rostral part the sternomastoid muscle. After emerging from the cord, the **spinal part of the accessory nerve** [XI] ascends in the spinal canal, passes through the foramen magnum, and joins the **cranial part of the accessory nerve** [XI]. The **accessory nerve**, now containing both spinal and cranial parts, leaves the skull through the jugular foramen. Inferior to the foramen the cranial part of the accessory nerve separates from the accessory nerve proper and is distributed with, and 'accessory' to, the vagal nerve [X].

The **lumbosacral nucleus**, also called the **central nucleus**, extends from L5 to S2 where it consists of a cluster of neurons lying between the lateral and medial divisions of the ventral horn. Though its peripheral distribution is uncertain, it may be responsible for the innervation of the pelvic diaphragm.

An unusual nuclear group along the ventral margin of the human ventral horn of the second sacral segment and extending from the inferior part of S1 to the upper part of S3 is the **nucleus of Onuf**. Fiber-stained sections of the human sacral cord reveal a sparsity of myelinated fibers in the neuropil surrounding this nucleus that facilitates its identification. This nucleus innervates striated muscles of the external anal and the external urethral sphincter in humans.

3.4. FUNCTIONAL CLASSES OF NEURONS

Columns of neurons characteristic of the human spinal cord give each level of the cord a characteristic shape with a possibility of four functional classes of neurons at any level. Since each category of neuron may provide

processes into a spinal nerve, any spinal nerve may be composed of from one to four functional types of fibers. Thus, the concept of the 'functional components of a spinal nerve' has as its basis the functional class of neuronal cell bodies in the spinal cord that contributes fibers to the spinal nerve.

3.4.1. Four Classes of Neurons in the Spinal Cord

The four classes of neurons in the spinal cord are **general somatic afferent** (GSA), **general visceral afferent** (GVA), **general somatic efferent** (GSE), and **general visceral efferent** (GVE). Sensory or **afferent neurons** conduct impulses toward their neuronal soma. The dorsal horn (alar lamina) receives sensory impulses and forms the sensory neuronal pool of the spinal cord. Motor or **efferent neurons** conduct impulses away from their cell body to muscles or glands to cause contraction or secretion. The ventral horn (basal lamina), with the lateral horn, forms the efferent neuronal pool of the spinal cord.

3.4.2. General Somatic vs. General Visceral Afferent Neurons

General **somatic** afferent neurons conduct impulses for pain, temperature, touch, pressure, and proprioception from areas of skin, connective tissue, muscles, tendons, and articular joints. General **visceral** afferent neurons conduct impulses from the large organs inside the body (viscera), including the heart, stomach, lungs and intestines and their associated blood vessels. Most dorsal horn neurons are secondary general somatic afferent neurons receiving sensory impulses from primary general somatic afferent neurons in the spinal ganglia. General visceral afferent neurons, on the lateral aspect of the base of the dorsal horn, form the **secondary visceral gray substance** from T1 to L1 or L2 and S2 to S4. These visceral neurons receive sensory impulses from primary general visceral afferent neurons in spinal ganglia.

3.4.3. General Somatic vs. General Visceral Efferent Neurons

Somatic efferent neurons supply striated, skeletal muscles; visceral efferent neurons supply nonstriated muscle, cardiac muscle, and glands. Much of the ventral horn consists of general somatic efferent neurons. General visceral efferent neurons in the lateral horn form the intermediolateral nucleus in segments T1 to

L1 or L2 (Table 3.1) or the **thoracolumbar** (sympathetic) **part of the autonomic nervous system** (ANS). The sacral parasympathetic nucleus, a collection of general visceral efferent neurons from S2 to S4, is also part of the lateral horn. Its neurons form the **sacral division of the craniosacral** (parasympathetic) **part of the autonomic nervous system**.

The dorsal horn includes general somatic afferent and general visceral afferent neurons, the lateral horn has only general visceral efferent neurons, and the ventral horn has only general somatic efferent neurons. These four neuronal groups provide the cell bodies of origin for all spinal nerves. Thus, a given **spinal nerve** (Fig. 3.6) may have from one to four kinds of functional fibers depending on the neurons of origin from which its fibers arise (usually efferent) or to which its fibers project (usually afferent). The overwhelming majority of neurons in the spinal gray matter are interneurons that do not belong to these specific functional categories.

3.4.4. Some Ventral Root Axons are Sensory

A basic concept of neurology, established in the early 1800s, is the **Bell–Magendie law**. This hypothesis claims that the dorsal spinal roots are sensory and the ventral roots are motor. The Bell–Magendie law would no longer be tenable based on the recognition that thinly myelinated efferents, from neuronal cell bodies in or near the lateral horn, leave the cord in the dorsal roots. In humans, about 27% of ventral root axons are nonmyelinated and probably sensory. Cell bodies of these fibers are in spinal ganglia. Dorsal horn sensory neurons may therefore receive impulses through nonmyelinated fibers that entered by the ventral roots. Perhaps a revised Bell–Magendie law could state that the dorsal horns of the spinal cord are sensory and the ventral horns are motor.

3.5. FUNICULI/FASCICULI/TRACTS – WHITE MATTER

Surrounding the gray matter of the spinal cord, with its distinct neuronal columns, are numerous longitudinally oriented myelinated fibers that form the **white matter** of the spinal cord. In humans, myelination takes place in spinal cord fibers until about the seventh prenatal month. The white matter of the spinal cord is divisible into three large bundles of fibers called **funiculi**. These are the **dorsal, ventral,** and **lateral funiculus** (Fig. 3.9). These three funiculi have smaller bundles called **fasciculi**

or **tracts**. By definition, a tract or fasciculus is a bundle of fibers in the CNS with a common origin and termination, such as the **fasciculus** gracilis (FG) or the lateral spinothalamic **tract** (LST) (see Fig. 7.1). A neuronal path, however, is a chain of synaptically linked neurons that serve a particular function (see Fig. 6.8). For example, the path for discriminatory tactile impulses from the toes has several components linked together: the fasciculus gracilis (a tract), the medial lemniscus (another tract), and thalamocortical projections (a third tract). Other paths carry impulses from peripheral body areas to the spinal cord, brain stem, thalamus, and cerebral cortex. Fibers from motor neurons in the cerebral cortex or from subcortical areas may pass without interruption to the spinal cord. Among the many ascending tracts in the spinal cord are those carrying impulses for pain, temperature, touch, pressure, proprioception, and vibration. Also ascending in the spinal cord are tracts that carry impulses to the cerebellum.

Descending in the spinal cord are descending motor tracts that influence skeletal muscles plus tracts carrying impulses for autonomic reflexes and the regulation of autonomic activity. Propriospinal fibers are long and short fibers that connect the cervical enlargement with the lumbosacral enlargement. In nonhuman primates, there are long descending propriospinal fibers arising from neurons in spinal laminae I, IV to VIII, and X of the cervical enlargement. Numerous, short, ascending and descending intersegmental tracts connect adjacent spinal segments for intrinsic and intersegmental activity and reflexes. The precisely arranged spinal gray matter along with the ascending and descending tracts impart a distinctive appearance to each segment of the spinal cord.

Tracts in funiculi that traverse the spinal cord have names that reflect their origin, position in the spinal cord, and site of termination. The **lateral spinothalamic tract** (LST), carrying impulses for pain and temperature, travels in the **lateral** funiculus, originates at **spinal** levels, and terminates in the **thalamus** (see Fig. 7.1). Other tracts carry the name of the investigator who first described the path or was associated with early studies of the path. In 1855, a German neurologist by the name of Heinrich Lissauer described the **dorsolateral tract,** which has the alternative name of the 'tract of Lissauer'. Those interested in the meaning of neuroanatomical eponyms are encouraged to consult the reference book by Lockard listed at the end of this chapter. Several tracts have names reflecting their shape or contour such as the wedge-shaped, fasciculus cuneatus [Latin, wedge].

A well-organized system of tracts allows the spinal cord to contribute to the functioning of the nervous system. Appreciation of the origin, course, and termination of certain tracts in the spinal cord is essential to understanding the organization of the central nervous system. Knowledge of a tract, its relation to specific spinal nuclei, and to other paths is of tremendous value in making an anatomical diagnosis of a neurological disease. Such information is useful in determining the cause of the disease and ultimately in determining a course of therapy.

3.6. SPINAL REFLEXES

Although the nervous system is a functional entity, it is also appropriate to speak of different 'levels' of function. Function occurs at cortical, subcortical, thalamic, brain stem and spinal cord levels. Although each level executes certain activities of its own, these activities are modifiable by, or correlated with, activities at other levels. The highest level of neural functioning, the cerebral cortex plays an essential role in modifying or regulating neural activity at all other levels.

When a spinal segment receives sensory impulses, it may respond by initiating a well-localized movement called a **spinal reflex**. Reflex response to a stimulus is the primary functional unit of activity at the spinal level. Figure 3.11 illustrates the structural elements in a spinal segment that provide for a simple reflex response. A sharp tap on the patellar tendon, with the knee slightly

flexed, causes contraction of the extensor muscle in the thigh causing the leg to kick forward. In the **knee jerk reflex** (also termed the patellar tendon or quadriceps reflex), tapping the patellar tendon is the initial stimulus suddenly stretching its muscle, activating a specialized receptor in that muscle. The receptor transforms this stimulus into a nerve impulse that travels in the peripheral process of a spinal ganglionic neuron, to that neuron and then toward the spinal cord in the central process of the same spinal ganglionic neuron. As the central process of the spinal ganglionic neuron enters the dorsal horn, it relays the impulse to a motoneuron in the ventral horn that then transmits the impulse to the quadriceps femoris, causing it to contract in response to the initial stimulus. Although only one spinal segment is illustrated, it is probable that neurons in the third and perhaps fourth lumbar segments are also involved in the quadriceps reflex.

Careful examination of spinal reflexes is basic to assessing the function of individual spinal segments. The quadriceps reflex and other spinal reflexes involve a receptor, sensory neuron, and motoneuron. If any part of the reflex is injured, interfering with the conduction of impulses, the reflex response lessens or is absent. Injury may involve the receptors, peripheral or central process of the spinal ganglion, the spinal ganglion itself, ventral horn motoneuron, ventral root, peripheral nerve, neuromuscular junction, or the muscle itself. Descending paths may influence spinal reflexes. Increased reflex activity may appear as repetitive muscle contractions with intervening relaxations. Exaggeration of such reflexes may result from a loss of inhibitory influence by higher levels of the nervous system. Table 3.2 lists a number of clinically useful deep and superficial spinal cord reflexes and their segmental level of representation.

Most of the spinal cord reflexes listed in Table 3.2 involve tapping of tendons to stimulate stretch receptors in associated muscles. When a muscle stretches beyond a certain point, the **muscle stretch** or **myotatic reflex** causes it to tighten and attempt to shorten. You may feel this type of tension during stretching exercises. Since receptors that receive the initial stimuli involved in myotatic reflexes are deep to the skin, the term **deep reflex** may be used. Myotatic reflexes involve only two neurons and a single synapse between them. Thus, they are also termed **monosynaptic reflexes**. **Superficial reflexes** involve stimulation of receptors in or near skin. Complex reflexes or reactions that involve the participation of receptors, sensory neurons, interneurons, and motoneurons are **polysynaptic**. Many responses, miscellaneous neurological signs, defense, postural and righting reflexes, plus other reactions are polysynaptic.

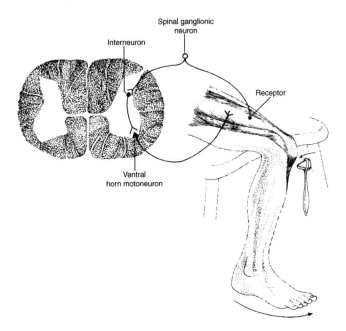

FIGURE 3.11. Connections in a segment of the human spinal cord that provide for a simple reflex response. As the patellar tendon is tapped, its muscle is stretched. From receptors in the muscle, this quadriceps reflex is then initiated causing extension of the leg.

TABLE 3.2. Certain useful human spinal cord reflexes and their segmental level of representation

Deep reflexes	Segments	Site of stimulus	Normal response
Biceps	C5–C6	Biceps tendon	Flexion of forearm or muscle twitch without movement
Brachioradialis	C5–C6	Brachioradialis	Flexion of forearm tendon with supination
Triceps	C6–C7	Triceps tendon	Extension of forearm or muscle twitch without movement
Quadriceps	L3–L4	Patellar tendon	Extension of leg
Triceps surae	L5–S1	Tendo calcaneus	Plantar flexion of foot or twitch of the triceps surae
Superficial reflexes	Segments	Site of stimulus	Normal response
Upper abdominal	T8–T9	Skin of upper abdominal quadrant	Contraction of abdominal muscles
Lower abdominal	T11–T12	Skin of lower abdominal quadrant	Contraction of abdominal muscles
Cremasteric	L1–L2	Skin on upper, medial thigh	Elevation of testis
Plantar	S1–S2	Plantar surface of the foot	Flexion of toes

Although there is overlap in these different categories, many are of diagnostic significance.

Spinal reflexes are part of a steady state of activity that underlies the normal functioning of the nervous system. Monosynaptic reflexes, essential to the control of muscle tone, act to prevent injury by overstretching tendons or muscles, and are essential in maintaining upright posture for walking, and in limb movements. They act as a feedback mechanism allowing for the establishment of a delicate balance between tendon stretch and muscle contraction while permitting the muscle to meet any demand by maintaining a normal level of tone and activity. Spinal reflexes thus provide the structural basis for all spinal cord motor activity.

All necessary elements of a functional spinal reflex (e.g. sensory and motoneurons, the muscle, and the connections between these structures) are completed at or before the eighth prenatal week. By the end of that week, the numbers of axodendritic synapses substantially increase. Axosomatic synapses rapidly proliferate between 10½ and 13 postovulatory weeks. In both instances, the appearance of these synapses coincides with the early appearance of behavioral changes in fetuses. Appearance of reflexes at certain fetal stages, their presence at birth, and their presence or disappearance at appropriate times in infancy and childhood, are useful indicators of normal growth and development of the nervous system.

3.7. SPINAL MENINGES AND RELATED SPACES

3.7.1. Spinal Dura Mater

Three membranes or **meninges** surround and protect the spinal cord. These include an external **spinal dura**

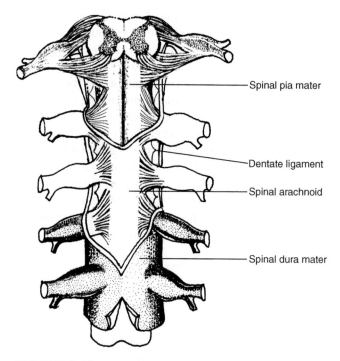

FIGURE 3.12. Ventral aspect of part of the human spinal cord. Three layers of meninges that cover the cord are shown including the denticulate ligaments.

Spinal pia mater

Dentate ligament

Spinal arachnoid

Spinal dura mater

mater (Fig. 3.12) of dense fibrous tissue continuous with the inner layer of the intracranial dura. Dura mater surrounds the spinal ganglia and nerves as they traverse the intervertebral foramina. The dura envelopes the spinal cord and ends as the **dural sac** at the level of sacral vertebra S2 or S3. Spinal dura mater is relatively avascular in contrast to cranial dura. Deep to the spinal dura mater is the **subdural space**. In the fetus, there is no separation of the spinal dura from the arachnoid. A subdural space appears after birth. An **epidural space**, external to the spinal dura mater, separates the cord and its coverings from the bony vertebral walls.

Filled by loose connective tissue, fat, and a plexiform venous bed, the internal vertebral venous plexuses communicate with veins in the cervical, thoracic, and abdominal regions.

3.7.2. Spinal Arachnoid

Inside the spinal dura mater, and in continuity with the cerebral arachnoid, is a delicate membrane called the **spinal arachnoid** (Fig. 3.12) consisting of fine elastic tissue with a smooth internal surface that loosely invests but does not adhere to the spinal surface. Extending from the inner surface of the arachnoid to the deepest membrane, the spinal pia mater, are delicate strands of connective tissue, the **arachnoid trabeculae**. Beneath the arachnoid, and between it and the spinal pia mater, is the **subarachnoid space** that contains **cerebrospinal fluid** (CSF). Cerebrospinal fluid in the spinal subarachnoid space communicates freely with that in the cranial subarachnoid space and that around the spinal roots. Studies of the flow of cerebrospinal fluid indicate considerable variations in spinal fluid dynamics.

Arachnoid villi occur on nearly every spinal root in the thoracic and lumbar regions. These small fingerlike protrusions of arachnoid tissue are internal to the spinal dura mater, extend into it, or penetrate it. Epidural venous plexuses are intimately associated with these arachnoid villi. Such a close relationship allows cerebrospinal fluid to leave the subarachnoid space through the arachnoid villi and enter the venous system.

3.7.3. Spinal Pia Mater

The deepest meningeal layer surrounding the spinal cord, the **spinal pia mater** (Fig. 3.12), is intimately bound to the spinal cord surface, follows its every contour, and has numerous blood vessels. These vessels are vulnerable to injury during the performance of a lumbar puncture. If these vessels are injured, blood accumulates in the subarachnoid space. Human neural crest cells take part in the early formation of the pia mater.

Denticulate or Dentate Ligaments

Condensations of pial tissue are on either side of the cervical and thoracic cord between the dorsal and ventral rootlets. These meningeal specializations, the **denticulate ligaments** (Fig. 3.12), serve to fix the lateral aspect of the cord to the inside of the spinal dura mater yet allow movement of the cord in an anterior-posterior direction. Maximum flexion of the head directs the spinal cord to the anterior part of the vertebral canal. The spinal cord, dura, and spinal roots move up and down in the vertebral canal in flexion or extension of the neck. Such movements are about 1.8 cm at the level of spinal roots C8–T5. Flexion causes the length of the vertebral canal to increase, stretching the spinal cord and spinal dura mater, especially between C2–T1. The dentate ligaments may transmit axial stress or tension between the cord and spinal dura mater. Surgical section of the dentate ligaments, in the face of adherence of the cord to the floor of the vertebral canal because of disease, is a method of surgical treatment to reduce the resulting tension.

3.8. SPINAL CORD INJURY

In the United States, some 220,000 to 285,000 people are living with spinal cord injuries with nearly 11,000 new cases every year. While there are a number of experimental models of the acute phase (the first 48 hours) of spinal cord injury and a great deal of neurophysiological, morphological, and biochemical data available on this topic, there is no satisfactory clinical treatment for this phase of spinal cord injury. Treatment may involve the use of corticosteroids, osmotic diuretics, antiadrenergic compounds, or other miscellaneous drugs, enzyme therapy, hyperbaric oxygenation, hypothermia, or fetal implants. Development of a successful treatment for spinal cord injury will bring relief to thousands of patients and their families.

3.8.1. Transverse Hemisection of the Spinal Cord (Brown-Séquard Syndrome)

Transverse hemisection of the spinal cord (injury to half of the spinal cord) (Fig. 3.13), is termed the Brown-Séquard syndrome (named for Charles Edouard Brown-Séquard (1817–1894), who is known for his work in delineating the sensory pathways in the spinal cord). This condition may arise from unilateral compression or penetrating trauma that leads to hemisection of the spinal cord. The principal structures affected include ascending and descending paths as well as motoneurons at the level of injury. This includes the descending lateral and ventral corticospinal paths (which have already decussated) and the ascending pain and temperature path (from the contralateral side). Also involved are the ascending gracile and cuneate fasciculi (carrying discriminative touch, pressure, proprioceptive and vibratory sensation from the site of the injury) and the ventral horn lower motor neurons at that level and on that site of injury. The accompanying symptoms include an ipsilateral motor and sensory loss (discriminative touch, pressure, proprioception, and vibration) on the site of the lesion along with a contralateral loss of pain and temperature.

3.8.2. Syringomyelia

The abnormal development of a cyst in the central part of the upper cervical cord (Fig. 3.14) is termed **syringomyelia**. This cyst, called a syrinx, is lined with glia and may expand and elongate over time destroying tissue and impinging on spinal cord gray and white matter. The central canal is not necessarily the primary site in all cases of syringomyelia. About 21,000 American men and women have syringomyelia. Because of the oblique arrangement of fibers in the dorsal funiculi according to the levels at which they arise and the kind of impulses carried by them, such an injury near

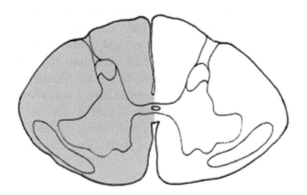

FIGURE 3.13. Transverse section of the human spinal cord depicting the location of a lesion (dark area of half the cord) leading to a Brown-Séquard syndrome.

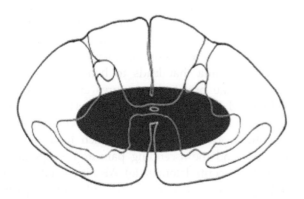

FIGURE 3.14. Transverse section of the upper cervical spinal cord depicting the location of the syrinx in syringomyelia (hatched area).

the central canal at upper cervical levels affects fibers transmitting vibratory impulses at several levels of the body, even though the injury is at a single spinal level. If the injury expands into the dorsal funiculi, fibers transmitting proprioceptive impulses become involved, as do fibers from additional levels of the body. Secondary fibers giving rise to the lateral spinothalamic tract and carrying painful and thermal impulses decussate in the ventral white commissure ventral to the central canal. Hence, in syringomyelia of the upper cervical cord, there may be loss of pain and temperature in both upper limbs in addition to dorsal funicular signs.

3.9. BLOOD SUPPLY TO THE SPINAL CORD

A thorough knowledge of the spinal cord blood supply is of paramount importance in understanding the signs and symptoms caused by vascular disease of the cord. More information on this subject is in Chapter 23.

FURTHER READING

Angevine JB Jr (1973) Clinically relevant embryology of the vertebral column and spinal cord. Clin Neurosurg 20:95–113.

Augustine JR, White JF (1986) The accessory nerve nucleus in the baboon. Anat Rec 214:312–320.

Barson AJ (1970) The vertebral level of termination of the spinal cord during normal and abnormal development. J Anat 106:489–497.

Barson AJ, Sands J (1977) Regional and segmental characteristics of the human adult spinal cord. J Anat 123:797–803.

Blinkov SM, Glezer II (1968) *The Human Brain in Figures and Tables: A Quantitative Handbook.* Basic Books, New York.

Carlstedt T, Cullheim S, Risling M (2004) Spinal cord in relation to the peripheral nervous system. In: Paxinos G, Mai JK, eds. *The Human Nervous System.* 2nd ed. Chapter 9, pp 250–263. Elsevier/Academic Press, Amsterdam.

Chambers WW, Liu C-N (1972) Anatomy of the spinal cord. In: Austin GM, ed. *The Spinal Cord: Basic Aspects and Surgical Considerations.* 2nd ed. pp 5–56. Thomas, Springfield.

Coggeshall RE (1979) Afferent fibers in the ventral root. Neurosurgery 4:443–448.

Crosby EC, Humphrey T, Lauer EW (1962) *Correlative Anatomy of the Nervous System.* Macmillan, New York.

DeArmond SJ, Fusco MM, Dewey MM (1989) *Structure of the Human Brain.* 3rd ed. Oxford University Press, New York.

Frigon A, Rossignol S (2006) Functional plasticity following spinal cord lesions. Prog Brain Res 157:231–260.

Kido DK, Gomez DG, Pavese AM Jr, Potts DG (1976) Human spinal arachnoid villi and granulations. Neuroradiology 11:221–228.

Rexed B (1964) Some aspects of the cytoarchitectonics and synaptology of the spinal cord. Prog Brain Res 11:58–92.

Riley HA (1943) *An Atlas of the Basal Ganglia, Brain Stem and Spinal Cord: Based on Myelin-Stained Material*. Williams & Wilkins, Baltimore.

Routal RV, Pal GP (2000) Location of the spinal nucleus of the accessory nerve in the human spinal cord. J Anat 196:263–268.

Schoenen J, Faull RLM (2004) Spinal cord: cyto- and chemoarchitecture. In: Paxinos G, Mai JK, eds. *The Human Nervous System*. 2nd ed. Chapter 7, pp 190–232. Elsevier/Academic Press, Amsterdam.

Schoenen J, Grant G (2004) Spinal Cord: Connections. In: Paxinos G, Mai JK, eds. *The Human Nervous System*. 2nd ed. Chapter 8, pp 233–249. Elsevier/Academic Press, Amsterdam.

Standring S (2004) Editor-in-chief. *Gray's Anatomy*. 39th ed, Section 2, Neuroanatomy: Chapter 18, Spinal Cord, pp 307–326. Elsevier/Churchill Livingstone, London.

Tatarek NE (2005) Variation in the human cervical neural canal. Spine J 5:623–631.

Vandenabeele F, Creemers J, Lambrichts I (1996) Ultrastructure of the human spinal arachnoid mater and dura mater. J Anat 189:417–430.

Wolpaw JR (2006) The education and re-education of the spinal cord. Prog Brain Res 157:261–280.

Within the brain stem are concentrated all the motor and sensory paths passing to and from the cerebral hemispheres, the nuclei of most of the cranial nerves, also centers controlling respiration, circulation, bladder and rectal sphincters and deglutition. Because of the great number of functions so closely grouped into such a small region, the localization of lesions to the brain stem is rarely difficult.

Walter E. Dandy (1886–1946)

The Brain Stem

Later in the third week of human development, three main divisions of the brain (encephalon) are recognizable in the neural folds: these include the **prosencephalon** (forebrain), **mesencephalon** (midbrain) and **rhombencephalon** (hindbrain). As these three main divisions change into five secondary brain subdivisions, the **mesencephalon** undergoes slight change and remains as the **midbrain** whereas the **rhombencephalon** (hindbrain) transforms into the **metencephalon** and the **myelencephalon**. The **metencephalon** becomes the **pons** and the **cerebellum**; the myelencephalon develops into the **medulla oblongata**. The **brain stem** comprises the **midbrain, pons,** and **medulla oblongata** (Fig. 4.1);

these constituents are identifiable by about five weeks of development.

4.1. EXTERNAL FEATURES

4.1.1. Medulla Oblongata

The most caudal part of the brain stem, the medulla oblongata, is continuous caudally with the spinal cord and rostrally with the pons. The pyramids, inferior olive, and the attachment of cranial nerves VIII, IX, X, XI, and XII are conspicuous external features of the medulla

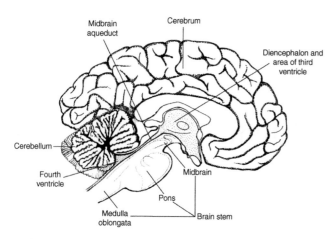

FIGURE 4.1. Medial surface of a human cerebral hemisphere including a median section of the diencephalon, brain stem, and cerebellum. The area of the fourth ventricle, aqueduct of the midbrain and third ventricle is lightly shaded. Note the midbrain interposed between the cerebral hemispheres and diencephalon superiorly and the remainder of the brain stem inferiorly.

oblongata. The **fourth ventricle** and part of the **central canal** are associated with the medulla oblongata.

Ventral Surface of the Medulla Oblongata

Since the brain stem is bilaterally symmetrical, structures on one side only are included in this description. A **ventral median fissure** (Fig. 4.2), defines the ventral surface of the medulla oblongata as it runs from the medulla/cord junction to the medulla/pons junction. At the caudal border of the pons, pontocerebellar fibers passing from the pons to the cerebellum delineate the lower pons from the upper medulla oblongata. The ventral median fissure continues caudally to the spinal cord where it ends, as fibers of the decussation of the pyramids (motor decussation) (Fig. 4.2) cross obliquely from one side of the brain stem to the other. The caudal limit of the medulla oblongata on the ventral surface of the brain stem is the middle of these decussating fibers of the motor decussation. The **pyramids** (Fig. 4.2) are protuberances lateral to the ventral median fissure, on either side, formed by underlying fiber bundles, the **pyramidal tracts**, originating in the cerebral cortex and ending in the spinal cord. Because of the **decussation of the pyramids** (motor decussation), one side of the brain is said to control or influence the opposite side of the body for voluntary motor activity. Each level of the brain stem has characteristic external structural features that result from the presence internally of specific nuclear groups or fiber bundles.

Adjoining each pyramid is an oval mass called the **inferior olive** (Fig. 4.2). The underlying inferior olivary nucleus gives rise to this conspicuous surface feature. A **ventrolateral sulcus,** normally occupied by the exiting

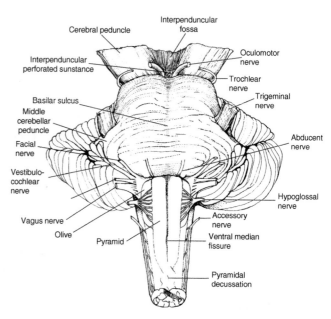

FIGURE 4.2. Ventral surface of the human brain stem with the cranial nerves labeled.

rootlets of the hypoglossal nerve [XII] (Fig. 4.2), separates the inferior olive from the pyramid. Rootlets of this cranial nerve appear as small fascicles of fibers leaving the brain stem. The lateral edge of the inferior olive is a shallow groove called the **retro-olivary groove**. A few millimeters lateral to this retro-olivary groove in the retro-olivary area, are the **rootlets of the glossopharyngeal nerve** [IX], **vagus** [X], and **cranial root of the accessory nerve** [XI] (Fig. 4.2) from superior to inferior. The **abducent nerve** [VI] emerges in line with, and rostral to, the hypoglossal rootlets at the pontomedullary junction (Fig. 4.2). Laterally, the **facial** [VII] and the **vestibulocochlear** [VIII] **nerves** lie in line with rootlets of cranial nerves IX, X, and XI (Fig. 4.2).

Dorsal Surface of the Medulla Oblongata

A dorsal view of the medulla oblongata, with the overlying cerebellum removed, reveals the medullary width expanding as it is followed rostrally (Fig. 4.3). On this dorsal or posterior brain stem surface, the medulla oblongata has an inferior, closed part, resembling the spinal cord, and a superior, open part that forms the **floor of the fourth ventricle**. The expansion of the central canal into the fourth ventricle causes this open part. The point at which the closed part of the medulla oblongata begins to open is called the **obex** (Fig. 4.3) – a frequently used neurosurgical landmark. The closed part of the medulla oblongata has a **dorsal median sulcus** (Fig. 4.3) in the median plane. On either side of it is

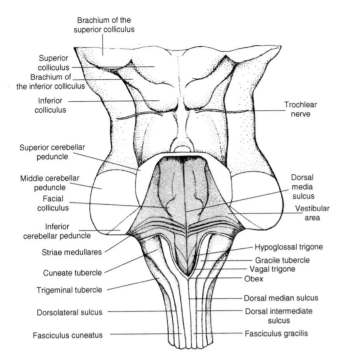

Brachium of the
superior colliculus

Superior
colliculus

Brachium of
the inferior colliculus

Inferior
colliculus

Superior cerebellar
peduncle

Middle cerebellar
peduncle

Facial
colliculus

Inferior
cerebellar peduncle

Striae medullares

Cuneate tubercle

Trigeminal tubercle

Dorsolateral sulcus

Fasciculus cuneatus

Trochlear
nerve

Dorsal
media
sulcus

Vestibular
area

Hypoglossal trigone

Gracile tubercle
Vagal trigone
Obex

Dorsal median sulcus

Dorsal intermediate
sulcus

Fasciculus gracilis

FIGURE 4.3. Dorsal surface of the human brain stem with the cerebellum removed. The area forming the floor of the fourth ventricle is lightly shaded.

an elongated eminence formed by an ascending tract, the **fasciculus gracilis** (Fig. 4.3). The rostral end of this tract is distinguished by an external swelling, the **gracile tubercle** (Fig. 4.3), formed by the underlying gracile nucleus and terminal fibers of its fasciculus. The lateral boundary of the gracile tubercle and fasciculus gracilis is the **dorsal intermediate sulcus** (Fig. 4.3). A second tract, the **fasciculus cuneatus** (Fig. 4.3), has at its rostral end a slight swelling, the **cuneate tubercle** (Fig. 4.3) marking the underlying cuneate nucleus. Fibers of the fasciculus cuneatus end in the cuneate nucleus and, in doing so, contribute to the formation of this swelling. A third elevation, lateral to the cuneate tubercle, is the **trigeminal tubercle** (Fig. 4.3) marking the underlying trigeminal spinal tract. Lateral to the trigeminal tubercle are entering or emerging rootlets of the glossopharyngeal nerve [IX], vagus [X], and cranial root (vagal part) of the accessory nerve [XI].

The dorsal surface of the open part of the medulla oblongata contributes to the floor of the fourth ventricle, a rhomboidal cavity at the rostral end of the medulla oblongata and the caudal end of the pons (Fig. 4.1). The area forming the fourth ventricle is lightly shaded in Figure 4.3. The area of the fourth ventricle remains hidden from view by the overlying cerebellum. A **dorsal median sulcus** divides the ventricular floor at this level (Fig. 4.3). On either side of this sulcus is the caudal

continuation of the **medial eminence** that includes, at this point, two small swellings, the hypoglossal and vagal trigones. The **hypoglossal trigone** (Fig. 4.3), adjacent to the dorsal median sulcus, marks the underlying **hypoglossal nucleus**. Lateral to the hypoglossal trigone is the **vagal trigone** (Fig. 4.3), formed by the underlying **dorsal vagal nucleus**. Lateral to the vagal trigone and separated from it by the sulcus limitans is the **vestibular area** (Fig. 4.3) covering the vestibular nuclei. The rostral boundary of the medulla oblongata corresponds to a band of fibers, the **striae medullares** (Fig. 4.3) that cross the floor of the fourth ventricle. If these striae are inconspicuous or absent, the rostral boundary corresponds to the widest part of the fourth ventricle or to the position of the lateral apertures of the fourth ventricle.

4.1.2. Pons

The other part of the metencephalon associated with the fourth ventricle, the **pons**, is continuous with the midbrain superiorly and the medulla oblongata inferiorly. In the pons are the abducent [VI], facial [VII], and vestibular [VIII] nuclei and part of the trigeminal [V] nuclear complex. Trigeminal roots (motor, sensory, and intermediate) enter or emerge at middle pontine levels; the abducent, facial, and vestibulocochlear nerves enter or emerge at the pontomedullary junction. Pontine sections reveal three levels: rostral, middle, and caudal.

Ventral Surface of the Pons

The ventral pontine surface in humans has prominent horizontally oriented fiber bundles crossing the median plane that project to the cerebellum. These **pontocerebellar fibers** form the prominent **middle cerebellar peduncle** (MCP) (Fig. 4.2). Roots of the **trigeminal nerve** [V] (Fig. 4.2) enter or emerge at pontine levels midway along the middle cerebellar peduncle and in line with the **facial nerve** [VII]. The trigeminal nerve divides the pons into pretrigeminal and post-trigeminal regions; two-thirds of the pons is post-trigeminal. The abducent [VI] and facial [VII] nerves enter or emerge at the pontomedullary junction (Fig. 4.2) from the bulbopontine sulcus. The ventral pontine surface shows a prominent depression, the **basilar sulcus** (Fig. 4.2), occupied by the basilar artery.

Dorsal Surface of the Pons

The dorsal pontine surface is clearly visible with the cerebellum removed. On either side of its dorsal median sulcus is the **medial eminence**, whose middle part often has a small swelling, the **facial colliculus** (Fig. 4.3) formed by the underlying abducent nucleus [VI] and intramedullary fibers of the facial nerve [VII] that arch over it. The caudal part of the medial eminence, formed by the hypoglossal and vagal trigones, extends into the medulla oblongata. A **vestibular area** occupies the remaining lateral ventricular floor caudally in the pons (Fig. 4.3), marking the position of the underlying vestibular nuclei. These nuclei are crossed by the transversely directed fibers of the **striae medullares** that pass to the cerebellum (Fig. 4.3) forming the rostral medullary boundary on the dorsal surface of the brain stem.

4.1.3. Midbrain

Many nuclei and tracts, particularly the oculomotor [III] and trochlear [IV] nuclei, occur in the **midbrain**, the most rostral part of the human brain stem. The midbrain has a central canal, the **aqueduct of the midbrain** (Fig. 4.1), which communicates rostrally with the third ventricle of the diencephalon and caudally with the fourth ventricle (Fig. 4.1) at the level of the pons. Sections of the midbrain reveal two levels: rostral and caudal.

Ventral Surface of the Midbrain

The ventral surface of the midbrain presents two **cerebral peduncles**, separated by the **interpeduncular fossa** (Fig. 4.2). Internally the cerebral peduncle or base of the midbrain consists of a fibrous part termed the **cerebral crus** and a cellular part termed the **substantia nigra**. Fibers in each cerebral crus descend in the pons and into the medulla oblongata where some of them form the medullary pyramids and then enter the decussation of the pyramids. The **oculomotor nerves** [III] emerge from the interpeduncular fossa as numerous blood vessels enter it (Fig. 4.2). Removal of these vessels from their points of entrance collectively delineates an area called the **posterior perforated substance** (Fig. 4.2). Each **trochlear nerve** [IV] emerges dorsally from the brain stem (Fig. 4.3), travels along the lateral surface of a cerebral peduncle, and appears on the ventral surface of the midbrain (Fig. 4.2).

Dorsal Surface of the Midbrain

The **trochlear nerve** [IV], smallest of the cranial nerves, is the only cranial nerve to emerge from the dorsal aspect of the brain stem. On the dorsal midbrain are four small eminences – two **superior** and two **inferior colliculi** (Fig. 4.3). Extending obliquely and rostrally from each inferior colliculus is a band of fibers, the **brachium of the inferior colliculus** (Fig. 4.3). A much smaller **brachium**

of the superior colliculus (Fig. 4.3) passes over the surface of the medial geniculate body. The **superior cerebellar peduncles** (Fig. 4.3), on the dorsal aspect of the brain stem, emerge from the cerebellum and continue to the midbrain to participate in the formation of the superior part of the roof of the fourth ventricle.

4.2. CEREBELLUM AND FOURTH VENTRICLE

The cerebellum is one of two specialized metencephalic structures. Though not a part of the brain stem, the cerebellum and fourth ventricle have a close relation to it.

4.2.1. Cerebellum

The **cerebellum**, a conspicuous structure attached to the dorsal aspect of the pons (Figs 4.1, 4.2), is composed of a **cerebellar vermis** (Fig. 4.4), a wormlike structure in the median plane with a vertical orientation, and greatly expanded lateral parts, the **cerebellar hemispheres** (Fig. 4.4). The cerebellar surface displays transverse **cerebellar folia** (Fig. 4.4) that give it a laminated appearance. Cerebellar fissures of varying depths (corresponding to the cerebral gyri) separate the cerebellar folia. Due to the extensive folding of the cerebellar cortex, some 85% of the cerebellar surface area remains buried in the depths of the cerebellar fissures.

Fissures divide the cerebellar hemispheres into **cerebellar lobules** and the lobules into **lobes**. The three cerebellar lobes are the **anterior, posterior,** (Fig. 4.4) and **flocculonodular**. Each lobe has vermian and hemispheric parts. The **primary fissure** (Fig. 4.4) separates the anterior cerebellar lobe from the posterior lobe; the **dorsolateral** and **postnodular fissures**, which are in essence continuous, separate the posterior lobe from the flocculonodular lobe.

The cerebellum has an external layer of gray matter, the **cerebellar cortex**, and several subcortical **cerebellar nuclei**. The **white matter of the cerebellum** consists of fiber bundles that intervene between the cerebellar cortex and these nuclei. On leaving the cerebellum, fibers from cell bodies in the cerebellar nuclei form three distinct **cerebellar peduncles** connecting the cerebellum with parts of the brain stem and spinal cord.

4.2.2. Fourth Ventricle

The **fourth ventricle** is a tent-shaped cavity whose roof has a superior and inferior part. The superior cerebellar

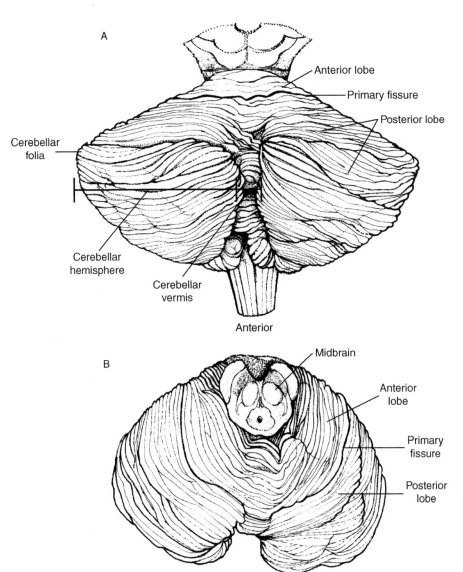

FIGURE 4.4. **A**, Dorsal surface of the human cerebellum. Note in the median plane, the vermis, and the large expanded cerebellar hemispheres on either side of the vermis. **B**, Superior surface of the cerebellum with the anterior lobe, primary fissure, and a part of the posterior lobe.

peduncles and the superior medullary velum form the superior part of the roof of the fourth ventricle whereas the inferior medullary velum and **choroid plexus** form the inferior part of the roof of the fourth ventricle. The latter consists of a rich capillary bed, pia mater, and single continuous layer of choroidal epithelial cells that participate in the formation of cerebrospinal fluid. Choroidal epithelial cells develop from the ependymal lining of the ventricles. The blood vessels and connective tissue core of the choroid plexus derives from a vascular fold of pia mater termed the **tela choroidea**. The dorsal pontine surface and the open part of the medulla oblongata (Fig. 4.1) form the floor of the fourth ventricle. The fourth ventricle narrows rostrally into the **aqueduct of the midbrain** and caudally into the **central canal of the medulla oblongata**. The fourth ventricle opens into the surrounding subarachnoid space by means of a single **median** as well as two **lateral apertures**.

4.3. ORGANIZATION OF BRAIN STEM NEURONAL COLUMNS

The organization and arrangement of brain stem neuronal groups are similar to those of the spinal cord. Their basic plans are similar – neuronal columns arranged longitudinally throughout their lengths. Special aspects of the brain stem and its development are noteworthy. First, the neural tube near the fourth ventricle widens like an opening book displacing the alar laminae laterally in the brain stem. Second, many special sense organs such as those related to audition, vestibular sensation and taste develop in relation to the brain stem. Third, some musculature associated with the brain stem has a special embryological origin from mesenchyme of the pharyngeal arches. These neurons are in a special column in the brain stem.

Based on these considerations, the adult brain stem has the same four general neuronal columns as the spinal cord together with three special neuronal columns. The four general neuronal columns of the brain stem are designated as **general somatic efferent** (GSE), **general somatic afferent** (GSA), **general visceral efferent** (GVE), and **general visceral afferent** (GVA) (Fig. 4.5). The three special neuronal columns unique to the brain stem are the **special somatic afferent** (SSA), **special visceral afferent** (SVA), and **special visceral efferent** (SVE) columns (Fig. 4.5).

Special somatic afferent (SSA) neurons of the brain stem receive special sensory impulses from the ear, an ectodermal derivative. **Special visceral afferent** (SVA) neurons receive special sensory impulses related to feeding. They supply the taste buds on the tongue. **Special visceral efferent** (SVE) neurons innervate muscles with a special origin from mesenchyme of the pharyngeal arches. These are the muscles of mastication (from pharyngeal arch 1), muscles of facial expression (from pharyngeal arch 2), muscles of the soft palate, pharynx, and larynx (from pharyngeal arches 3 and 4), as well as the sternomastoid and the superior (descending) part of the trapezius (from pharyngeal arch 5). The inferior (ascending) part of the trapezius receives its innervation from general somatic efferent neurons in the ventral horn of the spinal cord (from cervical somites).

Figure 4.5 illustrates the orderly arrangement of brain stem nuclei of cranial nerves into functional columns. Understanding this arrangement, coupled with an understanding of the position of functionally significant ascending and descending fiber paths, will provide a

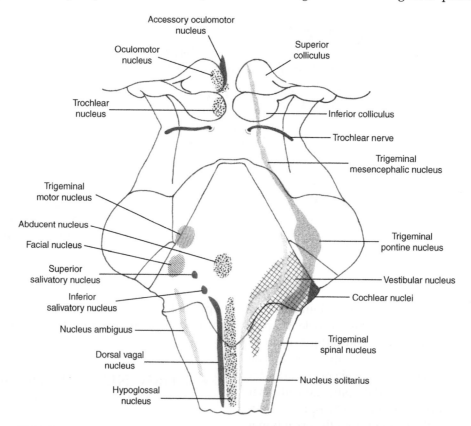

FIGURE 4.5. Cranial nerve nuclei projected on the human brain stem. Note the orderly arrangement of the brain stem nuclear groups into functional columns. Efferent nuclei are shown on the left: general somatic efferent (oculomotor, trochlear, abducent, and hypoglossal nuclei), general visceral efferent (accessory oculomotor, superior salivatory, inferior salivatory and dorsal vagal nuclei), and special visceral efferent (trigeminal motor nucleus, facial nucleus, and nucleus ambiguus). Afferent nuclei are shown on the right: general somatic afferent (trigeminal mesencephalic nucleus, trigeminal pontine nucleus, and trigeminal spinal nucleus), special somatic afferent (vestibular and cochlear nuclei), special visceral efferent and general visceral afferent (nucleus of the solitary tract) (after Crosby, Humphrey, and Lauer, 1962).

firm foundation for understanding the entire brain stem and the functional components of the cranial nerves.

4.3.1. Functional Components of the Cranial Nerves

Since there are seven functional types of nuclei of cranial nerves in the brain stem, it is theoretically possible for each cranial nerve to have one to seven functionally distinct fiber types. On this basis, we speak of 'functional components' or 'fiber types' in a cranial nerve. Designation of any cranial nerve as having one or more functional components is dependent upon the efferent neurons from which its motor fibers arise, or on the neurons on which its sensory fibers end. Fibers of the trochlear nerve [IV] are processes of neurons in the **trochlear nucleus**, a component of the somatic efferent column of the brain stem. Hence, the trochlear nerve is a somatic efferent (SE) nerve. The oculomotor nerve [III], however, includes fibers from two nuclear groups: the **oculomotor nucleus**, a somatic efferent nucleus, and the **accessory oculomotor nucleus**, a general visceral efferent nucleus. Therefore, the oculomotor nerve is designated as somatic efferent (SE) and general visceral efferent (GVE) nerve. Table 4.1 provides a summary of the functional components, nucleus of origin or reception and functions of the cranial nerves in humans.

4.3.2. Efferent Columns

General Somatic Efferent Column

The **general somatic efferent** (GSE) **column** of the brain stem is nearest the median plane (Fig. 4.5) resembling the somatic efferent column in the spinal cord. Its neurons occur in the midbrain, pons, and medulla oblongata (Fig. 4.5) and supply skeletal muscle. The GSE column includes the following nuclear groups: in the rostral part of the midbrain, the **oculomotor nucleus**, in the caudal part of the midbrain, the **trochlear nucleus**, in the lower pons, the **abducent nucleus**, and at medullary levels, the **hypoglossal nucleus** (Fig. 4.5). These four nuclei are in line with one another throughout their rostrocaudal extent in the brain stem, are classified as **general somatic efferent** (GSE) nuclei, and their respective cranial nerves each have a somatic efferent component (Table 4.1).

General Visceral Efferent Column

Lateral to the somatic efferent column (Fig. 4.5) is a column similar to the spinal cord general visceral efferent column. The brain stem **general visceral efferent** (GVE) **column** includes preganglionic parasympathetic neurons that supply nonstriated and cardiac muscle and exocrine glands and consists of the cranial division of the craniosacral (**parasympathetic**) part of the autonomic nervous system extending from the midbrain to medullary levels (Fig. 4.5). The general visceral efferent nuclear groups in this brain stem column are the **accessory oculomotor** (Edinger–Westphal) **nucleus** at rostral levels of the midbrain, the **superior salivatory nucleus** in the lower pons, the **inferior salivatory nucleus** and the **dorsal vagal nucleus** in the medulla oblongata (Fig. 4.5). Neurons in these four general visceral efferent nuclei contribute fibers to four or perhaps five cranial nerves (Table 4.1).

4.3.3. Afferent Columns

General Somatic Afferent Column

Laterally in the brain stem, this column (Fig. 4.5) resembles the somatic afferent column at spinal levels. The **general somatic afferent column** receives general sensory impulses from cutaneous areas of the face and head and consists of the **trigeminal nuclear complex** – a discontinuous column of neurons from the midbrain to rostral cervical levels (Fig. 4.5) where it overlaps the substantia gelatinosa and is divisible into three parts: the **mesencephalic**, **pontine**, and the **spinal trigeminal nuclei**. The **trigeminal pontine** and **trigeminal spinal nuclei** form an ice cream and cone-shaped structure (Fig. 4.5). Sitting atop the ice cream, like a dunce hat, is the trigeminal mesencephalic nucleus (Fig. 4.5). The **trigeminal mesencephalic nucleus** (Fig. 4.5) consists of pseudounipolar neurons whose peripheral processes carry proprioceptive impulses to the brain stem. The trigeminal nuclear complex, forming the cone, is a general somatic afferent nucleus. Hence, the trigeminal nerve [V] has a general somatic afferent component, as do the facial [VII], glossopharyngeal [IX], and vagal [X] nerves (Table 4.1).

Special Somatic Afferent Column

The brain stem **special somatic afferent column** is situated dorsolaterally (Fig. 4.5) and has no counterpart at spinal levels. The **special somatic afferent column** receives impulses from ectodermal structures, especially the inner ear. The **vestibular nuclei** are in the caudal part of the pons and the rostral part of the medulla whereas the **cochlear nuclei** (Fig. 4.5) are exclusively in the rostral part of the medulla. These nuclei together form the special somatic afferent column (Table 4.1).

TABLE 4.1. Functional components, nucleus of origin or reception, and functions of the cranial nerves in humans

Cranial nerve	Functional components	Brain stem nucleus of origin or reception	Functions
Olfactory	SVA	Olfactory epithelium of the nasal cavity	Sense of smell
Optic	SSA	Retinal ganglionic neurons	Visual sensation
Oculomotor	GSE	Oculomotor nucleus	Motor supply to levator palpebrae superioris and ocular muscles except superior oblique and lateral rectus
	GVE	Accessory oculomotor nucleus	Motor supply to sphincter pupillae and ciliary muscle
Trochlear	GSE	Trochlear nucleus	Motor supply to the superior oblique
Trigeminal	SVE	Trigeminal motor nucleus	Motor supply to muscles of mastication, as well as the mylohyoid, anterior belly of the digastric, tensor tympani and tensor veli palatini
	GSA	Trigeminal nuclear complex	Superficial and deep sensation from the face and head
Abducent	GSE	Abducent nucleus	Motor supply to the lateral rectus
Facial	SVE	Facial nucleus	Motor supply to facial muscles
	GVE	Superior salivatory nucleus	Secretions of submandibular, sublingual, lacrimal, oral, nasal and palatine glands
	GSA	Trigeminal spinal nucleus	Sensation from external ear and tympanic membrane
	GVA	Nucleus solitarius	To blood vessels of the face and some inside the skull
	SVA	Nucleus solitarius	Sense of taste from anterior two–thirds of the tongue
Vestibulocochlear	SSA	Vestibular nuclei	Vestibular sensation
	SSA	Cochlear nuclei	Auditory sensation
Glossopharyngeal	SVE	Nucleus ambiguus	Motor supply to stylopharyngeus muscle
	GVE	Inferior salivatory nucleus	Secretion of parotid gland and serous glands near vallate papillae
	GSA	Spinal trigeminal nucleus	Sensation from external and middle ear
	GVA	Nucleus solitarius	Sensation from back of mouth, tongue, oral- and nasopharynx, carotid bodies and carotid sinus
	SVA	Nucleus solitarius	Taste from posterior third of the tongue
Vagal	SVE	Nucleus ambiguus	Motor supply to constrictor muscles of the pharynx, and intrinsic muscles of the larynx
	GVE	Dorsal vagal nucleus	Motor supply to cardiac muscle, nonstriated muscle, glands of respiratory and gastro-intestinal system
	GSA	Trigeminal spinal nucleus	Sensation from back of ear and external auditory meatus
	GVA	Nucleus solitarius	Sensation from pharynx, larynx, thoracic and abdominal viscera
	SVA	Nucleus solitarius	Taste from epiglottis
Accessory	SVE	Nucleus ambiguus	Motor supply to the muscles of the soft palate with the exception of the tensor veli palatini (innervated by mandibular nerve)
Cranial root (vagal part)	GVE	Dorsal vagal nucleus	Motor supply to nonstriated muscle and glands of respiratory and gastrointestinal tract with vagus
Accessory Spinal root (spinal part)	GSE	Accessory nucleus	Motor supply to trapezius and sternomastoid muscles
Hypoglossal	GSE	Hypoglossal nucleus	Motor supply to extrinsic and intrinsic muscles of the tongue

GSA, general somatic afferent; GSE, general somatic efferent; GVA, general visceral afferent; GVE, general visceral efferent; SSA, special somatic afferent; SVA, special visceral afferent; SVE, special visceral efferent.

(The classification of the optic nerve [II] is also **special somatic afferent** though its neurons are not in the brain stem.)

General and Special Visceral Afferent Columns

These brain stem columns are indistinguishable from one another lateral to the general visceral efferent column (Fig. 4.5). No special visceral efferent column is at spinal levels. The **general visceral afferent and special visceral afferent** (GVA/SVA) **column** receives general visceral impulses from the back of the mouth, tongue, tonsil, oro- and naso-pharynx, from the carotid bodies and sinuses through the glossopharyngeal nerve [IX], and from the pharynx, larynx, thoracic and abdominal viscera by means of the vagal nerve [X] (Table 4.1). The GVA/SVA nuclear group also receives special visceral afferent impulses from taste buds on the anterior two-thirds of the tongue through the facial nerve [VII], posterior third of the tongue by means of the glossopharyngeal nerve [IX], and from epiglottic taste buds through the vagal nerve (Table 4.1). The nuclear group with both general and special visceral afferent neurons includes the **nucleus of the solitary tract** (Fig. 4.5), present in all medullary levels and extending into the pons (Fig. 4.5). The olfactory nerve [I] is a special visceral afferent (SVA) nerve but its neurons are not in the brain stem (Table 4.1).

4.4. INTERNAL FEATURES

Before describing characteristic internal features at representative transverse levels in the brain stem, two general comments are in order. The term **reticular formation** (RF) refers to areas at all brain stem levels, consisting of multipolar neurons of varying size embedded in masses of interweaving fibers. The reticular formation has indistinct boundaries and extends from through the medulla oblongata, pons, and midbrain with overlap into the spinal cord and diencephalon. The structural and functional organization of the reticular formation is the subject of Chapter 9.

4.4.1. Endogenous Substances

A unique feature of the organization of the nervous system is the presence in different locations throughout the nervous system of endogenous **monoamines** (catecholamines and indoleamines). **Norepinephrine**, **epinephrine**, and **dopamine** are examples of catecholamines; **serotonin** is an example of an indoleamine. These substances, thought to serve as neurotransmitters, are identifiable in specific brain regions in a variety of mammals. The brain stem of primates has nuclei or tracts with catecholamines and serotonin in neuronal cell bodies, axons, and axon terminals.

Epinephrine, a catecholamine in the adrenal glands of most species, may function as a chemical transmitter at synapses in the human brain stem. Its localization there corresponds to the localization of phenylethanolamine-N-methyltransferase, the enzyme for converting norepinephrine to epinephrine. The reticular formation of the human brain stem has 20 ng/g of epinephrine. In the human medulla oblongata, epinephrine is associated with the human hypoglossal and dorsal vagal nuclei (35–45 ng/g) of the medulla oblongata. Dense accumulations of catecholaminergic neurons are lateral and ventrolateral to the dorsal vagal nucleus of nonhuman primates. In the human midbrain, the superior colliculus has about 5 ng/g of epinephrine while the area of the oculomotor nuclei has about 15–40 ng/g.

The **locus coeruleus** is the largest collection of **norepinephrine-containing neurons** in the mammalian brain. There are a number of catecholamine cell groups and paths in the primate brain including a dorsal ascending pathway mainly associated with the locus coeruleus. There is a loss of neurons in the locus coeruleus with normal aging (25–60% by the ninth decade), in elderly patients with Alzheimer's disease, and in those with Parkinson's disease. In the human midbrain, the pars compacta of the **substantia nigra** has three catecholamines: dopamine (1000 ng/g), norepinephrine (200 ng/g), and epinephrine (about 2 ng/g). Dopaminergic neurons in the compact part of the substantia nigra transport their neurotransmitter to the putamen and globus pallidus over nigrostriate fibers. There is progressive cell loss of neurons in the substantia nigra of patients with Parkinson's disease. Dopaminergic neurons and neuronal processes are also in all primate retinae thus making dopamine the major catecholamine in the retina.

4.4.2. Medulla Oblongata

In this description of the internal features of the medulla oblongata, landmarks provided will help the reader in later descriptions of paths that traverse various levels of the medulla oblongata. Attention focuses on the relations between longitudinal tracts and transverse structures such as the cranial nuclei and their intramedullary fibers.

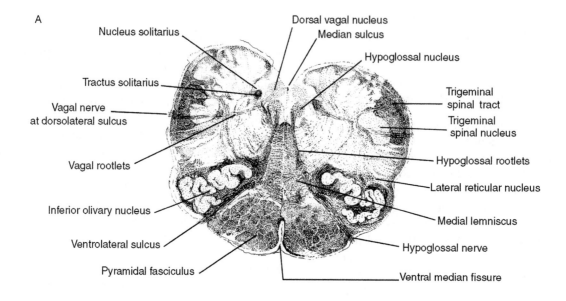

A

Nucleus solitarius

Tractus solitarius

Vagal nerve
at dorsolateral sulcus

Vagal rootlets

Inferior olivary nucleus

Ventrolateral sulcus

Pyramidal fasciculus

Dorsal vagal nucleus
Median sulcus

Hypoglossal nucleus

Trigeminal
spinal tract
Trigeminal
spinal nucleus

Hypoglossal rootlets

Lateral reticular nucleus

Medial lemniscus

Hypoglossal nerve

Ventral median fissure

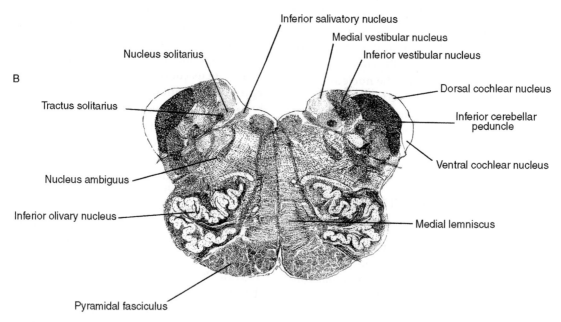

B

Nucleus solitarius

Tractus solitarius

Nucleus ambiguus

Inferior olivary nucleus

Pyramidal fasciculus

Inferior salivatory nucleus

Medial vestibular nucleus

Inferior vestibular nucleus

Dorsal cochlear nucleus

Inferior cerebellar
peduncle

Ventral cochlear nucleus

Medial lemniscus

FIGURE 4.6. Transverse sections of the adult human medulla oblongata. **A**, At the level of the hypoglossal, cranial root of the accessory, and vagal nuclei and nerves. **B**, At the level of the glossopharyngeal and vestibulocochlear nuclei and nerves. Toward the bottom of the figure is the ventral surface of the transverse section while the top of the figure corresponds to the dorsal surface of the section. Both drawings are magnified approximately four times.

Level of the Hypoglossal, Cranial Accessory, and Vagal Nuclei and Nerves

In this and all subsequent sections of the brain stem, the ventral surface of the transverse section is near the bottom of the figure and the dorsal surface is near the top. On the ventral surface in the median plane is the ventral median fissure. Lateral to it and on either side are **pyramidal tracts** (Fig. 4.6) composed of corticospinal fibers (from cerebral cortex to the spinal cord),

corticobulbar fibers (from cerebral cortex to cranial nerve nuclei of the brain stem or 'bulb'), and cortico-reticular fibers. Lateral to each pyramidal tract is the **inferior olivary nucleus** (Fig. 4.6) forming a prominent external landmark, the **olive**. Dorsal to each pyramidal tract and extending dorsally is a prominent longitudinal bundle, the **medial lemniscus** (Fig. 4.6). The medial lemnisci are interolivary in position and are part of a bilateral ascending sensory path. Along the superior margin of the pyramidal tracts and interspersed in the medial lemnisci and inferior olivary nuclei of the brain stem of nonhuman primates are neuronal cell bodies containing serotonin. Similar cell bodies are ventral in the **nuclei of the medullary raphé**.

In the median plane and on the dorsal surface of the medulla oblongata (Fig. 4.6), is the **dorsal median sulcus**. On either side of this sulcus is the continuation of the **medial eminence** that includes, at this point, two small swellings, the hypoglossal and vagal trigones. The **hypoglossal trigone** is nearest the dorsal median sulcus and marks the underlying **hypoglossal nucleus** (Fig. 4.6) – a collection of general somatic efferent neurons that innervate the intrinsic and extrinsic muscles of the tongue. Unilateral injury to one hypoglossal nucleus leads to an ipsilateral, flaccid paralysis of the tongue. Intramedullary fibers that form the **hypoglossal nerve** [XII] pass ventrally from each nucleus, between the pyramidal tract and inferior olivary nucleus, and emerge from the ventrolateral sulcus between the pyramid and olive (Fig. 4.6). Scattered neurons connect each hypoglossal nucleus with the spinal somatic efferent column from which it embryologically develops.

Structural information is useful in localizing injury to the CNS particularly if deficits involve two or more related structures such as a longitudinal tract along with a cranial nerve nucleus or cranial intramedullary fibers. A patient with paralysis of the limbs on the right side of the body (hemiplegia) as well as the left half of the tongue has an injury involving the left intramedullary hypoglossal fibers and the left pyramidal tract as they lie in relation to each other (Fig. 4.6). Since fibers in the left pyramidal tract decussate at the medulla–cord junction (for control of voluntary motor activity on the right side of the body), a left-sided injury rostral to the motor decussation causes a right-sided paralysis of the body. When this patient protrudes their tongue, it deviates to the left because of (1) the forward pull of the nonparalyzed genioglossus muscle; (2) elongation of the tongue by the intrinsic muscles on the nonparalyzed side; and (3) because of a lag produced by muscles in the paralyzed half. The site of this injury must be in the medulla oblongata where the longitudinally oriented pyramidal tract comes into relationship with the transversely oriented intramedullary hypoglossal fibers before the latter emerge from the brain stem. Knowledge of this relationship is clinically useful in localizing such an injury.

Dorsal and lateral to the **hypoglossal trigone** is the **vagal trigone**. These trigones contribute to the caudal part of the medial eminence in the medulla. The vagal trigone is superficial to the underlying **dorsal vagal nucleus** (Fig. 4.6). As a general visceral efferent nucleus,

the dorsal vagal nucleus is in line with the **intermediolateral nucleus** in the spinal cord and extends from the rostral cervical cord to the rostral part of the hypoglossal nucleus or superior to it (Fig. 4.5). Epinephrine is associated with the human hypoglossal and dorsal vagal nuclei (35–45 ng/g). Dense accumulations of catecholaminergic neurons are lateral and ventrolateral to the dorsal vagal nucleus of nonhuman primates. Progressive widening of the central canal at spinal levels into the medullary fourth ventricle causes the dorsal vagal nucleus to shift its position as it extends from the caudal into the rostral part of the medulla oblongata. At the rostral level of the medulla oblongata, the dorsal vagal nucleus remains lateral to the hypoglossal nucleus (Fig. 4.5). The **dorsal vagal nucleus** has two groups of neurons: those inferior to the open floor of the fourth ventricle innervating the heart and respiratory system, and those innervating abdominal viscera in the caudal half of the medulla oblongata. Each dorsal vagal nucleus in nonhuman primates supplies fibers to the ipsilateral and contralateral vagal nerve [X]. Consequently, there is bilateral representation in the dorsal vagal nucleus.

Thinly myelinated fibers from the dorsal vagal nuclei pass ventrolaterally and emerge from the brain stem in the **retro-olivary area** as rootlets of the vagal nerve [X]. Some of the fibers from the dorsal vagal nuclei may join the glossopharyngeal nerve [IX] and cranial root of the accessory nerve [XI]. These preganglionic parasympathetic fibers synapse with postganglionic neurons in ganglia that lie in, on, or near the heart, lungs, oesophagus, and gastrointestinal tract to the left colic flexure.

In the medulla oblongata is the **nucleus ambiguus**, a **special visceral efferent nucleus** that contributes fibers to the glossopharyngeal, vagal, and accessory nerves. Because the nucleus ambiguus consists of clusters of neurons connected by scattered neurons, this nucleus is satisfactorily identifiable only at certain medullary levels (hence its name). There is a somatotopic arrangement of neurons in the nucleus ambiguus that supply muscles derived from **pharyngeal (visceral) arch mesoderm**. Neurons in the caudal part of the nucleus ambiguus contribute fibers to the cranial root (vagal part) of the accessory nerve, those in the intermediate part (and perhaps greatest extent) of this nucleus provide fibers to the vagal nerve, and those in the rostral part (at rostral medullary levels) provide fibers to the glossopharyngeal nerve. Therefore, the nucleus ambiguus supplies motor nerves to the pharynx, larynx, and soft palate. Clinical and experimental observations reveal the following arrangement of neurons in the nucleus ambiguus: neurons supplying the stylopharyngeus are in the rostral part of the nucleus, followed by those supplying the pharyngeal and laryngeal muscles, and neurons supplying the palatal muscles in the caudal part of the nucleus. A rostrocaudal motor pattern occurs in the nucleus ambiguus of nonhuman primates for the laryngeal muscles. Neurons in the rostral part of the nucleus ambiguus supply the lateral crico-arytenoid muscle followed caudally by those for the remainder of the laryngeal adductors (which close the larynx). Neurons supplying the abductor (posterior crico-arytenoid – which opens the larynx) in nonhuman primates are between those for the crico-arytenoid and the rest of the adductors. A similar motor pattern in the human nucleus ambiguus would account for small injuries that cause impaired function of one or more pharyngeal, laryngeal, or soft palate muscles. The possibility of involving several muscles innervated by one cranial nerve without involving all the muscles innervated by that nerve holds for any motor nucleus of the brain stem that exhibits a motor pattern.

Lateral in this section, and at all medullary levels, are the **trigeminal spinal tract** and its accompanying **trigeminal spinal nucleus** (Fig. 4.6). In the caudal part of the medulla oblongata, these structures are superficial forming a trigeminal tubercle. The trigeminal spinal nucleus is a general somatic afferent nucleus concerned with sensations from the face, cornea, and oral mucosa, the tongue and dental pulp, the nasal mucosa, and anterior half of the scalp.

Also at this level are the **solitary tract** and the associated **nucleus of the solitary tract** (Fig. 4.6). The solitary tract includes descending general and special visceral afferent fibers that enter the brain stem in the vagal nerve [X] and descend to the medulla–cord junction. Surrounding the longitudinal extent of this descending tract is the **nucleus of the solitary tract** (Fig. 4.6). The nucleus and the tract form a continuous complex from the middle part of the pons to the caudal part of the medulla oblongata. General and special visceral afferents in the vagus [X] end in this complex, as do some fibers from the facial [VII] and glossopharyngeal nerves [IX] that contribute to the solitary tract. The special visceral afferent fibers in this tract and nucleus are gustatory in function. Catecholaminergic neurons occur in the solitary tract and at its ventrolateral border in various primates. Catecholaminergic fibers are scattered in and near the solitary tract of nonhuman primates.

The **lateral reticular nucleus**, a constituent of the reticular formation, is in the lateral medulla oblongata dorsal to the dorsolateral edge of the inferior olivary nucleus and ventral to the ventrolateral edge of the trigeminal spinal tract (Fig. 4.6). In primates, its catecholaminergic neurons form part of a column of catecholaminergic neurons and fibers extending rostral to the inferior olivary nucleus and into the rostral part of the pons.

Level of the Glossopharyngeal and Vestibulocochlear Nuclei and Nerves

At this level, the fourth ventricle is widest. The **dorsal and ventral cochlear nuclei** appear on the dorsal, lateral, and ventrolateral margins of the **inferior cerebellar peduncle** (ICP) (Figs 4.3, 4.6). These cochlear nuclei, classified as **special somatic afferent** neurons, are secondary neurons in the auditory path. A decrease in volume of this nucleus appearing beyond the fifth decade is due to the loss of non-neuronal elements. Medial to the inferior cerebellar peduncle is another special somatic afferent nucleus, the **vestibular nuclear complex** (Fig. 4.6). The vestibular part of the **vestibulocochlear nerve** [VIII] is concerned with the maintenance of body, head, and eye position, and orientation in space. Its nuclei are secondary neurons in the vestibular path. Two groups of vestibular neurons at this level are the medial and inferior vestibular nuclei (Fig. 4.6). The latter nucleus is medial to the inferior cerebellar peduncle and has noticeable vertically oriented fibers scattered among its neurons. The medial vestibular nucleus is medial to the inferior vestibular nucleus (Fig. 4.6). Entering vestibular fibers pass diagonally through the peduncle to reach the vestibular nuclei. Cochlear fibers pass along the lateral border of the inferior cerebellar peduncle and enter the cochlear nuclei where they end. These relations are clinically significant in that a single blood vessel (posterior inferior cerebellar artery) supplies all structures in the dorsolateral medulla oblongata at this level.

Autonomic fibers traveling in the glossopharyngeal nerve [IX] to the otic ganglion and then to the parotid gland, have their neurons of origin in the **inferior salivatory nucleus** (Fig. 4.6). At this level, the inferior salivatory nucleus is at the rostral pole of the dorsal vagal nucleus (Fig. 4.5). Also at this level (Fig. 4.6) is the nucleus ambiguus that provides special visceral efferent fibers by way of the glossopharyngeal nerve to the stylopharyngeus muscle.

The **glossopharyngeal nerve** [IX] transmits sensory impulses from the tongue, tonsils, oral and nasopharynx, carotid bodies and sinuses. These **general visceral afferent** fibers enter the brain stem in the glossopharyngeal nerve and contribute to the solitary tract (Fig. 4.6) before ending in the nucleus of the solitary tract (Fig. 4.6). Each glossopharyngeal nerve carries taste impulses (special visceral afferent – SVA) from the posterior third of the tongue. Central processes of these SVA fibers enter the brain stem as part of the glossopharyngeal nerve to enter the **solitary tract** (Fig. 4.6), particularly its ventricular part. The glossopharyngeal nerve and the vagus receive general somatic afferent impulses from a small part of the external and middle ear. These general somatic afferents enter the brain stem with the glossopharyngeal nerve then join the trigeminal spinal tract before ending in its nucleus (Fig. 4.6).

Relations between certain brain stem nuclei at this level underlie the gag (pharyngeal) reflex, perhaps in association with emesis and other accompanying responses. Vestibular connections to the dorsal vagal nucleus, nucleus of the solitary tract, nucleus ambiguus, inferior salivatory nuclei, and descending connections to spinal autonomic neurons provides the basis for motion sickness and such accompanying responses as reverse peristalsis, nausea, regurgitation, frequent swallowing, pallor, sweating, and salivation.

4.4.3. Pons

Sections of the pons reveal three levels: rostral, middle, and caudal (Figs 4.7, 4.8). Each level shares certain general internal features and each is divisible into a dorsal part or **pontine tegmentum** and a ventral part or **basilar pons**. Inspection of the basilar pons enables one to identify the appropriate pontine level. At all pontine levels the basilar pons has **longitudinal** and **transverse pontine fibers** that are intermingled with the **pontine nuclei** (Figs 4.7, 4.8). These nuclei (not to be confused with the trigeminal pontine nuclei) extend into the pontine tegmentum and serve as relay stations in a path from the cerebral cortex to the cerebellum. They are separable into secondary nuclear groups according to their position in relation to the longitudinal fiber bundles. Projection and intrinsic neurons occur in the pontine nuclei of nonhuman primates.

Transverse pontine fibers at this level are predominantly **pontocerebellar fibers** (Figs 4.7, 4.8) that enter and form the middle cerebellar peduncles (Fig. 4.7) as they approach the cerebellum. Among these fibers and the pontine nuclei are **longitudinal pontine fibers** that have a distinctive appearance in transverse sections (Fig. 4.7). Many of these longitudinal fibers – such as the corticospinal, corticobulbar, corticoreticular, and corticopontine fibers, originate in the cerebral cortex. The first two are components of the pyramidal system; the latter are part of a corticopontocerebellar system of fibers. The most common synaptic endings in the pontine nuclei are corticopontine endings. The **fourth ventricle** gradually narrows from the caudal part of the pons (Fig. 4.7) to the rostral part of the pons (Fig. 4.8) where it tapers off before continuing as the **aqueduct of the midbrain**.

Level of the Facial and Abducent Nuclei (Caudal Pons)

The **abducent nucleus** (Fig. 4.7) is in the pontine tegmentum beneath the facial colliculus (a small elevation on the floor of the fourth ventricle). No change in neuronal number occurs after birth in the human abducent nucleus though the nucleus nearly doubles in length as the brain grows. Fibers from this somatic efferent nucleus turn ventrocaudally to emerge at the caudal pontine border (in line with the hypoglossal rootlets) to supply the lateral rectus muscle of the same side. Abducent fibers pass caudally to leave the brain stem inferior to the level of the nucleus itself. In the basilar pons, the abducent fibers pass lateral to the longitudinally oriented corticospinal fibers before leaving the brain stem. Relations between the transverse abducent fibers and the longitudinal corticospinal fibers are of clinical importance. Destruction of the right abducent nucleus or the fibers arising from it will cause paralysis of the right lateral rectus muscle; the right eye deviates medially toward the nose (esotropia or strabismus) because of unopposed action of the right medial rectus. Injury to the basilar pons involving the right abducent fibers and right corticospinal fibers causes medial deviation of the right eye with paralysis of the left upper and lower limbs. Injury to both abducent nuclei or to both abducent nerves causes **bilateral internal strabismus**. Because of its long intracranial course, expanding masses such as tumors often injure the abducent nerve [VI].

The **facial nucleus**, in the caudal part of the pontine tegmentum (Fig. 4.7) medial to the trigeminal spinal tract and nucleus but separated from them by the emerging facial fibers (Fig. 4.7), is the origin of the **special visceral efferent** component of the facial nerve [VII] that supplies the facial muscles. The latter are derived embryologically from **pharyngeal arch 2**. The total number of human facial neurons on one side estimated using an optical fractionator is 10,470. Fibers innervating lower facial muscles arise from the dorsal part of the facial nucleus; fibers supplying upper facial muscles (for eyelid closure and wrinkling of the forehead) arise from neurons in the ventral part of the facial nucleus. Fibers from neurons in the facial nucleus have an unusual course, passing to the ventricular floor, medial to the abducent nucleus where they collect in a compact bundle beneath the fourth ventricular floor. This compact bundle of facial fibers is termed the **genu**

A

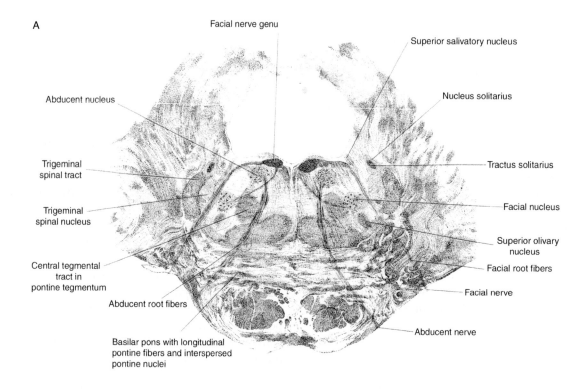

Facial nerve genu

Superior salivatory nucleus

Abducent nucleus

Nucleus solitarius

Trigeminal
spinal tract

Tractus solitarius

Trigeminal
spinal nucleus

Facial nucleus

Superior olivary
nucleus

Central tegmental
tract in
pontine tegmentum

Facial root fibers

Facial nerve

Abducent root fibers

Abducent nerve

Basilar pons with longitudinal
pontine fibers and interspersed
pontine nuclei

B

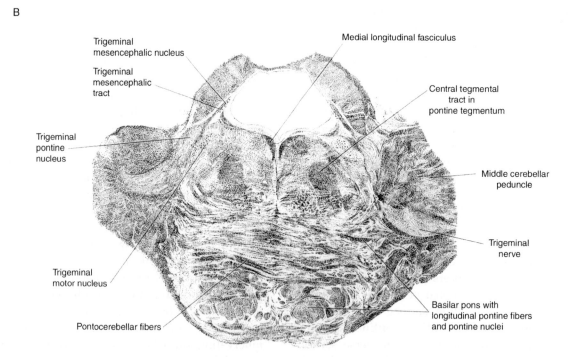

Trigeminal
mesencephalic nucleus

Medial longitudinal fasciculus

Trigeminal
mesencephalic
tract

Central tegmental
tract in
pontine tegmentum

Trigeminal
pontine
nucleus

Middle cerebellar
peduncle

Trigeminal
nerve

Trigeminal
motor nucleus

Basilar pons with
longitudinal pontine fibers
and pontine nuclei

Pontocerebellar fibers

FIGURE 4.7. **A**, A transverse section of the caudal third of the adult human pons at the level of the facial and abducent nuclei and nerves. **B**, Drawing of a transverse section of the middle third of the adult human pons at the level of the trigeminal nerve [V], pontine and trigeminal motor nuclei. Both drawings are magnified approximately three times.

of the facial nerve [VII] (Fig. 4.7). From this point, the genu of the facial nerve ascends and turns laterally to arch over the abducent nucleus from medial to lateral. These special visceral efferent fibers then pass ventro-laterally between the trigeminal spinal nucleus and the lateral aspect of the facial nucleus before emerging from the pontomedullary junction. These exiting facial nerve fibers are rostral to, and aligned with, exiting fibers of

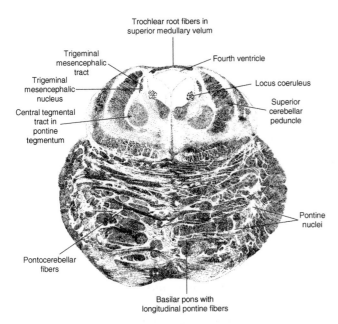

Labels on figure:
Trochlear root fibers in superior medullary velum
Trigeminal mesencephalic tract
Trigeminal mesencephalic nucleus
Central tegmental tract in pontine tegmentum
Pontocerebellar fibers
Fourth ventricle
Locus coeruleus
Superior cerebellar peduncle
Pontine nuclei
Basilar pons with longitudinal pontine fibers

FIGURE 4.8. A transverse section of the rostral third of the adult human pons at the level of the pontine isthmus (pretrigeminal level). Drawing is magnified approximately three times.

the glossopharyngeal [IX], vagal [X], and accessory [XI] nerves in that rostral to caudal order.

Injuries to the facial nucleus, its intramedullary fibers, or its fibers outside the brain stem causes paralysis of all facial muscles ipsilateral to the injury. In such cases, the patient cannot close one eye, wrinkle the forehead, or smile on that side. The palpebral fissure widens because of paralysis of the orbicularis oculi muscle, and the corner of the mouth sags on the injured side. Because the **special visceral efferent** facial fibers do not join the other facial components until after the facial genu, injury before the facial genu involves only this component without involving the others. In humans, some facial fibers join a layer of fibers beneath the floor of the fourth ventricle and contribute to the contralateral facial nerve without entering the facial genu.

The **superior salivatory nucleus**, in the caudal part of the pontine tegmentum, provides **general visceral efferent** fibers to the facial nerve. Neurons of this nucleus form small clusters, lateral to, and intermingled with, the emerging facial fibers (Fig. 4.7). Processes of superior salivatory neurons join the facial fibers and distribute to the submandibular ganglion that innervates the submandibular, sublingual, and other oral salivary glands. Destruction of a superior salivatory nucleus is clinically insignificant if other salivary glands are unaffected. A special group of neurons in or near the superior salivatory nucleus termed the **lacrimal nucleus** supply the lacrimal gland. Destruction of these lacrimal neurons or

their processes in the facial nerve eliminates lacrimal secretion and causes a dry eye – a serious symptom accompanying facial paralysis.

Special visceral afferent facial fibers that carry impulses from taste buds on the anterior two-thirds of the tongue enter the pons and separate from the incoming facial fibers to enter the **solitary tract** and then end in that part of the nucleus of the solitary tract on the ventricular side of the solitary tract. The solitary tract and the nucleus of the solitary tract are identifiable in the caudal part of the pons (Fig. 4.7). General somatic afferent facial fibers that carry impulses from a small part of the auricle and perhaps the external surface of the tympanic membrane join the trigeminal spinal tract on entering the caudal part of the pons.

The **superior olivary nucleus**, an essential link in the auditory path, is at caudal levels of the pons in the ventrolateral tegmentum ventral to the facial nucleus (Fig. 4.7A) and dorsoventral in its orientation. The human superior olive has a conspicuous medial segment and a more difficult to delineate lateral segment. In nonhuman primates, catecholaminergic neurons and fibers form the **ventral catecholaminergic path** that extends from the medullary **lateral reticular nucleus** to a region near the periphery of the superior olivary nucleus and medial to the intrapontine facial fibers. The **ventral catecholaminergic path** continues to rostral pontine levels along a course corresponding to the path of the **central tegmental tract** (Fig. 4.7A), a descending fiber bundle with various origins that ends in the inferior olivary nucleus of the medulla oblongata.

Level of the Entering Trigeminal Nerve (Middle Pons)

A transverse section at this level (Fig. 4.7B) reveals fibers of the **trigeminal nerve** [V]. Here at the middle pons level the basilar pons resembles that in the caudal part of the pons. Corticopontine fibers are greater in number at this level but in smaller bundles.

The anterior half of the scalp to the vertex (except for a small area at the mandibular angle), the cornea, oral mucosa, tongue, dental pulp, and nasal mucosa are innervated by general somatic afferents in the trigeminal nerve [V] that have neurons of origin in the trigeminal ganglion. Central processes of these ganglionic neurons pass to the middle pons as the **trigeminal sensory root**. As they enter the brain stem on the lateral margin of the pons, they mark the boundary between the pons and middle cerebellar peduncle (Fig. 4.7). Fibers continue medially into the pontine tegmentum. The **trigeminal pontine nucleus** is a **general somatic**

afferent nucleus in the lateral part of the pontine tegmentum of the middle pons dorsolateral to the entering trigeminal sensory roots (Fig. 4.7). The trigeminal pontine nucleus is about 4.5 mm in rostrocaudal length. It receives impulses through the trigeminal sensory root whose fibers encapsulate the trigeminal pontine nucleus before entering it.

Medial to trigeminal fibers in the dorsolateral pontine tegmentum is the **trigeminal motor nucleus** (Fig. 4.7). Although nearer the floor of the fourth ventricle, it is in line with other brain stem special visceral efferent nuclei. Its fibers leave the brain stem, form the **trigeminal motor root,** and innervate the muscles of mastication. A pattern of localization is present in the trigeminal motor nucleus for the muscles it innervates including the muscles of mastication, the mylohyoid, tensor veli palatini, tensor tympani, and anterior belly of the digastric. Sectioning the trigeminal motor root seriously impairs chewing. Because of the pull by the nonparalyzed contralateral pterygoid muscles, the resultant paralysis leads to mandibular deviation to the injured side. In addition to the trigeminal motor and sensory roots, small intermediate filaments are between these roots in humans. Such accessory or intermediate fibers are a component of the trigeminal motor root. Clarification of their neuronal origin, peripheral and central connections, may aid in understanding their function. About 27% of ventral root axons leaving the human spinal cord are nonmyelinated and are likely to be sensory in function. Nonmyelinated fibers make up about 12–20% (about 300–1000) of the fibers in the trigeminal motor root in humans. Their function is unclear.

The **trigeminal mesencephalic nucleus** (Fig. 4.7) has pseudounipolar neurons that resemble those in spinal ganglia. Unlike the latter, the trigeminal mesencephalic neurons did not migrate from the developing neural plate but remained in the brain stem. They appear in the lateral periaqueductal gray substance from trigeminal to rostral levels of the midbrain. At trigeminal levels, they serve as neurons of origin for general somatic afferent proprioceptive trigeminal fibers.

Peripheral processes of trigeminal mesencephalic neurons have specialized endings in the masticatory muscles. These fibers enter the trigeminal motor root but bypass the trigeminal motor and pontine nuclei to form the **trigeminal mesencephalic tract** (Fig. 4.7). Neurons of the trigeminal mesencephalic nucleus are scattered along the tract in the lateral part of the pontine periventricular gray substance. Fibers of the trigeminal mesencephalic tract in humans extend caudally to the level of the facial nerve. Some fibers of the facial nerve join the trigeminal mesencephalic tract. Such fibers may be the proprioceptive component of the facial nerve. The trigeminal mesencephalic tract continues into the rostral part of the midbrain (superior collicular level). Trigeminal mesencephalic neurons send their central processes caudalward to the trigeminal pontine or trigeminal motor nucleus. At pontine levels, catecholaminergic neurons surround the **trigeminal mesencephalic tract** of nonhuman primates. Finally, the nucleus of the solitary tract, whose rostral pole extends to about the middle of the trigeminal pontine nucleus, appears immediately dorsal to that nucleus. Surrounding fibers often obscure the rostral part of the nucleus of the solitary tract.

Level of the Pontine Isthmus (Rostral Pons)

The junction of the pons and midbrain, where the fourth ventricle narrows (Fig. 4.8) is termed the **pontine isthmus.** It is the rostral part of the pons, the narrowest part of the brain stem, and the narrowest part of the fourth ventricle. The basilar pons at this level resembles the basilar pons of the middle and caudal parts of the pons in that it has **longitudinal** and **transverse pontine fibers** that are intermingled with the **pontine nuclei** (Fig. 4.8). **Longitudinal pontine fibers** are numerous at this level and less compact than at other pontine levels.

The **superior medullary velum** (Fig. 4.8), a thin layer of tissue between both superior cerebellar peduncles (Fig. 4.8), forms a roof over the rostral part of the fourth ventricle. Each superior cerebellar peduncle has fibers that join the cerebellum with the midbrain and diencephalon. Catecholaminergic neurons and fibers adjoin, and occasionally are in, the superior cerebellar peduncle of nonhuman primates. No entering or emerging cranial nerves occur at the level of the pontine isthmus. Trochlear fibers are identifiable as they cross in the superior medullary velum (Fig. 4.8) before emerging from the dorsal surface of the pons-midbrain junction. Catecholaminergic neurons and fibers occur near or in decussating trochlear fibers in nonhuman primates. Neurons of origin for these trochlear fibers are at caudal levels of the midbrain.

The **medial longitudinal fasciculus** (MLF), a prominent coordinating path in the pons, extends rostrally into the midbrain (Fig. 4.8) and caudally into the medulla oblongata as a well-circumscribed fiber bundle near the median plane (Fig. 4.7B). Serotonergic neurons are among fascicles of the medial longitudinal fasciculus at the level of the isthmus in nonhuman primates. The trigeminal mesencephalic root and nucleus (Fig. 4.8) are identifiable parallel to the lateral wall of the progressively decreasing fourth ventricle. Intermingled in or medial and ventral to the trigeminal mesencephalic neurons are multipolar, pigmented neurons. These two groups of neurons are distinguishable in the lateral periventricular gray substance. In nonhuman primates, pigmented neurons consisting of catecholaminergic (presumably noradrenergic) neuronal cell bodies collectively form the **locus coeruleus** (Fig. 4.8). Occasionally they occur at this level in the lateral wall and roof of the fourth ventricle. Neuronal cell bodies in the human locus coeruleus have small, spherical, cytoplasmic, protein bodies, rich in free basic amino acids. Their presence serves as a marker for catecholamine-synthesizing neurons. A **dorsal catecholaminergic path** in nonhuman primates originates at isthmus levels from the locus coeruleus and continues into the midbrain.

4.4.4. Midbrain

The **superior and inferior colliculi** are rounded eminences on the dorsal surface of the midbrain (Figs 4.3, 4.9) while on the ventral surface of the midbrain are large fiber bundles termed the **cerebral peduncles**. Sections of the midbrain reveal rostral and caudal levels that correspond to levels of the inferior and superior colliculi, respectively. Internally from dorsal to ventral, both levels of the midbrain have a **tectum** [Latin, roof] formed by the colliculi, a **tegmentum** and a **base** (Table 4.2). Between the tectum and tegmentum of the midbrain is the **periaqueductal gray substance** consisting of neurons arranged around the **aqueduct of the midbrain** (Fig. 4.9). The **base of the midbrain** (basis pedunculi) includes a cellular part, the **substantia nigra,** and a fibrous part, the **cerebral crus**. The **tegmentum of the midbrain** (Fig. 4.9) is centrally located on either side of the median plane and extends from the **substantia nigra** to the **aqueduct of the midbrain** (Fig. 4.9). The **periaqueductal gray substance** is continuous caudally with the **periventricular gray substance** at pretrigeminal pontine levels and rostrally with the gray matter lateral to the third ventricle. Different neuronal types and three divisions of the periaqueductal gray substance are identifiable in humans: **nucleus medialis**, an inner ring around the aqueduct; **nucleus dorsalis**, a region of smaller neurons dorsal to the aqueduct; and **nucleus lateralis**, an outer layer dorsolateral in position.

The fibrous part of the basis pedunculi, the **cerebral crus**, has numerous longitudinally oriented fiber bundles that begin in the cerebral cortex, descend into the midbrain, pons, medulla oblongata, and spinal cord. The cellular part of the basis pedunculi, the **substantia**

TABLE 4.2. Internal organization of the human midbrain

Tectum of the midbrain
Roof formed by the colliculi – two levels
1. superior colliculi – defines rostral level of the midbrain
2. inferior colliculi – defines caudal level of the midbrain

Periaqueductal gray
Area of neurons arranged around the aqueduct of the midbrain
At both levels of the midbrain
Continuous rostrally with the periventricular gray

Tegmentum of the midbrain
Centrally located on either side of the median plane
At both levels of the midbrain
Extends from the substantia nigra to the aqueduct of the midbrain

Base of the midbrain
At both levels of the midbrain with cellular and fibrous parts
1. Substantia nigra – cellular part
   A. pars reticularis – contains many dendrites and synaptic endings and is considered the plexiform layer of the substantia nigra; composed mainly of GABAergic neurons
   B. pars compacta – cellular and dorsal, considered the ganglionic layer; dopaminergic neurons
2. Cerebral crus – fibrous part
   Corticospinal, corticobulbar, and corticopontine fibers here

nigra (Fig. 4.9), whose neurons have dark brown neuromelanin pigment, extends from the midbrain into the diencephalon. Neuromelanin is a waste product of catecholamine metabolism and a marker for catecholaminergic neurons. It makes its appearance in the substantia nigra about the third year of life reaching its adult degree of pigmentation by 10–18 years of age. There is a progressive increase in the pigmentation of the substantia nigra until age 60 with a gradual decline from 60 to 90 years. While most of the pigmented substantia nigra is the cellular **compact part**, it has a nonpigmented **reticular part** with many dendrites and synaptic endings as well as **lateral** and **retrorubral parts**. The reticular part in humans is a plexiform layer whose neurons are immunoreactive for parvalbumin. The compact part is a ganglionic layer. The substantia nigra has significant connections with the motor system. The characteristic pathological feature of Parkinson's disease is decreased pigmentation and, at times, reduction in the size of the substantia nigra. The neurons of the substantia nigra (the largest accumulation of neuromelanin containing neurons) and locus coeruleus (the densest accumulation of neuromelanin containing neurons) normally have a few spherical protein bodies that either form clusters among the melanin granules or are evenly dispersed in the neuronal cytoplasm and in proximal dendrites. A noticeable reduction in these protein bodies occurs in the

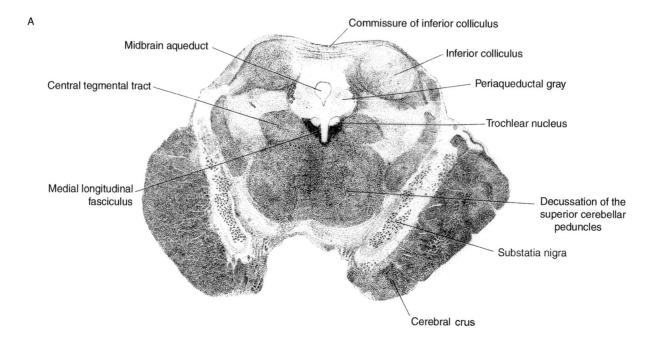

A

Commissure of inferior colliculus

Midbrain aqueduct

Inferior colliculus

Central tegmental tract

Periaqueductal gray

Trochlear nucleus

Medial longitudinal fasciculus

Decussation of the superior cerebellar peduncles

Substatia nigra

Cerebral crus

B

Superior colliculus

Midbrain aqueduct

Cerebral tegmental tract

Accessory oculomotor nucleus

Oculomotor nucleus

Cerebral crus

Medial longitudinal fasciculus

Substantia nigra

Red nucleus

Oculomotor root fibers

FIGURE 4.9. **A,** A transverse section of the caudal half of the adult human midbrain at the level of the inferior colliculus and trochlear nucleus. **B,** Drawing of a transverse section of the rostral half of the adult human midbrain at the level of the superior colliculus and the oculomotor nucleus and nerve [III]. Both drawings are magnified approximately three times.

brains of patients who have suffered from Parkinson's disease. In humans, the substantia nigra has three catecholamines: dopamine (1000 ng/g), norepinephrine (200 ng/g), and epinephrine (about 2 ng/g). Only dopaminergic neurons occur in the compact part.

Level of the Inferior Colliculus and Trochlear Nucleus (Caudal Midbrain)

The inferior colliculi form the tectum of the midbrain at the level of the trochlear nucleus (Fig. 4.9A). A layer of neurons and fibers surrounds each colliculus and interconnects them by way of the **commissure of the inferior colliculus** (Fig. 4.9A). Inside each inferior colliculus is the **nucleus of the inferior colliculus**, a relay station for auditory impulses ascending to higher levels. Deafness results from injury involving both inferior colliculi (such as a cyst compressing on these structures).

The **trochlear nucleus** (Fig. 4.9A), a somatic efferent nucleus, is ventral to the periaqueductal gray substance and the **aqueduct of the midbrain** near the median plane of the midbrain tegmentum. A capsule of fibers separating it from the periaqueductal gray substance identifies its rostral extent. Trochlear fibers proceed around the periaqueductal gray substance to the level of the caudal part of the inferior colliculi. Bundles of trochlear fibers curve dorsal to the **aqueduct of the midbrain**, cross through the superior medullary velum (forming the trochlear decussation), and emerge from the dorsal aspect of the junction of the pons and midbrain caudal to the opposite inferior colliculus. As the trochlear fibers follow this course, they delineate the path of migration of the trochlear neurons in development. Because of this decussation, left trochlear neurons innervate the right superior oblique muscle and vice versa. No significant age-related change in neuronal number occurs in the human trochlear nucleus. Positive immunostaining for choline acetyltransferase (ChAT) is demonstrable in the human trochlear nucleus. ChAT is a specific marker for mapping cholinergic neurons and their paths.

Dorsal to the trochlear nuclei, at inferior collicular levels, is the dorsal nucleus of the raphé or the supratrochlear nucleus. In, and adjoining, this nucleus in nonhuman primates, are densely packed, serotonergic neurons. The trigeminal mesencephalic nucleus and tract occur at the edge of the periaqueductal gray substance. Trigeminal mesencephalic neurons, at or near the trochlear nucleus, may contribute proprioceptive fibers to the trochlear nerve for the superior oblique muscle.

Catecholaminergic neurons are identifiable at this level near the ventrolateral border of the periaqueductal gray substance in nonhuman primates. Fibers of the dorsal catecholaminergic path in nonhuman primates are intermingled with these neurons as this path ascends from the locus coeruleus and into the rostral part of the midbrain (superior collicular level).

The **medial longitudinal fasciculus** is near the median plane indented by the trochlear nucleus that appears partly embedded in it. This fasciculus helps to coordinate activity of certain brain stem nuclei concerned with vestibular and oculomotor functions. The **decussation of the superior cerebellar peduncles**

(Fig. 4.9A), occurring at inferior collicular levels, is centrally located in the tegmentum of the midbrain ventral to the medial longitudinal fasciculus and dorsomedial to the substantia nigra. In general, the decussation of the superior peduncles carries cerebellar fibers to the rostral part of the midbrain and then to the thalamus. Serotonergic neurons occur among and lateral to the most rostral of the decussating fibers of the superior cerebellar peduncles in nonhuman primates.

Level of the Superior Colliculus and Oculomotor Nucleus (Rostral Midbrain)

The tectum at this level presents the **superior colliculus**, which shows alternating layers of neurons and fibers giving it a laminated pattern (Fig. 4.9B) suggesting a high degree of correlation. Seven layers occur in the superior colliculi in humans. A deep white layer is the efferent collicular region, a collection of fibers that border the periaqueductal gray substance. Nearly all efferent superior collicular fibers, destined for the brain stem and spinal cord, pass, in part, through this layer. Many of these efferent fibers cross in the commissure of the superior colliculi. The human superior colliculus has about 5 ng/g of epinephrine.

Innumerable tracts interconnecting regions rostral and caudal to the midbrain characterize the midbrain tegmentum at this level. Here also are paired **oculomotor nuclei** (Fig. 4.9B). They are somatic efferent and extend through the caudal three-fourths of the midbrain. Each nucleus is dorsal and medial to the medial longitudinal fasciculus (Fig. 4.9). These two nuclei fuse caudally and overlap the rostral ends of the trochlear nuclei. Fibers from each oculomotor nucleus pass in the medial longitudinal fasciculus, then ventrally through the medial part of the red nucleus, substantia nigra, and cerebral crus (Fig. 4.9). Catecholaminergic neurons are interspersed in the median plane among the oculomotor fibers before they emerge from the oculomotor sulcus in the interpeduncular fossa. There is a somatotopic arrangement of neurons in the oculomotor nucleus supplying different ocular muscles. Studies in the baboon indicate that neurons supplying the medial rectus, inferior rectus, inferior oblique, and most of the neurons supplying the levator palpebrae superior contribute fibers to the ipsilateral oculomotor nerve whereas superior rectus neuronal cell bodies contribute fibers to the contralateral oculomotor nerve. In addition, neurons supplying the levator palpebrae superioris overlap with medial rectus neuronal cell bodies in the oculomotor nucleus. There is also overlap among neurons that supply the inferior rectus, inferior oblique and superior rectus. The area of the oculomotor nuclei in humans has a high content of epinephrine (15–40 ng/g). All of the large neurons in the human oculomotor nucleus are cholinergic.

The **accessory oculomotor** (Edinger–Westphal) **nucleus** has rostral and caudal parts with different connections and functions. This nucleus is the source of general visceral efferent oculomotor [III] fibers (Fig. 4.9). The rostral part is involved in light reflexes, the caudal part with accommodation. Cholinergic neurons are present in the human accessory oculomotor nucleus.

The **trigeminal mesencephalic nucleus** and its accompanying **tract** are identifiable at superior collicular levels lateral to the periaqueductal gray substance. Processes from trigeminal mesencephalic neurons form the tract – a likely source of proprioceptive fibers for certain muscles innervated by the cranial nerves. Apart from the periaqueductal gray substance, the tegmentum at this level has large and small neurons with the larger referred to as tegmental nuclei. A conspicuous nucleus at rostral midbrain levels is the **red nucleus** (Fig. 4.9). In fresh brains, this nucleus has a reddish color. Fibers of the superior cerebellar peduncle and other paths encapsulate each red nucleus. Since the superior cerebellar peduncles decussate at inferior collicular levels, each half of the cerebellum influences the contralateral red nucleus. Because of its connections with motor structures, the red nucleus plays an essential role in motor activity.

Catecholaminergic fibers ascend from the medulla oblongata, pons, and caudal part of the midbrain to skirt the red nucleus and enter the diencephalon in nonhuman primates. Catecholaminergic neurons adjoin the dorsolateral border of the red nucleus and intermingle laterally with the medial lemniscus. The dorsal catecholaminergic path from the locus coeruleus in nonhuman primates continues through the midbrain and to the thalamus. At rostral midbrain levels, this path is in the ventrolateral part of the periaqueductal gray substance.

FURTHER READING

Alkan A, Sigirci A, Ozveren MF, Kutlu R, Altinok T, Onal C, Sarac K (2004) The cisternal segment of the abducens nerve in man: three-dimensional MR imaging. Eur J Radiol 51:218–222.

Blessing WW, Gai WP (1997) Caudal pons and medulla oblongata. In: *Handbook of Chemical Neuroanatomy.* Björklund A, Hökfelt T, eds. Vol 13: The Primate Nervous System, Part I, pp 139–186. Elsevier, Amsterdam.

Covenas R, Martin F, Salinas P, Rivada E, Smith V, Aguilar LA, Diaz-Cabiale Z, Narvaez JA, Tramu G (2004) An immunocytochemical mapping of methionine-enkephalin-Arg(6)-Gly(7)-Leu(8) in the human brainstem. Neuroscience 128:843–859.

Crosby EC, Humphrey T, Lauer EW (1962) *Correlative Anatomy of the Nervous System.* Macmillan, New York.

DeArmond SJ, Fusco MM, Dewey MM (1989) *Structure of the Human Brain.* 3rd ed. Oxford University Press, New York.

England MA, Wakely J (1991) *Color Atlas of the Brain & Spinal Cord: An Introduction to Normal Neuroanatomy.* Mosby Year Book, St. Louis.

Fedorow H, Tribl F, Halliday G, Gerlach M, Riederer P, Double KL (2005) Neuromelanin in human dopamine neurons: comparison with peripheral melanins and relevance to Parkinson's disease. Prog Neurobiol 75:109–124.

FitzGerald MJT (2004) Cerebellum. In: Standring S, Editor-in-Chief. *Gray's Anatomy.* 39th ed, pp 353–368. Elsevier/Churchill Livingstone, London.

Foote SL (1997) The primate locus coeruleus: the chemical neuroanatomy of the nucleus, its efferent projections, and its target receptors. In: *Handbook of Chemical Neuroanatomy.* Björklund A, Hökfelt T, eds. Vol 13: The Primate Nervous System, Part I, pp 187–215. Elsevier, Amsterdam.

Gluhbegovic N, Williams TH (1980) *The Human Brain: A Photographic Guide.* Harper & Row, Hagerstown.

Hardman CD, McRitchie DA, Halliday GM, Cartwright HR, Morris JG (1996) Substantia nigra pars reticulata neurons in Parkinson's disease. Neurodegeneration 5:49–55.

Hornung JP (2003) The human raphe nuclei and the serotonergic system. J Chem Neuroanat 26:331–343.

Itzev DE, Ovtscharoff WA, Marani E, Usunoff KG (2002) Neuromelanin-containing, catecholaminergic neurons in the human brain: ontogenetic aspects, development and aging. Biomed Rev 13:39–47.

Koutcherov Y, Huang X-F, Halliday G, Paxinos G (2004) Organization of Human Brain Stem Nuclei. In: Paxinos G, Mai JK, eds. *The Human Nervous System.* 2nd ed. Chapter 10, pp 267–320. Elsevier/Academic Press, Amsterdam.

Kulesza RJ Jr (2007) Cytoarchitecture of the human superior olivary complex: Medial and lateral superior olive. Hear Res 225:80–90.

Lachman N, Acland RD, Rosse C (2002) Anatomical evidence for the absence of a morphologically distinct cranial root of the accessory nerve in man. Clin Anat 15:4–10.

Lewis DA, Sesack SR (1997) Dopamine systems in the primate brain. In: *Handbook of Chemical Neuroanatomy.* Björklund A, Hökfelt T, eds. Vol 13: The Primate Nervous System, Part I, pp 263–375. Elsevier, Amsterdam.

Lorke DE, Kwong WH, Chan WY, Yew DT (2003) Development of catecholaminergic neurons in the human medulla oblongata. Life Sci 73:1315–1331.

Marani E, Usunoff KG (1998) The trigeminal motonucleus in man. Arch Physiol Biochem 106:346–354.

Nathan H, Goldhammer Y (1973) The rootlets of the trochlear nerve. Anatomical observations in human brains. Acta Anat (Basel) 84:590–596.

Olszewski J, Baxter D (1982) *Cytoarchitecture of the Human Brain Stem.* 2nd ed. Karger, Basel.

Paxinos G, Huang X-F (2004) *Atlas of the Human Brainstem.* Academic Press, San Diego.

Riley HA (1943) *An Atlas of the Basal Ganglia, Brain Stem and Spinal Cord: Based on Myelin-stained Material.* Williams & Wilkins, Baltimore.

Standring S (2004) Editor-in-Chief. *Gray's Anatomy.* 39th ed, Section 2, Neuroanatomy: Chapter 19, Brain Stem, pp 327–352. Elsevier/Churchill Livingstone, London.

Tracey I, Iannetti GD (2006) Brainstem functional imaging in humans. Suppl Clin Neurophysiol 58:52–67.

Tubbs RS, Oakes WJ (1998) Relationships of the cisternal segment of the trochlear nerve. J Neurosurg 89:1015–1019.

Usunoff KG, Marani E, Schoen JH (1997) The trigeminal system in man. Adv Anat Embryol Cell Biol 136:1–126.

Usunoff KG, Itzev DE, Ovtscharoff WA, Marani E (2002) Neuromelanin in the human brain: a review and atlas of pigmented cells in the substantia nigra. Arch Physiol Biochem 110:257–369.

Zec N, Filiano JJ, Kinney HC (1997) Anatomic relationships of the human arcuate nucleus of the medulla: a DiI-labeling study. J Neuropathol Exp Neurol 56:509–522.

Zec N, Kinney HC (2003) Anatomic relationships of the human nucleus of the solitary tract in the medulla oblongata: a DiI labeling study. Auton Neurosci 105:131–144.

In the last analysis, the cerebral lobes, the cerebellum and the quadrigeminal bodies, medulla oblongata, medulla spinalis, the nerves, all these essentially different parts of the nervous system, all have specific properties, proper functions and distinct effects. And in spite of this marvelous diversity of properties, of function and of effects they constitute nonetheless a unitary system.

Pierre Flourens (1794–1867)

5

The Forebrain

The **prosencephalon** (forebrain) differentiates into the **telencephalon** and **diencephalon** in the fourth week of development. In adults, the **telencephalon** is composed of the **telencephalon medium, cerebral hemispheres, part of the basal nuclei** (caudate and putamen), and the **rhinencephalon**. The adult **diencephalon** consists of the **epithalamus** (Fig. 5.1), **hypothalamus, dorsal thalamus**, and **subthalamus**. The dorsal thalamus is in general termed the 'thalamus'. Table 2.2 summarizes the adult derivatives of the human forebrain along with the associated ventricular cavity for each part of the adult brain.

Brain weight more than doubles in the first postnatal year, quadruples by about three years in males and between three and four years in females, and takes about 15 years to reach the mean highest value for young adults which is 1340 g for females and 1450 g for males. The male brain is greater in weight than the female with an average difference at all ages of about 9.8%. The structural and functional significance of this difference in brain weight is unclear although the greater height and weight of males may account for this difference. A progressive decline in brain weight begins at about 45 or 50 years and continues into advancing age, reaching its lowest value after 86 years, by which time mean brain weight has decreased by about 11%. Such a decrease in brain weight is most likely due to general neuronal wear and tear along with some age-related neuronal loss.

5.1. TELENCEPHALON

5.1.1. Telencephalon Medium

The middle of the telencephalon, the **telencephalon medium** (Fig. 5.1) is dorsal to the **optic chiasma** occupying the rostral part of the third ventricle. The telencephalon medium, a small part of the adult telencephalon, consists of the **preoptic area, anterior commissure**, and **lamina terminalis** (Fig. 5.1). The **preoptic area** lies on either side of the third ventricle, anterior to a line extending from the ipsilateral **interventricular foramen** to the anterior surface of the optic chiasma (Fig. 5.1). Although developmentally a part of the telecephalon medium, the preoptic area functionally

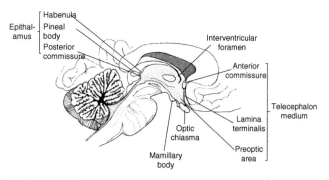

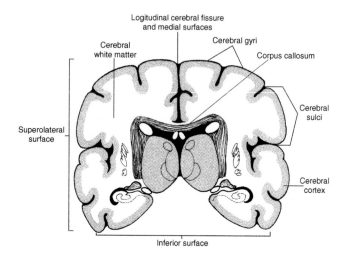

FIGURE 5.1. A median section of the human brain including two subdivisions of the forebrain, the telencephalon medium and the diencephalon.

FIGURE 5.2. A coronal section of the human cerebral hemispheres. Each cerebral hemisphere has a superior, lateral, inferior, and medial surface. The superior and lateral surfaces may be combined into the superolateral face or surface. The medial surfaces of the cerebral hemispheres adjoin each other in the median plane along the longitudinal cerebral fissure.

belongs with the hypothalamus (a diencephalic region). The **lamina terminalis** is a thin membrane forming the rostral wall of the third ventricle. The lamina terminalis interconnects the optic chiasma with the anterior commissure (Fig. 5.1) and the preoptic area of one side with that of the other. The **anterior commissure** is a conspicuous fasciculus composed of interconnections between the temporal lobes that cross the median plane in the upper lamina terminalis. The anterior commissure is 12% larger in females than in males even though the male brain is larger than the female brain. In homosexual men it is 18% larger than it is in heterosexual women and 34% larger than in heterosexual men. The preoptic area, anterior commissure, and lamina terminalis are obvious on the medial surface of each cerebral hemisphere (Fig. 5.1).

5.1.2. Cerebral Hemispheres

The human brain consists of a single cerebrum (one cerebrum) divided into two lateral telencephalic outgrowths that form the right and left **cerebral hemispheres** (two hemispheres) (Fig. 5.2). Each cerebral hemisphere has three poles (frontal, temporal, and occipital), four surfaces (superior, lateral, medial and inferior), and five lobes (frontal, parietal, temporal, occipital, and insular). The cerebral hemispheres completely cover the diencephalon and occupy the anterior and middle cranial fossae. The **lateral ventricles (ventricles 1 and 2)** are fluid-filled cavities in the cerebral hemispheres. As the lateral surface of each cerebral hemisphere balloons out, their medial surfaces meet in the median plane and flatten (Fig. 5.2). Separating the two cerebral hemispheres in the median plane is the **longitudinal cerebral fissure** (Fig. 5.2). Each hemisphere has four surfaces: **superior, lateral, medial**, and **inferior** distinguished by numerous convolutions or **cerebral gyri** (Fig. 5.2) separated from

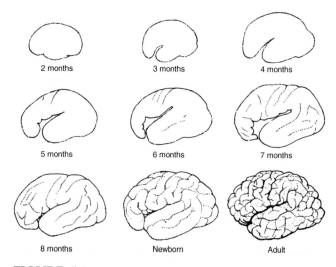

FIGURE 5.3. External appearance of the human left cerebral hemisphere at intervals from the second prenatal month until adult size is achieved (after Lockard, 1948 and Chi, Dooling and Gilles, 1977).

other gyri by grooves, the **cerebral sulci** (Fig. 5.2). Certain deeper sulci form boundaries for the largest divisions of each cerebral hemisphere, the five **cerebral lobes**, namely the **frontal, parietal, temporal, occipital** (see Fig. 5.4), and **insular lobes**. Figure 5.3 illustrates the external appearance of the cerebral hemispheres at monthly intervals. Initial fissuring appears at about 24 weeks and is completed shortly before birth at which time all folds, convolutions, sulci and gyri that characterize the adult brain are present. There is an asymmetry in the developing right and left cerebral hemispheres such

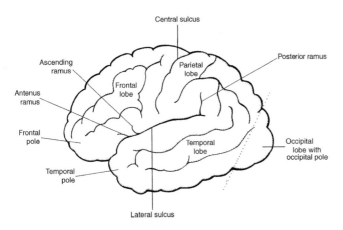

FIGURE 5.4. Superolateral surface of the left cerebral hemisphere including the boundaries and location of four of the five cerebral lobes.

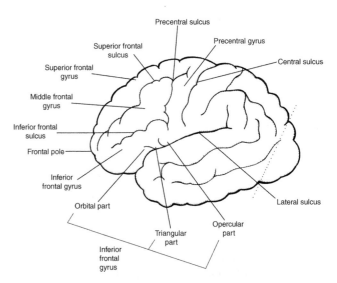

FIGURE 5.5. Superolateral surface of the left cerebral hemisphere including the boundaries, major sulci, and gyri of the frontal lobe.

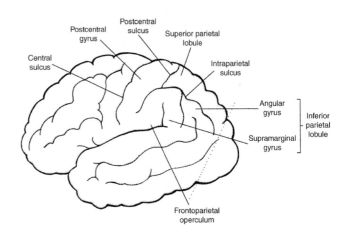

FIGURE 5.6. Superolateral surface of the left cerebral hemisphere including the boundaries, major sulci, and gyri of the parietal lobe.

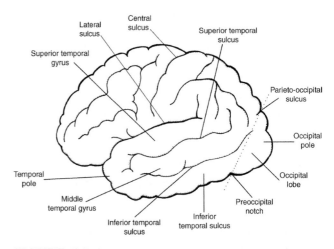

FIGURE 5.7. Superolateral surface of the left cerebral hemisphere including the boundaries, major sulci, and gyri of the temporal and occipital lobes.

that the initial frontal and temporal gyri in the right hemisphere are evident 1–2 weeks earlier than those in the left hemisphere are. Secondary gyri also appear earlier in the right cerebral hemisphere. Only the frontal, temporal, and occipital lobes have identifiable lobes. These poles are useful landmarks in the adult and can be easily identified and followed during development.

Superior and Lateral Surfaces (Superolateral face)

Although each hemisphere has recognizable superior and lateral surfaces, it is difficult to say where one ends and the other begins. A number of structures are continuous from one surface to the other. Thus, for simplicity of description these two surfaces are collectively the **superolateral surface** or face. The **central sulcus** and the **lateral sulcus** are two prominent sulci on the **superolateral surface** (Figs 5.2, 5.4–5.7) of each cerebral hemisphere. Each **central sulcus** makes a 70° angle as it ascends obliquely across the superolateral surface of each cerebral hemisphere separating the frontal lobe, anterior to it, from the parietal and occipital lobes posterior to it (Fig. 5.4). Each **lateral sulcus** (Fig. 5.4) arises from the base of the frontal lobe, passing posterior to separate the temporal lobe inferior to it from the frontal and parietal lobes superior to it. Posteriorly the lateral sulcus has a **posterior ramus** extending into the parietal lobe (Fig. 5.4). Anteriorly, it has an **anterior ramus** and an **ascending ramus**, both of which extend into the frontal lobe (Fig. 5.4). Numerous sulci separate the lobes of each cerebral hemisphere into smaller subdivisions termed **convolutions**.

The term **cortex** [Latin, bark] refers to the 'bark' or gray matter that covers the cerebral surface while the term **pallium** [Latin, cloak] refers to the **cerebral cortex** *and* the subjacent **cerebral white matter** (Fig. 5.2). The human cerebral cortex is thin (varying from 1.5 to 4.5 mm in thickness in different brain areas) and usually composed of six layers of neurons, axons, and dendrites (the number of layers varies in different brain areas). Folding of the cerebral cortex occurs such that only a third of its surface area is visible; two-thirds is invisible and folded, hidden in cerebral gyri (Fig. 5.2).

The **frontal lobe**, essential for voluntary movements of the limbs, body, head, and eyes, also participates in certain aspects of speech and is responsible for the exceedingly complex phenomena of intelligence and behavior in humans. The frontal lobe has a **frontal pole** (Figs 5.3, 5.5) and a characteristic pattern of sulci and gyri. The caudal boundary of the frontal lobe is the **central sulcus**. On the superolateral surface, the inferior boundary of the frontal lobe is the **lateral sulcus** (Fig. 5.4). Parallel to the central sulcus, but anterior to it, is the **precentral sulcus** (Fig. 5.5). Between these two sulci (central and precentral) is the **precentral gyrus**. Each frontal lobe also has **superior** and **inferior frontal sulci** (Fig. 5.5). A middle frontal sulcus is present in some brains. These two sulci begin in the precentral sulcus and extend to the frontal pole roughly parallel to the lateral sulcus (Fig. 5.5). They bound a **superior frontal gyrus**, dorsal to the superior sulcus, an **inferior frontal gyrus** ventral to the inferior sulcus, and a **middle frontal gyrus** between them (Fig. 5.5). Each inferior frontal gyrus has three parts, **orbital**, **triangular**, and **opercular**, bounded by the anterior and ascending rami of the lateral sulcus. The ascending ramus of the lateral sulcus (Fig. 5.4) is the anterior boundary of the **opercular part** of the inferior frontal gyrus (Fig. 5.5). The frontal **operculum** [Latin, lid] helps form a lid or covering for the insular lobe. The anterior ramus of the lateral sulcus (Fig. 5.4) is the posterior boundary of the **orbital part** of the inferior frontal gyrus (Fig. 5.5) while the **triangular part** (Fig. 5.5) is between the anterior and ascending rami of the lateral sulcus.

The **parietal lobe**, involved in recognition and discrimination of certain sensations, also associates various types of sensory information permitting the distinction of differences in textures, perception of small changes in temperature, awareness of position of body parts, and appreciation of position of two stimuli simultaneously applied close together (two-point discrimination or discriminative touch). The parietal lobe allows comparison of qualities of sensation, is involved in certain aspects of speech, and in automatic or following ocular movements (such as occur when the eyes automatically follow a tennis ball from one side of

the net to the other). Immediately behind the **central sulcus** is the **postcentral sulcus** (Fig. 5.6). Between the central and the postcentral sulci is the **postcentral gyrus** of the parietal lobe (Fig. 5.6). Both the **postcentral gyrus** and its sulcus parallel the central sulcus. An **intraparietal sulcus** (Fig. 5.6) begins in the middle part of the postcentral sulcus and runs posteriorly to separate the **superior** and **inferior parietal lobules** (Fig. 5.6). Parts of the inferior parietal lobule and the inferior frontal gyrus contribute to the upper lip of the lateral sulcus termed the **frontoparietal operculum** (Fig. 5.6). The lower part of the inferior parietal lobule includes the **supramarginal** and the **angular gyri** (Fig. 5.6). The former caps the posterior tip of the lateral sulcus (Fig. 5.6) and the latter caps the posterior tip of the superior temporal sulcus (Fig. 5.6). These two gyri are often indistinct, occurring instead as a region contiguous with their respective sulci, they are parts of the human brain network responsible for language.

The **occipital lobe** (Fig. 5.7), deep to the occipital bone of the skull, is concerned with the awareness of vision, including color, form and motion. The occipital lobe and parietal lobe participate in mediating following ocular movements. The posterior tip of each cerebral hemisphere in the occipital lobe is the **occipital pole** (Fig. 5.7). About 44 mm anterior to the occipital pole but on the inferior margin of the superolateral hemispheric surface is the **preoccipital notch** (Fig. 5.7). About 5 cm from the occipital pole, but on the superior margin of the superolateral surface, is an ascending continuation of the **parieto-occipital sulcus** (Fig. 5.7). The parieto-occipital sulcus is on the medial surface of each cerebral hemisphere. A line drawn on the superolateral surface from the parieto-occipital sulcus superiorly, to the preoccipital notch inferiorly is the rostral boundary of the occipital lobe (Fig. 5.7).

The **temporal lobe** (Fig. 5.7) is deep to the temporal bone of the skull and is involved in **audition**, the appreciation of sounds, and with the **vestibular sensation**, the awareness of equilibrium and orientation in space. The temporal lobe supplements voluntary movements normally carried out in the frontal lobe. Finally, the temporal lobe is involved in certain autonomic functions and with certain aspects of language.

The temporal lobe has a **temporal pole** (Fig. 5.7) at its anterior tip. Two sulci, the **superior** and **inferior temporal sulci** (Fig. 5.7), parallel the lateral sulcus and divide the temporal lobe into the **superior**, **middle**, and **inferior temporal gyri** (Fig. 5.7). That part of the superior temporal gyrus covering the insular lobe is the temporal operculum. **Transverse temporal gyri**, related to audition, extend from the superior temporal

TABLE 5.1. Terminology applied to the temporal gyri and sulci

Oldest term	Old term	Newest term	Explanation
First temporal sulcus	Superior temporal sulcus	Superior temporal sulcus	Sulcus on the lateral surface, between the superior temporal gyrus above and middle temporal gyrus below
Second temporal sulcus	Middle temporal sulcus	Inferior temporal sulcus	Sulcus on lateral surface, between middle temporal gyrus above and inferior temporal gyrus below
Third temporal sulcus	Inferior temporal sulcus	Occipitotemporal sulcus	Sulcus on inferior surface, between lateral and medial occipitotemporal gyri
Fourth temporal sulcus	Collateral sulcus	Collateral sulcus	Sulcus on inferior surface, between parahippocampal gyrus medially and medial occipitotemporal gyrus laterally

gyrus into the lateral sulcus. For this reason, these transversely oriented gyri are invisible from the superolateral surface but visible upon separation of the lips of the lateral sulcus. Only one transverse temporal gyrus is present in the left hemisphere but two are in the right hemisphere in most human brains. When present, the single, left transverse temporal gyrus is shorter and forms a more obtuse angle than the right gyri. The right gyri develop earlier, are larger, more often subdivided, but shorter in rostrocaudal length than the one on the left. When one or more transverse temporal gyri occur, they have named sulci forming their boundaries. The terminology applied to the temporal gyri and sulci are somewhat confusing in light of the changes in terminology applied to these structures. The interested reader might want to refer to Table 5.1 for more information comparing these changing terms.

The fifth lobe of the cerebral hemispheres, the **insular lobe** or **insula**, is in the lateral sulcus covered by the opercular parts of the frontal, parietal, and temporal lobes. Removal or displacement of the opercular parts of the frontal and parietal lobes reveals the **insular lobe** (Fig. 5.8) forming the floor of the lateral sulcus. The insular lobe supplements voluntary motor activity carried out in the frontal lobe and is concerned with the awareness of taste and other general visceral sensibilities. The insular lobe, bounded by the **circular insular sulcus** (Fig. 5.8), has several gyri (Fig. 5.8) namely, a **single long insular** and **several short insular gyri**. The **central insular sulcus** (Fig. 5.8) separates the long insular gyrus from the short insular gyri.

Medial Surface

The largest structure on the medial surface of each hemisphere is the **corpus callosum**, an enormous bundle of fibers interconnecting corresponding areas in the two cerebral hemispheres. The size and shape of the corpus callosum varies greatly from individual to individual and there are sex differences in the shape of callosum.

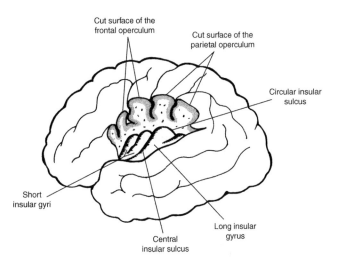

FIGURE 5.8. Superolateral surface of the left cerebral hemisphere. The frontoparietal operculum is dissected away to show the boundaries, major sulci, and gyri of the insular lobe.

In some cases, the corpus callosum is lacking, a condition termed agenisis of the corpus callosum. There are age related changes in the corpus callosum during childhood and adulthood. The whole of the corpus callosum or body of the corpus callosum is composed of a **rostrum, genu, trunk,** and **splenium**. The splenium is the posterior part of the corpus callosum (Fig. 5.9) and the genu is its anterior part with the **trunk of the corpus callosum** between them (Fig. 5.9). The splenium is more bulbous shaped in women with a greater maximum width but it is more tubular shaped in men. The **genu** of the corpus callosum turns inferiorly and posteriorly toward the **anterior commissure** (Fig. 5.9) as the **rostrum** of the corpus callosum. Inferior to the genu and rostrum is the **subcallosal area** (Fig. 5.9) – a collective term including the **subcallosal gyrus** and, inferior to it, the **septal area**. Callosal fibers radiating to the frontal lobe (radiations of the corpus callosum) from the genu form the **frontal forceps**; those radiating from the splenium to the occipital lobe form the **occipital forceps**.

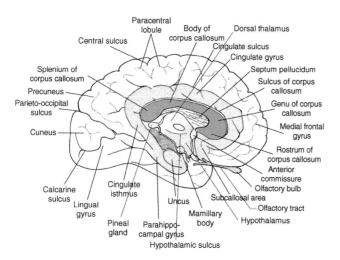

FIGURE 5.9. Medial surface of the left cerebral hemisphere with certain cerebral sulci and gyri.

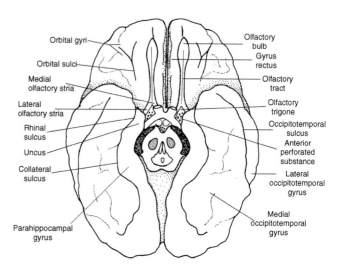

FIGURE 5.10. Inferior surface of the human cerebral hemispheres with certain cerebral sulci, cerebral gyri, and other structures.

Attached to the inferior border of the corpus callosum is the thin **septum pellucidum** (Fig. 5.9) – a transparent, membranous partition that separates the right and left cerebral ventricles and forms their medial wall. The septum pellucidum is bilaminar with a potential space between its layers (cavum septi pellucidi) that is normally absent at birth but may remain open as a median cleft. There is an enormous variability in the reported incidence of cavum septi pellucidi (from 2% to 71.5%). Enlargement of the cavum, though rare in healthy individuals, has been reported in patients with schizophrenia. Dorsal to, and following the contours of the corpus callosum, is the **cingulate gyrus** (Fig. 5.9) that then passes behind the splenium, narrows into the **cingulate isthmus** (Fig. 5.9) before becoming continuous with the **parahippocampal gyrus** (Fig. 5.9). Separating the cingulate gyrus from the corpus callosum is the **sulcus of the corpus callosum** (Fig. 5.9). The **cingulate gyrus** is continuous rostrally with the **subcallosal gyrus** (the dorsal part of the subcallosal area).

The **central sulcus** extends from the superolateral surface on to the medial surface for a short distance (Fig. 5.9). On either side of this sulcus on the medial surface of each cerebral hemisphere is the **paracentral lobule** (Fig. 5.9). It has a contribution from the frontal lobe, the precentral or anterior part of the paracentral lobule, and a contribution from the parietal lobe, the postcentral or posterior part of the paracentral lobule. The remainder of the medial surface of the frontal lobe consists of the **medial frontal gyrus** (Fig. 5.9). The **parieto-occipital sulcus** (Fig. 5.9), the rostral boundary of the occipital lobe, intersects at almost a right angle with the rostral end of the **calcarine sulcus** (Fig. 5.9) with the two sulci then continuing caudally as a common

groove. The calcarine sulcus divides the occipital lobe into the **cuneus** superiorly and the **lingual gyrus** inferiorly (Fig. 5.9). The **precuneus**, part of the parietal lobe, is anterior to the parieto-occipital sulcus (Fig. 5.9) but posterior to the paracentral lobule. The **cingulate sulcus** (Fig. 5.9) follows the contour of the cingulate gyrus and separates it in order from the medial frontal gyrus, the paracentral lobule, and the precuneus.

Inferior Surface

On the inferior surface of each frontal lobe is the **olfactory bulb** (Figs 5.9, 5.10). Extending from each bulb is an **olfactory tract** (Figs 5.9, 5.10) that passes posteriorly along the inferior surface of each cerebral hemisphere and divides into **medial** and **lateral striae** that bound a triangular area, the **olfactory trigone** (Fig. 5.10). Behind the trigone is an area perforated by numerous blood vessels, the **anterior perforated substance** (Fig. 5.10). Each olfactory bulb and tract lies in relation to the **olfactory sulcus** of the frontal lobe but the olfactory sulcus is itself hidden from view by the bulb and tract. Medial to the sulcus is the **gyrus rectus** (Fig. 5.10). Lateral to each bulb and tract are several **orbital sulci** arranged in the shape of a Y or H. These sulci separate the surface into various **orbital gyri** (Fig. 5.10).

Each cingulate gyrus passes beneath the splenium of the corpus callosum, enters the temporal lobe, and becomes continuous with the **parahippocampal gyrus**. On the rostromedial part of this gyrus is the hooklike **uncus** (Figs 5.9, 5.10). The **collateral sulcus** (Fig. 5.10) separates each parahippocampal gyrus from the remainder of the temporal lobe. The rostral end of the

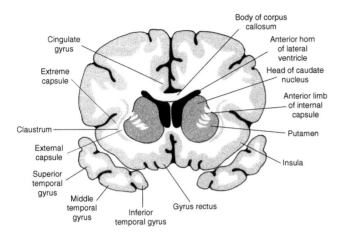

FIGURE 5.11. A coronal section of the rostral part of the human cerebral hemispheres at the level of the anterior limb of the internal capsule where it separates the caudate nucleus from the putamen.

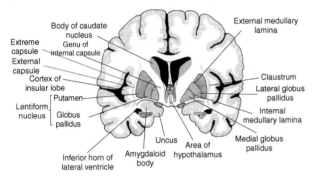

FIGURE 5.12. A coronal section of the human cerebral hemispheres at the level of the genu of the internal capsule.

collateral sulcus is the **rhinal sulcus** (Fig. 5.10). These collateral and rhinal sulci may not be continuous. An **occipitotemporal sulcus** may be present on the inferior surface of the occipital and temporal lobes (Fig. 5.10) separating the **lateral** and **medial occipitotemporal gyri** (Fig. 5.10).

5.1.3. Basal Nuclei

Anatomically there are five subcortical masses of gray matter buried in the depths of the cerebral hemispheres. These five nuclear groups, the **caudate nucleus**, **putamen**, **claustrum** (Figs 5.11, 5.12), **globus pallidus** and the **amygdaloid body (complex)** (Fig. 5.12), belong to the telencephalon and were formerly termed the deep telencephalic nuclei. They are most clearly visible on coronal sections through the cerebral hemispheres. The *Terminologia Anatomica* (1998) uses the term **basal nuclei** to refer to three of these five nuclei (namely the caudate nucleus, putamen, and globus pallidus) but lists the claustrum and amygdaloid body as constituents of the

TABLE 5.2 The basal nuclei

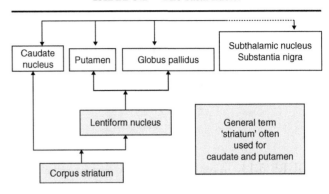

basal forebrain. The amygdaloid body is a component of the limbic system. In the following account, we will use the term **basal nuclei** more broadly to include the following five structures: the **caudate nucleus**, **putamen**, and **globus pallidus** along with the **substantia nigra** and **subthalamic nucleus** (parts of the mesencephalon and diencephalon, respectively) (Table 5.2). This affiliation of these five structures under the heading **basal nuclei** is a functional, not an anatomical or developmental, consideration. Indeed the caudate, putamen and globus pallidus are deep telencephalic structures whereas the substantia nigra is a mesencephalic structure and the subthalamic nucleus is a diencephalic structure. The basal nuclei are an important part of the extrapyramidal motor system along with extrapyramidal areas in the cerebral cortex.

A number of areas of the human nervous system present challenges with regard to terminology and nomenclature. One of these areas is the basal nuclei. It is standard practice among neuroanatomists to designate a collection of neurons in the peripheral nervous system as a **ganglion** and a collection of neurons in the central nervous system as a **nucleus**. For this reason, the use of the term basal nuclei appears justified for the five nuclear groups described above. However, the term **basal ganglia** is also used for many of these same structures and this term is sanctioned by usage and deeply engrained in the literature and in textbooks.

Two of the basal nuclei, the putamen and globus pallidus, are collectively termed the **lentiform nucleus** (Fig. 5.12) because of their shape [Latin, lens-shaped]. The term **corpus striatum** refers collectively to the caudate nucleus, putamen, and globus pallidus whereas the term **striatum** applies generally to the caudate and putamen only. Thus, in speaking of projections to the basal nuclei from various areas of the brain without specifying a particular nucleus of the basal nuclei, we

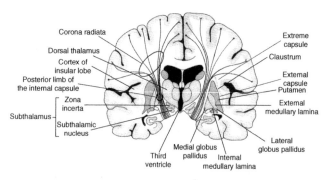

FIGURE 5.13. A coronal section of the human cerebral hemispheres at the level of the posterior limb of the internal capsule. The cortical origin and partial course of the projection fibers that form the internal capsule are illustrated including fibers of the corona radiata.

would use such terms as 'corticostriate,' 'thalamostriates,' or 'nigrostriate'.

Caudate Nucleus

The **caudate nucleus** is a constituent of the basal nuclei and one of several important C-shaped structures in the human brain. Each **caudate nucleus** has a relatively large head (Fig. 5.11) projecting from the lateral side into the **frontal (anterior) horn of the lateral ventricle** (Fig. 5.11) of each cerebral hemisphere. The **head of the caudate** narrows into the **body of the caudate** near the **interventricular foramina**. The **body of the caudate** (Fig. 5.12) narrows considerably as it turns into the temporal lobe to become the **tail of the caudate** that then curves along the roof of the ventricle and extends rostrally to fuse with the **amygdaloid body** (Fig. 5.12).

Putamen

The ventral part of the head of the caudate is continuous with the corresponding part of the **putamen** (Fig. 5.11). Functionally and structurally, the caudate nucleus and putamen represent a common mass incompletely separated in the human brain by a large band of fibers, the **anterior limb of the internal capsule** (Fig. 5.11). A fiber bundle, the **external** or **lateral medullary lamina** (Figs 5.12, 5.13), separates the putamen from the globus pallidus. As the putamen is followed in sections through each cerebral hemisphere (Figs 5.12, 5.13), it begins to disappear at caudal levels of the diencephalon.

Globus Pallidus

The **globus pallidus** [Latin, pale] – also called the pallidum – has myelinated fibers that give it a pale appearance and its name. A band of fibers, the **lateral or**

external medullary lamina (Figs 5.12, 5.13), demarcates the globus pallidus on its lateral aspect from the putamen. A **medial** or **internal medullary lamina** (Figs 5.12, 5.13) separates the globus pallidus into **lateral** or **external** (often abbreviated as GPe) and **medial or internal** (often abbreviated as GPi) segments (Figs 5.12, 5.13). There may be three segments of the human globus pallidus. Viewed together the putamen and globus pallidus are lens-shaped and therefore called the **lentiform nucleus** (Fig. 5.12). The caudate and lentiform nucleus (putamen plus the globus pallidus) collectively form the **corpus striatum**.

Subthalamic Nucleus

The **subthalamic nucleus**, shaped like a biconvex lens, lies along the medial border of the internal capsule, inferior to the **hypothalamic sulcus** (Fig. 5.9) and lateral to the hypothalamus and **third ventricle** (Fig. 5.13). It lies in close relation to the rostral continuation of the substantia nigra and has reciprocal connections with the globus pallidus by way of a fiber bundle, the **subthalamic fasciculus**. Structurally the **subthalamic nucleus**, **zona incerta** as well as the **nuclei of perizonal fields** (of Forel), designated as **H, H1**, and **H2 fields of Forel**, belong to the **subthalamus** (Fig. 5.13). Functionally the subthalamic nucleus is an important constituent of the basal nuclei.

Substantia Nigra

In the unstained human brain, neurons in the **substantia nigra** are identified by their dark brown neuromelanin pigment, giving this cell mass a distinctive black color and its name. Neuromelanin is a waste product of catecholamine metabolism. The substantia nigra, extending through the midbrain and into the diencephalon, consists of four parts: **compact, reticular, lateral**, and **retrorubral**. Neurons of the human substantia nigra contain three different catecholamines of which dopamine (1000 ng/g) is the most prominent. Dopamine synthesized in the pigmented cells of the compact part of the substantia nigra reaches the striatum. Based on functional grounds, the substantia nigra is an important constituent of the basal nuclei.

Claustrum

Immediately lateral to the putamen is a thin but distinct band of fibers, the **external capsule** that separates the **putamen** from the **claustrum** (Figs 5.11, 5.12, 5.13). Another band, the **extreme capsule**, is between the lateral surface of the claustrum and the cortex of the **insular lobe** (Figs 5.12, 5.13). A ventral extension of the claustrum continues into the temporal lobe approaching the amygdaloid body. Usually no sharp transition is apparent between these structures, as parts of the caudate, putamen, claustrum, and amygdaloid body are closely associated. The claustrum is a constituent of the basal forebrain but not the basal nuclei.

5.1.4. Rhinencephalon

The term **rhinencephalon** [Greek, nose brain] refers to cerebral structures related to olfaction, particularly in lower animals. In humans, this area of the brain consists of many structures on the medial and inferior surfaces of each cerebral hemisphere. One, the **anterior perforated substance**, is posterior to the **olfactory trigone** and best visualized on the inferior surface of the brain (Fig. 5.10). Another, the **subcallosal area,** is on the medial surface of each cerebral hemisphere inferior to the genu and rostrum of the corpus callosum. It includes a dorsally located **subcallosal gyrus** and a ventrally located **septal area**. As the cingulate gyrus turns into the subcallosal area (Fig. 5.9) it is continuous with the subcallosal gyrus that is inferior to the **rostrum of the corpus callosum** (Fig. 5.9).

5.2. DIENCEPHALON

The central part of the forebrain, the **diencephalon**, forms part of the walls in the middle and posterior parts of the third ventricle. The **epithalamus** (Fig. 5.1), **dorsal thalamus** (Figs 5.9, 5.13), **subthalamus** (Fig. 5.13), and **hypothalamus** (Figs 5.9, 5.12) are recognizable constituents of the diencephalon. The dorsal thalamus is generally termed the **thalamus**.

5.2.1. Epithalamus

The **epithalamus**, the smallest part of the diencephalon, is at the posterior end of the third ventricle. It includes the **pineal gland** and **habenula** (Fig. 5.1). The **pretectal area** and **pretectal nuclei** are parts of the epithalamus, as are three fiber bundles: the **posterior commissure** (Fig. 5.1), **habenular commissure**, and the **habenulo-interpeduncular tract**.

The pineal gland, weighing only 173 mg in humans, is a neuroendocrine transducer that releases a chemical substance in response to neural stimuli by means of autonomic nerves that stimulate neurons in the pineal gland to synthesize and release melatonin. In particular, postganglionic neurons from the superior cervical ganglia release norepinephrine that stimulates the $beta_1$-adrenergic receptors on the pinealocytes. No extrapineal production of melatonin is identifiable though it occurs in the human retina. High concentrations of melatonin occur in the pineal gland during menstruation with lower levels occurring during ovulation, suggesting that melatonin in humans is concerned with reproductive function. Melatonin and hence the pineal gland are of importance in metabolic regulation in humans. Since the pineal gland in adult humans is partially calcified, it becomes a useful landmark on plain skull X-rays. Its location, relative to the median plane, serves as a useful indicator of shifts involving intracranial contents.

The **habenula**, a small nucleus in the posterior diencephalon (Fig. 5.1) and another external feature of the epithalamus, is involved in correlating olfactory and gustatory sensations. Each posterior and habenular commissure has a few fibers. Their function in primates is uncertain.

5.2.2. Thalamus

The dorsal thalamus, generally referred to as the thalamus, makes up most of the human diencephalon (Figs 5.9, 5.13). Composed of at least 30 named nuclear groups (see Table 14.1) with various functions, the **thalamus** is a sensory relay station for auditory, vestibular, visual, and somatosensory impulses. Functionally the thalamus reinforces and regulates cortical activity. Some aspects of extremes of temperature and pain, general tactile impulses, vibration, and gustatory sensations, likely reach consciousness at thalamic levels in humans. Discriminative, well-localized sensations, however, require cortical participation. Complex correlations of sensory information that lead to feelings of well-being, pleasure, or displeasure depend on normal functioning of the thalamus. Certain thalamic nuclei are involved in motor activity. The thalamus is essential to the relay of information from all directions to specific destinations. The proximity of the thalamus, surrounding basal nuclei, internal capsule, and third ventricle is clinically significant.

5.2.3. Subthalamus

The **subthalamus** (ventral thalamus) has several nuclei that serve to relay impulses involved in motor activity. Impulses from the other constituents of the basal nuclei project to the subthalamus, an essential component of the motor system. The subthalamus consists of the **sub-thalamic nucleus**, **zona incerta**, and **nuclei of perizonal fields** (of Forel) designated as **H**, **H1**, and **H2**. **H** corresponds to the nucleus of the medial field, **H1** corresponds to the nuclei of the dorsal field along with fibers of the thalamic fasciculus, and **H2** corresponds to the nucleus of the ventral field along with the fibers of the lenticular fasciculus. The close proximity of the subthalamus to the thalamus and hypothalamus is clinically significant.

5.2.4. Hypothalamus

Although it is only four grams of tissue, the **hypothala-mus** consists of almost 30 named nuclei (see Table 19.1) along the walls of the third ventricle (Fig. 5.12). These nuclei are concerned in humans with diverse activities such as water balance, hunger, temperature regulation, sleep and wakefulness, emotional expression, activation and regulation of the autonomic nervous system, and neuroendocrine regulation. The junction of the midbrain and diencephalon traverses the epithalamic commissure dorsally and the caudal part of the **mamillary bodies** [from Latin, mamilla] of the hypothalamus ventrally (Figs 5.1, 5.9).

5.3. CEREBRAL WHITE MATTER

Numerous fiber bundles between the cerebral cortex and the basal nuclei form the white matter of the cerebral hemispheres (Fig. 5.2). Fibers in the white matter are of three types: **projection**, **commissural**, and **association fibers**.

Projection fibers arise in the cerebral cortex and end in various subcortical regions or vice-versa. The **internal capsule**, an example of projection fibers (Fig. 5.13), expands like an open fan of radiating fibers toward the cerebral cortex. Superior to the basal nuclei and thalamus, horizontal fibers of the corpus callosum interdigitate with the vertically radiating fiber bundles of the internal capsule. This interdigitation separates the internal capsule into smaller fascicles creating a crown of fibers called the **corona radiata** (Fig. 5.13). Inferiorly, internal capsular fibers collect and funnel into the cerebral crus at the base of the midbrain. Based on its relation to the basal nuclei and thalamus, the internal capsule is divisible into an **anterior limb** (Fig. 5.11), **genu** (Fig. 5.12), and a **posterior limb** (Fig. 5.13). Each **anterior limb** (Fig. 5.11) is between the head of the caudate nucleus and

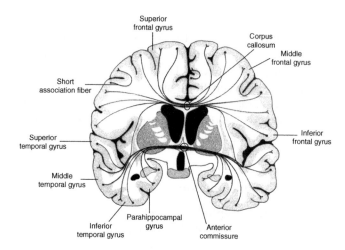

FIGURE 5.14.   A coronal section of the human cerebral hemispheres, at the level of the anterior commissure. Some commissural fibers that form the corpus callosum and the anterior commissure are illustrated.

the rostromedial border of the lentiform nucleus. A **genu** (Fig. 5.12) is at the level of the interventricular foramen and bordered laterally by the apex of the globus pallidus, rostrally by the caudate nucleus, and posteriorly by the thalamus. Each **posterior limb** has the lentiform nucleus caudolaterally and the thalamus caudomedially (Fig. 5.13). Each posterior limb has postlenticular and sublenticular parts reflecting the relation of the posterior limb to the lentiform nuclei. Injury to the internal capsule may involve the basal nuclei and parts of the thalamus.

Association fibers interconnect cortical areas in the same cerebral hemisphere. They are long or short, depending on the distance separating the areas they connect. **Short association fibers** (Fig. 5.14) interconnect adjacent gyri and remain intracortical or travel through the white matter. **Commissural fibers**, such as the **corpus callosum** and the **anterior commissure** (Fig. 5.14), interconnect the two cerebral hemispheres. Many fibers in the corpus callosum are collaterals of projection and association fibers.

The infinite complexity of brain function in humans is the result of rich and varied correlations that occur by means of projection, commissural, and association fibers. The number of these connections is enormous and the cortical associations they allow are exceedingly complex. Additional study of the functions of these fibers in humans would be beneficial.

FURTHER READING

Allen LS, Gorski RA (1991) Sexual dimorphism of the anterior commissure and massa intermedia of the human brain. J Comp Neurol 312:97–104.

Allen LS, Gorski RA (1992) Sexual orientation and the size of the anterior commissure in the human brain. Proc Natl Acad Sci USA 89:7199–7202.

Allen LS, Richey MF, Chai YM, Gorski RA (1991) Sex differences in the corpus callosum of the living human being. J Neurosci 11:933–942.

Chi JG, Dooling EC, Gilles FH (1977) Gyral development of the human brain. Ann Neurol 1:86–93.

Crosby EC, Humphrey T, Lauer EW (1962) *Correlative Anatomy of the Nervous System.* Macmillan, New York.

Crosby EC, Schnitzlein HN (1982) *Comparative Correlative Anatomy of the Vertebrate Telencephalon.* Macmillan, New York.

Crossman AR (2004) Basal ganglia. In: Standring S, Editor-in-Chief. *Gray's Anatomy.* 39th ed, pp 419–430. Elsevier/Churchill Livingstone, London.

Crossman AR (2004) Cerebral hemisphere. In: Standring S, Editor-in-Chief. *Gray's Anatomy.* 39th ed, pp 387–418. Elsevier/Churchill Livingstone, London.

Crossman AR (2004) Diencephalon. In: Standring S, Editor-in-Chief. *Gray's Anatomy.* 39th ed, pp 369–386. Elsevier/Churchill Livingstone, London.

DeArmond SJ, Fusco MM, Dewey MM (1989) *Structure of the Human Brain.* 3rd ed. Oxford University Press, New York.

Dekaban AS, Sadowsky D (1979) Changes in brain weights during the span of human life: relation of brain weights to body heights and body weights. Ann Neurol 4:345–356.

England MA, Wakely J (1991) *Color Atlas of the Brain and Spinal Cord: An Introduction to normal neuroanatomy.* Mosby Year Book, St. Louis.

Flores LP (2002) Occipital lobe morphological anatomy: anatomical and surgical aspects. Arq Neuropsiquiatr 60:566–571.

Gluhbegovic N, Williams TH (1980) *The Human Brain: A Photographic Guide.* Harper & Row, Hagerstown.

Iaria G, Petrides M (2007) Occipital sulci of the human brain: Variability and probability maps. J Comp Neurol 501:243–259.

Jackson GD, Duncan JS (1996) *MRI Neuroanatomy: A New Angle on the Brain.* Churchill Livingstone, New York.

Mai JK, Assheuer J, Paxinos G (2004) *Atlas of the Human Brain.* 2nd ed. Elsevier/Academic Press, Amsterdam.

Ono M, Kubic S, Abernathey CD (1990) *Atlas of the Cerebral Sulci.* Thieme Medical Publishers, New York.

Paxinos G, Huang X-F (2004) *Atlas of the Human Brainstem.* Academic Press, San Diego.

Rhoton AL Jr (2002) The cerebrum. Neurosurgery 51: [Suppl 1] 1–51.

Standring S (2004) Editor-in-Chief. *Gray's Anatomy.* 39th ed. Section 2, Neuroanatomy: Chapter 22. Cerebral Hemisphere, pp 387–418. Elsevier/Churchill Livingstone, London.

Young PA, Young PH (1997) *Basic Clinical Neuroanatomy.* Lippincott Williams & Wilkins, Philadelphia.

A multitude of tracks over which there is run a multitudinous Marathon. Ten to fifteen thousand million tracks with branch tracks are in the human brain alone. Messages are hurrying everywhere. Each helps to join the north of the body to the south, the east to the west, to blend the total into the affairs of planet, galaxies, universe, the affairs of the inanimate, the animate.

Gustav Eckstein, 1970

6

Introduction to Ascending Sensory Paths

6.1. RECEPTORS

The nervous system is a constellation of **neurons** and **neuroglial cells** that are concerned with the **reception** of stimuli, their **transduction** into **nerve impulses**, the **conduction** of such impulses to various parts of the nervous system, and the ultimate **association, correlation, appreciation,** or **interpretation** of these impulses. **Receptors** in skin, other tissues, and organs are continuous with nerves putting them in functional continuity with the nervous system. Individual nerve fibers show

a spectrum of diameters and it is possible to record activity from this entire spectrum in humans. Such experiments indicate a close correlation between activities in fibers of different sizes and different sensations. Nonmyelinated fibers constitute the majority in cutaneous nerves. Table 6.1 provides a functional classification of fibers in human peripheral nerves based on their diameter and conduction velocity.

Although skin in humans has abundant sensory nerves, the density of innervation varies. There is a dense innervation of the face and limbs but a sparse

TABLE 6.1.   Classification of peripheral nerve fibers by diameter and conduction velocity

Fiber type	Diameter, micrometers	Myelinated or nonmyelinated	Conduction velocity, m/sec	Structures innervated
Aα (Ia, Ib)	Largest diameter (13–22 μm)	Myelinated	Fastest conduction 70–120 m/sec	Alpha motoneurons; neuromuscular spindles (primary fibers for muscle length and rate of change of length) and neurotendinous spindles (muscle tension)
Aβ (II)	Large diameter (8–13 μm)	Myelinated	Fast conduction 69.1 ± 7.4 m/sec (range 25–75)	Neuromuscular spindles (secondary fibers), tactile disks (contact, pressure, form, texture = light touch), tactile corpuscles (contact, velocity, movement = discriminative touch), lamellar corpuscles (pressure, high frequency vibration, stretch), Ruffini corpuscles (contact and skin stretch = light touch) and nociceptors
Aγ	Medium diameter (4–8 μm)	Myelinated	Medium conduction 15–40 m/sec	Gamma motoneurons
Aδ (III)	Small diameter (1–4 μm)	Myelinated	Slow conduction 10.6 ± 2.1 m/sec (range 5–30)	Free nerve endings (sharpness, pinprick or first pain) and free nerve endings (coolness, 15–45°C)
B	Very small diameter (1–3 μm)	Lightly myelinated	Very slow conduction 3–14 m/sec	Preganglionic autonomic efferents and visceral afferents
C (IV)	Smallest diameter (0.4–1.2 μm)	Nonmyelinated	Slowest conduction 1.2 ± 0.2 m/sec (range 0.5–2.0)	Postganglionic autonomic, free nerve endings (slow, burning second pain, heat pain, cold pain), free nerve endings (warmth, 20–40°C), blood vessels, sweat glands, and arrector pili muscles

Modified from Hagbarth, 1979; Light and Perl, 1993; Tran et al., 2001; Djouhri and Lawson, 2004; Mann, 2004; Mountcastle, 2005.

innervation of the posterior surface of the trunk. Body parts that reach out to explore our sensory world, such as our hands and feet, have a dense innervation as does the face and genitalia but not the hairy skin of the body, though hairs themselves are generously innervated.

6.2. CLASSIFICATION OF RECEPTORS BY MODALITY

Using **modality** or type of sensation to which a given receptor is sensitive as a means of classifying receptors, there are several types of identifiable receptors. These include **mechanoreceptors, thermoreceptors, nociceptors, chemoreceptors, photoreceptors,** and **osmoreceptors.**

6.2.1. Mechanoreceptors

Mechanoreceptors respond to nonpainful mechanical displacements or forces applied to them. There are musculoskeletal, cutaneous, visceral, and serosal mechanoreceptors.

Musculoskeletal Mechanoreceptors

Musculoskeletal mechanoreceptors are in muscle, at musculotendinous junctions, and in joints. These receptors respond to changes in muscle length or tension, and movements of joints.

Cutaneous Mechanoreceptors

Cutaneous mechanoreceptors include a variety of structures; receptors responsive to general or light touch and pressure are an example. Brushing a wisp of cotton across the skin, pressing on the skin with the head of a pin, or lightly touching it with a fingertip stimulates them. Endings in follicles of hairs are an example of cutaneous mechanoreceptors serving general or light touch. In hairless (glabrous) skin there are four identifiable cutaneous mechanoreceptors related to tactile perception including tactile disks (of Merkel), tactile corpuscles (of Meissner), lamellar corpuscles (of Pacinia), and Ruffini corpuscles. Each responds to cutaneous motion and deformation in a different way.

Visceral Mechanoreceptors

Visceral mechanoreceptors occur throughout the thorax, abdomen, and pelvis. Carotid baroreceptors in the wall of the carotid sinus that slow heart rate and lower systolic arterial pressure subsequent to changes in blood pressure (via reflex action) are an example of visceral mechanoreceptors. The aorta and brachiocephalic arteries also have baroreceptors. Other visceral mechanoreceptors include lung irritant receptors and pulmonary stretch receptors in the lungs and airways as well as atrial volume receptors, ventricular pressure receptors, and epicardial and pericardial receptors in the heart. Visceral mechanoreceptors in nonstriated muscle of abdominal viscera respond to steady mechanical stimuli such as compressing, stroking, or stretching the abdominal viscera. We are consciously aware of stomach or bladder 'fullness' because of the transmission of impulses from the walls of these organs. Such receptors help to effect visceral reflex activity.

Serosal Mechanoreceptors

Serosal mechanoreceptors respond to traction on the mesentery and to compression or distention of the adjacent organs. Serosal mechanoreceptors at the root of the mesentery and along the branches of its arteries signal intestinal movement. Mucosal receptors, along the urethra, respond to the flow of fluid.

Mechanoreceptors in the internal ear receive sound waves by means of a complex sequence of movements of the tympanic membrane, auditory ossicles, and the associated fluid medium, causing stimulation of sensory hair cells in the cochlea. More information on these receptors is available in Chapter 10. Another example of mechanoreceptors are the vestibular receptors in the inner ear specialized for the sense of awareness of position or movement of the body caused by the effects of gravity, acceleration and deceleration (equilibrium and orientation).

6.2.2. Thermoreceptors

Thermoreceptors respond to small changes in skin temperature, especially cool and warm temperatures, but are insensitive to mechanical stimuli. A dual set of such receptors probably exists in human skin as separate 'cold' and 'warm' spots yielding cold or warm sensations with natural or electrical stimuli. Hence, fibers for cold and warm sensations occur in human cutaneous nerves. Thermal sensations result from appropriate stimuli over a temperature range of 13–45°C. Small diameter, slowly conducting, myelinated Aδ fibers innervate thermoreceptors for cold whereas receptors for warmth, active at normal skin temperature (about 32°C and above), are innervated by the smallest diameter, slowly conducting, nonmyelinated C fibers. The former are more numerous, more sensitive, and nearer the surface. Thermal sensation is evaluated using test tubes containing cold (15–20°C) or warm (40–45°C) water. 'Hot,' a distinct sensory quality results from stimulation of the human skin near 46°C. Temperatures from 47–49°C, applied to skin, result in superficial, well-localized, intense pain; temperatures of 11–13°C are painful but deeply localized.

6.2.3. Nociceptors

Noxious insults activate nociceptors and evoke pain (with possible tissue injury). An example of nociception would be driving a pin into or through the skin yielding tissue damage. There are several types of nociceptors including **thermal nociceptors**, which respond to extremes of temperature, **mechanical nociceptors**, which respond to mechanical force, **chemical nociceptors**, that respond to certain chemical irritants and **polymodal nociceptors**, which respond to a combination of thermal, mechanical, and chemical stimuli. If such stimuli are prolonged, tissue injury takes place with liberation of substances that stimulate these nociceptors.

Mechanical Nociceptors

Pinching, pricking, or scraping the skin stimulates mechanical nociceptors innervated by either nonmyelinated C fibers or myelinated, small diameter Aδ fibers. Nonmyelinated C fibers mediate poorly localized, burning pain, whereas the Aδ fibers serve superficial, local pain such as occurs with the prick of a pin (the sense of 'sharpness').

Thermal Nociceptors

Heating skin more than 46°C or cooling it below 11–13°C, will stimulate thermal nociceptors. Ordinary thermal receptors are distinguishable experimentally from thermal nociceptors. Nerves to thermal nociceptors are either nonmyelinated C fibers or small myelinated Aδ fibers.

Chemical Nociceptors

Located in skin, chemical nociceptors are stimulated by itching powder, a nettle leaf, acetic acid, or by other chemicals such as histamine, serotonin, bradykinin, prostaglandins, or substance P. These receptors also respond to the by-products of substances released from injured tissue. Nociceptors in viscera (including the heart, lungs, testes, and biliary system), muscle, and joints receive their innervation from thinly myelinated Aδ or nonmyelinated C fibers. Visceral pain results from the release of a chemical substance from injured tissue that stimulates visceral nociceptors. An example is cardiac pain (angina pectoris) owing to inadequate blood flow in the coronary arteries (acute coronary insufficiency) that leads to tissue ischemia. As the cardiac muscle continues to work, substances released from the ischemic tissue concentrate locally thus stimulating visceral nociceptors.

Visceral pain also results from urinary tract infections (e.g. cystitis), gastric ulcerations, gastric distension, and forms of gastroenteritis (e.g. colitis, appendicitis). In addition, visceral pain results from inflammation of visceral or parietal peritoneum, powerful contractions of abdominal organs (spasmodic colic), distention caused by stones in the gallbladder, biliary ducts, renal calyces or pelves, and in the urinary bladder, or by impacted feces in the colon, or from uterine contractions in labor. In these examples, there is either the production of a substance stimulating chemical nociceptors or excessive mechanical force (stretch or tension) directly stimulating mechanical nociceptors.

4. Polymodal Nociceptors

Nociceptors that respond to a combination of thermal, mechanical, or chemical stimulation are termed **polymodal nociceptors**. Areas of skin innervated by polymodal nociceptors often overlap such that a single noxious stimulus activates several nociceptors.

6.2.4. Chemoreceptors

Chemoreceptors (gustatory and olfactory receptors) are sensitive to substances in solution. The receptors for **gustatory sensation** in humans are modified epithelial structures (taste buds) on specialized papillae of the tongue and soft palate. Taste buds contain a number of taste cells (perhaps up to 40) and while each cell is sensitive to one kind of taste stimuli (salt, sweet, sour, bitter, or umami) most taste buds may respond to stimuli from two or three types of taste. More information on these receptors is available in Chapter 17.

Chemoreceptors for **olfactory sensation** occur in the olfactory epithelium. They are modified bipolar neurons with bulbous endings directed toward the nasal cavity. Interaction with odorant molecules takes place at the olfactory receptor cell membrane. Molecules of certain

shapes presumably fit into specific sites along receptor membranes yielding changes in the permeability of the membrane that leads to impulses in the olfactory nerves.

The largest collections of chemoreceptors in humans are the **carotid bodies** in the angles of the carotid bifurcation. Another collection of arterial chemoreceptors, the **aortic body**, is at the root of the aorta. Each carotid body is a distinct pale pink structure (small, flat, ovoid, weighing nearly 4 mg, and measuring about $2 \times 5 \times 6$ mm in humans). Chemoreceptors respond to changes in composition of their extracellular fluid. Alterations in pH, oxygen tension, and to a lesser extent, carbon dioxide tension in arterial blood, will stimulate them, increasing the rate and depth of respiration. A prolonged decrease in oxygen to the carotid body occurs in pulmonary emphysema, or at high altitudes, causing enlargement of the carotid bodies, or **carotid body hyperplasia**. Blood flow in a carotid body per unit of tissue may exceed that in any other body organ. These peripheral chemoreceptors are capable of rapidly detecting blood hypoxia and then adjust the central respiratory drive to motoneurons innervating the muscles of respiration. Abdominal chemoreceptors in the gastric mucosa are sensitive to acid and characteristic of all regions of the gastrointestinal tract.

6.2.5. Photoreceptors

These specialized receptors respond to electromagnetic waves in the visual spectrum of frequencies. They are in the retina and fully considered in Chapter 12.

6.2.6. Osmoreceptors

Another special type of receptor, the osmoreceptor responds to changes in osmolality of solutions or to changes in osmotic pressure. These receptors respond to intracarotid injections of hypertonic saline solutions and excess sodium ions in the third ventricle. Osmoreceptors occur near the hypothalamic nuclei along the walls of the third ventricle and in the walls of the carotid artery, especially the carotid sinus.

6.3. SHERRINGTON'S CLASSIFICATION OF RECEPTORS

Based on their distribution and function, Sherrington (1906) classified receptors into three groups – **exteroceptors**, **interoceptors**, and **proprioceptors**.

6.3.1. Exteroceptors

Responding to stimuli in the external environment, exteroceptors are at or near the surface of the body. Two categories of exteroceptors are general cutaneous receptors, including endings in hair follicles and endings related to touch. Special receptors in the nose for smell, in the mouth for taste, in the eye for vision, and in the ear for hearing are exteroceptors.

6.3.2. Interoceptors

Interoceptors respond to stretch, volume, or pressure in the walls of viscera and adventitia of blood vessels, and to excessive muscle contraction or to overstretching of a visceral organ. Interoceptors are essential in regulating blood flow and pressure in the cardiovascular system, in controlling respiration, and in general homeostasis. They are involved in monitoring blood gases, the osmolality of the blood as well as blood glucose. Craig (2002) suggests that interoception is the sense of the physiological condition of the entire body, not just the viscera.

6.3.3. Proprioceptors

These receptors respond to stimuli arising in muscles, tendons, joints, and musculotendinous junctions. Changes in muscle length, tension, or the force of muscle contractions, movements of joints, and changes in body and limb position are the appropriate stimuli for these receptors. Proprioceptors are essential to coordination of movements and maintenance of posture. Neuromuscular and neurotendinous spindles are two examples of proprioceptors. The vestibular nerve [VIII] receives proprioceptive input leading to sensations of position or movement of the body and head while maintaining our equilibrium and orientation.

6.4. STRUCTURAL CLASSIFICATION OF RECEPTORS

Based on their structure there are three categories of receptors: **free nerve endings**, **endings in hair follicles**, and **terminal corpuscles of nerves**.

6.4.1. Free Nerve Endings

Free nerve endings (Fig. 6.1) are nonmyelinated or finely myelinated fibers that lose myelin layers and neurilemmal

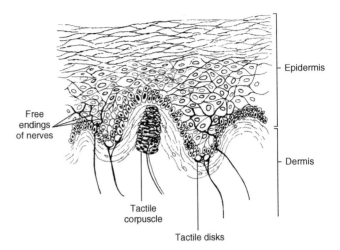

FIGURE 6.1. A vertical section of human skin with a thin epidermis. Note the free nerve endings interdigitated among the basal epidermal cells. A nonmyelinated fiber is visible ending as a tactile meniscus at the base of a tactile epithelioid cell and forming a tactile disk. A tactile corpuscle is beneath the epidermis throughout the depth of a dermal papilla.

cells as they approach the dermal–epidermal junction of the skin. These fibers end as fine, naked, axoplasmic filaments between cells in the basal layer of the human epidermis and papillary dermis. They are also in the epithelia of some mucous membranes, synovial membranes in joints, deep fascia, in joint capsules, tendons, ligaments, periosteum, bone, and viscera. These endings serve as **thermoreceptors** for the sense of cooling (range 15–45°C) and the sense of warming (20–40°C) and **nociceptors** for various aspects of pain including the sense of sharpness (pinprick or first pain), slow, burning pain, heat pain, and cold pain. They are slowly adapting and have a high threshold. Nonmyelinated fibers also end freely, deep to the epidermis of hairless (glabrous) skin where their function is unclear. A neural network of fine, wavy nerves with a horizontal orientation is a constituent of the human dermis and of mucosal surfaces. Fibers in this network end as a mesh of nonmyelinated fibers in a subepidermal position.

6.4.2. Endings in Hair Follicles

About 10 to 15 individual fibers innervate a single hair in humans (Fig. 6.2). They approach its follicle and encircle the hair before ending as **parallel endings** (Fig. 6.2). This collar of terminals arranged parallel to the root of the hair is itself surrounded by **circumferential fibers** (Fig. 6.2). An abundance of such receptors and the large amount of hairy skin in humans suggests that they are sensory. **Endings in hair follicles** (Fig. 6.2) serve as

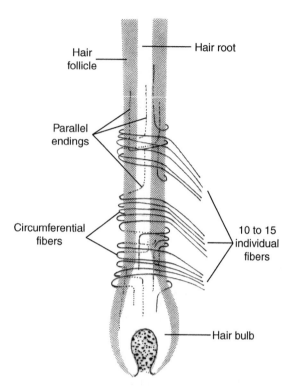

FIGURE 6.2. Many (10–15) fibers innervating a single human hair. Note the collar formed by the circumferentially-arranged fibers that end parallel to the hair (×40) (redrawn from Munger, 1971).

mechanoreceptors responding to any displacement of the hair. Consequently, they serve as receptors for light touch. These receptors are very sensitive and rapidly adapting.

6.4.3. Terminal Endings of Nerves

Nonencapsulated tactile disks, encapsulated nerve endings, neurotendinous spindles, and **neuromuscular spindles** are examples of **terminal endings of nerve.** Gradations occur in the appearance of many of these receptors and they are vulnerable to age-related changes.

Nonencapsulated Tactile Disks

Nonencapsulated tactile disks are concentrated in the basal epidermal layer of human digital skin, in nail beds, in gingival mucosa, and in the red border of the lips (Fig. 6.1). They consist of a disklike or crescent-shaped expansion (**meniscus**) of a slowly-adapting type 1 nonmyelinated nerve fiber that approximates the base of perhaps as many as 100 **tactile epithelioid cells** (of Merkel) thus covering an area of about 5 mm². These receptors are

termed **tactile disks** (Fig. 6.1). The tactile epithelioid cells are equal to or larger than keratinocytes, and may be derivatives of neural crest. In the palm of the hand, the **tactile disks** serve as mechanoreceptors. Arranged in groups or 'touch spots,' they function as receptors for contact, pressure, the two-dimensional form, and texture of objects which might be described collectively as light touch. Tactile disks are slowly adapting, have small receptive fields and low thresholds. They are sensitive to points, edges, and curvature, and readily discriminate surfaces with dots or ridges. (Epithelioid cells with a similar appearance but without contact with nerve terminals are part of a diffuse neuroendocrine system, are not mechanoreceptors, and are likely the origin of a primary skin tumor termed the Merkel cell carcinoma.)

Encapsulated Endings of Nerves

These receptors differ in size, shape, and name. Each has a large diameter sensory fiber enveloped by a distinct capsule or corpuscle. Receptors belonging to this group are **tactile corpuscles, lamellar corpuscles, corpuscles of Krause, corpuscles of Ruffini,** and **Golgi–Mazzoni corpuscles.**

Tactile corpuscles (of Meissner) are ellipsoid encapsulated receptors in the palm and tips of the fingers, forearms, lips, and in the plantar skin of the big toe. They are 40–70 μm in diameter and up to 150 μm in length. Each corpuscle is in a dermal papilla (Fig. 6.1) and supplied by two to six fibers (Fig. 6.3) one of which is a rapidly adapting afferent (Aβ) fiber while the others are likely to be C fibers. One afferent fiber supplies as many as 25–40 corpuscles over 2–5 dermal papilla. The myelin layer is lost before each afferent enters the corpuscle, which consists of horizontally arranged layers of cells apparently epithelial in nature, perhaps derived from neurilemmal cells. Between the cells of the corpuscle, naked nerve fibers ramify forming loops and spirals (Fig. 6.3). Therefore, this receptor typically shows parallel endings closely surrounded by layers of neurilemmal cells. **Tactile corpuscles** more densely innervate the skin than do the Merkel disks (150 per cm² vs. 100 cm² at the fingertips in humans). They respond to localized tactile stimuli particularly related to contact, velocity, and vibration (or flutter, best at 20–30 Hz) as well as motion across or into the skin and are able to distinguish between two similar stimuli simultaneously applied at different points (often termed two-point discrimination). The tactile receptors are specialized mechanoreceptors for **discriminative aspects of touch** and have a low threshold but are rapidly adapting. Recent studies suggest that these receptors and their afferents provide a

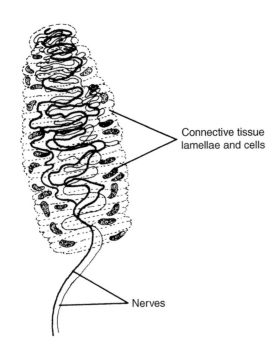

Connective tissue
lamellae and cells

Nerves

FIGURE 6.3.   A human tactile corpuscle deep to the epidermis. For orientation, see Fig. 6.1. Note the two types of fibers that leave the base of the corpuscle (×400) (material provided by Dr. James Hightower).

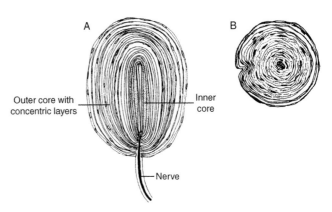

Outer core with
concentric layers

Inner
core

Nerve

FIGURE 6.4.   A human lamellar corpuscle. **A**, Cut parallel to the long axis of the corpuscle. **B**, Cut perpendicular to the long axis of the corpuscle. Note the concentric layers of cells and supporting elements resembling the cut face of an onion. Leaving the base of the corpuscle is a myelinated sensory fiber (×70) (material provided by Dr. James Hightower).

neural image of motion signals from the whole hand that are important for grip control and the motion of objects contacting the skin.

The sensation of discriminative touch varies in different parts of the body. Essential aspects of tactile innervation are the number and distribution of identifiable mechanoreceptors (presumably all four types of cutaneous mechanoreceptors) in various areas of skin. The density of innervation increases slightly from palm to fingers, with an appreciable increase in the fingertips. Such findings agree with the well-known increase in tactile acuity in a proximal-distal direction on the hand with particularly well developed acuity at the fingertips. The visually impaired often use the distal 12–15 mm of the fingers to recognize shapes of letters. A paper clip bent in the shape of an inverted 'V' and applied to hairless skin is a simple means of testing discriminative touch. The skin is simultaneously stimulated at two points by the ends of the bent paper clip.

Lamellar corpuscles (of Pacini) are large, ovoid mechanoreceptors that average 1 mm in length, with a diameter of 0.7 mm. They have an outer capsule of multiple layers of simple squamous epithelium (Fig. 6.4) resembling the cut surface of an onion that surrounds a smaller, inner core. About 70 layers occur in the capsule of the largest human corpuscles. These receptors occur between the dermis and subcutaneous tissue of the palm (about 800 in the each palm), fingers (about 350 in each finger), and soles, in connective tissue near tendons and joints, in periosteum, interosseous membranes of limbs, fascia covering muscles, pancreas, mesentery, serous membranes, and in the breast, penis, and clitoris. They are also in the human middle ear related to all three ossicles, and are associated with the tendons of the stapedius and tensor tympani. One or two thick myelinated fibers supply each lamellar corpuscle then branch, lose their myelin, and enter the inner core of the corpuscle. They are extremely sensitive to skin motion and respond to deformation in the nanometer range, deep pressure, high frequency vibration (best at 250 Hz), and to stretch. With age, they lose their ovoid shape, become larger, coiled, and irregular in shape. The lamellar corpuscles are rapidly adapting but have low thresholds. They allow us to detect vibration transmitted though an object or tool grasped in the hand as though our fingers were present at the working surface of the tool. Stimulation of the afferents to these lamellar corpuscles yields sensations of contact, movement, and high frequency vibration.

Corpuscles of Krause are in the deep layers of hairless skin, oral mucosa, pharyngeal wall, conjunctiva, and the external genitalia of both sexes. In the latter location, they are termed **genital corpuscles**. They have a well-defined capsule, are oval, 40 to 150 μm in diameter, and innervated by one or two thick myelinated fibers that lose myelin as they enter the inner core of the capsule. Genital corpuscles are thought to be structurally similar to tactile corpuscles by some observers but distinct entities by others. They likely serve as mechanoreceptors involved in touch.

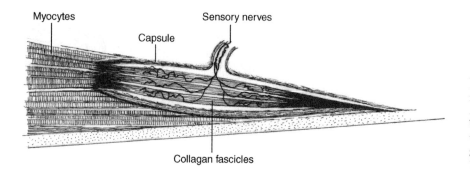

FIGURE 6.5. A human neurotendinous spindle in series with myocytes. As sensory nerves enter the spindle, they branch along the body of the spindle, lose their myelin layer, and end among the collagen fascicles of the tendon (after Bridgman, 1970).

Corpuscles of Ruffini are simple, dermal receptors with a thin capsule and a large subcapsular space. They occur in hairy and hairless (glabrous) skin, in deep fascia and capsules of joints, ligaments, tendons, and fibrous sheaths of flexor tendons. Corpuscles of Ruffini are mechanoreceptors that provide information about direction, magnitude, and rate of change of tension in skin, and between skin and deeper structures. They respond to stretching of skin, pressure, and joint movement perhaps providing a neural image of skin stretch over the whole hand.

Golgi–Mazzoni corpuscles are constituents of the deep fascia, joint capsules, ligaments, and periosteum of fingers in humans. They are circular or oval, from 100 to 150 μm in diameter, and 250 μm in length. A single myelinated fiber usually enters this presumed mechanoreceptor.

6.4.4. Neurotendinous Spindles

These receptors present at musculotendinous junctions and consist of fine tendinous (collagen) fascicles that are continuous with myocytes (muscle cells) but enclosed in a delicate capsule (Fig. 6.5). Tension exerted by a contracted muscle directly influences the tendinous fascicles extending the length of the neurotendinous spindle. Activation of sensory endings on the tendinous (collagen) fascicles (Fig. 6.5) causes reflex-induced muscle relaxation that aids in the production of smoothly coordinated contractions. Sensory fibers to the neurotendinous spindles belong to the Aα group (myelinated group Ib fibers) (Table 6.1).

6.4.5. Neuromuscular Spindles

These highly specialized, encapsulated, fusiform receptors called **neuromuscular spindles** (Fig. 6.6) have a complex structure with 2 to 14 specialized **intrafusal myocytes** in a distinct fibrous capsule (Fig. 6.6). The entire structure of the neuromuscular spindle is parallel to the **extrafusal myocytes** adjoining the spindle capsule (Fig. 6.6). There is a characteristic expanded central region of the spindle. Three kinds of intrafusal myocytes are identifiable including two types with central clear nuclei (**nuclear bag intrafusal myocytes** – about three or four of these occur in a spindle in humans) and a third type with a central chain of nuclei (**nuclear chain intrafusal myocytes**). Histochemical and ultrastructural variability occurs along the length of human intrafusal myocytes. These receptors are in all but a few skeletal muscles.

Usually one to three nerves, each containing five to twenty fibers, enters the capsule of the neuromuscular spindle and provide innervation to this complex region (Fig. 6.6). Both sensory and motor fibers are involved, as the structure is a sensory receptor with an important motor component. Large, thickly myelinated sensory fibers called **primary fibers** (Fig. 6.6) that are classified as myelinated group Ia fibers, enter the central region of the spindle and end on nuclear bag myocytes. These primary endings, called **annulospiral terminations**, wrap around the nuclear bag myocytes. Central terminations of primary endings make monosynaptic connections with spinal cord ventral horn motoneurons (alpha motoneurons). Smaller secondary fibers, classified as myelinated group II fibers, end on nuclear chain myocytes adjoining the central region of the spindle. These **secondary sensory endings** (Fig. 6.6) are longer than the primary, and are variously termed **flower-wreath**, **flowerlike**, or **flower-spray endings**.

Sensory information from neuromuscular spindles travels back to the CNS. Since neuromuscular spindles are parallel to the muscle, the state of the intrafusal myocytes gives an indication of the overall length of the muscle. Stretching the muscle stretches the intrafusal myocytes thus activating the sensory receptors in the spindle. When the muscle is relaxed tension on the spindle is relaxed and the appropriate signals sent to the CNS. Although neuromuscular spindles are also involved in awareness of position of joints and their

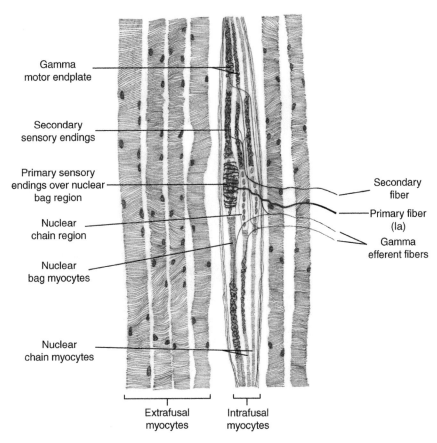

Gamma
motor endplate

Secondary
sensory endings

Primary sensory
endings over nuclear
bag region

Nuclear
chain region

Nuclear
bag myocytes

Nuclear
chain myocytes

Secondary
fiber

Primary fiber
(Ia)

Gamma
efferent fibers

Extrafusal
myocytes

Intrafusal
myocytes

FIGURE 6.6. Motor and sensory innervation of the human neuromuscular spindle. Primary sensory endings on the nuclear bag myocytes are in the equatorial region of the spindle. Secondary endings, on nuclear chain myocytes, are adjoining the equatorial region. Gamma motor end-plates are on both sides of the intrafusal myocyte equatorial region.

movement (but not awareness of individual muscles), they monitor muscle length and tension in normal posture and movement.

Gamma motoneurons in the ventral horn of the spinal cord allow neuromuscular spindles to regulate activity originating in muscle and influence the state of muscle contraction. These gamma motoneurons give rise to fine (1–3 μm) myelinated **gamma efferent fibers** (γ-efferent) that spread along the length of myocytes on both sides of the intrafusal equatorial region before terminating as gamma motor endplates. More information on the functional aspects of the various elements of the neuromuscular spindle is available in Chapter 15, The Motor System.

6.5. REFLEX CIRCUITS

6.5.1. The Monosynaptic Reflex

The simplest circuit mediating activity in the nervous system is a **monosynaptic reflex** (Fig. 6.7), in which a single interneuronal synapse exists between two neurons,

one sensory, and one motor. Incoming sensory information from a neuromuscular spindle travels over the **peripheral process** of a **primary sensory neuron** such as a spinal ganglionic cell (Fig. 6.7). The central process of this same primary sensory neuron carries the resulting impulse into the spinal cord (Fig. 6.7) to a **secondary neuron** such as a ventral horn motoneuron (Fig. 6.7). This motoneuron sends messages over its peripheral process to the same muscle from which the sensory information originated. The spinal ganglionic neuron in this instance is the **sensory component of the reflex** and the ventral horn motoneuron is the **motor component of the reflex**. The **myotatic** or **stretch reflex** is an example of a monosynaptic reflex participating in maintenance of intrinsic muscle tone. Passive stretching of the muscle results in a reflex response.

6.5.2. Complex Reflexes

Receptors also participate in **complex reflexes** that involve many synapses and several levels of the spinal cord or brain stem. In such instances, the first neuron is

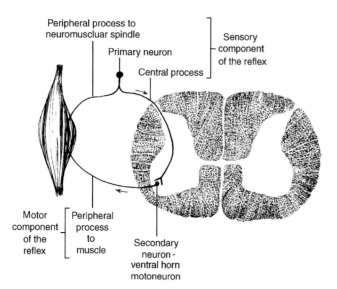

FIGURE 6.7. A monosynaptic reflex. Only a single inter-neuronal synapse may exist between the two neurons in the simplest reflex circuit.

in the ganglion of a spinal or cranial nerve. Its central process enters the CNS and synapses with one or more **interneurons**, which serve as a means of communication between the sensory and motor parts of the reflex. Interneurons usually send axons to several efferent neurons that supply motor units in several muscles. An example of a complex reflex, involving many neurons and several levels of the spinal cord, is the response to thermal pain on touching a hot stove. An initial stimulus, transmitted through many synapses and spinal levels, causes withdrawal of the hand from the hot stove (the efferent component of the reflex).

6.6. GENERAL SENSORY PATHS

Sensory paths are routes by which sensory information from peripheral receptors reaches all parts of the CNS. Therefore if skin of the arm is pinched, the stimulus is transformed by a receptor, transmitted into the CNS and on to the cerebral cortex where it is interpreted as a pinch, localized to the arm, and a decision is made regarding what to do about it (move away from the pincher?). There may be an accompanying reflex response. Sensory paths consist of neurons linked together at synapses and carrying specific information to its proper destination. A sensory path beginning at spinal levels may consist of a primary neuron in a spinal ganglion, a secondary neuron in the dorsal horn of the spinal cord whose central process contributes to an ascending tract ending in the thalamus,

and a tertiary neuron in the thalamus that projects its impulses on to a quaternary neuron in the cerebral cortex.

6.6.1. Classification of Sensory Paths by Function

General somatic afferent paths transmit impulses for pain, temperature, touch, pressure, vibration, and proprioception; **special somatic afferent** paths convey visual, auditory, and vestibular impulses. **General visceral afferent** paths conduct impulses from viscera and perhaps from walls of blood vessels, whereas **special visceral afferent** paths transmit olfactory and gustatory impulses. **General somatic efferent** paths convey impulses to striated muscles and **general visceral efferent** paths conduct impulses to nonstriated muscle, cardiac muscle, and to exocrine glands. Lastly, the **special visceral efferent** paths transmit impulses like those in general somatic efferent paths, but to striated muscles of specialized embryological origin such as the laryngeal muscles.

6.7. ORGANIZATION OF GENERAL SENSORY PATHS

6.7.1. Receptors

Chapters 7 and 8 include a discussion of several general sensory paths. Such paths share certain similarities, as shown in Fig. 6.8. The first link in a general sensory path is a specialized sensory **receptor** (Fig. 6.8). Specific receptors for different sensations such as pain, temperature, touch, pressure, proprioception or vibration are at the peripheral end of the **primary neuron** (Fig. 6.8) in a general sensory path.

6.7.2. Primary Neurons

The **primary neuron** (Fig. 6.8) in a general sensory path has its cell body in a spinal ganglion or in the trigeminal ganglion and is always on the same side as the receptor. Primary neurons have a **peripheral process** that reaches the skin and innervates the specialized sensory receptor and a **central process** that carries impulses from the peripheral receptor into the central nervous system (Fig. 6.8).

Peripheral Processes of Primary Neurons

Sensory nerves innervating the skin are **peripheral processes** of primary neurons made up of myelinated fibers from 1–14 μm in diameter and nonmyelinated

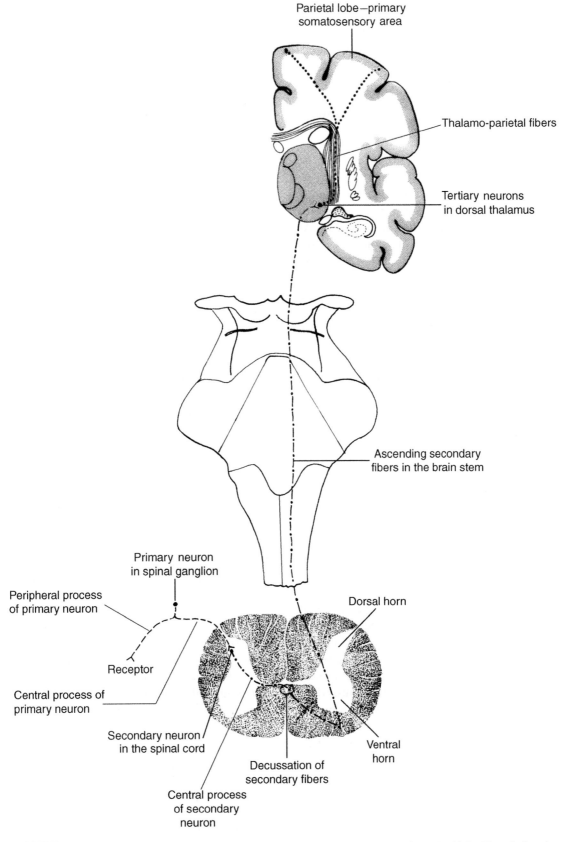

Parietal lobe—primary
somatosensory area

Thalamo-parietal fibers

Tertiary neurons
in dorsal thalamus

Ascending secondary
fibers in the brain stem

Primary neuron
in spinal ganglion

Dorsal horn

Peripheral process
of primary neuron

Receptor

Central process of
primary neuron

Secondary neuron
in the spinal cord

Ventral
horn

Decussation of
secondary fibers

Central process
of secondary
neuron

FIGURE 6.8. A general somatic afferent path from its receptor to its termination in the parietal lobe. Note the location of the neurons along the path and the decussation of fibers from secondary neurons.

fibers from 1–4 μm in diameter. Studies have shown that the smaller the diameter of the nerve fiber, the slower the conduction velocity in that fiber. Consequently, a mechanical nociceptor innervated by a small to medium-sized axon of 3–6 μm, or by a nonmyelinated axon of 1 μm or less, has a slow conduction velocity compared to a mechanoreceptor innervated by a myelinated axon of 6–12 μm.

Central Processes of Primary Neurons

Central processes of primary neurons in general somatic afferent paths enter the central nervous system (Fig. 6.8) and end on **secondary neurons** (Fig. 6.8). As central processes of primary neurons enter the spinal cord or brain stem, there is segregation of sensory fibers. In the spinal cord, thin fibers carrying painful and thermal impulses from the body or painful impulses from viscera enter the lateral part of each dorsal horn and take up a position at the tip of the dorsal horn. Fibers concerned with general or light touch, discriminative tactile sensation, and proprioception from muscles, tendons and joints, are larger in diameter and enter the medial division of the dorsal root.

6.7.3. Secondary Neurons

The neuronal cell bodies of **secondary neurons** (Fig. 6.8) in a general somatic afferent path are always in the CNS and on the same side as primary neurons (Fig. 6.8). These secondary neurons are in the dorsal horn at spinal levels, in the trigeminal nuclear complex, or in other nuclear groups of the spinal cord and brain stem.

Central Processes of Secondary Neurons

The central processes of most secondary neurons in general somatic afferent paths **decussate** (cross the median plane) and ascend on the contralateral side (Fig. 6.8). In so doing they form **ascending sensory paths** (also called 'long tracts') in the spinal cord and brain stem that end on **tertiary neurons** in the **thalamus** (Fig. 6.8). Decussation of fibers from secondary neurons is a characteristic of general somatic afferent paths.

Secondary neurons have a well-ordered layered or laminar arrangement. The functional significance of this arrangement is that other fibers (both local and descending) are able to influence them. Dorsal horn neurons are divisible into different functional classes depending on their response to differing input from cutaneous and muscle receptors.

Secondary neurons in general sensory paths are essential sites at which inhibitory mechanisms operate. Descending paths inhibit secondary neurons, either directly or through interneurons. Interneurons in the spinal cord and brain stem may directly inhibit secondary neurons. Although the details of such descending and local inhibition are not always clear, normal activity in sensory paths is likely the result of a delicate balance between excitation and inhibition, with descending and local inhibitory systems playing vital roles. A variety of complex interactions of both excitation and inhibition are possible resulting in the conduction of modified information over this second link in a general sensory path.

6.7.4. Thalamic Neurons

Tertiary neurons in general somatic afferent paths have their neuronal cell bodies in the thalamus (Fig. 6.8). General somatic afferent impulses likely enter consciousness in humans at the thalamic level. If localization ('where') and discrimination ('what') of these general sensations is necessary then the cerebral cortex must become involved. This would require that these impulses project from the thalamus to the cerebral cortex where such additional processing can take place.

6.7.5. Cortical Neurons

The final termination of general somatic afferent impulses is the parietal lobe. From thalamic neurons, such impulses travel by way of **thalamoparietal fibers** to the **primary somatosensory cortex** corresponding to **Brodmann's areas 3, 1,** and **2** along the **postcentral gyrus** and **posterior part of the paracentral lobule** (Fig. 6.8). Normally impulses travel in the sensory paths only in one direction. Such directional specificity of neuronal conduction is a manifestation of the principle of **polarization** whereby each neuron in a chain conducts impulses only from its receptor, through its dendrites, to its cell body, and out its axonal process. Impulses traveling in all neurons and along all sensory paths follow this principle of polarization.

6.7.6. Modulation of Sensory Paths

Any component of a sensory path including its receptor or the peripheral and central processes of its neurons may be functionally modified or **modulated** and its effectiveness diminished or enhanced by physiological, biochemical, pharmacological, and pathological influences.

Repetitive stimulation of an already excited and active receptor leads to a suppression of sensation. Receptor desensitization occurs such that its response to normal stimuli is reduced or limited. Receptor desensitization is a versatile mechanism for shaping sensory transmission in general sensory paths. A receptor may also display a period of enhanced excitation after an initial period of stimulation or perhaps an increased sensitivity to a specific stimulus. Another aspect of modulation is the possibility of inhibition by descending paths or by collaterals of other paths exerting local effects wherever synapses occur.

Certain polypeptides such as bradykinin, amines such as histamine and 5-hydroxytryptamine (serotonin), and prostaglandins function as modulators of sensory mechanisms. These substances alter receptor sensitivity or interact with naturally occurring substances along sensory paths serving as an intermediary step in the process of receptor excitation. They also interact with substances injected into the body. Pharmacologic agents may depress sensory paths, causing analgesia (freedom from pain). Pharmacological aspects of sensory inhibition are of obvious clinical relevance.

Complete transection of a nerve leads to a loss of sensation in the region served by that nerve. Such diminished sensation is termed **hypesthesia** (hypoesthesia). Inflammation, infections, expanding tumors, or toxic substances are examples of other pathological conditions that can lead to hypesthesia. The latter agents have access to components of sensory paths and consequently are able to influence activity throughout the path.

FURTHER READING

Bealer SL, Smith DV (1975) Multiple sensitivity to chemical stimuli in single human taste papillae. Physiol Behav 14:795–799.

Birder LA, Perl ER (1994) Cutaneous sensory receptors. J Clin Neurophysiol 11:534–552.

Bolton CF, Winkelmann RK, Dyck PJ (1966) A quantitative study of Meissner's corpuscles in man. Neurology 16:1–9.

Bridgman CF (1970) Comparisons in structure of tendon organs in the rat, cat and man. J Comp Neurol 138: 369–372.

Burke D (1978) The fusimotor innervation of muscle spindle endings in man. TINS 1:89–92.

Cauna N (1980) Fine morphological characteristics and microtopography of the free nerve endings of the human digital skin. Anat Rec 198:643–656.

Cervero F (1985) Visceral nociception: peripheral and central aspects of visceral nociceptive systems. Philos Trans R Soc Lond B Biol Sci 308:325–337.

Collins DF, Refshauge KM, Todd G, Gandevia SC (2005) Cutaneous receptors contribute to kinesthesia at the index finger, elbow, and knee. J Neurophysiol 94:1699–1706.

Craig AD (2002) How do you feel? Interoception: the sense of the physiological condition of the body. Nat Rev Neurosci 3:655–666.

Djouhri L, Lawson SN (2004) Aβ-fiber nociceptive primary afferent neurons: a review of incidence and properties in relation to other afferent A-fiber neurons in mammals. Brain Res Brain Res Rev 46:131–145.

Inglis JT, Kennedy PM, Wells C, Chua R (2002) The role of cutaneous receptors in the foot. Adv Exp Med Biol 508:111–117.

Johnson KO (2001) The roles and functions of cutaneous mechanoreceptors. Curr Opin Neurobiol 11:455–461.

Johnson KO, Yoshioka T, Vega-Bermudez F (2000) Tactile functions of mechanoreceptive afferents innervating the hand. J Clin Neurophysiol 17:539–558.

Just T, Stave J, Pau HW, Guthoff R (2005) In vivo observation of papillae of the human tongue using confocal laser scanning microscopy. ORL J Otorhinolaryngol Relat Spec 67:207–212.

Light AR, Perl E (1993) Peripheral sensory systems. In: *Peripheral Neuropathy*. Dyck PJ, Thomas PK, Griffin JW, Low PA, Poduslo JF, eds. Saunders, Philadelphia, pp 149–165.

Messlinger K (1996) Functional morphology of nociceptive and other fine sensory endings (free nerve endings) in different tissues. Prog Brain Res 113:273–298.

Mountcastle VB (2005) *The Sensory Hand: Neural Mechanisms of Somatic Sensation*. Harvard University Press, Cambridge.

Sherrington CS (1906) *The Integrative Action of the Nervous System*. Yale University Press, Newhaven.

Tran TD, Lam K, Hoshiyama M, Kakigi R (2001) A new method for measuring the conduction velocities of A beta-, A delta- and C-fibers following electric and CO_2 laser stimulation in humans. Neurosci Lett 301:187–190.

Vallbo AB, Olausson H, Wessberg J (1999) Unmyelinated afferents constitute a second system coding tactile stimuli of the human hairy skin. J Neurophysiol 81:2753–2763.

Vega JA, Haro JJ, Del Valle ME (1996) Immunohistochemistry of human cutaneous Meissner and Pacinian corpuscles. Microsc Res Tech 34:351–361.

But the truth is that the word Pain has two senses which must now be distinguished. a. A particular kind of sensation, probably conveyed by specialized nerve fibers and recognizable by the patient as that kind of sensation whether he dislikes it or not (e.g. the faint ache in my limbs would be recognized as an ache even if I didn't object to it). b. Any experience, whether physical or mental, which the patient dislikes. It will be noticed that all Pains in sense 'a' become Pains in sense 'b' if they are raised above a certain very low level of intensity, but that Pains in the 'b' sense need not be Pains in the 'a' sense. Pain in 'b' sense, in fact, is synonymous with 'suffering,' 'anguish,' 'tribulation,' 'adversity,' or 'trouble,' and it is about it that the problem of Pain arises.

C.S. Lewis, 1940

Paths for Pain and Temperature

In the human nervous system, there are **general somatic afferent** paths that carry impulses for somatic pain and temperature from peripheral receptors to the thalamus. From the thalamus, such impulses project to the cerebral cortex. In addition to these general somatic afferent paths, there is a general visceral afferent path carrying impulses for visceral pain to higher levels of the nervous system.

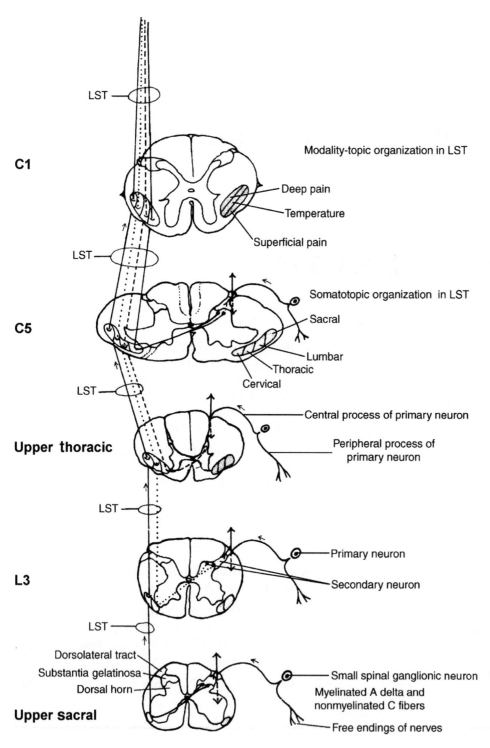

LST

C1

Modality-topic organization in LST

Deep pain

Temperature

Superficial pain

LST

C5

Somatotopic organization in LST

Sacral

Lumbar

Thoracic

Cervical

LST

Upper thoracic

Central process of primary neuron

Peripheral process of primary neuron

LST

L3

Primary neuron

Secondary neuron

LST

Dorsolateral tract

Substantia gelatinosa

Dorsal horn

Upper sacral

Small spinal ganglionic neuron

Myelinated A delta and nonmyelinated C fibers

Free endings of nerves

FIGURE 7.1. Lateral spinothalamic tract in humans as it originates in, and ascends through, the spinal cord. Note the probable pattern of localization for the lateral spinothalamic tract at C5. Also, note the pattern for localization of different types of impulses carried along lateral spinothalamic tract shown on C1. Further course of the lateral spinothalamic tract through the brain stem is illustrated in Fig. 7.2.

7.1. PATH FOR SUPERFICIAL PAIN AND TEMPERATURE FROM THE BODY

The primary path carrying impulses for sharp and well-localized somatic pain and temperature from superficial and certain deep tissues of the limbs, trunk,

neck, and back of the head through the spinal cord and brain stem to the thalamus is termed the **lateral spinothalamic tract**. As its name suggests, it is **lateral** in position throughout its course, begins at **spinal** levels, and ends in the **thalamus** (Figs 7.1, 7.2). Descriptive terminology applied to this and other sensory paths is

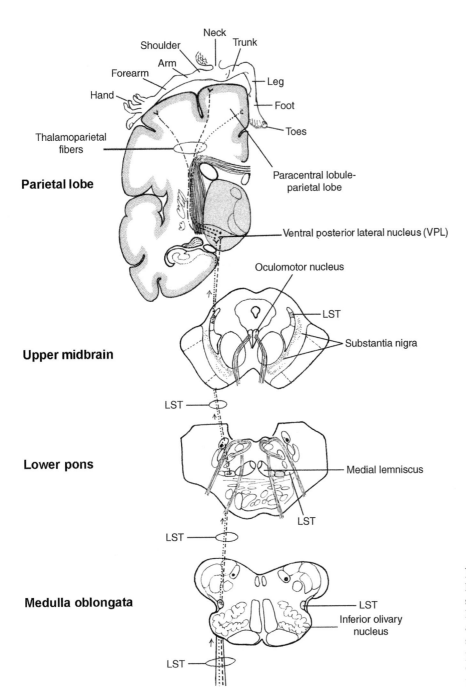

FIGURE 7.2. Course and position of the lateral spinothalamic tract through the brain stem, its termination in the ventral posterior lateral nucleus (VPL) of the dorsal thalamus, and its thalamoparietal projection. Note the sensory homunculus illustrated along the postcentral gyrus (hand, forearm, arm, shoulder, neck, trunk, and hip) and paracentral lobule (leg, foot, toes) of the parietal lobe.

often very helpful in understanding the anatomical aspects of such paths.

7.1.1. Modalities

Superficial or pricking pain and itch from skin, well-localized deep or somatic pain from joints, fascia, periosteum, tendons, and muscles, together with different degrees of thermal sensation such as warmth or heat and coolness or coldness are examples of the modalities of sensation served by the **lateral spinothalamic tract.**

Spinoreticular and spinotectal tracts, accompanying the lateral spinothalamic tract, aid in the transmission of painful impulses.

7.1.2. Receptors

Receptors involved in the path for pain and temperature from the body are **nociceptors** and **thermoreceptors** (see Table 6.1); nociceptors are probably free nerve endings. In humans, the duration of stimulus and size of the stimulated area influence the discrimination of pain. Areas of skin innervated by polymodal nociceptors (those excited by one or more category of stimuli, such as mechanical and thermal stimuli) often overlap in humans, so that a single noxious stimulus activates several nociceptors. Although the hot or cold character of a sensation is attributable to activity of specific thermoreceptors, their structure is unclear.

Lightly pricking the skin with a pin results in **pin-prick sensation** or more accurately, the **sense of sharpness**, and is clinically useful to test for **superficial pain**. (Driving a pin into or through the skin causing tissue damage would be an example of a painful or injurious sensation termed **nociception**.) Squeezing the tendo calcaneus or forearm muscles is a test of deep or **somatic pain**. Applying test tubes containing cold (15–20°C) and warm (40–45°C) water to the skin is one means of testing thermal sensation.

7.1.3. Primary Neurons

Peripheral Processes

Cell bodies of primary neurons in the path for superficial and somatic pain and temperature from the body are small, dark **spinal neurons in the spinal ganglia** at all levels of the spinal cord (Fig. 7.1). These primary neurons utilize glutamate as their neurotransmitter. The **peripheral processes** of these primary neurons synapse with thermoreceptors or nociceptors. Peripheral nerves carrying impulses for coolness are myelinated Aδ fibers, whereas those supplying receptors for warmth are nonmyelinated C fibers. These are active at normal skin temperature (32°C and upward). Small to medium-sized myelinated Aδ fibers and nonmyelinated C fibers transmit impulses for cutaneous pain (Table 6.1). In most mammals including primates, A-fiber nociceptors conducting in the Aβ conduction velocity range are mechano-heat units. While these peripheral processes function as dendrites, structurally they are axons.

Efforts to relieve certain painful conditions often focus on the pain-carrying fibers in peripheral nerves using injections of alcohol and local anesthetics. Other treatments use heat, massage, acupuncture, and electrical stimulation including **transcutaneous electrical nerve stimulation** (TENS) involving the surgical implantation of electrodes on, in or near peripheral nerves. Spinal cord simulation is a commonly used method of treatment for chronic low back pain.

Central Processes

Upon receiving appropriate stimuli, thermoreceptors or nociceptors transform such stimuli into nerve impulses

and transport them through the peripheral process of a **primary neuron** to its cell body of origin in a spinal ganglion. These sensory impulses then travel through the **central process** (Fig. 7.1) of the primary neuron and enter the spinal cord through the **lateral division** of each **dorsal root**. These central processes of primary neurons are thinner than their corresponding peripheral processes. Each dorsal root in humans divides on average into six rootlets. One method used in the surgical management of pain, called **selective dorsal rhizotomy** [Greek, rhiz (o) or root], requires cutting pain-carrying fibers located in the lateral division of the dorsal root at the cord-rootlet junction. Complete section of dorsal roots superior and inferior to the painful region is termed **dorsal rhizotomy**. On entering the spinal cord, fibers carrying impulses for pain and temperature bifurcate into ascending and descending branches. The longer branches ascend at least one spinal cord segment and in doing so contribute to the formation of the **dorsolateral tract** (of Lissauer). The dorsolateral tract (Fig. 7.1) is identifiable at the apices of the dorsal horns. Since fibers carrying impulses for pain and temperature enter the spinal cord at all spinal levels, this dorsolateral tract is at all spinal levels. About 80% of its fibers come from dorsal roots; the remaining 20%, called **propriospinal fibers**, are intrinsic to the cord. Seventy to eighty percent of dorsolateral fibers are nonmyelinated; the rest are thinly myelinated.

Congenital Insensitivity to Pain with Anhidrosis

The **dorsolateral tract** varies regionally in transverse spinal cord sections, being long and narrow at cervical enlargement (C4–T1) levels, but short and wide at lumbosacral enlargement (L2–S2) levels. The extensive surface area of the limbs and the corresponding extensive sensory innervation accounts for a greater size of this tract at these levels. Absence of pain is a symptom of several disorders, both congenital and acquired. One of these conditions, termed **congenital insensitivity to pain with anhidrosis** (absence of painful and thermal sensations and a lack of sweating), is distinguished by the absence of small neurons in the spinal ganglia and a deficiency of the associated small diameter myelinated and nonmyelinated pain-carrying fibers. Because of the absence of these pain-carrying fibers, there is also an absence of the dorsolateral tract. Free nerve endings in patients with congenital insensitivity to pain are often completely lacking in the preputial skin. Abnormal development of neural crest cells that should develop into small pain-carrying neurons in the spinal ganglia is a likely cause of this condition. Findings in those with this condition confirm the role of free nerve endings, small neurons in the spinal ganglia, associated small myelinated and nonmyelinated fibers, and the dorsolateral tract in transmission of painful impulses and the innervation of sweat glands. Since this condition involves abnormal and degenerated peripheral nerve fibers, it is termed a **neuropathy**. Increasingly the underlying genetic abnormality is the means used to classify the different types of hereditary sensory and autonomic neuropathies.

Throughout its length, the dorsolateral tract gives off collaterals to the dendrites and cell bodies of many neurons in the adjoining superficial dorsal horn. Near the tract is a thin layer of large, flat, marginal neurons and horizontal fibers collectively forming a **marginal zone** (spinal lamina I), ventral to which is a clear region of neuronal somata, the **substantia gelatinosa** (spinal lamina II) (see Fig. 3.10). The latter is divisible into a ventral small-celled inner region and a dorsal smaller-celled outer region. The dorsal part of the **nucleus proprius** (spinal lamina III) is ventral to spinal lamina II (Fig. 3.10). Almost all fine myelinated or nonmyelinated primary sensory fibers, entering through the lateral division of each dorsal root, end in spinal laminae of the superficial dorsal horn. The superficial dorsal horn receives a rich and functionally diverse sensory input. Surgical destruction of the **dorsal root zone of entry** (DREZ), including the dorsolateral tract and substantia gelatinosa, is performed to relieve chronic pain in the limbs caused by avulsion of the cervical dorsal roots. Falls or blows to the head or shoulder often tear these roots from the spinal cord.

7.1.4. Secondary Neurons

The cell bodies of secondary neurons in the path for superficial pain and temperature are in the **dorsal horn** (Fig. 7.1) of all spinal cord levels. The substantia gelatinosa is one of several nuclear groups where impulses related to pain and temperature sensation are subject to modulation. Sensory activity in the dorsal root afferents influences the substantia gelatinosa. In turn, the substantia gelatinosa receives descending inhibitory inputs, influences its own neurons, larger ones in the same or adjacent segments, and those in the brain stem. Consequently, the substantia gelatinosa has neurons that receive and relay impulses along with intrinsic neurons that interact with, modify, or integrate incoming sensory stimuli. Thus, the substantia gelatinosa serves as a transition zone and integrative region in the pain and temperature path.

Secondary neurons in the path for pain and temperature in nonhuman primates have their cell bodies predominately in the **marginal zone** (spinal lamina I) and in the **deeper layers of the dorsal horn** (spinal lamina IV–VI) (Fig. 3.10). Responses to noxious stimuli by spinothalamic tract cell bodies in the spinal cord of nonhuman primates are consistent with the view that these neurons are essential to the transmission of painful impulses. Three groups of these neurons are identifiable: those activated by low-intensity mechanical stimuli, those activated by nociceptive stimuli, and those with intermediate characteristics between the other two.

Central Processes of Secondary Neurons Form the Lateral Spinothalamic Tract

Myelinated and nonmyelinated fibers from these secondary neurons cross the median plane (decussate) to the opposite side of the spinal cord (Fig. 7.1). These decussated fibers swing ventromedially, pass ventral to the central canal through the **ventral white commissure**, and reach the ventral part of the opposite **lateral funiculus** (Fig. 7.1). Such fibers decussate in humans in the spinal segment containing their parent cell body although they receive sensory information from the level below the location of their parent cell body. In the contralateral lateral funiculus, they turn and ascend in the spinal cord and brain stem to reach the thalamus and in so doing form the **lateral spinothalamic tract** (Fig. 7.1). These ascending secondary fibers give off short collaterals that synapse with ventral horn neurons at the same level as the decussation of the secondary fibers and for several segments rostral to this. Fibers in the human lateral spinothalamic tract are less than 10 μm in diameter and have a conduction velocity of 40 m/s. Cutting these decussating fibers in the ventral white commissure, a procedure called **myelotomy**, was a method of pain management in the 1920s.

Somatotopic Organization of the Lateral Spinothalamic Tract

Fiber distribution in the lateral spinothalamic tract (LST) occurs according to the parts of the body they serve, an arrangement called **somatotopic organization**. At sacral levels, the LST has fibers only of sacral origin (Fig. 7.1). At lumbar levels, the ascending lumbar fibers join the ascending sacral fibers on their medial aspect, pushing the sacral fibers dorsolaterally (Fig. 7.1). At thoracic levels, the ascending thoracic fibers join the ascending lumbar fibers on their medial aspect with the sacral fibers ventromedial to the lumbar fibers (Fig. 7.1). Fibers from cervical levels join the LST in a similar somatotopic manner. Such a grouping of fibers and the curvature of the cord causes a layered arrangement of ascending fibers in the LST. At rostral cervical levels, the LST has sacral fibers dorsal and lateral, cervical fibers ventral and medial, with lumbar and thoracic fibers between them (Fig. 7.1). Knowledge of this arrangement is of value to neurosurgeons contemplating interruption of this path for the relief of pain, called **anterolateral cordotomy**. **Percutaneous cordotomy** employs a lateral approach to the cord at the C1–C2 vertebral interspace and a special electrode used to make a radio frequency heat lesion in the anterolateral quadrant.

Modality-topic Organization in the Lateral Spinothalamic Tract

In addition to the somatotopic arrangement of spinothalamic fibers, there is also a **modality-topic organization** (Fig. 7.1) of fibers in the LST. Spinothalamic fibers carrying impulses for cutaneous pain are superficial whereas fibers carrying impulses for well-localized deep pain from muscles, tendons, and joints are near the gray matter. Between these two groups are fibers transmitting impulses for temperature (Fig. 7.1). In the human spinal cord, the descending path for automatic respiration, the **ventrolateral reticulospinal tract**, is intermingled with, and deep to, fibers of the ventral part of the lateral spinothalamic tract. Although unilateral destruction of this tract by way of an anterolateral cordotomy causes insignificant respiratory loss, bilateral section of this descending path for automatic respiration causes severe respiratory embarrassment.

Stimulation at spinal levels of the lateral spinothalamic tract in humans evokes sharp pain localized to specific parts of the contralateral surface of the body, often described as 'burning' or 'hot pain' and as pain 'like needles'. Sensations of tingling, warmth, and coolness result from decreasing stimulus intensity. In another study, such stimulation resulted in sensations localized to both sides of the body 12% of the time. A clear-cut dorsoventral somatotopic organization in the human lateral spinothalamic tract is present as revealed by electrical stimulation.

7.1.5. Position of the Lateral Spinothalamic Tract in the Brain Stem

As the lateral spinothalamic tract ascends in the lateral part of the brain stem, it gives off collaterals to many named and unnamed nuclei in the medial reticular formation of the medulla oblongata and pons and to the periaqueductal gray substance of the midbrain. Through these collaterals, the LST influences brain stem nuclei and is able to modify sensory transmission.

In the **medulla oblongata**, the lateral spinothalamic tract is dorsolateral to the inferior olivary nucleus (Fig. 7.2). In the **pons**, the LST lies on the lateral aspect of the medial lemniscus (Fig. 7.2) where it is identifiable by stimulation. Surgical destruction of the LST at this level is a means of abolishing certain types of pain. In the **midbrain**, the LST is dorsal to the lateral part of the substantia nigra (Fig. 7.2) where it joins other ascending sensory paths as they approach the thalamus. Electrical stimulation of the lateral spinothalamic tract in the midbrain results in sharp sensations of pain, burning, heat, cold, or numbness on the body, contralateral to the stimulation. Low-threshold stimulation here evokes contralateral sensations of cool or warm tingling, burning, or rarely, pain; 10% of the responses are on the ipsilateral side of the body. The LST ascends from the midbrain to the thalamus where it may synapse with tertiary neurons (Fig. 7.2). In humans, less than a third of the fibers in the lateral spinothalamic tract reach the thalamus.

7.1.6. Thalamic Neurons

Fibers of the lateral spinothalamic tract terminate on tertiary neurons primarily in the **ventral posterior lateral nucleus** (VPL) of the thalamus (Fig. 7.2) as well as in the **intralaminar nuclei** particularly the central lateral (CL) nucleus. Spinothalamic fibers end somatotopically in the ventral posterior nucleus (VP) with fibers from the upper limb ending in the medial part of VPL whereas fibers from the lower limb end in the lateral part of VPL (Fig. 7.2). Some spinothalamic fibers reach the ventral and posterior part of VL_p where it lies close to the boundary of VPL. Presumably, there is also an orderly termination of spinothalamic fibers by modality in VPL. In nonhuman primates, impulses for superficial pain from cutaneous areas of limbs reach the caudal part of the ventral posterior lateral nucleus (VPL_c) whereas painful impulses from fascia, tendons, and joints project to its oral part (VPL_o). In anesthetized monkeys, many neurons in VPL_c receive inputs from nociceptors in the periphery, presumably from **thermal** and **mechanical nociceptors**. General aspects of pain and extremes of temperature in humans likely enter consciousness at the thalamic level. If localization ('where') and discrimination ('what') of these general sensations is necessary then the cerebral cortex must become involved. This would require the relay of these impulses from the thalamus to the cerebral cortex where additional processing can take place.

Thalamic Stimulation in Humans

Electrical stimulation of the ventral posterior lateral nucleus (VPL) in humans yields a contralateral sensation of numbness and tingling; painful sensations following stimulation of VPL are rare. Sensations described as warm and cool tingling, and occasionally as burning or painful do occur following VPL stimulation. Such responses are usually contralateral with a small percentage of ipsilateral responses. In a series of patients, sensations described as sharp pain, aching pain, or burning sensation and localized to the contralateral body followed VPL stimulation. Thalamic stimulation in humans reveals a discrete somatotopic organization in the ventral posterior lateral nucleus (VPL) with upper limb spinothalamic fibers ending in a dorsal position and those from the lower limb ventral, near the medial geniculate nucleus (MG).

Thalamic Targets in the Treatment of Painful Conditions

Thalamic stimulation as a part of **deep brain stimulation** (DBS) procedures, thalamic destruction and the interruption of thalamofrontal fibers are methods used to treat painful conditions. The ventral posterior nucleus (VP), some intralaminar nuclei including the parafascicular (PF) and centromedian nuclei (CM), and the medial nuclear group of the pulvinar (Pul) have all been thalamic targets in the treatment of painful conditions. Chapter 14 provides additional information on the nuclei of the thalamus in humans

7.1.7. Cortical Neurons

Primary Somatosensory Cortex (SI)

The final link in the pain and temperature path consists of fibers from tertiary neurons in the dorsal thalamus that project to the **anterior parietal cortex**. These **thalamo-parietal fibers** (Fig. 7.2) enter the **posterior limb of the internal capsule** and end in a precise order on neurons in the **postcentral gyrus** and the posterior part of the **paracentral lobule of the parietal lobe** (Fig. 7.2). This cortical region, also termed the **primary somatosensory cortex** or **SI**, corresponds to **areas 3**, **1** and **2** according to the cortical mapping scheme proposed by **Brodmann** (1909). Brodmann defined 52 discrete cortical areas based on the presences of individual histological elements, their layering, and their parcellation in the adult human brain. Brodmann's area 3 is divisible into cytoarchitectonic areas 3a and 3b in nonhuman primates and humans.

Thus, there are four strip-like somatosensory areas in the human primary somatosensory cortex. SI appears to participate in the sensory-discriminative aspects of pain in humans. In monkeys, VPL projects to all four subdivisions of SI (3a, 3b, 1 and 2) and to SII.

In addition to SI, recent imaging studies in humans suggest that the **secondary somatosensory cortex** or **SII** as well as the **cingulate gyrus**, **anterior insula**, and the **prefrontal cortex** are involved in pain processing. The cingulate gyrus as a constituent of the limbic system (Chapter 18) plays an important role in emotion and, through its many autonomic projections, in the expression of that emotion. Brain imaging studies have demonstrated that the suffering aspect of pain activates the anterior cingulate gyrus in humans. Thus, it is not surprising that pain and emotion come together in the cingulate gyrus. SII likely participates in the recognition, learning, and memory of painful events whereas the anterior insula may play a role in the autonomic reactions to painful stimuli. Studies using intra-cerebral recordings suggest that SII in humans also integrates nociceptive and non-nociceptive inputs. Additional evidence for this suggestion derives from the input of both pain and temperature impulses to SII via the lateral spinothalamic tract along with the input of touch and proprioception impulses to SII via the medial lemniscus. The posterior insula appears to be involved in tactile and innocuous temperature perception. Imaging studies in humans reveal the ability to experience another person's pain, characteristic of **empathy**, involves activation of the anterior insula and rostral anterior cingulate gyrus. Clearly, multiple brain regions are involved in the sensory perception of pain and temperature in humans. Table 7.1 summarizes the pain and temperature paths from the body and compares them with the paths from trigeminal levels.

The Sensory Homunculus

Electrical stimulation of the postcentral gyrus in conscious patients evokes sensations of numbness and tingling; rarely reported are sensations of pain or temperature. In 163 patients, with 800 responses, coldness was reported 13 times, pain 11 times, and heat twice. Stimulation along this gyrus and the posterior part of the paracentral lobule causes sensations on different areas of the contralateral body. Such stimulation reveals a pattern of representation of different parts of the body, from toes to tongue and oriented along the postcentral gyrus and paracentral lobule. This pattern of representation of body parts drawn across the postcentral gyrus and paracentral lobule body as a distorted

TABLE 7.1.   Summary of paths for pain and temperature from the body

Modalities	Superficial pain, deep pain, temperatures	Chronic, visceral pain
Receptors	Nociceptors and thermoreceptors	Visceral nociceptors
Primary neurons	Smaller neurons in all spinal ganglia	Smallest neurons in all spinal ganglia
1° peripheral processes	To receptors as A-delta and C fibers	To receptors as A-delta and C fibers
1° central processes	Via lateral division form dorsolateral tract	Enter lateral division may divide before ending
Secondary neurons	Lamina I lamina IV–VI	Secondary visceral gray substance
2° central processes	Decussate to form **lateral spinothalamic tract**	Crossed and uncrossed to form **spinoreticulothalamic tract**
Location 2° fibers	Contralateral	Bilateral path
Spinal cord	Ventral in lateral funiculus	Lateral part of proprius bundles
Medulla	Lateral above inferior olive	Interolivary position
Pons	Lateral to medial lemniscus	Between fibers of VI, VII
Midbrain	Dorsolateral to medial lemniscus	Crossed reticulothalamic fibers from midbrain
Dorsal thalamic neurons	Lateral part of ventral posterior nucleus (VPL) of dorsal thalamus	VPL, intralaminar, reticular nuclei; may enter consciousness here
Thalamoparietal fibers	Via posterior limb of the internal capsule	Perhaps via posterior limb of the internal capsule
Cortical areas	Parietal lobe, areas 3, 1, and 2 in the postcentral gyrus and paracentral lobule; primary somatosensory cortex (SI)	Precentral operculum of the frontal lobe, anterior insula, anterior cingulate cortex, perhaps prefrontal cortex
Comments	Note sensory homunculus; also somatotopic and modality-topic arrangement throughout	Emotional component to this type of pain called suffering; visceral pain may be referred

caricature of a little man or parts of the body is termed the **sensory homunculus** (Fig. 7.2).

Certain aspects of the human sensory homunculus are of special interest. The hand, wrist, forearm, arm, shoulder, and neck follow a continuous pattern along the homunculus as does the trunk, hip, leg, foot, and toes. This continuity is lost between the face and hand. The larger cortical representation given to the face, lips, tongue, as well as the hand, and digits is out of proportion to that given to other body parts but is in keeping with the greater peripheral sensitivity and the greater density of receptors in these parts. The toes, foot, ankle and leg have their representation on the posterior part of the paracentral lobule on the medial surface of the brain (Fig. 7.2). The representation of the thigh or hip occurs at the medial to superolateral surface junction followed in order by the trunk, neck, shoulder, arm, forearm, and hand (Fig. 7.2) across the superolateral brain surface. Variations occur in the location of the hip/trunk boundary. The genitalia and perineum likely have their representation on the medial surface of the human brain just ventral to the representation of the toes (though this representation is unclear). Recent recordings of cerebrocortical potentials evoked by stimulation of the dorsal nerve of the penis in humans suggest a representation of the penis with the hip and upper leg near the junction of the medial to superolateral surface. Additional studies should help to clarify the location of the cortical representation of the human genitalia and perineum. The cutaneous regions adjacent to the median plane in the trunk have a bilateral representation in SI. The representation of the fingers occurs along the posterior wall of the central sulcus. The anterior surface of the body faces the central sulcus.

The Multiple Representation Hypothesis

In nonhuman primates, each cytoarchitectural component of the **primary somatosensory cortex** corresponding to **areas 3a, 3b, 1,** and **2** of **Brodmann** is a separate map of the body, receiving inputs from different peripheral receptors. Areas 3b and 1 represent cutaneous regions and are essentially mirror images, yet differ enough to suggest that each has a distinct role in cutaneous sensation. Area 2 appears to represent predominantly deep receptors, including those signaling the position of joints and other deep body sensations, whereas area 3a represents sensory input from muscles. Cortical representation along the postcentral gyrus reflects a composite of somatotopically-organized areas instead of a continuous homunculus. Functional imaging studies of the human SI show that multiple digit representations occur in the four cytoarchitectonic subdivisions of SI (3a, 3b, 1 and 2) resembling the **multiple representation hypothesis** that is present in nonhuman primates. In addition to these four cytoarchitectonic subdivisions of SI, there is also a columnar cortical organization in the postcentral gyrus. The definition of these cytoarchitectural subdivisions during imaging studies in humans depends on approximations based on anatomical landmarks.

Thermal and Painful Auras

Loss of discriminative pain or thermal sensation from cortical injury is well known. Thermal and painful sensations may foreshadow the onset of an epileptic seizure, a phenomenon known as an **aura**. In 267 patients with focal seizures, only seven described painful sensations as a part of their seizure. A male patient has had seizures since nine years of age marked by sensations of excruciating genital pain.

Sensory Jacksonian Seizures

Sensory Jacksonian seizures consist of abnormal, localizable, cutaneous sensations without apparent prior stimulation. Such sensations spread or progress to adjacent cutaneous areas in the body or along a limb, reflecting propagation of an epileptic discharge – an abnormal firing of neurons in the postcentral gyrus. Because they are alert and aware of their seizures, such patients are able to describe what and how they feel during the seizures. The careful analysis of seizure patterns in 42 patients, involving subjective sensory experiences without objective signs, reveals a cortical sensory sequence with the thumb juxtaposed to the lower lip and corner of the mouth but not to the brow and corner of the eyelid as classically described (see Fig. 7.6). Destruction or ablation, including removal of the postcentral gyrus or parietal lobotomy was once a treatment of choice for the relief of pain.

The nervous system is capable of discriminating and comparing qualities of pain and slight variations in temperature. The ability to recognize, assess and localize pain and thermal sensations requires cortical participation. The frontal lobe compares current sensations with those previously experienced, thus providing a basis for immediate reactions and influencing intended behavior.

7.1.8. Modulation of Painful and Thermal Impulses

Painful and thermal impulses conducted along the pain and temperature path are subject to modification at any point along the entire path – at the initial receptors or at synapses in the spinal cord, brain stem, thalamus, and cerebral cortex. In nonhuman primates, a descending inhibitory influence on spinothalamic tract cell bodies in spinal lamina I results from stimulation of the somatosensory and motor cortices. Other sites are in or near the nuclei of the raphé in the medulla oblongata, caudal part of the pons, and reticular formation next to the periaqueductal gray substance of the midbrain. Stimulation of these nuclei inhibits noxious impulses

from entering the cord through nonmyelinated C fibers. The corticospinal tract, a descending motor path in nonhuman primates, carries fibers from the precentral gyrus of the frontal lobe and the postcentral gyrus of the parietal lobe that influence neurons in spinal laminae IV and V of the dorsal horn. Fibers containing catecholamine are in the marginal layer and substantia gelatinosa of the spinal cord in nonhuman primates. They may be terminals of descending fibers from the brain stem involved in modifying the transmission of impulses in the lateral spinothalamic tract.

Influence of Endogenous Substances on the Lateral Spinothalamic Tract

Natural (endogenous) substances in the brain influence neurons or synapses along the lateral spinothalamic tract, facilitating activity along the path. In human dorsal roots, fibers containing substance P resemble the nonmyelinated C fibers known to transmit painful and thermal impulses. Indirect evidence (from nonprimates) suggests that this polypeptide facilitates the transmission of painful impulses between these fibers and neurons of the dorsal horn. Substance P is found in terminals in the human substantia gelatinosa, in a few fibers of the gray commissures of the cord, in the lateral funiculus parallel to the dorsal horn, around ventral horn neurons and, rarely, in the ventral root.

An example of inhibitory effects is the depression of activity in spinothalamic neuronal cell bodies from application of serotonin near them. The origin, course, and termination of such descending inhibitory paths, and the substances or mechanisms involved in modifying activity in the lateral spinothalamic tract remains to be clarified in nonhuman primates and humans. Whatever the influence of a descending path on an ascending path, be it inhibition or excitation, their combined activities are an important part of normal sensory processing.

7.2. PATH FOR VISCERAL PAIN FROM THE BODY

Certain painful impulses arising from the viscera are persistent, poorly localized, predominantly bilateral, difficult to define, and often accompanied by an emotional component. Such pain, called **visceral pain**, is anatomically and physiologically distinct from the acute, somatic pain served by the lateral spinothalamic tract. The path responsible for the transmission of visceral pain is composed of neurons with bifurcating axons of varying

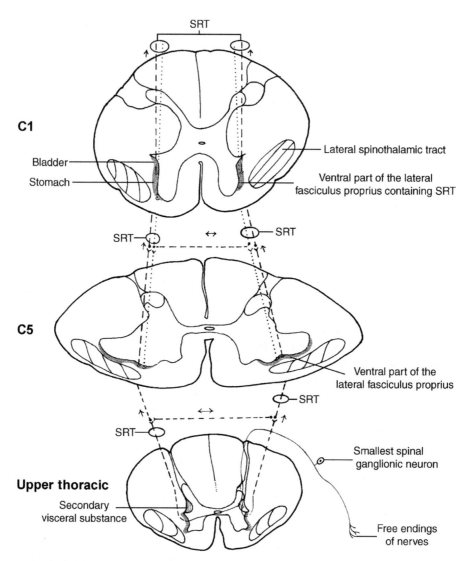

FIGURE 7.3. Origin and partial course of the human spinoreticulothalamic tract (SRT) from upper thoracic to C1 spinal levels. This bilateral, multisynaptic path carries chronic, visceral pain from spinal levels, ascends through the brain stem, and to the dorsal thalamus. Note the probable pattern of localization for fibers in the SRT carrying impulses from the bladder and stomach (adapted from Crosby, Humphrey, and Lauer, 1962).

length and many synapses extending from the spinal cord to the thalamus (Fig. 7.3). In keeping with the terminology first used for these connections in the early 1900s, and in light of its **spinal** origin, course in the **reticular** formation, and **thalamic** termination, this system is termed the **spinoreticulothalamic tract** (SRT) (Fig. 7.3). Many questions regarding its structure and function in nonhuman primates and humans remain unanswered.

7.2.1. Modalities and Receptors

The modalities carried in the spinoreticulothalamic tract include persistent aching, burning, unpleasant, disagreeable, or dull pain caused by carcinoma,

inflammatory states, chemical stimuli, ischemia, distention of hollow organs, or excessive muscle contractions. Visceral pain is diffuse and poorly localized. When localized to an area distant from the actual site of the pain-causing process, such pain is termed **referred pain**. Hence, one quality of visceral pain is its referral to a well-defined segmental level or to a distant region. There is often a delay between the initial stimulus and its conscious perception, and persistence of the sensation beyond the cessation of its stimulus. Structural and functional evidence for specific visceral pain receptors, called **visceral nociceptors**, is lacking; they are probably free nerve endings (Fig. 7.3). Nociceptors in visceral organs such as the heart, lungs, testes and biliary system as well as those in muscles and joints are responsive to extremes of temperature, mechanical force, and certain chemical irritants.

7.2.2. Primary Neurons

Primary neurons in the **spinoreticulothalamic tract** are cells in spinal ganglia with small cell bodies and peripheral processes (Fig. 7.3) that travel in somatic nerves, along blood vessels, in autonomic nerves, and perhaps traverse the sympathetic trunk and its ganglia. Fibers carrying visceral pain are thinly myelinated Aδ or nonmyelinated C fibers. Central processes of these primary neurons enter the lateral division of each dorsal root; a few fibers enter the ventral roots.

Because about 25% of the ventral root fibers are nonmyelinated (and presumably sensory), pain is often evoked on ventral root stimulation in conscious patients. Experiments indicate that about two-thirds of these ventral sensory fibers are visceral in nature. Additional studies will likely clarify the structural and functional aspects of these visceral sensory fibers in the ventral roots. Cutting only the dorsal roots to relieve pain (dorsal rhizotomy) leaves these visceral sensory fibers intact. Destruction of the neuronal cell bodies of origin of these sensory fibers prevents sensory impulses from entering the central nervous system through the ventral roots. Since their cell bodies are presumably in the spinal ganglia, a **multiple spinal ganglionectomy** would eliminate, at the source, all sensory fibers (and all sensory impulses through these fibers) from entering the spinal cord. Blocking spinal ganglionic neurons with alcohol or destroying them by extremes of temperature would also block the transmission of sensory impulses. As the thinly myelinated and nonmyelinated visceral pain-carrying fibers enter the spinal cord, they end in the segment of entrance, or bifurcate and ascend or descend to end superior or inferior to this segment on the cell bodies of secondary neurons.

7.2.3. Secondary Neurons

Impulses carried in central processes of primary visceral sensory neurons ultimately influence neurons in the dorsal horn. The precise relationship between these entering fibers and dorsal horn neurons is unclear. Nonmyelinated C fibers enter the lateral division of each dorsal root in nonhuman primates and end in the substantia gelatinosa (spinal lamina II). For this reason, the substantia gelatinosa is likely involved in visceral and somatic pain. The cell bodies of origin of some secondary neurons in this

path, found laterally at the base of the dorsal horn, form the **secondary visceral substance** (Fig. 7.3), that extends from T1 to L1 or L2 and from S2 to S4 spinal levels. The central processes of the secondary neurons in this path form an ascending visceral pain path called the **spinoreticulothalamic tract**.

Central Processes of Secondary Neurons Form the Spinoreticulothalamic Tract

Because of its partial origin from the general visceral afferent neurons in the secondary visceral substance, the **spinoreticulothalamic tract** may also be termed the **secondary ascending visceral tract**. Neurons contributing to sensations from muscle and viscera are identifiable in spinal laminae IV–VI of the dorsal horn of anesthetized nonhuman primates. Experiments in cats have implicated neurons in spinal laminae VI–VIII as the origin of the spinoreticulothalamic path. Two-thirds of responses from neurons in these spinal laminae were ipsilateral, a third were contralateral. Experimental studies suggest that the lateral spinothalamic and spinoreticulothalamic tracts are functionally different, most of their fibers having separate cell bodies of origin in the spinal cord.

A collection of fibers, called the **proprius bundles**, surround the entire spinal gray matter (except in the area near the dorsolateral tract) at the junction between the gray matter and fibers in the dorsal, ventral and lateral funiculi. Ascending axons in the visceral pain path arise from secondary cell bodies and enter the ventral part of the lateral funiculi on both sides of the cord. In this position, both crossed and uncrossed secondary visceral pain fibers contribute fibers that enter the proprius bundles. Fibers related to visceral pain and traveling in the spinoreticulothalamic tract (SRT) are in the **ventral parts of the lateral fasciculus proprius** (Fig. 7.3). The spinoreticulothalamic tract is a bilateral ascending path ideally organized to transmit diffuse, poorly localized, chronic visceral pain to brain stem levels. Surgical interruption of fibers in the dorsal funiculi (on either side of the dorsal median sulcus) carried out in a limited number of patients with visceral pain as the result of uterine or colon cancer suggests that some visceral pain fibers ascend in the dorsal funiculi. The relation of these fibers to the spinoreticulothalamic tract is unclear.

Somatotopic Organization in the Spinoreticulothalamic Tract

Although experimental evidence regarding the anatomic organization of this system is limited, clinical evidence suggests it is somatotopically organized. Ventrally in the lateral fasciculus proprius near the central canal, are ascending fibers for sensory impulses from the bladder. They are dorsal to the ascending fibers for sensory impulses from the stomach (Fig. 7.3). The ascending visceral fibers in the dorsal funiculi noted in the previous paragraph suggest a possible localization there for visceral sensory impulses from pelvic and abdominal areas. In nonhuman primates, most fine fibers of spinal origin that are deep in the ventrolateral funiculus ascend to the reticular formation and end in the ventromedial part of the medulla oblongata and pons. Coursing in the lateral funiculus of the cord and continuing perpendicular to the long axis of the brain stem, the bilateral spinoreticulothalamic tract conveys impulses to the brain stem reticular formation.

Position of the Spinoreticulothalamic Tract in the Brain Stem

The spinoreticular part of the visceral pain path is in the medial reticular formation of the medulla, pons, and midbrain where it contributes to the ascending reticular system. More on the brain stem reticular formation is available in Chapter 9.

At medullary levels, spinoreticular fibers are between or perhaps dorsal to the **inferior olives** and, in the pons, dorsomedial to the superior olive between the exiting abducent and facial intramedullary fibers. Many spinoreticular fibers end in the reticular formation at medullary and pontine levels, while the majority terminate at the junction of the pons and midbrain. The reticular formation of the midbrain appears to form the ascending continuation of the spinoreticulothalamic tract by providing crossed reticulothalamic fibers. The spinoreticulothalamic system extends from spinal levels ('spino' part), through the brain stem reticular formation ('reticulo' part), and into the thalamus ('thalamic' part).

A few **spinomesencephalic fibers** end in the dorsolateral periaqueductal gray substance, and some **spinotectal fibers** end in the tectum of the midbrain. Most of these fibers in humans are ipsilateral and most accompany the spinoreticulothalamic tract.

Involvement of the Periaqueductal and Periventricular Regions in Chronic Pain

The periaqueductal gray substance in humans exhibits high levels of binding of opiate receptors. **Opium** and its derivative **morphine** are plant products used because of their narcotic and analgesic effects. (**Analgesia** refers to freedom from pain in humans and **antinociception** for

the suppression of reflex behavior in experimental animals caused by noxious stimuli.) These substances attach to specialized **receptor sites** on neuronal membranes. Since the incidence of such **opiate receptor sites** varies in the brain, high levels of binding in the periaqueductal gray substance of the midbrain are noteworthy because this opiate-to-receptor binding either yields analgesia or initiates events that culminate in analgesia. In two regions of the nonhuman primate brain (the periaqueductal–periventricular region and the lateral periaqueductal gray substance medial to the lateral spinothalamic tract), injections of morphine produce antinociception. Based on the above, the junction between the midbrain and diencephalon in humans, particularly the **periaqueductal and periventricular regions** is likely involved in chronic pain.

Endorphins and Pain

The **endogenous opioid peptides** are hormone-like substances produced in the brain that bind to opiate receptors and may reduce sensitivity to pain and stress. The three major groups of these natural substances are the **endorphins**, the **enkephalins**, and the **dynorphins and neoendorphins**. The periaqueductal gray substance of the midbrain is one of many regions in which these opioid peptides are present. These observations suggest that the brain has the ability to furnish its own opiates that are in turn capable of producing analgesia.

Stimulation of the Periaqueductal and Periventricular Regions

Deep brain stimulation (DBS) of the periaqueductal gray substance inhibits the responsiveness of neurons in spinal lamina V to noxious visceral input in rats. Since direct spinal projections exist in nonhuman primates from the periaqueductal gray substance of the midbrain and surrounding reticular formation, this region likely influences spinal neurons involved in visceral pain. Electrical stimulation and therapeutic injuries to the human periaqueductal gray substance confirm this possibility. Sensations elicited by stimulation of neurons at the edge of the aqueduct and 5 mm laterally, in those undergoing neurosurgical treatment for chronic pain, are described as unpleasant and referred to the center of the body, deep in the head, neck, chest (perhaps near the heart), or abdomen. A patient complained of periumbilical sensations as well as sensations from the bladder and experienced the desire to urinate during periaqueductal gray stimulation. Painful sensations elicited by such stimulation have a diffuse and unpleasant quality often accompanied by autonomic and emotional reactions including feelings of fear. Complex sensory reactions to stimulation of the periaqueductal gray substance of the midbrain in humans are like spontaneous reactions in those undergoing an intense painful experience such as chronic pain. Small therapeutic lesions in this region of the midbrain calm the patient and relieve suffering, but do not yield analgesia – especially in those suffering from intractable pain caused by carcinoma in the head and neck. In another study, electrical stimulation of the peri-aqueductal region effectively reduced chronic pain with numerous side effects, including a sensation of smother-ing, abnormal ocular movements, nausea, and vertigo. Stimulation of the ventral periventricular/periaqueduc-tal gray substance in 15 awake humans undergoing deep brain stimulation for treatment of chronic pain caused a mean reduction in systolic blood pressure (in seven patients) whereas stimulation of the dorsal peri-ventricular/periaqueductal gray substance caused a mean increase in systolic blood pressure (in six patients). Collectively these observations indicate that while peri-aqueductal gray substance may not be the best target for stimulation designed to reduce or control chronic pain in humans there may well be a role for the periaqueductal gray substance in cardiovascular control in humans.

7.2.4. Thalamic Neurons

As ascending reticulothalamic fibers leave the midbrain to enter the thalamus, some enter the posterior hypothal-amus in a periventricular position near the posterior part of the third ventricle. Spinoreticulothalamic fibers end in the **intralaminar nuclei** especially the parafascicular (PF) and centromedian nuclei (CM), the **reticular nucleus (Rt)**, and parts of the **ventral posterior nucleus** (VP). The intralaminar nuclei are the site of conscious appreciation of diffuse, poorly localized, noxious stimuli. Experimen-tal injections of morphine into these nuclei produced antinociception. Stimulation or destruction of the intra-laminar region in humans, and the medial pulvinar, relieves pain without detectable sensory loss for 2–12 months. In another study, stimulation along the medial aspect of the parafascicular nucleus in the periventricu-lar region resulted in 'good-to-excellent' reduction of chronic pain with minimal side effects. These patients expressed feelings of relaxation and well-being without overwhelming emotional overtones. Reduction of pain was contralateral to the stimulation yet affected bilateral peripheral fields, especially in medial regions of the body. Other studies by these same investigators demon-strated effective relief with long-term implantation of electrodes in the periventricular region near the posterior

part of the third ventricle. Two-thirds of these patients noted reduction of pain on self-stimulation. Pain relief via this **stimulated analgesia** outlasts stimulation by many hours in some patients. Stimulated analgesia involving the periventricular region in humans relieves chronic pain with minimal complications (perhaps by activating and releasing **opioid peptides** – substances that appear in the cerebrospinal fluid after stimulation). Naloxone, an opiate antagonist, inhibits chronic pain produced by periventricular stimulation with implanted electrodes. Stimulation of the periventricular region appears effective in treating chronic pain in humans.

7.2.5. Cortical Neurons

Although impulses for visceral pain reach conscious-ness at a thalamic level, areas of the cerebral cortex are also involved in visceral sensation. Intralaminar nuclei, part of the 'diffuse thalamocortical projection system', relay visceral impulses to the cerebral cortex including the cingulate gyrus. The diffuseness of these cortical projections likely accounts for the poor localization of visceral pain at the cortical level. Cortical neurons for visceral sensation in humans are in the **precentral operculum of the frontal lobe**, in the **anterior insula**, the pregenual **anterior cingulate cortex,** the **anterior midcingulate cortex** and perhaps in the prefrontal cor-tices. That part of the precentral operculum associated with visceral sensations is outside the SII cortical region but between Brodmann's areas 43 and 44.

7.2.6. Suffering Accompanying Pain

The calming effect of medial lesions in the midbrain resemble the effects of frontal lobotomy for relief of pain. Such operations fail to abolish the perception of pain, but modify the emotional responses to pain. Their fear of pain seemingly disappears due to an alteration in the patient's attitude. Because of adverse changes in per-sonality and intelligence produced by this procedure, it is a method of therapeutic desperation. Modification of emotional response to pain after therapeutic lesions in the midbrain or frontal lobotomy emphasizes an affec-tive aspect of pain called **suffering** that accompanies certain types of pain and varies from individual to indi-vidual. Memories of pain and its anticipation influence the suffering accompanying painful experiences. Absence of pain in combat or in emotional stress accom-panying serious injuries, illustrates this aspect of pain. Suffering is probably independent of the cause or dura-tion of pain. The degree to which pain influences our daily lives, including the ability to carry out family responsibilities and to work, has a profound effect on the suffering that accompanies pain. Some patients use pain for secondary gain such as in compensation for injuries to avoid work or to gain sympathy from others. Neurons in the rostral part of the frontal lobe – referred to as the 'prefrontal cortex' – send axons to the dorsolateral quadrant of the periaqueductal gray substance in the midbrain of nonhuman primates. Activation of these prefrontal-periaqueductal fibers is likely to underlie emotional responses accompanying stimulation in this region of the midbrain in humans. Brain imaging studies have demonstrated that the suffering aspect of pain activates the **anterior cingulate cortex** in humans. The cingulate gyrus as a constituent of the limbic system (Chapter 18) plays an important role in emotion and, through its many autonomic projections, in the expression of that emotion.

7.2.7. Visceral Pain as Referred Pain

The perception of visceral pain is that it comes from a point distant from its origin. This phenomenon is termed **referred pain**. Pain arising from cervical skeletal muscle and surrounding structures, from the tongue, teeth, and tonsils of the oral cavity or from the temporomandibular joint (including chewing-gum earache in children) is often referred to the auricle. Pain from the nasopharynx and paralaryngeal region may be referred to the epigastrium and to the area around the xiphoid process in the median plane (particularly the area of the seventh to the ninth thoracic segments). Pain from the common bile duct and cardiac pain (angina pectoris) is often referred to the left shoulder, arm, and side of the neck. Dysmenorrheal pain (from hyperactivity of the uterine muscle and myometrial ischemia) is often referred to the lower back and abdomen, and pain in early labor (from stretch of the cervix) is likely felt between the umbilicus and pubic symphysis, laterally on the iliac crest, and posteriorly on the lower lumbar and upper sacral spines. With intense stimulation caused by contraction and distention of the uterus in labor, pain is referred to the umbilical region, upper thighs, gluteal region, and middle sacral area.

The distribution of intra-abdominal pain in normal subjects and in those suffering from disease of this region has received considerable attention. Inflating a balloon introduced during colonoscopy leads to pain at several sites along the colon. In those without primary pain, this produces lower, left-sided abdominal pain. Inflation in the ascending and transverse colon produces periumbilical pain. In similar experiments

involving control subjects, distention of the splenic flexure caused pain in the abdominal left upper quadrant at the site of distention. Distention at the splenic flexure in some persons evokes a sense of pain or discomfort in areas of the body above the diaphragm that may also be sites of referral of pain from angina pectoris. In another series of normal subjects, balloon distention of the sigmoid colon, and occasionally the upper rectum, resulted in pain referred to the suprapubic or left iliac region simulating intestinal colic (pain due to distention of the bowel by gas). Distention of the rectum caused a characteristic sensation of fullness referred to the rectum itself or the sacral region.

An appealing theory underlying referred pain is that visceral pain and cutaneous pain, arriving over separate paths, converge in the spinal cord or thalamus (**convergence theory**) on common neurons. Sensory fibers from visceral areas may converge on spinal neurons with a cutaneous receptive field. From this point, impulses travel along separate routes and are interpreted as though arising from two body locations – at the site of their visceral origin and at a cutaneous, referred site. In nonhuman primates, sensory impulses from the thoracic viscera, cardiopulmonary region, and from the skin and muscle of the chest and forearm converge on spinothalamic neurons. This may account for referred cardiac pain.

Cell bodies in the dorsal horn that respond to noxious stimuli and are excited by input from sensitive mechanoreceptors and nociceptors are concentrated in spinal laminae IV–VI in nonhuman primates. These neurons have large receptive fields with a sensitive central zone surrounded by a less sensitive periphery. Because such neurons respond to both tactile and noxious stimuli, they are termed '**wide dynamic range**' **neurons** in contrast to '**narrow dynamic range**' **neurons** that respond only to tactile stimuli. 'Wide dynamic range' neurons receive a convergent input from deep or visceral receptors and their activation likely accounts for referred pain. Nociceptive substances, produced in skin or subcutaneous tissue, stimulate the endings of spinal nerves that supply a given area of skin and thus evoke referred pain. Occasionally the pain of angina is referred to the site of previous somatic pain (including the ear, jaws, teeth, and oropharynx). The precise mechanism for this false localization is unclear.

Referred nonpainful cutaneous sensations elicited by a non-noxious stimulation, such as scraping fingernails on the scalp, plucking at hairs, toenails, or minor skin eruptions differ in many ways from referred pain. The resultant sensations, usually referred to a relatively distant cutaneous point, are never painful, but sharply localized and startling in their suddenness.

These nonpainful sensations may be accompanied by an itch that is relieved by rubbing the site of the referred sensation.

7.2.8. Transection of Fiber Bundles to Relieve Intractable Pain

Transection of the lateral spinothalamic tract (**anterolateral cordotomy**) to relieve intractable pain is often unsuccessful. One year after cordotomy, approximately half of those treated had neither relief from pain nor complete analgesia. Impulses for chronic, intractable pain travel to higher levels of the central nervous system by the spinoreticulothalamic tract. Without interruption of this path on both sides of the spinal cord, visceral pain will persist. The position of the lateral spinothalamic tract and the spinoreticulothalamic tract at spinal levels is fortunate, because fibers of the SRT (conducting visceral pain) lie medial to the lateral spinothalamic tract (Fig. 7.3). Therefore, transection in the ventrolateral quadrant involving the spinoreticulothalamic tract in the ventral part of the **lateral fasciculus proprius** will transect the path for visceral pain *and* that for somatic pain. Even so, because of the multisynaptic and bilateral nature of the spinoreticulothalamic tract, it is impossible to interrupt visceral impulses ascending in the cord unless bilateral transection occurs. Because the spinoreticulothalamic tract is separate from, and medial to, the lateral spinothalamic tract in the brain stem, a tractotomy in the pons or midbrain only interrupts the lateral spinothalamic tract. Although the periaqueductal gray substance in humans is unsuitable for stimulation designed to reduce or control chronic pain, the success of analgesia produced by stimulation of the diencephalic periventricular region is encouraging. These observations suggest the brain stem periaqueductal gray substance and adjacent periventricular region in the posterior third ventricle participate in modulating human pain. Details of the anatomy of the paths for pain in humans, an understanding of the different painful conditions occurring in humans, the precise roles of the ever-growing list of endogenous agents in the perception of pain, and the effectiveness of various therapeutic measures in relieving pain need further investigation.

7.3. THE TRIGEMINAL NUCLEAR COMPLEX

The **trigeminal nerve** [V] provides sensory innervation to the face, cornea, and oral mucosa, including the tongue, dental pulp, nasal mucosa, paranasal sinuses, and scalp back to the vertex of the skull. The trigeminal nerve and its branches carry impulses for pain and temperature from this extensive area to the central nervous system. Such sensory impulses travel through the brain stem to the thalamus by way of the **ventral trigeminothalamic tract**.

7.3.1. Organization of the Trigeminal Nuclear Complex

The trigeminal nuclear complex resembles an ice-cream cone in shape, extending from the middle pons into the spinal cord. The ice cream on the cone is the **trigeminal pontine nucleus**, some 4.5 mm in rostrocaudal length; the cone is the **trigeminal spinal nucleus** with its three subnuclei (Fig. 7.4). The trigeminal nuclear complex in humans develops in a cervicorostral direction (caudal part, interpolar part, then rostral part). There is a close correlation between development and activity in this complex: differentiation in the caudal part accompanies the earliest observed reflex response of human embryos to trigeminal stimulation. The **trigeminal pontine nucleus** is in the middle pons at the level of entrance of the trigeminal nerve (Fig. 7.4) and in the lateral part of the pontine tegmentum. Its neurons are small, some 25–28 μm in average diameter. The **trigeminal spinal nucleus** extends from the caudal end of the trigeminal pontine nucleus into the lower pons and medulla oblongata to reach the spinal cord (Fig. 7.4), where it intermingles with spinal laminae II and III of the upper cervical cord. Based on cytoarchitectural differences, the **trigeminal spinal nucleus** is divisible into **caudal**, **interpolar**, and **rostral parts** (Fig. 7.4). The **caudal part of the trigeminal spinal nucleus** (Fig. 7.4) extends from the obex of the medulla oblongata caudally to about C2 (and perhaps as low as C3 or C4) where it overlaps and intermingles with neurons of the dorsal horn, making this region of the spinal cord gray matter conspicuous in stained transverse sections. The cytoarchitecture of this **caudal part of the trigeminal spinal nucleus** resembles the apex of the dorsal horn in that it has three subnuclei from superficial to deep. The **interpolar part of the trigeminal spinal nucleus** (Fig. 7.4), about 6 mm in length, is evident at medullary levels through the caudal two-thirds of the **inferior olive** (corresponding to the level of the accessory cuneate nucleus). The **rostral part of the trigeminal spinal nucleus** (Fig. 7.4) continues from the rostral part of the **inferior olive** at medullary levels to the level of abducent [VI] and facial [VII] nuclei in the lower pons, lateral to the intrapontine facial fibers. It is then continuous rostrally with the

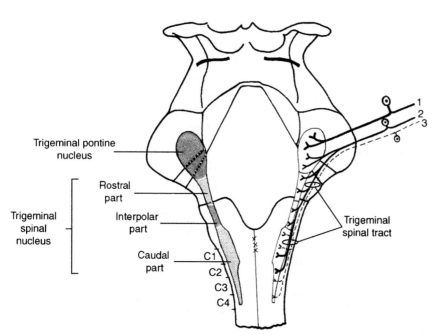

FIGURE 7.4. Organization of the trigeminal nuclear complex projected on the brain stem. Also shown are types of entering sensory fibers and their distribution to the trigeminal nuclear complex. **1,** Largest and most thickly myelinated fibers. **2,** Smaller and less thickly myelinated fibers. **3,** Smallest fibers with a slight amount of or no myelin (after Humphrey, 1969).

trigeminal pontine nucleus and caudally with the interpolar part of the trigeminal spinal nucleus.

7.3.2. Organization of Entering Trigeminal Sensory Fibers

General somatic afferents from the face, cornea, oral, and nasal mucosa, tongue, dental pulp, paranasal sinuses, and scalp to the vertex of the skull have their neuronal cell bodies of origin in the **trigeminal ganglion**. Each human trigeminal ganglion has some 27,400 neurons (range 20,000 to 35,400) with some reports of upwards of 80,000 neurons in each ganglion. Peripheral processes of these trigeminal ganglionic neurons pass to appropriate peripheral receptors. As their central processes approach the brain stem, they form the **trigeminal sensory root** before ending in a well-organized manner in the trigeminal complex. In baboons, about 45% of trigeminal sensory root fibers are nonmyelinated. Central processes of these primary neurons project onto the trigeminal nuclear complex in accordance with their function. The largest and most thickly myelinated fibers (Fig. 7.4) carry discriminative tactile information and end in the trigeminal pontine nucleus. Smaller and less thickly myelinated fibers (Fig. 7.4) carry general tactile information, and bifurcate on entry, sending one branch to the trigeminal pontine nucleus and another caudal branch into the **trigeminal spinal tract** (Fig. 7.4). Even smaller, thinly myelinated and some nonmyelinated fibers (Fig. 7.4) that carry impulses for general or light touch, pain, temperature (hot and cold),

itch, and possible 'tickle,' enter the pons, and at once descend in the trigeminal spinal tract to end in the trigeminal spinal nucleus (Fig. 7.4). As these fibers descend in the trigeminal spinal tract in nonhuman primates, they taper and lose their myelin. The course of the trigeminal spinal tract in the lower medulla is evident on the brain stem surface as the **trigeminal tubercle**. The trigeminal spinal tract overlaps the dorsolateral tract at cervical levels of the spinal cord.

In addition to these general somatic afferents in the trigeminal sensory root, there is also a population (about 20%) of small (0.3–0.5 µm) nonmyelinated fibers in the human **trigeminal motor root** similar to the non-myelinated sensory fibers in ventral spinal roots. Although the source and function of these nonmyelinated fibers is unclear, they are probably sensory. Several small **trigeminal accessory** or **intermediate fibers**, of uncertain function, exist between the trigeminal motor and sensory roots in humans.

7.4. PATH FOR SUPERFICIAL PAIN AND THERMAL EXTREMES FROM THE HEAD

7.4.1. Modalities and Receptors

The path carrying impulses for superficial pain (including itch), thermal extremes (hot and cold), possibly tickle, and general or light touch from areas of the head innervated by the trigeminal nerve (including the dental pulp) is the **ventral trigeminothalamic tract** (Fig. 7.5).

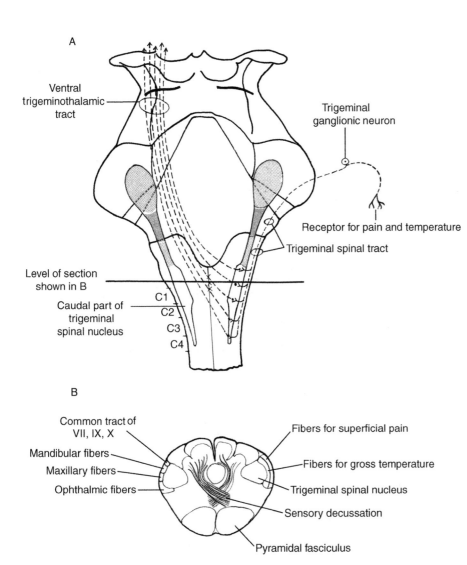

A

Ventral trigeminothalamic tract

Trigeminal ganglionic neuron

Receptor for pain and temperature

Trigeminal spinal tract

Level of section shown in B

C1
C2
C3
C4

Caudal part of trigeminal spinal nucleus

B

Common tract of VII, IX, X

Mandibular fibers

Maxillary fibers

Ophthalmic fibers

Fibers for superficial pain

Fibers for gross temperature

Trigeminal spinal nucleus

Sensory decussation

Pyramidal fasciculus

FIGURE 7.5. **A,** Entering trigeminal fibers involved in the transmission of impulses for superficial pain and extremes of temperature. Such impulses are transmitted by secondary neurons in the caudal part of the trigeminal spinal nucleus and are then relayed to higher levels in the ventral trigeminothalamic tract (VTT). **B,** Illustration of a transverse section of the medulla oblongata at the level in **A.** On the left is the dorsoventral lamination of mandibular, maxillary and ophthalmic fibers in the trigeminal spinal tract. Cutaneous components of the facial, glossopharyngeal, and vagal nerves descending in the trigeminal spinal tract on the dorsolateral aspect of the mandibular fibers are shown. On the right side is illustrated the probable position of superficial pain fibers (lateral) and the fibers for thermal extremes (medial) in the trigeminal spinal tract (after Humphrey, 1969).

Receptors involved are probably free nerve endings and thermoreceptors (Fig. 7.5). The human cornea contains only fine arborizations of nerves. Since the cornea responds to light touch and noxious stimuli, these fine nerves must be sensitive to both types of stimuli. One patient experienced pain in a localized part of her gums for five years. Following surgical excision of the area to relieve the pain, large numbers of free nerve endings were in the epithelium and lamina propria of the excised gingival tissue. These observations emphasize the relationship of free nerve endings to the transmission of painful impulses at trigeminal levels.

7.4.2. Primary Neurons

The **primary neurons** (Fig. 7.5) in the path for superficial pain and thermal extremes at cranial levels are small neurons found in the **trigeminal ganglia.**

Peripheral processes of these primary neurons are probably nonmyelinated C fibers and myelinated Aδ fibers that respond to innocuous and noxious thermal and mechanical stimuli with multimodal nociceptors and thermoreceptors, respectively, at their terminations. Separate 'cold' spots on the hairy skin of the face in nonhuman primates receive their innervation by Aδ myelinated fibers. Central processes of these primary neurons enter the central nervous system in the **trigeminal sensory root.** They enter the middle pons and turn to descend in the medulla oblongata and upper spinal cord as part of the **trigeminal spinal tract** (Figs 7.4, 7.5).

Primary fibers descending in this tract, and carrying impulses for pain and thermal extremes from the head, end in the caudal part of the **trigeminal spinal nucleus** (Fig. 7.5). In this location are secondary neuronal cell bodies in the path for pain and thermal extremes in the head. Transmission of impulses along a sensory path is

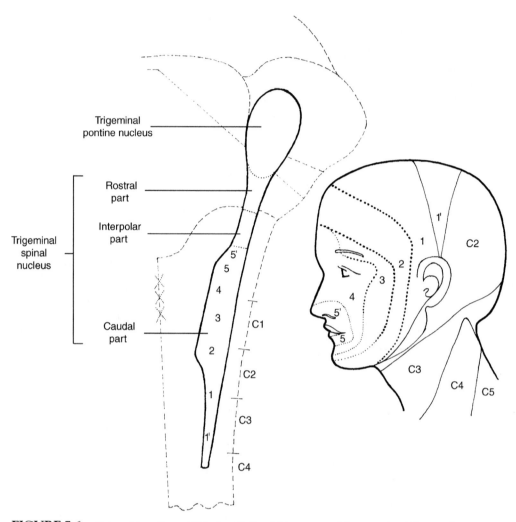

FIGURE 7.6. Onion-skin pattern of Déjerine. On the right are concentric zones about the mouth-nose area and extending to the ear. Levels of representation of these facial zones in the caudal part of the trigeminal spinal nucleus on the left are indicated by numbers on the face and on the nucleus. (After Humphrey, 1969.)

modifiable by descending paths and by endogenous substances. In the trigeminal complex, considerable modification of sensory information takes place in the caudal part of the trigeminal spinal nucleus.

Somatotopic Organization of Fibers in the Trigeminal Spinal Tract

Three peripheral trigeminal branches – mandibular, maxillary, and ophthalmic – bring impulses from their respective territories and send fibers to the trigeminal spinal tract where they descend as far as the second or third cervical segment. A dorsoventral, **somatotopic organization of fibers** is demonstrable in the trigeminal spinal tract, with mandibular fibers dorsal, ophthalmic fibers ventral, and maxillary fibers between them (Fig. 7.5B). Clinical evidence suggests that fibers carrying

painful impulses are lateral in the tract, whereas the fibers carrying impulses for thermal extremes are medial (Fig. 7.5B).

Somatotopic Organization of Fibers Distributing to the Caudal Part of the Trigeminal Spinal Nucleus

In addition to the arrangement of fibers in the trigeminal spinal tract according to the area innervated and modality served, there is also a layered arrangement of fibers associated with impulses for pain and thermal extremes, distributing to the caudal part of the trigeminal spinal nucleus. In this arrangement, called the 'onion-skin' pattern (of Déjerine), there are five facial zones (Fig. 7.6), beginning with a central circumoral zone, three intervening zones, and a peripheral facial zone. Sensory fibers from these facial zones have a rostrocaudal order as they descend along the **caudal part** of the trigeminal spinal nucleus. In particular, fibers supplying peripheral face areas descend to end in the **caudal part** of the trigeminal spinal nucleus in the spinal cord. Fibers supplying the circumoral zone end rostrally in the caudal part of the trigeminal spinal nucleus of the lower medulla oblongata near the obex (Fig. 7.6). Cutaneous branches of the cervical plexus (C2, C3) supply skin over the angle of the mandible, a region spared by unilateral injury to the trigeminal sensory root.

7.4.3. Secondary Neurons

Secondary neuronal cell bodies in the trigeminal path for pain and thermal extremes are in the **caudal part of the trigeminal spinal nucleus** and adjacent **lateral reticular formation**. Neurons with receptor terminals activated by noxious and thermal stimuli give off small, thinly myelinated, and nonmyelinated central processes that end in the substantia gelatinosa subdivision of the caudal part of the trigeminal spinal nucleus in nonhuman primates. At upper cervical levels sensory cervical spinal fibers and fibers of the descending trigeminal spinal tract overlap, as do secondary neurons in the lateral spinothalamic tract and the trigeminothalamic system. Painful impulses from this region of overlap may travel to higher levels in either the spinothalamic or the trigeminothalamic path. Axons containing the neurotransmitter serotonin synapse with neurons of the spinothalamic and trigeminothalamic tract in spinal lamina I and spinal lamina V in nonhuman primates.

Endogenous opiate-like substances are identifiable in human trigeminal neurons. Fibers, terminals, and neuronal cell bodies containing leu-enkephalin are in the trigeminal spinal nucleus in humans at caudal levels of the medulla (presumably in the caudal part of the trigeminal spinal nucleus). Since equivalent areas of the human spinal cord receive their innervation from fibers containing substance P, interaction of substance P and enkephalins in the processing of painful impulses is likely.

Central Processes of Secondary Neurons Form the Ventral Trigeminothalamic Tract

Axons of secondary neurons, with cell bodies in the zonal and deep magnocellular subdivisions of the caudal part of the trigeminal spinal nucleus and in the adjacent lateral reticular formation, accumulate in small fascicles, swing obliquely forward, and decussate in the brain stem. Rostral to their origins these fascicles are in the lateral brain stem where they help to form the **ventral trigeminothalamic tract (VTT)** (Fig. 7.5). A precise position for this tract in the medulla oblongata is difficult to define, because its fibers spread out as they leave the caudal part of the trigeminal spinal nucleus. In the lower pons (Fig. 7.7), VTT is a more compact bundle intermingling with the dorsal edge of the medial lemniscus, a prominent ascending sensory path. In the upper midbrain (Fig. 7.7), VTT remains on the dorsal aspect of the medial lemniscus before projecting forward to end in the thalamus (Fig. 7.7). Experiments in primates demonstrate descending fibers from the zonal subdivision of the caudal part of the trigeminal spinal nucleus; they are likely an indirect input to the ascending ventral trigeminothalamic tract.

7.4.4. Thalamic Neurons

The ventral trigeminothalamic tract, carrying impulses for pain and thermal extremes in the head, ends in the **ventral posterior medial nucleus** (VPM) (Fig. 7.7) of the thalamus. Some ventral trigeminothalamic fibers end in that part of the ventral posterior lateral nucleus (VPL) that is rostrolateral to the ventral posterior medial nucleus (VPM). A somatotopic organization exists in the ventral posterior nucleus (VP) with ascending sensory fibers from the body ending in the lateral part of this nucleus (VPL) and those from the face ending in the medial part (VPM). The representation of the body, face, and head across the ventral posterior nucleus forms a '**thalamic homunculus**'. A contralateral, medial to lateral arrangement of sensory responses is demonstrable in the ventral posterior nucleus (VP) beginning medially in the ventral posterior medial nucleus (VPM) with representation for the oral cavity, followed laterally by labial and facial representation, then that of the fingers, thumb, and rest of the hand. The representation of the forearm, arm, trunk, and leg follows in the lateral limit of the ventral posterior lateral nucleus (VPL). In humans, sensations of crude pain and thermal extremes such as those under discussion likely enter consciousness at the thalamic level. If localization ('where') and discrimination ('what') of these crude sensations is necessary, then the cerebral cortex must become involved. This would require the relay of these impulses from the thalamus to the cerebral cortex where additional processing can take place.

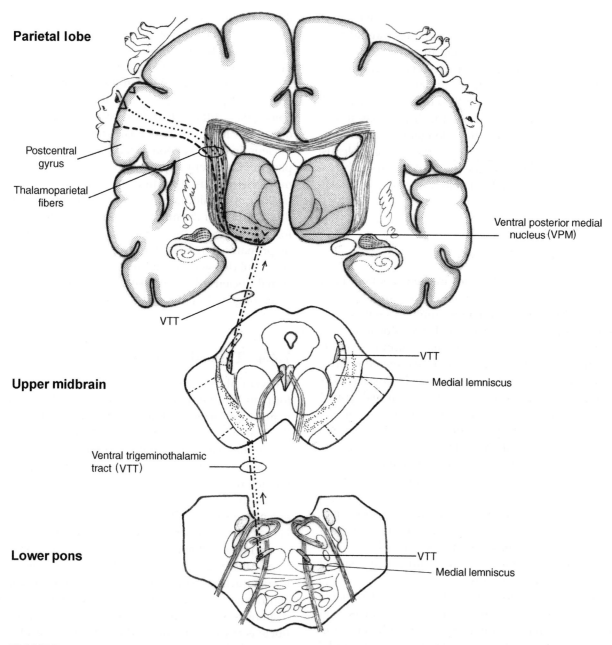

FIGURE 7.7. Course and position of the ventral trigeminothalamic tract (VTT) through the brain stem, its termination in the ventral posterior medial nucleus (VPM) of the dorsal thalamus, and its thalamoparietal projection. Note the face and hand parts of the sensory homunculus illustrated across the postcentral gyrus of the parietal lobe. On the left is the classical somatotopic arrangement of the face relative to the hand whereas on the right is depicted a recent pattern with the thumb juxtaposed to the lower lip and corner of the mouth.

7.5. PATH FOR THERMAL DISCRIMINATION FROM THE HEAD

7.5.1. Modality and Receptors

The forehead is the most sensitive to temperature of all cutaneous areas; it detects an increase of

$0.001°C/sec/cm^2$. Impulses for thermal discrimination (warmth versus coolness) from the face and head reach the trigeminal nuclear complex differently than impulses for thermal extremes. Both types of thermal impulses, however, project to higher levels through the **ventral trigeminothalamic tract** (Fig. 7.8). The structure of the involved **thermoreceptors** (Fig. 7.8) is unclear.

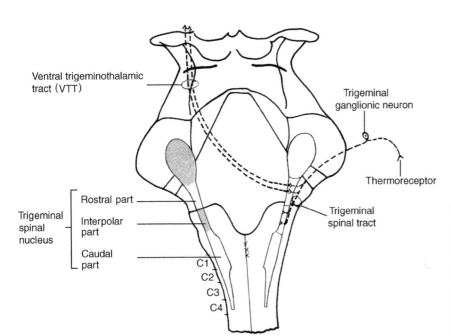

FIGURE 7.8. Entering trigeminal sensory fibers involved in the transmission of thermal discrimination. Such impulses are transmitted to secondary neurons in rostral and interpolar parts of the trigeminal spinal nucleus and then are relayed from the rostral part to higher levels over the ventral trigeminothalamic tract (VTT) as shown. See Fig. 7.7 for the course and position of this tract through the brain stem, its dorsal thalamic relay, and its ultimate termination.

7.5.2. Primary Neurons

Primary neurons in this path for thermal discrimination occur in the **trigeminal ganglia** (Fig. 7.8). Their peripheral processes in nonhuman primates are myelinated Aδ or nonmyelinated C fibers. Their receptors are sensitive to small (less than 1°C) increases or decreases in the temperature of the skin. Their central processes enter the pons in the trigeminal sensory root, descend in the **trigeminal spinal tract**, and end in the **rostral and interpolar parts** of the trigeminal spinal nucleus (Fig. 7.8).

7.5.3. Secondary Neurons

Secondary neurons in this path for thermal discrimination in the head occur in the **rostral part** of the trigeminal spinal nucleus in the lower pons. Fibers from these secondary neurons accumulate in small fascicles, swing obliquely forward and decussate to join the **ventral trigeminothalamic tract** (VTT) (Fig. 7.8) which continues to ascend through the pons and midbrain (Fig. 7.7). Thus, the ventral trigeminothalamic tract is made up of fibers carrying impulses for superficial pain, including itch, possibly tickle, light touch, thermal extremes, and thermal discrimination.

7.5.4. Thalamic Neurons

Impulses for thermal discrimination, relayed over fibers of the **ventral trigeminothalamic tract**, synapse with tertiary neurons in the **ventral posterior medial nucleus** (Fig. 7.7). **Thalamoparietal fibers** (Fig. 7.7) travel from here to the area of representation of the face in the postcentral gyrus of the parietal lobe (Fig. 7.7) where they terminate on fourth-order cortical neurons.

7.5.5. Cortical Neurons

Although sensations of thermal extremes and superficial pain enter consciousness at a thalamic level in humans, their accurate localization and the discrimination of fine thermal differences require cortical participation. Thalamoparietal fibers from the ventral posterior medial nucleus (VPM) reach that part of the postcentral gyrus of the parietal lobe designated as the **primary somatosensory cortex** or **SI** and corresponding to **Brodmann's areas 3, 1,** and **2.** Impulses from the head that reach the ventral posterior medial nucleus (VPM) do not project to the paracentral lobule because it receives sensory impulses from lower regions of the body by means of the lateral spinothalamic tract. Brodmann's area 3 is divisible into areas 3a and 3b in humans. Area 3a, the floor of the central sulcus, is a transition region between Brodmann's area 4 (corresponding to the **primary motor cortex**) along the precentral gyrus and area 3b on the posterior bank of the central sulcus. Areas 1 and 2 are on the lateral surface of the postcentral gyrus. Areas 3a, 3b, 1, and 2 in nonhuman primates and humans each constitute a separate map of the body and each receives input from a particular set of

TABLE 7.2.   Summary of paths for pain and temperature from the head

Modalities	Superficial pain, thermal extremes, light touch	Thermal discrimination (warmth vs. coolness)
Receptors	Nociceptors, thermoreceptors and mechanoreceptors	Thermoreceptors
Primary neurons	Smaller neurons in trigeminal ganglia	Trigeminal ganglia
1° peripheral processes	To receptors	A-delta or C fibers to receptors
1° central processes	Enter pons via trigeminal sensory root	Enter pons via trigeminal sensory root
Secondary neurons	Trigeminal spinal nucleus: note onion-skin pattern	Trigeminal spinal nucleus
2° central processes	Crossed to opposite side of the brain stem	Cross to the opposite side of the brain stem
Location 2° fibers	In **ventral trigeminothalamic tract**	In **ventral trigeminothalamic tract**
Spinal cord	Not in spinal cord	Not in spinal cord
Medulla	Not evident in medulla	Not evident in medulla
Pons	Dorsal to medial lemniscus	Dorsal to medial lemniscus
Midbrain	Medial aspect of medial lemniscus	Medial aspect of medial lemniscus
Dorsal thalamic neurons	Medial part of ventral posterior nucleus (VPM) of dorsal thalamus	Medial part of ventral posterior nucleus (VPM) of dorsal thalamus
Thalamoparietal fibers	Via posterior limb of the internal capsule	Via posterior limb of the internal capsule
Cortical areas	Facial area of postcentral gyrus in parietal lobe, areas 3, 1, and 2; primary soma (SI)	Facial area of postcentral gyrus in parietal lobe, areas 3, 1, and 2; primary somatosensory cortex (SI)
Comments	Note sensory homunculus; somatotopic and modality-topic organization throughout	Accurate localization ('where') and discrimination ('what') of fine thermal differences requires involvement of the cerebral cortex

receptors **(multiple representation hypothesis)**. Areas 3b and 1 are representations of cutaneous areas and are mirror images of each other, but differ enough to suggest distinct functions.

Experimental studies of the face area of SI in nonhuman primates reveal that oral soft tissues such as the gingiva and inner aspect of the lips have their representation near the surface of the postcentral gyrus whereas the dental pulp has its representation along the floor of the central sulcus, coinciding with area 3a. Stimulation of the dental pulp in humans causes painful sensations that are accurately localized suggesting involvement of the primary somatosensory cortex in dental pulp pain in humans.

The area of representation of the face, occupying most of the lower half of the postcentral gyrus of the parietal lobe (Fig. 7.7), receives trigeminal impulses from the face and mouth. Originally, the sensory homunculus depicted that part of the face area concerned with the upper face and head to be superior to that concerned with the lips and oral cavity. The back of the head has its representation caudally in the postcentral gyrus: the face orients toward the central sulcus. Sensory seizures, analyzed in many patients, indicate that a revision of the accepted homuncular pattern along the postcentral gyrus is in order. In particular, the thumb is adjacent to the lower lip and corner of the mouth (Fig. 7.7). Table 7.2 summarizes the location of neurons and fibers in the paths for pain and temperature from the head.

7.6. GENERAL SOMATIC AFFERENT COMPONENTS OF VII, IX AND X

Although the **trigeminal nerve** [V] is the sensory nerve of the head, the facial [VII], glossopharyngeal [IX], and vagal [X] nerves carry impulses for pain and temperature from restricted areas in and around the ear. Sensory fibers in the posterior auricular branch of the facial nerve [VII] join the auricular branch of the vagus [X] to supply skin on the back of the auricle, posterior part of the **external acoustic meatus**, and part of the lateral surface of the tympanic membrane. The glossopharyngeal nerve [IX] supplies the medial surface of the tympanic membrane, mastoid air cells, and auditory tube. A communicating branch joins the glossopharyngeal nerve with the vagal auricular branch. Cell bodies of primary general somatic afferent neurons are in the geniculate, superior vagal and inferior glossopharyngeal ganglia. Their central processes enter the brain stem in certain cranial nerves (VII, IX or X), descend in the trigeminal spinal tract dorsolateral to descending mandibular fibers (Fig. 7.5), form the **common tract of the facial, glossopharyngeal**, and **vagal nerves,** and then synapse with secondary neurons in the caudal part of the trigeminal spinal nucleus. Secondary fibers from these neurons decussate and ascend in the ventral trigeminothalamic tract.

7.7. TRIGEMINAL AND OTHER NEURALGIAS

Trigeminal neuralgia (or *tic douloureux*) is an excruciatingly painful affliction of the face consisting of sudden, recurring attacks of stabbing pain associated with facial contortions and grimaces. These attacks (from a few to hundreds per day) begin and end abruptly, and last a few seconds to a few minutes. Pain is often unilateral and typically involves the maxillary division of the trigeminal nerve (V^2). Trigeminal neuralgia is common in females and twice as common on the right side. Attacks often abate for weeks or months at a time. The lightest touch of a 'trigger zone' often near the angle of the mouth or nostrils is likely to set off an attack. Consequently, afflicted patients are reluctant to touch the face, eat, drink, talk, shave, wash their face, brush their teeth, or chew. Usually there is no accompanying objective sensory or motor loss.

7.7.1. Causes of Trigeminal Neuralgia

Several causes of trigeminal neuralgia are likely, including compression of the trigeminal sensory root by bony irregularities along the petrous temporal ridge, compression of the root of the nerve near the pons by arterial loops, tumors or other masses, or irritation by sclerotic plaques. Pathological findings in trigeminal neuralgia consist of degenerative changes in neuronal cell bodies of the trigeminal ganglion or its peripheral branches, including myelin fragmentation, segmental demyelination, and hypermyelination. Ectopic action potentials generated at the terminal myelinated segment of trigeminal fibers are another possible mechanism of trigeminal neuralgia. At such sites, the velocity of an action potential slows enough to re-excite the adjacent axonal membrane at a time when it is no longer refractory, thus giving rise to an ectopic or reflected action potential. Generation of ectopic action potentials is a property of normal trigeminal sensory fibers occurring where axonal size changes, at its terminal myelinated segment, or at the point at which a large myelinated fiber gives off a nonmyelinated collateral branch.

7.7.2. Methods of Treatment for Trigeminal Neuralgia

There are various medical and surgical methods for treating trigeminal neuralgia. Relief of pain, without sensory loss, often occurs following **vascular decompression** of the trigeminal sensory root by loops of the superior cerebellar artery, anterior inferior cerebellar artery, an unnamed artery, or by a small vein.

Partial section of the trigeminal sensory root at the pons causes permanent relief of pain because the transected fibers do not regenerate. There is an accompanying numbness in the corresponding facial region. Transection of the trigeminal sensory root near the pons for trigeminal neuralgia assumes that sensory impulses travel only in the sensory root. If there is postoperative preservation of sensation, it is likely due to nonmyelinated sensory fibers in the trigeminal motor root.

The injection of alcohol into the trigeminal sensory root may cause partial destruction of the fibers in that root. Although loss of light touch and cutaneous pain is likely to result, in one series tactile sensation returned in the previous area of neuralgia. If painful sensations return, trigeminal neuralgia is likely to recur. Alcohol and other chemicals injected into the trigeminal ganglion may be helpful in the treatment of trigeminal neuralgia. Difficulty in controlling the spread of such substances may lead to postoperative complications such as persistent anesthesia of the cornea leading to inflammation (keratitis).

Trigeminal neuralgia is treatable by thermocoagulation of the trigeminal ganglion or its peripheral branches. Small, myelinated $A\delta$ and nonmyelinated C fibers in the trigeminal peripheral branches are destroyed differentially using a thermal level that spares thickly myelinated, $A\alpha$ fibers carrying tactile information, thus eliminating the perception of pain but preserving tactile perception in a restricted facial zone. The effect is likely temporary because peripheral trigeminal fibers retain the ability to regenerate. Indeed, after thermocoagulation, the recurrence of trigeminal neuralgia tends to increase with the passage of time.

Another procedure, **medullary tractotomy**, involves transection of the trigeminal spinal tract, rostral to the motor decussation but caudal to the obex resulting in the elimination of pain with sparing of tactile sensation on the face; corneal responses are maintained. A modification of this procedure involves surgical removal of the trigeminal spinal tract. Excised tissue, examined with the aid of an electron microscope, reveals striking and regular changes in myelin layers. Degenerative hypermyelination is characteristic of trigeminal neuralgia.

Stereotactic trigeminal nucleotomy involves destruction of the caudal part of the trigeminal spinal nucleus. The 'onion-skin' pattern of distribution of trigeminal fibers to the caudal trigeminal spinal nucleus, described earlier in this chapter, is physiologically identifiable with electrical stimulation performed prior to nucleotomy for accurate identification. Many who suffer from trigeminal neuralgia think their pain is of dental origin and thus seek the services of a dental practitioner. Unfortunately, they often have repeated extractions without relief from pain before receiving appropriate medical or surgical therapy.

7.8. GLOSSOPHARYNGEAL NEURALGIA

Glossopharyngeal neuralgia resembles trigeminal neuralgia but is less common. The term **vagoglossopharyngeal neuralgia** (in recognition of a possible role of the vagal nerve in this painful condition) refers to this type of rare pain that is slightly more common in those over 50 years of age. Onset of severe pain in the throat or auricle (or in the sensory area innervated by either nerve) is abrupt and lasts a few seconds to a minute. Such pain is sharp, cutting, shooting, stabbing, or like an electric shock; others report a persistent, dull, aching, or burning pain often precipitated by swallowing, chewing, coughing, or other similar activities. Temporary relief from pain for one or two hours after application to the painful site of a 10% solution of cocaine or other surface anesthetic is a dependable aid in diagnosis of glossopharyngeal neuralgia. Patients with pain limited to the auricle often have a facial neuralgia involving facial somatic sensory fibers. Stimulation of the facial nerve [VII] with production of spontaneous pain will identify the involved nerve. Glossopharyngeal neuralgia results from trauma, vascular compression by a loop of a blood vessel, an abnormal styloid process or stylohyoid ligament, tumors, or local infection.

Surgical treatment of glossopharyngeal neuralgia involves transection of the glossopharyngeal nerve [IX] and the upper two or three vagal rootlets at the jugular foramen. Because of the proximity of the trigeminal spinal tract and common tract of the facial, glossopharyngeal, and vagal nerves, it is possible to eliminate pain in glossopharyngeal or vagal sensory areas by selective medullary tractotomy of the common tract. In those with recognizable vascular compression of glossopharyngeal or vagal nerves at their entrance into the brain stem, vascular decompression may be effective.

FURTHER READING

Apkarian AV, Hodge CJ (1989) Primate spinothalamic pathways: I. A quantitative study of the cells of origin of the spinothalamic pathway. J Comp Neurol 288:447–473.

Aziz Q, Schnitzler A, Enck P (2000) Functional neuroimaging of visceral sensation. J Clin Neurophysiol 17:604–612.

Bradley WE, Farrell DF, Ojemann GA (1998) Human cerebro-cortical potentials evoked by stimulation of the dorsal nerve of the penis. Somatosens Mot Res 15:118–127.

Bowsher D (2005) Pain. Pain 113:430.

Cervero F (1980) Deep and visceral pain. In: Kosterlitz HW, Terenius LY, eds. *Pain and Society*. Life Sciences Research Report 17, pp 263–282. Verlag Chemie, Weinheim.

Cervero F (1985) Visceral nociception: peripheral and central aspects of visceral nociceptive systems. Philos Trans R Soc Lond B Biol Sci 308:325–337.

Davis KD, Kwan CL, Crawley AP, Mikulis DJ (1998) Functional MRI study of thalamic and cortical activations evoked by cutaneous heat, cold, and tactile stimuli. J Neurophysiol 80:1533–1546.

Disbrow E, Roberts T, Krubitzer L (2000) Somatotopic organization of cortical fields in the lateral sulcus of Homo sapiens: evidence for SII and PV. J Comp Neurol 418:1–21.

Eickhoff SB, Lotze M, Wietek B, Amunts K, Enck P, Zilles K (2006) Segregation of visceral and somatosensory afferents: an fMRI and cytoarchitectonic mapping study. NeuroImage 31:1004–1014.

Feirabend HK, Choufoer H, Ploeger S, Holsheimer J, van Gool JD (2002) Morphometry of human superficial dorsal and dorsolateral column fibres: significance to spinal cord stimulation. Brain 125:1137–1149.

Frot M, Rambaud L, Guenot M, Mauguiere F (1999) Intracortical recordings of early pain-related CO_2-laser evoked potentials in the human second somatosensory (SII) area. Clin Neurophysiol 110:133–145.

Frot M, Garcia-Larrea L, Guenot M, Mauguiere F (2001) Responses of the supra-sylvian (SII) cortex in humans to painful and innocuous stimuli. A study using intracerebral recordings. Pain 94:65–73.

Frot M, Mauguiere F (2003) Dual representation of pain in the operculo-insular cortex in humans. Brain 126:438–450.

Gelnar PA, Krauss BR, Szeverenyi NM, Apkarian AV (1998) Fingertip representation in the human somatosensory cortex: an fMRI study. NeuroImage 7:261–283.

Graziano A, Jones EG (2004) Widespread thalamic terminations of fibers arising in the superficial medullary dorsal horn of monkeys and their relation to calbindin immunoreactivity. J Neurosci 24:248–256.

Green AL, Wang S, Owen SL, Xie K, Liu X, Paterson DJ, Stein JF, Bain PG, Aziz TZ (2005) Deep brain stimulation can regulate arterial blood pressure in awake humans. NeuroReport 16:1741–1745.

Hallin RG, Torebjörk HE (1974) Activity in unmyelinated nerve fibers in man. Adv Neurol 4:19–27.

Jones EG (2002) A pain in the thalamus. J Pain 3:102–104.

Karatas A, Caglar S, Savas A, Elhan A, Erdogan A (2005) Microsurgical anatomy of the dorsal cervical rootlets and dorsal root entry zones. Acta Neurochir (Wien) 147:195–199.

LaGuardia JJ, Cohrs RJ, Gilden DH (2000) Numbers of neurons and non-neuronal cells in human trigeminal ganglia. Neurol Res 22:565–566.

Nathan PW, Smith M, Deacon P (2001) The crossing of the spinothalamic tract. Brain 124:793–803.

Ostrowsky K, Magnin M, Ryvlin P, Isnard J, Guenot M, Mauguiere F (2002) Representation of pain and somatic sensation in the human insula: a study of responses to direct electrical cortical stimulation. Cereb Cortex 12:376–385.

Overduin AS, Servos P (2004) Distributed digit somatotopy in primary somatosensory cortex. NeuroImage 23:462–472.

Pappagallo M (2005) *The Neurological Basis of Pain*. McGraw-Hill, New York.

Pearce JMS (2006) Brodmann's cortical maps. J Neurol Neurosurg Psychiatry 76:259.

Pearce JM (2006) Glossopharyngeal neuralgia. Eur Neurol 55:49–52.

Ralston HJ 3rd (2005) Pain and the primate thalamus. Prog Brain Res 149:1–10.

Singer T, Seymour B, O'Doherty J, Kaube H, Dolan RJ, Frith CD (2004) Empathy for pain involves the affective but not sensory components of pain. Science 303:1157–1162.

Standring S (2004) Editor-in-chief. *Gray's Anatomy*. 39th ed. Section 2, Neuroanatomy: Chapter 18: Spinal Cord, pp 307–326. Elsevier/Churchill Livingstone, London.

Standring S (2004) Editor-in-chief. *Gray's Anatomy*. 39th ed. Section 2, Neuroanatomy: Chapter 19. Brain Stem, pp 327–352. Elsevier/Churchill Livingstone, London.

Standring S (2004) Editor-in-chief. *Gray's Anatomy*. 39th ed. Section 2, Neuroanatomy: Chapter 21, Diencephalon, pp 369–386. Elsevier/Churchill Livingstone, London.

Standring S (2004) Editor-in-chief. *Gray's Anatomy*. 39th ed. Section 2, Neuroanatomy: Chapter 22. Cerebral Hemisphere, pp 387–418. Elsevier/Churchill Livingstone, London.

Torebjörk HE (1979) Activity in C nociceptors and sensation. In: Kenshalo DR, ed. *Sensory Functions of the Skin of Humans*. 2nd Int Symp on Skin Senses, Fla State U, pp 313–321. New York, Plenum.

Treede RD, Kenshalo DR, Gracely RH, Jones AK (1999) The cortical representation of pain. Pain 79:105–111.

Trevino DL, Carstens E (1975) Confirmation of the location of spinothalamic neurons in the cat and monkey by the retrograde transport of horseradish peroxidase. Brain Res 98:177–182.

Usunoff KG, Marani E, Schoen JH (1997) The trigeminal system in man. Adv Anat Embryol Cell Biol 136:1–126.

Usunoff KG, Popratiloff A, Schmitt O, Wree A (2006) Functional neuroanatomy of pain. Adv Anat Embryol Cell Biol 184:1–115.

Vogt BA (2005) Pain and emotion interactions in subregions of the cingulate gyrus. Nat Rev Neurosci 6:533–544.

Weiss N, Lawson HC, Greenspan JD, Ohara S, Lenz FA (2005) Studies of the human ascending pain pathways. Thalamus Related Syst 3:71–86

White JC (1952) Conduction of visceral pain. New Engl J Med 246:686–691.

White JC (1954) Conduction of pain in man. Observations on its afferent pathways within the spinal cord and visceral nerves. AMA Arch Neurol Psychiatr 71:1–23.

Willis WD Jr, Zhang X, Honda CN, Giesler GJ Jr (2002) A critical review of the role of the proposed VMpo nucleus in pain. J Pain 3:79–94.

Willis WD, Westlund KN (1997) Neuroanatomy of the pain system and of the pathways that modulate pain. J Clin Neurophysiol 14:2–31.

Willis WD Jr, Westlund KN (2004) Pain system. In: Paxinos G, Mai JK, eds. *The Human Nervous System*. 2nd ed. Chapter 30, pp 1125–1170. Elsevier/Academic Press, Amsterdam.

Despite all the accumulated knowledge, all the tracing of skin paths, all the mapping of sensory cortexes, who in hell or heaven could account for the feeling of velvet rubbed against him in the dark?

Gustav Eckstein, 1970

8

Paths for Touch, Pressure, Proprioception, and Vibration

8.1. PATH FOR GENERAL TACTILE SENSATION FROM THE BODY

The path conveying general tactile impulses from the limbs, trunk, neck, and back of the head through the spinal cord and brain stem to the thalamus is the **ventral spinothalamic tract** (VST). Throughout its course, VST is ventral in position, begins at **spinal** levels, and ends in the **thalamus**. Many presumed spinothalamic tract neurons in the primate dorsal horn respond to weak forms of stimulation and hence contribute to the transmission of tactile information. Because details of the origin, course, and termination of the ventral spinothalamic tract in primates are uncertain, many authors combine the ventral and lateral spinothalamic tracts under the collective term **spinothalamic system**.

In the following account, the ventral spinothalamic tract is a distinct path carrying general tactile impulses.

8.1.1. Modalities and Receptors

The ventral spinothalamic tract carries **general tactile** sensation (also called **general** or **light touch**). Light stimuli (therefore the term 'light touch') without pressure: a wisp of cotton or the head of a pin applied to the skin, stroking of hairs, or blowing on the skin are examples of appropriate stimuli used to test general tactile sensation. **Endings in hair follicles** and **tactile disks** (of Merkel) are examples of cutaneous mechanoreceptors (Fig. 8.1) that respond to general tactile stimuli. Endings in hair follicles (Fig. 6.2) respond to displacement of a hair. Tactile disks (Fig. 6.1) occur in groups forming functional 'touch spots'. These disks are specialized complexes of nerve fibers with a disk-like ending that terminates on modified epithelial cells. Areas innervated by branches of individual axons overlap considerably. These general tactile receptors in humans are **slowly adapting** (SA-I) **units** with large receptive fields having obscure borders. These tactile disks are sensitive to edges, corners, and curvature.

8.1.2. Primary Neurons

Primary neurons serving light touch have large cell bodies and are in all spinal ganglia (Fig. 8.1). Peripheral processes of these primary neurons supply the appropriate receptors and their central processes pass through the **medial division** of each **spinal root** to enter the spinal cord (Fig. 8.1). As they enter the ipsilateral **dorsal funiculus** (Fig. 8.1), each fiber bifurcates into a longer ascending branch (Fig. 8.1) and a shorter descending branch. These afferent fibers terminate on secondary neurons in the dorsal horn over six to eight segments resulting in a great deal of overlap.

8.1.3. Secondary Neurons

Secondary neurons in this general tactile path occur in the **dorsomedial part** of **nucleus proprius** (spinal lamina III) at all spinal levels. Their axons swing ventromedial through the **ventral white commissure** (Fig. 8.1), decussate, accumulate along the ventrolateral margin of the spinal cord near the ventral roots, and ascend (Fig. 8.1). The resultant ascending bundle carrying general tactile information from spinal levels is the **ventral spinothalamic tract** (VST) (Fig. 8.1). A close relationship and overlap between this tract and the lateral spinothalamic tract in the cord is the basis for considering a

single spinothalamic system in humans serving pain and temperature as well as general tactile sensation. The fact that primary fibers from one spinal ganglionic cell related to general tactile sensation on entering the spinal cord may be distributed over six to eight cord segments precludes any sharp localization in the ventral spinothalamic tract.

Position of the Ventral Spinothalamic Tract in the Brain Stem

In the human brain stem, the VST is in the lateral medulla oblongata, medial to the lateral spinothalamic tract (Fig. 8.2). The close relationship at medullary levels of the lateral and ventral spinothalamic tracts supports the concept of a single spinothalamic system in humans. In the middle pons at the level of entry of trigeminal sensory fibers, the **ventral spinothalamic tract** is medial to the **lateral spinothalamic tract** and on the lateral edge of the **medial lemniscus** (Fig. 8.2) – a tract transmitting more discriminative tactile sensation from all levels of the body. In the midbrain, the **ventral spinothalamic tract** is dorsal and lateral to the **medial lemniscus** but ventromedial to the **lateral spinothalamic tract** (Fig. 8.2). The primate ventral spinothalamic tract provides fibers to a small nuclear mass in the midbrain, the nucleus of Darkschewitsch, and to the diencephalic periventricular region.

8.1.4. Thalamic Neurons

Tertiary neurons serving general tactile sensations occur in the **ventral posterior lateral nucleus** (VPL) (Fig. 8.2) and in the **caudal part of the intralaminar nuclei**. General tactile impulses enter consciousness at the level of the thalamus in humans. Accurate localization and discrimination of these impulses would require that they project onto the cerebral cortex. Table 8.1 summarizes the path for general tactile sensation from the body and compares it to the body path for discriminative touch, pressure, proprioception, and vibration.

8.2. PATH FOR TACTILE DISCRIMINATION, PRESSURE, PROPRIOCEPTION, AND VIBRATION FROM THE BODY

In addition to the ventral spinothalamic tract, there is a path carrying discriminative touch and other information from the limbs, trunk, neck, and back of the head

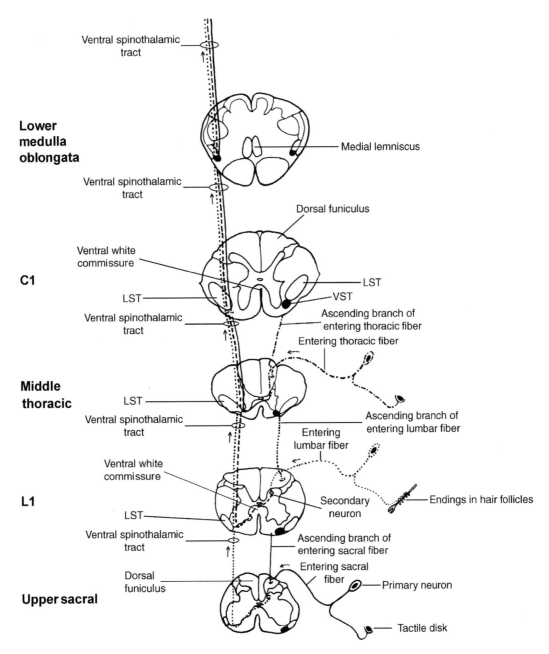

FIGURE 8.1. The human ventral spinothalamic tract (VST) including its origin and course in the spinal cord. Its course in the brain stem is illustrated in Fig. 8.2. As the central processes of the primary neurons enter the spinal cord they bifurcate in the dorsal funiculus into a longer ascending branch and a short descending branch that turns inferiorly, medial to the fasciculus gracilis.

to appropriate levels of the nervous system in humans. Such impulses travel in the **fasciculus gracilis** and **fasciculus cuneatus** (FGCF) of the spinal cord (Fig. 8.3) and relay through the brain stem in the **medial lemniscus** (ML) before reaching the thalamus. From the thalamus, such impulses project to the postcentral gyrus of the parietal lobe.

Observations in nonhuman primates and reports of patients with dorsal funicular lesions but intact

vibration and proprioception, or in whom no clinical deficit was observed, suggest the presence of another path in the lateral funiculus serving discriminative touch, pressure, proprioceptive, and vibratory sensation. Its existence in nonhuman primates and in humans is unclear. Participation of other paths in the transmission of discriminative touch, pressure, proprioceptive, and vibratory impulses, particularly in light of injury to the dorsal funiculi, requires additional investigation.

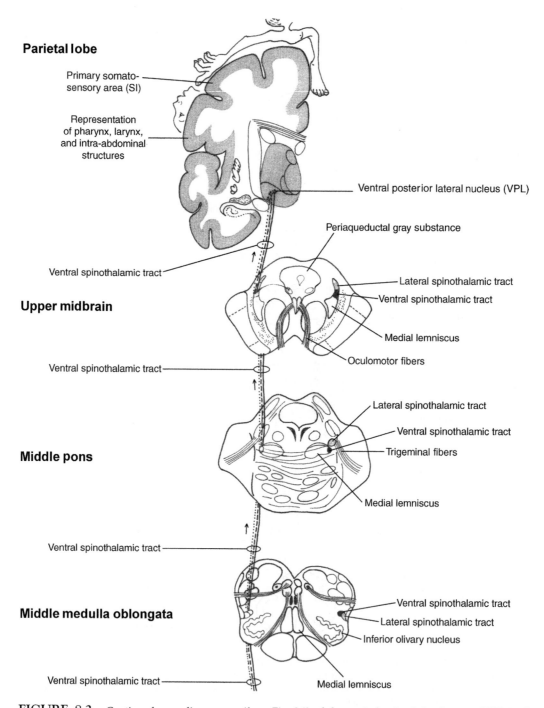

Parietal lobe

Primary somato-
sensory area (SI)

Representation
of pharynx, larynx,
and intra-abdominal
structures

Ventral posterior lateral nucleus (VPL)

Periaqueductal gray substance

Ventral spinothalamic tract

Upper midbrain

Lateral spinothalamic tract
Ventral spinothalamic tract

Medial lemniscus

Oculomotor fibers

Ventral spinothalamic tract

Lateral spinothalamic tract

Ventral spinothalamic tract

Middle pons

Trigeminal fibers

Medial lemniscus

Ventral spinothalamic tract

Middle medulla oblongata

Ventral spinothalamic tract
Lateral spinothalamic tract
Inferior olivary nucleus

Ventral spinothalamic tract

Medial lemniscus

FIGURE 8.2. Continued ascending course (from Fig. 8.1) of the ventral spinothalamic tract (VST) in the medulla, pons, and midbrain and its end in the ventral posterior lateral nucleus (VPL) of the dorsal thalamus. General tactile impulses transmitted over this path in humans enter consciousness at a dorsal thalamic level. Their projection to the cerebral cortex is not shown.

8.2.1. Modalities and Receptors

The **fasciculus gracilis-fasciculus cuneatus** and **medial lemnisci** transmit impulses for **tactile discrimination, pressure, proprioception,** and **vibration**. The term proprioception refers to the subjective awareness of the position and movement of joints, limbs, and other body parts, but not the direct awareness of the muscles in these parts. Proprioceptive sensation is essential for the coordination of movement and the maintenance of posture.

TABLE 8.1. Summary of paths for touch, pressure, proprioception and vibration from the body

Modalities	Light touch	Discriminative touch, pressure, proprioception, and vibration
Receptors	Endings in hair follicles; tactile disks	Neuromuscular and neurotendinous spindles, joint receptors, tactile corpuscles, lamellar corpuscles
Primary neurons	Fairly large neurons in all spinal ganglia	Largest spinal ganglionic neurons; mushroom shaped with coiled glomerulus
1° peripheral process	To receptors	To receptors in skin, muscles, tendons and joints
1° central processes	Via medial division enter dorsal funiculus, bifurcate, cover 6–8 segments	Thickly myelinated, enter medial division and ipsilateral dorsal funiculus, bifurcate here
Secondary neurons	Dorsomedial part of nucleus proprius	Nucleus gracilis and nucleus cuneatus
2° central processes	Decussate via ventral white commissure	Enter and contribute to internal arcuate system of fibers then decussate in sensory decussation
Location 2° fibers	In **ventral spinothalamic tract** – contralateral	Form **gracile and cuneate components of the medial lemniscus**
Spinal cord	Ventral funiculus adjoining LST	Secondary fibers not present in spinal cord
Medulla	Medial to LST	Dorsal to pyramidal fasciculi
Pons	Between ML and LST	In the pontine tegmentum
Midbrain	Ventromedial to LST	Superior to substantia nigra, lateral to red nucleus
Dorsal thalamic neurons	Lateral part of ventral posterior nucleus (VPL) also intralaminar nuclei	Caudal part of ventral posterolateral nucleus (VPL); somatotopic and modality topic arrangement; strictly contralateral representation
Thalamoparietal fibers	May enter consciousness in dorsal thalamus	In posterior third of posterior limb of the internal capsule
Cortical areas	Contralateral in SI, bilateral in SII, inferior parietal lobule (area 40)	Primary somatosensory cortex (SI); areas 3a, 3b, 1 and 2 of Brodmann; note sensory homunculus here; also multiple representation hypothesis

Two aspects of proprioceptive sensation are the awareness of **movement**, or **kinesthesia**, and **positional sense**. An example of the former would be detection of movement of one or two degrees at the interphalangeal joints. One method of testing proprioceptive sensation is by lightly grasping the finger or toe of the patient, without applying pressure, and then passively moving the digit, while avoiding contact between them. Recognition of the position of a digit, its direction, and range of movement are all part of the sense of proprioception.

Proprioceptors

These receptors respond to stimuli arising in muscles, tendons, joints, and at muscle-tendon junctions. **Neuromuscular spindles** (Fig. 6.6), **neurotendinous spindles** (Fig. 6.5), and **joint receptors** function as **proprioceptors**. These receptors are stimulated by changes in muscle length, tension, or movement, changes in the angle of joints, and by changes in limb or body position. Continuous input arises in these proprioceptors causing awareness of the relative position and movements of our body parts. The contribution of different proprioceptors to proprioception varies on different parts of the body such that patients with prostheses of finger joints retain normal positional sense postoperatively. Anesthetizing the skin and deep tissues of the fingers or toes in normal patients severely impairs or eliminates normal positional sense. Anesthetizing the knee joint capsule, skin of the joint, or both the skin and joint has no influence on positional sense of the knees. Therefore, positional sense of the knee is likely dependent on input from intramuscular receptors. A striking demonstration of the importance of such receptors for normal positional sense is the observation that their activation by vibrating tendons causes an illusion that a stationary limb is moving or is in a position different from its real position.

Tactile Corpuscles

Tactile corpuscles (Fig. 8.3) are receptors for **discriminative** or **two-point touch** – the ability to distinguish the sensations at each point when two points are simultaneously stimulated. Large fibers innervating tactile corpuscles branch repeatedly – in young adults, a single axon often innervates 20 receptors. Areas innervated by axonal branches overlap. Because of this, age-related reduction in receptors, or that caused by pathologic conditions, does not alter tactile discrimination.

Discriminative touch is testable with a bent paper clip applied to the skin in such a way as to avoid motion of hairs. Areas of hairless skin (the lips, the area around the nostrils, and the palmar surface of the fingers) are the best sites on which to demonstrate discriminative touch. The least distance for discriminative touch in humans is 1.7 mm on the ring finger in the third decade of life. Since the ability to distinguish two points as two points varies on different parts of the body, when testing discriminative

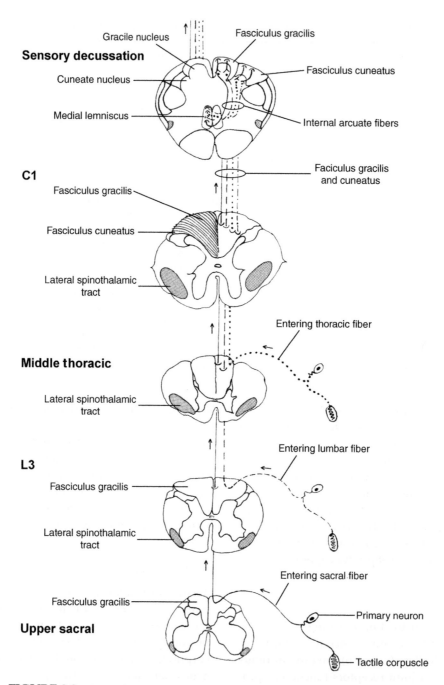

FIGURE 8.3. Fasciculus gracilis and fasciculus cuneatus in humans as they originate and then ascend in the spinal cord to enter the lower medulla. Each entering fiber divides into a short descending branch and a conspicuous branch ascending in the cord. These descending fibers are not shown. At lower medullary levels, the fibers of secondary neurons in this path cross the median plane in the sensory decussation to form the gracile and cuneate components of the medial lemniscus. Figures 8.5 and 8.6 illustrate the course of the medial lemniscus in the brain stem.

touch a comparison between similar areas on both sides of the body is in order.

Discriminative tactile receptors in humans are termed **rapidly adapting** (RA) **units** with small and well-defined cutaneous receptive fields of relatively

uniform sensitivity that tend to be smaller distally. The higher the tactile acuity of a cutaneous region, the higher will be the density of rapidly adapting units. An increase in concentration of tactile corpuscles from palm to fingertips in humans correlates with increased tactile acuity in the proximal-distal direction on the hand with the highest acuity at the fingertips. In the visually impaired, a limited area of the fingers (the distal 12–15 mm) participates in the recognition of letter shapes. Braille, a system of tactile reading for the visually impaired, uses characters made up of clusters of one to six raised dots with 64 possible patterns arranged in arrays of three by two. Six raised dots are the normal limit for discriminating separate points by a single digit (usually the index finger). These dots are set at a distance of 2.5 or 3.0 mm apart – within the normal threshold of tactile discrimination (normally about 2.0 mm on the fingertips). In reading a page of Braille, at an average rate of 80 to 100 words per minute, mechanoreceptors in the moving fingers receive information from some 2500 dots per minute, and relay that information to the cerebral cortex of the Braille reader in an orderly fashion where it is discriminated and interpreted.

Actions of a skilled Braille reader emphasize the importance of movement in tactile recognition. In this context, it is logical to refer to touch as the only 'active' sense. Although a stationary finger recognizes a cluster of raised dots, a horizontal sweeping movement of the finger aids in discriminating a series of characters aligned into words. Placing a cookie cutter in the shape of a number on the palm of the hand demonstrates the importance of movement in tactile discrimination. An accurate perception of the number is unlikely. Tracing the number on the palm with a blunt point makes it easily recognized and perceived. The difference in sensory discrimination by a stationary finger (*passive touch*) versus a moving finger (*active touch*) is the result of differences in receptor activation and accompanying cortical processing. When the finger is stationary, only general tactile receptors (slowly adapting units) are active and only the spatial aspects of sensory input comes in to the nervous system. As the finger moves, activation of the tactile corpuscles (rapidly adapting units) and lamellar corpuscles takes place and the brain receives the temporal and spatial aspects of this sensory input giving the nervous system a broader spectrum of sensory input to use in discriminating surface detail.

Tactile sensation is more complex than merely the ability to distinguish two points as two points. In addition to appreciating the **shape** of objects, tactile sensation encompasses hardness, roughness, smoothness, slipperiness, stickiness, 'rubberiness,' elasticity, and other properties perceived as **texture** (for example, distinguishing between a paper clip and a small safety pin or between two different coins). Though texture is a significant aspect of tactile sensation, other aspects of tactile sensation are an appreciation of movement of a stimulus across skin, the orientation of edges, corners, or solid surfaces pressed on skin and a perception of other qualities that help us appreciate the **shape and texture** of an object.

Lamellar Corpuscles

Lamellar corpuscles are mechanoreceptors (Fig. 6.3) that are sensitive to stimuli of pressure and vibration – especially vibrations of high frequency (100–400 Hz). They respond to joint movement and muscle contraction. A single fiber usually supplies a lamellar corpuscle or a small cluster of corpuscles. Primary endings of neuromuscular spindles in humans are also receptive to vibratory stimuli. Tactile corpuscles in nonhuman primates are responsive to low frequency vibrations (30 Hz) called **flutter**. A progressive decrease in sensitivity to vibration takes place with age (if tested on the thenar eminence) perhaps because of alterations in, or reduction of, these lamellar receptors.

Testing of vibratory sensation or **pallesthesia** involves the use of an oscillating tuning fork (128 Hz or 256 Hz) on any bony surface or prominence, especially the dorsomedial aspect of the first metatarsal bone, medial and lateral malleoli of the ankle, flat surface of the proximal tibia, or the dorsum of the metacarpal bone of the index finger. In addition, the ability to detect a tuning fork before it stops vibrating and the relative differences between vibratory sensations on both sides of the body are noted. When the result of the tuning fork test is ambiguous, an electromagnetic vibrator is a useful tool.

Recent evidence indicates that **genital corpuscles** are likely to be mechanoreceptors for touch. They respond to friction in the anal canal. **Corpuscles of Ruffini**, known to be particularly sensitive to stretching of skin, likely participate in pressure reception and respond to joint movement. Ruffini corpuscles in humans are slowly adapting (SA II) units.

8.2.2. Primary Neurons

The cell bodies of **primary neurons** in the path for tactile discrimination, pressure, proprioception, and vibration occur in all spinal ganglia (Fig. 8.3). They are large, mushroom-shaped, ganglionic neurons with a coiled or convoluted axon. Peripheral processes of these primary cell bodies supply receptors in skin, muscles, tendons, and joints. Their thickly myelinated central processes enter the **medial division** of each dorsal root. These primary fibers are dorsomedial in position at the junction of the spinal cord and rootlet. They then enter the ipsilateral **dorsal funiculus** (Fig. 8.3) of the spinal cord and bifurcate. Each entering fiber divides into a short descending branch that turns inferiorly, medial to the fasciculus gracilis and a conspicuous ascending branch (Fig. 8.3) that reaches the junction of the upper cord and lower medulla. Descending primary fibers from cervical and upper thoracic spinal levels form the fasciculus interfascicularis, whereas the descending primary fibers from lower thoracic, lumbar, and sacral levels collectively form the septomarginal fasciculus. These two fasciculi occur in the dorsal funiculi, medial to the fasciculus gracilis.

Ascending Central Processes of Primary Neurons

As primary fibers carrying discriminative touch, pressure, and proprioception ascend in the **dorsal funiculus**, they form two fiber bundles, the **fasciculus gracilis** and the **fasciculus cuneatus** (Fig. 8.3). Ascending fibers from all sacral, lumbar and some thoracic spinal levels form the **fasciculus gracilis** (Fig. 8.3). Primary fibers carrying general tactile impulses also ascend and descend in the fasciculus gracilis. Fibers of the fasciculus cuneatus originate from upper thoracic (T6 and above) and all cervical levels. Primary fibers for general tactile impulses ascend and descend in the fasciculus cuneatus. In addition to the large number of **primary sensory fibers** in the dorsal funiculi, with their cell bodies in spinal ganglia, there are nonprimary sensory fibers with their cell bodies of origin in spinal laminae IV and V that ascend in the primate dorsal funiculi. Large numbers of nonmyelinated fibers occur in the nonprimate dorsal funiculi. Their function in nonhuman primates and in humans is uncertain.

Somatotopic and Modality-topic Organization of Fibers in the Dorsal Funiculi

Ascending fibers of the fasciculi gracilis and cuneatus in the dorsal funiculi are **somatotopically organized** and obliquely oriented (Figs 8.3, 8.4). **Sacral fibers** enter the lower-most levels of the cord and ascend in a dorsomedial position in the dorsal funiculi near the dorsal median septum (Fig. 8.3); **cervical fibers** are ventral in position near the dorsal horns (Fig. 8.4). Between the sacral and cervical fibers in the dorsal funiculi are **lumbar** and **thoracic fibers** (Figs 8.3, 8.4). In addition to this somatotopic pattern of fibers in the dorsal funiculi, there is also a pattern impressed on the dorsal funiculi (Fig. 8.4) based on the type of impulses carried by these fibers (**'modality-topic pattern'**). Fibers carrying vibratory impulses are nearest the central canal whereas those carrying discriminative tactile impulses are superficial, with fibers carrying proprioceptive impulses between the other two (Fig. 8.4). Because of the oblique arrangement of fibers in the dorsal funiculi according to the levels at which they arise and the kind of impulses carried by them, injury near the central canal at upper cervical levels (Fig. 8.4) affects fibers transmitting vibratory impulses at several levels of the body, even though the injury is at a single spinal level. If the injury expands into the dorsal funiculi, fibers transmitting proprioceptive impulses become involved, as do fibers from additional levels of the body. Such an injury is characteristic of **syringomyelia**, an expansion of the central canal or a cavity near it at the level of the upper cervical cord and perhaps the lower end of the brain stem. If the latter is involved, the condition is termed **syringobulbia**.

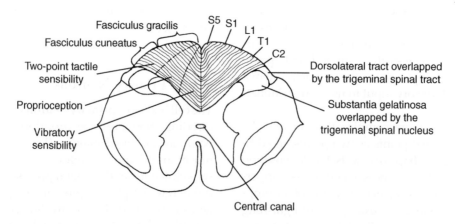

FIGURE 8.4.   Pattern of localization for body regions in the fasciculus gracilis and fasciculus cuneatus in humans at the level of C1. Also, note the probable pattern for localization of different types of impulses carried along these fasciculi (after Winkler, 1918 and Schneider, Kahn, Crosby and Taren, 1982).

Secondary fibers giving rise to the lateral spinothalamic tract and carrying painful and thermal impulses decussate in the ventral white commissure ventral to the central canal. Hence, in syringomyelia of the upper cervical cord, there may be loss of pain and temperature in both upper limbs in addition to dorsal funicular signs. In syringobulbia, there is likely loss of facial painful and thermal sensations because of involvement of the trigeminal spinal tract or its nucleus at medullary levels.

Injuries to the Ascending Fibers in the Dorsal Funiculi

Interruption of the dorsal funiculi at upper cervical spinal levels is usually accompanied by loss of one or more modalities, including vibration, proprioception, and discriminative tactile sensation. Even though the primary fibers carrying general tactile impulses ascend in the dorsal funiculus, general tactile discrimination carried in the ventral spinothalamic tract remains intact in dorsal funicular involvement because of the multiple synapses of its primary fibers. Loss of sensation after dorsal funicular injury likely involves the entire body below, and ipsilateral to, the injury. Sensory loss is usually **dissociated** which means there is impairment of some modalities but not others. Clinically movement and position sense are simultaneously tested. The patient closes their eyes with their hands (or foot) outstretched as the examiner lightly grasps the finger or toe of the patient, without applying pressure, and then passively moves the digit, while avoiding contact between digits. Recognition of the position of a digit, its direction, and range of movement are all part of the sense of proprioception.

The loss of proprioception due to dorsal funicular damage is termed **akinesthesia**. Dorsal funicular injury often yields an accompanying incoordination of movement termed **sensory ataxia** involving the lower limbs. Patients with sensory ataxia [Greek, without order or arrangement] have a loss of muscle coordination attributable to the loss of awareness of the position of their lower limbs. In an effort to compensate for this, they walk with a wide-based gait, feet thrown out, continually watch their feet, and may slap their feet on the floor. Their ataxic gait is irregular and jerky. Although vision compensates to a degree for the proprioceptive loss, with the eyes closed, the gait worsens and the patient is unable to walk (Romberg's sign). Ataxia is also likely to result following injury to the cerebellum (cerebellar ataxia).

One practical test for dorsal funicular function is the **numeral-tracing test** that involves tracing numerals (zero to nine) on the skin with a blunt point. Recognition of numerals or letters written on the skin is termed **graphesthesia**. Failure to do so, called **agraphesthesia** or **graphanesthesia** is an indication of dorsal funicular dysfunction. The level of injury is determined by noting areas of the body that have lost this ability compared with those in which the ability remains.

8.2.3. Secondary Neurons

Axons of primary neurons forming the fasciculi gracilis and cuneatus synapse with secondary neurons at C1 and lower medullary levels. Secondary neurons in the path for tactile discrimination, pressure, proprioception, and vibration occur in the **gracile nucleus** at C1 and lower medullary levels and the **cuneate nucleus** at lower medullary levels (Fig. 8.3). An orderly pattern exists in the termination of primary fibers on secondary neurons. Fibers from lower sacral levels traverse the **fasciculus gracilis** and end caudally in the **gracile nucleus**, whereas those from upper cervical levels traverse the fasciculus cuneatus and end rostrally in the cuneate nucleus. Many nonprimary sensory fibers from spinal laminae IV and VI ascend in the dorsal funiculi and end in the **gracile and cuneate nuclei**. These ascending fibers often modify activity in the **gracile** and **cuneate nuclei**.

The gracile and cuneate nuclei in primates each have two distinct cellular parts. One part receives input from fasciculi gracilis and cuneatus, and the other, called the reticular part, consists of multipolar neurons like those in the medullary reticular formation. Fibers from secondary neurons throughout both parts of these nuclei reach the thalamus.

Descending Fibers to the Gracile and Cuneate Nuclei

Studies in nonhuman primates reveal somatotopically organized **descending cortical projections** to the nuclei gracilis and cuneatus from areas of the limb and trunk represented along the postcentral gyrus of the parietal lobe, from Brodmann's area 5 in the superior parietal lobule, and from the Brodmann's area 4 along the precentral gyrus of the frontal lobe. These descending cortical connections occur in humans. The cerebral cortex often influences the transmission of impulses in the nuclei gracilis and cuneatus, in spinal lamina I neurons, and in the lateral reticular neurons of spinal laminae V and VI. Descending fibers from the ventral parts of these nuclei, at about the middle of their rostrocaudal extent, synapse with dorsal horn neurons in nonhuman primates known to contribute to spinothalamic, spinocervical, and cervicothalamic paths. The function of these descending corticogracile and corticocuneate projections is uncertain.

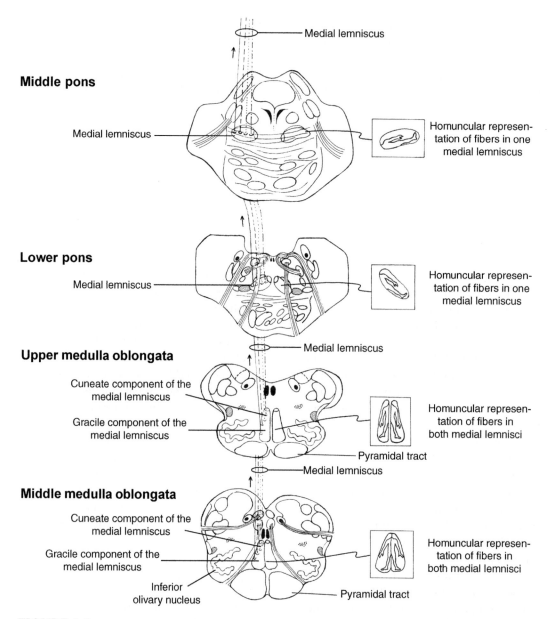

Middle pons

Medial lemniscus

Homuncular represen-
tation of fibers in one
medial lemniscus

Lower pons

Medial lemniscus

Homuncular represen-
tation of fibers in one
medial lemniscus

Medial lemniscus

Upper medulla oblongata

Cuneate component of the
medial lemniscus

Gracile component of the
medial lemniscus

Homuncular represen-
tation of fibers in
both medial lemnisci

Pyramidal tract

Medial lemniscus

Middle medulla oblongata

Cuneate component of the
medial lemniscus

Gracile component of the
medial lemniscus

Homuncular represen-
tation of fibers in
both medial lemnisci

Inferior
olivary nucleus

Pyramidal tract

FIGURE 8.5.   Continued course (from Fig. 8.3) of the gracile and cuneate components of the medial lemniscus (ML) in the medulla and into the pons. The somatotopic arrangement of fibers in the medial lemnisci at medullary levels is illustrated by a homunculus drawn out to the side of the two medullary levels. At the level of the pons, a homunculus representing one side of the body is drawn to the side of these levels. Each homunculus has a neck and posterior head that are not shown. Fig. 8.6 illustrates the continued course of this path into the midbrain ending in the dorsal thalamus.

The Sensory Decussation and Formation of the Medial Lemnisci

Axons of secondary neurons in the **gracile** and **cuneate nuclei** (Fig. 8.3) swing ventrally from these nuclei through the reticular formation of the medulla oblongata as part of the internal arcuate fibers (Fig. 8.3). The internal arcuate fibers curve or bend like a bow in the tegmentum of the medulla oblongata. After leaving the gracile and cuneate nuclei and entering the internal arcuate fibers, these secondary fibers then cross the median plane as the **decussation of the medial lemnisci** or the **sensory decussation**. The sensory decussation is actually the point in the median plane between the two medial lemnisci where the fibers from one side cross the median plane. These decussated fibers accumulate contralaterally in the brain stem, dorsal to each pyramidal tract as a compact bundle – the **gracile and cuneate components of the medial lemniscus**. The medial lemniscus (Fig. 8.3)

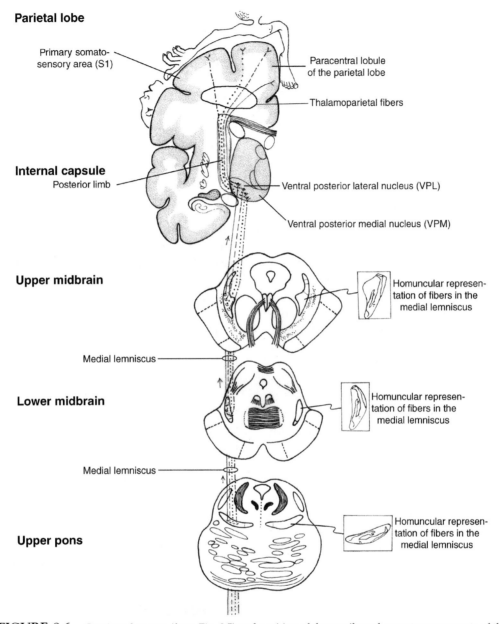

Parietal lobe

Primary somato-sensory area (S1)

Paracentral lobule of the parietal lobe

Thalamoparietal fibers

Internal capsule
Posterior limb

Ventral posterior lateral nucleus (VPL)

Ventral posterior medial nucleus (VPM)

Upper midbrain

Homuncular represen-tation of fibers in the medial lemniscus

Medial lemniscus

Lower midbrain

Homuncular represen-tation of fibers in the medial lemniscus

Medial lemniscus

Upper pons

Homuncular represen-tation of fibers in the medial lemniscus

FIGURE 8.6. Continued course (from Fig. 8.5) and position of the gracile and cuneate components of the medial lemniscus (ML) in the upper pons, midbrain, and its ending in the ventral posterior lateral nucleus (VPL) of the dorsal thalamus. Thalamoparietal projections from the ventral posterior lateral nucleus to the parietal post-central gyrus complete this path. The somatotopic organization of fibers in the upper pons and midbrain includes a neck and posterior head. Indicated across the postcentral gyrus and paracentral lobule is the sensory homuncu-lus for the representation of different parts of the body including the neck, face and posterior head. Inferior to the facial representation is a representation of the pharynx, larynx, and intra–abdominal structures.

is the next link in a path that began as the fasciculi gracilis and cuneatus in the spinal cord. Thus the **gracile and cuneate fasciculi/medial lemniscus system** is the ascending path from spinal levels to the thalamus for discriminative touch, pressure, proprioceptive, and vibratory impulses (Figs 8.3, 8.4, 8.5, 8.6).

There is a segmental arrangement of fibers of the fasciculi gracilis and cuneatus as they ascend through the cord to reach lower medullary levels where they synapse in the nuclei gracilis and cuneatus. As secondary fibers leave the **gracile and cuneate nuclei** to decussate, the gracile fibers are ventral in the medial lemniscus where they form the **gracile component of the medial lemniscus** (Fig. 8.5) whereas cuneate fibers (forming the **cuneate component of the medial lemniscus**) are dorsal in position in the medial lemniscus (Fig. 8.5). This dorsal-ventral arrangement of cuneate and gracile fibers leads to a pattern in each medial lemniscus with fibers from lower thoracic, lumbar, and sacral levels (gracile component) ventral near the pyramidal tract and fibers from the upper thoracic and cervical levels (cuneate component) in the dorsal part of the medial lemniscus (Fig. 8.5). The representation of the body in both medial lemnisci can be illustrated as a **homunculus** – a 'little man' or manikin standing with one foot on each pyramidal tract (Fig. 8.5). The segmental arrangement of fibers in the fasciculi gracilis and cuneatus of the human spinal cord and medulla oblongata below the **sensory decussation** changes to a somatotopic pattern in the medial lemniscus of the pons, midbrain, thalamus, and cerebral cortex.

Position of the Medial Lemniscus in the Brain Stem

The medial lemniscus on each side of the brain stem gradually shifts as it ascends through that structure. In the medulla oblongata, each medial lemniscus is dorsal to a pyramidal tract with their long axes in a dorsoventral orientation (Fig. 8.5). In the pons, however, the medial lemnisci separate, each representing the contralateral side of the body, with the long axis of each lemniscus parallel to the floor of the fourth ventricle (Figs 8.5, 8.6). A half-homunculus is now lying on either side of the pons with the head and neck directed medially and the foot laterally (Figs 8.5, 8.6). Each half-homunculus contains fibers from the contralateral side of the body. In the midbrain, each medial lemniscus shifts position and is dorsal to the substantia nigra and lateral to the red nucleus (Fig. 8.6). At this level, the half-homunculus on either side has its foot directed dorsolaterally and its neck, ventrolaterally. Therefore, the medial lemniscus changes orientation and rotates almost 180° as it travels through the brain stem.

8.2.4. Thalamic Neurons

Fibers of the medial lemniscus synapse with tertiary neurons in the caudal part of the **ventral posterior lateral nucleus** (VPLc) (Fig. 8.6). They end somatotopically with fibers from the **cuneate component of the medial lemniscus** (carrying impulses from thoracic and cervical levels) ending medially and rostrally, and fibers of the **gracile component of the medial lemniscus** (carrying impulses from lower thoracic, all lumbar and sacral levels) ending laterally and caudally in the caudal part of the ventral posterior lateral nucleus (VPL). Orientation of the body representation actually follows the shape of this thalamic nucleus. Mapping the human thalamus by stimulation reveals a representation of the thumb medial in the ventral posterior lateral nucleus (VPL) near that of

the mandible in the ventral posterior medial nucleus (VPM), followed in order by the fingers, palm, dorsum of the hand, forearm, arm, shoulder, trunk, and lower limb, with representation of the foot impinging on the internal capsule. Such a strictly contralateral representation is especially discrete for the hand, including the fingers. The lateral spinothalamic tract homunculus is dorsoventral, appearing to be standing on the medial geniculate nucleus (MG). When one combines the thalamic representations of the lateral spinothalamic tract with that of the medial lemniscus, a **double somatotopic representation** of the body is evident in the human thalamus.

Some fibers in the medial lemnisci end in the magnocellular (medial or internal) part of the **medial geniculate nucleus** (MG_{mc}), **pulvinar nuclei** (Pul), and **ventral part** of the **suprageniculate nucleus** (SGe). Stimulation in one human dorsal funiculus causes activity in the contralateral ventral posterior lateral nucleus (VPL) and perhaps in the contralateral intralaminar nuclei. Responses from these latter nuclei were diffuse and of longer latency suggesting involvement of elements of the spinoreticulothalamic tract.

Activation of thalamic neurons occurs in this system in conscious nonhuman primates by various mechanical stimuli including touch-pressure on the skin, mechanical stimulation of deep fascia or periosteum, or gentle rotation of a limb joint applied to a circumscribed, contralateral receptive field. An organization by modality is in the ventral posterior nucleus (VP) in humans. Activation of neurons in the posterior and inferior part of the ventral posterior nucleus (VP) results from cutaneous stimuli; stimuli related to movements, change of position of joints, and muscle contractions activate other neuronal somata, more rostral and dorsal.

Thalamoparietal Fibers

The caudal part of the ventral posterior lateral nucleus (VPL) projects impulses to the cerebral cortex of the parietal lobe through **thalamoparietal fibers** (Fig. 8.6) that pass in the **posterior third of the posterior limb of the internal capsule**. Sensory fibers in the internal capsule are somatotopically organized in a rostrocaudal, face-arm-leg sequence. As these sensory fibers travel in the internal capsule, they push adjoining motor fibers laterally. Reciprocal parietothalamic fibers occur in nonhuman primates and are likely in humans.

Unilateral vascular injury to the thalamus in humans near the caudal part of the ventral posterior lateral nucleus (VPL) often leads to a sensory deficit in the entire contralateral side of the face, arm, and leg. The deficit is either transient or there is persistent numbness and mild sensory loss in the contralateral hand, foot or both, spreading to the remainder of the involved side of the body, including the face. The limbs feel large and swollen with such terms as 'numb,' 'asleep,' 'tingling,' 'dead,' or 'frozen' used by patients to describe their affected limbs.

8.2.5. Cortical Neurons

Primary Somatosensory Cortex (SI)

All discriminative sensations, including discriminative touch and proprioception, require cortical participation for their conscious awareness. Vibratory and pressure sensations (as well as general tactile sensations) likely enter consciousness in humans at a thalamic level. Injury to the cerebral cortex in humans does not result in loss of vibratory sensation. Thalamoparietal fibers reach layer IV of the postcentral gyrus of the parietal lobe on the lateral surface of each cerebral hemisphere and the posterior paracentral lobule on the medial surface of each hemisphere. Collectively this region, on both lateral and medial surfaces of each cerebral hemisphere, is termed the **primary somatosensory cortex** or **SI** (Fig. 8.6) and is composed of four strip-like cytoarchitectural subdivisions identifiable in nonhuman primates and humans as **areas 3a, 3b, 1,** and **2**. Reciprocal and somatotopically organized projections are made from a single thalamic site to area 3b and to areas 1 and 2.

The Sensory Homunculus

A pattern of body representation, illustrated as a caricature drawn across SI, is termed the **sensory homunculus** (Fig. 8.6). The pattern across SI reflects the orderly arrangement of neurons along this path (gracile and cuneate fasciculi/medial lemniscus system) from the periphery to the cerebral cortex. Certain features of the human sensory homunculus are of interest. For example, the area representing the face and head is so large that it appears to disrupt the remaining sequence of body parts. The toes, foot, ankle and leg have their representation along the posterior part of the paracentral lobule on the medial surface of the brain (Fig. 8.6). The representation of the thigh or hip occurs at the medial to superolateral surface junction followed in order by the trunk, neck, shoulder, arm, forearm, and hand (Fig. 8.6) across the superolateral surface of the brain. The genitalia and perineum likely have their representation on the medial surface of the human brain just ventral to the representation of the toes (though this representation is unclear). Recent recordings of cerebrocortical potentials evoked

by stimulation of the dorsal nerve of the penis in humans suggest a representation of the penis with the hip and upper leg near the junction of the medial to superolateral surface. Additional studies should help to clarify the location of the cortical representation of the human genitalia and perineum. There is considerable variability in the position of the leg/body boundary from patient to patient. In humans, all five fingers have a separate medial to lateral representation along the wall of the central sulcus not on the lateral surface. The posterior body has its representation caudally in the postcentral gyrus; its anterior aspect faces the central sulcus. An intervening area for the upper limb separates representations of the face and trunk. Since cutaneous areas with greatest sensitivity have a correspondingly larger cortical representation, sizes of homuncular parts indicate the extent of primary somatosensory cortex devoted to those parts. Table 8.1 summarizes the path for tactile discrimination, pressure, proprioception, and vibration from the body and compares it to the body path for general tactile sensibility.

The Multiple Representation Hypothesis

In nonhuman primates, each cytoarchitectural component of the **primary somatosensory cortex** corresponding to **areas 3a, 3b, 1,** and **2 of Brodmann** is a separate map of the body, receiving inputs from different peripheral receptors. Areas 3b and 1 represent cutaneous regions and are essentially mirror images, yet differ enough to suggest that each has a distinct role in cutaneous sensation. Area 2 appears to represent predominantly deep receptors, including those signaling the position of joints and other deep body sensations, whereas area 3a represents sensory input from muscles. Cortical representation along the postcentral gyrus reflects a composite of somatotopically-organized areas instead of a continuous homunculus. Functional imaging studies of the human SI show that multiple digit representations occur in the four subdivisions of SI (3a, 3b, 1 and 3) resembling the **multiple representation hypothesis** that is present in nonhuman primates. The definition of these cytoarchitectural subdivisions during imaging studies in humans depends on approximations based on anatomical landmarks.

Secondary Somatosensory Cortex

Adjoining the **primary somatosensory cortex** or **SI** is the **secondary somatosensory cortex** or **SII**, on the superior bank of the lateral sulcus and continuing onto the parietal operculum, which is surprisingly deep in the

TABLE 8.2.　'What' vs. 'Where' in the human cerebral cortex

Cortical region	'What' – ventral path or stream	'Where' – dorsal path or stream
	Perception, identification, recognition	*Location and spatial relations*
Auditory cortex	Recognition of familiar sounds and intelligible speech	Spatial aspects of audition; activated earlier
	Primary auditory cortex → Wernicke's region (superior temporal gyrus) → Broca's region (inferior frontal gyrus) via the external capsule and uncinate fasciculus	Primary auditory cortex → Wernicke's region (superior temporal gyrus) → area 40 in the inferior parietal lobule → Broca's region (inferior frontal gyrus) via the arcuate fasciculus
Somatosensory cortex	Tactile object recognition	Tactile object localization
	Primary somatosensory cortex → secondary somatosensory cortex → inferior parietal lobule → frontal pole, left prefrontal region	Primary somatosensory cortex → secondary somatosensory cortex → superior parietal lobule (area 7) and precuneus (area 19)
Visual cortex	Object vision – form, color and features; faster	Spatial vision – visual scene and motion
	Primary visual cortex → secondary visual cortex → inferior temporal cortex → prefrontal cortex	Primary visual cortex → secondary visual cortex → superior parietal cortex → prefrontal cortex

human brain. Cytoarchitectonic mapping of this area reveals four distinct areas on the human parietal operculum termed **OP 1-4** (OP for operculum). OP 1 and OP 2 seem to belong to **Brodmann's area 40** in the inferior parietal lobule whereas OP 3 and OP 4 correspond to **Brodmann's area 43**. There are two somatotopic representations of the body on the lateral operculum in OP 1 and OP 4. A third somatotopic map is deep within the lateral sulcus in area OP 3. A diversity of stimuli may activate SII in humans including vibration, light touch, pain and visceral sensations. This multimodal input to SII suggests that it would appropriately be termed a **sensory association area**. Imaging studies in humans have confirmed that this area consists of at least two subdivisions, just as it does in nonhuman primates, with each subdivision containing a complete representation of the body. More than 10 somatosensory association areas are identifiable in the nonhuman primate brain. SII in nonhuman primates receives fibers from the ventral posterior medial nucleus (VPM) and the caudal part of the ventral posterior lateral nucleus (VPL). Reciprocal corticothalamic fibers occur from SII in nonhuman primates to the ventral posterior nucleus (VP). Output from SII is to SI, to the primary motor cortex (corresponding to Brodmann's area 4), and to the premotor cortex (non-primary motor cortex corresponding to Brodmann's area 6). SII neurons exhibit predominantly contralateral, moderate to well-defined receptive fields that are usually larger than the receptive fields of neurons in SI.

Fibers from SI to SII project to area 5 in the superior parietal lobule, to the **primary motor cortex** and to the **premotor cortex** in the frontal lobe. Only those parts of SI, SII and area 4 related to similar somatic regions are interconnected. SI and SII interconnect with one another

and with the **primary motor cortex** through intrinsic connections. Parts of SI representing regions of the body along the median plane make connections with the same areas in the contralateral cerebral hemisphere through commissural fibers. Indeed, each of the subdivisions of SI sends fibers to the analogous area of the opposite cerebral hemisphere and to the contralateral SII. There are also projections from SII to its counterpart in the opposite cerebral hemisphere and to the rostral extension of area 3 lying inferior to the **primary motor cortex**.

Unilateral injury to the primary somatosensory cortex in humans yields a severe and lasting contralateral sensory loss, with particular involvement of the distal limbs including impairment of pressure, discriminative touch, and positional sensation. Studies in nonhuman primates show separate and selective consequences of destruction in the various subdivisions of SI. These losses seem to emphasize that each cytoarchitectural subdivision makes a specific contribution to tactile discrimination. Unilateral or bilateral removal of monkey area SII causes a profound defect of tasks requiring tactile learning and retention.

'What' and 'Where' Processing in the Somatosensory Cortex

At the cortical level, the auditory and visual systems in primates are each organized into 'what' and 'where' paths (Table 8.2). This 'what' and 'where' model includes a ventral ('what') stream for the perception, identification, and recognition of objects and a dorsal ('where') stream for locating objects and appreciating the spatial relations among objects. It has long been proposed that at the cortical level the somatosensory system in nonhuman primates is organized into 'what' and 'where' paths. This concept suggests that somatosensory information travels to the primary somatosensory cortex, relays in serial and stepwise fashion to sensory association areas before ultimately reaching multimodal sensory areas. Based on imaging studies in humans, the ventral stream for tactile object recognition ('what') appears to involve the primary somatosensory cortex, secondary somatosensory cortex and the inferior parietal lobule (bilaterally) before reaching association areas in the frontal pole and left prefrontal region. The dorsal stream for tactile object localization ('where') appears to involve the primary and secondary somatosensory cortices and then association areas in the superior parietal lobule (area 7 of Brodmann) and the precuneus (area 19).

8.2.6. Spinal Cord Stimulation for the Relief of Pain

Transection of the fasciculi gracilis and cuneatus, called dorsal cordotomy, was popular in the late 1940s as a means of alleviating certain painful conditions. Some 60 years later surgical interruption of fibers in the dorsal funiculi (on either side of the dorsal median sulcus termed 'midline myelotomy') is being carried out in a limited number of patients with visceral pain as the result of uterine or colon cancer. Electrical stimulation of the dorsal funiculus, called '**dorsal column stimulation**' (DCS), or electroanalgesia, has been in use since the early 1970s to treat a variety of chronic, intractable painful conditions. The use of DCS has as its basis the presence of ascending fibers of the dorsal funiculi and their role in the **gate-control theory of pain**. This theory recognizes the presence and importance of three spinal cord elements in the transmission of painful impulses: neurons of the substantia gelatinosa, ascending fibers in the dorsal funiculi, and central transmission cells (T-cells) in the dorsal horn. Presumably, central transmission cells yield ascending pain-carrying paths while the substantia gelatinosa acts as a gate-control system, modifying sensory input from peripheral fibers before such input reaches the centrally located T-cells. Activity in fibers of the dorsal funiculus that reaches the cerebral cortex returns to the spinal cord to influence the gating mechanism in the substantia gelatinosa. T-cells monitor sensory input and ultimately transmit impulses that result in the response to, or the perception of, pain. The final output of this system depends on interaction among substantia gelatinosal somata, dorsal funicular fibers, and T-cells. Such interaction takes place against a background of activity from a variety of sources including the constant bombardment of the spinal cord by impulses (over small myelinated and nonmyelinated fibers) holding the gate open. There is a disproportionate increase in large versus small-fiber input taking place when a peripheral stimulus sets up an increase in activity in the large fibers. These large fibers influence the substantia gelatinosa by inhibiting the T-cells and closing the gate to the transmission of painful impulses; increased activity in small myelinated or nonmyelinated fibers presumably decreases the inhibition of the substantia gelatinosa on the T-cells and opens the gate. The total number of active fibers, the frequency of the impulses they transmit, and the balance of activity in small versus large fibers, are essential ingredients in the effectiveness of this system. Selectively influencing large-diameter fibers, decreasing the small-fiber input, and closing the gate is the mechanism that would lead to the therapeutic control of pain.

Electrical stimulation of large diameter peripheral fibers in the upper limb in humans does not abolish or diminish the quantity of pain induced by stimulation of small nonmyelinated C-fibers in the same limb. No reduction in pain is observable when all fibers are conducting. Electrical stimulation of large peripheral fibers reduces, and tends to obliterate, deep pain induced by squeezing the human tendo calcaneus. In certain conditions (Friedreich's ataxia, polyneuropathy of hypothyroidism, polyarteritis nodosa, prolonged phenytoin administration), however, there is a disproportionate destruction of large fibers with preservation of small ones, yet such conditions are usually painless. In other conditions, there is a decrease in small fibers with a great deal of pain. Collectively these observations question certain aspects of the gate-control theory of pain, but emphasize the often-overlooked distinction between experimentally produced pain and clinical pain. Despite these comments, spinal cord stimulation is successful in reducing pain but not alleviating it in about 30% of attempts (perhaps as high as 50–70% in some conditions). The goal of these procedures is to improve pain relief, increase patient activity levels, and reduce the use of narcotic medications.

Placing electrodes in the spinal dura mater or in a subdural, extra-arachnoid position (spinal electrode implantation) in contact with the spinal cord provides a means of stimulating the dorsal funiculi. Relief from pain is associated with a feeling of numbness or **paresthesia** usually referred to the painful region. Such relief occurs with chronic, visceral, or deep pain, or pain of central origin but not with sharp and acute to subacute pain in skin, muscle, or bone. Postoperatively, even as patients achieve relief for the first time because of dorsal column stimulation, they cannot suppress pain from their surgical wound. A patient was able to control his chronic pain, but was unable to obtain relief from an intercurrent fracture. Such stimulation acts selectively on mechanisms of chronic pain but not on those of acute pain. The benefits of this stimulation may result from **central inhibition in the thalamus**. Thalamic recordings while stimulating demonstrate a reduction in spontaneous activity in a small population of neurons. In anesthetized nonhuman primates, such stimulation inhibits pain transmission at spinal levels – especially spinothalamic neurons in spinal laminae I and V. Inhibition of pain transmission is likely caused by antidromic stimulation of collaterals of dorsal funicular fibers. Hence, the therapeutic success of this procedure is attributable, in part, to inhibition of spinal neurons. The presence of ipsilateral descending projections from gracile and cuneate nuclei in nonhuman primates, which synapse with dorsal horn neurons (whose central processes contribute to spinothalamic, and cervicothalamic tracts), is a possible explanation for the effectiveness of stimulation of the dorsal funiculi. The configuration of the field in the spinal cord influenced by stimulation of the dorsal funiculi, including the depth of penetration of the stimulus, is impossible to define. Cerebrospinal fluid may act as a conductive medium, transmitting electrical stimuli to the dorsal funiculi, dorsal horns, and to the spinothalamic and spinoreticulothalamic tracts.

In addition to the gate-control theory and central inhibition at the thalamic level there are other possible mechanisms of action that may underlie successful spinal cord stimulation in the face of painful conditions. These include γ-aminobutyric acid (GABA) and adenosine-mediated action on dorsal horn neuronal activity. Whatever the mechanism of action underlying spinal cord stimulation, it is estimated that some 14,000 new patients worldwide receive treatment for various painful conditions using spinal electrode implantation. With additional understanding of the underlying mechanisms involved in humans, improvements in equipment, surgical methods, and appropriate selection of patients, the future for spinal cord stimulation to improve pain relief, increase activity levels, and reduce the use of narcotic medications remains promising.

8.3. PATH FOR TACTILE DISCRIMINATION FROM THE HEAD

The **trigeminal nerve** [V] is the primary sensory nerve to the face, cornea, oral and nasal mucosa, tongue and dental pulp, paranasal sinuses, and scalp to its vertex. As such, this cranial nerve and its branches carry impulses for tactile discrimination, pressure, proprioception, and vibration from these regions to the CNS. The cutaneous region of the mandibular angle receives its innervation from superficial or cutaneous branches of the cervical plexus (C2, C3). Once inside the CNS, impulses from trigeminal areas relay through the brain stem to the thalamus in secondary ascending trigeminal paths including a **dorsal** and a **ventral trigeminothalamic tract**, each of which carries different modalities of sensation. A description of the trigeminal nuclear complex is available in Chapter 7.

8.3.1. Modalities and Receptors

Tactile corpuscles (Fig. 8.7), on the face and anterior half of the scalp, are receptors for discriminative touch – the most acute of human tactile sensations. Tactile sensitivity on the face is comparable to, but slightly less than, that on the fingertips and mucosal surfaces of the external genitalia. Chapter 6 contains a description of certain features of these receptors.

8.3.2. Primary Neurons

Primary neurons transmitting discriminative tactile sensations from trigeminal regions in the head are the largest neurons in the **trigeminal ganglion**. Their peripheral processes are thickly myelinated fibers that leave the trigeminal ganglion and synapse with **tactile corpuscles** (Fig. 8.7). Central processes of these primary neurons enter the middle third of the pons in the **trigeminal sensory root** and end in the ipsilateral **trigeminal pontine nucleus** (Fig. 8.7).

8.3.3. Secondary Neurons

Secondary neurons for tactile discrimination are dorsomedial in the trigeminal pontine nucleus. Fibers emerge from neurons in the rostral end of the trigeminal pontine nucleus and ascend ipsilaterally in the brain stem; some fibers decussate and ascend contralaterally. Therefore, a bilateral path of crossed and uncrossed fibers, called the **dorsal trigeminothalamic tract** (Fig. 8.7) is formed ascending through the upper pons, midbrain, and into the thalamus. At upper pontine levels, these tracts are in the center of the dorsal part of the pontine tegmentum. As they ascend into the midbrain, the dorsal trigeminothalamic tract is dorsal

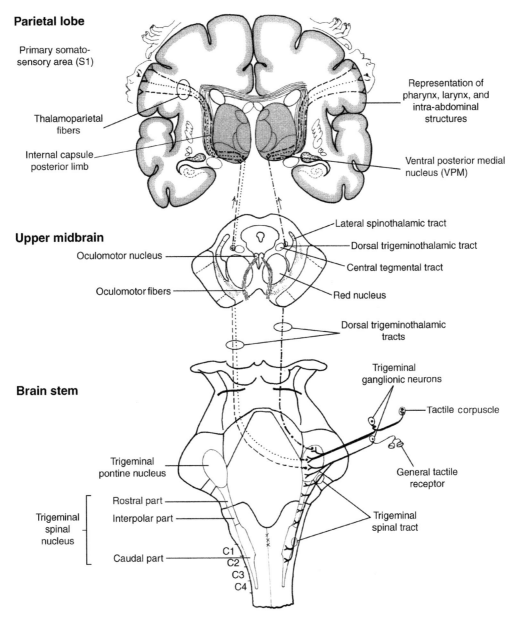

Parietal lobe

Primary somato-sensory area (S1)

Thalamoparietal fibers

Internal capsule posterior limb

Representation of pharynx, larynx, and intra-abdominal structures

Ventral posterior medial nucleus (VPM)

Upper midbrain

Oculomotor nucleus

Oculomotor fibers

Lateral spinothalamic tract

Dorsal trigeminothalamic tract

Central tegmental tract

Red nucleus

Dorsal trigeminothalamic tracts

Brain stem

Trigeminal ganglionic neurons

Tactile corpuscle

Trigeminal pontine nucleus

General tactile receptor

Trigeminal spinal nucleus

Rostral part

Interpolar part

Caudal part

C1
C2
C3
C4

Trigeminal spinal tract

FIGURE 8.7. Entering trigeminal fibers involved in the transmission of discriminative tactile, some general tactile and proprioceptive sensibilities. Such impulses are transmitted by secondary neurons in the trigeminal pontine nucleus and are then relayed to higher levels in the dorsal trigeminothalamic tract (DTT). This tract carries crossed and uncrossed fibers in the midbrain and to the ventral posterior medial nucleus (VPM) of the dorsal thalamus (after Humphrey, 1982). Note the sensory homunculus illustrated across the postcentral gyrus. The classical pattern of the sensory homunculus is shown on the left whereas a recent pattern with the thumb juxtaposed to the lower lip and corner of the mouth is shown on the right (after Lende and Popp, 1976).

to each **red nucleus** but lateral to the **central tegmental tract** (Fig. 8.7).

Experiments in nonhuman primates show somato-topically organized descending projections from the postcentral gyrus of the parietal lobe to all parts of the trigeminal nuclear complex. Similar connections occur in humans. Hence, the cerebral cortex likely influences the transmission of impulses in the trigeminal nuclear complex.

8.3.4. Thalamic Neurons

Each **dorsal trigeminothalamic tract** ends bilaterally in the **ventral posterior medial nucleus** (VPM) (Fig. 8.7). The ventral posterior nucleus (VP) in humans is divisible into a lateral part – the **ventral posterior lateral nucleus** (VPL) for representation of the body, and a medial part – the **ventral posterior medial nucleus** (VPM) for representation of the head. Thalamoparietal fibers from tertiary neurons in VPM reach fourth-order neurons in the postcentral gyrus of the parietal lobe.

8.3.5. Cortical Neurons

Primary Somatosensory Cortex (SI)

Though nondiscriminatory aspects of sensation in the head reach consciousness at a thalamic level, recognition of discriminative tactile sensations requires cortical participation in humans. Thalamoparietal fibers leaving VPM enter the internal capsule to reach the **primary somatosensory cortex** in the parietal lobe. The primary somatosensory cortex corresponds to **areas 3a, 3b, 1,** and **2 of Brodmann**. Each subdivision of SI contributes to the representation of a given territory of the head. The functional organization of the part of SI receiving its input from areas of the face and head is similar to those regions of SI receiving inputs from the body with

one exception. Impulses from the face and head reaching the ventral posterior medial nucleus (VPM) do not project to the paracentral lobule of the parietal lobe on the medial surface of the cerebral hemisphere. That part of the sensory homunculus devoted to the head is only on the lateral surface of each cerebral hemisphere, separated from representation of the trunk by an intervening area for the upper limb. The cortical representation for the face and head (Fig. 8.7) is so large it appears to disrupt the remaining sequence of the body. Since the dorsal trigeminothalamic tract is a bilateral path, unilateral cortical injury does not influence discriminative touch on the face. An arrangement of the human sensory homunculus, based on an analysis of sensory Jacksonian seizures in many patients, has the thumb juxtaposed to the lower lip and corner of the mouth (Fig. 8.7). This slight alteration in the classical sensory homunculus is in agreement with recent studies in nonhuman primates.

TABLE 8.3. Summary of paths for touch, pressure, proprioception and vibration from the head

Modalities	Discriminative touch	Light touch	Proprioception, pressure, vibratory sensation
Receptors	Tactile corpuscles	Ending in hair follicles tactile disks	Neuromuscular/tendinous spindles, pressure and joint receptors
Primary neurons	Trigeminal ganglion	Trigeminal ganglion	Trigeminal mesencephalic nucleus
1° peripheral processes	End on tactile corpuscles	Supply appropriate receptors	Supply appropriate receptors
1° central processes	Enter trigeminal sensory root	Enter sensory root of trigeminal nerve	Project to trigeminal pontine nucleus; form trigeminal mesencephalic tract
Secondary neurons	Dorsomedial part of trigeminal pontine nucleus	Most in trigeminal subnuclei; some in trigeminal pontine nucleus	Trigeminal pontine nucleus
2° central processes	Some contralateral, some ipsilateral	From pontine nuclei enter DTT, from trigeminal spinal nucleus enter VTT	Some contralateral some ipsilateral
Location 2° fibers	Dorsal trigeminothalamic tract DTT	Dorsal and ventral trigeminothalamic tracts	Dorsal trigeminothalamic tract
Spinal cord	Not in spinal cord	Not in spinal cord	Not in spinal cord
Medulla	Not in medulla	VTT not identifiable in medulla; DTT not in medulla	DTT not in medulla
Pons	Upper pons in pontine tegmentum	DTT near CTT; dorsal to medial lemniscus	DTT near CTT
Midbrain	Dorsal to red nucleus	Dorsal to red nucleus; VTT medial to medial lemniscus	Dorsal to red nucleus
Dorsal thalamic neurons	VPM	VPM	VPM
Thalamoparietal fibers	Discriminative tactile impulses project to parietal lobe – facial area		Proprioceptive sensations only project to parietal lobe – face area
Cortical termination	Primary somatosensory cortex (SI); areas 3a, 3b, 1, and 2		Face region of SI – areas 3a, 3b, 1, 2
Comments	Sensory homunculus in humans – **discriminative trigeminal paths only to face area not to the paracentral lobule.** Also note the thumb juxtaposed to lower lip.		

CTT, central tegmental tract; DTT, dorsal trigeminothalamic tract; SI, somatosensory cortex; VPM, ventral posterior medial nucleus; VTT, ventral trigeminothalamic tract.

Secondary Somatosensory Cortex

The human **secondary somatosensory cortex or SII** is a small region in the superior bank of the lateral sulcus and continuing into the parietal operculum. In conscious primates, cell bodies with trigeminal receptive fields are identifiable in the anterior part of SII. These are the only neurons in SII with a predominantly bilateral receptive field. Table 8.3 summarizes the path for tactile discrimination from the head and compares it with other paths from the head for light or general touch and the path for proprioception, pressure, and vibration.

8.4. PATH FOR GENERAL TACTILE SENSATION FROM THE HEAD

8.4.1. Modalities and Receptors

Impulses related to general tactile sensation from the face and head (tested by brushing, blowing, or light pressure on skin, or bending hairs) reaches thalamic levels by way of the **ventral trigeminothalamic tract** (VTT). The ventral trigeminothalamic tract has an origin, course, and termination different from that of the dorsal trigeminothalamic tract, which carries impulses for discriminative tactile sensation.

Tactile disks and endings in hair follicles serve general tactile sensation from the face and head. In the cornea, however, arborizations of nerves ending freely throughout this structure are likely to serve as the receptors for general tactile stimuli.

8.4.2. Primary Neurons

Primary neurons serving general tactile sensation from the face and head are small neurons in the trigeminal ganglion, with thinly myelinated fibers. Peripheral processes of these primary neurons supply the appropriate receptors (Fig. 8.8). Their central processes enter the middle third of the pons (Fig. 8.8) by way of the trigeminal sensory root. Upon entering the pons, these fibers bifurcate, and send one branch to the trigeminal pontine nucleus; the other branch descends into the trigeminal spinal tract (Fig. 8.8). These descending fibers synapse with neurons at all levels of the trigeminal spinal nucleus.

8.4.3. Secondary Neurons and Their Central Processes

Secondary neurons in the general tactile path from the face and head occur in all subnuclei of the trigeminal spinal nucleus, particularly the caudal half of its interpolar part, deeper layers of its caudal part, in the reticular

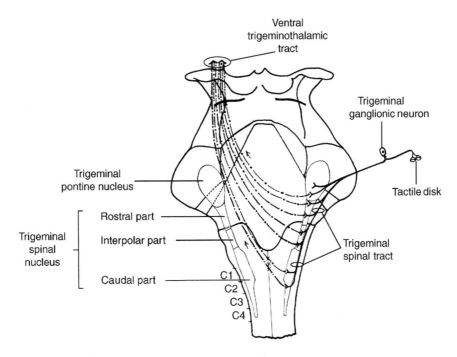

FIGURE 8.8. Entering trigeminal fibers involved in the transmission of general tactile sensation. Such impulses are relayed to secondary neurons in all levels of the sensory trigeminal nuclear complex and are then relayed to higher levels in the ventral trigeminothalamic tract. Continued course of this path is illustrated in Fig. 8.9. Some general tactile sensation is carried in the dorsal trigeminothalamic tract as shown in Fig. 8.7 (after Humphrey, 1982).

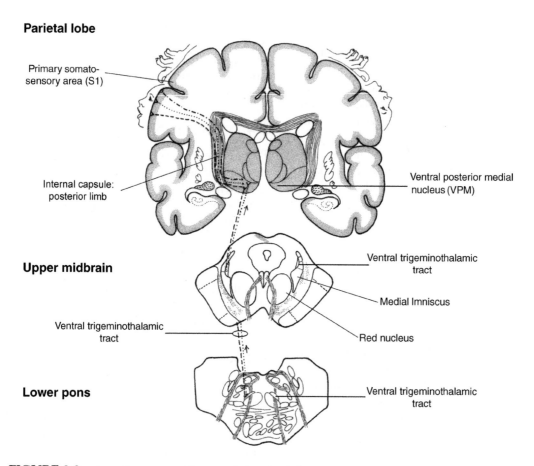

Parietal lobe

Primary somato-
sensory area (S1)

Internal capsule:
posterior limb

Ventral posterior medial
nucleus (VPM)

Upper midbrain

Ventral trigeminothalamic
tract

Medial lmniscus

Ventral trigeminothalamic
tract

Red nucleus

Lower pons

Ventral trigeminothalamic
tract

FIGURE 8.9. Ascending course of the ventral trigeminothalamic tract (from Fig. 8.8) in the lower pons and upper mid-brain. This tract ends in the ventral posterior medial nucleus (VPM) of the dorsal thalamus and its impulses are then relayed to the cerebral cortex through thalamoparietal fibers. Classical pattern for the facial part of the sensory homunculus is shown on the left whereas a recent pattern with the thumb juxtaposed to the lower lip and corner of the mouth is shown on the right (after Lende and Popp, 1976).

formation adjoining the caudal part of the trigeminal spinal nucleus, and in the trigeminal pontine nucleus. Fibers of these secondary neurons do one of two things: 1) those from the trigeminal spinal nucleus ascend obliquely, decussate, and enter the **ventral trigeminothalamic tract**, coursing with it through the brain stem to the thalamus (Figs 8.8, 8.9) whereas 2) axons of secondary neurons in the trigeminal pontine nucleus contribute to the bilateral **dorsal trigeminothalamic tract** (Fig. 8.7). As noted earlier, this path also carries discriminatory tactile stimuli from the face and head.

8.4.4. Thalamic Neurons

Tertiary neurons, serving general tactile sensation from the face and head, occur in the **ventral posterior medial nucleus** (VPM) (Fig. 8.9). Both the **ventral trigeminothalamic tract** (carrying general tactile impulses) and the **dorsal trigeminothalamic tract** (carrying some general tactile and discriminative tactile

impulses) end in the **ventral posterior medial nucleus** (VPM). General tactile impulses enter consciousness at this level in humans. If localization ('where') and discrimination ('what') of these general sensations is necessary then the cerebral cortex must become involved. This would require the relay of these impulses from the thalamus to the cerebral cortex where additional processing can take place.

Mapping the human thalamus by stimulation has demonstrated sensory responses in the medial part of the ventral posterior nucleus (presumably the ventral posterior medial nucleus) that represent impulses from the oral cavity including the gums, tongue, and pharynx. There is a generous representation of the lips and cheek in this area with a meager representation for the scalp, forehead, side, and back of the head. Representation of the thumb is medial in the ventral posterior lateral nucleus (VPL) adjoining the mandibular representation in the ventral posterior medial nucleus (VPM). Table 8.3 summarizes the path for general tactile sensation from the head and compares it with related paths for discriminative touch as well as paths for proprioception, pressure and vibration from the head.

8.5. PATH FOR PROPRIOCEPTION, PRESSURE, AND VIBRATION FROM THE HEAD

8.5.1. Modalities and Receptors

The **dorsal trigeminothalamic** tract, carrying general and discriminative tactile impulses, also transmits impulses for proprioception (position and movement of the mandible), superficial and deep pressure, vibration, and perhaps deep pain from muscles, tendons, and joints of the head. Neuromuscular spindles in the muscles of mastication (except the lateral pterygoid), tensor tympani, and tensor veli palatini, and pressure receptors in periodontal membranes around individual or groups of teeth, neurotendinous spindles, and receptors in joints, especially the temporomandibular joint, are the receptors serving proprioception, pressure and vibration in the head.

8.5.2. Primary Neurons

Primary neurons serving proprioception, pressure, and vibration in the face and head occur in the **trigeminal mesencephalic nucleus** (Fig. 8.10). These large (40–50 μm), pale neurons, of neural crest origin, distribute along the lateral border of the periaqueductal gray substance at, or rostral to, the middle pons and extending as far rostrally as the superior colliculus level in the upper midbrain – a distance of about 24–26 mm in length. Cells at caudal levels of the trigeminal mesencephalic nucleus are smaller (25–30 μm in diameter) than those in the rostral part of this nucleus. Trigeminal mesencephalic neurons have many features of spinal ganglionic cells, also of neural crest origin. Some primary neurons supplying neurotendinous spindles and joint receptors occur in the trigeminal ganglion.

Processes of these primary trigeminal mesencephalic neurons, as thin myelinated axons some 2–3 μm in diameter, continue caudally to the **trigeminal pontine nucleus**, and, in doing so, form the trigeminal mesencephalic tract (Fig. 8.10). Each process of a pseudounipolar trigeminal mesencephalic neuron bifurcates into a long branch that leaves the brain stem with efferent fibers of the trigeminal motor root and passes peripherally to an appropriate receptor. In addition to each long peripheral branch, a short central branch passes to the **trigeminal**

pontine nucleus (Fig. 8.10). Unilateral destruction of 80 to 100% of trigeminal mesencephalic nuclear cells, and the accompanying trigeminal mesencephalic tract, causes a strong, but never complete, preference for chewing on the contralateral side. There is a transient clumsiness in manipulating food. Muscles on the side of the injury seem to behave normally during mastication when tested immediately after injury.

8.5.3. Secondary Neurons

Secondary neurons in the path for proprioception, pressure, and vibration occur in the **trigeminal pontine nucleus** (Fig. 8.10). The major projection from these secondary neurons includes a crossed bundle of fibers and a smaller projection of fibers from the dorsal third of the ipsilateral trigeminal pontine nucleus. The resulting bilateral path, the **dorsal trigeminothalamic tract** (DTT), also carries general and discriminative tactile impulses from the face and head (Fig. 8.10). At upper pontine levels, the dorsal trigeminothalamic tract is in the dorsal aspect of the pons in the center of the tegmentum. As it ascends into the midbrain, it is dorsal to the red nucleus (Fig. 8.10).

8.5.4. Thalamic Neurons

Tertiary neurons in the path for proprioception, pressure, and vibration from the face and head (Fig. 8.10) are in the **ventral posterior medial nucleus** (Fig. 8.10). Impulses for pressure, vibration, and general tactile sensibility enter consciousness at a thalamic level in humans. However, proprioceptive and discriminative tactile impulses must reach the cerebral cortex in order for them to come into consciousness. For that reason, the discriminatory aspects of sensation such as proprioception and discriminative touch that reach the thalamus in the dorsal trigeminothalamic tract project over thalamoparietal fibers to the postcentral gyrus of the parietal lobe.

8.5.5. Cortical Neurons

Impulses for discriminative touch and proprioception from the face and head traveling in the **dorsal trigeminothalamic tract** require the participation of the **postcentral gyrus** of the parietal lobe for their conscious appreciation. Impulses from the **ventral posterior medial nucleus** (VPM) project over thalamoparietal fibers to the area of representation of the face in the primary somatosensory cortex (SI) on the lateral surface

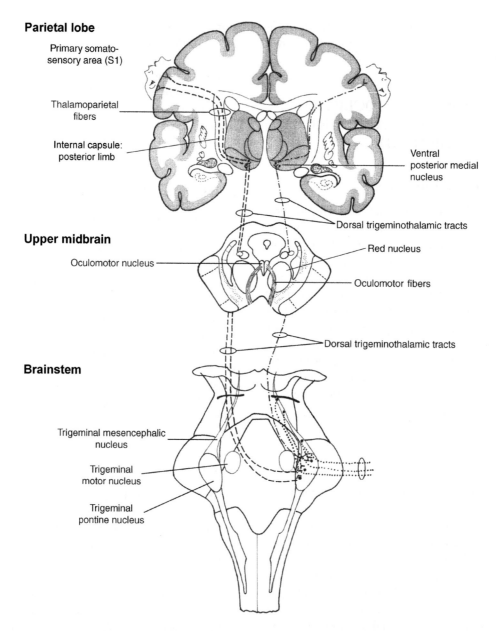

Parietal lobe

Primary somato-
sensory area (S1)

Thalamoparietal
fibers

Internal capsule:
posterior limb

Ventral
posterior medial
nucleus

Dorsal trigeminothalamic tracts

Upper midbrain

Oculomotor nucleus

Red nucleus

Oculomotor fibers

Dorsal trigeminothalamic tracts

Brainstem

Trigeminal mesencephalic
nucleus

Trigeminal
motor nucleus

Trigeminal
pontine nucleus

FIGURE 8.10.   Relations of the trigeminal motor nucleus, trigeminal mesencephalic nucleus and tract, and the dorsal trigeminothalamic tract (DTT). This bilateral path from the trigeminal pontine nucleus courses in the midbrain and ends in the ventral posterior medial nucleus (VPM) of the dorsal thalamus. Thalamoparietal fibers from this dorsal thalamic nucleus to the face area of the primary somatosensory cortex (SI) are also shown (after Humphrey, 1982).

(Fig. 8.10) of the cerebral hemisphere. Cytoarchitectural subdivisions of SI (areas 3a, 3b, 1, and 2) provide somatotopic and modality-topic representations of different parts of the head. Table 8.3 gives a summary of the path for proprioception, pressure, and vibration from the head and compares it with related paths from the head for discriminative touch and for light touch.

8.6.   TRIGEMINAL MOTOR COMPONENT

Neurons of the **special visceral efferent** component of the trigeminal nerve are large (35–45 μm in diameter), multipolar neurons of the **trigeminal motor nucleus**

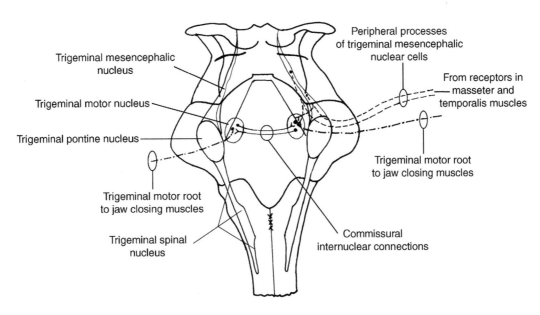

FIGURE 8.11. Connections mediating the jaw–closing reflex (after Humphrey, 1982).

(Fig. 8.10). The human trigeminal motor nucleus is medial to the **trigeminal pontine nucleus** in the middle third of the pons (Fig. 8.10) and some 4–4.5 mm in rostrocaudal length. Its cells feature a dense accumulation of chromatophil (Nissl) substance that extends into the proximal dendrites. The human trigeminal motor nucleus includes a group of larger neurons 35–45 μm in diameter, a second group of smaller neurons less than 25 μm in diameter (some of which are likely to be γ motoneurons), and a third group of sparsely-distributed small neurons in the range of 10–15 μm. These latter small cells are likely interneurons.

Axons of these motoneurons emerge from the brain stem to form the **trigeminal motor root**. The trigeminal motor fibers continue to the muscles of mastication (masseter, temporalis, medial and lateral pterygoids), mylohyoid, anterior belly of the digastric, tensor tympani, and tensor veli palatini. A pattern of localization is present in nonhuman primates in the trigeminal motor nucleus for the various muscles it supplies.

Almost 20% of the trigeminal motor root fibers in humans are nonmyelinated. Painful impulses may enter the CNS in such fibers in the trigeminal motor root in a manner similar to afferents entering the ventral roots of the spinal cord. This provides a plausible explanation for the failure of relief to occur from trigeminal neuralgia after rhizotomy of the trigeminal sensory root. Unilateral injury to the trigeminal motor root causes deviation of the jaw to the side of the injury because of weakness or paralysis of the pterygoids ipsilateral to the lesion. Normally each medial pterygoid acts together with the

lateral pterygoid on that side to protrude the mandible, moving it forward and toward the opposite side. If the pterygoids on one side are weak or paralyzed, they will lag behind while the contralateral pterygoids are protruding the mandible on their own pushing it toward the weakened side. Thus, the jaw deviates toward the side of the weakness or lesion side.

8.7. CERTAIN TRIGEMINAL REFLEXES

8.7.1. Mandibular, Masseter, or 'Jaw-Closing' Reflex

To demonstrate the jaw-closing reflex, the examiner places their index finger over the middle of the mandible with the mouth slightly open, tapping on the index finger with a reflex hammer, and thereby stretching the masseter and temporalis muscles. The normal response is sudden closure of the jaws caused by contraction of the stretched muscles. Clinically this reflex tests the integrity of the receptors, neurons, and neuronal processes that underlie it. Receptors are innervated by **peripheral processes** of primary neurons in the **trigeminal mesencephalic nucleus** (Fig. 8.11) whose central processes are short branches going directly to the trigeminal motor nucleus. The efferents for this reflex are secondary neurons in the trigeminal motor nuclei (Fig. 8.11) on the two sides of the brain stem that have **commissural internuclear connections** (Fig. 8.11) between them. Axons of these secondary neurons leave the brain stem in the

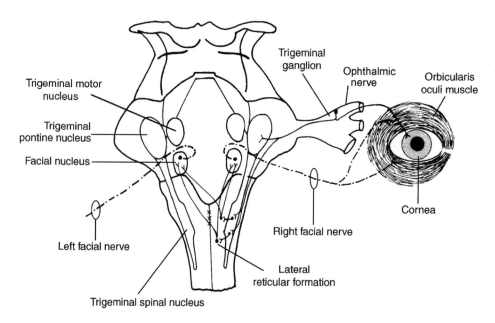

FIGURE 8.12. Connections forming the basis of the corneal reflex (after Humphrey, 1982 and Ongerboer de Visser, 1980).

trigeminal motor root (Fig. 8.11) to reach the masseter and temporalis muscles. Commissural internuclear connections between trigeminal motor nuclei permit closure of the mouth on stimulation of the appropriate receptors.

8.7.2. Corneal Reflex

To demonstrate the **corneal reflex**, the examiner lightly touches the cornea with a piece of tissue paper, wisp of cotton, puff of air, or other light mechanical stimulus. The result of such stimulation leads to closure of both eyelids caused by contraction of the orbicularis oculi of the stimulated eye (direct corneal response) and contraction of the contralateral muscle (consensual corneal response). The corneal reflex helps protect the eyes from injury caused by foreign particles.

Receptors for the corneal reflex are arborizations of fine nerves that end freely in the cornea. Primary neurons for the afferent limb of this reflex are in the trigeminal ganglia (Fig. 8.12). Peripheral processes of these **primary neurons** reach the cornea through the **ophthalmic nerve** (Fig. 8.12), nasociliary nerves, and long ciliary nerves. Central processes of these **primary neurons** enter the middle third of the pons in the trigeminal sensory root. They then bifurcate, yielding a branch ending in the **trigeminal pontine nucleus** and another branch that descends in the trigeminal spinal tract. The latter descending fibers synapse with secondary neurons in the **trigeminal spinal nucleus** (Fig. 8.12).

Secondary neurons for the corneal reflex are at all levels of the **trigeminal spinal nucleus**. Axons of these secondary neurons continue as internuclear fibers through the **lateral reticular formation** to both **facial nuclei** (Fig. 8.12). The efferent side of this reflex consists of fibers from each **facial nucleus** to its corresponding **orbicularis oculi muscle** (Fig. 8.12). Therefore, stimulation of one cornea causes contraction of both eyelids provided both the afferent and the efferent limbs underlying this reflex are intact.

Injuries to the trigeminal nerve [V] result in a diminished corneal reflex that may be detectable before any other sensory deficit. Injury to the trigeminal ophthalmic division interrupts small myelinated and nonmyelinated fibers from the cornea while sparing a sufficient number of the larger cutaneous fibers to the forehead to maintain normal facial sensation. Examination of the corneal reflex is a simple means of assessing injuries to the trigeminal nerve and its central connections with the facial nuclei.

FURTHER READING

Accornero N, Berardelli A, Bini G, Cruccu G, and Manfredi M (1980) Corneal reflex elicited by electrical stimulation of the human cornea. Neurology 30:782–785.

Apkarian AV, Hodge CJ (1989) Primate spinothalamic pathways: II. The cells of origin of the dorsolateral and ventral spinothalamic pathways. J Comp Neurol 288:474–492.

Apkarian AV, Hodge CJ (1989) Primate spinothalamic pathways: III. Thalamic terminations of the dorsolateral and ventral spinothalamic pathways. J Comp Neurol 288:493–511.

Bradley WE, Farrell DF, Ojemann GA (1998) Human cerebro-cortical potentials evoked by stimulation of the dorsal nerve of the penis. Somatosens Mot Res 15:118–127.

Burton H, Fabri M, Alloway K (1995) Cortical areas within the lateral sulcus connected to cutaneous representations in areas 3b and 1: a revised interpretation of the second somatosensory area in macaque monkeys. J Comp Neurol 355:539–562.

Celesia GG (1979) Somatosensory evoked potentials recorded directly from human thalamus and Sm I cortical area. Arch Neurol 36:399–405.

Disbrow E, Roberts T, Krubitzer L (2000) Somatotopic organization of cortical fields in the lateral sulcus of Homo sapiens: evidence for SII and PV. J Comp Neurol 418:1–21.

Eickhoff SB, Amunts K, Mohlberg H, Zilles K (2006) The human parietal operculum. II. Stereotaxic maps and correlation with functional imaging results. Cereb Cortex 16:268–279.

Eickhoff SB, Grefkes C, Zilles K, Fink GR (2007) The somatotopic organization of cytoarchitectonic areas on the human parietal operculum. Cereb Cortex 17:1800–1811.

Eickhoff SB, Schleicher A, Zilles K, Amunts K (2006) The human parietal operculum. I. Cytoarchitectonic mapping of subdivisions. Cereb Cortex 16:254–267.

Fabri M, Polonara G, Salvolini U, Manzoni T (2005) Bilateral cortical representation of the trunk midline in human first somatic sensory area. Hum Brain Mapp 25:287–296.

Gelnar PA, Krauss BR, Szeverenyi NM, Apkarian AV (1998) Fingertip representation in the human somatosensory cortex: an fMRI study. NeuroImage 7:261–283.

Gildenberg PL, Murthy KSK (1980) Influence of dorsal column stimulation upon human thalamic somatosensory-evoked potential. Appl Neurophysiol 43:8–17.

Hardy TL, Bertrand G, Thompson CJ (1980) Organization and topography of sensory responses in the internal capsule and nucleus ventralis caudalis found during stereotactic surgery. Appl Neurophysiol 42:335–351.

Humphrey T (1982) The central relations of the trigeminal nerve. In: Schneider RC, Kahn EA, Crosby EC, Taren JA, eds. Correlative Neurosurgery. 3rd ed, 2:1518–1532. Thomas, Springfield.

Kaas JH (2004) Somatosensory System. In: Paxinos G, Mai JK, eds. The Human Nervous System. 2nd ed. Chapter 28, pp 1059–1092. Elsevier/Academic Press, Amsterdam.

Kerr FWL (1975) The ventral spinothalamic tract and other ascending systems of the ventral funiculus of the spinal cord. J Comp Neurol 159:335–356.

Krubitzer L, Clarey J, Tweedale R, Elston G, Calford MA (1995) Redefinition of somatosensory areas in the lateral sulcus of macaque monkeys. J Neurosci 15:3821.3839.

Lee SH, Kim DE, Song EC, Roh JK (2001) Sensory dermatomal representation in the medial lemniscus. Arch Neurol 58(4):649–651.

Lockard BI, Kempe LG (1988) Position sense in the lateral funiculus? Neurol Res 10:81–86.

Marani E, Schoen JH (2005) A reappraisal of the ascending systems in man, with emphasis on the medial lemniscus. Adv Anat Embryol Cell Biol 179:1–74.

Marani E, Usunoff KG (1998) The trigeminal motonucleus in man. Arch Physiol Biochem 106:346–354.

Meyerson BA, Linderoth B (2000) Mechanisms of spinal cord stimulation in neuropathic pain. Neurol Res 22:285–292.

Meyerson BA, Linderoth B (2006) Mode of action of spinal cord stimulation in neuropathic pain. J Pain Symptom Manage 31(Suppl 4):S6–12.

Mishkin M (1979) Analogous neural models for tactual and visual learning. Neuropsychologia 17:139–151.

Mountcastle VB (2005) The Sensory Hand: Neural Mechanisms of Somatic Sensation. Harvard University Press, Cambridge.

Ongerboer de Visser BW (1980) The corneal reflex: electrophysiological and anatomical data in man. Prog Neurobiol 15:71–83.

Overduin AS, Servos P (2004) Distributed digit somatotopy in primary somatosensory cortex. NeuroImage 23:462–472.

Pearce JMS (2006) Brodmann's cortical maps. J Neurol Neurosurg Psychiat 76:259.

Ralston HJ 3rd (2005) Pain and the primate thalamus. Prog Brain Res 149:1–10.

Reed CL, Klatzky RL, Halgren E (2005) What vs. where in touch: an fMRI study. NeuroImage 25:718–726.

Standring S (2004) Editor–in–chief. Gray's Anatomy. 39th ed, Section 2, Neuroanatomy: Chapter 18, Spinal Cord. Chapter 19, Brain Stem. Chapter 21, Diencephalon. Chapter 22, Cerebral Hemispheres, pp 307–352 and 369–418. Elsevier/Churchill Livingstone, London.

Stojanovic MP, Abdi S (2002) Spinal cord stimulation. Pain Physician 5:156–166.

Ungerleider LG, Haxby JV (1994) 'What' and 'where' in the human brain. Curr Opin Neurobiol 4:157–165.

Usunoff KG, Marani E, Schoen JH (1997) The trigeminal system in man. Adv Anat Embryol Cell Biol 136:1–126.

Waite PME, Ashwell KWS (2004) Trigeminal sensory system. In: Paxinos G, Mai JK, eds. The Human Nervous System. 2nd ed. Chapter 29, pp 1093–1124. Elsevier/Academic Press, Amsterdam.

These nonspecific mechanisms are distributed widely through the central core of the brain stem and, as spokes radiate from the hub of a wheel to its peripheral working rim, so functional influences of these central systems can be exerted in a number of directions: caudally upon spinal levels . . . rostrally and ventrally upon hypothalamic and pituitary mechanisms . . . and more cephalically and dorsally still, upon the cortex in the cerebral hemispheres. . . . Just as all spokes move together in the turning of a wheel, though they may bear weight sequentially, so the variously directed influences of these nonspecific reticular systems are closely interrelated in normal function.

Horace W. Magoun, 1963

9

The Reticular Formation

The human reticular formation (RF) includes reticular nuclei, reticulothalamic fibers, diffuse thalamic projections to the cortex, ascending cholinergic projections, descending noncholinergic projections, and descending reticulospinal projections. The well-defined series of reticular nuclei are in the central core of the brain stem (Fig. 9.1) embedded in masses of ascending multisynaptic fibers. The ascending reticulothalamic fibers originate in the brain stem reticular nuclei, then reach the dorsal thalamus and as a series of diffuse thalamic projections relay information to the cerebral cortex. The ascending cholinergic projections and ascending noncholinergic projections reach the cerebral cortex. The descending reticulospinal projections allow the reticular nuclei to influence spinal cord motoneurons. While the reticular nuclei may be well defined, their connectivity, neurochemistry, and neuropharmacology in nonhuman primates and in humans remains unclear. In addition to its role in attention, arousal, and consciousness, other diverse functions such as respiration, cardiovascular activity, visceral reflexes, and involuntary motor activity are often attributable to the reticular formation.

9.1. STRUCTURAL ASPECTS

Reticular neurons exhibit a breadth of sizes and shapes, some with long ascending or long descending axons while other reticular neurons exhibit long radiating dendrites that allow for substantial overlapping of dendritic fields. Examination of stained preparations of this region reveals long, radiating dendrites that give a reticular or 'netlike' appearance to this area of the brain stem and give this part of the brain stem its name. These dendrites may extend hundreds of microns in the dorsoventral as well as the mediolateral plane.

Paxinos and Huang (1995), in their atlas of the human brain stem, have delineated in some detail the reticular nuclei. Table 9.1 summarizes the nuclear groups belonging to the reticular formation in humans based on the terminology and abbreviations employed by Paxinos and Huang along with the recommendations of the *Terminologia Anatomica* (1998). Figures accompanying this description are from the Paxinos and Huang (1995) atlas. In the human brain stem, some

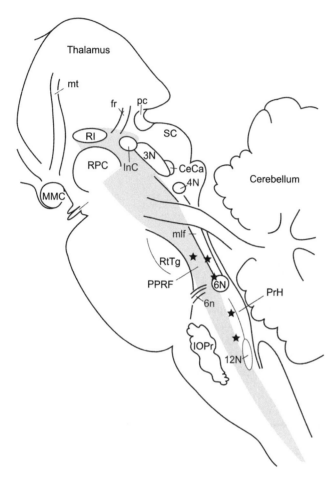

FIGURE 9.1. Median section of the human brain stem showing the location of the reticular formation (shaded area). (Modified from Büttner-Ennever and Horn in Paxinos and Mai, 2004.) **Abbreviations:** 3N, oculomotor nucleus; 4N, trochlear nucleus; 6N, abducent nucleus; 6n, abducent nerve; 12N, hypoglossal nucleus; mlf, medial longitudinal fasciculus; RPC, red nucleus; SC, superior colliculus.

14 reticular nuclei or subnuclei are identifiable in the medulla oblongata, another six reticular nuclei are identifiable in the pons, and three reticular nuclei are in the midbrain. Thus, there are some 23 named reticular nuclei or subnuclei identifiable in the human brain stem. Table 9.2 attempts to depict the relationship of these nuclei to one another and their location at different levels of the brain stem in humans. In addition to these named reticular nuclei, it may be appropriate as the details of their connectivity and chemical neuroanatomy become available to include under the collective term 'reticular formation' certain tegmental nuclei (lateral dorsal tegmental nucleus and ventral tegmental nucleus), the locus coeruleus, the nuclei of the raphé, and certain nuclei in the basal forebrain.

In the introduction to the first edition of their human brain stem atlas (1954), Olszewski and Baxter noted the

obvious fundamental physiological significance of the reticular formation. They also noted a lack of detailed information about the anatomy of the reticular formation. Our knowledge of the reticular nuclei in the human brain stem has as its basis the work of Olszewski and Baxter in their 1952 atlas and as well he efforts of Paxinos and Huang in their atlas of the human brain stem (1995). As noted above some 23 named nuclei are clearly identifiable in the human brain stem. There is growing understanding of the chemical neuroanatomy of these nuclei. However, our understanding of the detailed afferent and efferent connections of these nuclei and the role these connections play in the physiology of the reticular formation is lacking in nonhuman primates and humans. The very nature of the reticular formation – multisynaptic and diffuse – presents a great challenge to closing the gap between physiological observations and anatomical studies. Rather than abandon the term 'reticular formation' we should renew our efforts to elucidate its anatomy and physiology using combined microanatomical, biochemical, pharmacological and behavioral approaches, particularly in nonhuman primates and humans.

9.1.1. Reticular Nuclei in the Medulla

The spinal cord–medulla oblongata transition occurring about 11 mm below the obex (Fig. 9.2) includes both spinal cord structures (ventral horn and spinal accessory nucleus) and medullary structures (pyramidal decussation, parts of the lateral corticospinal tract, the fasciculi gracilis and cuneatus and the trigeminal spinal tract and its nucleus). Identifiable at this transition (Fig. 9.2) are the **dorsal reticular nucleus** (DRt) which extends some 17 mm in length, the **ventral reticular nucleus** (VRt) some 18 mm in length, the **medial reticular nucleus** (MRt) some 16 mm in length, and the **intermediate reticular zone** (IRz). Since the intermediate reticular zone is some 29 mm in length, it extends from the medulla oblongata into the lower pons (Table 9.2). These four reticular nuclei are between the trigeminal spinal nucleus and the fibers of the motor decussation that are crossing the median plane at this lower medullary level. The A1 and C1 catecholamine groups are in the ventrolateral part of the human intermediate reticular zone.

At slightly more rostral medullary levels, and about 5 mm below the obex (Fig. 9.3), the dorsal, ventral, medial and intermediate reticular nuclei are still present in the same relative position between the fibers of the motor decussation and the trigeminal spinal nucleus. Now, however, all the reticular nuclei are

TABLE 9.1.   Nuclear groups of the reticular formation in humans

Brain stem level	Named nuclear group	Abbreviation	Levels*
Medulla			
	Ventral reticular nucleus	VRt	4–21
	Dorsal reticular nucleus	DRt	5–21
	Medial reticular nucleus	MRt	6–21
	Intermediate reticular zone	IRt	5–33
	Lateral reticular nucleus, proper	LRt	12–27
	Lateral reticular nucleus, subtrigeminal part	LRtS5	15–28
	Lateral reticular nucleus, parvicellular part	LRtPC	
	Epiolivary lateral reticular nucleus	EO	16–23
	Interfascicular nucleus	IFH	17–23, 25, 26
	Lateral paragigantocellular nucleus	LPGi	21–32
	Dorsal paragigantocellular nucleus	DPGi	23–32
	Gigantocellular reticular nucleus	Gi	22–33
	Gigantocellular reticular nucleus, ventral part	GiV	22–27
	Gigantocellular reticular nucleus, alpha part	Giα	30–33
	Parvicellular reticular nucleus	PCRt	22–32
Pons			
	Parvicellular reticular nucleus, alpha part	PCRtα	33–36
	Pontine reticular nucleus, caudal part	PnC	33–36
	Reticulotegmental nucleus of the pons	RtTg	34–45
	Pontine reticular nucleus, oral part	PnO	37–50
	Pedunculopontine tegmental nucleus, diffuse part	PPTgD	48–52
	Pedunculopontine tegmental nucleus, compact part	PPTgC	50–53
Midbrain			
	Cuneiform nucleus	CnF	50–56
	Mesencephalic reticular fields	MRF	56–62
	Parapenduncular nucleus	PaP	59–62

Modified from *Terminologia Anatomica*, 1998; Paxinos and Huang, 1995.
*The extent of these nuclei at various levels of the brain stem are shown in Table 9.2.

larger and accompanied by the **lateral reticular nucleus** (LRt) which makes its appearance at this level (Fig. 9.3) and extends for some 16 mm in length. In the midst of these reticular nuclei are neurons of the **nucleus ambiguus**. The inferior olivary nucleus has yet to make its appearance at this level of the medulla oblongata.

At the level of the hypoglossal nucleus, dorsal vagal nucleus, inferior olivary nucleus, and some 3 mm above the obex (Fig. 9.4), the reticular nuclei continue to occupy the central core of the brain stem. In addition to the five reticular nuclei previously identified (**dorsal, ventral, medial, lateral** and **intermediate reticular nuclei**), this more rostral level of the medulla oblongata presents three additional reticular nuclei. First is the **lateral reticular nucleus, subtrigeminal division** (LRtS5) some 14 mm in length and just inferior to the trigeminal spinal nucleus (Fig. 9.4). Another is the **lateral reticular nucleus, epiolivary division** (EO) some 8 mm in length between the intermediate reticular nucleus and the inferior olive (LRtEO). Finally, about 10 mm in rostrocaudal

length, is the **interfascicular hypoglossal nucleus** (IFH) found medial to the ventral reticular nucleus (Fig. 9.4).

The reticular nuclei continue to form the central core of the brain stem at more rostral levels of the medulla oblongata. At this rostral level the fourth ventricle is completely open and the vestibular nuclei evident, about 11 mm above the obex (Fig. 9.5). Four reticular nuclei appear at this level for the first time along with parts of two other previously described reticular nuclei. The **inferior reticular nucleus** (IRt) is still evident (Fig. 9.5) as is the rostral tip of the **lateral reticular nucleus, subtrigeminal division** (LRtS5). Medial to the inferior reticular nucleus from superior to inferior on this section (Fig. 9.5) is the **dorsal paragigantocellular nucleus** (DPGi) some 10 mm in length, the **gigantocellular reticular nucleus** (Gi) some 12 mm in length, and the **lateral paragigantocellular nucleus** (LPGi) some 12 mm in length. Lateral to the inferior reticular nucleus at this level is the **parvicellular reticular nucleus** (PCRt) which measures some 11 mm in length.

TABLE 9.2. Relationship of reticular nuclei and their location in the brain stem

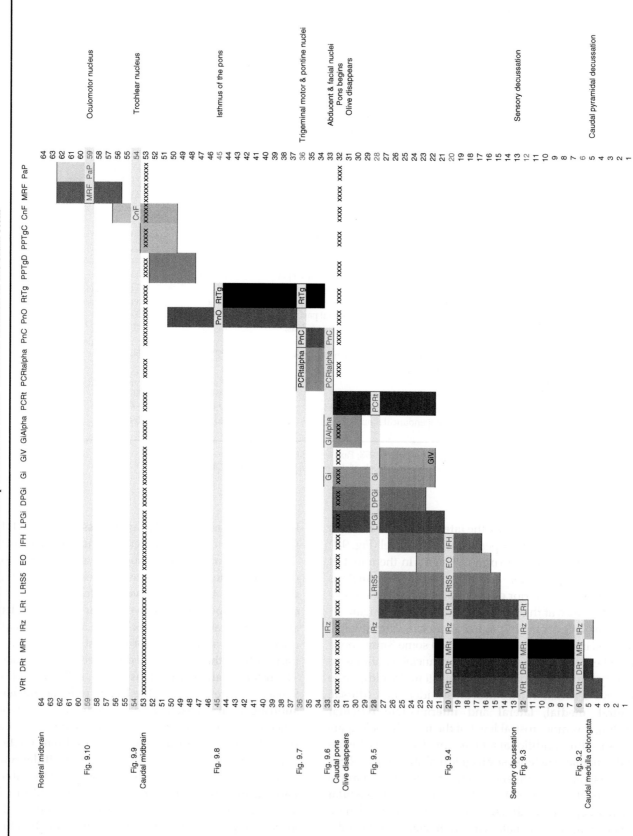

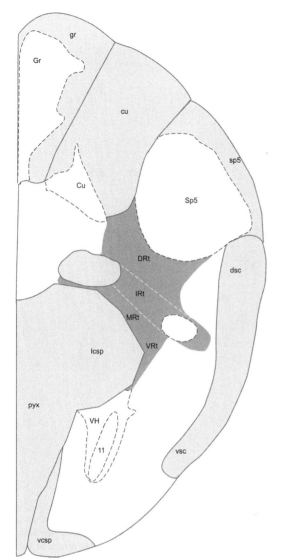

FIGURE 9.2. Transverse section through the human medulla oblongata about 11 mm below the obex. The reticular nuclei at this level are lightly shaded. (Modified from Figure 6 of Paxinos and Huang, 1995.) **Abbreviations for Figs 9.2–9.10:** 3, oculomotor nucleus; 3n, oculomotor nerve or its root; 4, trochlear nucleus; 4n, trochlear nerve or its root; 4V, 4th ventricle; 6, abducent nucleus or spinal cord layer 6; 6n, root of abducent nerve; 7, facial nucleus; 7n, facial nerve or its root; 10, dorsal vagal nucleus; 11, accessory nucleus; 11n, accessory nerve; 12, hypoglossal nucleus; 12n, root of hypoglossal nerve; AON, accessory oculomotor nucleus; Aq, midbrain aqueduct; bic, brachium of the inferior colliculus; bsc, brachium of the superior colliculus; CnF, cuneiform nucleus; cp, cerebral peduncle; ctg, central tegmental tract; Cu, cuneate nucleus; cu, cuneate fasciculus; DC, dorsal cochlear nucleus; DPGi, dorsal paragigantocellular nucleus; DRt, dorsal reticular nucleus; dsc, dorsal spinocerebellar tract; DT, dorsal tegmental nucleus; g7, genu of the facial nerve; Gi, gigantocellular reticular nucleus; Gr, gracile nucleus; gr, fasciculus gracilis; IC, inferior colliculus; icp, inferior cerebellar peduncle; IOA, inferior olive; IRt, intermediate reticular zone; LC, locus coeruleus; lcsp, lateral corticospinal tract; ll, lateral lemniscus; LPGi, lateral paragigantocellular nucleus; LRt, lateral reticular nucleus; LSO, lateral superior olive; LVe, lateral vestibular nucleus; m5, trigeminal motor root; Me5, trigeminal mesencephalic nucleus; me5, trigeminal mesencephalic tract; MG, medial geniculate nucleus; ml, medial lemniscus; mlf, medial longitudinal fasciculus; MRF, mesencephalic reticular fields; MRt, medial reticular nucleus; MSO, medial superior olive; MVe, medial vestibular nucleus; P7, perifacial zone; Pa6, paraabducent nucleus; PAG, periaqueductal gray; pc, posterior commissure; PCRt, parvicellular reticular nucleus; Pn, pontine nuclei; PnC, pontine reticular nucleus, caudal part; PnO, pontine reticular nucleus, oral part; Pr5, principal sensory trigeminal nucleus; py, pyramidal tract; pyx, pyramidal decussation; R, raphé; RN, red nucleus; RtT, reticulotegmental nucleus; SC, superior colliculus; scp, superior cerebellar peduncle; SNC, substantia nigra; sol, solitary tract; Sol, solitary nucleus; sp5, trigeminal spinal tract; Sp5, trigeminal spinal nucleus; SVE, spinal vestibular nucleus; SpVe, superior vestibular nucleus; its, tectospinal tract; tth, trigeminothalamic tract; Tz, nucleus of the trapezoid body; tz, trapezoid body; VC, ventral cochlear nucleus; vcsp, ventral corticospinal tract; VH, ventral horn; VRt, ventral reticular nucleus; vsc, ventral spinocerebellar tract; VT, ventral tegmental nucleus; VTA, ventral tegmental area.

Not evident on the levels described above is the **gigantocellular reticular nucleus, ventral part** (GiV) measuring some 6 mm in length and the **gigantocellular reticular nucleus, alpha part** (Giα) just 4 mm in length. The gigantocellular reticular nucleus, ventral part (GiV) is at levels between Figs 9.4 and 9.5 just ventral to the gigantocellular reticular nucleus and dorsal to the inferior olive, dorsal nucleus. The gigantocellular reticular nucleus, alpha part (Giα) lies rostral to Fig. 9.5 and lateral to nucleus raphé magnus. It extends from that level rostrally into the lower pons.

9.1.2. Reticular Nuclei in the Pons

Several medullary reticular nuclei extend for a millimeter or two into the lowest levels of the pons (the level of the abducent and facial nuclei), about 16 mm above the obex (Fig. 9.6). These include the intermediate reticular zone, the gigantocellular reticular nucleus, and the gigantocellular reticular nucleus, alpha part. The **pontine reticular nucleus, caudal part** (PnC) is directly ventral to the abducent and para-abducent nuclei but dorsal to the central tegmental tract. The latter tract contains many ascending reticulothalamic fibers originating in the reticular nuclei of the medulla oblongata and pons and destined for the thalamus. The PnC extends rostrally for about 4 mm accompanied by a nucleus of about the same length, the **parvicellular reticular nucleus, alpha part** (PCRtα) which is on either side of the intrapontine facial root fibers (Fig. 9.6). The reticular nuclei form a sizable part of the tegmentum of the pons at this level.

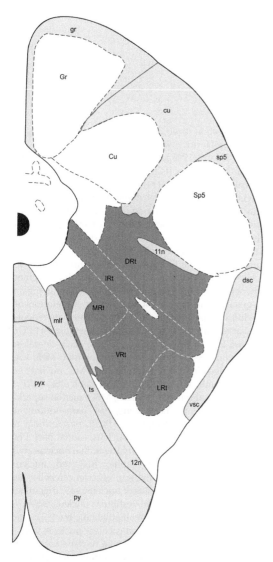

FIGURE 9.3. Transverse section through the human medulla oblongata about 5 mm below the obex. The reticular nuclei at this level are lightly shaded (modified from Figure 12 of Paxinos and Huang, 1995). **Abbreviations:** see Fig. 9.2.

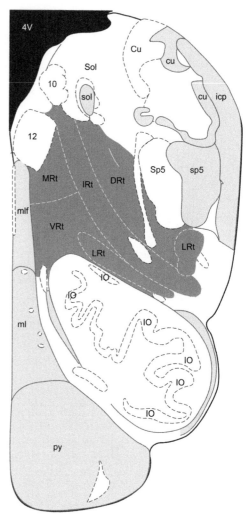

FIGURE 9.4. Transverse section through the human medulla oblongata about 3 mm above the obex. The reticular nuclei at this level are lightly shaded (modified from Figure 20 of Paxinos and Huang, 1995). **Abbreviations:** see Fig. 9.2.

At the level of the trigeminal motor and trigeminal pontine nuclei, some 19 mm above the obex (Fig. 9.7) and in the middle pons, the two previous pontine reticular nuclei (PnC and PCRtα) are still evident. Accompanying these two nuclei is the **reticulotegmental nucleus of the pons** (RtTg) that is medial and dorsal to the medial lemniscus and has a rostrocaudal length of about 12 mm.

At the isthmus level of the pons, some 28 mm above the obex (Fig. 9.8), the **pontine reticular nucleus, oral part** (PnO) makes its appearance lateral to the paramedian raphé nucleus. Ventral to the paramedian raphé nucleus is the rostral extent of the **reticulotegmental nucleus of the pons** (RtTg). The reticular nuclei make up a decreasing area within the pontine tegmentum at this level. The

reticulotegmental nucleus of the pons disappears at more rostral levels of the pons but the pontine reticular nucleus, oral part, with an overall length of 14 mm extends rostrally another five millimeters. At that point, three other pontine reticular nuclei join it, namely: the **pedunculopontine tegmental nucleus, diffuse part** (PPTgD), the **pedunculopontine tegmental nucleus, compact part** (PPTgC), and the **cuneiform nucleus** (CnF). The latter nucleus, some seven millimeters in length, is identifiable in the upper pons and the lower midbrain.

9.1.3. Reticular Nuclei in the Midbrain

Representative levels of the midbrain in humans include the level of the trochlear nucleus (inferior collicular

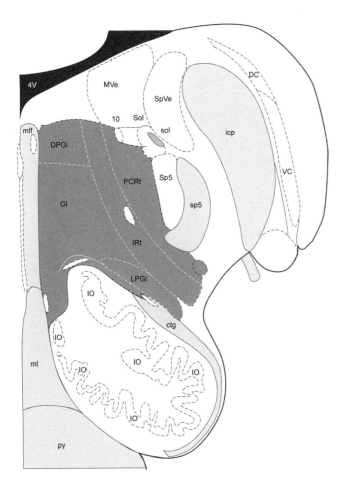

FIGURE 9.5. Transverse section through the human medulla oblongata about 11 mm above the obex. The reticular nuclei at this level are lightly shaded (modified from Figure 28 of Paxinos and Huang, 1995). **Abbreviations:** see Fig. 9.2.

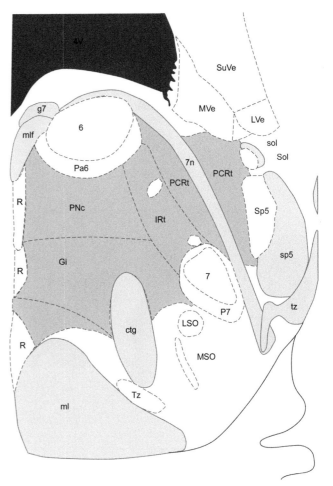

FIGURE 9.6. Transverse section through the lower third of the human pons about 16 mm above the obex. The reticular nuclei at this level are lightly shaded (modified from Figure 33 of Paxinos and Huang, 1995). **Abbreviations:** see Fig. 9.2.

level) and the level of the oculomotor nucleus (superior collicular level). At trochlear levels, some 37 mm above the obex (Fig. 9.9), only one representative of the reticular nuclei, the **cuneiform nucleus** (CnF) is evident. CnF is ventral to the nuclei of the inferior colliculus. More rostrally, some 41 mm above the obex, at the level of the oculomotor nucleus (Fig. 9.10), the **mesencephalic reticular field** (MRF) is identifiable ventral to the medial lemniscus. One millimeter higher, the very small **parpeduncular nucleus** (PaP) makes an appearance. Even more rostrally, PaP remains small whereas the mesencephalic reticular fields are somewhat larger.

9.2. ASCENDING RETICULAR SYSTEM

Moruizzi and Magoun (1944) discovered that stimulation of the brain stem reticular formation in cats

resulted in activation of the electroencephalogram (EEG) in the ipsilateral cerebral hemisphere. Such stimulation eliminated high-amplitude slow waves and led to the emergence of fast, low-amplitude electrical activity in the brain typically seen during waking. These EEG changes were elicitable following stimulation of the medullary reticular formation, the pontine and midbrain tegmentum and the dorsal hypothalamus. These effects occur by a series of reticular relays that activate the cerebral cortex by way of the **diffuse thalamic projection system**. This paper brought into focus the role of the **ascending reticular activating system** (ARAS) and led to greater understanding of the physiology of arousal and consciousness. Activation in the midbrain reticular formation and intralaminar nuclei has been demonstrated using positron emission tomography in humans who went from a relaxed awake state to an attention-demanding reaction-time test. This study

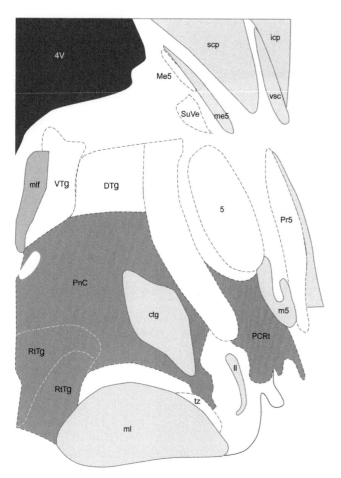

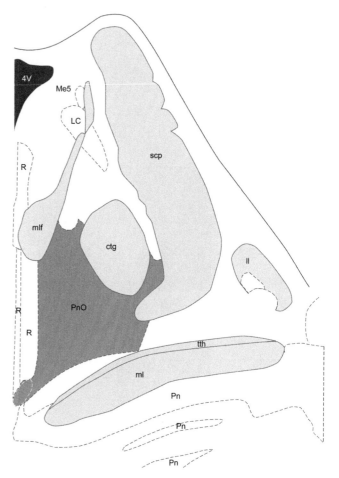

FIGURE 9.7. Transverse section through the middle third of the human pons about 19 mm above the obex. The reticular nuclei at this level are lightly shaded (modified from Figure 36 of Paxinos and Huang, 1995). **Abbreviations:** see Fig. 9.2.

FIGURE 9.8. Transverse section through the isthmus of the human pons about 28 mm above the obex. The reticular nuclei at this level are lightly shaded (modified from Figure 45 of Paxinos and Huang, 1995). **Abbreviations:** see Fig. 9.2.

confirms the role of these areas in arousal and vigilance in humans. It seems likely that this region of the midbrain reticular formation and dorsal thalamus might control not only the transition from sleep to an alert state but also from a relaxed wakefulness to a high level of general attention.

The contributions of various neurotransmitters to this process are receiving a great deal of interest. Observations in nonhuman primates and humans have led to the suggestion by Mesualm (1995) that in addition to these reticulothalamic projections the concept of the ARAS should include two sources of ascending cholinergic projections. These include a traditional projection in the upper brain stem arising from cholinergic cell groups Ch5–Ch6, that is, the pedunculopontine nucleus (especially its compact part) and the laterodorsal tegmental nucleus, respectively. The pedunculopontine tegmental nucleus, compact part, is one of the named reticular nuclei in humans (Table 9.1). A second projection arises

from cholinergic cell groups Ch1–Ch4 in the basal forebrain, that is, from cholinergic cells associated with the medial septal nucleus (Ch1), with the vertical nucleus of the diagonal band (Ch2), with the horizontal limb of the diagonal band nucleus (Ch3), and with the nucleus basalis (of Meynert) designated as Ch4. Also included in this scheme would be noncholinergic regulatory paths arising from the hypothalamus (histaminergic), ventral tegmental area (dopaminergic), locus coeruleus (adrenergic) and the nuclei of the raphé (serotonergic) of the brain stem. Fibers from these areas in turn send widespread projections to the thalamus and cerebral cortex that influence attention, arousal, and the level of consciousness. This suggestion would greatly expand the structures traditionally considered under the umbrella of the reticular formation. Additional studies of the connectivity and chemical neuroanatomy of these cell groups and their relationship to defined reticular nuclei of the brain stem (Table 9.1) would be helpful.

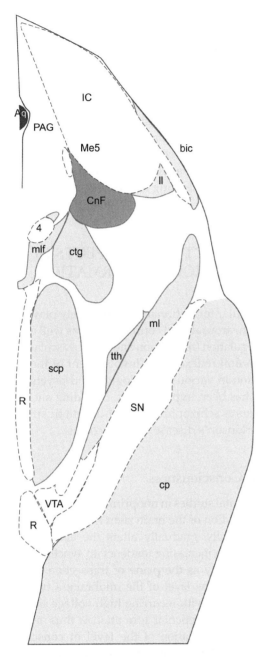

FIGURE 9.9. Transverse section through the trochlear nucleus of the human midbrain about 37 mm above the obex. The reticular nuclei at this level are lightly shaded (modified from Figure 54, Paxinos and Huang, 1995). **Abbreviations:** see Fig. 9.2.

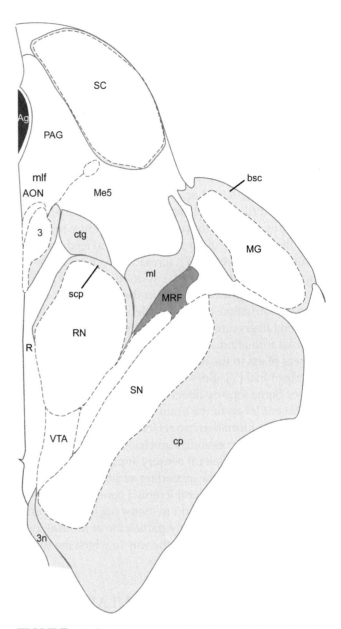

FIGURE 9.10. Transverse section through the oculomotor nucleus of the human midbrain about 41 mm above the obex. The reticular nuclei at this level are lightly shaded (modified from Figure 58 of Paxinos and Huang, 1995). **Abbreviations:** see Fig. 9.2.

The primary origin of the ascending reticular system is the caudal part of the **gigantocellular reticular nucleus** (Gi) in the medulla oblongata and the **caudal part of the pontine reticular nucleus** (PnC). Many of these fibers ascend in the **central tegmental tract**. The principal destination of fibers of the ascending reticular system is the reticular nuclei, intralaminar nuclei, and median nuclei of the thalamus. Impulses from

these thalamic nuclei project to the cerebral cortex and increase its excitability. Other ascending reticular fibers do not pass beyond the reticular formation of the midbrain. Multisynaptic fibers from reticular nuclei in the midbrain reach the hypothalamus, preoptic area, and septal region.

The term **central receptive pool** (*centrum receptorium*) used in reference to the reticular formation emphasizes the enormous convergence of somatic and visceral sensory information, both collateral and terminal, on the reticular formation and recognizes the possibility of integrating information from a wide variety of sources. This includes input from all known sensory systems innervating cutaneous areas, muscles, joints, cardiovascular and lung stretch receptors and chemoreceptors. The precise way in which this integration occurs is unclear in nonhuman primates and humans. It does appear that these converging impulses drive the neurons of the reticular formation with their widespread and extensive dendritic and axonal processes. One example is the lateral spinothalamic tract. In humans, only one third of the fibers in this path reach the thalamus. Thus, the path for pain and temperature impulses from the body has many opportunities to feed into and influence the reticular formation. Other sensory impulses enter the reticular formation via ascending spinoreticular fibers that end in the medial part of the reticular formation. Sensory nuclei of the cranial nerves also send fibers directly or indirectly as collaterals to the reticular formation. There is evidence for some primary afferent fibers to the reticular formation in and through the trigeminal [V], glossopharyngeal [IX], and vagal [X] nerves. Some sensory fibers of these cranial nerves enter at different levels of the brain stem and pass directly to the reticular formation in a region near the **solitary tract** and the lateral vestibular nuclei. A most interesting aspect of such collateral sensory input to the reticular formation is that of the ascending auditory path, the lateral lemniscus. The lateral lemnisci contribute fibers to the reticular formation and to motor nuclei in the brain stem. Such collaterals likely participate in arousal, providing a structural basis for the way in which the sound of an alarm clock acts to alert us.

9.3. DESCENDING RETICULAR SYSTEM

Clinical studies have demonstrated that the reticular formation is capable of influencing somatic and autonomic neurons at spinal levels via descending **reticulospinal fibers**. The medial reticular formation of both the medulla oblongata and pons gives rise to long descending axons that reach caudal levels of the spinal cord. Those fibers originating from neurons in medullary parts of the medial reticular formation send fibers bilaterally into the spinal cord whereas descending fibers originating from reticular neurons in the pons send descending fibers to the ipsilateral parts of the spinal cord. The lateral reticular formation projects fibers medially to influence descending reticular neurons in the medial reticular formation. Afferent fibers from various regions of the nervous system influence

the reticular nuclei of the brain stem by way of these descending fibers. There are descending reticulospinal fibers involved in respiration. Primary somatosensory cortical areas 3, 1, and 2 and premotor area 6 send projections to that part of the medial reticular formation of the medulla oblongata and pons that gives rise to reticulospinal fibers. The basal nuclei and the cerebellum (particularly the fastigial nucleus and perhaps the dentate nucleus of the cerebellum), send afferents to the descending reticular formation. Thus, there is a wide spectrum of descending influences on the reticular formation of both a motor and a sensory nature.

9.4. FUNCTIONAL ASPECTS OF THE RETICULAR FORMATION

The reticular formation in humans likely plays a role in attention, arousal, and consciousness as well as homeostatic regulation (respiration and cardiovascular activity), and visceral reflexes. The involvement of the reticular formation in various neurological and psychiatric disorders has been hypothesized, including such diverse disorders as schizophrenia, post-traumatic stress disorder, Parkinson's disease, and narcolepsy.

9.4.1. Consciousness

Experimental studies in nonprimates have demonstrated that transection of the brain stem from lower medullary levels rostrally, gradually alters the electroencephalogram with an increasing tendency to synchronization of cortical activity as the plane of transection reaches the midbrain. At the level of the midbrain, a typical sleep rhythm results with recurring high voltage slow waves. The brain stem reticular formation is thus an essential area for the regulation of the **level of consciousness**. While the **content of consciousness** depends on the cerebral cortex, crude alertness, arousal or consciousness per se depends on activity in the reticular formation and its involvement in the sleep–wakefulness cycle. The brain stem reticular formation preserves wakefulness by offering a constant background of impulses that are destined for the cerebral cortex. These impulses profoundly influence arousal and the electrical activity of the cerebral cortex. There are significant regional differences. A complete description of levels of consciousness requires that other ascending and descending fiber paths receive sufficient consideration. The level of consciousness may be increased (activated, aroused) or decreased (inhibited) leading to sleep.

A considerable amount of the cerebral cortex is removable from any lobe of the brain in humans without loss of consciousness. However, injuries at the junction of the midbrain and diencephalon produce immediate clinical and electroencephalographic manifestations of coma or other altered states of consciousness from which the patient cannot be aroused. Injuries in other parts of the brain stem disturb the level of consciousness with accompanying signs that vary with the site of the injury. These accompanying signs are of localizing value. Injuries to the pons may result in the loss of consciousness with respiratory and cardiac abnormalities and hypertonicity of the limbs. Impaction of the medulla oblongata as the result of cerebellar tonsilar herniation may lead to respiratory arrest and circulatory changes resulting in a loss of consciousness. Digital compression of the human medulla oblongata may yield unconsciousness, stuporous breathing, and bradycardia.

9.4.2. Homeostatic Regulation

Respiration

Respiration is an important aspect of homeostatic regulation as well as a complex, rhythmic, motor activity involving the coordinated contraction of the various respiratory muscles. These include the diaphragm, the accessory respiratory muscles (external and internal intercostals, scalene muscles, sternocleidomastoid, levatores costarum, and abdominal muscles), and the upper airway muscles (laryngeal, pharyngeal, genioglossus muscles). These muscles receive their innervation from lower motoneurons in the brain stem or spinal cord. Respiratory neurons in the reticular formation of the brain stem rhythmically drive these respiratory-related lower motoneurons. Transneuronal tracing studies by means of the rabies virus in the mouse indicate that many structures in the brain stem and beyond (cerebellum, hypothalamus, thalamus, and cerebral cortex) have descending projections to brain stem respiratory neurons. Much of our understanding of the structure and function of the brain stem respiratory network comes from experimental studies in the adult cat. There is much disagreement regarding the extent and topographic organization of the respiratory neurons in the human brain stem and their relationship to respiratory groups found in the cat and rat.

Inspiratory and expiratory neurons are found in the caudal half of the medulla oblongata in primates and presumably also in humans. The area of the expiratory responses lies in the dorsal part of the reticular formation of the medulla oblongata including the parvicellular

reticular nucleus and gigantocellular reticular nucleus, particularly its dorsal part. The **expiratory region** surrounds the inspiratory field. The **inspiratory region** is medial and dorsal to the rostral half of the inferior olive just ventral and lateral to, and cupped over by, the expiratory region. This would include the lateral and intermediate reticular nuclei near the nucleus ambiguus. Inspiratory neurons stimulate expiratory neurons as well as excite other inspiratory neurons. Neurons in these respiratory fields project into the ventrolateral part of the spinal cord as a **descending ventrolateral reticulospinal path** that ends on motoneurons of the muscles of respiration. Presumably, some of these descending projections excite inspiratory motoneurons while others may inhibit phrenic motoneurons during expiration. Normally the medullary respiratory regions are under the influence of pontine, hypothalamic, thalamic, and cortical control. Thus, these respiratory fields in the reticular formation of the medulla oblongata are a minimum substrate for the maintenance and regulation of respiratory activity.

Structures in the reticular formation of the pons including a **pneumotaxic region** and an **apneustic region** play a fundamental role in normal breathing. The **pneumotaxic region** is in the rostral part of the pontine tegmentum and thought to be involved in the inhibition of respiratory activity. Apneusis is a sustained inspiratory pause. The **apneustic region** is in the tegmentum of the middle and caudal parts of the pons where it exerts a strong tonic effect on the medullary inspiratory region. Abnormal periodic breathing may occur with expanding injuries of the reticular formation of the brain stem or even respiratory arrest if these respiratory regions of the brain stem reticular formation are injured. Long periods of respiratory arrest may occur with maximal respiratory depth following stimulation of this apneustic region. Recent studies, however, suggest the pons may have only a minor role in the control of normal breathing.

The intermediate reticular zone, shown to have a role in respiration in the cat and rat, may have a similar role in humans. This zone in these experimental animals is a 'gasping center' having the capacity to restore breathing through gasping when quiet breathing fails due to noxious stimuli or due to the failure of higher levels to regulate respiration. The finding of α2-adrenergic receptor binding in the intermediate reticular zone in the human fetus and infant suggests that this receptor may play a role in human gasping responses in early life.

Cardiovascular Activity

Another important function of the reticular formation in relation to homeostatic regulation is the involvement of the reticular formation in **cardiovascular activity**. Both pressor and depressor areas are present in the medullary reticular formation. The depressor area probably is tonically active with continuous inhibition of spinal cardiovascular neurons. There is an acceleratory and an inhibitory cardiac region in the medulla. The inhibitory region connects with the vagal nerve [X] and the excitatory region of the brain stem reticular formation connects with sympathetic nerves. Stimulation of the ventral periventricular/periaqueductal gray substance in 15 awake humans undergoing deep brain stimulation for treatment of chronic pain caused a mean reduction in systolic blood pressure (in seven patients) whereas stimulation of the dorsal periventricular/periaqueductal gray substance caused a mean increase in systolic blood pressure (in six patients). Collectively these observations indicate that there may well be a role for the periaqueductal gray substance in cardiovascular control in humans. The relationship of this periventricular/periaqueductal gray region to the pressor and depressor areas in the medullary reticular formation is unclear.

9.4.3. Visceral Reflexes

Gag Reflex

Another aspect of the function of the reticular formation is the participation of its neurons in various reflex arcs at brain stem levels. A good example is the gag or pharyngeal reflex. If a tongue blade is touched to the back of the tongue in the tonsilar area, there will be elevation and constriction of the pharyngeal musculature with retraction of the tongue. The afferent limb of this reflex is the glossopharyngeal nerve [IX]. The primary neurons in this path are general visceral afferent sensory neurons in the inferior (petrosal) ganglion of the glossopharyngeal nerve at the jugular foramen. The central processes of these primary neurons pass directly to the **solitary tract** at brain stem levels. Secondary neurons in this path are in the **nucleus of the solitary tract**. There is bilateral discharge to efferent neurons from the nucleus of the solitary tract. The efferent limb of this pharyngeal reflex begins in the nucleus ambiguus, the dorsal vagal nucleus, and the adjacent reticular neurons of the medulla. The existence of crossed and uncrossed discharge via the vagal nerve [X] to the constrictors of the pharynx and to muscles that constrict the pharynx and close the epiglottis results in a gag response when the back of the human tongue is stimulated. There may be some discharge from medullary reticular neurons for connections to the hypoglossal nucleus for retraction of

the tongue as well. The pharyngeal reflex functions to initiate deglutition or swallowing.

Vomiting

The gag reflex will lead to vomiting if preganglionic neurons in the brain stem and spinal cord are involved. Thus with widespread autonomic discharge from the brain stem reticular formation there will be an accompanying dilatation of the pupils, sweating, pallor of the face, and an irregular heartbeat. These signs require connections from the dorsal vagal nucleus to preganglionic neurons of the myenteric plexuses of the esophagus and stomach and postganglionic fibers to the smooth musculature that yields a reverse peristalsis and emesis. There are literally hundreds of other reflexes, many are muscle stretch reflexes, but many, such as the swallowing, sucking, coughing, hiccupping, yawning, pupillary, and respiratory reflexes, use neurons of the reticular formation as a constituent of their reflex arc. Many important reflexes related to gastric motility and sensations involve neurons of the reticular formation.

9.4.4. Motor Function

While the corticospinal path is the primary descending motor path for fine skilled, voluntary movements of the upper and lower limbs, there are other descending motor paths. These include the so-called common discharge paths from neurons in the tegmentum of the midbrain and brain stem reticular formation whose fibers descend and influence lower motoneurons of the brain stem and spinal cord. Under the heading of common discharge paths are the rubrospinal, associated tegmentospinal, and reticulospinal paths. Associated tegmentospinal fibers originate in the tegmentum of the midbrain and descend on both sides of the brain stem ending in the lower spinal cord. These fibers supplement the rubrospinal fibers. Neurons contributing to these common discharge paths receive input from diverse areas of the nervous system involved in motor activity including extrapyramidal motor areas, the basal nuclei, and cerebellum. Output from these neurons over the common discharge paths is the result of considerable integration and correlation of impulses arriving by way of corticotegmental fibers, the ansa system, cerebellorubral, and cerebellotegmental fibers. These common discharge paths are partly facilitatory and partly inhibitory.

In the human spinal cord, the descending path for automatic respiration, the ventrolateral reticulospinal tract, is intermingled with, and deep to, fibers of the ventral part of the lateral spinothalamic tract. Although unilateral destruction of this tract by way of an antero-lateral cordotomy causes insignificant respiratory loss, bilateral section of this descending path for automatic respiration causes severe respiratory embarrassment.

FURTHER READING

Blessing WW (1997) *The Lower Brainstem and Bodily Homeostasis*. Oxford University Press, New York.

Blessing WW, Gai WP (1997) Caudal pons and medulla oblongata. In: *Handbook of Chemical Neuroanatomy*. Björklund A, Hökfelt T, eds. Vol 13: The Primate Nervous System, Part I, pp 139–186. Elsevier, Amsterdam.

Foote SL (1997) The primate locus coeruleus: the chemical neuroanatomy of the nucleus, its efferent projections, and its target receptors. In: *Handbook of Chemical Neuroanatomy*. Björklund A, Hökfelt T, eds. Vol 13: The Primate Nervous System, Part I, pp 187–215. Elsevier, Amsterdam.

Garcia-Rill E (1997) Disorders of the reticular activating system. Med Hypotheses 49:379–387.

Green AL, Wang S, Owen SL, Xie K, Liu X, Paterson DJ, Stein JF, Bain PG, Aziz TZ (2005) Deep brain stimulation can regulate arterial blood pressure in awake humans. NeuroReport 16:1741–1745.

Heidel KM, Benarroch EE, Gene R, Klein F, Meli F, Saadia D, Nogues MA (2002) Cardiovascular and respiratory consequences of bilateral involvement of the medullary intermediate reticular formation in syringobulbia. Clin Auton Res 12:450–456.

Hilaire G, Pasaro R (2003) Genesis and control of the respiratory rhythm in adult mammals. News Physiol Sci 18:23–28.

Huang XF, Paxinos G (1995) Human intermediate reticular zone: a cyto- and chemoarchitectonic study. J Comp Neurol 360:571–588.

Kinomura S, Larsson J, Gulyas B, Roland PE (1996) Activation by attention of the human reticular formation and thalamic intralaminar nuclei. Science 271:512–515.

McCormick DA (1992) Neurotransmitter actions in the thalamus and cerebral cortex and their role in neuromodulation of thalamocortical activity. Prog Neurobiol 39:337–388.

Mesulam MM (1995) Cholinergic pathways and the ascending reticular activating system of the human brain. Ann NY Acad Sci 757:169–179.

Mesulam MM, Geula C, Bothwell MA, Hersh LB (1989) Human reticular formation: cholinergic neurons of the pedunculopontine and laterodorsal tegmental nuclei and some cytochemical comparisons to forebrain cholinergic neurons. J Comp Neurol 283:611–633.

Moruzzi G, Magoun, HW (1949) Brain stem reticular formation and activation of the EEG. Electroenceph Clin Neurophysiol 1:455–473.

Nathan PW, Smith MC (1982) The rubrospinal and central tegmental tracts in man. Brain 105:223–269.

Olszewski J, Baxter D (1982) *Cytoarchitecture of the Human Brain Stem*, 2nd ed. Karger, Basel.

Paus T (2000) Functional anatomy of arousal and attention systems in the human brain. Prog Brain Res 126:65–77.

Pearce JMS (2003) The nucleus of Theodor Meynert (1833–1892). J Neurol Neurosurg Psychiatry 74:1358.

Plum F, Posner JB (1982) The pathologic physiology of signs and symptoms of coma. In: Plum F, Posner JB, *The Diagnosis of Stupor and Coma*, pp 1–86. FA Davis Co, Philadelphia.

Richter DW (1996) Neural regulation of respiration: rhythmogenesis and afferent control. In: Greger R, Windhorst U, *Comprehensive Human Physiology*, pp 2079–2095. Springer–Verlag, Berlin.

Steriade M, Glenn LL (1982) Neocortical and caudate projections of intralaminar thalamic neurons and their synaptic excitation from midbrain reticular core. Neurophysiol 48:352–371.

Steriade M (1996) Arousal: revisiting the reticular activating system. Science 27:225–226.

St John WM, Paton JF (2004) Role of pontile mechanisms in the neurogenesis of eupnea. Respir Physiol Neurobiol 143:321–332.

Usunoff KG, Itzev DE, Lolov SR, Wree A (2003) Pedunculopontine tegmental nucleus. Part 1: Cytoarchitecture, transmitters, development and connections. Biomed Rev 14:95–120.

Van der Kooy D (1987) The reticular core of the brain-stem and its descending pathways: anatomy and function. In: Fromm GH, Faingold C, Browning RA, Burnham WM, eds. *Epilepsy and the Reticular Formation: the Role of the Reticular Core in Convulsive Seizures*. Neurology and Neurobiology 27:9–23. Alan R. Liss Inc, New York.

Zec N, Kinney HC (2001) Anatomic relationships of the human nucleus paragigantocellularis lateralis: a DiI labeling study. Auton Neurosci 89:110–124.

The problems of deafness are deeper and more complex, if not more important, than those of blindness. Deafness is a much worse misfortune. For it means a loss of the most vital stimulus – the sound of the voice – that brings language, sets thoughts astir, and keeps one in the intellectual company of man.

Helen Keller (1880–1968)

10

The Auditory System

Hearing occurs by way of the **cochlear nerve**, the auditory part of the **vestibulocochlear nerve** [VIII]. The cochlear nerve is functionally classified as **special somatic afferent**, being responsible for a special exteroceptive sensation relating us to our environment. Hearing may well be the primary sense for human communication.

10.1. GROSS ANATOMY

10.1.1. External Ear

The external ear consists of the **auricle** and **external acoustic meatus** (Fig. 10.1). The **auricle**, a large, convoluted piece of elastic cartilage covered by skin, conducts sound waves along the external acoustic meatus and aids in the localization of the source of sound. The external acoustic meatus, about 25.4 mm in length, has a lateral cartilaginous part and a medial, slightly longer, bony part. At the medial end of the meatus is the **tympanic membrane** (Fig. 10.1) of the middle ear.

10.1.2. Middle Ear

The tympanic membrane is a nearly circular, tense membrane forming a partition between the external and middle ear. With the body erect, it faces laterally, forward and downward. The trigeminal nerve [V] is the main

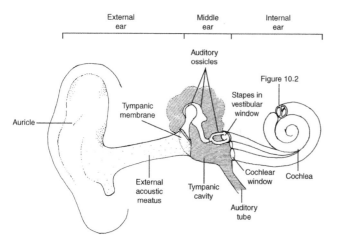

FIGURE 10.1. Three parts of the ear: (1) the external ear consisting of the auricle, external acoustic meatus, and its medial boundary the tympanic membrane; (2) the middle ear, conveniently divided into a tympanic cavity (containing the auditory ossicles), mastoid adnexa, and pharyngotympanic tube; and (3) the cochlear part of the internal ear. A section is made of one turn of the cochlea. It is shown in detail in Fig. 10.2.

sensory nerve to this membrane. Tympanic stimulation in humans may cause a reflex cough or an unpleasant or painful sensation. Free nerve endings in the human tympanic membrane are like those in the cornea. The tympanic membrane is the lateral boundary of the **middle ear** composed of the **tympanic cavity**, **auditory ossicles**, including their joints and muscles, and the **auditory (pharyngotympanic) tube**. The **tympanic cavity**, an air space housed in the temporal bone (Fig. 10.1), houses the three **auditory ossicles** called the **malleus, incus**, and **stapes**. The manubrium and lateral process of the malleus attach to the medial aspect of the tympanic membrane. The tympanic cavity communicates with mastoid air cells and, by the **auditory tube** (Fig. 10.1), with the nasopharynx. The auditory tube opens in swallowing, equalizing air pressure between middle ear and environment. The medial wall of the cavity presents the oval or **vestibular window** (Fig. 10.1), normally closed by the base of the **stapes** (Fig. 10.1). The conversion of sound waves reaching the tympanic membrane into vibrations of the ossicles reaches the vestibular window, relays to the cochlear fluids, and then to sensory structures in the internal ear. The medial wall also has a small niche containing a round opening – the **cochlear window**. A trilaminar membrane normally closes this portal of entry from the middle to the inner ear. In humans, it has an outer squamous epithelial layer with an underlying basement membrane, a middle fibrous layer of collagen, elastin, fibroblasts, vessels, and nerves, and an inner layer of mesothelial cells. Mucosal membranous

'veils' often cover the cochlear membrane forming 'false' membranes. These veils likely protect the cochlear membrane and influence its permeability.

Muscles of the Auditory Ossicles and the Acoustic Reflex

Two muscles, the tensor tympani and the stapedius, control the auditory ossicles. The tendon of the **tensor tympani** is attached to the manubrium of the malleus and displaces the tympanic membrane medially when it contracts, pushing the base of the stapes into the vestibular window. The **stapedius** attaches to the neck of the stapes such that when it contracts, it pulls the base of the stapes laterally and slightly out of the vestibular window. With gradually increasing sound or sudden loud sounds, both the tensor and stapedius contract. In doing so, they simultaneously oppose each other and increase the rigidity of the auditory ossicles. This **acoustic response** or **reflex** dampens the vibrations of the auditory ossicles and protects sensitive receptors in the internal ear. Both muscles contract before vocalization. Some individuals are able to contract these muscles voluntarily. Examination of the functional condition of the middle ear, including the mobility of the tympanic membrane and auditory ossicles, and the state of intra-aural muscle reflexes requires special procedures including **impedance audiometry**.

10.1.3. Internal Ear

Cochlea, Cochlear Labyrinth, and Cochlear Duct

The human internal ear has an intricate series of continuous, fluid-filled cavities called the **bony labyrinth** divisible into several parts: three **semicircular canals**, a centrally placed and ovoid **vestibule**, and a conical **cochlea** (see Fig. 11.1). The **modiolus** is a central cone of trabecular bone supporting the cochlea. The **cochlea**, about 33.2 mm in length on average, based on five separate studies (with extreme values of 25.3 and 40.1 mm), is shaped like the spiral shell of a snail with two and a half turns around its axis (Fig. 10.1). The auditory or **cochlear labyrinth** is in the petrous temporal bone. Sections of the cochlea reveal it is divisible along its entire length into three fluid-filled, membranous compartments (Fig. 10.2) including a middle compartment that is roughly triangular in section and called the **cochlear duct**, an upper compartment, the **scala vestibuli**, and a lower compartment, the **scala tympani** (Fig. 10.2). The two scalae communicate at the apex of

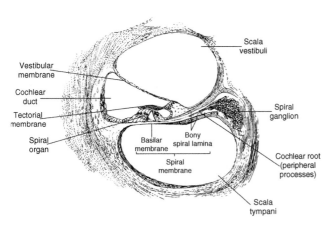

FIGURE 10.2. A section of one turn of the cochlea (For orientation see Fig. 10.1). Its three compartments – scala vestibuli, scala tympani, and cochlear duct, are visible. Note the spiral organ perched on the basilar membrane part of the spiral membrane in the cochlear duct. Details of the spiral organ are shown in Figs 10.3 and 10.4 (after Ballenger, 1977).

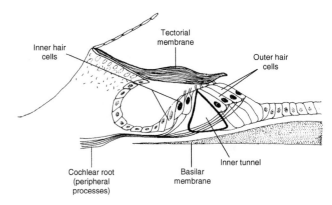

FIGURE 10.3. Spiral organ on the spiral membrane in the cochlear duct. Microvilli of the outer hair cells are firmly attached to the undersurface of the tectorial membrane into shallow pits called imprints. For orientation, see Fig. 10.2 (after Ballenger, 1977).

the cochlea at a region known as the **helicotrema**. The roof of the cochlear duct is the **vestibular membrane** (Fig. 10.2) while its floor is the **spiral membrane**. The spiral membrane in turn, is composed of a bony shelf, the **spiral lamina**, and another membrane, the **basilar membrane** (Fig. 10.2). The bony spiral lamina extends from the medial side of the cochlea about halfway across its cavity like the thread on a screw. The basilar membrane extends laterally from the edge of the spiral lamina to complete the floor of the cochlear duct. The lateral wall of the cochlear duct is a well-vascularized, multilaminar tissue called the **stria vascularis**.

Cochlear Fluids

The scalae vestibuli and tympani contain **perilymph**, a cochlear fluid similar to cerebrospinal fluid, with a high concentration of protein and sodium but a low concentration of potassium. Indeed, perilymph in the scala tympani communicates with the cerebrospinal fluid of the subarachnoid space by means of the cochlear aqueduct. Perilymph in the auditory part of the internal ear is continuous with that surrounding its vestibular part. Each cochlear duct has another fluid, **endolymph**, that has a large D.C. (direct current) potential (polarized by approximately 80 mV), low protein and sodium, but high potassium concentrations. Endolymph in the cochlear duct is continuous with that in the saccule of the membranous vestibular apparatus of the internal ear by means of the narrow **ductus reuniens**. Cochlear fluids transport nutrients from the blood stream to cochlear cells, transmit vibrations from the vestibular window to

the cochlear duct, and provide an appropriate ionic environment for the specialized auditory epithelium.

The Spiral Organ

The **spiral organ** is the specialized sensory epithelium of the cochlea (Fig. 10.2) on the **basilar part** of the **spiral membrane** in the cochlear duct (Fig. 10.2). Composed of **outer and inner hair cells**, the **tectorial membrane, supporting cells**, and **nerves**, the **spiral organ** (Figs 10.2, 10.3) varies in structure throughout its length – a reflection of its functional specialization. Supporting cells stabilize the cochlear hair cells and are likely absorptive. Nerves in the spiral organ consist of a group of afferents (peripheral processes of bipolar neurons in the spiral ganglia – Fig. 10.2) that conduct impulses from the spiral organ to auditory nuclei in the brain stem, and a smaller group of efferents carrying impulses from the brain stem back to the spiral organ.

Hair Cells of the Spiral Organ

Outer and inner hair cells (Figs 10.3, 10.4) are set at an angle to each other and named by their position relative to the modiolus. These non-neural hair cells, modified for the reception of auditory stimuli, extend the length of the cochlea. Each **spiral organ** (Fig. 10.3) in humans has five rows of tall columnar **outer hair cells** (an average of 12,350 based on three studies, with extreme values of 10,945 and 16,000) and a single row of bulkier, flask-shaped **inner hair cells** (an average of 3280 based on three studies with extreme values of 2695 and 4400) (Fig. 10.3).

Inner and outer hair cells of the cochlea differ in structure and in their innervation. As the name implies,

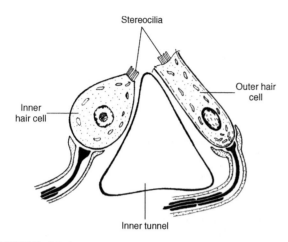

FIGURE 10.4. Inner and outer hair cells of the spiral organ aligned on either side of the inner tunnel. For orientation, see Fig. 10.3. Efferent fibers (dotted) and afferent fibers (black) are shown synapsing with the base of both types of cochlear hair cells (after Engström and Ades, 1973; Firbas, 1978).

their apical surfaces have numerous, stiff, **stereocilia** (Fig. 10.4). In mammalian development, each cell has a single **cilium**; the remaining cilia are **stereocilia**. The latter are cylindrical, 0.2–0.8 μm in diameter, and composed of parallel actin microfilaments in longitudinal bundles. In adults, the cilium disappears but its basal body remains. Stereocilia of inner hair cells are longer, fewer, and arranged linearly; stereocilia of outer hair cells are numerous, 2.5–7.2 μm in length, and arranged in 'W' or 'V' patterns. Stereocilia in humans show a gradation in height, being short at the cochlear base (4.2 μm) but tall at its apex (7.2 μm).

The Tectorial Membrane

Stereocilia of the tallest outer hair cells in adults are firmly attached to the undersurface of the **tectorial membrane** – a gelatinous plate of varying thickness (Fig. 10.3) formed by an extracellular matrix of fine fibrils. These stereocilia insert into shallow pits called **imprints** on the tectorial undersurface. The tallest row of outer hair cell stereocilia touches the tectorial membrane. In humans, imprints on the tectorial undersurface form two or three rows of small holes distributed throughout all turns of the cochlea.

The Spiral Ganglia

In the modiolus (the central cone of trabecular bone supporting the cochlea) are cell bodies of bipolar neurons, whose fibers form the **cochlear nerve**. Collectively these bipolar neurons (29–38,000 in each human cochlea) form the **spiral ganglion** (Fig. 10.2). **Peripheral processes** of

the spiral ganglionic neurons take an intricate course passing in the modiolus (Fig. 10.2) to synapse at the base of the cochlear hair cells along the spiral membrane (Fig. 10.3). Inner hair cells receive large myelinated fibers, one fiber to each hair cell. Outer hair cells receive small, nonmyelinated fibers that branch to innervate as many as 10 outer hair cells.

10.2. THE ASCENDING AUDITORY PATH

10.2.1. Modality and Receptors

The modality under consideration is **hearing**. The hearing of sounds requires their reception by specialized mechanoreceptors (the cochlear hair cells), transformation into impulses that travel in the cochlear nerve and conduction to the cochlear nuclei, through the brain stem and on to the thalamus. At thalamic levels, sound enters consciousness. Fibers from the auditory thalamus then relay auditory impulses to the cerebral cortex, in particular, to the **primary auditory area** in the temporal lobe; here sounds are associated, interpreted, and become meaningful.

Reception of Auditory Stimuli

A complex sequence of events enables cochlear hair cells to receive auditory stimuli. As a first step, sound, in the form of small changes in air pressure, strikes the **tympanic membrane** (Fig. 10.1) causing it to vibrate in a complex manner depending on the frequency of the sound. Because the manubrium of the malleus attaches to the tympanic membrane, vibrations of this membrane set the **auditory ossicles** (Fig. 10.1) in motion. Vibrations in the ossicular chain reach the base of the stapes covering the **vestibular window** (Fig. 10.1). A piston-like motion of the stapes in the vestibular window transmits the ossicular vibrations to the perilymph behind the window. Waves in the perilymph quickly travel through the scala vestibuli (Fig. 10.2) to the helicotrema and through the scala tympani (Fig. 10.2) to the cochlear window (Fig. 10.1). These traveling waves in the perilymph influence **vestibular** and **basilar membranes** and the contents of the **cochlear duct** (Fig. 10.2), including the endolymph in it and the **spiral organ** on the **basilar membrane**.

Transduction of Auditory Stimuli

Vibrations of the spiral membrane, relative to the tectorial membrane, deform the subtectorial space and cause endolymphatic fluid movement, bending or displacing stereocilia at the apices of the **cochlear hair cells** (Figs 10.3, 10.4). This displacement excites and then depolarizes these receptors releasing a neurotransmitter at the receptor base that activates the **peripheral ends of the cochlear fibers** (Fig. 10.4). Hence, the cochlear hair cells act as transducers, converting afferent mechanoacoustic energy into a chemoelectric form that elicits action potentials along the cochlear nerve. Though the identity of the neurotransmitter concentrated in presynaptic cochlear terminals is uncertain, both glutamate and aspartate are likely candidates.

Tuning of the Basilar Membrane

Each basilar membrane is nearly a hundred times stiffer at the cochlear base than at its apex, with fibers that span its width like the strings on a musical instrument. The width of the basilar membrane increases progressively from the cochlear base to its apex, but its thickness, and therefore stiffness, decreases from base to apex. Hair cells are concentrated at the cochlear base. These structural differences reflect a functional specialization along the cochlea; sounds of differing frequencies influence different points along the cochlea. High frequencies have the greatest effect at the basal end and low frequencies apically along the cochlea in a precise order. Thus, a high degree of 'tuning' exists in each part of the basilar membrane along the cochlear length.

Conduction of Sound

Sound often travels to the cochlear labyrinth through different routes: the usual route is through vibrations of the tympanic membrane and auditory ossicles that cause waves in the cochlear fluids. If the ossicles are absent, however, sound must traverse the middle ear by **air conduction**. If the stapes remains fixed in the vestibular window, as happens in otosclerosis, there is dampening of waves in the cochlea, decreasing an already inefficient means of conduction. Perception of a tuning fork applied directly to the mastoid process (Schwabach test) bypasses mechanical conduction in the middle ear, and causes structures in the cochlear duct to vibrate. These events are an example of **bone conduction** that takes place by vibration in bones of the skull.

10.2.2. Primary Neurons

Primary neurons in the auditory path (Fig. 10.5) are bipolar neurons derived from neural crest that form the **spiral ganglia**. Bipolar cell types in humans consist of large type I cells, forming about 94% of spiral ganglionic cells, and some smaller, type II cells, forming the remaining 6%. A maximum of 2% of all spiral ganglionic cells, both type I and type II, are myelinated in humans. Peripheral processes of these primary neurons synapse with the base of the cochlear hair cells. About 78–85% of afferent fibers in the cochlear nerve innervate inner hair cells; the remaining 15–22% supply outer hair cells. Different types of endings occur in humans at the base of the outer hair cells. A single terminal often possesses two types of synaptic specializations of opposite polarity; the neurons of origin for such terminals are uncertain. The existence of these reciprocal synapses raises the possibility of a local feedback influencing synaptic activity.

Central Processes of Primary Neurons –
the Cochlear Nerve

Central processes of the primary neurons in a spiral ganglia form a **cochlear nerve** (Fig. 10.5). These nerve fibers are often indistinguishable until they reach the brain stem where the cochlear nerve is caudal to the vestibular nerve. About 50,000 myelinated fibers are in each cochlear nerve as it skirts the pons entering (at a slightly obtuse angle) at the junction of the pons, medulla oblongata, and cerebellum (a three-dimensional recess), along the dorsal surface and lateral border of the **inferior cerebellar peduncle** (Fig. 10.5). Fibers in the cochlear nerve of nonprimates spiral as they pass to the brain stem in a direction corresponding to that of the cochlear spiral. Entering cochlear fibers bifurcate into ascending and descending branches and disperse in an orderly manner to the **dorsal** and **ventral cochlear nuclei** (Fig. 10.5).

10.2.3. Secondary Neurons

Secondary neurons in the auditory path are the **dorsal and ventral cochlear nuclei** (Fig. 10.5). These secondary neurons number about 96,000 in humans; the dorsal cochlear nucleus has a volume half or a third that of the ventral cochlear nucleus (ratio of 1:2.8). The **ventral cochlear nucleus** in humans is compact, 4.0–4.5 mm in rostrocaudal length and divisible into superior and inferior parts; the **dorsal cochlear nucleus** is a single, lunate structure, 2.5–3.0 mm in rostrocaudal width and superficial to the superior part of the inferior cerebellar peduncle. The majority of the dorsal cochlear nucleus is in the lateral recess of the fourth ventricle but the ventral cochlear nucleus is outside the lateral recess. The inferior

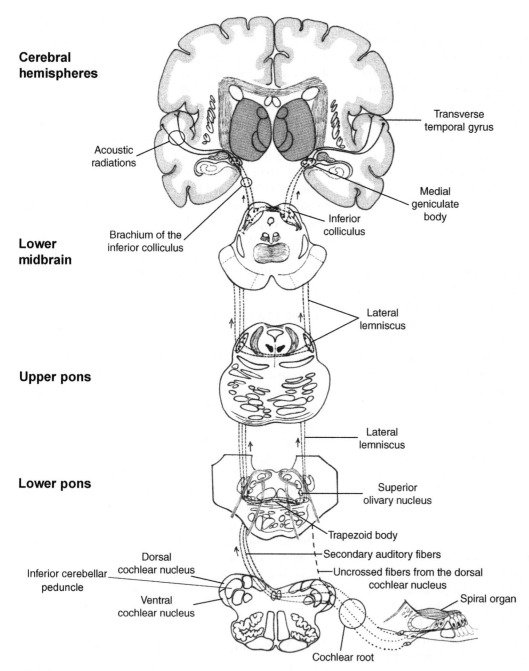

Cerebral hemispheres

Acoustic radiations

Transverse temporal gyrus

Medial geniculate body

Lower midbrain

Brachium of the inferior colliculus

Inferior colliculus

Lateral lemniscus

Upper pons

Lateral lemniscus

Lower pons

Superior olivary nucleus

Trapezoid body

Secondary auditory fibers

Dorsal cochlear nucleus

Uncrossed fibers from the dorsal cochlear nucleus

Inferior cerebellar peduncle

Spiral organ

Ventral cochlear nucleus

Cochlear root

FIGURE 10.5. Auditory path as it ascends from receptors in the spiral organ through the brain stem and diencephalon and onto the primary auditory cortex on the transverse temporal gyri of the temporal lobe.

medullary velum attaches to the surface of the cochlear nuclear complex. Both parts of the cochlear nuclei are at the pontomedullary junction lateral to the brain stem entry of the vestibulocochlear nerve [VIII]. The cochlear nuclei in humans are composed of seven neuronal types arranged in an orderly manner. Large, round or ovoid neurons of particular importance in hearing are the most numerous of the seven neuronal types.

Most cochlear fibers enter the ventral cochlear nucleus, bifurcate, and then terminate in that nucleus giving off a few terminals that reach the dorsal cochlear

nucleus. A few cochlear fibers bypass the cochlear nuclei to end in other brain stem nuclei. The neurotransmitter between the cochlear afferents and secondary cochlear neurons is probably glutamate.

Processes of Secondary Neurons – the Acoustic Stria and Lateral Lemniscus

Processes of secondary neurons form three bundles, collectively called **secondary auditory fibers** (Fig. 10.5). A ventral bundle ascends and spreads through the medial lemniscus to reach the opposite side of the brain stem, contributing to the **trapezoid body**, a trapezoid-shaped area of intermingled secondary auditory fibers, neurons, and fibers of the medial lemniscus (Fig. 10.5) in the lower pons. A small intermediate bundle in humans crosses the inferior cerebellar peduncle and the median plane. Fibers from the dorsal cochlear nucleus form a dorsal bundle that decussates near the ventricular floor. These bundles of secondary auditory fibers, (the **ventral, intermediate,** and **dorsal acoustic striae**), reach the opposite side of the brain stem, turn and ascend as the **lateral lemniscus (LL)** (Fig. 10.5).

10.2.4. Tertiary Neurons

Though some secondary auditory fibers decussate and turn to ascend as the lateral lemniscus (Fig. 10.5), others synapse with several groups of tertiary neurons such as the **superior olivary nucleus, nuclei of the lateral lemnisci** (scattered medially and laterally in them), and **nuclei of the trapezoid body** (intermingled with it). The **superior olivary nucleus** (Fig. 10.5) in humans is in the ventrolateral part of the lower pontine tegmentum and ventral to the facial nucleus. It is divisible into two segments – a prominent and vertically oriented **medial superior olive (MSO)** and a more circular **lateral superior olive (LSO)** found dorsal and lateral to the tip of the MSO (Fig. 9.6). The superior olivary nucleus, a tertiary group in the auditory path, is a relay and a reflex nucleus connected with the facial and trigeminal motor nuclei. Through these connections, loud sounds activate the stapedius and tensor tympani muscles of the middle ear as a part of the acoustic reflex. Each of the tertiary nuclei in the auditory path likely contributes fibers to the lateral lemniscus.

The Ascending Auditory Path – the Lateral Lemniscus

The human ascending auditory path is the **lateral lemniscus** (Fig. 10.5). The lateral lemniscus has fibers from secondary and tertiary neurons, crossed and uncrossed fibers from the dorsal cochlear nucleus, but predominately crossed fibers from the ventral cochlear nuclei. It also has fibers from both superior olivary nuclei, the nuclei of the trapezoid body, and from nuclei of the lateral lemniscus on both sides of the brain stem. The left lateral lemniscus has about 200,000 fibers while the right has about 184,000 (a 9% difference between the sides). Each is fully myelinated at birth; infants of one

month demonstrate responsiveness to, and discrimination of, speech sounds.

Stimulation of the Human Auditory Path

The auditory path in humans is sharply defined with electrical stimulation. Such stimulation of this path in the upper brain stem causes a smooth, synchronous, low-pitched note (buzzing like a bee or humming), sounds with a musical quality (ringing like a bell), a thin, tinny sound (ticking, sizzling, swishing, clicking, or cricket sounds), or a deep note (a roaring or rumbling sound). There is a contralateral preponderance of responses in humans. Presumably, the ipsilateral component is too small to be successfully stimulated.

10.2.5. Inferior Collicular Neurons

Quaternary (fourth-order) neurons in the ascending auditory path are the **nuclei of the inferior colliculi** (Fig. 10.5) in the lower midbrain. The inferior colliculi are multilaminar and richly interconnected through the **commissure of the inferior colliculus**. Each lateral lemniscus distributes fibers to the ipsilateral inferior colliculus. The major relay from these nuclei is by means of the **brachium of the inferior colliculus** (Fig. 10.5) to the thalamus. The inferior colliculus is an essential link in the ascending auditory path. Stimulation of it or its brachium in humans causes a 'loud noise' in the ipsilateral ear and a similar but fainter sound in the contralateral ear.

10.2.6. Thalamic Neurons

Quinary (fifth-order) neurons in the ascending auditory path are in the **medial geniculate nuclei (MG)** of the thalamus (Fig. 10.5). Externally these nuclei form the medial geniculate body, a small eminence about 5 mm wide, 4 mm deep, and 4–5 mm long, protruding from the posterior aspect of the diencephalon. Each human medial geniculate body has **dorsal** (MG_d), **medial** (internal or magnocellular) (MG_{mc}), and **ventral** (MG_v) parts. The ventral (principal or parvocellular) part is the largest subdivision occupying the ventrolateral quarter, the medial part occupies the ventromedial quarter, and the dorsal part forms a tier extending the length and breadth of the medial geniculate body. Major ascending projections of auditory neurons in the inferior colliculi reach the ventral part of the medial geniculate nucleus (MG_v). Fibers from each inferior colliculus enter the **brachium of the inferior colliculus** to reach their geniculate termination. A few fibers of the lateral lemniscus bypass the inferior colliculi to synapse with medial geniculate neurons. In the elderly, there is a reduction in the number and size of the fibers of the brachium of the inferior colliculus.

Tones of Different Pitch Reach Consciousness at Thalamic Levels

Tones of different pitch reach consciousness at the geniculate level in humans. Studies in primates indicate that the discrimination of sound localization, loudness, and pitch do not require involvement of the **primary auditory cortex** (corresponding to **Brodmann's area 41**) in the temporal lobe. Ablation of the primary auditory and periauditory cortical areas in humans does not influence the discrimination of tone sequences. Stimulation of the human medial geniculate nucleus (MG) causes a ringing, referred to the center of the head, or to a buzzing heard bilaterally but mainly in the contralateral ear.

10.2.7. Cortical Neurons

Primary Auditory Cortex (AI)

Neurons in each medial geniculate nucleus (MG) send axons to layers IIIb and IV of the **primary auditory cortex** or **AI** in the **anterior transverse temporal gyrus** (Fig. 10.6) corresponding to **Brodmann's area 41**. The anterior transverse and the posterior transverse temporal gyri are two parallel gyri that are part of the **superior temporal gyrus** as it extends into the lateral sulcus. By removing the overlying opercular parts of the frontal and parietal lobe or widely opening the lateral sulcus, the transversely oriented temporal gyri are visible. Substance P, a neuropeptide, is immunocytochemically localizable to small, round neurons in layers IV and V of the human auditory cortex. Length and surface area of the left primary auditory cortex is greater than the right in newborns and adults. In addition to the left-right asymmetry, this region also demonstrates an asymmetry of its cytoarchitectonic areas. Orderly projections from neurons of the medial geniculate nucleus (MG) to the primary auditory cortex, known as the **acoustic radiations** (Fig. 10.5), reach the temporal lobe through the **sublenticular part of the internal capsule**.

Extent of the Primary Auditory Cortex

Each human primary auditory cortex extends from the transverse temporal gyri laterally and onto the parietal operculum. In support of this is the observation that

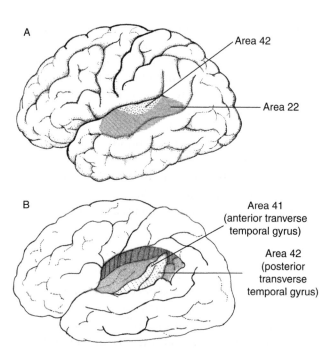

FIGURE 10.6. **A**, Part of primary auditory area 42 extending from the posterior transverse temporal gyrus is lightly dotted. Also illustrated is part of primary auditory association area 22 (horizontal lines). **B**, Removal of parts of the frontal and parietal opercula expose the primary auditory cortex often designated as AI and corresponding to Brodmann's areas 41 and 42 (lightly dotted).

patients with injuries limited to the parietal operculum are likely to be unable to hear loud and unpleasant noises. Evoked recordings are elicitable from the lateral surface of the human cerebral cortex, including the posterior two-thirds of the superior temporal gyrus and the frontal and parietal opercula. The posterior two-thirds of the superior temporal gyrus is that part of area 42 extending from the transverse temporal gyri to the lateral surface for a short distance (Fig. 10.6). Most of the superior temporal gyrus on the lateral surface of the brain, however, corresponds to Brodmann's **area 22** (Fig. 10.6) (and perhaps a small part of **area 52**). **Wernicke's region** corresponds with areas 42 and 22 of Brodmann and is an auditory association area.

Connections of the Primary Auditory Cortex

Auditory impulses reach the **primary auditory cortex**, corresponding to Brodmann's **area 41**. They also reach adjacent areas 42, area 22, and perhaps area 52. In addition to these connections, auditory callosal fibers are identifiable in the posterior third of the trunk of the corpus callosum by computerized tomography. These fibers were identifiable in a patient with disturbed auditory function and a discrete injury in that area.

Auditory Association Cortex

More than a dozen auditory association areas are identifiable in the nonhuman primate brain. One of these is on the posterior transverse temporal gyrus and on the opercular surface of the temporal lobe within the lateral sulcus corresponding to **area 42** of Brodmann. The lateral aspect of the superior temporal gyrus, corresponding to **area 22** (Fig. 10.6) and perhaps **area 52** of Brodmann on the opercular surface of the temporal lobe between area 41 and the insula are also regarded as auditory association areas. Localization, recognition, discrimination, and interpretation of sounds, as well as the understanding of language, takes place only at cortical levels. Discrimination of speech requires analysis of a complex spectrum of sounds that vary in sequence and extend over a large range of frequencies. The interpretation of sounds often requires the comparison of current sounds to previous auditory experiences. Though the primary auditory cortex of one hemisphere distinguishes a sound, the auditory cortices in both hemispheres are essential for its localization. Such localization requires comparison between the right and left auditory cortices of the direction, speed of transmission, and loudness of the sound. Thus, the exquisite auditory appreciation of sounds and language in humans requires not only the primary auditory cortex but also auditory association areas some of which are unimodal but others of which are multimodal. In addition to this complex serial processing, the combined and coordinated functioning of auditory areas in both hemispheres is necessary for the appreciation of sounds and language in humans.

Stimulation of the Auditory Association Cortex

Stimulation of **auditory association cortex** corresponding to Brodmann's areas 42, 22 and perhaps 52 leads to the subjective sensation of simple sounds called **audenes**; humans report a continuous buzzing, humming, rumbling, ringing, or hissing sound or they may report knocking, 'cricket-like,' or wavering sounds. They often complain of deafness or a change in the character of sounds at the time of stimulation. Such unusual sounds or slight deafness is often referable to the contralateral ear. Deafness caused by stimulation of the primary auditory cortex is perhaps analogous to numbness (not tingling) caused by stimulation of the primary somatosensory cortex, or shadows (not lights), when the primary visual cortex is stimulated.

Sounds heard during cortical stimulation belong to one part or another of the frequency spectrum. They have a degree of pitch, are not pure tones, and are lacking in fine pitch discrimination. Such observations emphasize the functional specificity of the primary auditory cortex or AI compared with the auditory association cortex. Simple sounds resulting from stimulation reflect capabilities of the primary auditory cortex. Stimulation at a distance from the primary auditory cortex is likely to introduce elements of the interpretation of sound. The fine discrimination of sounds, their interpretation, and their meaning requires involvement of auditory association areas. Stimulation in the anterior half of the right area 22 yields musical hallucinations; injury to this region often leads to a loss of ability to appreciate music.

'What' and 'Where' Processing in the Auditory Cortex

At the cortical level, the somatosensory and visual systems in primates are each organized into 'what' and 'where' paths (Table 8.2). This 'what' and 'where' model includes a ventral ('what') stream for the perception, identification, and recognition of objects and a dorsal ('where') stream for locating objects and appreciating the spatial relations among objects. At the cortical level the auditory system in nonhuman primates is also thought to be organized into ventral 'what' and dorsal 'where' paths. Functional imaging studies have confirmed this 'what' and 'where' functional organization for the human auditory system as well. Auditory information travels first to the primary auditory cortex and is then relayed in serial fashion to the auditory association cortex and then to other cortical association areas. In the case of the auditory system, the **dorsal 'where' stream** related to spatial aspects of audition becomes active significantly earlier than the ventral stream. This dorsal stream in humans involves the middle temporal gyrus and the superior temporal gyrus including its posterior third corresponding to Wernicke's area. Impulses then project to the supramarginal gyrus, corresponding to Brodmann's area 40 in the inferior parietal lobule, and via the arcuate fasciculus reach Broca's region for motor speech in the inferior frontal gyrus. The **ventral 'what' stream**, related to the recognition of familiar sounds and intelligible speech (but only in the left hemisphere for intelligible speech), projects to the anterior superior temporal gyrus and then on to the posterior planum polare. In addition, there are two streams projecting to functionally distinct areas of the frontal lobe in nonhuman primates related to auditory spatial and non-spatial processing, respectively. Although the auditory system is a bilateral system with two sets of ears and a bilateral path in the brain stem and beyond to the thalamus and cortex, only one ear is necessary in order to identify a sound but both ears are necessary to locate that sound in three-dimensional space. Because the human auditory system is involved in complex sounds related to language as well as simple to complex sounds, it seems likely that there are additional auditory processing streams beyond those described above.

Tonotopic Organization in AI

In the primary auditory cortex (AI), tones of different frequencies have their representation along specific parts of AI – a phenomenon called **tonotopic organization**. Positron emission tomography (PET) has demonstrated tonotopic organization as a feature of the human primary auditory cortex. Observations in humans demonstrate a point-to-point projection from sites along the cochlea to sites along the primary auditory cortex. Thus, AI has a complete and orderly representation of the audible frequency spectrum, with low tones represented rostrolaterally and higher frequencies caudomedially along its length.

Bilateral Nature of the Human Auditory Path

The human auditory path is a bilateral system with the lateral lemnisci of the two sides richly interconnected. Auditory information on one side of the brain reaches the other side through various commissural paths including the trapezoid body, commissure of the inferior colliculus, and corpus callosum. The only decussation of auditory information rostral to the inferior colliculi takes place through interconnections between the right and left auditory cortices. Because of this communication, each human cochlea has representation in the primary auditory cortex of both cerebral hemispheres. Stimulation studies, however, indicate a predominance of contralateral cochlear representation in humans. Normal volunteers subjected to the presentation of auditory stimuli (a tape-recorded story) through earphones to only one ear (three right and three left) displayed a focal increase in glucose metabolism in the temporal cortex contralateral to the stimulated ear that could be measured by a metabolic mapping method using the [18F]flurodeoxyglucose technique and positron emission transaxial tomographs. These observations and the results of stimulation of the auditory path in humans in the midbrain lead to the conclusion that, though this path is bilateral in structure, it is physiologically contralateral based on current physiological measuring methods.

Processing Self-Produced Vocalizations

The auditory cortex in each hemisphere likely receives information during self-produced vocalization that differs from that caused by the vocalizations of others. Cerebral areas activated by phonation may inhibit parts of the auditory cortex. These data suggest much processing in cortical areas responsible for auditory communications.

10.2.8. Comments

Presentation of a pure tone to one ear causes maximal stimulation of a single locus along the cochlea as a function of frequency of stimulus or pitch. Each spiral membrane in the basal cochlea vibrates primarily at high frequencies; its apical part vibrates predominately at low frequencies. Individual cochlear fibers innervate a restricted area of the cochlea and accumulate in an orderly manner in the cochlear nerve such that fibers serving tones of lower pitch (apical cochlear fibers) are near the center of the cochlear nerve; at its periphery are fibers carrying tones of higher pitch from basal cochlear regions. A general principle of neural organization is that peripheral receptors have an orderly representation at each step along an ascending path. The well-organized cochlea, 'represented' several times along the auditory path, is ultimately impressed on the primary auditory cortex (AI) in a tonotopic manner.

In humans, there is likely an orderly pattern of synapses of cochlear fibers with cochlear nuclei, cochlear nuclear fibers with inferior collicular neurons, and inferior collicular fibers with medial geniculate neurons. The pattern of termination of collicular fibers within the medial geniculate nucleus (MG) corresponds to an unrolled cochlea. Lastly, there is a topographic pattern of termination of medial geniculate fibers along the **primary auditory cortex** in and around the fourth cortical layer.

The lateral lemnisci contribute fibers to the reticular formation and to motor nuclei in the brain stem. Such collaterals likely participate in arousal, providing a structural basis for the way in which an alarm clock acts.

10.3. DESCENDING AUDITORY CONNECTIONS

Fibers from auditory areas in the cerebral cortex descend in the medial part of the lateral lemniscus to synapse with auditory structures in the brain stem. Therefore, the major auditory path has both ascending and descending fibers. Nuclei of the medial geniculate body, inferior colliculi, superior olive, and the ventral cochlear nuclei form links in this multisynaptic descending corticofugal path to lower auditory regions. Efferent fibers from these regions reach the base of the cochlear hair cells. There is a well-defined projection to the cochlea, called the **olivocochlear bundle**, containing some 600 efferent fibers. Large efferent cochlear terminals end on neuronal cell bodies of outer hair cells in the human spiral organ. Most of these endings at the base of the outer hair cells are efferent with the majority in the basal turn; their number gradually decreases near the apex.

In the human cochlea, GABA-like immunostaining is present in inner spiral fibers, tunnel spiral fibers, tunnel-crossing fibers, and at efferent endings synapsing with outer hair cells. In squirrel monkeys, efferent cochlear cell bodies of origin are identifiable by their immunoreactivity to antisera for choline acetyltransferase (ChAT). These neurons are in the hilus of the lateral superior olive, along the lateral border of the superior olivary complex with a few in the superior part of the ventral trapezoid nucleus. Some cochlear efferent neurons and fibers in this species are immunoreactive for leucine enkephalin (L-ENK). Since acetylcholinesterase activity is localizable to efferent fibers of the human cochlea, the probable efferent cochlear neurotransmitter is acetylcholine. Efferents of cerebellar origin end in the cochlear nuclei of nonhuman primates.

10.3.1. Electrical Stimulation of Cochlear Efferents

Electrical stimulation of cochlear efferents in nonprimates inhibits activity in cochlear hair cells and cochlear afferents. In normal speech, these efferents likely prevent humans from being deafened by their own voice. These efferents probably participate in screening out extraneous sounds in a noisy environment (background competition), allowing one to 'tune in' to sounds one wants to hear.

Deafness during stimulation of the primary auditory cortex presumably results from excitation of descending fibers to auditory structures at brain stem levels. Patients undergoing stimulation of auditory areas in the cerebral cortex often report that their own voice sounds strange to them. As soon as stimulation ceases, this effect stops.

10.3.2. Autonomic Fibers to the Cochlea

Autonomic fibers to the cochlea likely travel in the nervus intermedius of the facial nerve [VII]. Fibers that synapse with both nonmyelinated and small myelinated cochlear ganglionic neurons are identifiable in humans, perhaps lending support to the concept of parasympathetic fibers and neurons in human cochleae. Such fibers may be vasomotor to the stria vascularis, a collection of small blood vessels and capillaries in the upper part of the bony spiral lamina; they doubtless are part of the nonautonomic efferent auditory connections. Efferent fibers of the olivocochlear bundle likely influence small spiral ganglionic neurons.

10.4. INJURY TO THE AUDITORY PATH

According to the National Institute on Deafness and Other Communication Disorders (NIDCD), approximately 28 million Americans have a hearing impairment. Hearing loss affects approximately 17 in 1000 children under age 18. The incidence of hearing loss increases with age such that approximately 314 in 1000 people over age 65 have hearing loss and 40 to 50 percent of people 75 and older have a hearing loss. About two to three out of every 1000 children in the United States are born deaf or hard-of-hearing. Ten million Americans have suffered irreversible noise induced hearing loss, and 30 million more are exposed to dangerous noise levels each day. Hearing loss can be hereditary, or it can result from disease, trauma, or long-term exposure to damaging noise or medications.

10.4.1. Congenital Loss of Hearing

Impairment of hearing at birth is due to genetic and nongenetic causes. Genetic abnormalities likely lead to developmental defects of any part of the external, middle, or internal ear. Teratogens or acquired intrauterine 'insults' such as prenatal or perinatal maternal viral infections are examples of nongenetic abnormalities. In any event, whether congenital or acquired, the result is often impairment of hearing. Congenital loss of hearing is either conductive or sensorineural.

Congenital Conductive Loss of Hearing

Congenital conductive loss of hearing, also called obstructive or transmissive loss, indicates a loss of air conduction in the external or middle ear. Absence or pathologic closure of the external meatus, called **external meatus atresia**, often exists alone or combined with abnormalities in the middle ear, such as fused, deformed, or absent auditory ossicles. Other defects involving the ossicles, muscles, windows, and vascular structures in the middle ear may cause congenital conductive loss of hearing. The treatment of conductive hearing loss depends on the cause. A bacterial infection of the middle ear is treatable with antibiotics and it is often possible remove blockages of the external and middle ears. If damage occurs to the tympanic membrane it can often be surgically repaired. If damage occurs to the auditory ossicles by otosclerosis, they are replaceable with artificial bones.

Congenital Sensorineural Loss of Hearing

Congenital sensorineural hearing loss is either **sensory** (caused by injury of cochlear hair cells), **neural** (because of abnormalities of the cochlear nerve or nuclei, or of central auditory paths), or both. Cochlear nerves may be absent, atrophic, or compressed by blood vessels; the cochlea may have fewer turns than normal, its modiolus may be hypoplastic, or the spiral organ may be underdeveloped. Sensory hair cell loss or atrophy is a likely cause of sensorineural hearing loss as are abnormalities of the stria vascularis or vestibular membrane. The most common cause of severe to profound sensorineural deafness is **cochleosaccular aplasia**. Structural changes are usually confined to the cochlea and saccule, with atrophy of the stria vascularis (causing impaired production of endolymph), degeneration of the spiral organ, and retraction or rolling up of the tectorial membrane. The maculae sacculi and otoconia are usually degenerated, with collapse of the saccular wall. Neural and spiral ganglionic atrophy occurs, secondary to loss of cochlear hair cells and supporting cells. Some causes of sensorineural hearing loss are treatable. For example, an acoustic neuroma that impinges on the cochlear part of the vestibulocochlear nerve [VIII] is often removable. Hearing aids may be useful in improving the hearing and speech comprehension of people with sensorineural hearing loss.

10.4.2. Decoupling of Stereocilia

Since stereocilia of outer hair cells are firmly in contact with the tectorial membrane, any abnormalities of stereocilia, including stiffness, is likely to lead to a disruption or **decoupling** of cochlear hair cells from the tectorial membrane. Decoupling probably leads to decreased transmission of auditory signals, causing

loss of auditory sensitivity, recruitment (an exclusive sign of sensory hair cell dysfunction), poor speech discrimination, and **tinnitus** – a common complaint associated with cochlear disorders.

10.4.3. Tinnitus

According to the National Institute on Deafness and Other Communication Disorders (NIDCD), approximately 12 million Americans experience tinnitus, the conscious perception of sounds that appear to originate inside the head. Of these, at least one million experience it so severely that it interferes with their daily activities. Tinnitus affects almost 12% of men who are 65–74 years of age. Noises (ringing, buzzing, roaring, or whistling), tones (especially high-pitched tones), or tones superimposed on noises are examples of tinnitus. Such sounds are likely to be steady or pulsating, change daily in quality, yet are severe and debilitating in 20% of those afflicted (6.4% of the adult population). Usually the loudness is small (less than 10 dB SL), but some report tinnitus as loud as a jet aircraft taking off. Efforts to measure the loudness of tinnitus by a nontraditional method have equated tinnitus with a sound as loud as a freight train passing 30.5 m away. The debilitating nature of tinnitus is probably the result of its constant and unremitting presence, not its loudness. Tinnitus, in this special sensory system, may be analogous with chronic pain in the general sensory system.

Though tinnitus is usually a symptom accompanying loss of hearing (in about 75% of those who are deaf), in a study of 200 patients with tinnitus 30% had normal hearing. Partial decoupling of cochlear hair cells from the tectorial membrane probably increases the noise level at the sensory hair cell, giving rise to tinnitus of cochlear origin. Since individual cochlear hair cells are 'tuned' according to their position on the spiral membrane, the nature of the perceived noise in tinnitus of cochlear origin derives from the location and extent of the decoupled stereocilia. With decoupling of hair cells in the apical cochlea, a 'roaring' tinnitus is often experienced. Other sites of decoupling lead to a 'hissing' or perhaps a 'whooshing' sensation, particularly when speech frequencies are involved and speech frequency is affected. Vascular compression of the cochlear nerve may also lead to tinnitus.

10.4.4. Noise-Induced Loss of Hearing

Studies in primates of noise-induced loss of hearing show a correlation between the location of lost outer hair cells along the cochlea and the frequency at which hearing is impaired. Temporary or permanent loss of hearing, particularly for high tones, common in patients with infections of the middle ear, is probably the result of injury to hair cells of the lower basal cochlear turn. In about a third of individuals with injuries to the head, there is loss of hearing. Such loss can result from excessive pressure waves that travel along bones of the skull to the internal ear fluids causing injury to the delicate cochlear hair cells.

10.4.5. Aging and the Loss of Hearing

With aging parts of the human internal ear may degenerate leading to a loss of outer and inner hair cells, and loss of nerve fibers, especially in the basal coil of the cochlea. Aging may also lead to fused or modified stereocilia of cochlear hair cells, and accumulations of pigment in the upper part of supporting cells in the spiral organ and in cells of the stria vascularis. As degeneration continues, a gradual and progressive impairment of hearing is likely to follow. The pattern of degeneration in this case differs from that in the basal cochlear turn in those exposed to industrial noise or gunfire. In those over 50, there is a decrease in number of spiral ganglionic neurons and a decrease in volume of the ventral cochlear nuclei. There is often no cure for age-related hearing loss. In such cases, a hearing aid for one or both ears may prove helpful. Many different types of hearing aids are available.

10.4.6. Unilateral Loss of Hearing

The auditory path in the brain stem is bilateral, interconnected by way of fibers of the trapezoid body and the commissure of the inferior colliculus. Unilateral brain stem injury, involving one lateral lemniscus, is not often clinically detectable. Peripheral injury in the cochlea, spiral ganglion, or cochlear nerve may lead to unilateral loss of hearing.

10.4.7. Injury to the Inferior Colliculi

Selective impairment of hearing caused by injury to both inferior colliculi, though rarely reported, is often due to a cyst of the inferior colliculi or to a large pinealoma compressing both inferior colliculi (without involvement of the superior colliculi). A junior medical student complained of difficulty understanding speech, though he heard speech sounds. With conventional audiometry, his hearing was within normal limits. Psychiatric referral and treatment for 'an anxiety

neurosis, agitated depression, and typical medical student hypochondriasis' was unsuccessful. Subsequent Békésy audiometer and speech tests revealed a possible bilateral auditory injury. Re-examination revealed a massive pinealoma compressing the inferior colliculi.

10.4.8. Unilateral Injury to the Medial Geniculate Body or Auditory Cortex

Destruction of only one human medial geniculate body leads to no appreciable loss of hearing, though there is slight loss of acuity in the auditory field on the opposite side. Unilateral injury to the human auditory cortex possibly leads to a loss of ability to localize sources of sounds in the contralateral ear, including the direction of the sound and estimation of its distance. Routine tests with tuning forks, conventional audiometry of pure tones, and discrimination tests of undistorted speech are incapable of detecting subtle auditory deficits accompanied by patterns of oral speech and marked by an excessive rate of production with normal rhythm and melody.

Effects of unilateral injury to the temporal lobe, involving the **primary auditory cortex**, are subtle and not demonstrable by routine audiometric methods. No effect on audiograms for pure tones is evident, and intelligibility of signals of undistorted speech is usually good. With distortion of the verbal signal, however, there is a relative deficit in the ear contralateral to the injury.

10.4.9. Bilateral Injury to the Primary Auditory Cortex

After bilateral injury to the **primary auditory cortex**, nonhuman primates are initially unable to respond to sound but display some recovery for low and very high frequencies. However, they suffer a permanent loss of hearing in the middle frequency range. Though rare, bilateral injuries in humans result in a sudden onset of severe to total deafness, with some partial recovery, and a residual frequency-dependent loss of hearing at middle or high frequencies. Though such patients likely have substantial loss of hearing by routine audiogram examination, they are able to hear environmental sounds of low frequencies. Bilateral injuries profoundly impair the ability to understand speech, including single words or sentences. The size and depth of the injury determines the magnitude of the loss, in monkeys and humans. A patient suffered, on separate occasions, injury to the primary auditory cortex in both hemispheres. The second cerebral injury occurred as the patient slept. When he awoke, he was immediately aware that the world about him was strangely silent. He awoke 30 minutes after his alarm clock should have started ringing. He had not heard the alarm clock, or the sound of his electric razor, the running water, or traffic on the street. He heard nothing when others spoke to him.

10.4.10. Auditory Seizures – Audenes

During seizures involving the auditory cortex, patients likely experience **audenes**, including buzzing or drumming sounds that are similar to the simple sounds caused by cortical stimulation. These simple sounds are referred to the contralateral ear.

10.5. COCHLEAR IMPLANTS

Many devices are available to help improve the hearing of those with sensorineural loss of hearing. These include hearing aids and cochlear implants. Approximately 59,000 people worldwide have received cochlear implants including some 13,000 adults and nearly 10,000 children in the United States. These implants consist of single or multiple stimulating electrodes placed at various points along the auditory path such as in the cochlear window, scala tympani, and lateral wall of the first and second turns of the cochlea, directly into the cochlear nerve or into the cochlear nuclei. Another approach involves implanting electrodes in the primary auditory cortex; the primary advantage of this latter method is its application to patients who have suffered injury to the cochlea, cochlear nerve, or even brain stem auditory paths, provided the primary auditory cortex remained uninjured.

Usually these devices are very useful used in combination with lip-reading and have allowed those with impaired hearing to react to auditory stimuli in the environment. The sounds heard tend to be of a buzzing or electronic nature (similar to the audenes described above). Such devices are likely to help a patient to monitor the intensity of their own voice and make appropriate adjustments in speaking which may make conversation easier. Presently the efficiency of cochlear prostheses is low for restoring useful auditory perception of speech and enhancing speech communication. In the best of circumstances, some individuals are able to achieve not only sound awareness but also good speech comprehension such that they are able to converse on the phone. Cochlear implants can be particularly valuable for hearing impaired children when

implanted around the time language skills are developing fastest (age two or three) or during a sensitive period of 3.5 years in early childhood when the central auditory system is highly plastic.

10.6. AUDITORY BRAIN STEM IMPLANTS

Bilateral surgical removal of both cochlear nerves due to acoustic neuromas, bilateral acoustic neuroma growth that impinges on both cochlear nerves, trauma to both nerves, or atresia of the internal auditory meatus on both sides all result in hearing loss. Such patients cannot benefit from cochlear implants. One option is an auditory brain stem implant that uses a multichannel design and an array of electrodes placed over the surface of the dorsal and ventral cochlear nuclei in the lateral recess of the fourth ventricle. The success of this method depends on the underlying pathological conditions.

The auditory midbrain implant provides stimulation to the human inferior colliculus. This recently developed prosthesis is also suitable for those who cannot benefit from cochlear implants and who have bilateral injuries to their cochlear nerves.

FURTHER READING

Abe H, Rhoton AL Jr (2006) Microsurgical anatomy of the cochlear nuclei. Neurosurgery 58:728–739.

Ahveninen J, Jaaskelainen IP, Raij T, Bonmassar G, Devore S, Hamalainen M, Levanen S, Lin FH, Sams M, Shinn-Cunningham BG, Witzel T, Belliveau JW (2006) Task-modulated 'what' and 'where' pathways in human auditory cortex. Proc Natl Acad Sci 103:14608–14613.

Alain C, Arnott SR, Hevenor S, Graham S, Grady CL (2001) 'What' and 'where' in the human auditory system. Proc Natl Acad Sci 98:12301–12306.

Brugge JF, Volkov IO, Garell PC, Reale RA, Howard MA (2003) Functional connections between auditory cortex on Heschl's gyrus and on the lateral superior temporal gyrus in humans. J Neurophysiol 90:3750–3763.

Celesia GG (1976) Organization of auditory cortical areas in man. Brain 99:403–414.

Colletti V, Carner M, Miorelli V, Guida M, Colletti L, Fiorino F (2005) Auditory brainstem implant (ABI): new frontiers in adults and children. Otolaryngol Head Neck Surg 133:126–138.

Fex J, Altschuler RA (1986) Neurotransmitter-related immunocytochemistry of the organ of Corti. Hear Res 22:249–263.

Firszt JB, Ulmer JL, Gaggl W (2006) Differential representation of speech sounds in the human cerebral hemispheres. Anat Rec A Discov Mol Cell Evol Biol 288:345–357.

Glueckert R, Pfaller K, Kinnefors A, Schrott-Fischer A, Rask-Andersen H (2005) High resolution scanning electron microscopy of the human organ of Corti. A study using freshly fixed surgical specimens. Hear Res 199:40–56.

Glueckert R, Pfaller K, Kinnefors A, Rask-Andersen H, Schrott-Fischer A (2005) Ultrastructure of the normal human organ of Corti. New anatomical findings in surgical specimens. Acta Otolaryngol 125:534–539.

Howard MA, Volkov IO, Mirsky R, Garell PC, Noh MD, Granner M, Damasio H, Steinschneider M, Reale RA, Hind JE, Brugge JF (2000) Auditory cortex on the human posterior superior temporal gyrus. J Comp Neurol 416:79–92.

Hudspeth AJ (2005) How the ear's works work: mechano-electrical transduction and amplification by hair cells. C R Biol 328:155–162.

Kaas JH, Hackett TA (1999) 'What' and 'where' processing in auditory cortex. Nat Neurosci 2:1045–1047.

Khalfa S, Bougeard R, Morand N, Veuillet E, Isnard J, Guenot M, Ryvlin P, Fischer C, Collet L (2001) Evidence of peripheral auditory activity modulation by the auditory cortex in humans. Neuroscience 104:347–358.

Kulesza RJ Jr (2007) Cytoarchitecture of the human superior olivary complex: Medial and lateral superior olive. Hear Res 225: 80–90.

Lenarz T, Lim HH, Reuter G, Patrick JF, Lenarz M (2006) The auditory midbrain implant: a new auditory prosthesis for neural deafness – concept and device description. Otol Neurotol 27:838–843.

Lim HH, Anderson DJ (2006) Auditory cortical responses to electrical stimulation of the inferior colliculus: implications for an auditory midbrain implant. J Neurophysiol 96:975–988.

Moller AR (2006) Physiological basis for cochlear and auditory brainstem implants. Adv Otorhinolaryngol 64:206–223.

Moller AR (2006) History of cochlear implants and auditory brainstem implants. Adv Otorhinolaryngol 64:1–10.

Parker GJM, Luzzi S, Alexander DC, Wheeler-Kingshott CAM, Ciccarelli O, Lambon Ralph MA (2004) Lateralization of ventral and dorsal auditory-language pathways in the human brain. NeuroImage 24:656–666.

Perrot X, Ryvlin P, Isnard J, Guenot M, Catenoix H, Fischer C, Mauguiere F, Collet L (2006) Evidence for corticofugal modulation of peripheral auditory activity in humans. Cereb Cortex 16:941–948.

Pamulova L, Linder B, Rask-Andersen H (2006) Innervation of the apical turn of the human cochlea: a light microscopic and transmission electron microscopic investigation. Otol Neurotol 27:270–275.

Philibert B, Veuillet E, Collet L (1998) Functional asymmetries of crossed and uncrossed medial olivocochlear efferent pathways in humans. Neurosci Lett 253:99–102.

Romanski LM, Tian B, Fritz J, Mishkin M, Goldman-Rakic PS, Rauschecker JP Dual streams of auditory afferents target multiple domains in the primate prefrontal cortex. Nat Neurosci 2:1131–1136.

Schrott-Fischer A, Kammen-Jolly K, Scholtz AW, Gluckert R, Eybalin M (2002) Patterns of GABA-like immunoreactivity in efferent fibers of the human cochlea. Hear Res 174:75–85.

Ungerleider LG, Haxby JV (1994) 'What' and 'where' in the human brain. Curr Opin Neurobiol 4:157–165.

. . . the whole vestibular system is a physiologic unit that finds itself in constant activity, and according to its functional state the afferent impulses set up reflexes of determined pattern, because they find open only a limited number of the extremely numerous anatomic paths.

R. Lorente de Nó (1902–1990)

The Vestibular System

The **vestibular nerve**, one of two parts of the **vestibulocochlear nerve** [VIII] carries impulses related to **vestibular sensation**, the awareness of position or movement of our body caused by the effects of gravity and acceleration or deceleration. The vestibular system is a special somatic afferent, proprioceptive system that maintains our equilibrium and orientation. The descending or **efferent component** of the vestibular nerve influences skeletal muscles and regulates posture and ocular movements.

11.1. GROSS ANATOMY

11.1.1. Internal Ear

Bony Labyrinth

Information on the external and middle ear is available in Chapter 10 (Figs 10.1, 11.1). The **internal ear** (Fig. 11.1) is a continuous series of fluid-filled cavities, the **bony labyrinth**, in the petrous part of the temporal bone.

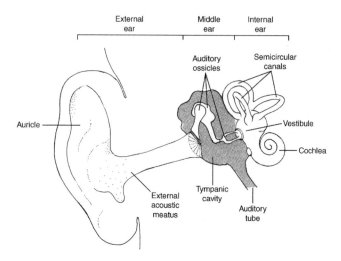

FIGURE 11.1. Osseous labyrinth of the internal ear consisting of the semicircular canals, vestibule, and cochlea. The membranous labyrinth is shown in detail in Fig. 11.2.

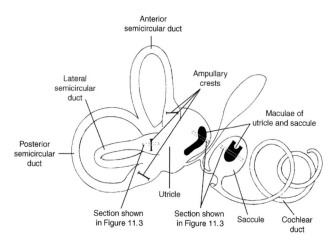

FIGURE 11.2. Vestibular components of the membranous labyrinth (semicircular ducts, utricle and saccule). Locations of the specialized sensory vestibular neuroepithelium (ampullary crests and maculae) are indicated. A section is made of each of the maculae and of one crest. They are shown in detail in Fig. 11.3 (modified from Sobatta, 1957 and Gardner, Gray and O'Rahilly, 1975).

These cavities have a thin periosteal lining with a delicate epithelioid layer and contain **perilymph**, a fluid that resembles extracellular fluid with a high concentration of protein. Perilymph is continuous between the vestibular and auditory parts of the internal ear. The **bony labyrinth** (Fig. 11.1) is composed of an auditory part, the **cochlea,** a vestibular part formed by three **semicircular canals** (anterior, posterior, and lateral) and a **vestibule**. The latter measures some 4 mm in diameter and receives the ends of the semicircular canals. Each canal forms about two-thirds of a circle and each is at right angles to the other two, an arrangement resembling the three adjoining sides at each corner of a cube. On average, semicircular canals are about 1 mm in diameter, but expand on one end to about twice its size as a **bony ampulla.**

Membranous Labyrinth

In the bony labyrinth is a continuous series of epithelial spaces and tubes, collectively called the **membranous labyrinth** (Fig. 11.2), filled with a fluid called **endolymph.** Endolymph has a large DC potential (relative to perilymph), low concentration of proteins and sodium but is high in potassium. Polysaccharides dissolved in the endolymph give it a greater viscosity than water – a matter of functional significance. Endolymph in the **cochlear duct** (an auditory structure) is continuous with endolymph in the membranous vestibular labyrinth. The **semicircular ducts** (**anterior, posterior, lateral**) are membranous parts in the bony semicircular canals (Fig. 11.2). Each semicircular duct has the same arrangement and name as the corresponding canal in

which it lies. The average diameter of a semicircular duct is 0.23 mm.

Membranous Ampullae and the Cristae Ampullaris

One end of each duct expands as a **membranous ampulla**, about 1.55 mm in height, 1.28 mm in width, and 1.94 mm in length. The ampullae are anterior, posterior, and lateral in name and position. Each has a saddle-shaped **ampullary crest** (Fig. 11.2) that has non-neuronal cells specialized for reception of vestibular stimuli.

Utricle and Saccule and their Associated Maculae

The **utricle** and **saccule** are connected parts of the membranous labyrinth in the bony vestibule (Fig. 11.2). The **utricle** in humans is ovoid, slightly flattened (Fig. 11.2) and measures 4.2 mm long and 6.5 mm in circumference; it communicates through five openings with the semicircular ducts and with the saccule which then communicates with the cochlear duct. The **saccule** in humans is smaller than the utricle, ovoid (Fig. 11.2) measuring 2.6 mm in length, 2.2 mm in width and 5.4 mm at its greatest circumference. Each utricle and saccule has a thickened area, the **macula** (Fig. 11.2) that has non-neuronal cells specialized for reception of vestibular stimuli. The **utricular macula** (Fig. 11.2) is anterolateral with the largest collection of vestibular receptors and a surface area approximately twice the saccular macula. The **utricular macula** (Fig. 11.2) is 2.3 mm long and 2.1 mm wide; the **saccular macula** is 2.2 mm long and 1.2 mm wide.

11.2. THE ASCENDING VESTIBULAR PATH

11.2.1. Modalities and Receptors

Vestibular receptors (see Fig. 11.6) are innervated by the **vestibular nerve** – the proprioceptive part of the vestibulocochlear nerve [VIII], serving vestibular sensation and classified as **special somatic afferent**. Human vestibular receptors can be stimulated without involving proprioceptors in muscles, tendons, and joints or in visual receptors. Hence, it is difficult to evaluate the role of vestibular receptors independent of these nonvestibular receptors. Other than **vertigo** (the abnormal sensation of motion of one's self or of external objects often described as 'dizziness'), there is no easily definable, discrete vestibular sensation. Instead, it appears that vestibular stimuli integrate with visual and somatosensory modalities. In the process of this integration, vestibular stimuli seem to lose their original, pure character. Overlap of vestibular and visual sensations with other somatosensory modalities often allows each to compensate for a deficiency in the other.

The Vestibular Epithelium – Vestibular Hair Cells and Supporting Cells

The vestibular epithelium, consisting of **vestibular hair cells** and **supporting cells** in the ampullary crests of each membranous ampulla and in both macula (Fig. 11.3), rests on subepithelial tissue containing afferent nerves (Fig. 11.3) and blood vessels. At the apex of each vestibular hair cell are **sensory hairs** (Fig. 11.3) consisting of many **stereocilia** and one **kinocilium** (Fig. 11.4). At their base, the supporting cells enclose nerves; at the epithelial surface, they surround the vestibular hair cells. Vestibular supporting cells have microvilli on their surface. Structural and functional differences exist between the vestibular epithelium of the cristae and maculae. Vestibular hair cells in the cristae and maculae consist of two cell types. **Type I vestibular hair cells** (with chalice-like synapses) are goblet-shaped with a narrow neck and a round base that has a neurite surrounding it in a cup-like manner (Fig. 11.4). **Type II vestibular hair cells** (with disseminated synapses) are cylindrical and innervated by several afferent and efferent fibers (Fig. 11.4). In addition to these two types of vestibular hair cells, the cristae and the maculae also have various supporting cells. The saccular macula in humans has about 18,800 vestibular hair cells – the utricular macula about 33,100. Each crista has about 7600 cells.

Age-Related Changes in Vestibular Hair Cells

With aging, there are structural changes manifested as a reduction in vestibular function. Disequilibrium of aging (appearing after 55) may be due to vestibular degeneration. Quantitative analyses in humans of hair cells in the vestibular epithelia show a reduction in hair cell number in those over 40, accompanied by a simultaneous reduction in number and quality of vestibular axons. Decrease in vestibular hair cells seemed to precede that for neurons in these subjects. A clear reduction in diameter of vestibular fibers also takes place in the elderly. A 40% reduction in number of vestibular hair cells in the cristae occurs in those over 70. A moderate but significant reduction (21%) of vestibular hair cell population in the macula of the utricle takes place in those between ages 70 and 95; in the macula of the saccule, the degree of this degeneration is about 24%. Large inclusions, identical with the appearance of lysosomes and lipofuscin granules, are in the vestibular epithelia of the elderly. Loss or derangement, increased fragility, or clumping of stereocilia and the formation of giant stereocilia also occur in the elderly. Age-related extracellular inclusions may be present under the basement membrane of entering nerves.

Cupula and the Cristae

The apices of vestibular hair cells in the cristae contain one **kinocilium** and 50 to 110 stiff **stereocilia** (Fig. 11.4) projecting into a gelatinous substance or **cupula** (Fig. 11.3C). The domed cupula blocks the lumen of the semicircular duct. Each duct, plus the cavity of the utricle related to it, forms a circle (Fig. 11.5). With rotation of the head, the semirigid endolymph in the semicircular ducts (Fig. 11.5) remains stationary, providing resistance against which the cupula and stereocilia projecting into it is deflected. The **deflection of the cupula** (Fig. 11.5) is in the opposite direction of the initial rotation of the head. Bending of stereocilia, caused by shearing motion of the cupula across their tips, excites vestibular hair cells and activates peripheral vestibular fibers innervating the hair cells. Ampullary receptors are sensitive to angular acceleration occurring with circular motion; their primary role, however, is to detect rotation of the head. Since each of the canals, ducts, and membranous ampullae is in a different plane, movement of the head in any plane stimulates receptors on one, two, or all three cristae.

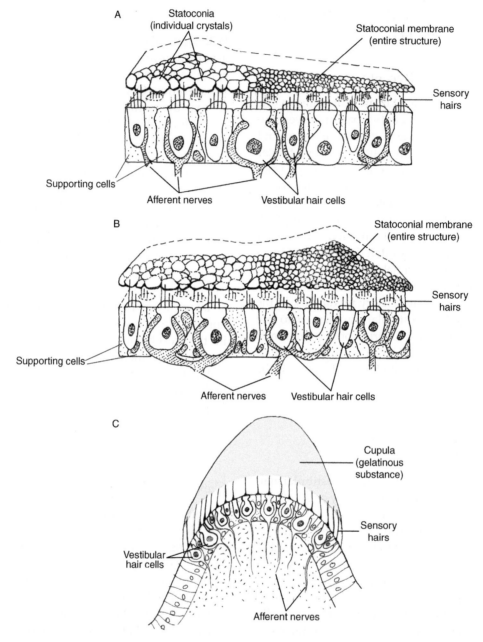

FIGURE 11.3. Sections of the specialized vestibular neuroepithelium at three different locations of the membranous labyrinth. (For orientation, see Fig. 11.2.) **A**, the structural organization of the macula utriculi and **B**, that of the macula sacculi (modified from Lindeman, 1973). **C**, Drawing of the crista ampullaris containing specialized vestibular hair cells (modified from Wersäll, 1972). Details of the vestibular hair cells are shown in Fig. 11.4.

Statoconia of the Maculae

A **statoconial membrane** (Figs 11.3A, 11.3B) covers stereocilia of vestibular hair cells in the maculae of the saccule and utricle. Stereocilia loosely attach to this membrane. The statoconial membrane consists of mucopolysaccharides, has the consistency of fine, packed sand, and is covered by a separate ridge or crystalline

layer made up of **statoconia** (Figs 11.3A, 11.3B). This membrane appears to make contact with the free surfaces of vestibular supporting cells. Statoconia are made of **calcite** or calcium carbonate, crystallized in trigonal form. Calcite differs from bones and teeth, which are made of a complex calcium phosphate compound called hydroxyapatite. A crystalline ridge of statoconia covers

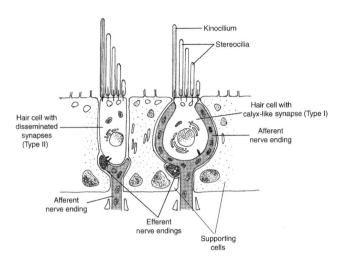

FIGURE 11.4. Two types of vestibular hair cells: type I cells with chalice-like synapses and type II cells with disseminated synapses. Fine efferent terminals are shown at the base of both types of cells (modified form Wersäll, 1972 and Wersäll and Bagger-Sjöbäck, 1974).

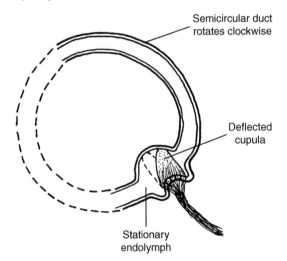

FIGURE 11.5. Each semicircular duct forms a full circle when the cavity of the utricle is included with it. When the head rotates in a clockwise direction in the plane of the page, the endolymph in the duct, with the consistency of a semi-rigid gel, remains stationary providing resistance against which the cupula and the stereocilia protruding into it must move or be deflected. Because of this resistance, the cupula and the stereocilia protruding into it moves in the opposite direction from the initial rotation of the head (modified from Roberts, 1976).

that area of the vestibular epithelium having the greatest concentration of cells with chalice-like synapses. Decreased production and loss of statoconia takes place with age.

Statoconia in humans are cylindrical with pointed ends, and are 3–19 µm in length. Delicate strands of organic substance interconnect them. Those in the saccular macula are uniform in size and often twice as large

as statoconia of the utricular macula. Statoconial surfaces have fine serrations often displaying shallow furrows. Saccular statoconial destruction increases with age; the number of utricular statoconia often decreases drastically. Remaining utricular statoconia usually show no sign of degeneration. Thickening of statoconial membranes is part of the aging process.

Statoconia are dense (2.95 g/cm^3 or about three times the density of water) and responsive to linear acceleration, including the effect of gravity, tilting of the head, and perhaps also to vibration. There are probably differences in response and, therefore, in function of statoconia in both maculae. Because of the loose attachment of the stereocilia of vestibular hair cells in the maculae to the statoconial membrane, motion of the head and maculae in a particular direction causes bending of the stereocilia in the opposite direction, thereby stimulating or inhibiting them. Movement of the head and macula in one direction with bending of stereocilia in the other is analogous to the backward movement of bristles on a brush as the body of the brush moves forward. In this example the 'brush' (vestibular hair cell) is upside down, with its bristles (stereocilia) projecting into the statoconial membrane (or into the cupula for ampullary receptors). With the head erect, the saccular macula is vertical, that of the utricle is horizontal. With these different orientations, the maculae probably serve a dual role, monitoring static position or acceleration of the head.

Positional and Functional Polarization of Vestibular Hair Cells

Positional polarization of vestibular hair cells refers to the eccentric position of the kinocilium on the vestibular hair cells. Its position in relation to the stereocilia determines the direction of polarization. **Functional polarization** refers to directional sensitivity of the sensory cells. Displacement of stereocilia in one direction yields an increase in discharge rate in vestibular nerve fibers (an excitatory effect). Displacement in the opposite direction has an inhibitory effect as the result of a decrease in discharge rate in fibers of the vestibular nerve. By determining the stereocilia-kinocilium relationship (**positional polarization**), the directional sensitivity (**functional polarization**) of the vestibular hair cells is inferred. Maculae in humans differ fundamentally in polarization pattern. Stimulation of sensory hairs on the vestibular hair cells in the maculae or ampullae causes current flow in the hair cells from terminal to basal surfaces that then depolarizes receptors and leads to release of a neurotransmitter at the synapse between the hair cell and its afferent terminal. Movement of the sensory hairs is likely to result in

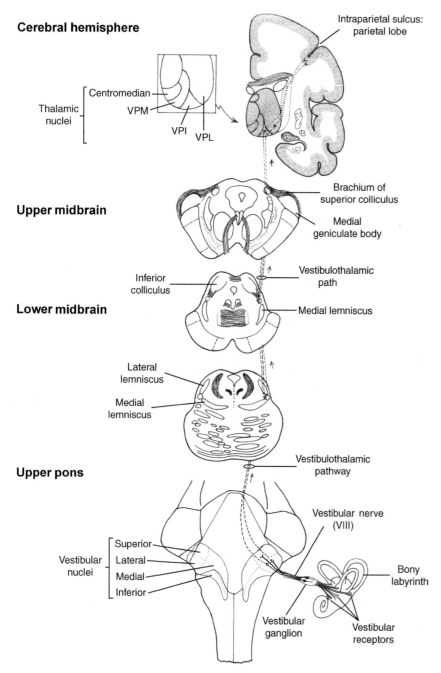

Cerebral hemisphere

Intraparietal sulcus:
parietal lobe

Thalamic nuclei
- Centromedian
- VPM
- VPI
- VPL

Upper midbrain

Brachium of
superior colliculus

Medial
geniculate body

Inferior
colliculus

Vestibulothalamic
path

Lower midbrain

Medial lemniscus

Lateral
lemniscus

Medial
lemniscus

Upper pons

Vestibulothalamic
pathway

Vestibular nerve
(VIII)

Vestibular nuclei
- Superior
- Lateral
- Medial
- Inferior

Bony
labyrinth

Vestibular
ganglion

Vestibular
receptors

FIGURE 11.6. Path for vestibular impulses from receptors in the peripheral vestibular apparatus to the primary vestibular cortex of the parietal lobe. This path is presumably bilateral in that there is never any direction of movement of the head that does not stimulate both sets of vestibular receptors. The thalamic termination of this path includes the oral part of the ventral posterior lateral nucleus (VPL_o), caudal part of the ventral lateral nucleus (VL_c), and dorsal part of the ventral posterior inferior nucleus (VPI_d). Some authors include VPI in the borders of the ventral posterior lateral nucleus (VPL) of the dorsal thalamus. The few primary fibers that reach the cerebellum are not shown.

increased neurotransmitter release; movement of the same hairs in the opposite direction causes decreased neurotransmitter release.

11.2.2. Primary Neurons

Primary neurons in the vestibular system are **bipolar neurons** with their cell bodies in the **vestibular ganglion** (Fig. 11.6). These bipolar neurons in humans are ovoid and vary from 15–37 µm in diameter. Each ganglion, located inferolaterally in the **internal acoustic**

meatus, has two groups of neurons interconnected by a narrow isthmus and collected in a larger **superior part** and a smaller **inferior part**. Some 18,000+ neurons occupy each ganglion, with a corresponding number of fibers both central and peripheral to it. In nonhuman primates, there is a high rate of resting discharge in primary vestibular neurons – they respond to semicircular canal acceleration in one direction with an increase in discharge (excitation) and to accelerations in the opposite direction with a decrease in discharge (inhibition). This phenomenon is functional polarization.

Efferent Fibers to the Vestibular Epithelium

Efferent fibers to the vestibular epithelium, though few in number, have collaterals that form a network at the base of the receptors. There are probably adrenergic, autonomic fibers innervating the vestibular epithelium.

Peripheral Processes of Primary Neurons

Peripheral processes of these primary neurons form two divisions of the vestibular nerve, with many branches supplying vestibular hair cells in the cristae and maculae. A **superior division**, arising from the superior part of the ganglion, supplies vestibular hair cells in the utricular macula, a small part of the saccular macula, and the cristae of the anterior and lateral semicircular ducts. An **inferior division**, arising from the inferior part of the ganglion, innervates vestibular hair cells in the main part of the saccular macula and on the crista of the posterior semicircular duct. Depending on the location of an electrode and length of stimulus, direct electrical stimulation of the inferior division to the saccule causes subjective or objective brief turning of the head (with short stimuli of 100 msec) or subjective or objective tilt of the body (using longer stimuli beyond 0.5 sec). The vestibular nerve lies next to the facial nerve [VII] in the **internal acoustic meatus**.

Acoustic Neuromas

Neurilemmal cells of the vestibular nerve or abnormal cells in the vestibular ganglion often give rise to **acoustic neuromas**, a common intracranial tumor. Such tumors often arise in the inferior division of the nerve and then assume a peripheral position as they grow. Because of the position in the internal acoustic meatus of the vestibular and cochlear roots of the nerve, the initial symptom of an acoustic neuroma is unilateral loss of hearing, gradual in onset for tones above 1 KHz. A unilateral deafness usually results within two years. Most patients, however, are unaware of the significance of their initial symptoms and fail to seek immediate attention. There is often a delay of a year or more from the onset of symptoms until a diagnosis is established.

In addition to involvement of the cochlear nerve, acoustic neuromas often impinge on the trigeminal nerve [V] at the pons causing an abnormal corneal reflex on the side of the tumor, or numbness of the face. Less often, the facial nerve [VII] is involved, with weakness of facial muscles, numbness of the tongue, or a change in taste sensation. In a series of 205 patients with acoustic neuromas, 15.1% (31) had ipsilateral symptoms of trigeminal neuralgia because of the effects of the tumor on the trigeminal nerve [V]. Rarely, trigeminal neuralgia is a symptom of a contralateral acoustic neuroma. In these cases, there is often compression, distortion, and rotation of the brain stem, causing a vessel to impinge contralaterally on the trigeminal nerve or causing stretch on the contralateral trigeminal nerve. These examples emphasize relations of the vestibulocochlear [VIII], facial [VII], and trigeminal [V] nerves with surrounding vessels.

Peripheral Vestibular Fibers

Peripheral vestibular fibers in primates may be thick or thin, myelinated or nonmyelinated, each with a characteristic discharge. Thin fibers supply cylindrical vestibular hair cells with disseminated synapses, are tonic receptors, and have regular-spaced action potentials. Thick vestibular fibers supply the type I vestibular hair cells with chalice-like synapses, are more phasic in behavior, and have irregular-spaced action potentials. These thick fibers are more sensitive to stimuli during movements of the head. In either case, whether thick or thin, vestibular fibers have a resting discharge that provides a constant tonic input to the central nervous system. Input from each labyrinth provides a source of excitation to secondary vestibular neurons.

Number of Vestibular Nerve Fibers and Changes with Age

Peripheral to the vestibular ganglion, the total number of myelinated fibers (between 1 day and 35 years of age) averages 18,346, with two-thirds in the superior division and one-third in the inferior division. The number of myelinated fibers approximates the number of neurons in each vestibular ganglion (18,439). Numbers of myelinated fibers in an older group (ages 75–85) averaged 11,506. There is a quantitative reduction (averaging some 37%) in vestibular fibers with age. A noticeable decrease in vestibular ganglionic cells, beginning about 40 years of age, is in keeping with a moderate but significant reduction in vestibular hair cell population in the maculae found in those between ages 70 and 95.

Central Processes of Primary Neurons

Central processes of primary vestibular neurons form the **vestibular nerve** (Fig. 11.6). A distinctive arrangement of fibers exists in the vestibular nerve of primates, fibers from the superior part of the vestibular ganglion are grouped rostrolaterally; those from the inferior part are caudomedial in the nerve. Fibers from the superior part take up approximately two-thirds of the vestibular nerve; those from the inferior part form the remaining third. Fibers from both parts of the ganglion have a precise arrangement in the vestibular nerve according to their peripheral distribution to the cristae and maculae. Fibers from the superior part are rostrolateral in the vestibular nerve with those innervating the cristae of the anterior (anterior ampullary nerve) and lateral (lateral ampullary nerve) canals being rostral to those to the macula of the utricle (utricular nerve). Fibers from the inferior part are caudomedial in the vestibular nerve with those innervating the crista of the posterior canal (posterior ampullary nerve) being rostral to those from the macula of the saccule (saccular nerve).

Although a few central processes of primary vestibular neurons are primary **vestibulocerebellar fibers** reaching parts of the cerebellum and concerned with equilibrium, most project to the **vestibular nuclei** (secondary vestibular gray) medial to the inferior cerebellar peduncle at medullary and pontine levels (Fig. 11.6). The vestibular nuclei form a surface feature on the floor of the fourth ventricle called the **vestibular area**.

11.2.3. Secondary Neurons

Secondary neurons in the vestibular path are the **superior**, **inferior**, **medial**, or **lateral vestibular nuclei** (Fig. 11.6). In humans, the vestibular nuclei containing about 200,000 neurons have a collective volume of 177 mm^3, with individual volumes of 22.4 mm^3 (lateral), 45.9 mm^3 (inferior), and 72.8 mm^3 (medial). The inferior vestibular nucleus is only in the medulla oblongata, the medial vestibular nucleus in the pons and medulla oblongata whereas the superior and lateral vestibular nuclei are only in the pons.

Spontaneous Activity in the Vestibular Nuclei

A relatively high level of spontaneous activity takes place in the vestibular nuclei of the monkey at rest, even if the semicircular canals have not been stimulated and are parallel to the plane of stimulation. A high rate of resting discharge in primary neurons leads to a substantial input to the secondary neurons. Labyrinthine injury may reduce or abolish the resting discharge of secondary neurons. Visual and vestibular stimuli influence the spontaneous discharge in secondary vestibular neurons in primates.

Superior Vestibular Nucleus

The presence in it of vertical bundles of myelinated cerebellovestibular fibers clearly distinguishes the superior vestibular nucleus. Caudally it extends from inferior to the rostral pole of the trigeminal pontine nucleus in the middle pons to the rostral third of the abducent nucleus where it overlaps the rostral pole of the medial vestibular nucleus. The rostrocaudal extent of the superior nucleus (some 4 mm) is exclusively at pontine levels; it is ventromedial to the superior cerebellar peduncle and the trigeminal mesencephalic nucleus and dorsal to the trigeminal pontine and motor nuclei in the dorsolateral corner of the pontine tegmentum.

Medial Vestibular Nucleus

The medial vestibular nucleus in humans is the longest, most voluminous and cellular of the vestibular nuclei measuring about 9 mm in length and extending from the lower third of the pons into the medulla. Its rostral pole, at the caudal tip of the abducent nucleus, overlaps the caudal end of the superior vestibular nucleus. The caudal extent of the medial vestibular nucleus, in the upper medulla oblongata, is about 1 mm caudal to the rostral tip of the hypoglossal nucleus. Throughout its extent, the medial vestibular nucleus is beneath the widest part of the floor of the fourth ventricle. In cross sections, it is a triangular, densely packed collection of neurons (more so than the other vestibular nuclei), with bundles of presumed cerebellovestibular fibers running in it.

Unilateral injuries (nondestructive but irritative) of the medial nucleus in primates cause many symptoms such as a **horizontal nystagmus** – an involuntary, rhythmic, side-to-side movement of the eyes, with a clearly defined fast and slow phase. There may also be an inability to coordinate voluntary movements while walking or standing, causing postural abnormalities, falling, or rotating to the side of the injury, a condition called **vestibular ataxia**. Lastly, a variable and perhaps transient diminution of deep reflexes, grasping responses, and righting reflexes may be present. Medial vestibular injuries in squirrel monkeys also result in tilt of the head to the side of the injury, ipsilateral limb flexion, and contralateral limb extension.

Lateral Vestibular Nucleus

The lateral vestibular nucleus is primarily at pontine levels, near the caudal pole of the trigeminal motor nucleus and contiguous with the caudal part of the superior vestibular nucleus. Caudally it extends some 4 mm to the caudal pole of the facial nucleus where it is bordered medially by the medial vestibular nucleus (especially at caudal levels) and laterally by the inferior cerebellar peduncle. Numerous descending fibers, presumably cerebellovestibular in nature, along with the presence of large neurons distinguish the lateral vestibular nucleus and divide it into medial and lateral parts. The medial part extends more rostrally than the lateral, is more cellular, and its place caudally is taken by the inferior vestibular nucleus.

Inferior Vestibular Nucleus

A close companion of the medial vestibular nucleus, except most rostrally, is the inferior vestibular nucleus. Some 5 mm in length, its rostral boundary is in the lower third of the pons at the caudal pole of the facial nucleus where it is continuous with the caudal pole of the lateral vestibular nucleus. Caudally, the inferior vestibular nucleus is at the rostral pole of the lateral cuneate nucleus. Throughout its course, except in its caudal extent, it is medial to the inferior cerebellar peduncle and lateral to the medial vestibular nucleus. Though the border between both nuclei is indistinct, a lower density of neurons and the presence of fibers among its cells easily differentiate the inferior vestibular nucleus from the medial. Vertical fibers among the neurons of the inferior vestibular neurons consist of descending branches of primary vestibular, vestibulospinal, and cerebellovestibular fibers.

Distribution of Primary Fibers to the Vestibular Nuclei

Primary fibers to the vestibular nuclei enter the brain stem between the inferior cerebellar peduncle and the trigeminal spinal tract. Though there is a distinctive pattern of distribution of primary fibers to the vestibular nuclei, and areas of common projection, there are regions in all four nuclei that fail to receive primary fibers. Primary fibers ascend, descend, or pass medially to the vestibular nuclei. Few, if any, entering vestibular fibers reach the dorsal part of the lateral vestibular nucleus in primates. Primary neurons in the vestibular ganglia, with peripheral processes supplying receptors in the cristae ampullares of the lateral and anterior semicircular canals, also yield central processes (primary fibers) that supply the rostral and lateral parts of the superior nucleus and the dorsal and lateral parts of the medial nucleus at, and above, entering vestibular fibers. Primary fibers, whose peripheral processes supply vestibular receptors in the crista of the posterior semicircular canal, end in medial and caudal parts of the superior vestibular nucleus. Vestibular ganglionic cells, whose peripheral processes innervate the macula of the utricle, have central processes (primary fibers), that traverse and give collaterals to the ventral part of the lateral vestibular nucleus then descend to give collaterals to the medial nucleus before ending in

the inferior nucleus. Finally, central processes of primary neurons supplying vestibular receptors in the macula of the saccule of primates enter the dorsolateral inferior vestibular nucleus, descend in it, and end throughout its rostrocaudal length.

Central Processes of Secondary Neurons – the Vestibulothalamic Path

Secondary neurons, projecting to the thalamus for relay to the cerebral cortex, provide the anatomical basis for the conscious perception of motion (including falls or losing our balance) and orientation in space. Fibers from each lateral vestibular nucleus contribute to an ipsilateral, **vestibulothalamic path** (Fig. 11.6); contralateral fibers originate in the superior vestibular nucleus. The location of this ascending path in the brain stem of primates is uncertain; it is probably between the medial and lateral lemnisci (perhaps nearer the lateral). Electrophysiological evidence from studies in rhesus monkeys confirms this localization and suggests a monosynaptic connection between the vestibular nuclei (secondary neurons) and **thalamic nuclei** (Fig. 11.6).

At inferior collicular levels of the human midbrain, vestibular responses are identifiable between 5 and 12 mm from the median plane (corresponding to a narrow zone between the medial and lateral lemnisci). Some researchers contend that vestibulothalamic fibers do not form a compact bundle but spread out in the brain stem, forming a bilateral system of fibers. Another view emphasizes the presence of two brain stem paths: one ascends with or near the lateral lemniscus; the other is ventral in the brain stem. At the junction of the midbrain and diencephalon, the latter passes lateral to the red nucleus and dorsal to the subthalamic nucleus. A double ascending system is in keeping with the presence of vestibular areas in both the parietal and temporal lobes. Future studies probably will define the brain stem location of the vestibulothalamic path in primates – be it crossed, uncrossed, or bilateral.

11.2.4. Thalamic Neurons

The ascending vestibulothalamic path of primates projects to various **thalamic nuclei** (Fig. 11.6) including bilaterally to neurons in the oral part of the **ventral posterior lateral nucleus** (VPL$_o$), caudal part of the **ventral lateral nucleus** (VL$_c$), and dorsal part of the **ventral posterior inferior nucleus** (VPI$_d$).

Several aspects of these thalamic nuclei are noteworthy. First, VPI$_d$ in rhesus monkeys approximates the magnocellular part of the medial geniculate body, wedged between two somatosensory relay nuclei of the thalamus, namely the lateral and medial parts of the ventral posterior nucleus. In this location, VPI_d is mediodorsal to the perioral representation of the face in VPM and laterodorsal to the hand (thumb) representation in VPL (VPI_d is considered as part of VPL by some authors). Second, fibers of the medial lemniscus carrying discriminatory tactile, pressure, proprioceptive, and vibratory stimuli synapse with neurons in VPL_c that in turn send their axons to the **primary somatosensory cortex** corresponding to **areas 3a, 3b, 1,** and **2** of Brodmann. Vestibular projections to VPL_o are scattered in small patches over this nucleus. The distribution of vestibular fibers in the thalamus permits interaction in VPL of vestibular and somatosensory (especially proprioceptive) inputs. Thus, vestibular and somatosensory nuclei in the thalamus of rhesus monkeys have a topographical and functional relationship. Third, studies in primates reveal the vestibular nuclear projection to these thalamic neurons is a sparse but definite bilateral projection. Finally, these thalamic neurons are activated in the alert monkey by vestibular stimulation including angular acceleration, rotation of an optokinetic cylinder, rotation of the visual surround, and rotation of the animal itself about a vertical axis (both to the ipsilateral and to the contralateral side) following appropriate proprioceptive stimuli. Discharge patterns of these thalamic neurons are unrelated to ocular movements. Other thalamic nuclei, identified as vestibular relay nuclei, include the **posterior nuclei** (PLi) and the magnocellular part of the **medial geniculate nucleus** (MG_{mc}). Vestibular impulses often merely pass through this area on their way to thalamic nuclei.

Based on these observations, VPL_o, VL_c, VPI_d, PLi, MG_{mc} are **vestibular relay nuclei.** Based on its projections to **primary motor cortex (area 4)**, however, VPL_o and its extension VL_c in monkeys are likely involved in vestibular aspects of motor coordination. VL_c in monkeys corresponds to the dorsal part of VL_p in humans. VPI and PLi with projections to the posterior part of the postcentral gyrus of the parietal lobe at the base of the intraparietal sulcus, corresponding to the **primary vestibular cortex** or Brodmann's **area 2v**, are probably involved in the conscious appreciation of vestibular sensation. The anterior part of the human intraparietal sulcus is involved in visually guided grasping.

Recordings from the Monkey Thalamus

Recordings made in the posterior part of the thalamus in rhesus monkeys have localized the vestibulothalamic tract between the medial lemniscus and brachium of the inferior colliculus. These recorded responses suggest that the projection between vestibular nuclei and thalamus is monosynaptic.

Stimulation of the Human Thalamus

Stimulation of the human thalamus in the superolateral part of the medial geniculate body and in the brachium of the inferior colliculus between 10 and 17 mm from the median plane causes vertigo and related phenomena including the feelings of clockwise or counter-clockwise rotation, rising or falling, floating, whole body displacement, fainting, or nausea. Such responses occur from thalamic stimulation of the region anterior to that giving rise to auditory responses. Stimulation of VPI in conscious humans evokes vestibular perceptions such as being tilted, whirled, or falling, and sensations of vertigo and of body movements. Less discriminative aspects of vestibular sensation are likely to become conscious at the thalamic level in humans.

These data, from nonhuman primates and humans, suggest that the ascending vestibulothalamic tract is a bilateral primary sensory projection system at brain stem levels between the medial and lateral lemnisci (perhaps nearer the lateral). Vestibular and proprioceptive impulses, integrated at thalamic levels, reach the cerebral cortex, where the perception of body position and movements, generally referred to as vestibular sensation, enters consciousness.

11.2.5. Cortical Neurons

Primary Vestibular Cortex (2v)

Experimental studies have identified the **primary vestibular cortex** of primates in the anterior part of the inferior lip of the **intraparietal sulcus** (Figs 11.6, 11.7) where it adjoins, and is continuous with the posterior part of the **postcentral gyrus**, corresponding to Brodmann's area 2. The **primary vestibular cortex** on the superior parietal lobule is therefore designated **area 2v** to distinguish it from that part of area 2 related to the **primary somatosensory cortex.**

Vestibular responses recorded from neurons in **area 2v** are predominantly contralateral. Single unit analysis of these neurons in rhesus monkeys has shown that they are not modality-specific to a single sensation as classically described for the auditory, visual, and somatosensory neurons. Units in the **primary vestibular cortex** or **area 2v** of primates probably integrate vestibular and proprioceptive inputs (muscle, pressure, or joint movements) that then converge in the vestibular nuclei and thalamus. Conscious appreciation of body position in space and its changes cannot rely exclusively on vestibular

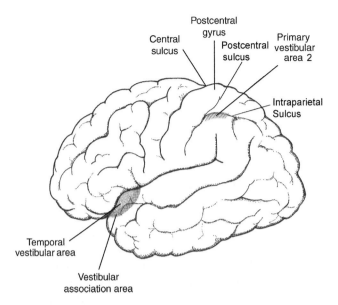

Postcentral
gyrus
Central
sulcus
Postcentral
sulcus
Primary
vestibular
area 2
Intraparietal
Sulcus
Temporal
vestibular area
Vestibular
association area

FIGURE 11.7. Lateral surface of the human cerebral cortex. At the base of the intraparietal sulcus, on its inferior lip, is the primary vestibular cortex (diagonal lines). The temporal vestibular cortex and adjacent vestibular association area along the rostral part of the temporal operculum are illustrated.

stimuli because the head moves with respect to the remainder of the body. Vestibular and visual inputs participate in the assessment of body position and motion. The presence of **primary vestibular cortex (area 2v)** in the parietal lobe is not surprising as both visual and proprioceptive stimuli activate these vestibular neurons in the cerebral cortex. An appreciation of the position and movement (kinesthesia) of joints reaches consciousness in the **primary somatosensory cortex**. These sensations (and perhaps visual input as well) probably combine with vestibular impulses related to body position in space for the ultimate recognition and appreciation of the motion and orientation of body parts, that is, for the conscious awareness of body orientation.

Vestibular and somatosensory areas in the parietal lobe of rhesus monkeys have both a topographical and functional relation in agreement with similar relations between vestibular and somatosensory nuclei in the thalamus. Anterior to the **primary vestibular cortex** are perioral neurons representing the face; anteromedially are those representing the hand (thumb).

Another parietal vestibular field is identifiable in squirrel monkeys and some lower mammals. Its location corresponds to area 3a, in the depths of the central sulcus. Area 3a in rhesus monkeys has a thin layer IV. In nonhuman primates and humans, it is a region of overlap anteriorly with the **primary motor cortex (area 4)** along the precentral gyrus. Connections to area 3a from vestibular relay nuclei (VPL$_o$, VL$_c$, VPI$_d$, PLi, and MG$_{mc}$) are lacking; area 3a, receiving its input from

VPL$_c$ of the thalamus, seems to serve group Ia afferents from neuromuscular spindles. These receptors probably receive inputs that reflect position of limbs and velocity of passive limb displacement. Further studies in nonhuman primates and humans would help define the role of area 3a in vestibular sensation.

Stimulation in the depth of the intraparietal sulcus in a patient with a brain tumor (meningioma) resulted in a detailed account of the perception of rolling in one direction. Electrical stimulation in the vestibular cortex of the parietal lobe in humans will evoke a sense of rotating or body displacement while the world around the patient remains stationary; seizure discharge here often yields an epileptic aura of that type. In another patient, stimulation of the parieto-occipital junction on the lateral surface of the cerebral hemisphere caused vertigo. Everything appeared to be moving in a clockwise direction. These sensations resembled those that accompanied her previous seizures. In another patient, a right parieto-occipital tumor initially irritated the **primary vestibular cortex (area 2v)**, causing a sensation of spinning. At times, the patient saw himself upside down. As the tumor continued to grow, there was disorientation in space and inversion of body image. **Vertigo**, the abnormal perception of movement or orientation, is likely to be secondary to injury to (pathological dysfunction) or stimulation of the primary vestibular cortex.

Input arises from two cortical fields to reach the **primary vestibular cortex (area 2v)** in rhesus monkeys. These two cortical fields consist of a longitudinal strip of area 18 on the posterior bank of the lunate sulcus (an inconstant feature of the human brain) and the neck-arm region of **primary motor cortex (area 4)**. The functional significance of these inputs is uncertain.

Finally, it is worth repeating that since there is integration of vestibular input with other sensory modalities (proprioceptive and visual) at cortical levels, it is reasonable to conclude that spatial skills functionally related to the parietal lobe such as right-left discrimination, directional sense, and body image, among others, are to a degree dependent on vestibular input and vestibulo-visual-somatosensory integration.

Temporal Vestibular Cortex

Vestibular sensations occur during temporal lobe stimulation in patients undergoing exploratory craniotomy for focal epilepsy. The **temporal vestibular cortex** (Fig. 11.7) in humans is on the superior temporal convolution along the rostral part of the temporal operculum. An expanding intracranial tumor that stimulated a vestibular area on the medial aspect of the temporal operculum and the insular lobe led to dizziness in a patient. As the tumor grew and destroyed this region, these vestibular symptoms disappeared. Irritative injury to this temporal lobe region led to the subjective feeling of whirling or the patient is likely to rotate in a quiet environment or feel as though they are rotating. Whether of parietal or temporal lobe origin, vestibular sensation is a complex phenomenon requiring the correlation, integration, and association of many types of sensations from various cortical areas. Rich interplay and complex association of vestibular-related information probably takes place in that area of the temporal lobe adjacent to the temporal vestibular cortex. This area likely functions as a **vestibular association area** (Fig. 11.7).

11.3. OTHER VESTIBULAR CONNECTIONS

In addition to the **ascending vestibulothalamic path** (Fig. 11.6), the vestibular system has significant connections with other regions of the nervous system. Entering primary vestibular fibers that bypass the vestibular nuclei and end in the cerebellum (primary vestibulocerebellar fibers) are an example of such connections. Secondary neurons in the vestibular nuclei send fibers to the spinal cord, brain stem motor nuclei innervating the extraocular muscles, reticular formation, and to contralateral vestibular nuclei. Elucidation of the functions of the vestibular system requires a consideration of its bilateral organization, including the bilateral labyrinths and their associated receptors with their particular orientation and arrangement, the bilateral vestibular nuclear complexes, and the variety of connections of these nuclei with diverse areas of the nervous system.

11.3.1. Primary Vestibulocerebellar Fibers

A few primary vestibular fibers end in the cerebellum. In nonhuman primates, these entering vestibular fibers ascend to traverse the superior vestibular nucleus, and then enter the cerebellum in a small inner part of the inferior cerebellar peduncle termed the **juxtarestiform body**. All parts of the vestibular ganglion send axons to the cerebellar cortex in the ipsilateral vermis (particularly to all parts of the nodulus, ventral folia of the uvula, and a few to the lingula). Another bundle of fibers, from ganglionic neurons innervating cristae of the semicircular ducts and maculae of the saccule and utricle, reach regions of the cerebellar flocculus. A third group of fibers enters the cerebellum to project ipsilaterally to folia of lobules V and VI on both sides of the primary fissure.

Fibers of the vestibular nerve to the flocculus are identifiable in the cerebellum of the human fetus. Those cerebellar regions that receive primary vestibulocerebellar fibers are termed the '**vestibulocerebellum**'. Stimulation of the cerebellar flocculus, nodulus, and uvula in rhesus monkeys causes **nystagmus** – an involuntary, rhythmic, horizontal or vertical movement of the eyes, having clearly defined fast and slow phases.

11.3.2. Vestibular Nuclear Projections to the Cerebellum

In the monkey, a few fibers from the medial vestibular nucleus reach the flocculus and a few fibers from the inferior vestibular nucleus reach the vermis of the anterior lobe, uvula, and the nodulus of the flocculonodular lobe. The vestibular nuclei receive more cerebellar fibers than they send to it.

11.3.3. Vestibular Nuclear Projections to the Spinal Cord

Humans have a narrow base for standing and a high center of gravity. Though vision helps indicate falling and stretch reflexes act during normal sway, the vestibular system and its spinal projections complement them in locomotion and standing, to help maintain equilibrium. Through a series of complex righting movements, requiring the shifting of body parts, including the head and neck, the vestibular system coordinates and adjusts the tone of extensor muscles by direct influence on alpha motoneurons and by acting on the gamma loop. Therefore, smooth and appropriate muscle responses occur with maintenance of balance and posture. Indeed, stimulation of, or injury to, these vestibulospinal projections is likely to cause a postural imbalance that leads to an incoordination of movement, or **vestibular ataxia**. Irritative injuries of these projections often lead to an objective tilting or falling in affected patients.

Medial Vestibulospinal Tract

Fibers arising from the **medial vestibular nucleus** (Fig. 11.8) course medially, decussate, and descend in the medial longitudinal fasciculus (MLF). In the cervical cord, these fibers are medial to the ventral horn in the ventral funiculus where they form the **medial vestibulospinal tract** (Fig. 11.8). Some uncrossed fibers in this tract are in the ipsilateral medial longitudinal fasciculus. The medial vestibulospinal tract influences muscles of the neck and upper limb and does not descend below

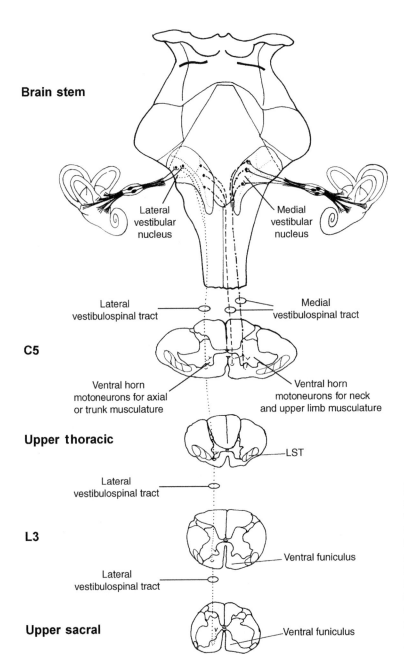

Brain stem

Lateral vestibular nucleus

Medial vestibular nucleus

Lateral vestibulospinal tract

Medial vestibulospinal tract

C5

Ventral horn motoneurons for axial or trunk musculature

Ventral horn motoneurons for neck and upper limb musculature

Upper thoracic

Lateral vestibulospinal tract

LST

L3

Lateral vestibulospinal tract

Ventral funiculus

Upper sacral

Ventral funiculus

FIGURE 11.8. Vestibular projections to the spinal cord. On the left are ipsilateral fibers from the lateral vestibular nucleus that contribute to the formation of the lateral vestibulospinal tract. A somatotopic arrangement of these vestibulospinal projections from the lateral vestibular nucleus exists. From the left medial vestibular nucleus are the crossed medial vestibulospinal fibers that course through the ventral funiculus to end at middle cervical levels with a few fibers extending into the lowermost cervical cord. From the right medial vestibular nucleus is an ipsilateral group of fibers that also contributes to the lateral vestibulospinal tract. Most of these fibers end in the upper cervical region with a few fibers to the middle cervical cord.

midthoracic levels. Because fibers of the medial vestibulospinal tract run in the ventral funiculus, this tract is also termed the **ventral vestibulospinal tract**. The medial vestibulospinal tract appears to function in relation to vestibular reflexes involving the neck.

Lateral Vestibulospinal Tract

Neurons in the **lateral part** of the **lateral vestibular nucleus** (Fig. 11.8) make a substantial contribution to the **lateral vestibulospinal tract** (Fig. 11.8) by way of ipsilateral fibers that descend below C7 to influence muscles of the trunk and lower limbs. Neurons in the **medial part**

of the lateral vestibular nucleus supply upper cervical spinal cord segments for muscles of the neck and upper limbs. Studies in the rhesus monkey, using retrograde axonal transport methods, have confirmed these projections and emphasized a somatotopic organization of fibers descending from the lateral vestibular nucleus. In primates, the medial vestibular nucleus also provides a contingent of descending vestibulospinal fibers that contribute to the lateral vestibulospinal tract. In the medulla oblongata, these ipsilateral fibers are lateral to the medial longitudinal fasciculus. Descending into the cervical spinal cord, they are scattered between the ventral surface of the cord and the ventrolateral border of the ventral horn. Most end in the upper cervical region, but a few reach the middle cervical level. Thus, in non-human primates and humans, the lateral vestibu-lospinal tract probably has ipsilateral descending fibers from medial and lateral vestibular nuclei. These descending fibers likely influence ventral horn motoneurons at all spinal levels. After injury to this path, the body, head, and chin may show deviations – the body to the opposite side and head and chin to the side of injury. Details of vestibulospinal fibers in humans and their relation to those in nonhuman primates warrant additional study.

Stimulation of Fibers from the Semicircular Canals

In nonhuman primates, stereotyped movements of the head in different planes accompany stimulation of fibers from the semicircular canals. The direction of these movements coincides with the plane of the canal stimulated. Impulses mediating these movements reach ventral horn motoneurons of the cervical cord innervating appropriate muscles of the neck by way of descending vestibulospinal fibers that also influence ventral horn motoneurons innervating muscles of the upper limb.

Labyrinthine Injuries

Patients with bilateral destruction of vestibular labyrinths initially show impaired posture and locomo-tion. After a period of compensation, posture and loco-motion becomes normal if such patients utilize visual and proprioceptive inputs. Because of enhanced propri-oceptive reflexes, there is increase in tone in the extensor muscles of the legs with a characteristic forward and backward body sway in the upright position. Patients with unilateral labyrinthine injury often show increased tone in the extensor muscles in the contralateral leg. Unilateral injury leads to an imbalance between vestibu-lar nuclei on both sides, with inhibition of the injured side and excitation of the other resulting from reduced contralateral inhibition by way of commissural inter-nuclear fibers.

Section of the Vestibular Nerve

Unilateral section of the vestibular nerve in rhesus monkeys causes an ipsilateral tilt of the head without postural asymmetry, but in baboons it affects spinal reflexes. Such studies suggest that depression of spinal reflexes on the side of injury is due to vestibulospinal inhibition of alpha and gamma motoneurons with opposite effects contralaterally.

11.3.4. Vestibular Nuclear Projections to Nuclei of the Extraocular Muscles

Intimate structural and functional relations exist between the vestibular nuclei and brain stem nuclei that innervate extraocular muscles. By means of the vestibu-lar connections with these nuclei, and internuclear con-nections among the extraocular motor nuclei, the vestibular system influences ocular movements in all directions and maintains gaze in a given direction despite movements of the body and head. Stimulation of vestibular nuclei in the alert or lightly anesthetized rhe-sus monkey with high cervical cord transection causes **nystagmus**, (horizontal, vertical, or rotatory) or ocular movements in different planes similar to those following stimulation of nerves from semicircular canals.

Recording from single neurons in the vestibular nuclei of conscious primates, reveals a large repertoire of discharge patterns related to almost every aspect of voluntary and compensatory ocular movements. Such studies emphasize the role of vestibular nuclei in coor-dinating vestibular and oculomotor inputs mediating movements in all directions. These vestibulo-ocular connections provide a structural basis for nystagmus.

Complete examination of the vestibular system must involve investigation of spontaneous or induced nystagmus (caused by vestibular stimulation through caloric, rotational, postural, galvanic, or optokinetic means). Details of the vestibular projections to the extraocular motor nuclei, certain vestibulo-ocular reflexes, their internuclear connections, and their functional sig-nificance are available in Chapter 13.

11.3.5. Vestibular Nuclear Projections to the Reticular Formation

The medial vestibular nucleus gives short ascending and descending fibers to various medullary nuclei directly or indirectly via the reticular formation. Such connections presumably explain the visceral symptoms that accom-pany vertigo, induced by external stimulation of the vestibular system, a condition called **motion sickness** with typical symptoms of pallor, cold sweating, nausea, and vomiting. Though the exact connections mediating such symptoms are unclear, they may involve several sets of connections. For example, they may involve direct pro-jections to the ipsilateral dorsal vagal nucleus from the medial vestibular nucleus, influencing nonstriated mus-cle of the stomach in reverse peristalsis. Other possible connections would be those to the nucleus of the solitary tract for the nausea, and to the nucleus ambiguus (inner-vating the pharynx and larynx) for the frequent swallow-ing and regurgitation. A last set of connections might be direct connections to the intermediolateral nucleus, supplying blood vessels and sweat glands, for pallor and cold sweating that occasionally accompany visceral reflexes in motion sickness. Reticulospinal projections to the phrenic nucleus may provide the necessary diaphragmatic movements that accompany vomiting. Finally, projections from the medial vestibular nuclear complex, to both superior and inferior salivatory nuclei, provide for the excessive salivation in motion sickness.

11.3.6. Vestibular Projections to the Contralateral Vestibular Nuclei

Neurons ventromedial in the medial vestibular nucleus of nonhuman primates send commissural fibers to the same part of the contralateral nucleus. In the fetus, intramedullary fibers of the vestibular nerve decussate to form the **vestibular decussation** or commissure; it is unclear whether secondary fibers from the vestibular nucleus accompany these commissural fibers. Experiments in rhesus monkeys suggest that this commissural path is inhibitory; increase in extensor tone in the muscles of the leg contralateral to a unilateral labyrinthine disturbance is likely to be a release phenomenon caused by interruption of inhibitory proprioceptive reflexes.

11.4. THE EFFERENT COMPONENT OF THE VESTIBULAR SYSTEM

Vestibular projections to vestibular receptors form the efferent component of this system. These connections resemble the efferent component of the auditory system. Vestibular efferents arise from a dense collection of neurons that are dorsal to the intrapontine part of the facial nerve [VII] and interposed between the abducent and superior vestibular nuclei. A bilateral neuronal column, referred to as **group e**, extends from rostral to middle levels of the abducent nucleus. Its neurons are acetylcholinesterase-positive. Efferents emerge from **group e**, enter the vestibular nerve, and synapse with vestibular hair cells. Each vestibular apparatus receives a bilateral innervation of vestibular efferents. The path, taken by both auditory and vestibular efferent fibers, is the **cochlear-vestibular efferent bundle** (CVEB). The number of vestibular efferents is few compared with the vestibular afferents. There is sparse topographic specificity between individual efferents and afferents whose discharges they modify. Nevertheless, these few efferent fibers, by means of multisynaptic contacts, probably

exert an influence on almost every afferent vestibular fiber. In nonhuman primates, vestibular efferents are predominantly excitatory and likely active during the large accelerations accompanying voluntary movement of the head. These efferent synapses are cholinergic.

11.5. AFFERENT PROJECTIONS TO THE VESTIBULAR NUCLEI

Inputs to the vestibular nuclei, from a variety of sources, play a vital role in normal functioning of the vestibular system. Sources of these afferents include the spinal cord, cerebellar cortex, deep cerebellar nuclei, contralateral vestibular nuclei, and perhaps the brain stem reticular formation. In humans, ascending spinovestibular fibers arise at lower spinal levels and end in the lateral vestibular nucleus. Massive bundles stream out of the cerebellum to reach the lateral vestibular nucleus.

Studies in primates have identified Purkinje cells in the central zone (zone I) of all ten folia of the cerebellar flocculus that send axons to the ipsilateral medial vestibular nucleus. Parallel zones on either side of the central zone (designated medial zone I and lateral zone III of the flocculus in the cerebellar vermis) send axons to the ipsilateral superior vestibular nucleus. These medial to lateral zones span all ten folia of the flocculus, occupying only their central regions with extrazonal areas of cerebellar cortex on either side.

Parts of the vestibular nuclei in primates also receive projections from the cerebellar fastigial nucleus. Autoradiographic tracing techniques have identified neurons in each fastigial nucleus that send axons bilaterally and symmetrically to ventral parts of the lateral and inferior vestibular nuclei. Ipsilateral, asymmetrical projections also arise from rostral parts of the fastigial nucleus to small regions of the ventrolateral medial vestibular nucleus. Commissural connections between the vestibular nuclei are likely both afferent and efferent. Neurons in the ventromedial medial vestibular nucleus send efferents to the same part of the contralateral nucleus.

11.6. VERTIGO

Vertigo or dizziness is an abnormal sensation of motion or orientation of oneself or external objects. Because vestibular, visual, and somatosensory stimuli are integrated, it is possible to produce vertigo by stimulation or injuries involving any of these modalities. Other manifestations likely to accompany vertigo include **vestibular**

ataxia caused by involvement of the vestibulospinal projections, **vestibular nystagmus** caused by vestibular nuclear injury, or **caloric nystagmus** caused by thermal stimulation of vestibular receptors. Certain **visceral symptoms** (pallor, cold sweating, nausea, vomiting) often accompany vertigo and are caused by stimulation of vestibular nuclear connections with the reticular formation of the brain stem.

11.6.1. Physiological Vertigo

Physiological vertigo and related phenomena result from stimulation in conscious humans of the brachium of the inferior colliculus, the superolateral part of the medial geniculate body, the ventral posterior inferior nucleus (VPI), and from stimulation of the **parietal and temporal vestibular cortical areas**.

Motion Sickness

Physiological vertigo, and its accompanying manifestations, often occur while ascending in elevators, riding in cars, roller coasters, ships at sea, and in aircraft or spacecraft. Such vertigo, called **motion sickness**, results from **sensory mismatch**, in which one set of receptors is receiving information indicating a certain motion, position, or orientation, while another set is receiving different information about motion, position, or orientation. In the closed cabin of a ship, visual stimuli are likely to indicate that the body is fixed and stable (with respect to the cabin) while vestibular receptors indicate a great deal of motion. One aspect of such 'sea sickness' is the observation, originally made in the early 1900s, that sailors gradually lose their sea sickness, a phenomenon called **habituation**. Ballet dancers, ice skaters, aerospace and marine personnel undergo a degree of occupational vestibular habituation.

Motion sickness involves patterns of acceleration different from normal environmental conditions that often occur in carnival rides and high-speed vehicles (land-based or air-borne). In these examples, receptors for motion continue to function as before, but generate sensations and reflex adjustments that are no longer appropriate to the unusual environmental conditions.

Motion Sickness in Space

Motion sickness in space involves intravestibular mismatch – a mismatch between information from statoconia versus that from cristae ampullares. In the absence of forces supporting gravity, as found in an orbiting

spacecraft, the statoconia are no longer responsive. Habituation takes place in three to six days and the motion sickness thereafter abates. On return to earth, however, transient vertigo appears and lasts hours to days. Some 15–20% of the American astronauts and Soviet cosmonauts experienced motion sickness in space.

Motion sickness in space, accompanied by inversion of body image, formed visual hallucinations, or unformed visual images such as flashes of light, probably requires involvement of the parietal, temporal, and occipital cortices. Reports of symptoms of motion sickness associated with flight in space and those experienced by patients with expanding injuries involving the **vestibular cortex** in the **parietal or temporal lobes** (or involvement of the underlying fiber bundles that may discharge to these and to remote areas of the brain) are similar. Motion sickness in space with cortical signs and symptoms might also be secondary to a transient vascular insufficiency to the primary vestibular cortex. One Soviet cosmonaut experienced a feeling of being upside down in his visually upright spacecraft – a feeling that persisted until re-entry.

Experiments carried out in future manned spacecraft ventures will help us understand the capacity in humans to adapt to rearranged sensory stimuli and understand the ability of our brain to compensate for disturbances in orientation and motion. Motion sickness is a serious problem that may severely impede the performance of an individual, whether weightless and orbiting the earth or land-locked and trying to get about the house.

Other Types of Physiological Vertigo

Other types of physiological vertigo include **vertigo due to height**. This condition may occur when the distance between an individual and visible stationary objects in the environment become critically large. **Visual vertigo** may result from viewing a motion picture sequence of an automobile chase (the visual sensation of movement takes place in the absence of a simultaneous somatosensory and vestibular input).

11.6.2. Pathological Vertigo

Vestibular Vertigo

Vestibular vertigo is a type of pathological vertigo that usually results from irritative or destructive injuries at some point in the vestibular system. Because of the interrelationship of vestibular, visual, and somatosensory modalities in vestibular sensation, it is also possible to have vertigo caused by disturbances of the visual and the somatosensory systems.

Rotational or Linear Vertigo

Unilateral injury to the vestibular labyrinth often leads to severe **rotational or linear vertigo** associated with vestibular nystagmus, vestibular ataxia, and nausea. Patients are likely to experience a subjective sense of motion in the same direction as the nystagmus. They may also experience a compensatory vestibulospinal reflex response that results in falling to the side of the injury. Vertigo with nystagmus is likely to occur after injury to the vestibulocochlear nerve [VIII] in the internal acoustic meatus or at its brain stem entry.

Positional or Postural Vertigo

Vertigo can result from changes in position or posture of the body or head. Such **positional or benign postural vertigo** is initiated by briskly turning head so that one ear is facing down, by lying back in bed, arising from bed, or bending forward. **Benign paroxysmal postural vertigo** usually does not occur when the head is in a normal position nor while seated. In a series of 10 patients, transection of the fibers to the posterior semicircular canal led to relief from such vertigo, indicating that this canal or its contents are involved in the pathophysiology of this disorder.

Alcohol Related Positional Vertigo with Nystagmus

A common cause of motion sickness with accompanying nystagmus is alcohol ingestion. When appropriate blood levels of alcohol are reached (40 mg per deciliter), alcohol diffuses into the cupulae of the semicircular ducts making them temporarily lighter than endolymph and sensitizing these ampullary receptors to forces of gravity. Vertigo accompanied by nystagmus results when the intoxicated individual lies down. As time goes on the alcohol then diffuses into the endolymph therefore canceling out its initial effect on the cupula. A quiet period ensues in which vertigo is absent (this is usually some 3½–5 hours after cessation of alcohol ingestion). Alcohol diffuses out of the cupula before leaving the endolymph, again setting up the imbalance in specific gravity between the cupula and the endolymph. Positional vertigo with nystagmus continues, beginning 5–10 hours after cessation of alcohol ingestion, and continues until all alcohol leaves the endolymph. It remains until the cupula and the endolymph regain their similar specific gravity – probably many hours after blood alcohol levels return to zero. Treatment of a hangover and its accompanying motion sickness with a morning-after drink has its basis in attempts to equalize the specific gravity of the cupula and the endolymph.

Vertigo Due to Vestibular Neuritis

Vestibular neuritis is a disorder of the vestibular system distinguished by sudden unilateral loss of vestibular function in an otherwise healthy patient without auditory involvement or other disease of the central nervous system. Single or multiple attacks of vertigo, perceived as turning, whirling, or dizziness, often associated with milder periodic or constant unsteadiness (vestibular ataxia) may occur in this condition. Current evidence suggests that the disease involves atrophy of one or both trunks of the vestibular nerve, with or without involvement of their associated receptors. Clinical and pathological aspects of vestibular neuritis suggest that a virus often causes this disorder.

FURTHER READING

Baloh RW, Halmagyi GM (1996) *Disorders of the Vestibular System*. Oxford University Press, New York.

Baloh RW, Honrubia V (2001) *Clinical Neurophysiology of the Vestibular System*. 3rd ed. Oxford University Press, New York.

Bergström BL (1973) Morphology of the vestibular nerve. I. Anatomical studies of the vestibular nerve in man. Acta Otolaryngol 76:162–172.

Bergström B (1973) Morphology of the vestibular nerve. II. The number of myelinated vestibular nerve fibers in man at various ages. Acta Otolaryngol 76:173–179.

Black FO, Simmons FB, Wall C III (1980) Human vestibule-spinal responses to direct electrical eighth nerve stimulation. Acta Otolaryngol 90:86–92.

Brandt T, Dieterich M (1999) The vestibular cortex. Its locations, functions, and disorders. Ann NY Acad Sci 871:293–312.

Brandt T, Strupp M (2005) General vestibular testing. Clin Neurophysiol 116:406–426.

Day BL, Fitzpatrick RC (2005) The vestibular system. Curr Biol 5:R583–R586.

de Waele C, Baudonniere PM, Lepecq JC, Tran Ba Huy P, Vidal PP (2001) Vestibular projections in the human cortex. Exp Brain Res 141:541–551.

Dieterich M (2004) Dizziness. Neurologist 10:154–164.

Eickhoff SB, Weiss PH, Amunts K, Fink GR, Zilles K (2006) Identifying human parieto-insular vestibular cortex using fMRI and cytoarchitectonic mapping. Hum Brain Mapp 27:611–621.

Fasold O, von Brevern M, Kuhberg M, Ploner CJ, Villringer A, Lempert T, Wenzel R (2002) Human vestibular cortex as identified with caloric stimulation in functional magnetic resonance imaging. NeuroImage 17:1384–1393.

Hawrylyshyn PA, Rubin AM, Tasker RR, Organ LW, Fredrickson JM (1978) Vestibulothalamic projections in man – a sixth primary sensory pathway. J Neurophysiol 41:394–401.

Lóken AC, Brodal A (1970) A somatotopical pattern in the human lateral vestibular nucleus. Arch Neurol 23:350–357.

Miyamoto T, Fukushima K, Takada T, De Waele C, Vidal PP (2005) Saccular projections in the human cerebral cortex. Ann NY Acad Sci 1039:124–131.

Richter E (1980) Quantitative study of human Scarpa's ganglion and vestibular sensory epithelia. Acta Otolaryngol 90:199–208.

Rosenhall U (1972) Vestibular macular mapping in man. Ann Otol Rhino Laryngol 81:339–351.

Rosenhall U (1973) Degenerative patterns in the aging human vestibular neuron-epithelia. Acta Otolaryngol 76:208–220.

Rosenhall U, Rubin W (1975) Degenerative changes in the human vestibular sensory epithelia. Acta Otolaryngol 79:67–80.

Ross MD, Johnsson L-G, Peacor D, Allard LF (1976) Observations on normal and degenerating human otoconia. Ann Otol Rhino Laryngol 85:310–326.

Sadjadpour K, Brodal A (1968) The vestibular nuclei in man: a morphological study in the light of experimental findings in the cat. J Hirnforsch 10:299–323.

Schneider RC, Crosby EC (1980) Motion sickness: Part I – A theory. Aviat Space Environ Med 51:61–64.

Schneider RC, Crosby EC (1980) Motion sickness: Part II – A clinical study based on surgery of cerebral hemisphere lesions. Aviat Space Environ Med 51:65–73.

Schneider RC, Crosby EC (1980) Motion sickness: Part III – A clinical study based on surgery of posterior fossa tumors. Aviat Space Environ Med 51:74–85.

Ylikoski J (1982) Morphologic features of the normal and 'pathologic' vestibular nerve of man. Am J Otol 3:270–273.

No video system or computerized camera, no matter how sophisticated, can match the ability of the human visual system to make sense of an infinite variety of images. That ability is made possible by the brain's capacity to process huge amounts of information simultaneously.

Margaret S. Livingstone, 1988

12

The Visual System

Vision, including the appreciation of the color, form (size, shape, and orientation), and motion of objects as well as their depth, is a **special somatic afferent** sensation served by the visual apparatus including the retinae, optic nerves, optic chiasma, lateral geniculate nuclei, optic tracts, optic radiations, and visual areas in the cerebral cortex.

12.1. RETINA

The photoreceptive part of the visual system, the **retina**, is part of the inner tunic of the eye. The retina has ten layers, that can be divided into an outermost, single layer of pigmented cells (Layer 1), the **pigmented layer**, and a neural part, the **neural layer** (Layers 2–9).

12.1.1. Pigment Layer[1]

The **pigmented layer**[1] [*In this chapter, the layers of the retina are indicated as superscripts in the text*] is formed by the retinal pigmented epithelium (RPE), a simple cuboidal epithelium with cytoplasmic granules of **melanin**. Age-related decrease and regional variations in melanin concentration in the pigmented layer[1] occur in humans. The pigmented layer[1] (Fig. 12.1) adjoins a basement membrane adjoining choroidal connective tissue. The free surfaces of these pigmented cells are adjacent to the tips of the outer segments of specialized neurons modified to serve as **photoreceptors**. One pigmented epithelial cell may contact some 30 photoreceptors in the primate retina. Outer segments of one type of photoreceptor, the **rods**, are cylindrical while the outer segments of the other type, the **cones**, are tapering. By absorbing

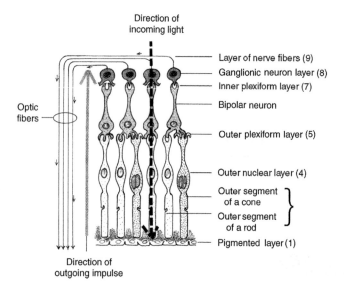

Direction of
incoming light

Layer of nerve fibers (9)
Ganglionic neuron layer (8)
Inner plexiform layer (7)
Bipolar neuron
Outer plexiform layer (5)
Outer nuclear layer (4)
Outer segment
of a cone
Outer segment
of a rod
Pigmented layer (1)

Optic
fibers

Direction of
outgoing impulse

FIGURE 12.1. Neuronal organization of the retina in humans. The direction of incoming light is shown. This stimulates the rods and cones that carry the resulting impulses in the opposite direction to bipolar neurons and then to ganglionic neurons. Axons of ganglionic neurons form the optic nerve [II] that carries visual impulses to the central nervous system (modified from Sjöstrand, 1961 and Gardner, Gray and O'Rahilly, 1969).

light and heat energy, pigmented cells protect photoreceptors from excess light. They also carry out resynthesis and isomerization of visual pigments that reach the outer segments of retinal photoreceptors. Pigmented cells demonstrate phagocytic activity, engulfing the apical tips of outer segments of retinal rods detached in the process of renewal. Age-related accumulation of lipofuscin granules takes place in the epithelial cells throughout the pigmented layer[1].

12.1.2. Neural Layer

The **neural layer** corresponds to the remaining nine layers of the retina (layers 2–10) illustrated, in part, in Fig. 12.1. Layer 2 is the **layer of inner and outer segments**[2] of cones, adjoining the pigmented layer[1]. Layer 3 is the **outer limiting layer**[3] and layer 4 is the **outer nuclear layer**[4]. Layer 5 is the **outer plexiform layer**[5] and layer 6 is the **inner nuclear layer**[6]. Layer 7 is the **inner plexiform layer**[7] and layer 8 is the **ganglionic layer**[8]. Layer 9 is the **layer of nerve fibers**[9] and layer 10 is the **inner limiting layer**[10]. Several types of retinal neurons (Fig. 12.1), interneurons, supporting cells and neuroglial cells occur in these nine layers. Most synapses in the retina occur in the outer[5] and inner plexiform[7] layers (Fig. 12.1). Such synapses in humans are chemical synapses.

The **layer of nerve fibers**[9] (Fig. 12.1) is identifiable with an ophthalmoscope as a series of fine striations near the inner surface of the retina. Such striations represent bundles of individual axons. Recognition of this normal pattern of striations often aids in early diagnosis of certain injuries. **Retinal astrocytes**, a neuroglial element, also occur in the **layer of nerve fibers**[9].

12.1.3. Other Retinal Elements

Other retinal elements include two types of interneurons, **horizontal** and **amacrine neurons**, as well as certain supporting cells, the **radial gliocytes** (Müller cells). Though their appearance in the retina of primates suggests a neuronal function, neither amacrine nor horizontal cells are 'typical' neurons considering their unusual synaptic organization and electrical responses. Processes of horizontal neurons, with cell bodies in the inner nuclear layer[6], extend into the outer plexiform layer[5] and synapse with dendrites of bipolar neurons.

Horizontal Neurons

Two types of horizontal neurons occur in humans: one synapses with cones, the other with rods. Synapses between horizontal neurons and rods and cones underlie the process of **retinal adaptation** – the mechanism by which the retina is able to change sensitivity as light intensities vary under natural conditions. Retinal adaptation probably involves two processes – **photochemical adaptation** by the photoreceptors and **neuronal adaptation** by retinal neurons (including horizontal neurons) and photoreceptors.

Amacrine Neurons

Amacrine [Greek, without long fibers] **neurons** are peculiar in having no axon. Their somata, occurring in the inner nuclear layer[6], exhibit a selective accumulation of the inhibitory neurotransmitter glycine. Amacrine neurons in humans also contain the inhibitory amino acid, γ-amino butyric acid (GABA), and several peptides including substance P, vasoactive intestinal peptide (VIP), somatostatin (SOM), neuropeptide Y (NPY), and peptide histidine-isoleucine (PHI). Substance P, VIP, and PHI occur in neuronal cell bodies in the inner plexiform layer[7] whereas GABA, substance P, VIP, SOM, and NPY occur in cell bodies in the ganglionic layer[8]. These peptidergic neurons are either displaced or interstitial amacrine neurons. Many amacrine neurons synapse with processes of other amacrine neurons in the **inner plexiform layer**[7].

In humans, this layer shows an unusual diversity of neurotransmitters, including GABA and fibers immunoreactive for substance P that may be processes of amacrine neurons. The inner plexiform layer[7] also features diffuse glycine labeling of processes of amacrine neurons, peptide immunoreactive fibers (presumably processes of amacrine neurons), and a density of high affinity [3H] muscimol binding sites that label high affinity GABA receptors. There are benzodiazepine receptors, [3H] strychine binding presumably to glycine receptors, dopamine receptor binding and dopaminergic nerve terminals, and a high density of muscarinic cholinergic receptors, but low levels of β-adrenergic receptors in the inner plexiform layer[7].

Radial Gliocytes

Radial gliocytes are specially differentiated supporting cells in various retinal layers that provide paths for metabolites to and from retinal neurons. Radial gliocytic processes separate photoreceptors from each other near the outer limiting layer[3]. As retinal neurons diminish near the retinal periphery, radial gliocytes replace them, showing a structural modification based on their location as well as a functional differentiation. GABA-like immunoreactiviy is demonstrable in radial gliocytes of the human retina.

Dopaminergic Retinal Neurons

Neurons that accumulate and those that contain dopamine and their processes are identifiable in the primate retina, therefore making dopamine the major catecholamine in the retina. One group of dopaminergic neurons, with many characteristics of amacrine neurons, called **dopaminergic amacrine neurons**, has their cell bodies in the inner nuclear layer[6]. Their dendrites arborize predominately in the outer part of the inner plexiform layer[7]. Here they synapse with other amacrine neurons, and hence are likely to be interamacrine neurons. A second group of dopaminergic neurons probably exists in humans, with cell bodies in the inner nuclear layer[6] and processes extending to both plexiform layers[5, 7]. Consequently, these neurons are termed **interplexiform dopaminergic neurons**. Perhaps they participate in the flow of impulses from inner[7] to outer plexiform layer[5]. Studies of content, uptake, localization, synthesis, and release of dopamine in the retina have helped to substantiate its neurotransmitter role in the human retina. These dopaminergic neurons are **light sensitive** and inhibitory. Peptidergic interplexiform neurons occur in the human retina. The presence of

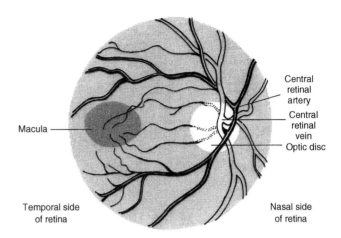

FIGURE 12.2. Normal fundus of the left eye as viewed by the examiner. Notice the pale optic disc on the nasal side with retinal vessels radiating from the disc and over the surface of the retina. Approximately three millimeters on the temporal side of the optic disc is the darker, oval macula that is only slightly larger than the optic disc. The center of the macula, the foveola, has only cones and is a region of acute vision.

acetylcholinergic receptors in the human retina indicates cholinergic neurotransmission takes place here. Though some properties of neurotransmitters exist at birth in humans, significant maturation of these properties takes place postnatally.

12.1.4. Special Retinal Regions

The Macula

The **macula** (Fig. 12.2) is a small region about 4.5 mm in diameter near the center of the whole retina but on the **temporal** or lateral side (Fig. 12.2). A concentration of yellow pigment consisting of a mixture of carotenoids, lutein, and zeaxanthin characterizes the macula. This pigment protects the retina from short-wave visible light and influences color vision and visual acuity (clarity or clearness of vision) by filtering blue light. After 10 years of age, there is much individual variation in macular pigmentation, but this variation is not age-related.

The Fovea Centralis and Foveola

The macula has a central depression about 1.5 mm in diameter called the **fovea centralis** [Latin, central depression or pit], where visual resolution is most acute and pigmented cells most densely packed. Visual acuity declines by about 50% just two degrees from the fovea. The adjoining choroid nourishes the avascular fovea. The central area of the fovea, the **foveola**, is thin, lacks at least four retinal layers (inner nuclear[6], inner

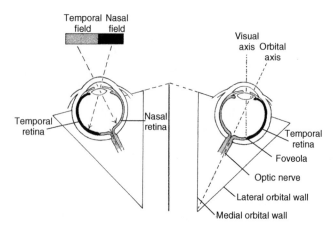

FIGURE 12.3. Anatomical relationship of the visual and orbital axes. The triangular-shaped walls of the orbit are shown. Bisecting the angle formed by the medial and lateral orbital walls on the right is a dashed line representing the longitudinal or orbital axis. The visual or optic axis passes from the object viewed to the foveola. The left visual field as viewed by the left retina is also shown. The visual image is inverted and reversed by the lens of the eye (modified from Gardner, Gray and O'Rahilly, 1969, after Whitnall, 1932).

plexiform[7], ganglionic[8], and layer of nerve fibers[9]), and is about 100–200 μm in diameter. The foveola does have cones, a few rods, and modified radial gliocytes. Density of cones is greatest in the foveola, with a peak density of 161,900 cones per square millimeter in one study. The foveal slope is termed the **clivus**.

The Fovea and the Visual Axis

The fovea is specialized for fixation, acuity, and discrimination of depth. A line joining an object in the visual field and the foveola is the **visual** (optic) **axis** (Fig. 12.3). Misalignment between the visual axes of the two eyes leads to **diplopia** (double vision) that severely disrupts visual acuity. The **orbital axis** extends from the center of the optic foramen (apex of the orbit), travels through the center of the optic disc, and divides the bony orbit into equal halves (Fig. 12.3).

Developmental Aspects of the Retina

The retina appears to be sensitive to light as early as the seventh prenatal month. Poorly developed at birth with a paucity of cones, **foveal photoreceptors** permit fixation on light about the fourth postnatal month. They remain immature for the first 15 months of life, becoming mature by 45 months coinciding with the observation that visual perception reaches the adult level at four years. While the visual capabilities of infants seem to be considerable, only elements in the peripheral retina are

fully functional a few days after birth continuing to develop for several months thereafter. Visual acuity, as measured by the ability to see shapes of objects, such as symbols or letters on a chart, develops rapidly after birth, reaches adult levels at six months, and shows a steady level until 60 years, after which it declines. With age, visual acuity for a moving target is poor compared to that for a stationary one. The foveal cones at 11 months are slim and elongated, like those in adults.

The Optic Disc

About 3 mm to the nasal side of the macula is the **optic disc** (Fig. 12.2). Processes of retinal ganglionic neurons accumulate here as they leave the retina and form the **optic nerve** [II]. Since there are no photoreceptors here, this area is not in vision but is physiologically a **blind spot**. The optic disc is paler than the rest of the retina, 1.5 mm in diameter and appears pink with its circumference or **disc margins**, slightly elevated. The center of the optic disc has a slight depression, the **physiological cup**, pierced by the **central retinal artery and vein** (Fig. 12.2). Since the retinal vessels go around, not across the macula, visual stimuli do not have to travel through blood vessels to reach photoreceptors in the macula. The optic disc is easily visible with an ophthalmoscope and therefore of commanding interest in certain diseases. In the face of disease, it is elevated, flat, excavated, or discolored – pale or white rather than pink.

12.1.5. Retinal Areas

Because the fovea is slightly eccentric, a vertical line through it divides the retina into unequal parts – the **hemiretinae**. That part of the retina on the temporal side of the fovea – the **temporal retina**, is slightly smaller than the **nasal retina** (Fig. 12.3). A horizontal line through the fovea divides the retina into **superior and the inferior retinal areas**. Combining superior and inferior retinal areas with temporal and nasal areas leads to four retinal quadrants in each eye. These are **superior temporal, inferior temporal, superior nasal,** and **inferior nasal quadrants**. About 41% of the retinal area in humans belongs to the temporal retina.

12.1.6. Visual Fields

The **visual field** (Fig. 12.4) is the visual space in which objects are simultaneously visible to one eye when that eye fixes on a point in that field. Since visual acuity is greatest near the visual axis (fovea), objects nearest

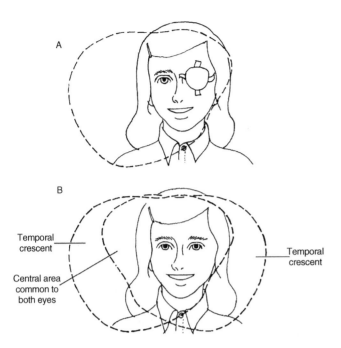

FIGURE 12.4. **A**, Uniocular visual field as visualized by the right retina. Because the lens inverts the visual image and reverses it, the inferior half of the retina views the superior half of the visual field, whereas the temporal part of the retina views the nasal part of the visual field. **B**, Visual fields of both eyes (binocular field). The central area common to both eyes is bounded on either side by the temporal crescents. In this central part, visual acuity is slightly greater than in the same area of either field separately.

to this point are clearest while objects farther from it are fainter with small objects becoming almost invisible. Differences in visual acuity are a reflection of differences in retinal sensitivity. The retina as a whole is most sensitive in its center (at the fovea) with sensitivity decreasing at its circular periphery.

Uniocular and Binocular Visual Fields

The **uniocular visual field** (Fig. 12.4) is that region visualized by one retina extending 60° superiorly, 70–75° inferiorly, 60° nasally, and 100–110° laterally from the fovea. The uniocular visual fields of each retina in humans overlap such that a **binocular visual field** is formed (Fig. 12.4). Though binocular interaction (the interaction between both eyes) does not occur in the newborn, this phenomenon appears by two to four months of age. By about 12 months, the binocular visual field reaches adult size – especially its superior part. The binocular field includes a **central part**, common to both retinae and almost circular in diameter, extending within a 30° radius of the visual axis (Fig. 12.4), and a **peripheral part** or **temporal crescent** (Fig. 12.4) visualized by one retina. The temporal crescent extends

between 60° and 100° from the median plane (visual axis).

Quadrants of the Visual Fields

Each uniocular field is divisible into quadrants: **superior and inferior nasal visual fields** and **superior and inferior temporal visual fields**. Though the temporal retinal area is smaller than the nasal retinal area, the temporal visual field is larger than the nasal visual field (Fig. 12.3). The difference in size of the retinal areas versus the visual fields is because of the transparent lens system and the inverse relation that exists between the position of any point in the visual field and its corresponding image on the retina.

Examination of the Visual Fields

Testing retinal function by examining the visual fields is an essential part of a neurological examination. Defects are likely to be age-related or result from cerebrovascular disease, tumors, or infections. The visual fields can be tested using colors or the fingers of the examiner. If the latter are used, the examiner faces or 'confronts' the patient (hence the term **confrontational visual field examination**) at a distance of about 3 feet and introduces his or her fingers or hand-held colored objects into the visual field of the patient from the periphery. The border of the visual field is the outer point at which the patient is aware of a finger or colored object. The confrontational method of examining the visual fields is useful in determining large or prominent defects in visual fields. Representation of the visual fields on a **visual field chart** (Fig. 12.5), which uses a coordinate system for specifying retinal location analogous to that in the visual field, provides a more precise physiological method of depicting the visual fields. The primary axis of this system is through the fovea. The horizontal meridian at zero degrees passes through the optic disc of the right eye and the 180° meridian passes through the optic disc of the left eye. The superior vertical meridian of both retinae is at 90° and the inferior vertical meridian at 270°. The macula has a diameter of 6°30′ when plotted on a visual field chart; the **fovea centralis**, its central depression, has a diameter of about 1°.

Sensitivity to light and volume of visual fields remain constant into the 37th year, after which they decrease linearly. Age-related decreases in retinal sensitivity influence the superior half of the visual field more than the inferior half, and peripheral and central visual field more than the pericentral region. Such changes are likely attributable to age-related changes

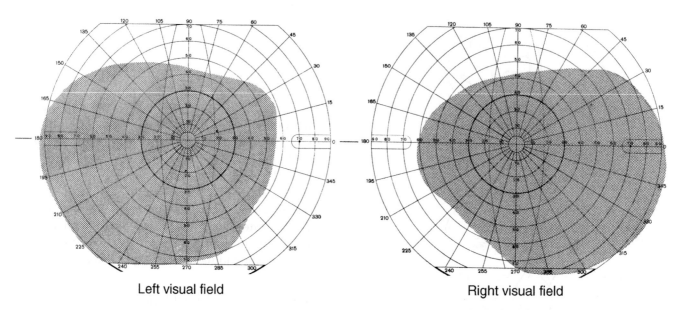

Left visual field   **Right visual field**

FIG. 12.5. Normal visual fields as recorded on a visual field chart. The field of the right eye is placed on the right and that of the left eye on the left of the chart. This is the physiological representation of the visual fields (chart provided by James G. Ferguson Jr., MD).

in photoreceptors, ganglionic neurons, and fibers in the primary visual cortex.

12.2. VISUAL PATH

12.2.1. Receptors

Rods and cones are neurons modified to respond to intensity and wavelength, thereby serving as the **receptors** in the visual path not as the primary neurons. The human visual system responds to light of wavelength from 400–700 nm. Each neuronal type, with their cell bodies in the outer nuclear layer[4], has an **outer segment** (whose shape gives the cell its name) in layer 2, an **inner segment**, and a **synaptic ending** that puts these photoreceptors in contact with dendrites of retinal **bipolar neurons** (Fig. 12.1) in the **outer plexiform layer**[5]. A loss of photoreceptors from the outer nuclear layer[4] with a concomitant loss of photoreceptors in the macula is observable in retinae of patients over 40.

Processes of horizontal neurons also synapse with bipolar dendrites in the outer plexiform layer[5]. Some 111–125 million rods and about 6.3–6.8 million cones tightly pack the plate-like retina in humans. Receptive surfaces of rods and cones face away from incoming light that must then pass through all other retinal layers before reaching the outer segments of the rods and cones (Fig. 12.1). Such an arrangement protects the photoreceptors from overload by excess stimuli. An image in the visual field reaches the retina as light rays that stimulate

the photosensitive pigments in the outer segments of rods and cones. Ultrastructural studies of rods in those over 40 reveal elongation and convolutions in the outer segments of individual rods, with some 10 to 20% affected by the seventh decade. These changes represent an aging phenomenon.

Visual Pigments

A visual pigment, **rhodopsin**, exists in the outer segment of rods. Retinal rods in humans have a mean wavelength near 496.3 ± 2.3 nm. Different light-absorbing pigments in the outer segments of cones permit the identification of three classes of cones in humans. Each class absorbs light of a certain wavelength in the visible spectrum. These include cones sensitive to light of long wavelength (with a mean of 558.4 ± 5.2 nm) that are 'red sensitive', cones sensitive to light of middle wavelength (with a mean at 530.8 ± 3.5 nm) that are 'green sensitive', and cones sensitive to light of short wavelength (with a mean of 419.0 ± 3.6 nm) that are 'blue sensitive'. Our ability to appreciate color requires the proper functioning of these classes of cones and the ability of the brain to compare impulses from them. There are likely congenital dysfunctions of these cones leading to disorders of color vision.

Melatonin, synthesized and released by the pineal gland, is identifiable in the human retina on a wet weight basis. A melanin-synthesizing enzyme, hydroxyindole-O-methyltransferase (HIOMT), is in the human retina. Cytoplasm of rods and cones has HIOMT-like immunoreactivity, suggesting that these cells are involved in synthesizing melatonin. Perhaps melatonin regulates the amount of light reaching the photoreceptors.

Visual Pigments and Phototransduction

The initial step in conversion of light into neural impulses, a process called **phototransduction**, requires photosensitive pigments to undergo a change in molecular arrangement. Each retinal photoreceptor absorbs light from some point on the visual image and then generates an appropriate action potential that encodes the quantity of light absorbed by that photoreceptor. Action potentials thus generated are carried to the **bipolar neurons** and then to the **ganglionic neurons** (Fig. 12.1), in a direction opposite to that of incoming visual stimuli.

Scotopic and Photopic Vision

Rods are active in starlight and dim light at the lower end of the visible spectrum (**scotopic vision**). These same rods are overwhelmed in ordinary daylight or if lights in a darkened room suddenly brighten. With only one type of rod, it is not possible to compare different wavelengths of light in dim light or starlight. Under such conditions, humans are completely color-blind. Cones function in bright light and daylight (**photopic vision**) and are especially involved in color vision with high acuity. Such attributes are characteristic of the fovea, where the density of cones is greatest. Optimal foveal sensitivity, as measured by one investigator, occurred along the visible spectrum at a wavelength of about 562 nm, resembling the absorbance of long-wave 'red' cones. Density of cones falls sharply peripheral to the fovea though it is higher in the nasal than in the temporal retina.

Retinal Photoreceptors are Directionally Transmitting and Directionally Sensitive

Retinal rods and cones are directionally transmitting and directionally sensitive, qualities based on many structural features of photoreceptors and their surroundings. Photoreceptors are transparent, with a high index of refraction. Near the retinal pigment epithelium[1], processes of pigmented cells separate photoreceptors from each other whereas near the outer limiting layer[3] processes of radial glial cells separate photoreceptors. Interstitial spaces around photoreceptors, created by these processes, have a low index of refraction. The combination of transparent cells with a high index of refraction, and an environment distinguished by a low index of refraction, creates a bundle of fiber optic elements. The system of photoreceptors and fiber optics effectively and efficiently receives appropriate visual stimuli and then guides light to the photosensitive pigment in their outer segments.

12.2.2. Primary Retinal Neurons

Bipolar neurons, whose cell bodies occur in the inner nuclear layer[6] together with amacrine neurons, are **primary neurons** in the visual path. Bipolar and amacrine neurons display selective accumulation of glycine and are likely interconnected, allowing feedback between them, which is possibly significant in retinal adaptation or other aspects of visual processing. Retinal bipolar neurons are comparable to bipolar neurons in the spiral ganglia that serve as primary auditory neurons and those in the vestibular ganglia that serve as primary vestibular neurons. Terminals of **rods and cones** synapse with dendrites of **bipolar neurons** (Fig. 12.1) in the outer plexiform layer[5]. Cone terminals (pedicles) in primates are larger than rod terminals (spherules). Rods synapse with **rod bipolar neurons**; each cone synapses with a **midget** and a **flat bipolar neuron**. Though a midget bipolar neuron synapses with one cone, a flat bipolar neuron often synapses with up to seven cones. Midget bipolar neurons seem color-coded; flat bipolar neurons are probably concerned with brightness or luminosity. As many as 10–50 rods synapse with a single rod bipolar neuron. A neurotransmitter, either glutamate or aspartate, links the photoreceptors with bipolar neurons. Terminals of primary bipolar neurons (and processes of many amacrine neurons) synapse with dendrites of retinal ganglionic neurons and with many amacrine neurons in the **inner plexiform layer**[7] (Fig. 12.1). Therefore, bipolar neurons carry visual impulses from the outer[5] to the inner plexiform layer[7] (Fig. 12.1). In the primate inner plexiform layer[7], at least 35% of synapses are bipolar synapses. Since the remaining synapses are with amacrine neurons, the latter neurons likely have a role in processing visual information.

12.2.3. Secondary Retinal Neurons

Retinal ganglionic neurons with cell bodies in the **ganglionic layer**[8] (also containing displaced amacrine neurons) are secondary neurons in the visual path. There is a sparse synaptic plexus in the layer of nerve fibers[9] where it adjoins the ganglionic layer[8]. Some synapses in this zone stain positively for γ-amino butyric acid (GABA) in humans. These contacts are from displaced amacrine neurons.

Type I Retinal Ganglionic Neurons

At least three types of ganglionic neurons are identifiable in the human retina. **Type I ganglionic neurons**, also called 'giant' or 'very large' ganglionic neurons, have laterally-directed dendrites that ramify forming large dendritic fields in the inner plexiform layer[7] These large neurons usually have somal diameters between 26 and 40 μm (called 'J-cells'), some are up to 55 μm (called 'S-cells').

Type II Retinal Ganglionic Neurons

Type II ganglionic neurons (also called **parasol cells or M-cells**) have large cell bodies (20–30 μm or more in diameter) with a bushy dendritic field and axons that are thicker than axons of type III ganglionic neurons. Type II ganglionic neurons numbering no more than 10% of retinal ganglionic neurons send processes to tertiary neurons in **magnocellular layers** of the lateral geniculate nucleus (LG). Hence, type II or parasol cells are also called **M-cells**. They are not selective to wavelength, have large receptive fields, and are sensitive to the fine details needed for **pattern vision**.

Type III Retinal Ganglionic Neurons

The most numerous retinal ganglionic neurons (80%) are **type III ganglionic neurons** (also called **midget cells or P-cells**) with small cell bodies (10.5 to 30 μm) and smaller dendritic fields. Since they send processes to tertiary visual neurons in **parvocellular layers** of the dorsal lateral geniculate nucleus (LGd) they are termed P-cells. They have small receptive fields, are selective to wavelength (they respond selectively to one wavelength more than to others), and are specialized for **color vision**. In all primates, there are likely two types of P-cells: those near the retinal center participating in the full range of color vision and those outside the retinal center that are red cone dominated. In addition to type II and type III neurons, retinae of nonhuman primates contain another class of ganglionic neurons – **primate γ-cells**, which send axons to the midbrain. Further study will aid in determining the role of various retinal ganglionic neurons in processing visual stimuli and in visual perception. In the visual systems of primates, with their great visual ability, at least two mechanisms exist – one for fine detail (needed for pattern vision) and the other for color.

General Features of Retinal Ganglionic Neurons

Ganglionic neurons in the fovea centralis are smaller than ganglionic neurons in the peripheral part of the retina.

Their dendrites synapse with terminals of primary bipolar neurons and with many amacrine neurons in the inner plexiform layer[7]. The type of retinal bipolar neuron (rod, flat, or midget) that synapses with a retinal ganglionic neuron is uncertain. Although both rods and cones likely influence the same retinal ganglionic neuron, it responds to only one type of photoreceptor at any particular time, with some responding exclusively to stimulation by cones. Central processes of ganglionic neurons, along with processes of retinal astrocyte and radial gliocytes, collectively form the retinal **layer of nerve fibers**[9] that eventually becomes the **optic nerve** [II]. Radial gliocytes separate axons in the layer of nerve fibers[9] into discrete bundles. Convergence of 130 photoreceptors on to a single ganglionic neuron may take place.

Receptive Fields of Retinal Ganglionic Neurons

The **receptive field** of a retinal neuron is the area in the retina or visual field where stimulation by changes in illumination causes a significant modification of the activity in that neuron (excitatory or inhibitory). If explored experimentally, receptive fields of retinal ganglionic neurons are circular and have a **center-surround organization**, with functionally distinct **central** (center) and **peripheral** (surround) zones. The response to light in the center of the receptive field may be excitatory or inhibitory. If stimulation in the central zone yields excitation, it is an **ON ganglionic** or **'on-center' cell**. If central zone stimulation yields inhibition, it is an **OFF ganglionic** or **'off-center' cell**. The ON cells detect bright areas on a dark background and the OFF cells detect a dark area on a bright background. In general, stimulation in the surround tends to inhibit effects of central zone stimulation – a phenomenon called **opponent surround**. Some neurons likely show an on-center, off-surround organization or vice-versa. A center-surround organization is present in tertiary visual neurons in the lateral geniculate body and in neurons of the visual cortex. This 'on' and 'off' arrangement of ganglionic cells is a feature of bipolar cells whose cell bodies occur in the inner nuclear layer[6] of the retina.

From the peripheral retina toward the fovea, sizes of the centers of receptive fields gradually decrease. Overall size of a receptive field, including center plus periphery, does not vary across the retina. The center of a receptive field seems to be served by rods or cones to bipolar neurons and to ganglionic neurons but its peripheral zone includes connections from rods or cones to bipolar neurons, to retinal interneurons (horizontal and amacrine neurons), and then to ganglionic neurons. Terminals of cones synapse with dendrites of bipolar neurons in the outer plexiform layer[5] whereas terminals of primary bipolar neurons synapse with dendrites of retinal ganglionic neurons in the inner plexiform layer[7]. Therefore, bipolar neurons carry visual impulses from the outer[5] to the inner plexiform layer[7]. There is likely a 1:1 relation between a foveal cone and a ganglionic neuron. The receptive fields of such ganglionic neurons, which are probably involved in the perception of small details, have small centers (perhaps the diameter of a retinal cone). Many rods and cones influence ganglionic neurons with large receptive fields. These neurons integrate incoming light from photoreceptors and are sensitive to moving objects and objects at low levels of light without much detail.

12.2.4. Optic Nerve [II]

Central processes of retinal ganglionic neurons along with processes of retinal astrocytes and radial gliocytes collectively form the retinal **layer of nerve fibers**[9] that eventually becomes the **optic nerve** [II]. The **optic nerve** [II] has several parts including an intraocular, intraorbital, intracanalicular, and intracranial parts.

Intraocular Part of the Optic Nerve

Optic fibers in the eyeball form the **intraocular part**. Here the fibers are nonmyelinated and the nerve is narrow in comparison with the intraorbital part. As these fibers traverse the outer layers of the retina, then the choroid, and finally the sclera, they are termed the retinal, choroidal, and scleral parts of the intraocular optic nerve. Ultrastructurally the optic nerve resembles central white matter not peripheral nerve even though it is one of the 12 cranial nerves.

Intraorbital Part of the Optic Nerve

As the nonmyelinated intraocular optic fibers leave the eyeball, they pass through the lamina cribrosa sclerae (the perforated part of the sclera) to become the **intraorbital part** of the optic nerve [II]. Myelinated optic fibers begin posterior to the lamina cribrosa of the sclera. At birth, few fibers near the globe are myelinated. After birth and continuing for about two years, this process of myelination increases dramatically. As a developmental anomaly, myelination often extends from the lamina cribrosa into the intraocular optic nerve and is continuous with the retina. Using an ophthalmoscope, clusters of myelinated fibers appear as dense gray or white striated patches. The intraorbital part of the optic nerve is ensheathed by three meningeal layers: pia mater, arachnoid, and dura mater. Anteriorly these sheaths blend into the outer scleral layers. Here the subarachnoid and the potential subdural

space end. They do not communicate with the eyeball or intraocular cavity. As the optic nerves leave the orbit posteriorly via the optic canal, these meningeal sheaths are continuous with their intracranial counterparts. Therefore, there is continuity between the cerebrospinal fluid of the intracranial subarachnoid space and that in the thin subarachnoid space that extends by way of the optic canal, surrounds the intraorbital optic nerve, and ends at the lamina cribrosa. Along the course of the intraorbital part of the optic nerve, the inner surface of cranial pia mater extends into the optic nerve as longitudinal septa incompletely separating fibers into bundles. These septa probably provide some support for the optic nerve.

Each optic nerve [II] has about 1.1 million fibers (range 0.8 to 1.6 million) with variability between nerves. Most optic fibers (about 92%) are about 2 μm or less in diameter and myelinated, averaging 1–1.2 μm in diameter. A small, but statistically significant, age-related decrease in axonal number and density occurs in the human optic nerve. Substance P is localizable to the human optic nerves from 13–14 to 37 prenatal weeks.

Fibers from the macula travel together as the **papillomacular bundle** on the lateral side of the orbital part of the optic nerve immediately behind the eyeball (Fig. 12.6); small axons of small ganglionic neurons in the fovea centralis predominate in this bundle. Here the papillomacular bundle is especially vulnerable to trauma or to a tumor that impinges on the lateral aspect of the optic nerve. Fibers in the papillomacular bundle shift into the center of the optic nerve as they approach the optic chiasma (Fig. 12.6). At this point, fibers from retinal areas surrounding the macula and forming the **paramacular fibers** travel together; the remaining **peripheral fibers** from peripheral retinal areas are grouped together peripheral to the paramacular fibers.

Intracanalicular Part of the Optic Nerve

After traversing the orbit, intraorbital optic fibers enter the optic canal with the ophthalmic artery, as the **intracanalicular part of the optic nerve**. Meningeal layers on the superior aspect of this part of the nerve fuse with the periosteum of the canal superficial to the nerve, fixing it in place, preventing anteroposterior movement, and obliterating the subarachnoid and subdural spaces superior to it.

Intracranial Part of the Optic Nerve

The **optic nerve** [II] enters the middle cranial fossa as the **intracranial part of the optic nerve**, which measures about 17.1 mm in length, 5 mm in breadth, and 3.2 mm in height. From the optic canal, this part of the

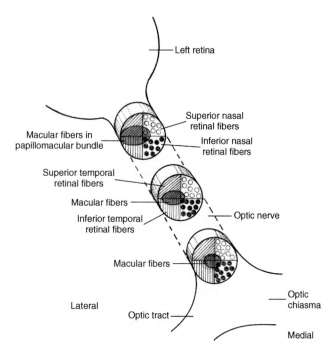

FIGURE 12.6. Course of optic fibers from the posterior aspect of the globe to the optic chiasm. Fibers from the macula are lateral in the optic nerve immediately behind the globe where they are vulnerable to injury. The macular fibers move to the center of the optic nerve as it approaches the chiasm. In this position, the macular fibers are surrounded and protected by fibers from paramacular parts of the retina (modified from Fig. 37 in Scott, 1957).

optic nerve then inclines with its fibers in a plane 45° from the horizontal. Intracranial parts of each optic nerve join to form the **optic chiasma** (Figs 12.6, 12.7).

Small efferent fibers traverse the optic nerve, retinal layer of nerve fibers[9] and bypass the retinal ganglionic neurons before synapsing with amacrine neurons in the inner nuclear layer[6]. About 10% of the fibers in the human optic disc are efferent. They probably excite amacrine neurons that then inhibit the ganglionic neurons. The many synapses of amacrine neurons with retinal ganglionic neurons allow a few efferents to influence many retinal ganglionic neurons.

Retinotopic Organization

Fibers from specific retinal areas maintain a definite position throughout the visual path, from retina to the primary visual cortex in the occipital lobe. Ample evidence, both clinical and experimental, of this **retinotopic organization** is present in primates. Experimental studies have emphasized such organization in the layer of nerve fibers[9] and in the optic disc, an arrangement continuing as central processes of almost all retinal ganglionic neurons enter the optic nerves. Fibers from retinal ganglionic neurons in the superior or inferior temporal retina are

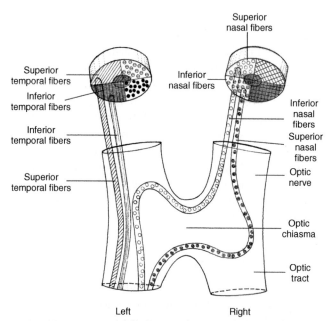

FIGURE 12.7. Course of fibers in the optic chiasma viewed from above. Fibers from the temporal half of the left retina have vertical (inferior temporal retina) or diagonal (superior temporal retina) lines through them. Fibers from the temporal retina do not cross in the chiasm. Fibers from the nasal half of the right retina have open (superior nasal retina) or closed (inferior nasal retina) circles in them. Fibers from the inferior retinal quadrant of each optic nerve cross in the anterior part of the chiasma and loop into the termination of the contralateral optic nerve before passing to the medial side of the tract. Fibers from the superior retinal quadrant of each optic nerve arch into the beginning of the optic tract ipsilaterally before crossing in the posterior part of the chiasma to reach the medial side of the contralateral optic tract (modified from Williams and Warwick, 1975).

superior or inferior in the optic nerve (Fig. 12.6); nasal retinal fibers are medial in the optic nerve.

12.2.5. Optic Chiasma – the Union of Both Intracranial Optic Nerves

Union of both intracranial optic nerves takes place in the **optic chiasma** (Fig. 12.7), a flattened, oblong structure measuring about 12 mm transversely, 8 mm anteroposteriorly, and 4 mm thick. Bathed by cerebrospinal fluid in the **chiasmatic cistern** of the subarachnoid space, the optic chiasma forms a convex elevation that indents the anteroinferior wall of the third ventricle. Since the intracranial optic nerves ascend from the optic canal, the chiasma tilts upward and its anterior margin is directed anteroinferiorly to the chiasmatic sulcus of the sphenoid bone; its posterior margin is directed posterosuperior.

The optic chiasma has decussating nasal retinal fibers from each optic nerve and nondecussating temporal retinal fibers from each optic nerve. Because of this decussation, axons of ganglionic neurons in the left hemiretina of each eye (temporal retina of the left eye and nasal retina of the right eye) will eventually enter the left optic tract (Fig. 12.7). Axons of ganglionic neurons in the right hemiretina of each eye (nasal retina of the left eye and temporal retina of the right eye) enter the right optic tract. Each optic tract therefore transmits impulses from the contralateral visual field. About 53% of fibers in each optic nerve (nasal retinal fibers) decussate in the chiasma: 47% (from each temporal retina) do not cross. These percentages reflect the nasal retina being slightly larger than the temporal retina and thus the temporal visual field is slightly larger than the nasal retinal field. Decussating fibers appear in the optic chiasma during the eighth week of development; uncrossed fibers begin to appear about the 11th week. The adult pattern of partial decussation in the chiasma appears by week 13.

The **anterior chiasmal angle**, between the optic nerves, narrows as the developing eyes approach the median plane. Fibers in the optic nerve and the anterior chiasmal margin are compressed and anteriorly displaced. Because of the breadth of the anterior chiasmal margin, some fibers arch into the optic nerves (Fig. 12.7). The narrower the angle, the more marked the arching. Crossed nasal fibers from ipsilateral and contralateral optic nerves and uncrossed fibers from ipsilateral nerves (temporal retinal fibers) are involved in this arching. In the posterior chiasma, with a wider angle, there is sparse arching of fibers.

In the primate chiasma, macular fibers are surrounded by those from paramacular retinal areas, fibers from superior retinal quadrants being dorsal and those from inferior retinal quadrants ventral in the chiasma. Fibers from peripheral and central superior retinal areas descend from the superior rim of the optic nerve and undergo inversion in the chiasma to enter each optic tract inferomedially. As noted earlier, about 10% of the fibers in the optic disc are efferents. Many authors suggest the presence of these efferents in the human optic nerve and chiasma. Their origin, course posterior to the chiasma, and function are unclear.

12.2.6. Optic Tract

The **optic tract** (Figs 12.7, 12.8) has fibers from both retina – contralateral nasal fibers and ipsilateral temporal fibers. The **right optic tract** has fibers from the right temporal and left nasal retina or, described in another way, fibers from the right hemiretina of each eye. The **left optic tract** has fibers from the left temporal and right nasal retina. Most secondary fibers in the optic tracts synapse with the cell bodies of tertiary neurons

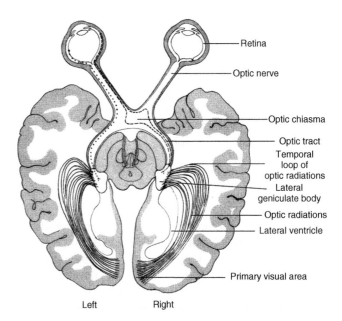

FIGURE 12.8. Retinal origin of optic fibers in humans, their decussation in the optic chiasm, course in the optic tracts, and termination in the lateral geniculate bodies. Note that only fibers from the nasal half of the retina, shown on the left, cross in the optic chiasma to enter the contralateral optic tract. From the lateral geniculate body, the optic radiations pass to the occipital lobe to end in the primary visual area 17.

in the thalamus: a few enter the superior colliculi of the midbrain. The arrangement of fibers in the optic tracts is retinotopic with macular fibers dorsal, those from the superior retina medial, and those from the inferior retinal quadrants lateral.

12.2.7. Thalamic Neurons

Tertiary neurons in the visual path are in the **dorsal part of the lateral geniculate nucleus** (LG$_d$) of the lateral geniculate body (Fig. 12.8) of the dorsal thalamus. Some retinal ganglionic neurons in monkeys also project fibers to the lateral dorsal nucleus (LD) and anterodorsal nucleus (AD). An almost one-to-one ratio exists between optic tract fibers and lateral geniculate somata such that practically all the retinal ganglionic neurons synapse with lateral geniculate somata. There is a direct, bilateral projection from retina to the **pretectal complex** (consisting of five small nuclei) in the diencephalon and a direct, retinal projection to the superior colliculus in humans. There is more information on these nongeniculate retinal connections in Chapter 13.

The Lateral Geniculate Body

Each human **lateral geniculate body** (LGB) is triangular and tilted about 45° with a hilum on its ventromedial surface. Fibers of the optic tract enter on its anterior, convex surface. The horizontal meridian of the visual field corresponds to the long axis of each lateral geniculate body, from hilum to convex surface. The fovea is represented in the posterior pole of the lateral geniculate with the upper quadrant of the visual field represented anterolaterally and the lower quadrant anteromedially in the lateral geniculate nucleus (LG).

Layers of the Lateral Geniculate Nucleus

Sections through the grossly visible lateral geniculate body reveal the microscopically visible lateral geniculate nucleus. The **lateral geniculate nucleus** is surprisingly variable in structure, with several segments: one with two layers, another with four, and one in the caudal half with six parallel layers. The six-layered part has two large-celled layers (an outer **magnocellular layer** ventral to an inner magnocellular layer) and four small-celled layers (an inner, outer, and two superficial **parvocellular layers**). A poorly developed **S-region** is ventral to the magnocellular region in humans. Neurons in the parvocellular layers display rapid growth that ends about six months after birth. Parvocellular neurons reach adult size near the end of the first year; those in magnocellular layers continue to grow rapidly for a year after birth, reaching adult size by the end of the second year. Reduction in mean diameter (and consequently cell volume), is observable in lateral geniculate neurons in patients with severe visual impairment (blindness). There was reduced cytoplasmic RNA, nucleolar volume, and tetraploid nuclei in glial cells.

In primates, each adjacent pair of geniculate neuronal layers is functionally distinct – one activated by the ipsilateral eye, the other by the contralateral eye. In magnocellular and parvocellular layers of the lateral geniculate nuclei in the monkey, the number of neurons receiving impulses from the contralateral eye is greater than that from the ipsilateral eye. During development and before birth, geniculate neurons segregate into layers for the left and right eyes. All type III or midget retinal ganglionic neurons in primates send axons to parvocellular layers (thus they are also named **P-cells**), whereas the larger type II or parasol ganglionic neurons of peripheral retina, and some paramacular retinal neurons provide fast-conducting, large fibers that synapse with magnocellular geniculate cell bodies (thus they are named **M-cells**). Because of this arrangement, representation of each retinal half occurs three times in each lateral geniculate nucleus – in a magnocellular layer and in two parvocellular layers.

Neurophysiological Studies of Lateral Geniculate Neurons

Neurophysiological studies in nonhuman primates reveal that some geniculate neurons are concerned with form, others with color, but most handle both aspects of visual stimuli simultaneously. The 'center-surround' organization characteristic of retinal ganglionic neurons is present in the monkey lateral geniculate nuclei. In one study, lateral geniculate neurons were selectively sensitive to blue, green, yellow, and red. Magnocellular neurons have transient discharge patterns and high contrast gain, whereas the parvicellular neurons have sustained discharge patterns and low contrast gain. Between these cellular layers are thin zones of fibers.

Termination of Retinal Fibers in the Lateral Geniculate Nuclei

Superior retinal fibers end medially in the lateral geniculate nucleus (LG), as inferior fibers end laterally. As macular fibers end in the nucleus, they form a central cone, its apex directed to the hilus of this nucleus. Nasal retinal fibers decussate in the chiasma and end in geniculate nuclear layers 1, 4, and 6, temporal retinal fibers do not decussate in the chiasma but end in layers 2, 3, and 5. In prenatal humans, fibers immunoreactive to substance P occur in the optic nerve and reach the lateral geniculate nuclei. Each superior colliculus in primates receives many processes of ganglionic neurons from both retinae though the contralateral eye appears to receive the strongest input; inputs have a band-like arrangement in collicular zones representing peripheral regions of the binocular visual fields.

Amblyopia and the Lateral Geniculate Nucleus (LG)

Reduction in vision, called **amblyopia** or 'lazy eye', results from disuse of an eye. If the eyes differ in refractive power (called **anisometropia**) and if this condition remains uncorrected, amblyopia often results. Anisometropic amblyopia will result in a decrease in neuronal size in the dorsal lateral geniculate (LG_d) parvocellular layers connected with the 'lazy' eye.

12.2.8. Optic Radiations

Tertiary visual neurons, with their cell bodies in the lateral geniculate body, send axons as **optic radiations** (geniculocalcarine fibers) (Fig. 12.8) to the **primary visual cortex**, corresponding to **Brodmann's area 17** on

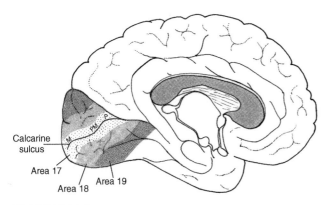

Calcarine
sulcus

Area 17

Area 18   Area 19

FIGURE 12.9. Medial surface of the left cerebral hemisphere to show the location of the primary visual area 17. This region is on the superior and inferior lips, banks, and depths of the calcarine sulcus. On the superior lip, the projection of the macular (M), paramacular (PM) and peripheral (P) parts of the contralateral superior nasal and ipsilateral superior temporal retinal quadrants is shown. Corresponding parts of the inferior retinal quadrants are projected on the inferior lip of the calcarine sulcus. Adjoining area 17 is Brodmann's area 18 that is bordered by area 19 as shown. Part of area 19 is in the parietal lobe anterior to the parieto-occipital sulcus. This parietal part of area 19 is the preoccipital area. Areas 18 and 19 are secondary visual areas.

the superior and inferior lips of the **calcarine sulcus** (Fig. 12.9) of the occipital lobe. Axons from the medial half of the dorsal lateral geniculate nucleus (LG_d) (carrying impulses from the superior retinal quadrants) pass posteriorly to the superior lip of the calcarine sulcus. Many axons from the lateral half of the dorsal lateral geniculate nucleus (LG_d) (carrying impulses from the inferior retinal quadrants) arch into the rostral part of the temporal region as far forward as 0.5 to 1 cm lateral to the tip of the temporal horn and the deeply located amygdaloid body (near the plane of the uncus). They then reach the inferior lip of the calcarine sulcus. These arching fibers from the inferior retina, with a few macular fibers, form the **temporal loop** (of Meyer) **of the optic radiations** (Fig. 12.8). In general, fibers in the optic radiations have a dorsoventral arrangement into three bundles: those from the superior peripheral retina, a central group from the macula, and a ventral one from the inferior retina. Though these fibers have a retinotopic organization, as they course in the temporal lobe, there is considerable variation in their position in the temporal lobe and an asymmetry in arrangement between the two lobes. Collaterals of optic radiations often enter the ipsilateral parahippocampal gyrus.

Termination of the Optic Radiations

The optic radiations end in an orderly manner in the primary visual cortex (Fig. 12.8) of the occipital lobe,

specifically in the superior and inferior lips of the calcarine sulcus (Fig. 12.9). Fibers carrying impulses from the macula (Fig. 12.9) end most posteriorly (1 to 3 cm rostral to the occipital pole), those from the paramacular retina (Fig. 12.9) adjoin them, and those from the unpaired, peripheral retina (Fig. 12.9) end most anteriorly along the calcarine sulcus (Fig. 12.9). The area of macular projection along the primary visual cortex is larger than the area of macular projection on the dorsal lateral geniculate nucleus (LGd). The latter area is larger than the retinal macular area. In the rhesus monkey, the central 5% of the visual field occupies 20% of layer 6 in the lateral geniculate body and 42% of the primary visual cortex. A few fibers of the optic radiations reach the lateral surface of the human cerebral hemisphere. Such projections show individual variation and, where present, often extend 1–1.5 cm onto the lateral surface. In nonhuman primates, most lateral geniculate neurons send axons to the primary visual cortex, with a sparse projection from neurons between the lateral geniculate laminae (interlaminar zones) to visual association areas of the cortex.

12.2.9. Cortical Neurons

Primary Visual Cortex (V1)

At the cortical level, there is reception, identification, and interpretation of visual impulses. The **primary visual cortex,** on the superior and inferior lips, banks, and depths of the **calcarine sulcus** (Fig. 12.9) on the medial surface of the occipital lobe, corresponds to **Brodmann's area 17.** About two-thirds of the primary visual cortex is in the calcarine sulcus, hidden from view. The primary visual cortex, extending from the occipital pole posteriorly to the parieto-occipital sulcus anteriorly, is designated **visual area 1, V1,** the **striate area** or **striate cortex.** Myelinated fibers of the visual radiations enter area 17 and end in its layer IV (the stria of the internal granular layer or stripe of Gennari) forming a visibly evident stripe of fibers that give the primary visual cortex a striated appearance and hence give rise to the term striate area or striate cortex. The primary visual cortex contains a direct representation of retinal activation and carries out low-level feature processing.

Surrounding primary visual area 17 are a number of secondary or **'extrastriate' visual areas** designated **visual area 2** or **V2** and corresponding to **Brodmann's areas 18 and 19** (Fig. 12.9). Areas 18 and 19 do not have a visible stripe of fibers in layer IV. Part of area 19 is in the parietal lobe anterior to the parieto-occipital sulcus.

Parts of areas 18 and 19 are on the lateral surface of the occipital lobe near the occipital pole. These secondary visual areas are **visual association areas**. These extrastriate areas participate in further processing and more advanced analysis of visual information that comes from the primary visual area. Fibers from area 17 end in layers III and IV of area 18, whereas fibers from area 18 end in upper (layers I, II, III) and lower (V and VI) layers of area 17.

The retinotopic organization of the human visual cortex is identifiable by positron-emission tomography. Impulses from the macula project most caudally near the occipital pole but do not extend onto the lateral surface, while peripheral areas of the retina project most rostrally along the calcarine sulcus. Paramacular regions project their impulses between these two. The superior retina projects on the superior lip of the calcarine sulcus while the inferior retina projects on the inferior lip of the calcarine sulcus.

Layers of the Primary Visual Cortex

The **primary visual cortex** (V1/Area 17) is thin, averages 1.8 mm in thickness, and amounts to about 3% (range 2–4%) of the entire cerebral cortex. Although it resembles other cortical areas, being arranged in six layers (layers I–VI), extensive quantitative analyses and correlation studies in humans have identified at least ten layers in the primary visual cortex: layers I, II, III, IVa, IVb, IVc, Va, Vb, VIa, and VIb. The primary visual cortex occupies about 21 cm^2 in each cerebral hemisphere. Area 17 in young adults has about 35,000 neurons per mm^3, alternately cell-sparse and cell-rich horizontal laminae with a conspicuous fibrous layer IV (stria of the inner granular layer), a thin, cell-poor layer V, and a thin, cell-rich layer VI. Layer IV has several subdivisions designated IVa, IVb, and IVc while layer IVc, in turn, is divisible into IVc-α and IVc-β. Neurons in each layer have a distinctive size, shape, density, and response to visual stimuli. Those in layer IV show the simplest response to visual stimuli and reveal an intermingled input from both eyes. Neurons in layers I–III and V–VI are complex in responses and usually driven by both eyes. Neurons in layer IV of the striate cortex send axons to neurons in layers II and III whereas neurons in layers II and III send axons to other cortical areas. Neurons in layer V send axons to the superior colliculus whereas neurons in layer VI send axons back to the lateral geniculate nuclei.

About 67% of the primary visual cortex is not visibly evident on the cortical surface but rather is in the calcarine sulcus, its branches, or accessory sulci. As myelinated fibers of the optic radiations enter area 17 to end in layer IV, they add to the thickly myelinated intracortical fibers there, forming a broad and visible layer, the **stria of the internal granular layer** (layer IVb). Thus, the **primary visual cortex** is termed the **striate cortex**. Layer IV of the primary visual cortex occupies about 33% of the total cortical thickness. About 20% or less of the synapses in layer IV occur on processes of neurons from the lateral geniculate nuclei. Hence, the intrinsic input to this layer, structurally and perhaps functionally, is dominant. A great deal of thalamic and intrinsic input converges on visual neurons in the cerebral cortex. There is a gradual reduction in the myelin in this stria, beginning in the third decade of life. This is likely the result of normal aging but also due to blindness, Alzheimer's disease, or multiple-infarct dementia. The human **primary visual cortex** is responsible for **conscious vision** but not visual interpretation. No appreciable visual consciousness is demonstrable at thalamic levels in humans. Recent studies have suggested that V1 neurons in nonhuman primates report color contrast and therefore may participate in color vision along with neurons outside of V1.

Extrastriate Visual Areas

Secondary Visual Cortex (V2/Area 18)

Area 18, the secondary visual cortex or area V2, surrounds V1, connects with it, and lacks a specialized layer IV. It is termed 'extrastriate' as it is outside or beyond the striate cortex. More than 30 extrastriate visual association areas are identifiable in the temporal, parietal, and occipital lobes of primates. Some of these are secondary visual association areas, some are tertiary and others are likely to be multimodal association areas. Primary visual area 17 sends many fibers to extrastriate visual areas 18 and area 19 that have an especially well-differentiated system of intracortical and myelinated fibers. Area 18 in turn has reciprocal connections with other extrastriate areas.

Extrastriate Area 19

Extrastriate area 19 is the most rostral part of the visual cortex in the occipital lobe. This latter area is not homogeneous but is divisible into a number of visual areas. It is likely a tertiary visual area.

Visual Area V4

This visual area receives input from the parvicellular layers of the lateral geniculate in nonhuman primates. In humans, this extrastriate visual area is specialized for different aspects of object recognition including color and shape. Patients with lesions in this area have an inability to see color, a condition called **achromatopsia**. Visual area V4 is in the collateral sulcus or lingual gyrus of the occipital lobe.

Visual Area V5

Another extrastriate visual area is visual area V5 in the ascending limb of the inferior temporal sulcus that is involved in the perception of motion in humans, including both speed and direction. This area in nonhuman primates receives input from the magnocellular layers of the lateral geniculate nucleus by way of the primary visual cortex. There may be direct projections from V1 to V5 or indirect to V5 through V2 or V3. This motion pathway likely extends beyond the middle temporal area to the medial superior temporal area, the parietal lobe, and the frontal eye fields. Patients with lesions in this area may have a selective disturbance of movement vision such as visual tracking. This visual area in humans is comparable in many ways to area MT in nonhuman primates.

Magno and Parvo Paths from Retina to Visual Cortex

The types of retinal ganglionic neurons (type II or type M cells and type III or type P cells), and their relation to different layers in the dorsal lateral geniculate nuclei (magnocellular and parvocellular) define two parallel paths from retinal ganglionic neurons to the visual cortex. These structural divisions ('**magno**' and '**parvo**') differ in color, acuity, speed, and contrast sensitivity. At cortical levels, these two divisions are probably selective for form, color, movement, and stereopsis.

'What' and 'Where' Processing in the Visual Cortex

At the cortical level the somatosensory, auditory and visual systems in primates are each organized into 'what' and 'where' paths (Table 8.2). Under this concept, information travels first to the primary visual cortex and then relays in serial fashion through a series of increasingly complex visual association areas (the extrastriate visual area). This 'what' and 'where' model of vision in nonhuman primates includes a **ventral stream** ('**what**' path), the occipito-temporo-prefrontal path for perception, identification, and recognition of visually presented objects (object vision, for example, faces and words) based on features like color, texture and contours. The **dorsal stream** ('**where**' path), or occipito-parieto-prefrontal path participates in the appreciation of the

spatial relations among objects (spatial vision) as well as for the visual guidance of movements toward objects in visual space. Examples of objects would be faces, buildings and letters. The occipitotemporal cortex includes Brodmann's areas 19 and 37 whereas the occipitoparietal cortex includes parts of Brodmann's area 19 and area 7 in the superior parietal lobule. The 'prefrontal part' of these paths includes parts of the inferior frontal gyrus corresponding to Brodmann's areas 45 and 47 as well as the dorsal part of premotor area 6. Both these paths in the end send information related to identity and location to the same areas of the prefrontal cortex so that this is not a completely segregated system. There seems to be some left hemisphere specialization or dominance for visual form in the ventral stream. Finally, there is much more to this story including the possibility of additional functional streams or even 'streams within streams'. The myriad of extrastriate visual areas makes this highly probable.

Developmental Aspects of the Visual Cortex

Some differentiation of neurons and dendritic growth takes place in the primary visual cortex in humans in the first few postnatal months with a regular decrease in neuronal density from 21 prenatal weeks until about the fourth postnatal month. However, most developmental changes in neuronal structure and connections in the human visual system take place in the absence of visual experience. Synaptic development in the human primary visual cortex covers a period from the third trimester prenatally to the eighth month postnatally by which time synaptic density and number is maximal. Adult levels of synaptic density occur at 11 years, being 40% less than at eight months. Synaptic density is probably lower in the human primary visual cortex when compared with other cortical areas. Postnatal synaptic reduction occurs in monkeys, with the total number of synapses being greater at six months than in adults. Neuronal differentiation, dendritic growth, changes in neuronal density, synaptogenesis, and synapse elimination in the human primary visual cortex provide excellent examples of plasticity in the central nervous system. The timing and sequence of these events coincides with the development of certain visual functions. When synaptogenesis is rapid (four to five postnatal months), there is a sudden increase in visual abilities including binocular interactions. The apparent excess production of synapses and their eventual elimination is probably a manifestation of activation of certain cortical circuits (neuronal somata, processes, and synapses) that are in use, stabilize, and persist. Nonactivated elements of this circuit often regress and disappear.

Studies of the primary visual cortex in humans suggest that it has an overabundance of synapses that are non-specific or labile from the fourth to the eighth postnatal months, regression and stabilization follow between the eighth month and 11th year, followed by a persistent, stable period throughout adulthood. By analogy, what starts out as a large mass of clay (the developing primary visual cortex with neurons, processes, and synapses) is 'sculptured' (neuronal differentiation, dendritic growth, changes in neuronal density, synaptic elimination) during development until a final form results, that is, the formation of the adult primary visual cortex. No evidence exists for age-related neuronal loss in the human primary visual cortex.

12.3. INJURIES TO THE VISUAL SYSTEM

12.3.1. Retinal Injuries

Depending on the nature, location, and size of the injury, changes in visual acuity, visual fields, and perhaps abnormal visual sensations may occur in humans. The most frequent cause of **retinal injury** is generalized vascular disease. Involvement of both retinae results in complete blindness. A small injury to the retina often leads to a visual field defect corresponding to the position, shape, and extent of the retinal injury. Blindness in the visual field corresponding to the macular retinal area with sparing of the peripheral field is a **central scotoma**. In such cases, vision is lost in a central area surrounded by an area of normal vision, like the hole in a doughnut, with the hole representing the scotoma. Patients often describe visual field defects as spots, glares, shades, veils, or blank areas of vision. If the injury involves fibers in the layer of nerve fibers[9], the visual field defect conforms to the retinal area represented by those fibers. Therefore, a small injury to the macular fibers, or to the optic disc, has a drastic effect. Degeneration of retinal ganglionic neurons was present in the retinas of eight of 10 patients with Alzheimer's disease.

Separation of the pigmented layer of the retina from the neural layers results in a condition called **retinal detachment**. This is likely due to one or more holes in the retina that permit fluid to enter between the pigmented and neural layers. Photocoagulation, cryotherapy, and diathermy are useful methods of repairing these holes and correcting the detachment.

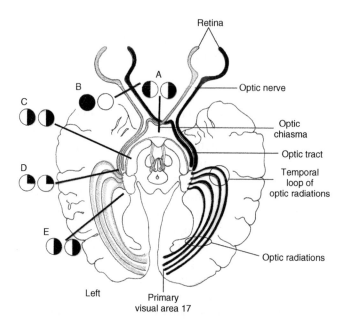

FIGURE 12.10. Visual field deficits caused by interruption or transection of fibers at certain points along the visual path. **A,** Section of the optic chiasma with a resulting bitemporal hemianopia (loss of vision in the temporal parts of both right and left visual fields). **B,** Section of the left optic nerve with blindness in the left visual field and a normal right visual field. **C,** Section of the optic tract causing a contralateral homonymous hemianopia. **D,** Section of the optic radiations in the temporal lobe with an incongruous visual field defect. The temporal part of the right visual field is affected, as is the superior nasal quadrant of the left visual field – a superior quadrantanopia. **E,** Section of the optic radiations in the parietal lobe with a resulting contralateral homonymous hemianopia (modified from Harrington, 1981).

12.3.2. Injury to the Optic Nerve

Injury to one **optic nerve** [II] by inflammation, demyelination or vascular disease may lead to **complete blindness** the uniocular visual field of that eye (Fig. 12.10B). Injury to the lateral part of the optic nerve as the nerve leaves the eyeball often involves the papillomacular bundle. The affected patient will have impaired vision in the macular part of the visual field of that eye, with normal peripheral vision. This condition is termed a **central scotoma**. **Optic neuropathy** is a functional disturbance or pathological change in the optic nerve. Impairment of brightness is a consistent finding with optic neuropathy. Objects and surfaces appear as shades of gray with an absence of color that persists in the face of changes in ambient illumination and accompanying changes in reflected light. Gray levels of an object or surface normalize over a broad range of illumination – a phenomenon called **brightness constancy**.

Swelling of the optic disc, called **papilledema**, may result from a space-occupying, intracranial tumor or as an indirect result of a swollen brain. Papilledema can occur without impairment of vision. In one series, the optic nerves in eight of 10 patients with Alzheimer's disease exhibited widespread axonal degeneration, including sparse packing of axons and considerable glial replacement. Radiation therapy for pituitary tumors and craniopharyngiomas often causes necrosis of fibers in the optic nerve and chiasma.

12.3.3. Chiasmal Injuries

Fibers in the optic chiasma may be flattened or stretched and their vascular supply interrupted by trauma, vascular disease, or tumors of the hypophysial or parasellar region, causing visual impairment. Transection of the chiasma by a gunshot wound in the temple will lead to blindness. If a hypophysial tumor expands beyond the sella (suprasellar extension), it can elevate and flatten the optic nerves and chiasma causing injury to only those fibers from the inferior retina. The result may be a symmetrical, superior temporal visual field defect called **bitemporal superior quadrantanopia**. If the tumor continues to expand and impinge on the optic chiasma and its decussating fibers from each nasal hemiretina, a visual field defect results with loss of vision in both temporal visual fields – a defect called **bitemporal hemianopia** (Fig. 12.10A). Hemianopia (also hemianopia) means 'half without vision' and the term bitemporal refers to the affected visual fields (both temporal crescents). The anatomical basis of bitemporal hemianopia is injury of chiasmal decussating nasal retinal fibers (Fig. 12.10), causing a sharply defined temporal field defect.

Though it seems easy to correlate visual field defects with the arrangement of fibers in the optic chiasma, the invasive character of chiasmal injuries and their effects on its vascular supply often result in visual field defects that defy any correlation with chiasmal anatomy. Examination of visual fields using confrontation with colors may help detect early chiasmal injuries.

12.3.4. Injuries to the Optic Tract

Ganglionic neurons in the left hemiretina of each eye (temporal retinal fibers of the left eye and nasal retinal fibers of the right) send axons to the left cerebral hemisphere, whereas ganglionic neurons in the right hemiretina of each eye (nasal retinal fibers of the left eye and temporal retinal fibers of the right) send axons to the right cerebral hemisphere. Injury to the left optic tract damages fibers from the temporal hemiretina of the left

eye and fibers from the nasal hemiretina of the right eye as they pass to the primary visual cortex, causing a defect in the right half of each uniocular visual field. The resulting condition is termed **homonymous hemianopia** (Fig. 12.10C). 'Homonymous' means that the defect is in the same or similar half of each uniocular visual field whereas 'hemianopia' means that half of each visual field is injured. The optic tract is short, small in diameter, and closely related to the oculomotor nerve, cerebral peduncle, uncus, and posterior cerebral artery. Compression of the optic tract against adjacent structures may follow increased intracranial pressure or injuries in the cranial cavity. Because some fibers in the optic tract transmit impulses for pupillary reflexes, an **afferent pupillary defect** (described in Chapter 13) is likely contralateral to optic tract injury. As pupillomotor fibers in the optic tract are absent from the optic radiations, a complete homonymous hemianopia with an afferent pupillary defect distinguishes injury in an optic tract from one in the optic radiations. Injury to the optic tract causes atrophy in the retinae and optic nerves after about six weeks.

Visual field defects from postchiasmal injuries are substantial and most often of vascular origin. They are detectable with confrontation techniques using the fingers to delineate the visual fields. Such homonymous defects usually have a slight chance of spontaneous recovery, though there is often some improvement within 48 hours of the cortical injury.

12.3.5. Injury to the Lateral Geniculate Body

Nonvascular injuries such as tumors, which infiltrate or compress the lateral geniculate, cause incongruent field defects (the fields are not superimposable). If the injury is limited to the lateral aspect of the lateral geniculate nucleus (LG), where inferior retinal fibers end, a defect in the superior nasal fields (superior quadrantanopia) results.

The lateral geniculate nucleus (LG) receives blood from two sources. The **anterior choroidal artery** normally arises as a single trunk from the supraclinoid part of the internal carotid artery several millimeters distal to the posterior communicating artery. It then makes an anterior approach to the lateral geniculate body along the optic tract (passing from the lateral to the medial side of the tract) before entering the choroidal fissure to end in the **choroid plexus** of the temporal horn. With regard to the visual system, the anterior choroidal artery sends branches to the optic tract and lateral geniculate body (anterior hilum and anterolateral half of this nucleus) and supplies the optic radiations in the retrolenticular part of the posterior limb of the internal capsule. Because of this, a typical anterior choroidal artery infarction causes a congruent defect in the superior and inferior quadrants of the same half of each visual field (a contralateral homonymous hemianopia).

One or more of the **posterior choroidal rami** of the posterior cerebral artery (see Figs 22.2, 22.3, 22.9) supply the posteromedial parts of the lateral geniculate nucleus (LG) on their way to the **choroid plexus** of the lateral ventricle. Injury to the medial aspect of this nucleus, where superior retinal fibers end – the territory of the posterior choroidal artery – causes a defect in the inferior visual fields without involvement of the macular area. Macular fibers form a central cone in the lateral geniculate nucleus (LG), with its apex directed to the nuclear hilus.

12.3.6. Injuries to the Optic Radiations

Owing to their length, the optic radiations are more often subject to injury than the optic tract or the lateral geniculate nucleus (LG). Injury may occur in the internal capsule or in the temporal lobe as the optic radiations travel through them to reach the occipital lobe. The resulting visual field loss is termed a **contralateral homonymous hemianopia**. Here the defect is in the **contralateral half** (hemianopia) of the visual field of each eye (Fig. 12.10E), that is, on the side of the visual field of each eye that is contralateral to the side of injury. The same or **homonymous** half of each uniocular field is involved. Injury to the optic radiations in the temporal lobe may damage a variable number of fibers that arch into the temporal lobe as part of the **temporal loop of the optic radiations**. Fibers from the ipsilateral inferior temporal retina are more anterior and ventral in the temporal loop than the crossed inferior nasal retinal fibers and therefore more vulnerable to injury involving the temporal lobe or small surgical resections of the temporal lobe. The resulting visual field defect in this instance is a **superior nasal quadrantanopia** (Fig. 12.10D), depending on the number of fibers involved. A field defect caused by injury to the optic radiations depends on the nature, extent, and rate of development of the injury, and whether the fibers involved are in the temporal, parietal, or occipital lobe. Ischemic injury to the optic radiations causes decreased glucose metabolism in the appropriate part of the primary visual cortex when examined with positron emission tomography in conjunction with [18]FDG.

The degree to which patients with homonymous hemianopia are aware of their visual deficit varies from complete awareness to complete unawareness. Analysis of computerized tomographic scans of 41 patients demonstrated smaller injuries in the occipital lobe in those patients who were aware of their defect. Patients unaware of their visual defect had extensive, anteriorly located injuries in the parietal lobe.

12.3.7. Injuries to the Visual Cortex

Injuries to the inferior lip of both visual cortices will lead to blindness in the superior half of both visual fields. If, however, the inferior lip on only one side is affected, the loss will be in the superior quadrant on the opposite side and the resulting deficit will be a **contralateral superior quadratic anopia**. Patients often describe the visual field defect caused by a cortical injury as a mist or a haze. If the left **primary visual cortex** is injured, a **contralateral** (right-sided) **homonymous hemianopia** will occur in the right half of each uniocular visual field. Patients with visual field defects learn to look with their good eye into the area not well seen by the other eye. Patients easily and unknowingly carry out compensation for visual field defects. Rehabilitation in patients with visual field deficits attributable to injuries to the primary visual cortex has proven unsuccessful to date. Ischemic injuries to the human visual cortex, causing visual field defects such as homonymous hemianopia, are demonstrable by metabolic mapping. Such methods reveal low glucose utilization in parts of the striate cortex consistent with the visual field loss. Glucose utilization in the adjacent extrastriate cortex is also lower in such instances.

Severe damage to the superior and inferior lips of the calcarine sulcus (the entire **primary visual cortex**) and the optic radiations may occur during occipital lobectomy, performed to remove tumors. In such cases, there is a contralateral homonymous hemianopia with distinct sparing of vision along a narrow strip about 1–3° from the foveal center. With other types of postchiasmal injuries, there is often a contralateral homonymous hemianopia with foveal sparing. Since the macula has a diameter of 6°30′ on a visual field chart, this 1–3° of sparing is foveal not macular in nature. Therefore, the term **foveal sparing** is appropriate for this phenomenon. With removal of the occipital lobe with preservation of the foveal area of the visual field contralateral to the injury, it is evident that there must be bilateral representation of the fovea centralis in the occipital lobes. The anatomical basis for foveal sparing, based on studies in nonhuman primates, is the presence of a 1° vertical strip along the retinal median plane (except in the fovea, where it widens to about 3°), where retinal ganglionic neurons from the nasal retina only intermingle and then yield ipsilateral and contralateral fibers that enter the optic nerves. These observations suggest that the central retinal region on either side of the fovea has a double representation in each cerebral hemisphere.

Injuries to the visual cortex in children do not show a uniform degree of sparing or recovery. Sparing, which does occur after such injury neonatally or in early childhood, often results from subcortical areas becoming proficient in functions that later are carried out primarily by the striate cortex. Altitudinal hemianopia is a visual field defect caused by bilateral injury to the occipital lobes. If the superior lips of both calcarine sulci are injured, an inferior altitudinal defect will result. Selective involvement of the inferior lips of both calcarine sulci with sparing of the superior lips causes a superior altitudinal defect. Though rare, these altitudinal field defects emphasize the representation of the superior visual fields along the inferior lip of the calcarine sulcus and the inferior visual fields along the superior lip of the calcarine sulcus.

FURTHER READING

Boothe RG, Dobson V, Teller DY (1985) Postnatal development of vision in human and nonhuman primates. Ann Rev Neurosci 8:495–545.

Bowmaker JK, Dartnall HJA (1980) Visual pigments of rods and cones in a human retina. J Physiol (Lond) 298:501–511.

Burkhalter A, Bernardo KL (1989) Organization of cortico-cortical connections in human visual cortex. Proc Natl Acad Sci 80:1071–1075.

DeYoe EA, Felleman DJ, Van Essen DC, McClendon E (1994) Multiple processing streams in occipitotemporal visual cortex. Nature 371:151–154.

Fox PT, Miezin FM, Allman JM, Van Essen DC, Raichle ME (1987) Retinotopic organization of human visual cortex mapped with positron-emission tomography. J Neurosci 7:913–922.

Glickstein M (1988) The discovery of the visual cortex. Sci Am 256:118–127.

Glickstein M, Whitteridge D (1987) Tatsuji Inouye and the mapping of the visual fields on the human cerebral cortex. TINS 10:350–353.

Goodale MA, Milner AD (1992) Separate visual pathways for perception and action. Trends Neurosci 15:20–25.

Goodale MA, Westwood DA (2004) An evolving view of duplex vision: separate but interacting cortical pathways for perception and action. Curr Opin Neurobiol 14:203–211.

Huk AC, Dougherty RF, Heeger DJ (2002) Retinotopy and functional subdivision of human areas MT and MST. J Neurosci 22:7195–7205.

Hurlbert A (2003) Colour vision: primary visual cortex shows its influence. Curr Biol 13:R270–R272.

Ishai A, Ungerleider LG, Martin A, Haxby JV (2000) The representation of objects in the human occipital and temporal cortex. J Cogn Neurosci 12 Suppl 2:35–51.

Lennie P (2003) Receptive fields. Curr Biol 13:R216–R219.

Livingstone MS, Hubel DH (1984) Specificity of intrinsic connections in primate primary visual cortex. J Neurosci 4:2830–2835.

Masland RH (2001) The fundamental plan of the retina. Nat Neurosci 4:877–886.

Massey SC (2006) Functional anatomy of the mammalian retina. In: Retina. Ryan SJ, Editor-in-chief. 4th ed. Chapter 4, 1:43–82. Elsevier/Mosby, Philadelphia.

Mishkin M (1979) Analogous neural models for tactual and visual learning. Neuropsychologia 17:139–151.

Neves G, Lagnado L (1999) The retina. Curr Biol 9:R674–R677.

Reh TA, Moshiri A (2006) The development of the retina. In: Retina. Ryan SJ, Editor-in-chief. 4th ed. Chapter 1, 1:2–21. Elsevier/Mosby, Philadelphia.

Rubino PA, Rhoton AL Jr, Tong X, Oliveira E (2005) Three-dimensional relationships of the optic radiation. Neurosurgery 57:219.227.

Schneider KA, Richter MC, Kastner S (2004) Retinotopic organization and functional subdivisions of the human lateral geniculate nucleus: a high-resolution functional magnetic resonance imaging study. J Neurosci 24:8975–8985.

Stensaas SS, Eddington DK, Dobelle WH (1974) The topography and variability of the primary visual cortex in man. J Neurosurg 40:747–755.

Stone J, Johnston E (1981) The topography of primate retina: a study of the human, bushbaby, and new- and old-world monkeys. J Comp Neurol 196:205–223.

Ungerleider LG, Haxby JV (1994) 'What' and 'where' in the human brain. Curr Opin Neurobiol 4:157–165.

Zeki S, Watson JD, Lueck CJ, Friston KJ, Kennard C, Frackowiak RS (1991) A direct demonstration of functional specialization in human visual cortex. J Neurosci 11:641–649.

Zilles K, Werners R, Büsching U, Schleicher A (1986) Ontogenesis of the laminar structure in areas 17 and 18 of the human visual cortex. A quantitative study. Anat Embryol 174:339–353.

Zilles K (1995) Is the length of the calcarine sulcus associated with the size of the human visual cortex? A morphometric study with magnetic resonance tomography. J Hirnforsch 36:451–459.

Zilles K (2004) Architecture of the human cerebral cortex. Regional and laminar organization. In: The Human Nervous System. Paxinos G, Mai JK, eds. 2nd ed. Chapter 27, pp 997–1055. Elsevier/Academic Press, Amsterdam.

The study of eye movements is a source of valuable information to both basic scientists and clinicians. To the neurobiologist, the study of the control of eye movements provides a unique opportunity to understand the workings of the brain. To neurologists and ophthalmologists, abnormalities of ocular motility are frequently the clue to the localization of a disease process.

R. John Leigh and David S. Zee, 2006

13

Ocular Movements and Visual Reflexes

13.1. OCULAR MOVEMENTS

13.1.1. Primary Position of the Eyes

Normally our eyes look straight ahead and steadily fixate on objects in the visual field. This is the **primary position** (Figs 12.3, 13.1) of the eyes. In this position, the visual axes of the two eyes are parallel and each vertical corneal meridian is parallel to the median plane of the head. The primary position is also termed the **position of fixation** or **ocular fixation**. The **position of rest** for the eyes exists in sleep when the eyelids are closed. In the newborn, the eyes often move separately. Ocular fixation and coordination of ocular movements takes place by about three months of age.

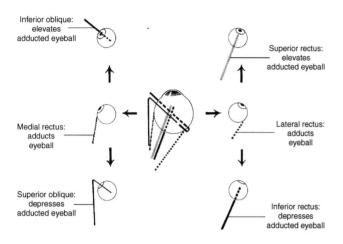

FIGURE 13.1. Certain actions of the muscles of the right eye. In the center, the eye is shown in its primary position with its six muscles also indicated. To the left of center the eye is adducted by the medial rectus. Left and above, the adducted eye is elevated by the inferior oblique. Left and below, the adducted eye is depressed by the superior oblique. To the right of center the eye is abducted by the lateral rectus. Right and above, the abducted eye is elevated by the superior rectus. Right and below, the abducted eye is depressed by the inferior rectus (modified from Gardner, Gray and O'Rahilly, 1975).

13.2. CONJUGATE OCULAR MOVEMENTS

Moving our eyes, head, and body increases our range of vision. Under normal circumstances, both eyes move in unison (yoked together or conjoined) and in the same direction. There are several types of such movements termed **conjugate ocular movements**. These include: (1) **miniature ocular movements**; (2) **saccades**; (3) **pursuit movements**; and (4) **vestibular movements**. The eyes move in opposite directions, independent of each other but with equal magnitude when both eyes turn medially to a common point such as during convergence of the eyes. Such **nonconjugate ocular movements** are termed **vergence movements**.

13.2.1. Miniature Ocular Movements

Because of a continuous stream of impulses to the extraocular muscles from many sources, the eyes are constantly in motion making as many as 33 back and forth **miniature ocular movements** per second. These movements occur while we are conscious, have our eyes in the primary position, and our eyelids are open. We are unaware of these movements in that they are smaller than voluntary ocular movements and occur during efforts to stabilize the eyes and maintain them in

the primary position. These miniature eye movements enhance the clarity of our vision. The **arc minute** is a unit of angular measurement that corresponds to one-sixtieth of a degree. Each arc minute is divisible into 60 arc seconds. During these miniature ocular movements, the eyes never travel far from their primary position – only about 2 to 5 minutes of arc on the horizontal or vertical meridian. The retinal image of the target remains centered on a few receptors in the fovea where visual acuity is best and relatively uniform. Miniature ocular movements encompass several types of movements. These include **flicks** (small, rapid changes in eye position, 1–3/sec, and about 6 minutes of arc), **drifts** (occurring over an arc of about 5 minutes), and **physiological nystagmus** (consisting of high frequency tremors of the order of 50–100 Hz with an average amplitude of less than 1 minute of arc – 5 to 30 seconds is normal).

13.2.2. Saccades

In addition to miniature ocular movements, two other types of voluntary ocular movements are recognized. **Saccades** (scanning or rapid ocular movements) are high velocity movements (angular velocity of 400–600°/sec) that direct the fovea from object to object in the shortest possible time. Saccades occur when we read or as the eyes move from one point of interest to another in the field of vision. While reading, the eyes move from word to word between periods of fixation. These periods of fixation may last 200–300 ms. The large saccade that changes fixation from the end of one line to the beginning of the next is termed the **return sweep**. Humans make thousands of saccades daily that are seldom larger than 5° and take about 40–50 ms. In normal reading such movements are probably 2° or less and take about 30 ms. Thus, saccades are fast, brief, and accurate movements brought about by a large burst of activity in the agonistic muscle (lateral rectus), with simultaneous and complete inhibition or silencing in the antagonistic muscle (medial rectus). Another burst of neural activity then steadily fixes the eye in its new position. The eye comes to rest at the end of a saccade not by the braking action of the antagonistic muscle but rather due to the viscous drag and elastic forces imposed by the surrounding orbital tissues. When larger changes are necessary beyond the normal range of a saccade, movement of the head is required. Saccades are rarely repetitive, rapid, and consistent in performance regardless of the demands on them. It is possible to alter saccadic amplitude voluntarily but not saccadic velocity. The ventral layers of the superior colliculus of the midbrain play an important role in the initiation and speed of saccades as well as the selection of saccade targets. Areas of the human cerebral cortex thought to be involved in the paths for saccades include the intraparietal cortex, frontal eye fields, and supplementary eye fields. There is an age-related increase in visually guided saccade latency.

13.2.3. Smooth Pursuit Movements

Another type of conjugate ocular movement is the **smooth pursuit** or **tracking movements** that occur when there is fixation of the fovea on a moving target. This fixation on the fovea throughout the movement ensures that our vision of the moving object remains clear during the movement. The amplitude and velocity for such tracking movements depends on the speed of the moving target – up to a rate of 30°/sec. Without the moving visual target, such movements do not take place. Many of the same cortical areas noted above as being involved in the paths for saccades (the intraparietal cortex, the frontal eye fields, and the supplementary eye fields) are involved in pursuit movements along with the middle temporal and medial superior temporal areas. Apparently, these overlapping areas have separate subregions for the two types of movements. There is an age-related decline in smooth pursuit movements such that eye velocity is lower than target velocity.

13.2.4. Vestibular Movements

The vestibular system also influences ocular movements. Movement of the head is required when larger changes in ocular movements are necessary beyond the size of normal saccades. The eyes turn and remain fixed on their target but as the head moves to the target, the eyes then move in a direction opposite to that of the head. Stimulation of vestibular receptors provides input to the vestibular nuclei that signals the velocity of the head needed and provides a burst of impulses causing ocular movements that are opposite to those of the head (thus moving the eyes back to the primary position). The brain stem reflex responsible for these movements is termed the **vestibulo-ocular reflex** (VOR). Such movements are termed **compensatory ocular movements** because they are compensating for the movement of the head and moving the eyes back to the primary position.

13.3. EXTRAOCULAR MUSCLES

Regardless of the type of ocular movement, the extraocular muscles, nerves and their nuclei, and the internuclear

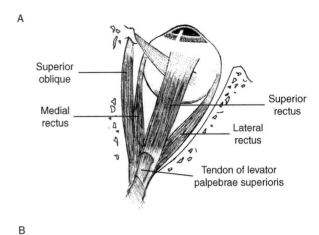

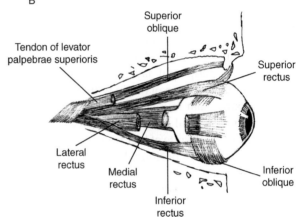

FIGURE 13.2. **A,** The muscles of the right eye as seen from above. The levator palpebrae superioris muscle is resected but its tendon of origin remains. **B,** The muscles of the right eye as seen from the lateral aspect. The levator is resected except for its tendon of origin and the middle of the lateral rectus is removed (modified from Gardner, Gray and O'Rahilly, 1975).

connections among them, all participate in ocular movements. The **extraocular** or **extrinsic eye muscles** include the **medial**, **lateral**, **superior**, and **inferior recti** and the **superior** and **inferior obliques** (Figs 13.1, 13.2). Except for the inferior oblique, all other extraocular muscles arise from the **common tendinous ring**, a fibrous ring that surrounds the margins of the optic canal. The extraocular muscles prevent ocular protrusion, help maintain their primary position, and permit conjugate ocular movements to occur.

Human extraocular muscles contain extrafusal (motor) and intrafusal (spindle) muscle fibers or myocytes. The extrafusal myocytes include at least two populations of myocytes and nerve terminals. Peripheral myocytes that are small in diameter, red, oxidative, and well suited for sustained contraction or tonus are termed '**slow**' or **tonic myocytes**. These tonic myocytes receive their innervation from nerves that discharge continuously, are involved in slower movements, and maintain the primary position of the eyes. Indeed, extraocular muscles seldom show signs of fatigue in that they work against a constant and relatively light load at all times. There are no slow myocytes in the levator palpebrae superioris. The inner core of large extraocular myocytes have 'fast', phasic, or twitch myocytes that are nonoxidative in metabolism, and better suited for larger, rapid movements. This inner core of large extraocular myocytes receives its innervation through large diameter nerves that are active for a short time. Cholinesterase positive 'en plaque' endings and 'en grappe' endings are on both types of myocytes. The 'en grappe' endings are somatic motor terminals that are smaller, lighter stained clusters or chains along a single myocyte.

Sections of human extraocular muscles reveal muscle spindles in the peripheral layers of small diameter myocytes near their tendon of origin with about 50 spindles in each extraocular muscle. Extraocular muscles are richly innervated skeletal muscles compared with other muscles in the body. In spite of this, humans have no conscious perception of eye position. Each spindle has 2–10 small diameter intrafusal myocytes enclosed in a delicate capsule. Nerves enter the capsule and synapse with the intrafusal myocytes. Age-related changes in human extraocular muscles include degeneration, loss of myocytes with muscle mass, and increase of fibrous tissue occurring before middle age and with increasing frequency thereafter. These findings probably account for age-related alterations in ocular movements, contraction and relaxation phenomenon, excursions, ptosis, limitation of eyelid elevation, and convergence insufficiency.

All extraocular muscles participate in all ocular movements maintaining smooth, coordinated ocular movements at all times. Under normal circumstances, no extraocular muscle acts alone nor is any extraocular muscle allowed to act fully hiding the cornea. Movement in any direction is under the influence of the antagonist extraocular muscles that actively participate in ending a saccade by serving as a brake. In some rare individuals the eyes can be voluntarily 'turned up' with open lids until the cornea are completely hidden.

The eyelids are closed in sleep and while blinking – an involuntary reflex involving brief (0.13–0.2 sec) eyelid closure that does not interrupt vision because the duration of the retinal after-image exceeds that of the act of blinking. In young infants rate of eye blinking is low, about eight blinks per minute, but this steadily increases over time to an adult rate of 15–20 blinks per minute.

Bilateral eyelid closure takes place in the corneal reflex (described in Chapter 8), on sudden exposure to intense illumination (dazzle reflex), by an unexpected and threatening object that moves into the visual field near the eyes (menace reflex), or by corneal irritants such as tobacco smoke. Application of a local anesthetic to the cornea does not interrupt blinking as it does in the congenitally blind and in those who have lost their sight after birth. Figure 13.1 illustrates actions of the extraocular muscles. Because of the complexity of the interactions among the extraocular muscles, it is best to examine them in isolation.

13.4. INNERVATION OF THE EXTRAOCULAR MUSCLES

The six extraocular muscles and the levator of the upper eyelid (levator palpebrae superioris) receive their innervation by three cranial nerves: the oculomotor, trochlear, and abducent. The extraocular muscles receive a constant barrage of nerve impulses even when the eyes are in the primary position. Impulses provided to the extraocular muscles allow the eyes to remain in the primary position or move in any direction of gaze. Ocular movements take place by increase in activity in one set of muscles (the agonists) and a simultaneous decrease in activity in the antagonistic muscles. The eyeball moves if the agonist contracts, the antagonist relaxes, or if both vary their activity together. Therefore, in the control of ocular movements, activity by the antagonists is as significant as activity of the agonists.

The abducent nerve [VI] or sixth cranial nerve, innervates the lateral rectus. The designation LR_6 indicates the lateral rectus innervation. The trochlear nerve [IV] or fourth cranial nerve innervates the superior oblique. The designation SO_4 indicates the superior oblique innervation. The remaining extraocular muscles and the levator palpebrae superioris receive their innervation through the oculomotor nerve [III], the third cranial nerve, for which the designation R_3 indicates the pattern of innervation.

If an extraocular muscle or its nerve is injured, certain signs will appear. First, there will be limitation of ocular movement in the direction of action of the injured muscle. Second, the patient visualizes two images that separated maximally when attempting to use the injured muscle. The resulting condition, called diplopia or double vision, results because of a disruption in parallelism of the visual axes. The images are likely to be horizontal (side by side) or vertical (one over the other) depending on which ocular muscle, nerve, or nucleus is injured.

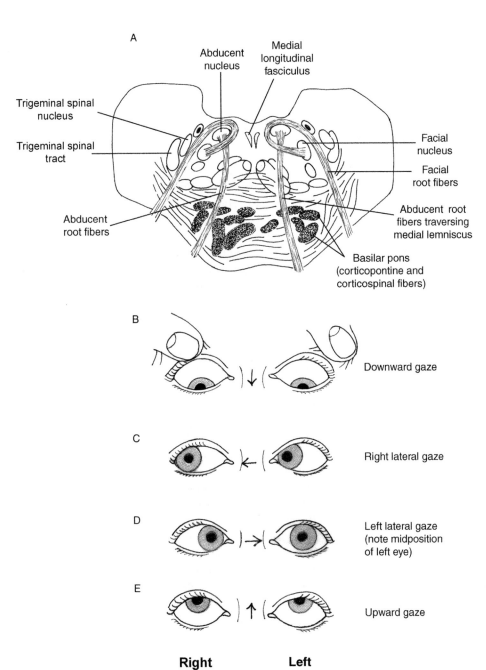

FIGURE 13.3. **A**, A transverse section of the lower pons showing the abducent and facial nuclei, their fibers and their relation to other structures at this level. **B**, **C**, **D**, **E**, The effects on ocular movements of a unilateral left abducent injury. Ocular movements are normal except for abduction of the left eye on left lateral gaze (**D**). The pupils are equal and reactive to light during all movements (modified from Spillane, 1975).

13.4.1. Abducent Nucleus and Nerve

The **abducent nerve** [VI] supplies the **lateral rectus muscle** (Figs 13.1, 13.2). Its nuclear origin, the **abducent nucleus**, is in the lower pons, lateral to the **medial longitudinal fasciculus** (MLF), and beneath the facial colliculus on the floor of the fourth ventricle (Fig. 13.3). The abducent axons leave the nucleus, cross the **medial lemniscus** and pontocerebellar fibers lying near the descending corticospinal fibers as they spread throughout the **basilar pons** (Fig. 13.3). These intra-axial relations of the abducent fibers are clinically significant.

Abducent axons emerge from the brain stem caudal to their nuclear level, at the pontomedullary junction where they collectively form the abducent nerve. Individual abducent cell bodies participate in all types of ocular movements, none of which are under exclusive control of a special subset of abducent somata.

Injury to the Abducent Nerve

The abducent nerve is frequently injured and has a long intracranial course in which it comes near many other structures. Thus, in addition to lateral rectus paralysis, other neurological signs are necessary to localize abducent injury. Isolated abducent injury is likely to be the only manifestation of a disease process for a considerable period. With unilateral abducent or lateral rectus injury, a patient will be unable to abduct the eye on the injured side (Fig. 13.3). Because of the unopposed medial rectus muscle, the eye on the injured side turns toward the nose, a condition called **unilateral internal** (convergent) **strabismus.** Double vision with images side-by-side, called **horizontal diplopia**, results when attempting to look laterally. Weakness of one lateral rectus muscle leads to a lack of parallelism in the visual axis of both eyes. Since the injured lateral rectus is not working properly, the paralyzed eye will not function in conjunction with the contralateral uninjured eye. Injury to the abducent nuclei or the abducent nerves will cause a **bilateral internal** (convergent) **strabismus** with paralysis of lateral movement of each eye and both eyes drawn to the nose. Often this is due to abducent involvement in or near the ventral pontine surface where both nerves leave the brain stem. In one series of abducent injuries, the cause was uncertain in 30% of the instances, due to head trauma in 17%, had a vascular cause in 17%, or was due to a tumor in 15% of those examined. Other common causes of abducent injury include increased intracranial pressure, infections, and diabetes.

13.4.2. Trochlear Nucleus and Nerve

The **trochlear nerve** [IV] innervates the **superior oblique muscle** (Fig. 13.2). Its cell bodies of origin are in the **trochlear nucleus** embedded in the dorsal border of the **medial longitudinal fasciculus** in the upper pons at the level of the trochlear decussation (Fig. 13.4). The rostral pole of the trochlear nucleus overlaps the caudal pole of the oculomotor nucleus. Fibers of the trochlear nerve originate in the trochlear nucleus, travel dorsolaterally around the lateral edge of the periaqueductal gray substance, and decussate at the rostral end of the superior medullary velum before emerging from the brain stem contralateral to their origin and caudal to the inferior colliculus as the **trochlear nerve** [IV]. The human trochlear nerve has about 1200 fibers ranging in diameter from 4 to 19 micrometers. Upon emerging from the brain stem, the trochlear nerve passes near the cerebral peduncles and then travels to the orbit. As they course in the brain stem from their origin to their emergence, trochlear fibers are unrelated to any intra-axial structures. The trochlear nerve is slender, has a long intracranial course and is the only cranial nerve that originates from the dorsal brain stem surface. The trochlear nerve

is the only cranial nerve all of whose fibers decussate before leaving the brain stem. Thus, the left trochlear nucleus supplies the right superior oblique muscle.

Injury to the Trochlear Nerve

Unilateral injury to the trochlear nerve causes limitation of movement of that eye and a **vertical diplopia** evident to the patient as two images, one over the other (not side-by-side as is found with abducent or oculomotor injury). Those with unilateral trochlear injury often complain of difficulty reading or going down stairs. Such injury is demonstrable if the patient looks downward when there is adduction of the injured eye. To compensate for a unilateral trochlear injury, some patients adopt a **compensatory head tilt** (Fig. 13.4B). With a right superior oblique paresis, the head may tilt to the left, the face to the right, and the chin down (Fig. 13.4B). In such instances, old photographs and a careful history may reveal a long-standing trochlear injury.

If the oculomotor nerve is injured and only the abducent and trochlear nerves are intact, the eye is deviated laterally, not laterally and downward, even though the superior oblique is unopposed by the paralyzed inferior oblique and superior rectus. In patients with unilateral oculomotor and abducent injury, sparing only the superior oblique innervation, the eye remains in its primary position. Superior oblique contraction (alone or in combination with the inferior rectus) does not cause rotation of the vertical corneal meridian (called **ocular intorsion**). Therefore, the function of the superior oblique is likely that of ocular stabilization, working with the inferior oblique and the superior and inferior recti in producing vertical ocular movements.

Because trochlear nerve fibers decussate at upper pontine levels before emerging from the brain stem, an injury here often damages both trochlear nerves. In 90% of the cases of vertical diplopia, the trochlear nerve is involved. The trochlear nerve is less commonly subject to injury than the abducent or oculomotor nerves. The list of causes of trochlear nerve paralysis is extensive including trauma (automobile or motorcycle accident with orbital, frontal or oblique blows to the head), vascular disease and diabetes with small vessel disease in the peripheral part of the nerve, and tumors.

Bilateral trochlear nerve injury likely results from severe injury to the head in which the patient loses consciousness and experiences coma for some time. The diplopia is usually permanent. The most likely site of bilateral fourth nerve injury is the superior medullary velum where the nerves decussate and the velum is thin such that decussating trochlear fibers are easily detached.

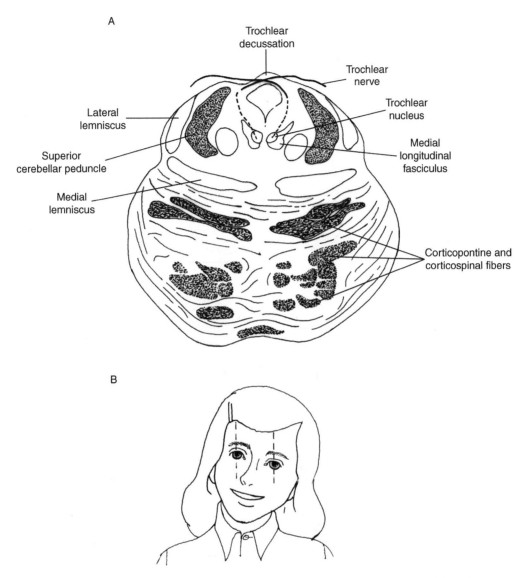

FIGURE 13.4. **A**, A transverse section of the upper pons at the level of the trochlear decussation. Although the trochlear nuclei lie rostral to this level, they are illustrated to emphasize the trochlear fibers leaving the brain stem (indicated by dashed lines). The effects of a unilateral trochlear nerve injury on ocular movements are illustrated in Fig. 13.5. **B**, A patient with a unilateral right trochlear nerve injury may manifest a compensatory tilt of the head to the left to reduce the vertical diplopia caused by a unilateral trochlear nerve lesion.

13.4.3. Oculomotor Nucleus and Nerve

The oculomotor nerve [III], innervating the remainder (R3) of the extraocular muscles, has its cells of origin in the **oculomotor nucleus** at the superior collicular level of the midbrain (Fig. 13.5). About 5 mm in length, the oculomotor nucleus extends to the caudal three-fourths of the superior colliculus. Throughout its length, it is dorsal and medial to the **medial longitudinal fasciculus** but ventral to the **aqueduct of the midbrain** (Fig. 13.5). At their caudal extent, the oculomotor nuclei fuse and overlap with the rostral part of the trochlear nuclei.

Various patterns of localization are identifiable in the oculomotor nucleus. In the baboon and presumably in humans, the inferior oblique, inferior rectus, medial rectus, and levator palpebrae superioris muscles receive their innervation from neurons in the ipsilateral oculomotor nucleus whereas the superior rectus receives fibers from neurons in the contralateral oculomotor nucleus. Functional neuronal groups in the baboon oculomotor nucleus intermingle with each other and do not remain segregated into distinct subnuclei. From the oculomotor nucleus, axons arise and cross the medial part

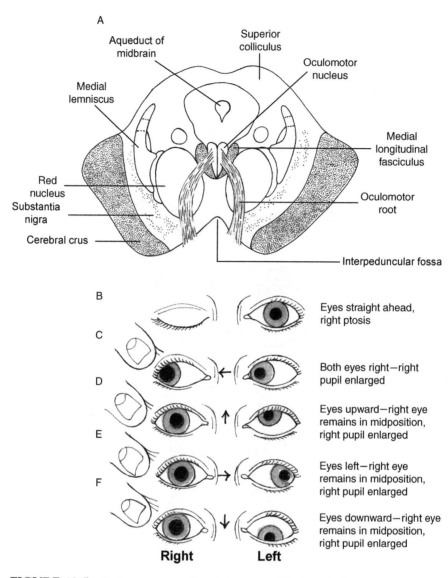

FIGURE 13.5. **A,** A transverse section of the upper midbrain at the level of the oculomotor nucleus and the emerging oculomotor fibers. The relation of these fibers to the medial longitudinal fasciculus, red nucleus, and to the medial part of the cerebral crus is significant. **B–F,** Effect on ocular movements and pupillary size of a unilateral right oculomotor nerve injury. There is a complete ptosis in **B**. In **C–F,** the ptosis is overcome by the examiner's finger. There is a dilated right pupil in **C–F** and intact movement of the right lateral rectus in **D**. In **D–F** the right eye is fixed and will not move up (**D**), medially (**E**), or down (**F**) (modified from Spillane, 1975).

of the **red nucleus** as well as the **substantia nigra** and **cerebral crus** (Fig. 13.5). These fibers then emerge from the **interpeduncular fossa** (Fig. 13.5). Once outside the brain stem, each nerve passes between a posterior cerebral and a superior cerebellar artery and then continues in the interpeduncular cistern of the subarachnoid space. In course, the oculomotor nerve is on the lateral aspect of the posterior communicating artery traversing the cavernous sinus before it enters the orbital cavity.

A significant number of ganglionic cells are scattered or clustered in the rootlets of the human oculomotor nerve. In addition, afferent fibers with neuronal cell bodies in the trigeminal ganglia are identifiable in the oculomotor nerve in humans. On entering the orbit in the lower part of the superior orbital fissure, the oculomotor nerve divides into a superior branch that innervates the superior rectus and the levator palpebrae superioris and an inferior branch that travels to innervate the inferior rectus, medial rectus, and the inferior oblique. Because of this method of branching, injuries that involve one branch while sparing the other often occur.

Unit activity recorded in the oculomotor and abducent nuclei of alert monkeys reveals that the behavior of these motoneurons is determined by eye position and velocity and not by the type of movement that established the position or velocity. Regardless of the source of stimulation, be it vestibular or visual, these motoneurons function in a consistent manner.

Injury to the Oculomotor Nerve

Unilateral injury to the oculomotor nerve leads to **ptosis**, abduction of the eye, limitation of movement, diplopia, and pupillary dilatation (Fig. 13.5). Ptosis [Greek, fall], caused by weakness or paralysis of the levator palpebrae superioris, exists if the lid covers more than half of the cornea including complete closure of the palpebral fissure. A mild or partial ptosis with the upper lid covering a third or less of the cornea may result from injury to the tarsal or palpebral muscle (of Müller) in the upper eyelid or with injury to the innervation of this muscle. The tarsal muscle is smooth muscle that has a sympathetic innervation and elevates the lid for approximately 2 mm. After injury to both oculomotor nuclei or to both nerves, loss of all ocular movements and the upper eyelids results with **double ptosis**. Abduction following unilateral oculomotor injury is likely due to the unopposed action of the lateral rectus causing **external strabismus** and the inability to turn that eye medially. The abducted eye is turned outward but not outward and downward even though the superior oblique is unopposed by the paralyzed inferior oblique (and perhaps the superior rectus). Pupillary dilatation may result from injury to the preganglionic parasympathetic fibers in the oculomotor nerve. These autonomic (pupillomotor) fibers arise from neurons in the **accessory oculomotor** (Edinger–Westphal) **nucleus** a compact neuronal mass on either side of the median plane through the rostral third (and slightly above) of the oculomotor nucleus. These preganglionic parasympathetic neurons are smaller than oculomotor neurons. Each neuronal mass is composed of rostral and caudal parts. With an expanding intracranial mass and compression or distortion of the oculomotor nerve, the ipsilateral pupil is frequently dilated, a condition called **paralytic mydriasis**, without any detectable impairment of the extraocular muscles. In one series, most oculomotor nerve injuries were of uncertain origin, 20.7% were vascular in nature, 16% caused by trauma, 13.8% due to aneurysms, and 12% resulted from tumors. In the same study, 48.3% of those with signs of oculomotor injury recovered.

A common misconception is that the eyes rotate in one direction following the head in the other direction, keeping the vertical corneal meridian perpendicular to the horizon. In humans and nonhuman primates, the eyes either do not undergo such rotation or do so only to a very small degree inadequate to compensate for the tilt of the head.

13.5. ANATOMICAL BASIS OF CONJUGATE OCULAR MOVEMENTS

Under normal conditions, ocular movements in the horizontal plane are dominant over those in other planes in primates. Disrupting the vertical muscles in nonhuman primates by severing their tendons, has no effect on the stability of horizontal movements. In all horizontal movements, it appears that the lateral rectus leads the way and determines the direction of movement. As the right eye turns laterally in a horizontal plane, the left eye turns medially. Movements of both eyes in a given direction and in the same plane are termed **conjugate ocular movements**. During such movements, the eyes move together (yoked, paired or joined) as their muscles work in unison with the ipsilateral lateral rectus and the contralateral medial rectus contracting simultaneously as their opposing muscles relax. Since motoneurons innervating the lateral rectus are in the lower pons and those innervating the medial rectus are in the upper midbrain, there must be a connection between these nuclear groups if they are to function in concert with one another.

Abducent neurons supply the ipsilateral lateral rectus. Adjoining the inferior aspect of the **abducent nucleus** (Fig. 13.6) is the crescent-shaped **para-abducent nucleus**. Fibers arise from the para-abducent nucleus, immediately decussate, and as internuclear fibers ascend in the **contralateral medial longitudinal fasciculus** (Fig. 13.6) to synapse with medial rectus neuronal cell bodies in the **oculomotor nucleus**. The anatomical basis for **horizontal conjugate ocular movements** involving the simultaneous contraction of the ipsilateral lateral rectus and the contralateral medial rectus depends on these connections. Connections exist allowing the opposing (antagonistic) muscles to relax as the agonist muscles contract. Abducent neurons use acetylcholine as their neurotransmitter whereas the neurons of the para-abducent nucleus use glutamate and aspartate as neurotransmitters. In addition to these cranial nerve ocular motor nuclei, there are premotor excitatory burst neurons that reside rostral to the abducent nucleus, inhibitory burst neurons that reside caudal to the abducent nucleus, and omnipause neurons near the median raphé at the level of the abducent nucleus. All three of these neuronal groups (excitatory,

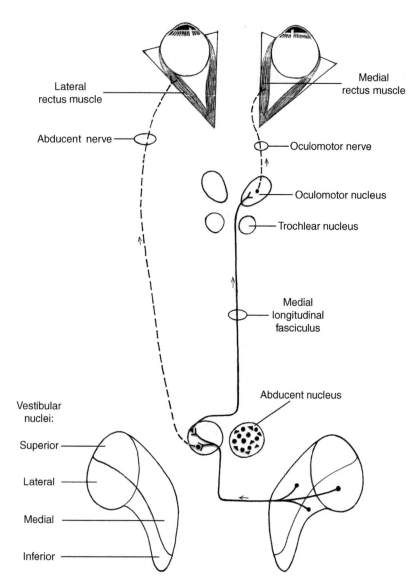

Lateral
rectus muscle

Medial
rectus muscle

Abducent nerve

Oculomotor nerve

Oculomotor nucleus

Trochlear nucleus

Medial
longitudinal
fasciculus

Abducent nucleus

Vestibular
nuclei:

Superior

Lateral

Medial

Inferior

FIGURE 13.6. Connections between the vestibular nuclei of the medulla, the abducent nuclei of the lower pons and the trochlear and oculomotor nuclei of the midbrain that underlie horizontal ocular movements from vestibular stimulation (after Crosby and Calhoun, 1965).

inhibitory, and omnipause) and their connections with abducent neurons are essential for horizontal ocular movements. Collectively these three neuronal groups form a physiological entity termed the **paramedian pontine reticular formation** (PPRF). Perhaps a better term for this group of neurons could be one that recognizes their anatomical relationship to named reticular nuclei in the human rostral medulla and pons as well as their function.

13.6. MEDIAL LONGITUDINAL FASCICULUS

The medial longitudinal fasciculus is a prominent bundle of fibers in the brain stem that participates in coordinating activity of several neuronal populations.

Well-circumscribed, near the median plane, and beneath the periaqueductal gray substance (Fig. 13.5), this bundle is overlapped dorsally and medially by the oculomotor nucleus at the superior collicular level (Fig. 13.5) and indented by the trochlear nucleus at upper pons levels (Fig. 13.4). In the lower pons, it is on the medial aspect of the abducent nucleus (Fig. 13.3). Therefore these three nuclear groups, related to ocular movements, form a column from the superior colliculus to the lower pons and all adjoin the medial longitudinal fasciculus. There is a large burst of activity in the agonistic muscle (lateral recti), with simultaneous and complete inhibition in the ipsilateral antagonistic muscle (medial recti). This occurs because there are fibers connecting neurons innervating the lateral rectus of one eye and the neurons innervating the medial rectus of the other eye as a basis for horizontal conjugate ocular movements.

These fibers form the **internuclear component of the medial longitudinal fasciculus** (Fig. 13.6). Recordings from single fibers in the medial longitudinal fasciculus of nonhuman primates reveal two physiologically identifiable types of fibers that discharge in relation to horizontal or vertical ocular movements indicating that two different groups of brain stem neurons are the source of inputs to these fiber types in the medial longitudinal fasciculus. The trigeminal motor, facial, and hypoglossal nuclei as well as the nucleus ambiguus have internuclear fibers interconnecting them through the medial longitudinal fasciculus as well. These internuclear fibers permit coordinated speech, chewing, and swallowing. Connections also exist in the medial longitudinal fasciculus that permit opening and closing of the eyelids while allowing the vestibular nuclei to influence ocular motor nuclei.

13.7. VESTIBULAR CONNECTIONS RELATED TO OCULAR MOVEMENTS

In addition to ocular movements in the horizontal plane induced by stimulation of the abducent nerves and nuclei and the medial longitudinal fasciculus, stimulation of many other parts of the nervous system such as the pontine reticular formation, vestibular receptors, nerves, and nuclei, the cerebellum, and the cerebral cortex often result in ocular movements in the horizontal plane. Indeed, the vestibular system probably influences ocular movements in all directions of gaze.

13.7.1. Vestibular Connections Related to Horizontal Ocular Movements

Receptors in this path are the vestibular hair cells on the ampullary crest in the lateral semicircular duct. Their primary neurons, in the **vestibular ganglia**, have peripheral processes that innervate these receptors and central processes that pass to the **vestibular nuclei** (Fig. 13.6) to synapse with **secondary neurons**. The secondary vestibular gray matter at medullary levels, perhaps the medial, rostral third of the inferior, and the caudal two-thirds of the lateral vestibular nuclei, are involved in this path for horizontal ocular movements. Axons of these secondary neurons proceed to the median plane, decussate and ascend in the contralateral **medial longitudinal fasciculus** (Fig. 13.6). These secondary fibers synapse with lateral rectus motoneurons in the **abducent nucleus** and with neurons in the para-abducent nucleus. Physiologically the vestibular nuclear complex influences

the contralateral abducent nucleus that innervates the lateral rectus muscle. Such connections between these ocular motor nuclei occur through the medial longitudinal fasciculus and are the same connections as those that underlie horizontal conjugate ocular movements.

A **secondary relay system** for reciprocal inhibition connects the vestibular nuclei with the ipsilateral abducent and para-abducent nuclei whose fibers innervate the contralateral oculomotor nucleus. It is by way of this secondary relay system in the medial longitudinal fasciculus (Fig. 13.6) that impulses for the inhibition of antagonistic muscles influence these muscles to relax as the agonist muscles contract, permitting smooth, coordinated, conjugate ocular movements.

By maintaining fixation despite movements of the body and head, the **vestibulo-ocular reflex** minimizes motion of an image on the retina as movements of the head occur. (If the reader rapidly shakes their head from side to side while reading these words, the words remain stationary and in focus.) Movements of the head increase activity in the already tonically active vestibular nerves. This increased neuronal activity relays to the ocular motor nuclei. The connections underlying the vestibulo-ocular reflex in the horizontal plane are the same as those that underlie horizontal conjugate ocular movements. Ocular position at any moment is the result of a balance of impulses from vestibular receptors and nuclei on one side of the brain stem versus impulses coming to the contralateral structures.

13.7.2. Vestibular Nystagmus

Vestibular nuclear stimulation in an alert or lightly anesthetized rhesus monkey with high cervical cord transection causes a succession of rhythmic, side-to-side ocular movements with quick and slow components. This centrally produced phenomenon is termed **vestibular nystagmus**. The eyes slowly deviate from their original position and away from the stimulus, followed by a rapid movement in the opposite direction toward the primary position. This **compensatory ocular movement** returns the eyes to the median plane in an effort to maintain the primary ocular position. Compensatory eye movements constitute a central vestibular reflex. Their initiation is not by vision, neck proprioception, or feedback from vestibular receptors. Recording from single neurons in the vestibular nuclei of conscious primates reveals a large repertoire of discharge patterns related to almost every aspect of voluntary and compensatory ocular movements. Such studies emphasize the role of the vestibular nuclei in coordinating vestibular and oculomotor inputs mediating movements in all directions.

Stimulation of the vestibular nerves to the semicircular canals in monkeys causes movements of the head and eyes. If the lateral canal nerve is stimulated, horizontal conjugate ocular movements occur whereas if the anterior or posterior semicircular canal nerve is stimulated, dysconjugate ocular movements occur.

13.7.3. Doll's Ocular Movements

Compensatory ocular movements that occur with changes in position of the head are under the influence of vestibular stimuli without influence from visual stimuli. Turning the head briskly in different directions in a newborn or a comatose patient with intact brain stem function leads to these reflexive, compensatory, or **doll's head** or **doll's ocular movements** (also referred to as **proprioceptive head turning**). When the eyes of a newborn are looking straight ahead and the is head extended,

the eyes will turn down involuntarily; flexing the head causes the eyes to turn up involuntarily. Turning the head to the right causes the eyes to turn to the left until they reach the primary position. Beyond one month of life, visual stimuli override this reflexive response and the response is no longer demonstrable. Motion of the head stimulates the appropriate vestibular receptors with connections from them to the vestibular nuclei and on to the abducent nuclei through the medial longitudinal fasciculus causing the eyes to move in the direction opposite the stimulus. With bilateral injury to the medial longitudinal fasciculi below the abducent nucleus, there will be no reflexive ocular movements when the head turns laterally because impulses from the vestibular receptors to the vestibular nuclei will have no way of reaching the abducent nuclei. After injury rostral to the abducent nucleus, the patient will have **dysconjugate ocular movements** or **bilateral internuclear ophthalmoplegia** so that when the head rotates to either side the lateral rectus on the side opposite the direction of rotation

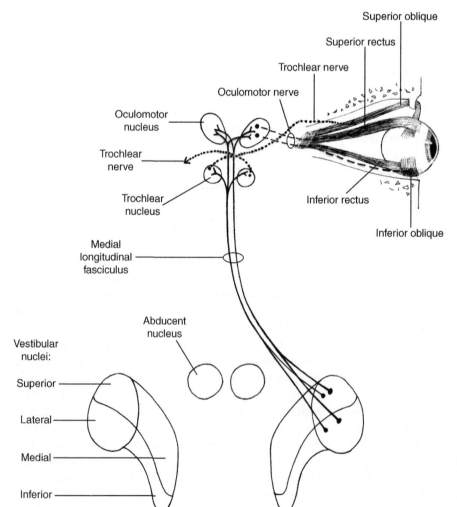

FIGURE 13.7. Connections between the pontine vestibular nuclei and the trochlear and oculomotor nuclei of the midbrain that underlie vertical ocular movements from vestibular stimulation (after Kahn, Crosby, Schneider and Taren, 1969).

will contract but the contralateral medial rectus with which it is connected, does not contract. Such individuals retain the ability to converge their eyes because the medial recti motoneurons in the oculomotor nuclei are intact. The absence of a response in infants or comatose patients suggests injury somewhere along this path.

13.7.4. Vestibular Connections Related to Vertical Ocular Movements

The receptors related to ocular movements in the vertical plane (Fig. 13.7) are probably vestibular hair cells on the superior ampullary crest at the peripheral end of the primary neurons in the vestibular ganglion. Central processes of these primary neurons synapse with secondary neurons in the vestibular nuclear complex. In the monkey, neurons in the superior vestibular nuclear complex (and perhaps in the rostral part of the lateral vestibular nucleus) have axons that proceed to the median plane to ascend exclusively in the ipsilateral medial longitudinal fasciculus. A few fibers enter the abducent nucleus but the majority synapse with trochlear and oculomotor neurons. These connections supply motor nuclei related to vertical and perhaps oblique ocular movements. A secondary relay system for reciprocal inhibition of the antagonistic muscles is involved in ocular movements in the vertical plane. In principle, this secondary system resembles a similar secondary relay system described for ocular movements in the horizontal plane.

13.8. INJURY TO THE MEDIAL LONGITUDINAL FASCICULUS

Injury to both medial longitudinal fasciculi between the oculomotor nucleus and the abducent nucleus causes a lack of coordinated, voluntary, ocular movements in either direction called **dysconjugate ocular movements**. In these instances, there is medial rectus paralysis on attempted horizontal conjugate ocular movement such that the patient can look laterally with either eye but in neither case will the contralateral eye turn medially. The contralateral eye remains in the primary position. Both eyes are able to turn medially or converge, as there is preservation of medial rectus function. This condition is termed **ophthalmoplegia** or 'eye stroke'. If there is **bilateral** injury to the **internuclear** fibers in the medial longitudinal fasciculi between the abducent and

oculomotor nuclei, the condition is termed **bilateral internuclear** (external) **ophthalmoplegia**. If only one MLF is injured, a **unilateral internuclear ophthalmoplegia** results. A patient with a long history of intermittent and progressive CNS symptoms with bilateral internuclear ophthalmoplegia is likely to have multiple sclerosis. Other causes include tumors or occlusive vascular brain stem disease.

13.9. INJURY TO THE VESTIBULAR NUCLEI

The vestibular nuclei receive a continuous stream of impulses from the vestibular receptors. If these impulses are excitatory, they increase the impulse frequency in the vestibular nerve above resting levels. If they are inhibitory, they decrease impulse frequency below resting levels. There are intimate and extensive interconnections between the vestibular nuclei and the ocular motor nuclei. Thus, any injury, or stimulation of the vestibular nuclei or nerves, will influence ocular movements. Irritative injury or experimental vestibular nuclear stimulation at upper medullary levels (medial or inferior nuclei) forces the eyes to the opposite side perhaps along with head deviation. The head and eyes turn *away* from the stimulus and may remain in that position. Vestibular nuclear destruction at medullary levels forces the eyes to the same side (toward the stimulus). In both of these instances, an imbalance exists in the discharge from the vestibular nuclei on either side. If the injury is not sufficiently irritative nor does it destroy the vestibular nuclei, the eyes will slowly turn to the contralateral side and then quickly return to the primary position. This is followed by a succession of rhythmic, side-to-side ocular movements characterized by a slow movement away from the stimulus followed by a quick return to the primary position, a phenomenon called **vestibular nystagmus** or more completely, **horizontal vestibular nystagmus with a quick component to the injured side**. The slow or **vestibular component** depends on the vestibular nuclei and is often difficult to see. Since this quick return or **compensatory component** is easier to see it is common practice to describe nystagmus by the direction of the quick component – an active return to the primary position. The compensatory, return, or quick component of vestibular nystagmus requires the participation of the brain stem reticular formation. The quick component of nystagmus is associated with an increase in frequency among reticular neurons. Therefore, vestibular nystagmus is dependent upon the interaction between vestibular and reticular nuclei. The concept of interaction is significant because there can be no quick component without the slow component. In any event, these ocular movements, be they forced or nystagmoid, represent an imbalance in the vestibular nuclear discharges on both sides of the brain stem.

Vertical and rotatory ocular movements may occur following superior vestibular nuclear stimulation or destruction in nonhuman primates. Injury to the vestibular nuclear complex at pontine levels involving the superior vestibular nucleus and perhaps the rostral part of the lateral vestibular nucleus will have a different result. The eyes will be forced into position – up or down or there is an upward rotatory nystagmus. If the injury involves considerable parts of the vestibular nuclear complex at pontine and medullary levels, an **oblique** or **rotatory nystagmus** often results depending on the specific vestibular nuclei involved. In the course of a progressive pathological disease process, there is likely to be a shift from an irritative to a destructive injury that upsets the balance between the vestibular areas on both sides. At the onset, nystagmus is likely present with a quick component to one side caused by an irritative injury. Later on in the disease, after destruction of the vestibular nuclei, the nystagmus reverses its direction with a quick component in the opposite direction.

A horizontal or vertical nystagmus may result from injury to upper cervical cord levels (C4 and above). Such a nystagmus is likely due to involvement of spinovestibular fibers in the **lateral** or **ventrolateral vestibulospinal tract**. This primarily uncrossed path supplies trunk and axial musculature. Vestibulospinal fibers often bring proprioceptive impulses from the spinal cord to the inferior vestibular nucleus. If these fibers are irritated, a horizontal nystagmus may result.

13.10. THE RETICULAR FORMATION AND OCULAR MOVEMENTS

Horizontal conjugate ocular movements can be induced in nonhuman primates by electrical stimulation of the medial nucleus reticularis magnocellularis of the pontine reticular formation which corresponds to the human pontine reticular nucleus, oral part (PnO) (Fig. 9.8) and the pontine reticular nucleus, caudal part (PnC) (Figs 9.6, 9.7). This area extends from the oculomotor and trochlear nuclei to the abducent nuclei where it is ventral to the medial longitudinal fasciculi, lateral to the median raphé, and dorsal to the trapezoid body. Two projections from this **paramedian pontine reticular formation** occur

in nonhuman primates: an ascending group of fibers through the ipsilateral oculomotor nucleus and a descending connection to the ipsilateral abducent nucleus. Electrical activity in this area precedes saccades whereas unilateral injury causes paralysis of conjugate gaze to the ipsilateral side. Unit activity recorded from this area in the monkey, followed by microstimulation of the recording site, resulted in the identification of three main categories of discharge pattern including burst units in association with saccades, tonic units with continuous activity related to position during fixation, and pause units that fired continuously during fixation but stopped during saccades.

Depending on stimulus parameters, medial pontine reticular formation stimulation causes horizontal ocular movements of constant velocity resembling the slow component of nystagmus, pursuit movements resembling the quick component of nystagmus, and saccades. Pupillary dilatation often accompanied stimulations. In nonhuman primates, horizontal saccades and the quick component of horizontal vestibular nystagmus likely have their origin in the medial pontine reticular formation. Activation of the ipsilateral lateral rectus and the contralateral medial rectus muscles occurs by medial pontine reticular stimulation through the descending connections from this region to the ipsilateral abducent nucleus. The path from the medial pontine reticular formation to the contralateral medial rectus has not more than two synapses. No vertical ocular movements are elicitable from this area. The finding of head and circling movements, if the animals were unrestrained, and pupillary dilatation accompanying medial pontine reticular stimulation, suggests that this region is not an exclusive integrator of neural activity responsible for ocular movements but a generalized **extrapyramidal motor area** involved in head, eye, and body movements. The role of the medial pontine reticular formation in human ocular movements is unclear.

The **prerubral field** (H field of Forel), an area of the tegmentum of the midbrain rostral to the red nucleus, has direct and indirect connections with the oculomotor nucleus in the monkey. The role of these reticulo-oculomotor connections in human ocular movements is also unclear.

13.11. CONGENITAL NYSTAGMUS

In addition to physiological nystagmus and vestibular nystagmus, some individuals are born with **congenital nystagmus**. In such cases, there is reduction in visual acuity because the image remains on the fovea and its receptors for a reduced period causing a drop in resolution.

While conjugate ocular movements occur by moving the eyes in the same direction, the vergence system maintains both eyes on an approaching or receding object by moving the eyes in opposite directions. However, convergence usually reduces or stops nystagmus: in some individuals, nystagmus results when they look at near targets with both eyes. Such **convergence-evoked nystagmus** is congenital or acquired.

13.12. OCULAR BOBBING

This distinctive, abnormal ocular movement involves abrupt, spontaneous, conjugate downward movement of the eyes followed by a slow return to their primary position with a frequency of 2–12/min. The eyes often remain downward for as long as 10 sec then drift upward. Horizontal conjugate ocular movements are absent with only bobbing movements remaining as the patient is typically comatose. Ocular bobbing differs from downward nystagmus in that the latter has an initial slow movement downward followed by a quick return to the primary position – the reverse of the rapid-slow sequence in ocular bobbing. Extensive, intrapontine injury is the most frequent cause of this phenomenon though cerebellar hemorrhage is another cause.

13.13. EXAMINATION OF THE VESTIBULAR SYSTEM

The vestibulo-ocular reflex and the integrity of the vestibular connections mediating it are testable in the normal conscious patient by using caloric stimulation and producing **caloric nystagmus**. Since this test permits examination of each vestibular apparatus separately, it detects unilateral peripheral vestibular injury. With the patient supine, eyes open in darkness, and the head elevated to 30° above the horizontal, 10–15 ml of warm water (about 40°C), cool to cold water (30°C), or less than 1 cc of ice water is slowly introduced into the external acoustic meatus. In this position, the lateral semicircular duct, responsible for lateral ocular movements, will be in a vertical plane (Fig. 13.8). In the normal, conscious patient, the use of warm water will result in a slow ocular movement away from the irrigated ear followed by a quick return to the primary position (Fig. 13.9). This induced back and forth ocular movement is termed **caloric nystagmus**.

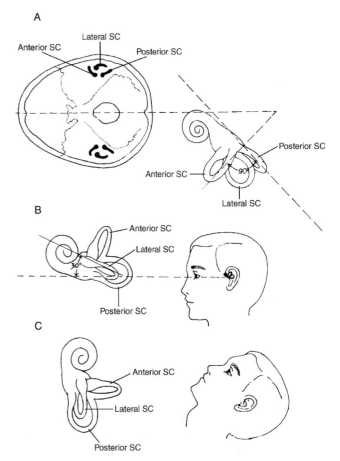

FIGURE 13.8. Anatomic position of the semicircular canals (bony labyrinth) in the skull base. **A,** The expanded ends of the semicircular canals (SC) are the ampullae that contain the vestibular receptors. No matter what direction the head is moved, complementary canals on the opposite sides of the head will always be stimulated. **B,** The bony labyrinth in anatomical position. Note that in this position the lateral semicircular canal is 30° above the horizontal. **C,** With the head tilted backward at an angle of 60°, the lateral semicircular canal is in a vertical position where it may be maximally stimulated during the caloric test (modified from Webb and Haymaker, 1969).

The **slow component,** away from the irrigated ear, is the **vestibular component** whereas the **quick component,** representing the **compensatory component,** is toward the primary position (the irrigated side). The quick component of caloric induced nystagmus is slightly slower than saccades. Caloric induced nystagmus is regular, rhythmic and lasts two to three minutes. The mnemonic **COWS** indicates the direction of the quick component of the response: 'CO' refers to '**cold opposite**' whereas 'WS' refers to '**warm same**'. When cold water is used, the quick component is away from the irrigated ear or to the opposite side, i.e. 'cold opposite'. When warm water is used, the quick component is to the same side as the irrigated ear, i.e. 'warm same'. The classification of nystagmus is in

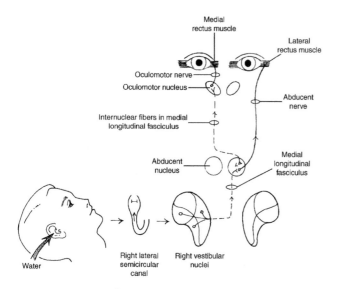

FIGURE 13.9. Connections that underlie the caloric test. With the head tilted backward at an angle of 60°, the lateral semicircular canal will be in a vertical position with its ampulla and vestibular receptors placed superiorly.

accordance with the direction of the quick component because the quick component is easily recognized.

An explanation of the caloric response (Fig. 13.9) is that the water placed in the external acoustic meatus sets up temperature gradients in the temporal bone that result in changes in endolymph density and activation of vestibular receptors (cupula deflection). Cold stimuli result in an endolymphatic current that moves away from the vestibular receptors whereas warm stimuli cause an upward endolymphatic current toward the vestibular receptors causing receptor stimulation (cupular deflection) and an increase in vestibular nerve activity on that side (Fig. 13.9). Since the vestibular nerve is tonically active at rest, warm water leads to an increase in impulses in the vestibular nerve to the vestibular nuclei on the stimulated side. Cold caloric stimulation has an opposite effect **decreasing the frequency of discharge** below the resting level on the irrigated side. This distorts the balance of neuronal activity between both vestibular nerves. The vestibular nerve and nuclei on the opposite side of the cold-water irrigation predominates and the eyes slowly turn toward the irrigated ear then quickly return to the primary position. Therefore, the nystagmus with cold water has its quick component opposite or away from the irrigated ear.

The simultaneous examination of the vestibular system on both sides involves the use of a Bårnåy chair. In this test, the patient sits quietly in a chair that rotates about a vertical axis. After some 30 sec of smooth, constant rotation, the patient, with eyes closed, will report

that they have no sensation of turning. If the chair is then suddenly brought to a halt (deceleration), the cupula (that gelatinous substance associated with the apices of vestibular hair cells in the cristae and into which the stereocilia project) will be deflected in the direction opposite to that of the rotation. This deflection provides stimuli and sensory discharges that the patient interprets as sensations of motion even though the patient is no longer rotating. Cupular deflection generates ocular movements. The eyes slowly turn and then quickly return to their primary position. This slow rotation, quick return pattern characteristic of **nystagmus** continues as long as the vestibular receptors are stimulated. The caloric test is reliable for demonstrating the presence of an acoustic neuroma. In one series, there was significantly reduced caloric response on the affected side in 94% of those patients tested who presented with symptoms of an acoustic neuroma.

13.14. VISUAL REFLEXES

The **iris** is a circular, pigmented diaphragm in front of the lens and behind the cornea. Its central border is free and bounds an aperture known as the **pupil** that normally appears black (because of reflected light from the retina). The pupils are normally round, regular, equal in diameter, centered in the iris, and usually 3–4 mm in diameter (range 2–7 mm). **Anisocoria** is a condition in which the pupils are unequal in size. Usually no pathological significance exists if the difference between the pupils is 1 mm or less. Some 15–20% of normal individuals show inequality of pupils on a congenital basis.

The pupils are small, react poorly at birth, and in early infancy, but are larger in younger individuals (perhaps 4 mm and perfectly round in adolescents, 3.5 mm in middle age, 3 mm or less in old age but slightly irregular). Though many factors influence pupillary size, **intensity of illumination** reaching the retina is most significant. Under ordinary illumination, the pupils are constantly moving with a certain amount of fluctuation in pupillary size, a condition that is termed **pupillary unrest**.

A **miotic pupil** is a pupil 2 mm or less in diameter. Causes of small pupils include alcoholism, arteriosclerosis, brain stem injuries, deep coma, diabetes, increased intracranial pressure, drug intoxications (morphine, other opium derivatives), syphilis, sleep (in which size decreases), and senility. **Mydriasis** is a condition in which the pupils are dilated more than 5 mm in diameter. Anxiety, cardiac arrest, fears, cerebral anoxia, pain, hyperthyroidism, injuries to the midbrain, and drug

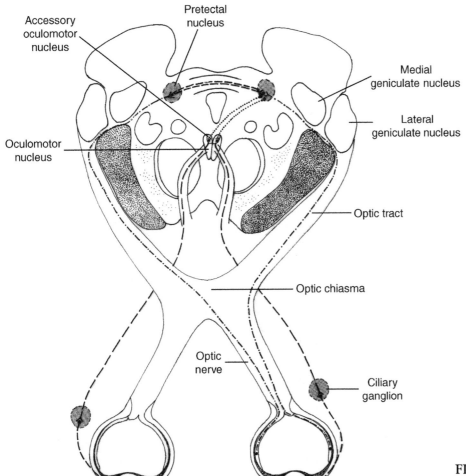

FIGURE 13.10. The light reflex pathway (modified from Crosby, Humphrey and Lauer, 1962).

intoxications such as cocaine and amphetamines may be the underlying cause of pupillary dilation. Pupillary dilation may exist during coma. The drug atropine is useful to dilate the pupils for diagnostic purposes. Though some gifted individuals can voluntarily produce pupillary dilatation, it may be passive in type due to paralysis of the sphincter mechanism or active in type due to direct stimulation of the dilator pupillae or the nerves that innervate that muscle.

13.14.1. The Light Reflex

If you shine a small penlight into the right eye and shade the other, both pupils will constrict – a phenomenon called **miosis**. The response in the stimulated eye is the **direct response** – that in the nonstimulated eye is the **consensual response** (crossed response). The delay of this response is a condition termed the Piltz–Westphal syndrome.

Anatomic Connections Mediating the Light Reflex

Both rods and cones are receptors for the light reflex. The primary neurons in this reflex path are retinal bipolar neurons and the secondary neurons are retinal ganglionic neurons. The appropriate impulses follow the visual path from bipolar to ganglionic neurons with central processes of the latter neurons contributing fibers to the optic nerve, optic chiasma, and optic tract (Fig. 13.10). Fibers for the light reflex separate from the optic tract to join the brachium of the superior colliculus. From here, they pass to the superior colliculus, and synapse with tertiary neurons in the **pretectal nuclear complex** on both sides (Fig. 13.10) of the diencephalon, rostral and ventral to the laminated part of the superior colliculus (and therefore, 'pretectal'). Central processes of these tertiary neurons (pretecto-oculomotor fibers) project bilaterally as to quaternary (fourth-order) neurons in this path in the rostral part of both **accessory oculomotor nuclei** (Fig. 13.10). This preganglionic parasympathetic nucleus lying rostral, dorsal, and dorsomedial to the oculomotor nucleus, sends its axons into the oculomotor nerve [III]. In the interpeduncular fossa, these fibers are superficial on the dorsomedial and medial aspect of the oculomotor nerve. They have a descending course as they travel from their brain stem emergence to their dural entry beneath the epineurium of the nerve. At their orbital entrance, these preganglionic fibers join the **inferior division of the oculomotor nerve**, and synapse with quinary (fifth-order) neurons in the ipsilateral **ciliary ganglion**. From each ciliary ganglion, postganglionic parasympathetic fibers enter the short ciliary nerves and pass to the **sphincter pupillae** of the iris. The sphincter pupillae is nonstriated muscle that develops from ectoderm. Retinal stimulation with a small penlight therefore causes contraction of both sphincter pupillae and constriction of both pupils.

13.14.2. The Near Reflex

On looking from a distant to a near object, **pupillary constriction** takes place in association with **ocular convergence** and **accommodation** of the lens. Ocular convergence refers to adduction of both eyes through medial recti contraction whereas accommodation refers to a modification in the power of the refraction of the lens caused by changes in the shape of the lens due to ciliary body movement. As the ciliary body moves anteriorly, decreased tension results on fibers of the ciliary zonule of the lens capsule and the lens becomes fatter. Alteration of the lens curvature results as its front surface moves toward the corneal vertex. Therefore, the lens thickens when near objects are viewed and the eye forms sharp images on the retina of objects that are at different distances from the eye.

Anatomic Connections Mediating the Near Reflex

The exact sequence of events, the appropriate stimulus, and the connections involved in this reflex are still a matter of question. Proprioceptive impulses from the converging muscles may serve as the necessary stimulus for accommodation and constriction or accommodation occurring simultaneously with convergence. The site of an object often provides the stimulus for the resulting constriction. Another possibility, because all three components of this reflex are obtainable by preoccipital cortical stimulation in humans, is that cortical areas are involved in initiating this reflex response.

Fibers of retinal origin separate from the optic tract to enter the superior colliculus. Both superior colliculi are interconnected and each discharges to the caudal part of the **accessory oculomotor nucleus** by way of **colliculo-oculomotor fibers** (tecto-oculomotor fibers). As with the light reflex, preganglionic parasympathetic fibers travel from their origin in the caudal part of the accessory oculomotor nucleus, enter the oculomotor nerve, and travel in it to the **ciliary ganglion**. Some fibers bypass the ciliary ganglion to synapse in the **episcleral ganglia** (a small collection of ganglionic cells in the sclera). Postganglionic parasympathetic fibers from the episcleral ganglion travel in the short ciliary nerves to supply the ciliaris whereas postganglionic fibers from the ciliary ganglion innervate the sphincter pupillae. Hence, in addition to pupillary constriction by way of the sphincter pupillae contraction, contraction of the ciliary muscles permits the ciliary body to move forward, decreasing tension on the lens. The increased curvature of the lens allows the eye to focus on near objects.

The rostral part of the **accessory oculomotor nucleus**, connected with the pretectal nuclear complex over **pretecto-oculomotor fibers**, participates in the **light reflex** whereas the caudal part of the accessory oculomotor nucleus participates in the **near reflex**. The caudal part of the accessory oculomotor nucleus connects with the superior colliculi over **colliculo-oculomotor fibers**. Since fibers to the respective parts of the accessory oculomotor nucleus do not pass through the same level of the midbrain, it is possible to injure one set of fibers (pretecto-oculomotor to the rostral part of the AON) and preserve the other (colliculo-oculomotor to the caudal part of the AON). Absence of pupillary constriction in the light reflex (direct and consensual response) with preservation of constriction in the near reflex is termed an **Argyll–Robertson pupil**. Causes of this condition include syphilis, diabetes, multiple sclerosis, alcoholic encephalopathy, and encephalitis.

Inactive pupils (Adie's pupils) do not respond to light or accommodation. This condition may be the result of a single circumscribed injury involving both accessory oculomotor nuclei in the rostral part of the midbrain or two small injuries, one injury involving each accessory oculomotor nucleus.

13.14.3. Pupillary Dilatation

The dilator pupillae muscles consist of nonstriated fibers derived from myoepithelial cells that form part of the underlying pigmented epithelium and hence are ectodermal in origin (in front of pigmented epithelium on the back of the iris) constituting the iridial part of the retina. Sympathetic fibers originating in neurons of the intermediolateral cell column in spinal segments T1 and

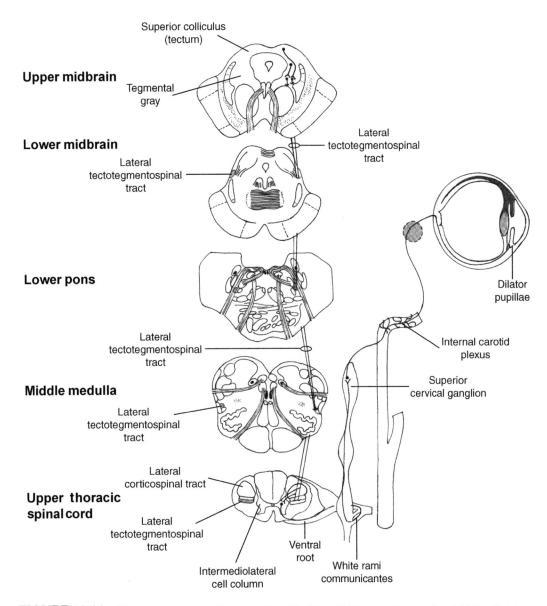

FIGURE 13.11. The origin, course, and termination of the lateral tectotegmentospinal tract. This path terminates on preganglionic neurons in the intermediolateral cell column at T1 and T2 cord levels. From these preganglionic neurons, fibers arise and exit the ventral roots from C8–T4 spinal cord levels to enter the sympathetic trunk through the white rami communicantes. These preganglionic fibers synapse in the superior cervical ganglion. Postganglionic fibers from this ganglion accompany the internal carotid artery as the internal carotid plexus. This plexus gives fibers that pass through the ciliary ganglion and short ciliary nerves to supply the dilator pupillae muscle (modified from DeJong, 1979).

T2 innervate the dilator pupillae. These neurons are termed the **ciliospinal nucleus** (or center of Budge). Preganglionic fibers leave the spinal cord in the C8–T2 ventral roots and enter the sympathetic trunk to synapse in the superior cervical ganglia. Postganglionic sympathetic fibers travel in the internal carotid plexus, enter the ophthalmic nerve [V$_1$] and reach the orbit by way of the nasociliary nerve. From here, they enter the long ciliary branches of the nasociliary nerve to reach the dilator pupillae and the tarsal or palpebral muscle (of Müller).

13.14.4. The Lateral Tectotegmentospinal Tract

Cells of the intermediolateral nucleus in spinal segments T1 and T2 supply sympathetic fibers to the dilator pupillae and are under the influence of a path that originates at upper levels of the midbrain (Fig. 13.11). From both superior colliculi (tectum) and the underlying tegmental gray matter of the midbrain, fibers accumulate, turn caudally, and descend in the lateral ipsilateral field of the brain stem. This path, the **lateral tectotegmen-tospinal tract** (Fig. 13.11), descends into the midbrain, pons, medulla oblongata, and spinal cord where it is ventral to the lateral corticospinal tract in the lateral funiculus. The termination of this path is in the interme-diolateral nucleus at T1 and T2. Destruction of this path leads to a partial ptosis and a small pupil ipsilaterally which does not dilate in response to light or to its absence.

13.14.5. Pupillary Pain Reflex and the Spinotectal Tract

Pupillary dilatation may result from a painful, cutaneous stimulus. In comatose patients, a **pupillary pain reflex** is elicitable by applying a painful stimulus on the cheek, below the orbit. Painful impulses reach the superior col-liculus (tectum) in the spinotectal tract as follows: pri-mary neurons in the trigeminal or certain spinal ganglia give off peripheral processes that have the appropriate nociceptors at their termination. Central processes of pri-mary neurons end in the substantia gelatinosa and the dorsal funicular gray. Fibers of secondary neurons pass ventrolaterally and decussate through the ventral white commissure taking up a position on the medial border of the lateral spinothalamic tract. This neither large nor well-myelinated **spinotectal path** ascends through the cord and into the brain stem. As it ascends, it gradually shifts to a position dorsal to the lateral spinothalamic tract at the uppermost tip of the medial lemniscus. The spinotectal path ends in the superior colliculus (which forms the tectum of the midbrain). Ventral trigeminothalamic fibers presumably also continue to the intermediate gray matter of the superior colliculus. Ascending painful impulses from the body in the spinotectal path and from the head in the ventral trigeminothalamic tract therefore reach the superior col-liculus. Here they are associated with the collicular regions that contribute to the lateral tectotegmentospinal tract. Hence, increase in pupillary size is likely a direct response to painful stimuli that travel in these paths.

13.14.6. The Afferent Pupillary Defect (Marcus Gunn Pupillary Sign)

In unilateral retinal or optic nerve disease, it is possible to observe pupillary constriction followed by dilata-tion on the affected side using the swinging flashlight test. In such cases, the examiner moves a small flash-light rapidly from one eye to the other and back again – every 2–3 sec. As the light moves from the good eye to the injured eye, there is an initial failure of immediate constriction of the injured pupil followed by dilatation. Removal of light from the normal side causes dilatation in the injured eye and is a normal consensual response to the absence of light in the normal eye. The normal consensual dilatation to darkness masks the impairment of the light reflex in the injured eye. The pupil on the unaffected side constricts normally. This **afferent pupillary defect** is also termed a **paradoxical reaction**, the **Marcus Gunn pupillary sign**, or the **swinging flashlight sign**. This sign is often the earliest indicator of optic nerve injury.

FURTHER READING

Augustine JR, DesChamps EG, Ferguson JG Jr (1981) Functional organization of the oculomotor nucleus in the baboon. Am J Anat 161:393–403.

Bahill AT, Adler D, Stark L (1975) Most naturally occurring human saccades have magnitudes of 15 degrees or less. Invest Ophthalmol 14:468–469.

Bortolami R, Veggetti A, Callegari E, Lucchi ML, Palmieri G (1977) Afferent fibers and sensory ganglion cells within the oculomotor nerve in some mammals and man. I. Anatomical investigations. Arch Ital Biol 115:355–385.

Burger LJ, Kalvin NH, Smith JL (1970) Acquired lesions of the fourth cranial nerve. Brain 93:567–574.

Büttner-Ennever JA, Henn V (1976) An autoradiographic study of the pathways from the pontine reticular formation involved in horizontal eye movements. Brain Res 108: 155–164.

Cohen B, Komatsuzaki A (1972) Eye movements induced by stimulation of the pontine reticular formation: evidence for integration in oculomotor pathways. Exp Neurol 36:101–117.

Dietert SE (1965) The demonstration of different types of muscle fibers in human extraocular muscle by electron microscopy and cholinesterase staining. Invest Ophthalmol 4:51–63.

Hall AJ (1936) Some observations on the acts of closing and opening the eyes. Br J Ophthalmol 20:257–295.

Henn V, Cohen B (1972) Eye muscle motor neurons with dif-ferent functional characteristics. Brain Res 45:561–568.

Henn V, Cohen B (1976) Coding of information about rapid eye movements in the pontine reticular formation of alert monkeys. Brain Res 108:307–325.

Horn AKE, Büttner-Ennever JA, Suzuki Y, Henn V (1995) Histological identification of premotor neurons for hori-zontal saccades in monkey and man by parvalbumin immunostaining. J Comp Neurol 359:350–363.

Jampel RS (1975) Ocular torsion and the function of the verti-cal extraocular muscles. Am J Ophthalmol 79:292–304.

Keller EL (1974) Participation of medial pontine reticular formation in eye movement generation in monkey. J Neurophysiol 37:316–332.

Keller EL, Robinson DA (1972) Abducens unit behavior in the monkey during vergence movements. Vision Res 12:369–382.

King WM, Lisberger SG, Fuchs AF (1976) Responses of fibers in medial longitudinal fasciculus (MLF) of alert monkeys during horizontal and vertical conjugate eye movements evoked by vestibular or visual stimuli. J Neurophysiol 39:1135–1149.

Leigh RJ, Zee DS (2006) The neurology of eye movements. In: *Contemporary Neurology Series*, 4th ed. Vol 70. Oxford University Press, New York.

Leisman G, Schwartz J (1977) Ocular-motor function and information processing: implications for the reading process. Int J Neurosci 8:7–15.

Leisman G, Schwartz J (1977) Directional control of eye movement in reading: the return sweep. Int J Neurosci 8:17–21.

Luschei ES, Fuchs AF (1972) Activity of brain stem neurons during eye movements of alert monkeys. J Neurophysiol 35:445–461.

McCrary JA III (1977) Light reflex anatomy and the afferent pupil defect. Trans Am Acad Ophthalmol Otolaryngol 83:820–826.

Pearson AA (1944) The oculomotor nucleus in the human fetus. J Comp Neurol 80:47–63.

Sharpe JA, Hoyt WF, Rosenberg MA (1975) Convergence-evoked nystagmus. Congenital and acquired forms. Arch Neurol 32:191–194.

Skvenski AA, Robinson DA (1973) Role of abducens neurons in vestibuloocular reflex. J Neurophysiol 36:724–738.

Weidman TA, Sohal GS (1977) Cell and fiber composition of the trochlear nerve. Brain Res 125:340–344.

Younge BR, Sutula F (1977) Analysis of trochlear nerve palsies: diagnosis, etiology, and treatment. Mayo Clin Proc 52:11–18.

Zahn JR (1978) Incidence and characteristics of voluntary nystagmus. J Neurol Neurosurg Psychiatry 41:617–623.

We are inclined to think of the thalamus as central to all cortical functions and to believe that a better understanding of the thalamus will lead to a fuller appreciation of cortical function . . . we suggest that cerebral cortex, without thalamus is rather like a great church organ without an organist: fascinating, but useless.

S. Murray Sherman and R.W. Guillery, 2001

14

The Thalamus

14.1. INTRODUCTION

The major part of the diencephalon in humans is the **dorsal thalamus**, generally referred to as the '**thalamus**'. The thalamus is in the median plane of each cerebral hemisphere and presents symmetrical right and left halves each with some 120 nuclear groups in humans. Along with immense structural complexity and functional significance, the thalamus is the site of convergence of impulses from a variety of sources permitting a great deal of integration, correlation, and association of impulses.

In general, the thalamus corresponds to those structures that bound the third ventricle. Its caudal boundary is the junction between the midbrain and diencephalon and its rostral boundary is roughly the anterior commissure. In some 70–80% of normal human brains, both halves of the thalamus meet in the median plane. When they do, a structure in the median plane, called

the **interthalamic adhesion** connects both halves. Although there is considerable variability among individuals in this structure, when present one or two small nuclei (belonging to the median nuclear group) and a few commissural fibers form the interthalamic adhesion in humans. The interthalamic adhesion is more often present in women than in men (in one study 68% of the males and 78% of the females had an interthalamic adhesion). When present it is 53% larger in females than in males, despite the fact that the male brain is larger than the female brain. The lateral boundary of the thalamus is a prominent fiber bundle, the posterior limb of the **internal capsule** (ic). The ventral boundary of the thalamus is the **hypothalamic sulcus** (Fig. 5.9), a surface feature on the wall of the third ventricle best visualized on the medial surface of the brain. The thalamus is superior to this sulcus whereas the **subthalamus** is inferior to the sulcus but lateral to the hypothalamus.

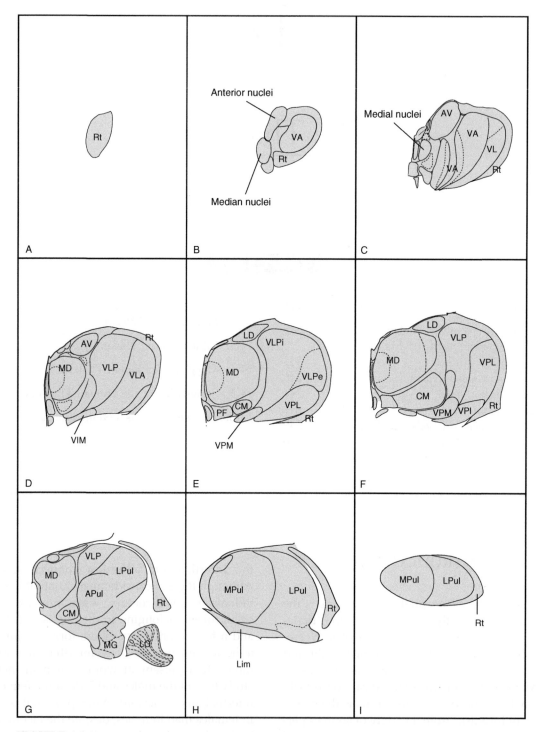

FIGURE 14.1. A series of coronal sections through the human brain posterior to the center of the anterior commissure to demonstrate the changing shape of the thalamic nuclei from rostral to caudal. (**A, B, C, D, E, F, G, H, I** are based on sections 25–49 respectively in Mai, Assheuer and Paxinos, 2004.)

Grossly the thalamus is roughly egg-shaped. Coronal sections through only the thalamus absent surrounding structures (Fig. 14.1) revealing the changing shape of the thalamus from rostral to caudal (Fig. 14.1) and its internal complexity with many named nuclei.

Coronal sections through the thalamus along with adjacent structures of the cerebral hemisphere reveal the presence of the **internal capsule** (ic) on the medial aspect of the thalamus (Figs 14.2 to 14.8) at almost all levels of the thalamus. The thalamus continues to

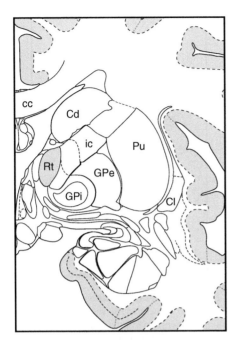

FIGURE 14.2. Coronal section through the human brain about 4 mm posterior to the center of the anterior commissure. The reticular thalamic nuclei (Rt) are the only thalamic nuclei present at this level. They are lightly shaded (modified from section 25 in Mai, Assheuer and Paxinos, 2004).

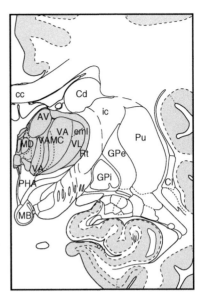

FIGURE 14.4. Coronal section through the human brain about 12 mm posterior to the center of the anterior commissure. The anteroventral (AV), medial dorsal (MD), reticular (Rt), ventral anterior (VA), and ventral lateral (VL) thalamic nuclei are present at this level and are lightly shaded (modified from section 31 in Mai, Assheuer and Paxinos, 2004).

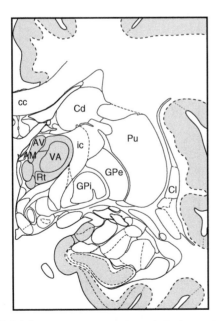

FIGURE 14.3. Coronal section through the human brain about 8 mm posterior to the center of the anterior commissure. The anteromedial (AM), anteroventral (AV) and ventral anterior (VA) thalamic nuclei are present at this level and are lightly shaded (modified from section 28 in Mai, Assheuer and Paxinos, 2004).

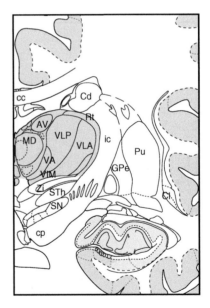

FIGURE 14.5. Coronal section through the human brain about 16 mm posterior to the center of the anterior commissure. The anteroventral (AV), medial dorsal (MD), reticular (Rt), ventral anterior (VA), and two parts of the ventral lateral (VLA and VLP) thalamic nuclei are present at this level and are lightly shaded (modified from section 34 in Mai, Assheuer and Paxinos, 2004).

increase in size as it extends posteriorly, always maintaining a medial relationship with the **internal capsule** (ic) (Figs 14.2 to 14.8) until the internal capsule disappears (Figs 14.9, 14.10). Internally the thalamus is divisible

into medial and lateral parts by a narrow band of myelinated fibers, the **internal medullary lamina** (iml, Figs 14.1, 14.3 to 14.8). On the lateral side of the thalamus is another myelinated band, the **external**

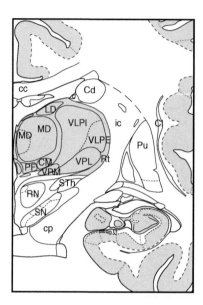

FIGURE 14.6. Coronal section through the human brain about 19.9 mm posterior to the center of the anterior commissure. The medial dorsal (MD), lateral dorsal (LD), reticular (Rt), parts of ventral lateral (VL), ventral posterior lateral (VPL), ventral posterior medial (VPM), centromedian (CM), and parafascicular (PF) thalamic nuclei are present at this level and are lightly shaded (modified from section 37 in Mai, Assheuer and Paxinos, 2004).

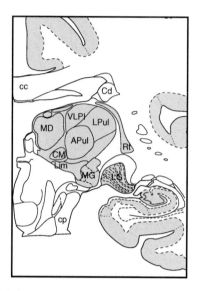

FIGURE 14.8. Coronal section through the human brain about 27.8 mm posterior to the center of the anterior commissure. The medial dorsal (MD), centromedian (CM), lateral posterior (LP), ventral posterior lateral (VPL), reticular (Rt), medial (MG) and lateral geniculate (LG), and the anterior pulvinar (APul) thalamic nuclei are present at this level and are lightly shaded (modified from section 43 in Mai, Assheuer and Paxinos, 2004).

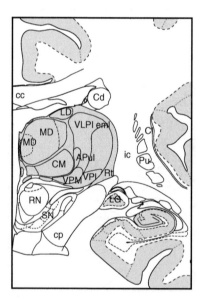

FIGURE 14.7. Coronal section through the human brain about 23.9 mm posterior to the center of the anterior commissure. The medial dorsal (MD), lateral dorsal (LD), centromedian (CM), ventral posterior medial (VPM), ventral posterior lateral (VPL), parts of ventral lateral (VLPI), reticular (Rt), and parafascicular (PF) thalamic nuclei are present at this level and are lightly shaded (modified from section 40 in Mai, Assheuer and Paxinos, 2004).

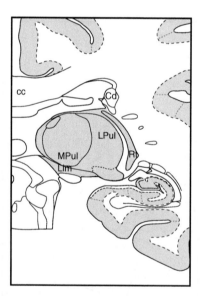

FIGURE 14.9. Coronal section through the human brain about 31.9 mm posterior to the center of the anterior commissure. The pulvinar (Pul) and reticular (Rt) thalamic nuclei are present at this level and are lightly shaded (modified from section 46 in Mai, Assheuer and Paxinos, 2004).

medullary lamina (eml, Fig. 14.1, Figs 14.3 to 14.9) separating the thalamus from the internal capsule (ic). Interposed between the external medullary lamina (eml) and the internal capsule (ic) are the **reticular** **nuclei** (Rt, Figs 14.3 to 14.10). Between the internal medullary lamina (iml) and the external medullary lamina (eml) are the ventral nuclei and the pulvinar nuclei (Pul) (Figs 14.8 to 14.10).

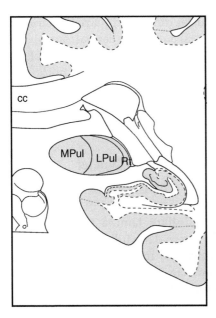

FIGURE 14.10. Coronal section through the human brain about 36 mm posterior to the center of the anterior commissure. The medial (MPul) and lateral (LPul) pulvinar and the reticular (Rt) thalamic nuclei are present at this level and are lightly shaded (modified from section 49 in Mai, Assheuer and Paxinos, 2004).

14.2. NUCLEAR GROUPS OF THE THALAMUS

The thalamus is 4 cm in length, with an anterior and posterior end and four surfaces: medial, lateral, ventral, and dorsal. About half of the 120 nuclei in the thalamus send fibers to the cerebral cortex and the other half send fibers to subcortical areas exclusively or in addition to a collateral cortical projection. Terminology related to the thalamic nuclei is extremely complex and varies from author to author even in the human brain. Nuclei of the thalamus are divisible into two functional groups: **relay nuclei** and **association nuclei**. For the sake of convenience, we will separate the thalamus into nine nuclear groups based on the work by Hirai and Jones, 1989 and Jones, 1997 along with the terminology found in *Terminologia Anatomica*, 1998 and used in the atlas of the human brain by Mai, Assheuer and Paxinos, 2004. These nine nuclear groups, listed in Table 14.1, include: **(1) anterior nuclei and lateral dorsal nucleus; (2) intralaminar nuclei; (3) medial nuclei; (4) median nuclei; (5) metathalamic nuclei; (6) posterior nuclear complex; (7) pulvinar nuclei and lateral posterior nucleus; (8) reticular nucleus; and (9) ventral nuclei.**

TABLE 14.1. Nuclei of the human thalamus

Terms in English	Abbreviation
1. Anterior nuclei and lateral dorsal nucleus	
a. Anterodorsal thalamic nucleus	AD
b. Anteromedial thalamic nucleus	AM
c. Anteroventral thalamic nucleus	AV
d. Lateral dorsal thalamic nucleus	LD
2. Intralaminar nuclei	
a. Central lateral thalamic nucleus	CenL
b. Central medial thalamic nucleus	CenM
c. Centromedian nucleus	CM
d. Paracentral thalamic nucleus	PC
e. Parafascicular thalamic nucleus	PF
3. Medial nuclei	
a. Medial dorsal thalamic nucleus (dorsomedial)	MD
b. Medial ventral nucleus	MV
4. Median nuclei	
a. Nucleus reuniens	Re
b. Paratenial thalamic nucleus	PT
c. Paraventricular thalamic nuclei	PV
d. Rhomboid thalamic nucleus	Rh
5. Metathalamic nuclei	
a. Medial geniculate nuclei Dorsal (MG_d), Medial (MG_m), Ventral (MG_v)	MG
b. Lateral geniculate nucleus	LG
6. Posterior nuclear complex	
a. Limitans thalamic nucleus	Lim
b. Posterior limitans nuclei	PLi
c. Suprageniculate thalamic nucleus	SG
7. Pulvinar nuclei and lateral posterior nucleus	
a. Lateral posterior nucleus	LP
b. Pulvinar nuclei Anterior (APul), Medical (MPul) Lateral (LPul), and Inferior (IPul) nuclei	Pul
8. Reticular nucleus	Ret
9. Ventral nuclei	
a. Ventral anterior thalamic nucleus Magnocellular (VA_{mc}) part	VA
b. Ventral lateral thalamic nucleus Anterior (VL_a) and posterior (VL_p) parts	VL
c. Ventral medial thalamic nucleus	VM
d. Ventral posterior nucleus	VP
Ventral posterior lateral thalamic nucleus Anterior (VPL_a) and posterior (VPL_p) parts	VPL
Ventral posterior medial thalamic nucleus	VPM
Ventral posterior inferior thalamic nucleus	VPI

Modified from Hirai and Jones, 1989; Jones, 1997; *Terminologia Anatomica*, 1998; Mai, Assheuer and Paxinos, 2004.

14.2.1. Anterior Nuclei and the Lateral Dorsal Nucleus

Anterior Nuclei

The triangular-shaped **anterior nuclei**, at the rostral end of the thalamus (Fig. 14.1B), are divisible into **anterodorsal**

(AD), **anteromedial** (AM), and **anteroventral** (AV) **nuclei** (AV, Figs 14.4, 14.5). These nuclei are easily identifiable because of the many fibers that surround them. As the anteroventral nucleus (AV) tapers posteriorly to a narrow tail, it blends into the **lateral dorsal nucleus** (LD, Figs 14.6, 14.7). The junction between AV and LD occurs at about the midpoint of the rostrocaudal extent of the human thalamus. The anterior nuclei and the lateral dorsal nucleus (LD) expand across the dorsal surface of the medial dorsal nucleus (MD) but are separable from it by fibers of the internal medullary lamina that encapsulate the medial dorsal nucleus on its anterior, posterior, lateral and ventral surfaces (and to a lesser extent on its medial and dorsal surfaces). The anterior nuclei receive impulses from the ipsilateral hypothalamus especially its mamillary body. The anterior nuclei relay these impulses to the cingulate gyrus of the limbic system. fMRI assessment of the connections of the human anterior nuclei reveals functional connectivity of this nucleus with the anterior cingulate cortex. Thus, the anterior nuclei play a role as a **relay nucleus in the limbic system** connecting the hypothalamus with the limbic areas of the cerebral cortex. Based on its connections with a variety of limbic structures, the anterior nuclei are often termed **limbic nuclei**. Functionally the anterior nuclei participate in learning and memory acquisition. Neuronal loss in the anterior thalamic nuclei occurs in alcoholic Korsakoff's psychosis. This neurodegeneration may be the neural substrate that underlies the amnesia observed in Korsakoff's patients. Focal ischemic damage to the human anterior nuclei or the major efferent path from these nuclei (the **mamillothalamic tract**) results in memory-related deficits. Damage to the mamillothalamic tract in humans is a necessary condition for the development of amnesia after thalamic injury. Neuronal number in the medial dorsal nuclei decreases by 24% to 35% in the brains of schizophrenics but by 16% in the AV/AM nuclei. Bilateral stimulation of the anterior nuclei through implantable electrodes resulted in clinically and statistically significant improvement in four of five patients with intractable partial epilepsy.

Lateral Dorsal Nucleus (LD)

The lateral dorsal nucleus (LD) and the anterior nuclei of the thalamus find themselves in the same group because the connections and functions of the lateral dorsal nucleus in humans are similar in part to those of the anterior nuclei. The internal medullary lamina (iml) splits around the lateral dorsal nucleus (LD, Figs 14.6, 14.7) as it does around the anterior nuclei. As one follows the anterior nuclei though the thalamus, the anterior nuclei diminish in size posteriorly and their position is taken by the lateral dorsal nucleus (LD) (Figs 14.6, 14.7). Some authors place LD in a 'dorsal nuclear group' with the pulvinar nuclei (Pul) and the lateral posterior nucleus (LP). The subicular cortex projects to the anterior nuclei, lateral dorsal nuclei, and the median nuclei in the monkey.

14.2.2. Intralaminar Nuclei

The thalamus is divisible into medial and lateral subdivisions by a narrow band of myelinated fibers, the **internal medullary lamina** (iml, Figs 14.3 to 14.8). Within the internal medullary lamina is a collection of nuclei termed the **intralaminar nuclei.** The nuclei in the intralaminar group are identifiable by their intense acetylcholinesterase staining and are divisible into a rostral and caudal group. In the human brain each group demonstrates a characteristic pattern of calcium-binding protein immunoreactivity. More rostrally in the internal medullary lamina are the **central lateral nucleus** (CenL), **central medial nucleus** (CenM), and the **paracentral nucleus** (PC). The rostral group of the intralaminar nuclei (CenM, CenL, PC) project to the anterior and posterior cingulate cortices and to the entorhinal cortex in the monkey. The central lateral nucleus in the monkey also projects to the superior temporal sulcus and frontal eye field.

More caudally in the internal medullary lamina (iml) are the **centromedian nucleus** (CM, Figs 14.6 to 14.8) and **parafascicular nucleus** (PF). The most conspicuous nucleus in the caudal group is the **centromedian nucleus** (Figs 14.6 to 14.8) – easily recognized in the human brain by its pale appearance when compared with adjacent thalamic nuclei. This pale appearance is probably the result of its rich fiber network and therefore its elevated myelin content. The centromedian nucleus has a large-celled part (magnocellular) and a homogenous population of densely packed cells that make up the small-celled part (parvocellular). Expansion and development of the centromedian nucleus is characteristic of nonhuman primates and humans.

The intralaminar nuclei receive impulses from widespread regions of the cerebral cortex (including premotor area 6 and the somatosensory cortex), have reciprocal connections with the primary motor area 4 (corticothalamic projections), and receive projections from other thalamic nuclei (interthalamic projections). Many fibers of the ascending reticular system end in the intralaminar nuclei as do fibers from the caudate, putamen, and internal segment of the globus pallidus (as part of the ansa lenticularis).

Activation of the midbrain reticular formation and of the thalamic intralaminar nuclei occurs in humans as they go from a relaxed awake state to an attention-demanding reaction-time task. These results confirm the role of the intralaminar nuclei in alertness and arousal as a part of the ascending reticular system.

Fibers from the medial lemniscus and some fibers in the trigeminothalamic paths terminate in the intralaminar nuclei. Tertiary neurons serving general tactile sensations are in the **caudal part of the intralaminar nuclei**. General tactile impulses likely enter consciousness at the level of the thalamus in humans. If localization ('where') and discrimination ('what') of these general sensations is necessary then the cerebral cortex must become involved. This would require that these impulses project from the thalamus to the cerebral cortex where such additional processing can take place.

Fibers of the lateral spinothalamic tract (body pain and temperature) as well as fibers of the spinoreticulothalamic tract (carrying visceral pain impulses) end in part in the **centromedian** and **parafascicular nuclei**. Pain-sensitive neurons are physiologically identifiable in the intralaminar nuclei of nonhuman primates and humans. Efferent connections of the parafascicular nuclei in the squirrel monkey reach both parts of the globus pallidus, as well as the pars compacta of the substantia nigra and the ventral tegmental area. This nucleus also projects to the hypothalamus, subthalamus, rostral intralaminar nuclei, medial dorsal thalamic nuclei and the median thalamic nuclei.

There appear to be substantial intralaminar projections to the striatum in highly specific functional circuits with the centromedian nucleus projecting to the putamen and the parafascicular nucleus to the caudate nucleus (thalamostriate projections). From the centromedian nucleus, there are diffuse and widespread fiber projections to the frontal cortex, including that of the precentral gyrus, anterior cingulate and dorsolateral cortex as well as to the piriform region (the anterior part of the temporal lobe medial to the rhinal sulcus). These diffuse thalamocortical projections are part of the so-called '**nonspecific or diffuse thalamocortical activating system**' which can be demonstrated by low-frequency ipsilateral stimulation of the intralaminar nuclei. Such stimulation results in recruiting responses throughout the cerebral cortex thus playing an important role in arousal and attention. The centromedian nucleus also projects to other thalamic nuclei particularly the ventral lateral nucleus.

Jones (1998, 2001, 2002 and elsewhere) has made the case for the presence of a **matrix** of thalamocortical neurons without nuclear borders found throughout the thalamus that project to superficial layers of the cerebral cortex over relatively wide areas that are not limited to architectonic boundaries. There is also a **core** of thalamocortical neurons concentrated in certain thalamic nuclei and projecting in a highly ordered manner to middle cortical layers. These projections follow specific architectonic borders. The matrix neurons receive subcortical projections that lack the somatotopic and modality topic order of the ascending sensory paths. The core neurons receive more ordered and precise subcortical inputs that have identifiable physiological properties. These core and matrix neurons are definable based on their staining characteristics using calcium-binding proteins. This viewpoint deserves further study and elucidation.

Conscious appreciation of diffuse, poorly localized, noxious stimuli takes place in the intralaminar nuclei of the thalamus. The diffuseness of cortical projections from the intralaminar nuclei may account for this poor localization. Stimulation of the human intralaminar nuclei leads to variable responses: contractures and clonus, a diffuse burning pain on the contralateral body, a feeling of warmth on the contralateral body, or a tingling in the hand. In patients with intractable pain, stimulation of the intralaminar nuclei may exacerbate the patient's spontaneous pain. Based on its connections and these stimulation-related observations, the intralaminar nuclei are a likely target during **deep brain stimulation** (DBS) for the neurosurgical relief of intractable pain. The reason for the success of this procedure is likely the interruption of some of spinoreticulothalamic and spinothalamic fibers that reach the intralaminar nuclei. Relief from pain for many months may occur after neurosurgical lesions of the centromedian nucleus (CM). Over a prolonged period, the pain often begins to recur in some patients. What is not clear in all of this is whether the beneficial effects of intralaminar nuclear stimulation or ablation result in changes in the patient's actual perception of pain or whether there is alteration in the patient's attitude towards their pain. The centromedian nucleus has also been the target for electrical stimulation in cases of difficult-to-control seizures.

Microelectrode recordings of neurons in the caudal group of the intralaminar nuclei (centromedian and parafascicular nuclei) in patients undergoing neurosurgical procedures for spasmodic torticollis provide evidence that these nuclei are involved in the neuronal mechanisms of selective attention.

14.2.3. Medial Nuclei

The **internal medullary lamina** is a band of fibers in the midst of the thalamus that divides the **medial nuclei** from more lateral nuclei (iml, Fig. 14.1). The most conspicuous representative of the medial nuclei is the **medial dorsal** (dorsomedial) **nucleus** (MD) (Figs 14.5 to 14.8). A **medial ventral nucleus** (MV) is part of the medial nuclei as well. The medial ventral nucleus in the left cerebral hemisphere may merge with the medial ventral nucleus in the right cerebral hemisphere at the interthalamic adhesion. When the medial ventral nuclei on the two sides of the brain merge with one another medially, they together form a **nucleus reuniens**. The medial nuclear group extends about two-thirds the length of the entire thalamus. Coronal sections of the human thalamus reveal the internal medullary lamina almost encircles the medial dorsal nucleus (MD, Figs 14.5 to 14.8). The ependyma of the lateral wall of the third ventricle, along with fibers of the internal medullary lamina, form the medial border of the medial dorsal nucleus (iml, MD, Figs 14.5 to 14.8). Posteriorly its lateral border is the internal medullary lamina and anteriorly its lateral border is the mamillothalamic tract. The anterior nuclear group and the lateral dorsal nucleus lie on the dorsal surface of the medial dorsal nucleus whereas its ventral surface is difficult to identify. Posteriorly the medial dorsal nucleus blends with the pulvinar. The human medial dorsal nucleus includes magnocellular, multiform, and parvocellular parts.

The medial nuclei receive impulses from almost all other thalamic nuclei and from the hypothalamus over the periventricular system of fibers. The medial nuclei have extensive connections with the prefrontal cortex, ventral surface of the frontal lobe, and olfactory structures. In addition to receiving impulses from these diverse regions of the nervous system, the medial nuclei also relay impulses back to these same areas. Finally, there is a projection of fibers in nonhuman primates from the medial dorsal nucleus to the amygdaloid body. Interestingly, this projection is not reciprocal. fMRI assessment of the connections of the human medial dorsal nucleus reveals functional connectivity of the MD nucleus with the dorsolateral prefrontal cortex. There were correlations with the left superior temporal, parietal, posterior frontal, and occipital regions as well.

Because of the complex interrelations with a variety of brain regions, associations or interrelations occurring in the medial nuclei are such that the source and quality of stimuli reaching these nuclei is lost. Various impulses that reach the medial nuclei discharge to the prefrontal cortex through the **anterior thalamic radiations**. Here such impulses enter consciousness as feelings of well-being, pleasure, displeasure, apprehension, or fear, often referred to as 'affective tone'. Such complex associations in the medial nuclei are an important part of our human personality. These feelings, correlated with many other

impulses of a visceral and a somatic nature, influence the responses of an individual. Although prefrontal lobotomies alter unfavorable feelings or affective tones such procedures also cause intellectual and personality changes. Imaging studies and postmortem studies of patients with schizophrenia reveal the medial dorsal nucleus and the pulvinar to be significantly smaller in comparison with controls.

An injury involving the medial dorsal nucleus in humans causes changes in motivational drive (abulia) including apathy, loss of the ability to solve problems, alterations in levels of consciousness, and often failure to inhibit inappropriate behavior. In one extensively studied patient who sustained a stab wound to that part of the left thalamus presumed to correspond to the medial dorsal nucleus, there was a severe verbal memory deficit. Brain imaging confirms the location of his thalamic injury.

Infusing Xylocaine through chemodes into the human thalamus has a depressant effect on the central nervous system. Teflurane gas, passed into the medial thalamus in a patient with terminal cancer and chronic pain, resulted in temporary suppression of thalamic activity. The long-term clinical significance of such efforts is unclear.

14.2.4. Median Nuclei

The median nuclei (Fig. 14.1B) include the **nucleus reuniens** (Re), the **paratenial nucleus** (PT), the **paraventricular thalamic nuclei** (PV), and the **rhomboid nucleus**. The nucleus reuniens is a term used for the fused medial ventral (MV) nuclei (representatives of the medial nuclei not the median nuclei) when they meet in the interthalamic adhesion. The paratenial nucleus is small and visible at the ventromedial edge of the medial dorsal nucleus. The paraventricular thalamic nucleus in monkeys projects to the subiculum, entorhinal cortex, and anterior cingulate cortex whereas the paratenial nucleus also projects to the anterior cingulate cortex in monkeys. The entire length of the superior temporal gyrus in monkeys receives afferents from nucleus reuniens along with the anterior cingulate cortex. A most interesting observation is that of dopaminergic axons profusely targeting the thalamus in nonhuman primates and humans. The highest density of this innervation was in the median thalamic nuclei followed by the medial dorsal nuclei, lateral median thalamic nuclei posterior nucleus and the ventral lateral (VL) motor nucleus. Autoradiographic and PET studies in humans have identified dopamine D_2-like receptors with relatively high densities in the median nuclei as well as in the intralaminar nuclei. The role of these dopaminergic projections to the thalamus may have important implications for a variety of neurological conditions (Parkinson's disease, schizophrenia, and drug addiction).

14.2.5. Metathalamic Body and Nuclei

These nuclei include the **lateral geniculate nucleus** (LG, Fig. 14.8) and the **medial geniculate nucleus** (MG, Fig. 14.8). These metathalamic nuclei are beneath the pulvinar at the transition between the midbrain and the diencephalon.

Lateral Geniculate Body and Nucleus

Each human **lateral geniculate body** is triangular and tilted about 45° with a hilum on its ventromedial surface. Sections of the lateral geniculate body reveal a **lateral geniculate nucleus** (LG), the thalamic relay nucleus for the visual system. Injury to the lateral geniculate nucleus on one side will lead to a contralateral homonymous hemianopia. The lateral geniculate nucleus in humans includes a laminated and horseshoe-shaped **dorsal part (LG$_d$)** and a small or absent **ventral part (LG$_v$)**. The horizontal meridian of the visual field corresponds to the long axis of each lateral geniculate body, from hilum to convex surface. The fovea is represented in the posterior pole of the lateral geniculate with the upper quadrant of the visual field represented anterolaterally and the lower quadrant anteromedially.

Nuclei and Layers of the Lateral Geniculate Body

The lateral geniculate body is surprisingly variable in structure, with several segments: one with two layers, another with four, and one in the caudal half with six parallel layers. The six-layered part has two large-celled layers – an outer magnocellular layer ventral to an internal magnocellular layer and four small-celled layers – an internal, outer, and two superficial parvocellular layers. Between cellular layers are thin zones of fibers. A poorly developed S-region is ventral to the magnocellular region in humans. Neurons in parvocellular layers display rapid growth that ends about six months after birth, reaching adult size near the end of the first year. Neurons in the magnocellular layers continue to grow rapidly for a year after birth, reaching adult size by the end of the second year. Reduction in mean diameter, (and consequently neuronal volume), was observed in lateral geniculate neurons in 24 patients with severe visual impairment (blindness). There was reduced cytoplasmic RNA, nucleolar volume, and tetraploid nuclei in neuroglia.

In primates, each adjacent pair of geniculate neuronal layers is functionally distinct – one activated by the ipsilateral eye, the other by the contralateral eye. In magnocellular and parvocellular layers of the lateral geniculate in the monkey, the number of neurons receiving impulses from the contralateral eye is greater than that from the ipsilateral eye. In development, geniculate neurons segregate into layers for the left and right eyes before birth. All type III or midget retinal ganglionic neurons in primates send axons to parvocellular layers (called P cells), whereas the larger type II or parasol ganglionic neurons of peripheral retina, and some paramacular retinal neurons, provide fast-conducting, large fibers that synapse with magnocellular geniculate neurons (called M-cells). Therefore, each half of the retina has three representations in each lateral geniculate nucleus (LG) – in a magnocellular layer and in two parvocellular layers.

Neurophysiological Studies of Lateral Geniculate Neurons

Neurophysiological studies in nonhuman primates reveal that some geniculate neurons are concerned with form, others with color, but most handle both simultaneously. The 'center-surround' organization characteristic of retinal ganglionic neurons occurs in monkey lateral geniculate nuclei. Lateral geniculate neurons are selectively sensitive to blue, green, yellow, and red.

Termination of Retinal Fibers in the Lateral Geniculate

Superior retinal fibers end medially as inferior fibers end laterally in the dorsal lateral geniculate nucleus (LG$_d$). As macular fibers end in the nucleus, they form a central cone, its apex directed to the hilus of this nucleus. Nasal retinal fibers decussate in the chiasma and end in the contralateral lateral geniculate in nuclear layers 1, 4, and 6, temporal retinal fibers do not decussate in the chiasma but end in the ipsilateral lateral geniculate in nuclear layers 2, 3, and 5. In prenatal humans, fibers immunoreactive to substance P occur in the optic nerve and reach the lateral geniculate nuclei; their function is uncertain. In primates, each superior colliculus receives many processes of ganglionic neurons from both retinae though the contralateral eye appears to be more strongly represented; inputs are arranged in bands in collicular zones representing more peripheral regions of the binocular visual fields.

Amblyopia and the Lateral Geniculate Nucleus

Reduction in vision caused by disuse of an eye is termed **amblyopia** (commonly called 'lazy eye'). If the eyes differ in refractive power (called **anisometropia**) and this condition remains uncorrected, amblyopia often results. Examination of the brain of a patient with anisometropic amblyopia showed a decrease in neuronal size in dorsal lateral geniculate parvocellular layers connected with the 'lazy' eye.

The Medial Geniculate Body and Nuclei

Each **medial geniculate body** (Fig. 10.5) is a small eminence about 5 mm wide, 4 mm deep, and 4–5 mm long, protruding from the posterior aspect of the diencephalon and containing a **medial geniculate nucleus** (MG, Fig. 14.8), the thalamic relay nucleus for the auditory system. Each medial geniculate nucleus contains a **dorsal part** (MG_d), a **ventral part** (MG_{mc}), and a **medial part** (MG_m). The ventral part, also termed the principal or parvocellular part is the largest part of the MG occupying the ventrolateral quarter, the medial part (also termed the internal or magnocellular part) occupies the ventromedial quarter, and the dorsal part forms a tier extending the length and breadth of the medial geniculate nucleus (MG). Ascending projections of auditory neurons in the inferior colliculi, carrying auditory impulses from both ears, enter the **brachium of the inferior colliculus** to reach the **ventral part of the medial geniculate nucleus** (MG_v). A few fibers of the lateral lemniscus bypass the inferior colliculi to synapse with medial geniculate neurons. Fibers of the brachium are fewer in number and size in the elderly.

Auditory impulses reaching the ventral part of the medial geniculate relay to the **primary auditory cortex** that correspond to **Brodmann's area 41** in the temporal lobe. This relay is over fibers of the **auditory radiations** that pass in the **sublenticular part of the internal capsule** on their way to the temporal lobe. An injury to one side of the cerebral hemisphere will not lead to an appreciable loss of hearing. This is because each medial geniculate nucleus (MG) receives fibers from both ears.

In addition to auditory impulses reaching the medial geniculate, the lateral spinothalamic tract and the medial lemniscus send some fibers to synapse with tertiary neurons in the **medial** or **magnocellular part of the medial geniculate nucleus**. The lateral spinothalamic tract carries impulses related to superficial pain and itch from skin, well-localized deep pain from joints, fascia, periosteum, tendons and muscles, together with different degrees of thermal sensation such as warmth or heat and coolness or coldness. The medial lemniscus carries discriminatory tactile, pressure, proprioceptive, and vibratory stimuli from the body and limbs.

Tones of Different Pitch Reach Consciousness in Humans at the Geniculate Level

Tones of different pitch likely enter consciousness at the geniculate level. Studies in primates indicate that the discrimination of sound localization, loudness, and pitch do not require involvement of the **primary auditory cortex** (area 41) in the temporal lobe. Ablation of the primary auditory and periauditory areas in humans does not influence the discrimination of tone sequences.

14.2.6. Posterior Nuclear Complex

The posterior nuclear complex of the human thalamus includes the nucleus limitans (Lim), the posterior nuclei (PLi), and the suprageniculate nucleus (SGe). These nuclei project to the cortex in and around the insula. Jones (2007) suggests that future studies are likely to relegate this group of nuclei to other nuclear complexes including the intralaminar nuclei, pulvinar nuclei, and medical geniculate nuclei.

14.2.7. Pulvinar Nuclei and Lateral Posterior Nucleus

This thalamic group includes the **pulvinar nuclei** (Pul) as well as the **lateral posterior nucleus** (LP). The latter nucleus is lateral to the **internal medullary lamina** and is therefore part of a 'lateral group' of nuclei according to some authors. Others include the lateral dorsal nucleus along with the pulvinar nuclei and the lateral posterior nucleus in the 'dorsal nuclei' of the thalamus. In the present discussion, the pulvinar nuclei and lateral posterior nucleus are collectively one of nine named nuclear groups. Because of their close relationship and similar connections, the lateral dorsal nucleus and the anterior nuclei collectively form another of these nine named nuclear groups.

Pulvinar Nuclei (Pul)

The large pulvinar nuclei (Pul) form the posterior pole of the thalamus in humans. Some authors have described anterior, medial, lateral, and inferior nuclei of the human pulvinar (Pul). fMRI assessment of the connections of the human pulvinar reveals functional connectivity of this nucleus with Brodmann's area 39 in the inferior parietal lobule. The **medial pulvinar** in nonhuman primates has widespread connections with the cerebral cortex of the frontal lobe (prefrontal cortex), inferior parietal lobule, superior temporal gyrus and sulcus, temporal pole, and the insular lobe as well as the cingulate cortex, amygdaloid body, and parahippocampal cortex. It also receives input from the superior colliculus of the midbrain. The medial pulvinar likely functions as both a thalamic relay and thalamic association nucleus related to visuospatial and somatosensory processing as well as directed attention. The **lateral and inferior nuclei** of the primate pulvinar have reciprocal connections with striate and extrastriate cortices. The lateral pulvinar also projects to the posterior parietal cortex and to the superior temporal cortex in nonhuman primates. The lateral and inferior nuclei of the human pulvinar may play an important role in visual processing. The pulvinar may also be an anatomical substrate for speech in light of the fibers of passage that travel through it to reach the centromedian nucleus and the medial dorsal nucleus. Most input then to the pulvinar in primates comes from the cerebral cortex and its output is back to the cerebral cortex. Based on these corticopulvinar and pulvinar-cortical connections the pulvinar in primates likely serves as a thalamic association nucleus. It is of interest to note that postmortem studies and imaging studies of patients with schizophrenia reveal the medial dorsal nucleus and the pulvinar to be significantly smaller in comparison with controls.

Lateral Posterior Nucleus (LP)

As its name suggests, the **lateral posterior nucleus** is lateral to the internal medullary lamina and extends posteriorly where the pulvinar replaces LP. The posterior parietal cortex and the extrastriate visual areas in nonhuman primates project to the lateral posterior nucleus as do the posterior cingulate and parahippocampal cortex. Little data is available regarding the function of the lateral posterior nucleus (LP).

14.2.8. Reticular Nucleus (Rt)

The **reticular nucleus** (Rt) is a thin sheet of neurons along the medial aspect of the internal capsule (ic) and along the entire lateral aspect of the thalamus. Neurons in this nucleus resemble those of the adjacent thalamic nucleus. In humans it contains two types of large, sparsely branched, long-dendrite, reticular, aspiny neurons. It also has neurons with spines that are densely branched. The reticular nucleus (Rt) is said to have a unique position in thalamic circuitry. It receives fibers from a variety of sources and is a region of termination for many fibers in the ascending reticular system. As thalamo-cortical fibers pass through this thin sheet of reticular neurons on the lateral aspect of the thalamus they provide collaterals

to the reticular neurons. GABAergic cells in the reticular nucleus then give rise to fibers that project back onto the same thalamic nuclear cells that gave rise to these thalamocortical fibers. Fibers coming from the cerebral cortex that are destined to terminate on cells in specific thalamic nuclei also provide collaterals to the reticular neurons. Again, GABAergic neurons in the reticular nucleus then give rise to fibers that project back to the particular thalamic nucleus on which these corticothalamic fibers terminated. The corticothalamic terminals predominate in this scheme of thalamic circuitry. The reticular nucleus relays impulses to other thalamic nuclei and is a region of relay for various thalamocortical systems. The reticular nucleus (Rt) has a widespread influence on cortical activity as a part of the '**nonspecific or diffuse thalamocortical activating system**' playing an important role in arousal, attention and conscious awareness. The reticular nucleus is thought to play a crucial role in the pathophysiology of human epilepsy and related disorders.

14.2.9. Ventral Nuclei

The ventral nuclei are divisible into four subdivisions: ventral anterior (VA), ventral lateral (VL), ventral medial (VM), and ventral posterior (VP) nuclei.

Ventral Anterior Nucleus

The **ventral anterior nucleus** (VA, Figs 14.3, 14.4) in nonhuman primates may be a thalamic relay nucleus for impulses from the substantia nigra. It receives afferents from the pars reticularis of the substantia nigra and from the internal segment of the globus pallidus (GPi) in nonhuman primates. The ventral anterior nucleus in primates projects its impulses to the cingulate, premotor, supplementary motor, and prefrontal cortex.

Ventral Lateral Nucleus

The **ventral lateral nucleus** (VL, Fig. 14.4) is posterior to the ventral anterior nucleus (VA) and occupies most of the anterior half of the ventral part of the thalamus. The ventral lateral nucleus has both anterior (VL_a) and posterior (VL_p) parts (Fig. 14.5). The posterior part is very large, occupying almost half of the volume of the ventral nuclear complex. In nonhuman primates, the posterior part of the ventral lateral nucleus (VL_p) is the site of termination of a dense projection of fibers from the dentate nucleus of the cerebellum. Such dentatothalamic fibers pass through the ipsilateral superior cerebellar peduncle and cross to the contralateral ventral lateral nucleus (VL_p). This nuclear group relays its impulses to the **primary motor area 4** in the frontal lobe. The posterior part of the ventral lateral nucleus (VL$_p$) as used in the present account (Table 14.1) corresponds to the ventral intermedial nucleus (Vim) as used in some schemes of human thalamic terminology (Hassler, 1982). This nucleus shows rich, spontaneous, rhythmic and nonrhythmic discharges related to tremor generation. For this reason, stereotaxic lesions of the posterior part of the ventral lateral nucleus (VL) in humans may be a useful target in the treatment of certain disorders of posture and movement such as limb rigidity, involuntary movement disorders, and the tremor of Parkinson's disease, essential tremor and perhaps tremor of other origins.

The **anterior part of the ventral lateral nucleus** (VL$_a$) receives fibers from the basal nuclei particularly the internal segment of the globus pallidus. This nuclear group then relays impulses to the **premotor area 6** in the frontal lobe. The ventral lateral nucleus (VL) (including its anterior and posterior parts), is therefore a **relay nucleus in the motor system** through which the cerebellum and the globus pallidus influence motor areas in the cerebral cortex. The ventral lateral nucleus (VL) is likely to modulate inputs from the cerebellum and basal nuclei to the primary motor cortex. Because of this, the ventral lateral nucleus (VL) is part of the '**motor thalamus**'.

Thalamic neurons in the ascending vestibulothalamic path of primates are identifiable in various nuclei of the thalamus including bilateral projections to neurons in the oral part of the ventral posterior lateral nucleus (VPL$_o$), caudal part of the ventral lateral nucleus (VL$_c$) (corresponding to the dorsal part of the human VL$_p$) and dorsal part of the ventral posterior inferior nucleus (VPI$_d$).

Ventral Medial Nucleus

The **ventral medial nucleus** (VM) occupies a transition zone between the ventral lateral nucleus (VL) and the ventral posterior nucleus (VP). Some authorities suggest that this nuclear subdivision belongs to VL – others suggest that it is a distinct subdivision. In addition to the controversy about where this nuclear subdivision belongs, there is also disagreement about the connections of this nuclear group. The ventral medial nucleus (perhaps along with the ventral anterior nucleus) may be a thalamic relay for impulses from the substantia nigra in nonhuman primates. The basal ventral medial nucleus (VM$_b$) in humans corresponding to the parvocellular part of the ventral posterior medial nucleus (VPM$_{pc}$) in monkeys is a gustatory relay nucleus.

Ventral Posterior Nucleus

The **ventral posterior nucleus** (VP), the largest subdivision of the ventral nuclear group, occupies the caudal half of the thalamus. This nuclear subdivision receives many ascending sensory systems from both the body and head including the termination of the medial lemniscus, dorsal and ventral trigeminothalamic tracts, and the lateral and ventral spinothalamic tracts. It serves as an end-station processing nondiscriminative sensations and as a relay station for more discriminative sensations that reach the cerebral cortex.

The ventral posterior nucleus (VP) is divisible into three subnuclei. These include the **ventral posterior lateral** (VPL, Figs 14.6, 14.7), **ventral posterior medial** (VPM, Figs 14.6, 14.7), and **ventral posterior inferior** (VPI, Fig. 14.7) **nuclei**. In humans, the ventral posterior lateral nucleus (VPL) may be divisible into **anterior** (VPL$_a$) and **posterior** (VPL$_p$) subnuclei in keeping with electrophysiological divisions of VPL in nonhuman primates (VPL$_o$ and VPL$_c$). A pattern for localization of body parts in the ventral posterior nucleus (VP) exists with sensory impulses from regions of the face and head ending in the medial part of the ventral posterior nucleus termed the **ventral posterior medial nucleus** (VPM). There may be some segregation of sensory inputs from cutaneous versus deep tissues of the face and head within VPM. Sensory impulses from regions of the body, limbs, and trunk end in the lateral part of the ventral posterior nucleus termed the **ventral posterior lateral nucleus** (VPL). A pattern is also impressed on the ventral posterior nucleus for the type of impulse entering this nucleus in nonhuman primates. Painful impulses from cutaneous areas reach the most caudal part of the ventral posterior nucleus (VPL$_p$) and project to postcentral areas 3b and 1 whereas proprioceptive impulses from deep tissues (muscle and joint inputs) which project to areas 3a and 2 in the postcentral gyrus reach the rostral parts or central core of this nucleus (VPL$_a$). Injury to this nuclear complex is rare but when it does occur, there is often interference with discriminative somatosensory sensations including the loss of pain and temperature on the contralateral side of the body. Paradoxically, however, if such injury is of a vascular nature involving the ventral posterior nucleus there is likely to be the development of the **thalamic pain syndrome** (of Déjerine–Roussy). Central pain is a conspicuous symptom of this syndrome.

Ventral Posterior Lateral Nucleus

The **lateral spinothalamic tract** carries impulses related to superficial pain and itch from skin, well-localized deep pain from joints, fascia, periosteum, tendons and muscles, together with different degrees of thermal sensation such as warmth or heat and coolness or coldness. Fibers of the lateral spinothalamic tract synapse with tertiary neurons in the **ventral posterior lateral nucleus** (VPL) of the thalamus (Fig. 7.2), in the **intralaminar nuclei** particularly the central lateral (CL) and central medial nuclei (CeMe), in **nucleus limitans** (Lim) of the posterior nuclear complex, and in the **magnocellular part of the medial geniculate nucleus** (MG_{mc}). Spinothalamic fibers end somatotopically in the ventral posterior nucleus (VP) such that fibers from the upper limb end medial and ventromedial to those from the lower limb (Fig. 7.2). Presumably, there is also an orderly termination of spinothalamic fibers by modality. In nonhuman primates, impulses for superficial pain from cutaneous areas of limbs relay to the caudal part of the ventral posterior lateral nucleus (VPL_c) whereas painful impulses from fascia, tendons, and joints project to its oral part (VPL_o). In anesthetized monkeys, many neurons in VPL_c receive inputs from nociceptors in the periphery, presumably from **thermal** and **mechanical nociceptors**. General aspects of pain and extremes of temperature in humans likely enter consciousness at the thalamic level.

Tertiary neurons serving **general tactile sensations** are in the **ventral posterior lateral nucleus** (VPL) (Fig. 8.2) and in the caudal part of the **intralaminar nuclei**. General tactile impulses enter consciousness at this level of the thalamus in humans. If localization ('where') and discrimination ('what') of these general sensations is necessary then the cerebral cortex must become involved. This would require that these impulses project from the thalamus to the cerebral cortex where such additional processing can take place.

Fibers of the medial lemniscus carrying discriminatory tactile, pressure, proprioceptive, and vibratory stimuli synapse with tertiary neurons in the **caudal part of the ventral posterior lateral nucleus** (VPL_c). They end in somatotopic manner with fibers from the cuneate component of the medial lemniscus (carrying impulses from thoracic and cervical levels) ending medially and rostrally in the caudal part of the ventral posterior lateral nucleus (VPL). Fibers of the **gracile component of the medial lemniscus** (carrying impulses from lower thoracic, all lumbar and sacral levels) end laterally and caudally in the caudal part of the ventral posterior lateral nucleus (VPL). Orientation of the body representation actually follows the shape of this thalamic nucleus. Neurons in VPL_c send fibers to **primary somatosensory areas 3a, 3b, 1**, and **2**. Some medial lemniscal fibers end in the **magnocellular part** of the medial geniculate

nucleus (MG_{mc}), pulvinar nuclei, and ventral part of the **suprageniculate nucleus** (SGe).

Thalamic neurons in the ascending vestibulothalamic path of primates are identifiable in various nuclei of the thalamus including bilateral projections to neurons in the **oral part of the ventral posterior lateral nucleus** (VPL_o) caudal part of the ventral lateral nucleus (VL_c), and dorsal part of the ventral posterior inferior nucleus (VPI_d). Vestibular projections to VPL_o are scattered in small patches over extensive areas of this nucleus. This distribution of vestibular fibers in the thalamus permits interaction in VPL of vestibular and somatosensory (especially proprioceptive) inputs. Thus, there is a topographical and functional relationship between vestibular and somatosensory nuclei in the thalamus of rhesus monkeys. Studies in primates reveal that the vestibular nuclear projection to these thalamic neurons is a sparse but definite bilateral projection. Based on these observations, VPL_o, VL_c, and VPI_d are **vestibular relay nuclei**. Based on its projections to the primary motor cortex, however, VPL_o and its extension VL_c in monkeys are probably involved in vestibular aspects of motor coordination.

Recordings made in the posterior part of the thalamus in rhesus monkeys have localized the vestibulothalamic tract lateral to the medial lemniscus. This is likely between the medial lemniscus and brachium of the inferior colliculus. These recorded responses suggest that the projection between the brain stem vestibular nuclei and the thalamus is monosynaptic.

Ventral Posterior Medial Nucleus

The ventral trigeminothalamic tract, carrying impulses for pain, thermal discrimination, thermal extremes, and general tactile sensations in the head, ends in the **ventral posterior medial nucleus** (VPM) (Fig. 7.7). Some ventral trigeminothalamic fibers end in that part of the ventral posterior lateral nucleus (VPL) that is rostrolateral to the ventral posterior medial nucleus (VPM). A contralateral, discrete, somatotopic representation of the body and head occurs in the ventral posterior nucleus in humans with ascending sensory fibers from the body ending in VPL and those from the face ending in VPM. This representation of body, face, and head across the ventral posterior nucleus is depicted by a '**thalamic homunculus**' that is similar to the motor homunculus that can be depicted across the human **primary motor cortex** and the sensory homunculus that can be depicted across the **primary somatosensory cortex**. A contralateral, medial to lateral arrangement of sensory responses in the ventral posterior nucleus exists beginning medially (in the ventral posterior medial nucleus) with the representation of the oral cavity, followed laterally by labial and facial representation, then representation of the fingers, thumb, and rest of the hand. Following the rest of the hand, in the ventral posterior lateral nucleus (VPL), is the representation of the forearm, arm, trunk, and leg to the lateral limit of this nucleus. A pattern probably exists in the ventral posterior nucleus in humans with regard to pain and temperature impulses. In humans, sensations of crude pain and thermal extremes enter consciousness at this level of the thalamus and for that reason need not project on to the cerebral cortex for their conscious appreciation. If localization ('where') and discrimination ('what') of these general sensations is necessary then the cerebral cortex must become involved. This would require that these impulses project from the thalamus to the cerebral cortex where such additional processing can take place.

Impulses for thermal discrimination, relayed over fibers of the **ventral trigeminothalamic tract**, synapse with tertiary neurons in the **ventral posterior medial nucleus** (Fig. 7.7). **Thalamoparietal fibers** (Fig. 7.7) travel from here to the area of representation of the face in the postcentral gyrus of the parietal lobe (Fig. 7.7) where they terminate on fourth-order cortical neurons.

The dorsal trigeminothalamic tract, carrying general tactile, discriminative touch, proprioceptive, pressure, and vibratory impulses from the head, ends bilaterally in the **ventral posterior medial nucleus** (Fig. 8.7). Thalamoparietal fibers from tertiary neurons in the ventral posterior medial nucleus reach the postcentral gyrus of the parietal lobe. In particular, proprioceptive and discriminative tactile impulses in the dorsal trigeminothalamic tract relay via thalamoparietal fibers to the cortex in the parietal lobe. General tactile as well as impulses for pressure and vibration, however, enter consciousness at the level of ventral posterior medial nucleus of the thalamus in humans and do not need to reach the cerebral cortex. If localization ('where') and discrimination ('what') of these general sensations is necessary then the cerebral cortex must become involved. This would require that these impulses project from the thalamus to the cerebral cortex where such additional processing can take place.

Ventral Posterior Inferior Nucleus

The dorsal part of the **ventral posterior inferior nucleus** (VPI$_d$) in rhesus monkeys approximates the medial or magnocellular part of the medial geniculate nucleus (MG), wedged between the lateral and medial parts of the ventral posterior nucleus (VP). In this location, VPI$_d$ is mediodorsal to the perioral representation

in VPM and laterodorsal to the hand (thumb) representation in VPL. (Some authors include VPI$_d$ as part of VPL).

Thalamic neurons in the ascending vestibulothalamic path of primates are identifiable in various nuclei of the thalamus including bilateral projections to neurons in the oral part of the ventral posterior lateral nucleus (VPL$_o$), the caudal part of the ventral lateral nucleus (VL$_c$), and the dorsal part of the **ventral posterior inferior nucleus** (VPI$_d$). The ventral posterior inferior nucleus and the posterior nuclei (PLi) relay vestibular impulses to the posterior part of the postcentral gyrus at the base of the intraparietal sulcus. This region corresponds to the **primary vestibular cortex** or area 2v. These thalamic nuclei are probably involved in the conscious appreciation of vestibular sensation.

14.3. INJURIES TO THE THALAMUS

Unilateral vascular injury to the thalamus near the caudal part of ventral posterior lateral nucleus (VPL$_c$) often causes a sensory deficit on the contralateral side involving the face, arm, and leg. This may include a transient or persistent numbness and mild sensory loss appearing in the contralateral hand or foot or both, spreading to the remainder of the involved side of the body, including the face. The limbs feel large and swollen with such adjectives as numb, asleep, tingling, dead, or frozen used in descriptions by affected patients.

A review of the literature on the consequences of thalamic infarctions in humans reveals that damage to the mamillothalamic tract is a necessary condition for the development of the amnesia syndrome provided there is already damage to the thalamus. Involvement of the median nuclei, intralaminar nuclei and the medial dorsal nucleus in thalamic infarction influences frontal lobe functioning yielding memory and disturbances of executive function.

Unilateral abnormal movement disorders are also a feature of localized thalamic lesions. In particular, myoclonus and dystonia restricted to the distal upper limb with a characteristic hand posture (flexion of the metacarophalangeal joints and extension of the interphalangeal joints) termed the 'thalamic hand' along with slow, pseudo-athetoid movements result from lesions to Vim and Vc. According to the present terminology (Table 14.1), Vim corresponds to the posterior part of the ventral lateral nucleus (VL$_p$) whereas Vc corresponds to the ventral posterior nucleus (VP). Postural and kinetic (action) tremors result from lesions to Vim presumably due to interruption of cerebellothalamic paths.

14.4. MAPPING THE HUMAN THALAMUS

Mapping the thalamus in humans by stimulation has demonstrated sensory responses in what is presumably the ventral posterior medial nucleus (VPM). These responses represent impulses from the oral cavity including the gums, tongue, and pharynx. The lips and cheek have a generous representation with a meager representation for the scalp, forehead, and side and back of the head. The mandibular representation is in the most lateral part of the ventral posterior medial nucleus adjoining the representation of the thumb in the medial part of the ventral posterior lateral nucleus.

Mapping the thalamus in humans demonstrates a pattern in the ventral posterior lateral nucleus beginning with the representation of the thumb in the medial part of the ventral posterior lateral nucleus. The representation of the fingers, palm, and dorsum of the hand, forearm, arm, shoulder, trunk, and lower limb follow in order, with representation of the foot impinging on the internal capsule. This strictly contralateral representation is especially discrete for the hand, including the fingers. The lateral spinothalamic tract homunculus is dorsoventral in orientation and appears to stand on the medial geniculate nucleus (MG). When one combines the thalamic representations of the lateral spinothalamic tract with that of the medial lemniscus, a **double somatotopic representation** of the body is evident in the human thalamus.

14.5. STIMULATION OF THE HUMAN THALAMUS

Stimulation of the **medial geniculate nucleus** (MG) causes a ringing, referred to the center of the head, or to a buzzing heard bilaterally but mainly in the contralateral ear. Physiological vertigo and related phenomena often result from stimulation in conscious humans of the brachium of the inferior colliculus, superolateral part of the medial geniculate body or in the ventral posterior inferior nucleus.

Electrical stimulation of the **ventral posterior lateral nucleus** (VPL) in humans yields contralateral sensations of numbness and tingling. Sensations described as warm and cool tingling, and occasionally as burning or painful do occur following stimulation. Such responses are usually contralateral with a small percentage of ipsilateral responses. In a third series of patients, sensations evoked by thalamic stimulation were localized to the contralateral body and described as sharp pains, aching, or burning sensations. Thalamic stimulation reveals a discrete somatotopic organization in the ventral posterior lateral (VPL) nucleus with upper limb spinothalamic fibers ending in a dorsal position and those from the lower limb ventral, near the medial geniculate nucleus (MG).

Stimulation of the **dorsal funiculi** in humans causes activity in the **ventral posterior lateral nucleus** and perhaps in the **intralaminar nuclei**. Responses from the latter nuclei were diffuse and of longer latency suggesting involvement of elements of the spinoreticulothalamic tract. The effect of stimulation of the dorsal funiculi on thalamic neuronal activity suggests that such stimulation reduces spontaneous activity of thalamic neurons.

Activation of thalamic neurons occurs in the **fasciculus gracilis/fasciculus cuneatus-medial lemniscal system** in conscious nonhuman primates by various mechanical stimuli including touch-pressure on skin, mechanical stimulation of deep fascia or periosteum, or gentle rotation of a limb joint applied to a circumscribed, contralateral receptive field. A modality topic organization also occurs in the human ventral posterior nucleus. Activation of neurons in the posterior and inferior part of the ventral posterior nucleus results from cutaneous stimuli whereas activation of neurons more rostral and dorsal occurs following movements, change of position of joints, and muscle contractions.

Vestibular stimulation activates thalamic neurons in the **ventral posterior lateral nucleus** (VPL$_o$), **caudal part of the ventral lateral nucleus** (VL$_c$), and dorsal part of the **ventral posterior inferior nucleus** (VPI$_d$) in the alert monkey. This may include angular acceleration, rotation of an optokinetic cylinder, rotation of the visual surround, rotation of the animal itself about a vertical axis (both to the ipsilateral and to the contralateral side), and appropriate proprioceptive stimuli. Discharge patterns of these thalamic neurons are unrelated to ocular movements. Stimulation of the **ventral posterior inferior nucleus** in conscious humans evokes vestibular perceptions such as being tilted, whirled, or falling, and sensations of vertigo and of body movements. Nondiscriminative aspects of vestibular sensation often become conscious at the thalamic level in humans.

Stimulation of the superolateral part of the medial geniculate body and in the brachium of the inferior colliculus between 10 and 17 mm from the median plane causes vertigo and related phenomena including feelings of clockwise or counter-clockwise rotation, rising or falling, floating, whole body displacement, fainting, or nausea. Such responses occur from thalamic stimulation of the region anterior to that giving rise to auditory responses.

Finally, as noted earlier, thalamic stimulation as a part of **deep brain stimulation** (DBS) procedures, thalamic destruction and the interruption of thalamofrontal fibers are methods used to treat painful conditions. The ventral posterior nucleus (VP), some intralaminar nuclei including the parafascicular (PF) and centromedian nuclei (CM), and the medial nuclear group of the pulvinar (Pul) have all been thalamic targets in the treatment of painful conditions.

14.6. THE THALAMUS AS A NEUROSURGICAL TARGET

Thalamic stimulation or destruction, and interruption of thalamofrontal fibers may be useful in the treatment of painful conditions. Thalamic targets include the ventral posterior nucleus, some intralaminar nuclei, and the medial nuclear group of the pulvinar. Stimulation along the medial aspect of the parafascicular nucleus in the periventricular region resulted in 'good-to-excellent' reduction of chronic pain with minimal side effects. These patients expressed feelings of relaxation and well-being without overwhelming emotional overtones. Reduction of pain was contralateral to the stimulation yet effected bilateral peripheral fields, especially in medial regions of the body. Other studies by these same investigators demonstrated effective relief with long-term implantation of electrodes in the periventricular region near the posterior part of the third ventricle. Two-thirds of such patients noted reduction of pain on self-stimulation. Relief outlasted stimulation by many hours in some patients. This method of **stimulated analgesia**, involving the periventricular region in humans, relieves chronic pain with minimal complications probably by activating and releasing opioid peptides that appear in the cerebrospinal fluid after stimulation. Periventricular stimulation through implanted electrodes is successful in inhibiting chronic pain. Naloxone, an opiate antagonist, abolishes the effect of such stimulation. Thus, stimulation of the periventricular region appears effective in treating chronic pain in humans. Stimulation or destruction of the intralaminar region in humans, and the medial pulvinar, relieves pain without detectable sensory loss for 2–12 months.

FURTHER READING

Aggleton JP, Desimone R, Mishkin M (1986) The origin, course, and termination of the hippocampothalamic projections in the macaque. J Comp Neurol 243:409–421.

Allen LS, Gorski RA (1991) Sexual dimorphism of the anterior commissure and massa intermedia of the human brain. J Comp Neurol 312:97–104.

Balercia G, Kultas-Ilinsky K, Bentivoglio M, Ilinsky IA (1996) Neuronal and synaptic organization of the centromedian nucleus of the monkey thalamus: a quantitative ultrastructural study, with tract tracing and immunohistochemical observations. J Neurocytol 25:267–288.

Berezhnaya LA (2006) Neuronal organization of the reticular nucleus of the thalamus in adult humans. Neurosci Behav Physiol 36:519–525.

Buchsbaum MS, Buchsbaum BR, Chokron S, Tang C, Wei TC, Byne W (2006) Thalamocortical circuits: fMRI assessment of the pulvinar and medial dorsal nucleus in normal volunteers. Neurosci Lett 404:282–287.

Engelborghs S, Marien P, Martin JJ, De Deyn PP (1998) Functional anatomy, vascularisation and pathology of the human thalamus. Acta Neurol Belg 98:252–265.

Fortin M, Asselin MC, Gould PV, Parent A (1998) Calretinin-immunoreactive neurons in the human thalamus. Neuroscience 84:537–548.

Garcia-Cabezas MA, Rico B, Sanchez-Gonzalez MA, Cavada C (2006) Distribution of the dopamine innervation in the macaque and human thalamus. NeuroImage 34:965–984.

Harding A, Halliday G, Caine D, Kril J (2000) Degeneration of anterior thalamic nuclei differentiates alcoholics with amnesia. Brain 123:141–154.

Hassler R (1982) Architectonic organization of the thalamic nuclei. In: Schaltenbrand G, Walker AE, eds. *Stereotaxy of the Human Brain*. 2nd ed. pp 140–180. Thieme, Stuttgart.

Hassler R (1959) Anatomy of the thalamus. In: Schaltenbrand G, Bailey P, eds. *Introduction to Stereotaxis with an Atlas of the Human Brain*, pp 230–290. Thieme, Stuttgart.

Herrero MT, Barcia C, Navarro JM (2002) Functional anatomy of thalamus and basal ganglia. Childs Nerv Syst 18:386–404.

Hirai T, Jones EG (1989) A new parcellation of the human thalamus on the basis of histochemical staining. Brain Res Rev 14:1–34.

Jones EG (1997) A description of the human thalamus. In: Steriade M, Jones EG, McCormick DA, eds. *The Thalamus*. Vol II, Chapter 9, pp 425–500. Elsevier Science, Amsterdam.

Jones EG (1998) The thalamus of primates. In: Bloom FE, Björklund A, Hökfelt T, eds. *Handbook of Chemical Neuroanatomy. Vol 14: The Primate Nervous System, Part II*, pp 1–298. Elsevier, Amsterdam.

Jones EG (1998) Viewpoint: the core and matrix of thalamic organization. Neuroscience 85:331–345.

Jones EG (1998) A new view of specific and nonspecific thalamocortical connections. Adv Neurol 77:49–71.

Jones EG (2001) The thalamic matrix and thalamocortical synchrony. Trends Neurosci 24:595–601.

Jones EG (2002) A pain in the thalamus. J Pain 3:102–104.

Jones EG (2002) Thalamic circuitry and thalamocortical synchrony. Philos Trans R Soc Lond B Biol Sci 357:1659–1673.

Jones EG (2002) Thalamic organization and function after Cajal. Prog Brain Res 136:333–357.

Jones EG (2007) The human thalamus. In: Jones EG. *The thalamus*. 2nd ed. Vol I, Chapter 18, pp. 1396–1447. Cambridge University Press, Cambridge.

Kemether EM, Buchsbaum MS, Byne W, Hazlett EA, Haznedar M, Brickman AM, Platholi J, Bloom R (2003) Magnetic resonance imaging of mediodorsal, pulvinar, and centromedian nuclei of the thalamus in patients with schizophrenia. Arch Gen Psychiat 60:983–991.

Kinomura S, Larsson J, Gulyas B, Roland PE (1996) Activation by attention of the human reticular formation and thalamic intralaminar nuclei. Science 271:512–515.

Lehericy S, Grand S, Pollak P, Poupon F, Le Bas JF, Limousin P, Jedynak P, Marsault C, Agid Y, Vidailhet M (2001) Clinical characteristics and topography of lesions in movement disorders due to thalamic lesions. Neurology 57:1055–1066.

Macchi G, Jones EG (1997) Toward an agreement on terminology of nuclear and subnuclear divisions of the motor thalamus. J Neurosurg 86:670–685.

Mai JK, Assheuer J, Paxinos G (2004) *Atlas of the Human Brain*. 2nd ed. Elsevier/Academic Press, Amsterdam.

Morel A, Magnin M, Jeanmonod D (1997) Multiarchitectonic and stereotactic atlas of the human thalamus. J Comp Neurol 387:588–630.

Munkle MC, Waldvogel HJ, Faull RL (1999) Calcium-binding protein immunoreactivity delineates the intralaminar nuclei of the thalamus in the human brain. Neuroscience 90:485–491.

Munkle MC, Waldvogel HJ, Faull RL (2000) The distribution of calbindin, calretinin and parvalbumin immunoreactivity in the human thalamus. J Chem Neuroanat 19:155–173.

Raeva SN (2006) The role of the parafascicular complex (CM-Pf) of the human thalamus in the neuronal mechanisms of selective attention. Neurosci Behav Physiol 36:287–295.

Ralston HJ 3rd (2005) Pain and the primate thalamus. Prog Brain Res 149:1–10.

Rieck RW, Ansari MS, Whetsell WO Jr, Deutch AY, Kessler RM (2004) Distribution of dopamine D2-like receptors in the human thalamus: autoradiographic and PET studies. Neuropsychopharmacology 29:362–372.

Sanchez-Gonzalez MA, Garcia-Cabezas MA, Rico B, Cavada C. The primate thalamus is a key target for brain dopamine. J Neurosci 25:6076–6083.

Schmahmann JD (2003) Vascular syndromes of the thalamus. Stroke 34:2264–2278.

Schneider KA, Richter MC, Kastner S (2004) Retinotopic organization and functional subdivisions of the human lateral geniculate nucleus: a high-resolution functional magnetic resonance imaging study. J Neurosci 24: 8975–8985.

Sherman SM, Guillery RW (2001) *Exploring the Thalamus*. Academic Press, San Diego.

Shipp S (2003) The functional logic of cortico-pulvinar connections. Philos Trans R Soc Lond B Biol Sci 358:1605–1624.

Steriade M, Jones EG, McCormick DA (1997) *The Thalamus*. 2 Vols. Elsevier Science, Amsterdam.

Van Buren JM, Borke RC (1972) *Variations and Connections of the Human Thalamus*. Springer, Berlin.

Van der Werf YD, Witter MP, Uylings HB, Jolles J (2000) Neuropsychology of infarctions in the thalamus: a review. Neuropsychologia 38:613–627.

Van der Werf YD, Witter MP, Groenewegen HJ (2002) The intralaminar and midline nuclei of the thalamus. Anatomical and functional evidence for participation in processes of arousal and awareness. Brain Res Rev 39: 107–140.

Winer JA (1984) The human medial geniculate body. Hear Res 15:225–247.

We have found the precentral convolution excitable over its free width, and continuously round into and to the bottom of the sulcus centralis. The 'motor' area extends also into the depth of other fissures besides the Rolandic, as can be described in a fuller communication than the present. The hidden part of the excitable area probably equals, perhaps exceeds, in extent that contributing to the free surface of the hemisphere. We have in some individuals found the deeper part of the posterior wall of the sulcus centralis to contribute to the 'motor' area.

A.S.F. Grünbaum and C.S. Sherrington, 1902

15

The Motor System: Part 1 – Lower Motoneurons and the Pyramidal System

15.1. REGIONS INVOLVED IN MOTOR ACTIVITY

Movement results from interaction among many areas of the nervous system. During motor activity, higher levels of the motor system, such as the cerebral cortex, regulate lower levels of the motor system, such as the brain stem and spinal cord. Regions involved in motor activity (Table 15.1) include the pyramidal system, the extrapyramidal system, the cerebellum, and the lower motoneurons in the brain stem and spinal cord. The **pyramidal system** or upper motoneurons, includes cortical areas and descending paths from neurons in these areas. The **extrapyramidal system** includes extrapyramidal cortical areas, the basal nuclei and related structures along with descending paths from neurons in these areas.

TABLE 15.1.   Regions involved in motor activity

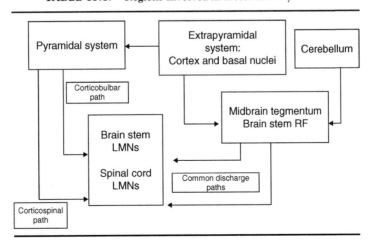

15.2. LOWER MOTONEURONS

Lower motoneurons are the final neuron in the motor system. They are geographically or positionally the 'lowest' neurons participating in motor activity. Sherrington called lower motoneurons the 'ultimate conductive link to a muscle'. Since this last link is termed the 'final path common to all impulses arising from any source', lower motoneurons are often termed the **final common path**. All parts of the motor system, in the end, control skeletal muscles through their influence on lower motoneurons.

15.2.1. Terms Related to Motor Activity

A **motor unit** includes one motoneuron, its axon, and the myocytes innervated by that motoneuron. Most motoneurons innervate several hundred myocytes. The term **neuromuscular unit**, referring to the same structures, emphasizes the functional inseparability of motoneurons and the myocytes they innervate. A **muscle unit** is the set of skeletal myocytes innervated by a single motoneuron. **Innervation ratio** is a term referring to the relationship between a single motoneuron and the myocytes it innervates and is expressed as motoneuron/myocytes. A relation exists between the required finesse of a movement and this ratio. Laryngeal muscles have the lowest innervation ratio of 1:2 or 1:3, the ocular muscles have a 1:13 to 1:20 ratio, the first lumbrical and dorsal interossei have a 1:100 to 1:340 ratio, and the gastrocnemius has a 1:600 to 1:1700 ratio (as much as 1 to several thousand).

15.2.2. Lower Motoneurons in the Spinal Cord

Lower motoneurons in the ventral horn of the spinal cord include **alpha motoneurons, gamma motoneurons**, and **interneurons**. With large axons, ranging from 10–16 μm in diameter, **alpha motoneurons** innervate **extrafusal myocytes** in skeletal muscles (Fig. 6.6) and normally exhibit a regular steady state activity. Descending motor paths can directly influence **alpha motoneurons** or do so indirectly through interneurons. **Gamma motoneurons** have axons that innervate **intrafusal myocytes** in the neuromuscular spindles (Fig. 6.6). These neurons, with small axons from 1–3 μm in diameter, constitute about a third of all ventral root fibers. **Interneurons** are the smallest elements yet they constitute the majority of neurons in the ventral horn outnumbering other motoneurons about 30 to one. These neurons are not passive elements. They receive incoming sensory impulses over dorsal root fibers actively transmitting such impulses to alpha motoneurons.

Structural and Functional Arrangement of Alpha Motoneurons

When viewed in transverse section the ventral horn of the spinal cord is divisible into **medial, lateral**, and **central** divisions that themselves are divisible into neuronal clusters concerned with the innervation of small groups of muscles. There is a functional pattern of arrangement of these subdivisions. Ventral horn alpha motoneurons have large, multipolar cell bodies with coarse chromatophil (Nissl) substance characteristic of efferent neurons. Axons of these cell bodies enter the ventral roots.

The **medial division of the ventral horn** is at almost all levels of the cord (perhaps not at C1, L5, or S1). When viewed in transverse section at regions other than the enlargements (particularly in the thoracic region), the medial division occupies almost the entire ventral horn. At enlargement levels, the medial subdivision occupies only the medial parts of the gray matter. This division is concerned with innervation of the neck, trunk, intercostal, and abdominal muscles.

The **lateral division of the ventral horn** is only at enlargement levels appearing in transverse section as a lateral extension of the ventral horn. (This is not the same structure as the lateral horn.) This division innervates the limbs and includes three nuclei, each of which are present at both enlargement levels. (1) Half of the **ventrolateral nucleus**, present from C4 to C8 innervates muscles of the shoulder girdle and upper arm while its other half present from L2 to S2 innervates the muscles of the hip and thigh. (2) Half of the **dorsolateral nucleus**, extending from C4 to T1 innervates muscles of the forearm and hand while the other half of the dorsolateral nucleus extending from L2 to S2 innervates muscles of the leg and foot. (3) Half of the **retrodorsolateral nucleus** consisting of large neurons from C8 to T1 innervates the intrinsic muscles of the hand while its other half, consisting of large neurons from S1 to S3, innervates the corresponding small muscles of the foot that move the toes.

The **central division of the ventral horn** has three neuronal groups that lie in the middle part of the ventral horn. These groups include the **phrenic, accessory, and lumbosacral nuclei**. The **phrenic nucleus** is conspicuous from C4 to C6 though there is a difference of opinion about its exact extent in humans. The phrenic nucleus innervates the diaphragm. The **accessory nucleus** extends from C1 to about C5 or C6. The caudal part of this nucleus innervates the trapezius and the rostral part, the sternomastoid muscle. The **lumbo-sacral nucleus** extends from about L5 to S2. When viewed in transverse section it is a cluster of neurons lying between the lateral and medial divisions of the ventral horn. This nucleus is likely responsible for the innervation of the pelvic diaphragm (levator ani and coccygeus).

An unusual nuclear group in the human ventral horn of the first, second, and third sacral segments is **Onuf's nucleus**. Fiber-stained sections of the human sacral cord reveal a paucity of myelinated fibers in the neuropil surrounding this nucleus that facilitates its identification. The similarities of this nucleus in humans to that in the cat, dog, rat, and squirrel monkey suggest that it probably innervates the perineal striated muscles in humans as it does in these experimental animals.

15.2.3. Activation of Motoneurons

Each neuronal type in the ventral horn (alpha motoneurons, gamma motoneurons, and interneurons) participates in motor activity. Such activity results through: (1) direct excitation or inhibition of alpha motoneurons; (2) indirect excitation or inhibition of alpha motoneurons through interneurons; or (3) through direct activation of gamma motoneurons that then indirectly activate alpha motoneurons. More information on **neuromuscular spindles** and their relationship to these neuronal types is available in Chapter 6.

All coordinated motor activity depends on the interplay between descending paths and their resulting effect on alpha, gamma or interneurons of the final common path. Lower motoneurons of the spinal cord are an **integrating center** receiving many influences and responding to these influences. The neuronal membrane and the dendrites of the lower motoneuron integrate the program, hold it and release it in a spiked discharge.

Studies of spinal motoneuronal function suggest that muscle contraction results from coactivation of both alpha and gamma motoneurons. Alpha motoneurons control muscle length and tension whereas the gamma motoneurons prevent spindle receptors from unloading during shortening of the muscle. Discharge of the efferents accompanies alpha motoneuron activation. Spindle afferents show an increase instead of a decrease in discharge rate during contraction. The cerebellum plays a major role in this facilitatory influence on γ efferents providing a background discharge of the gamma efferents.

15.2.4. Lower Motoneurons in the Brain Stem

Lower motoneurons in the spinal cord correspond to neuronal elements in the ventral horn; lower motoneurons in the brain stem correspond to the cranial nerve motor nuclei.

Oculomotor Nucleus

The **oculomotor nucleus**, the **somatic efferent** component of the third cranial nerve, is at superior collicular levels of the midbrain (Fig. 13.5). Each oculomotor nerve [III] receives fibers from both oculomotor nuclei. A pattern of localization is in the oculomotor nucleus for the individual muscles innervated by neurons in this complex. Studies in the baboon indicate that the neurons innervating the medial rectus, inferior rectus, inferior oblique, and most of the neurons innervating the levator palpebrae superioris contribute fibers to the ipsilateral oculomotor nerve whereas superior rectus neurons and some neurons for the levator contribute fibers to the contralateral oculomotor nerve. Neurons for the levator palpebrae superioris overlap with medial rectus neurons in the oculomotor nucleus. An overlap takes place among neurons that innervate the inferior rectus, inferior oblique and superior rectus.

Trochlear Nucleus

The **trochlear nucleus** is a **somatic efferent** nucleus that innervates only one muscle – the superior oblique. This nucleus is at inferior collicular levels of the midbrain (Fig. 13.5). Because of the decussation of trochlear fibers as they leave the dorsal aspect of the brain stem, each trochlear nucleus innervates the contralateral superior oblique muscle.

Trigeminal Motor Nucleus

The **trigeminal motor nucleus** is the **special visceral efferent** component of the trigeminal nerve [V]. This nuclear group, in the middle third of the pons (Fig. 4.7), displays a pattern of localization for the individual muscles it innervates. The trigeminal motor nucleus innervates the muscles of mastication, anterior belly of the digastric, mylohyoid, tensor tympani, and tensor veli palatini. Section of the trigeminal motor root seriously impairs chewing. Because of the pull by the nonpara-lyzed pterygoid muscles, the resultant paralysis leads to mandibular deviation to the side of the injury. Each trigeminal nerve probably receives fibers from both trigeminal motor nuclei.

Abducent Nucleus

The **abducent nucleus** is the **somatic efferent** compo-nent of the sixth cranial nerve located in the lower third of the pons beneath the facial colliculus (Figs 4.7, 13.3). This nucleus innervates one muscle – the lateral rectus. Its fibers pass caudally to leave the brain stem inferior to the level of the nucleus.

Facial Nucleus

The **facial nucleus**, in the lower third of the pons, is the **special visceral efferent** component of the **facial nerve** [VII]. A pattern of localization exists in this motor nucleus such that fibers innervating lower facial mus-cles arise from the dorsal part of the facial nucleus; fibers innervating upper facial muscles (for closure of the eyelids and wrinkling of the forehead) arise from neu-rons in the ventral part of the facial nucleus. Fibers from neurons in the facial nucleus have an unusual course, passing to the ventricular floor, medial to the abducent nucleus where they collect beneath the floor of the fourth ventricle as the **genu of the facial nerve** (Figs 4.7, 13.3). This bundle then ascends and turns laterally to arch over the abducent nucleus from medial to lateral. These special visceral efferent fibers then pass ventrolaterally between the trigeminal spinal nucleus and the lateral surface of the facial nucleus before emerging from the pontomedullary junction. The facial nerve fibers emerging at this point align with fibers of the glos-sopharyngeal, vagal, and accessory nerves that emerge below the facial fibers in a rostral to caudal pattern.

Injuries to the facial nucleus or its fibers cause paraly-sis of all facial muscles ipsilateral to the injury (Bell's palsy). An afflicted patient cannot close one eye, wrinkle the forehead, or smile on that side. The palpebral fissure widens (because of paralysis of the orbicularis oculi muscle) and the corner of the mouth sags on the injured side. Because the special visceral efferent facial fibers do not join the other facial components until after the facial genu, injury between the nucleus and the facial genu involves only the motor component without involving other components. In humans, some facial fibers join a layer of fibers beneath the floor of the fourth ventricle and contribute to the contralateral intrapontine part of the facial nerve without entering the facial genu.

Nucleus Ambiguus

The **nucleus ambiguus** is a **special visceral efferent** nucleus in the medulla oblongata (Fig. 4.6) that con-tributes fibers to the glossopharyngeal [IX], vagal [X], and accessory [XI] nerves. Neurons of the nucleus ambiguus innervate muscles derived from pharyngeal (visceral) arch mesoderm. Because the nucleus ambiguus is a series of neuronal clusters connected by scattered neurons, this nucleus is satisfactorily defin-able only at certain medullary levels (hence its name). Neurons in the caudal part of the nucleus ambiguus contribute fibers to the cranial root (vagal part) of the accessory nerve, those in the intermediate part (and perhaps its greatest extent) provide fibers for the vagal nerve, and those rostrally (at the rostral part of the medulla) provide fibers for the glossopharyngeal nerve [IX]. Therefore, the nucleus ambiguus provides motor nerves to the pharynx, larynx, and soft palate.

Clinical and experimental observations reveal a motor pattern in the nucleus ambiguus as follows: neurons innervating the stylopharyngeus are rostral, followed by those innervating the pharyngeal and laryngeal muscles, and most caudally, neurons innervating the palatal mus-cles. A rostrocaudal motor pattern occurs in the nucleus ambiguus of nonhuman primates for the laryngeal mus-cles. Rostrally located neurons innervate the cricoary-tenoid muscle followed caudally by those for the remainder of the laryngeal adductors. Neurons innervat-ing the abductor (posterior cricoarytenoideus) in nonhuman primates are between those for the cricoary-tenoideus and the rest of the adductors. A similar motor pattern in the human nucleus ambiguus accounts for small injuries that cause impaired function of one or more pharyngeal, laryngeal, or soft palatal muscles. The possibility of involving several muscles innervated by one cranial nerve, without involving all the muscles innervated by that nerve, holds for any motor nucleus of the brain stem that exhibits a motor pattern.

Hypoglossal Nucleus

The **hypoglossal nucleus** is a **somatic efferent** nucleus extending through most of the medulla. The human hypoglossal nucleus is a series of neuronal groups that innervate the intrinsic and extrinsic muscles of the tongue. Unilateral injury to one hypoglossal nucleus leads to a flaccid paralysis of the ipsilateral half of the tongue. Intramedullary fibers of the **hypoglossal nerve** [XII] pass ventrally from each nucleus, between the pyramidal tract and inferior olivary nucleus, to emerge from the ventrolateral sulcus between the pyramid and the **inferior olive** (Fig. 4.6). Each hypoglossal nucleus has connections by means of scattered neurons with other neurons in the spinal somatic efferent column that share a similar embryological origin.

15.2.5. Injury to Lower Motoneurons

After injury to lower motoneurons or their processes, there are motor but no sensory or cognitive disturbances. Manifestations of lower motoneuron injury include (1) **loss of motor power**; (2) **diminution or loss of reflexes**; (3) **loss of muscle tone** (therefore paralysis is described as flaccid); (4) **denervation atrophy of muscles**; and (5) **fasciculations**. **Loss of motor power** refers to the presence of muscle weakness or paralysis that is either focal or segmental involving only those muscles innervated by the injured lower motoneurons. The degree of such paralysis is dependent upon the number of neurons injured and is accompanied by **diminution** or **loss of reflexes**. In addition, there is loss of the ability to contract the affected muscles voluntarily, that is, a loss of muscle tone (hypotonicity), and weakness of all movements in which the affected muscles participate. Because of this loss of muscle tone, paralysis, if present, is a **flaccid paralysis**. No abnormal reflexes are associated with lower motoneuron injury. All muscle fibers innervated by injured lower motoneurons will lose their innervation and undergo **denervation atrophy** that begins in a few days and is progressive until loss of volume in the affected muscles takes place. Some 70–80% of the entire muscle mass is likely lost in three months.

A final manifestation of lower motoneuron injury is that the affected muscle fibers display sporadic and spontaneous action potentials. Although such activity is unseen initially, with electromyography this electrical activity is visible as **muscle fibrillations**. These fibrillations are often visible through intact skin as random **fasciculations**. Such fasciculations, which appear from 1–4 weeks after the initial injury and persist in sleep, are not tremors or abnormal movements. Normal individuals normally sense muscle cramps accompanied by fasciculations. Patients with lower motoneuron fasciculations do not usually feel them though their spouses or friends notice them.

Fasciculations accompanying lower motoneuron injury may be the result of increased irritability of muscle fibers to acetylcholine. Another suggestion is that these fasciculations originate distally at the junction of myelinated and nonmyelinated parts of the terminal axon of the injured lower motoneuron. The injured neuronal cell body apparently does not generate them.

Fasciculations probably represent a lack of the normal 'inhibitory influence' from the neuronal cell body over the distal axon. Two patients with a lower motoneuron disease, **amyotrophic lateral sclerosis** (ALS), had fasciculations that persisted and **increased** after section of the nerve, then stopped seven days later. Presumably, there was adequate time for degeneration to reach the distal terminals. Fasciculations are not evident accompanying peripheral neuropathy where axons are dying back toward the soma. Tongue fasciculations may be an early sign of multiple sclerosis.

15.2.6. Examples of Lower Motoneuron Disorders

Amyotrophic Lateral Sclerosis (ALS)

Amyotrophic lateral sclerosis (ALS) is a lower motoneuron disorder in which the primary structural change is selective neuronal loss. Other findings are secondary. In this disease, there is severe involvement of ventral horn alpha and gamma motoneurons; also cranial nerve nuclei especially the oculomotor, trigeminal, facial, and hypoglossal nuclei and the nucleus ambiguus. Although shrinkage and atrophy of neurons are evident, there is no chromatolysis. A glial scar subsequently replaces affected ventral horn motoneurons.

Poliomyelitis

Another example of a lower motoneuron disorder is poliomyelitis, an acute infectious disease particularly in children, caused by the human poliovirus. According to the World Health Organization (WHO), this disease once crippled 350,000 children in 125 countries (1988) but in 2002, there were only 1919 reported cases in seven countries worldwide. The virus invades the nervous system and replicates in the spinal cord leading to selective destruction of ventral horn lower motoneurons with rapidly progressive manifestations of lower motoneuron disease.

15.3. PYRAMIDAL SYSTEM

15.3.1. Corticospinal Component

The **corticospinal component of the pyramidal system** is synonymous with **corticospinal fibers** whose cell bodies of origin lie in the **cerebral cortex** and whose axons descend into the **spinal cord**. Since these fibers travel in the medullary **pyramids**, Turck (1851) termed the path the **pyramidal tract**. In each human medullary pyramid are about one million corticospinal fibers.

Cortical Origin

Some 60% (and perhaps as many as 80%) of the corticospinal fibers (Fig. 15.1) originate from neurons in the **frontal lobe** particularly in **areas 4**, **6**, and **8. Area 4**,

which corresponds to the precentral gyrus, is the **primary motor cortex. Area 6** is designated the **premotor cortex** in that it parallels and is anterior to or comes before, the primary motor cortex (hence, it is a 'pre' motor area but also a non-primary motor area). **Primary somatosensory areas 3a, 3b, 1**, and **2**, as well as **area 5** in the **parietal lobe** provide about 40% (or perhaps only 20%) of the corticospinal fibers. The anterior part

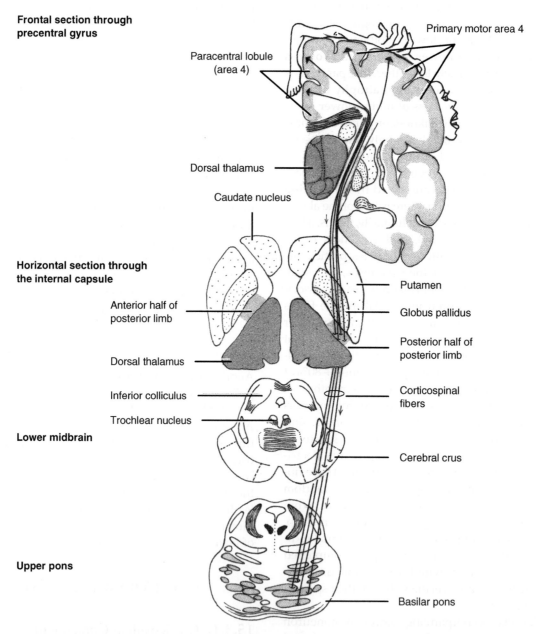

Frontal section through precentral gyrus

Paracentral lobule (area 4)

Primary motor area 4

Dorsal thalamus

Caudate nucleus

Horizontal section through the internal capsule

Anterior half of posterior limb

Dorsal thalamus

Inferior colliculus

Trochlear nucleus

Lower midbrain

Putamen

Globus pallidus

Posterior half of posterior limb

Corticospinal fibers

Cerebral crus

Upper pons

Basilar pons

FIGURE 15.1. Cortical (upper motoneuron) origin and course (into the upper pons) of the descending corticospinal fibers. Note the difference in position of the corticospinal fibers as they descend through different rostral-caudal levels of the internal capsule. Rostrally and on the left, the corticospinal fibers occur in the anterior half of the posterior limb. Caudally and on the right, corticospinal fibers occur in the posterior half of the posterior limb. Corticospinal fibers occupy the middle three-fifths of the fibrous part of the base of the midbrain (cerebral crus) before descending into the basilar pons. The remaining course and termination of these corticospinal fibers is shown in Fig. 15.2.

of the **paracentral lobule**, corresponding to that part of area 4 on the medial surface of each cerebral hemisphere, also gives rise to a small percentage of corticospinal fibers, particularly those for the lower limbs. Because neurons giving rise to corticospinal fibers originate in the cerebral cortex – the highest or uppermost level of the nervous system, the term **upper motoneuron** is used for these neurons of origin and applied to the pyramidal tract. In nonhuman primates, and presumably also in humans, there is a differential distribution of corticospinal fibers from each of these cytoarchitectonic fields to the different regions of the dorsal and ventral spinal cord gray matter.

The giant pyramidal neurons of Betz, totaling less than 40,000 per cerebral hemisphere in humans, make up a small percentage (2–3%) of the neurons in layer V of the **primary motor cortex**. Seventy-five percent of the Betz neurons occur in the leg region, 17.9% occur in the arm region, and 6.6% occur in the head region of area 4. This suggests a correlation between Betz neurons, the mass of the muscles and their antigravity status, or both. Muscles supplied by Betz neurons carry out the major antigravity activities of bipeds. Betz neurons are especially vulnerable to aging with 75% of them showing age-related changes by the eighth decade compared to 30% of the non-Betz cells. Age-related joint stiffness with associated slowing in motor performance may be the result of loss of Betz neurons and loss of extensor inhibition with inability to relax the antigravity muscles across weight-bearing joints. In addition to these Betz cells in layer V of area 4 in humans, there are relatively large pyramidal cells in sublayer IIIc and a low packing density of cells throughout all areas in the primary motor cortex in humans.

Motor Homunculus

Electrical stimulation of primary motor area 4 along the precentral gyrus in conscious patients evokes movements of contralateral parts of the body involving muscles of the head, trunk, and limbs. Stimulation along this gyrus and the anterior part of the paracentral lobule causes movements of different parts of the contralateral body. Plotting these areas of the body along the precentral gyrus and into the paracentral lobule results in a pattern of representation of the body, from toes to tongue, that is oriented along the precentral gyrus and anterior paracentral lobule. A distorted caricature, the **motor homunculus** (Fig. 15.1) illustrates this pattern. The size of the parts of the homunculus reflects the extent of the primary motor cortex devoted to individual parts of the body.

This motor pattern, based on the distribution of peripheral nerve fibers to appropriate muscles, exists across the precentral gyrus, among the corticospinal fibers, and at their termination. Experimentally the pattern is alterable demonstrating that it is not an inherent quality of the cerebral cortex. As an example, shoulder movements occur on stimulation of the shoulder area of the motor homunculus. If the accessory nerve [XI] supplying the sternomastoid and trapezius muscles is cut, and its proximal end sutured to the distal end of the facial nerve (allowing sufficient time for fibers from the accessory nerve to grow back along the route of the facial nerve) and then the same shoulder area of the cerebral cortex is restimulated, facial movements result. This experimental procedure would thus establish a 'new' cortical pattern.

Movements evoked from stimulation of the cerebral cortex reflect what muscles should contract and what body part should move rather than the position the limb is to obtain. The functioning of the cerebral cortex depends on the complete neuronal arch from the cerebral cortex to the periphery.

Double Representation Hypothesis

Studies in primates have demonstrated two spatially separate motor representations of the arm and hand in area 4. The long axes of these representations are oriented approximately parallel to the central sulcus. Stimulation in both areas causes the same movements and activates the same muscles at similar current strengths. This phenomenon is termed the **double representation hypothesis**. These two different areas or subareas of area 4 differ with regard to their somatosensory input. Movements that use tactile feedback for their execution have their representation in the caudal area whereas movements that use proprioceptive feedback for their execution have their representation in the rostral area. These findings along with cytoarchitectural observations and putative distributions of transmitter-binding sites have led to the suggestion that area 4 in humans includes **area 4 anterior (4a)** and **area 4 posterior (4p).** In addition to the motor representation across area 4, tactile and proprioceptive maps occur in 4a and 4p as well. Thus, the primary motor area in humans can be subdivided based on its anatomy, neurochemistry, and function. The thumb, index and middle fingers have representation in 4a and 4p in humans.

Position in the Internal Capsule

Corticospinal fibers pass through the white matter of the cerebral hemispheres in an orderly manner interdigitating with fibers of the corpus callosum. They then form part of the projection system of fibers of the cerebral hemispheres. As the corticospinal fibers descend, they funnel into the **internal capsule** – a large bundle of fibers between the basal nuclei and thalamus (Fig. 15.1). The internal capsule is a solid mass of fibers radiating out to the cerebral cortex. These radiating fibers on both sides of the brain form a crown of fibers termed the **corona radiata** (Fig. 5.13). Corticospinal fibers lie in the **anterior half of the posterior limb of the internal capsule**. However, studies indicate that corticospinal fibers shift their position as they descend from rostral to caudal levels of the internal capsule. Rostrally in the internal capsule, corticospinal fibers occur in the anterior half of the posterior limb whereas caudally the corticospinal fibers occur in the posterior half of the posterior limb. Corticospinal fibers intermingle in the internal capsule with fibers from the ventral lateral nucleus (VL).

Position in the Midbrain

Corticospinal fibers enter the fibrous part of the midbrain termed the **cerebral crus** [Latin, leg]. In particular, corticospinal fibers descend in the middle three-fifths of the cerebral crus (Fig. 15.1). On either side of the corticospinal fibers, both laterally and medially, are corticopontine fibers. A pattern is impressed on the corticospinal fibers such that fibers destined to innervate muscles of the feet have a more lateral location in the base of the midbrain (cerebral crus).

Position in the Pons

As the corticospinal fibers continue to descend, they enter the basilar pons (Fig. 15.1) where they are longitudinal in orientation as they intermingle with the pontine nuclei and with the transversely-oriented pontocerebellar fibers. A somatotopic pattern is impressed on these fibers with the feet oriented ventrolaterally.

Position in the Medulla Oblongata

As the corticospinal fibers enter the medulla oblongata, they descend in its ventral part forming compact **medullary pyramids** (Fig. 15.2). The medullary pyramids contain fibers other than corticospinal fibers. A pattern is in the pyramids such that corticospinal fibers for the upper limb are dorsal and medial whereas those for the lower limb are ventral and lateral. The spectrum of diameters among corticospinal fibers suggests that

in all mammals these fibers form a **slow conducting path**. Ninety-two percent of its fibers measure less than $4\,\mu m$, 8.4% of fibers measure less than $2\,\mu m$, and 2.6% have a diameter greater than $6\,\mu m$.

Decussation of the Pyramids (Motor Decussation)

At the lower medullary/upper cord transition region, some 75% of corticospinal fibers (range 70–90%) in the medullary pyramids take part in the **decussation of the pyramids** (motor decussation) (Fig. 15.2). Corticospinal fibers cross in large bundles alternately from right to left, then left to right, with fibers destined to innervate upper limb musculature crossing more rostrally in the motor decussation. At lower medullary levels the uncrossed fibers destined to innervate lower limb musculature are more dorsal and lateral and therefore less susceptible to superficial compression. Rarely, the decussation of the pyramids is lacking.

Cruciate Paralysis
Nielsen (1941) described two patients with bilateral upper limb paralysis from (1) a fractured odontoid process with backward projection of the upper end of the dens; and (2) tuberculosis of the odontoid process with medullary compression. Years later Bell (1970) described three additional patients with a similar paralysis resulting from (1) fracture dislocation of the odontoid process; (2) syringomyelia; and (3) a posterior fossa meningioma. In each of these patients, the injury involved the decussation of the pyramids at the lower medulla oblongata and only decussating upper limb fibers (Fig. 15.3).

Crossed Monoplegia
Wallenberg (1901) described several patients with injuries to the lateral aspect of the **motor decussation** involving fibers to the upper limb (which had already decussated) along with fibers to the lower limb (which had not yet decussated). Such injuries are likely the result of a hemorrhage, occlusion of the paramedian branches of the anterior spinal artery, or to trauma, such as a fall on the back of the head. The resulting condition is termed a **crossed monoplegia** because it involves the upper limb on the same side as the injury and the lower limb on the contralateral side (Fig. 15.3).

Lateral Corticospinal Tract

As the fibers that participate in the motor decussation enter the spinal cord, they take up a position in the **lateral**

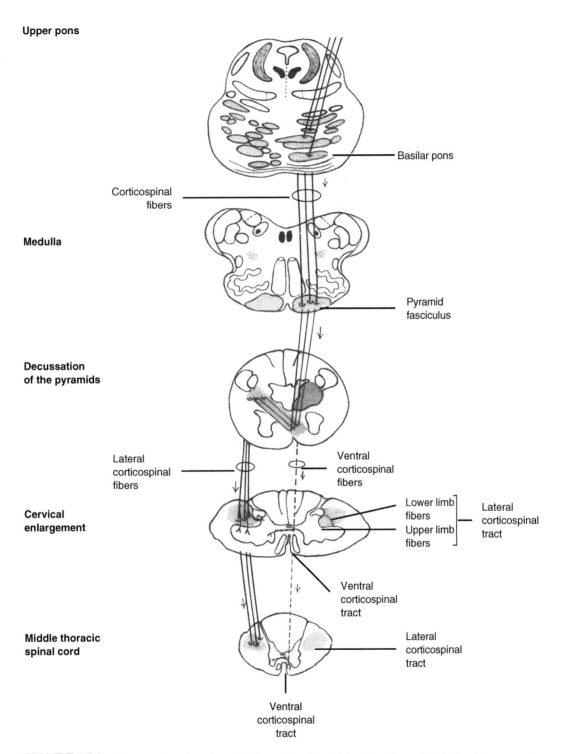

FIGURE 15.2.   Continuation of corticospinal fibers (from Fig. 15.1) as they descend in the basilar pons, enter the medullary pyramids, and decussate (75% of the corticospinal fibers) at lower medullary-upper spinal levels in the pyramidal decussation. At spinal levels, these crossed fibers form the lateral corticospinal tract whereas the uncrossed fibers (25%) form the ventral corticospinal tract. Many of the ventral corticospinal fibers do cross just before they terminate to end on contralateral spinal lower motoneurons. In the end some 90% of the corticospinal fibers cross either in the decussation or just before they terminate. About 10% of the corticospinal fibers do not cross at any point.

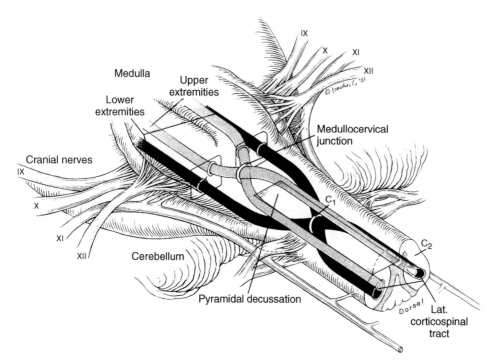

FIGURE 15.3. The pyramidal decussation in relation to other brain stem structures at the medulla oblongata-spinal cord junction. The upper limb fibers decussate rostral to those destined for the lower limbs. The characteristic lamination of the arm and leg fibers in the lateral corticospinal tract is reached at the C2 level of the spinal cord (modified from Dumitru and Lang, 1986).

funiculus as the **lateral corticospinal tract** (Fig. 15.2). Above L1, the lateral corticospinal tract is deep to, and thus protected by, the dorsal spinocerebellar tract. Below L1, however, the lateral corticospinal tract is superficial and therefore vulnerable to injury.

Ventral Corticospinal Tract

Those corticospinal fibers that do not participate in the **decussation of the pyramids** – some 25% of all corticospinal fibers (range 10–30%) – enter the **ventral funiculus** as the **ventral corticospinal tract** (Fig. 15.2). These uncrossed fibers are destined to influence motoneurons that innervate neck and trunk musculature. Although fibers destined to form the ventral corticospinal tract do not cross in the **motor decussation**, it is probable that many do cross in the ventral white commissure before they end (probably 15% of all corticospinal neurons decussate at this point). This leaves as many as 10% of all corticospinal fibers that may not ever cross – in either the motor decussation or prior to their termination in the spinal cord. These uncrossed fibers may innervate, in part, muscles in the shoulder.

Termination of the Lateral Corticospinal Tract

The majority of corticospinal fibers (55%) end in the cervical region innervating the upper limb, 25% end in the lumbosacral region innervating the lower limb, and the remaining 20% end in thoracic regions. In nonhuman primates, and presumably also in humans, there is a differential distribution of projections from neurons in various cortical areas that yield corticospinal fibers to different regions of the dorsal and ventral spinal cord gray matter. Some corticospinal fibers end directly on alpha and gamma motoneurons in the ventral horn. Processes of neurons in the **primary motor cortex** and those in the adjacent transitional sensory field (area 3a); terminate on interneurons at the base of the dorsal horn and in the intermediate zone. In this manner, they indirectly influence ventral horn alpha motoneurons.

Voluntary Control of the Bladder

Neurons in **paracentral area 4** yield fibers involved in the voluntary control of the external sphincter of the urethra. This '**bladder area**' in humans is in the paracentral lobule near the foot and ankle regions of the motor homunculus. Processes of these neurons accompany corticospinal fibers through the brain stem and into the spinal cord occupying a position on the ventral aspect of the lateral corticospinal tract in the lateral funiculus. They end at S2 to S4 levels to influence the external sphincter of the urethra and permit initiation of micturition. Bilateral injuries to the bladder area due to a falx meningioma may lead to the inability to start micturition with dribbling of urine from the overflowing bladder.

Are there Ascending Fibers in the Corticospinal Tract?

Ascending **spinocortical fibers** are identifiable among corticospinal fibers in humans. Whether or not these fibers occur in all primates is unclear. They likely play an inhibitory role over corticospinal discharges to permit selectivity of motor behavior and provide a method of checks and balances over the output of the pyramidal system.

Corticospinal Fibers from the Postcentral Gyrus

Corticospinal fibers originating from neurons in the postcentral gyrus of the parietal lobe reach sensory nuclei in the brain stem and spinal cord (e.g. the gracile nucleus, cuneate nucleus, the trigeminal nuclear complex, base of the dorsal horn, and the thoracic nucleus). Such fibers appear to exert some degree of regulatory control over these sensory nuclei.

Functional Aspects of the Lateral Corticospinal Tract

Corticospinal fibers regulate lower motoneurons and modify spinal reflexes. There is an increase in the knee jerk response with lateral corticospinal tract destruction because this tract is no longer actively regulating the spinal reflex. Corticospinal fibers either excite or inhibit lower motoneurons causing an appropriate response depending on the given situation. One either lightly kicks a football or kicks it as hard as possible. To do so requires corticospinal regulation of spinal cord lower motoneurons.

Continuous discharge or 'tonic activity' is characteristic of the entire pyramidal system. Corticospinal neurons are always active – even when alpha motoneurons are quiet. Such activity decreases during sleep. Before any muscle activity, there is firing in corticospinal neurons specific for an impending movement. This 'intention to move' activity can be recorded in corticospinal neurons. The **primary motor cortex** provides a readiness potential that reflects a preparatory event. Corticospinal neurons of origin are able to excite, inhibit, or reinforce lower motoneurons and therefore influence impending movements as well as those in progress.

Injury to the Pyramidal System (Upper Motoneuron Injury)

The signs and symptoms that result from injury to the pyramidal system at the cortical level or to its corticospinal fibers as they descend, depends on the extent, location, and rate of progression of the injury. The term upper motoneuron injury is applied to damage to the upper motoneurons in the cerebral cortex or the descending fibers from these neurons (pyramidal tract fibers). Characteristic upper motoneuron signs and symptoms are: (1) **loss of distal limb movements**; (2) **exaggeration of muscle stretch reflexes**; (3) **increased muscle tone**; (4) the appearance of **pyramidal reflexes such as the Babinski response**; (5) **no localized muscle atrophy**; (6) **no abnormal movements**; and (7) **no fasciculations**. The **loss of distal limb movements** involves a loss of dexterity and manipulative skill especially for fine, skilled, voluntary movements of the distal limbs. The resultant deficit can range from a weakness that is transient or permanent (**hemiparesis**) to a complete and lasting paralysis (**hemiplegia**). The affected limb is initially limp presumably because of involvement of those corticospinal fibers that end directly on alpha motoneurons. If the injury is above the motor decussation, the resulting deficit is on the contralateral side of the body whereas an injury below the decussation would lead to an ipsilateral deficit. When upper limb weakness is present it preferentially involves the wrist, finger and elbow extensors, supinators and external rotators and abductors of the shoulder with relative sparing of the flexor, pronator, and internal rotation muscles. When lower limb weakness is present it preferentially involves the foot and toe dorsiflexors, knee flexors, and flexors and internal rotators of the hips. Relative sparing of the extensors, external rotators, and plantar flexors.

The absence of corticospinal regulation on reflex activity leads to an **exaggeration of muscle stretch reflexes**. Initially the affected muscle is electrically silent at rest because of the release of spinal lower motoneurons from the influence of corticospinal fibers. Reflexes are hyperactive and **ankle clonus** is present. Ankle clonus involves passive dorsiflexion of the ankle causing alternate contraction and relaxation of muscles at the ankle joint. Examination of reflexes and their comparison on both sides of the body will aid in the analysis of the degree of impairment of reflex activity.

A third manifestation of upper motoneuron injury is **increased muscle tone** in the affected muscles. Normal muscle tone is the feeling of resistance or rigidity in the limbs when passively moving each limb at several joints. Such increased muscle tone depends on the degree of muscle contraction and the mechanical properties of the muscle. Muscle tone assessed by observing the position of the limbs at rest, by palpating the muscle and by determining the resistance to passive stretch and movement. Because of this increase in muscle tone, the resulting weakness or paralysis accompanying upper motoneuron injury is termed a **spastic paralysis**.

Upper motoneuron reflexes are evident because of involvement of corticospinal fibers that end on interneurons. The **Babinski response** is an excellent example of an upper motoneuron reflex. With stimulation of the lateral aspect of the plantar surface of the foot, there is normally a flexor plantar response. With upper motoneuron injury, there is an extensor plantar response of the big toe with fanning of the remaining toes. This is a most subtle and reliable index of UMN impairment. The Babinski response appears in nonwalking infants whose lateral corticospinal tract is not fully myelinated and in adults under hypnosis. Though there is no **localized muscle atrophy** with injury to the pyramidal system, there is likely to be some loss of volume because of disuse. There are **no abnormal movements** and no **fasciculations** with pyramidal system injury. Table 15.2 compares upper versus lower motoneuron injury in terms of the presence or absence of motor weakness, reflexes, muscle tone, muscle atrophy, and fasciculations.

There is always the possibility of substitution of function by other areas of the nervous system after upper motoneuron injury such that there is likely to be a partial recovery of useful motor activity. Such patients often learn to walk again using proximal musculature but usually do not see the return of finer movements involving the musculature of the hands or feet. The degree of functional recovery relates to the degree of initial injury, the age of the patient, their level of motivation and extent of postinjury therapeutic measures that are available. Recent evidence suggests the possibility of impairment on the ipsilateral side of the body even if the injury is limited to the contralateral hemisphere. This suggestion holds out the possibility for the potential involvement of neurons in the noninjured hemisphere to become involved in the recovery of motor function poststroke.

Todd's Paralysis

Postconvulsive depression or inhibition of function is distinguished by symptoms exactly like those caused by a structural injury to the pyramidal system. Maximum depression of movement may be present immediately after a severe convulsion. There is likely to be a flaccid paralysis with depressed stretch reflexes at onset because of the sudden removal of this excitatory path. Patients usually recover with proximal movement and strength returning first.

TABLE 15.2.   Comparison of UMN vs LMN injury

	UMN	LMN
Weakness	Yes	Yes
Reflexes	Increase	Decreased
Muscle tone	Increased	Decreased
Atrophy	No	Yes
Fasciculations	No	Yes

15.3.2. Corticobulbar Component

The **corticobulbar component** of the pyramidal system is a collection of fibers that accompany the corticospinal fibers of the pyramidal system and innervate lower motoneurons of the brain stem (Fig. 15.4).

Cortical Origin

Fibers of the **corticobulbar component** of the pyramidal system (Fig. 15.4) take their origin from the lower part of precentral gyrus, from the posterior part of the middle frontal gyrus, and adjoining part of the inferior frontal gyrus. Fibers from these cortical regions descend with corticospinal fibers until they reach the level of a cranial nerve motor nucleus. Hence, the alternative term for these fibers is corticonuclear fibers. They then distribute, often with partial decussation, to the lower motoneurons in the brain stem cranial nerve nuclei in which they end. Some corticobulbar fibers distribute ipsilaterally, others distribute bilaterally, but the largest number distributes contralaterally to several cranial nerve nuclei (Fig. 15.4). Corticobulbar fibers end directly or perhaps through interneurons, on lower motoneurons of the brain stem cranial nerve nuclei. This unusual distribution of corticobulbar fibers to brain stem cranial nerve nuclei underlies the unusual signs and symptoms that follow injuries to these fibers.

Position in the Internal Capsule

At the level of the internal capsule, corticobulbar fibers occur in the **genu of the internal capsule**. For this reason, the corticobulbar fibers were termed the **geniculate tract**. Corticobulbar fibers occur in the **genu of the internal capsu**le at rostral levels but at caudal levels of

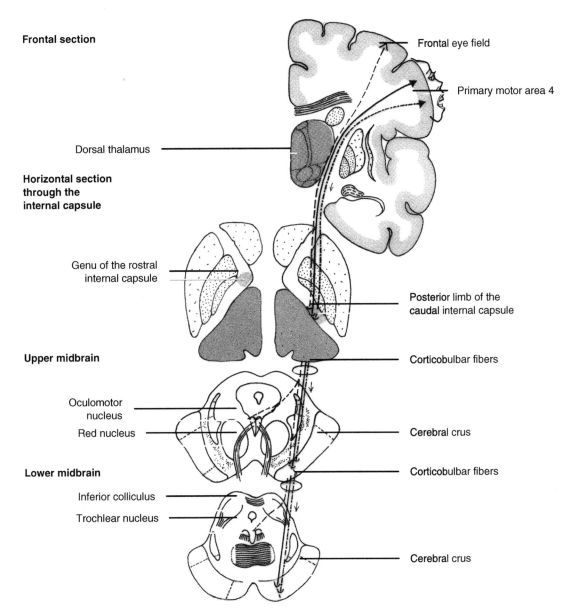

FIGURE 15.4.   Cortical origin, course, and mesencephalic termination of the corticobulbar fibers of the pyramidal tract. On the left at rostral levels of the internal capsule, corticobulbar fibers occur in the genu. As these fibers descend into the caudal internal capsule, depicted on the right, they occur in the posterior half of the posterior limb. Note the bilateral distribution of corticobulbar fibers to the oculomotor nuclei and the ipsilateral distribution to the trochlear nucleus. Fig. 15.5 illustrates the continuation and distribution of these corticobulbar fibers.

the internal capsule, they are with corticospinal fibers in the third quarter of the **posterior limb of the internal capsule** (Fig. 15.4).

Position in the Midbrain

Corticobulbar fibers descend through the internal capsule to enter the ventral part of each cerebral crus in the base of the midbrain. In this location, the corticobulbar fibers are ventral and ventromedial to corticospinal fibers (Fig. 15.4).

Position in the Pons and Medulla Oblongata

After leaving the midbrain, corticobulbar fibers enter the pons and take up a position in the dorsal part of the basilar pons beneath the medial lemnisci (Fig. 15.5). As the corticobulbar fibers enter the medulla oblongata, they remain inferior to the medial lemnisci in the medial and dorsal part of the medullary pyramids (Fig. 15.5).

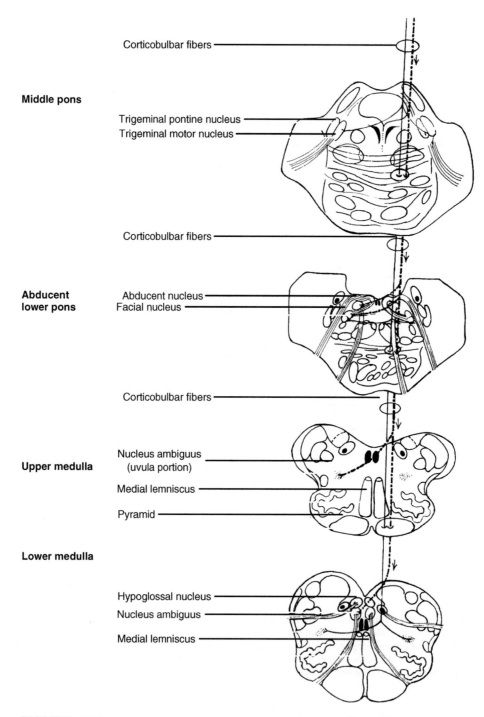

Corticobulbar fibers

Middle pons

Trigeminal pontine nucleus
Trigeminal motor nucleus

Corticobulbar fibers

**Abducent
lower pons**

Abducent nucleus
Facial nucleus

Corticobulbar fibers

Upper medulla

Nucleus ambiguus
(uvula portion)

Medial lemniscus

Pyramid

Lower medulla

Hypoglossal nucleus

Nucleus ambiguus

Medial lemniscus

FIGURE 15.5. Continuation of corticobulbar fibers and their distribution to pontine cranial nerve nuclei. Such fibers distribute bilaterally to the trigeminal motor nuclei, contralaterally to the abducent nucleus and the lower face part of the facial nucleus but bilaterally to the upper face part of the facial nucleus. Corticobulbar fibers continue into the medulla where they are distributed contralaterally to the uvula part of nucleus ambiguus, bilaterally to the remainder of nucleus ambiguus, but contralaterally to the hypoglossal nucleus.

Bilateral Distribution of Corticobulbar Fibers

Corticobulbar fibers distribute bilaterally to the oculomotor nuclei, trigeminal motor nuclei, greater part of the nucleus ambiguus, and to the ventromedial part of the facial nucleus (which innervates musculature in the upper part of the face) (Figs 15.4 and 15.5). The contralateral projection to the trigeminal motor nucleus is dominant. These corticobulbar fibers distribute to α-motoneurons in these nuclei in humans.

Contralateral Distribution of Corticobulbar Fibers

There is a contralateral distribution of corticobulbar fibers to the abducent nucleus, dorsolateral part of the facial nucleus (innervating musculature of the lower part of the face), that part of the nucleus ambiguus that innervates the uvula, and a contralateral distribution of corticobulbar fibers to the hypoglossal nucleus (Fig. 15.5).

Ipsilateral Distribution of Corticobulbar Fibers

There is an ipsilateral distribution of corticobulbar fibers to the trochlear nucleus. Fibers from the trochlear nucleus destined to form the trochlear nerve [IV] decussate before leaving the brain stem. Although the corticobulbar innervation of the trochlear nucleus is ipsilateral, its influence is on the contralateral superior oblique because of this decussation of the trochlear fibers (Fig. 15.4).

Peripheral Facial Weakness versus a Central Facial Weakness

A complete injury to the facial nucleus, intrapontine facial nerve fibers, or the facial nerve [VII] in its peripheral course causes paralysis of all facial muscles ipsilateral to the injury (Bell's palsy). An afflicted patient cannot close one eye, wrinkle the forehead, or smile on the side of the injury. The palpebral fissure widens (because of paralysis of the orbicularis oculi muscle) and the corner of the mouth sags on the injured side (Fig. 15.6). Because the special visceral efferent facial fibers do not join the other facial components until after the facial genu, injury before the genu involves only this component without involving the others. In humans, some facial fibers join a layer of fibers beneath the floor of the fourth ventricle and contribute to the contralateral facial nerve without entering the facial genu. The corticobulbar innervation of the facial nucleus is contralateral to that part of the facial nucleus that innervates only the lower half of the face but bilateral to that part of the facial nucleus that innervates the upper half of the face. Thus, a unilateral injury to the corticobulbar fibers destined for the contralateral facial nucleus will result in weakness or paralysis of the contralateral lower half of the face. This latter injury is likely to be anywhere along the course of the corticobulbar fibers before the facial nucleus.

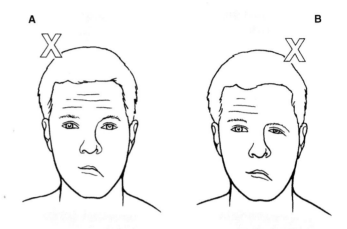

FIGURE 15.6. Appearance of the face following a lesion of corticobulbar fibers to the facial nucleus (**A**), versus a lesion of the root of the facial nerve. (**B**), The large X indicates the side of the lesion (from Haines, 1987).

Aberrant Corticobulbar Fibers

Another group of fibers, the **aberrant** (indirect) **corticobulbar fibers**, travel with the main group of corticobulbar fibers for part of their course then separate from them to enter the medial lemniscus. This separation takes place well above their levels of distribution to the cranial nerve nuclei. In the medial lemniscus, these aberrant corticobulbar fibers form the **descending component of the medial lemniscus**. Such fibers remain in, and descend with, the medial lemniscus until they reach the levels of the nuclei to which they distribute.

Three sets of aberrant corticobulbar fibers are present in the human brain stem. The first is a **midbrain aberrant path** that influences the oculomotor, abducent, and hypoglossal nuclei. This path turns off above the level of the oculomotor nucleus to enter the medial lemniscus and from there its fibers distribute bilaterally to the oculomotor nucleus, to the contralateral abducent nucleus, and bilaterally to the nucleus ambiguus. Such fibers end directly or by interneurons on their respective nuclei. An injury to the basilar pons at the level of the pontine isthmus destroys the corticobulbar fibers and the corticospinal fibers but does not involve the more deeply located midbrain aberrant path in the medial lemniscus. An individual with this injury is quadriplegic, unable to speak, smile, wrinkle their forehead, swallow voluntarily, or protrude their tongue. However, the patient is able to turn their eyes in all directions of gaze and shrug their shoulders. Such patients can interact with their environment, recognize objects and people, think, remember, and experience feelings and emotions.

The **Monte Cristo Syndrome** is so-called because a character in the story of *'The Count of Monte Cristo'*, M. Noirtier, suffered what was called apoplexy (a stroke) and became paralyzed except for ability to raise and lower his eyelids. M. Noirtier retained the ability to communicate with others through his lid and ocular movements. Plum and Posner (1966) proposed the term **locked-in syndrome** to refer to this neurological condition. In such cases, the midbrain aberrant corticobulbar fibers are intact but there is complete damage to the corticospinal and corticobulbar paths. Locked-in patients are said to be in a 'de-efferented' state or 'psuedocoma'.

The following description is from the back cover of a wonderful book entitled *'The Diving Bell and the Butterfly'*. It is the autobiography of Jean-Dominique Bauby, a French journalist and editor of the magazine *ELLE*. In December 1995, at the age of 43, Bauby suffered a sudden brain stem vascular accident and emerged from a coma several weeks later to find himself locked-in. He had lost movements of his trunk, limbs, and neck and was only able to move his left eyelid. The following summer he composed and dictated a series of reflections on his condition along with excursions into the realms of his memory, imagination, and dreams. The composition of this book was an extraordinary feat in itself. Bauby composed each passage in his mind and then dictated it, letter by letter, to an amanuensis who painstakingly recited a frequency-ordered alphabet until Bauby chose a letter by blinking his left eyelid once to signify 'yes'. In what was likely another heroic act of will, Bauby survived just long enough to see his memoir published in the spring of 1997.

15.3.3. Clinical Neuroanatomical Correlation

An excellent example of clinical neuroanatomy is a case involving injury to the genu and posterior limb of the internal capsule. Such an injury may result in the following signs: (1) **Hemiparesis** or **hemiplegia** of the contralateral limbs. (2) This hemiparesis is **spastic** in nature due to the accompanying increase in muscle tone. (3) There is a contralateral paralysis of the muscles of the lower face and perhaps those of the uvula and tongue, such that the patient was unable to smile voluntarily though they were able to wrinkle their forehead and close their eyes. (4) The patient would be unable to turn their eyes voluntarily toward the contralateral side.

FURTHER READING

Amunts K, Jancke L, Mohlberg H, Steinmetz H, Zilles K (2000) Interhemispheric asymmetry of the human motor cortex related to handedness and gender. Neuropsychologia 38:304–312.

Amunts K, Schlaug G, Schleicher A, Steinmetz H, Dabringhaus A, Roland PE, Zilles K (1996) Asymmetry in the human motor cortex and handedness. NeuroImage 4:216–222.

Binkofski F, Fink GR, Geyer S, Buccino G, Gruber O, Shah NJ, Taylor JG, Seitz RJ, Zilles K, Freund HJ (2002) Neural activity in human primary motor cortex areas 4a and 4p is modulated differentially by attention to action. J Neurophysiol 88:514–519.

Ehrsson HH, Naito E, Geyer S, Amunts K, Zilles K, Forssberg H, Roland PE (2000) Simultaneous movements of upper and lower limbs are coordinated by motor representations that are shared by both limbs: a PET study. Eur J Neurosci 12:3385–3398.

Geyer S, Ledberg A, Schleicher A, Kinomura S, Schormann T, Burgel U, Klingberg T, Larsson J, Zilles K, Roland PE (1996) Two different areas within the primary motor cortex of man. Nature 382:805–807.

Inoue T, Shimizu H, Yoshimoto T, Kabasawa H (2001) Spatial functional distribution in the corticospinal tract at the corona radiata: a three-dimensional anisotropy contrast study. Neurol Med Chir (Tokyo) 41:293–298.

Jankowska E, Edgley SA (2006) How can corticospinal tract neurons contribute to ipsilateral movements? A question with implications for recovery of motor functions. Neuroscientist 12:67–79.

Kim SH, Pohl PS, Luchies CW, Stylianou AP, Won Y (2003) Ipsilateral deficits of targeted movements after stroke. Arch Phys Med Rehabil 84:719–724.

Lacroix S, Havton LA, McKay H, Yang H, Brant A, Roberts J, Tuszynski MH (2004) Bilateral corticospinal projections arise from each motor cortex in the macaque monkey: a quantitative study. J Comp Neurol 473:147–161.

Macchi G, Jones EG (1997) Toward an agreement on terminology of nuclear and subnuclear divisions of the motor thalamus. J Neurosurg 86:670–685.

Marani E, Usunoff KG (1998) The trigeminal motonucleus in man. Arch Physiol Biochem 106:346–354.

Marx JJ, Iannetti GD, Thomke F, Fitzek S, Urban PP, Stoeter P, Cruccu G, Dieterich M, Hopf HC (2005) Somatotopic organization of the corticospinal tract in the human brainstem: a MRI-based mapping analysis. Ann Neurol 57:824–831.

Nathan PW, Smith MC, Deacon P (1990) The corticospinal tracts in man. Course and location of fibres at different segmental levels. Brain 113:303–324.

Penfield W, Rasmussen T (1950) *The Cerebral Cortex of Man. A Clinical Study of Localization of Function.* Macmillan, New York.

Picard N, Strick PL (1996) Motor areas of the medial wall: a review of their location and functional activation. Cereb Cortex 6:342–353.

Porter R, Lemon R (1993) *Corticospinal Function and Voluntary Movement.* Oxford University Press, New York.

Rademacher J, Burgel U, Geyer S, Schormann T, Schleicher A, Freund HJ, Zilles K (2001) Variability and asymmetry in the human precentral motor system. A cytoarchitectonic and myeloarchitectonic brain mapping study. Brain 124:2232–2258.

Roland PE, Zilles K (1996) Functions and structures of the motor cortices in humans. Curr Opin Neurobiol 6:773–781.

Toni I, Shah NJ, Fink GR, Thoenissen D, Passingham RE, Zilles K (2002) Multiple movement representations in the human brain: an event-related fMRI study. J Cogn Neurosci 14:769–784.

Urban PP, Wicht S, Vucorevic G, Fitzek S, Marx J, Thomke F, Mika-Gruttner A, Fitzek C, Stoeter P, Hopf HC (2001) The course of corticofacial projections in the human brainstem. Brain 124:1866–1886.

Woolsey CN, Erickson TC, Gilson WE (1979) Localization in somatic sensory and motor areas of human cerebral cortex as determined by direct recording of evoked potentials and electrical stimulation. J Neurosurg 51:476–506.

Yarosh CA, Hoffman DS, Strick PL (2004) Deficits in movements of the wrist ipsilateral to a stroke in hemiparetic subjects. J Neurophysiol 92:3276–3285.

. . . the human cerebellum is an enormously impressive mechanism. First of all, it contains more nerve cells (neurons) than all the rest of the brain combined. Second, it is a more rapidly acting mechanism than any other part of the brain, and therefore it can process quickly whatever information it receives from other parts of the brain. Third, it receives an enormous amount of information from the highest level of the human brain (the cerebral cortex), which is connected to the human cerebellum by approximately 40 million nerve fibers.

Henrietta C. Leiner and Alan L. Leiner, 1997

16

The Motor System: Part 2 – The Extrapyramidal System and Cerebellum

In addition to the **pyramidal system** and **lower motoneurons** of the brain stem and spinal cord, the **extrapyramidal system** and **cerebellum** (Table 15.1) are also involved in motor activity. In this chapter, we will describe these two components of the motor system and provide an overview of the effects of injuries that involve these two parts of the motor system.

16.1. EXTRAPYRAMIDAL SYSTEM

The **extrapyramidal system** is a general term for cortical and subcortical areas and the paths from them participating in motor activity by supplementing activities of the pyramidal system. These areas are 'extra' **pyramidal motor areas** in that they are in addition to or supplement the **primary motor cortex**. Since descending extrapyramidal fibers do not travel in the medullary pyramids, they are 'extrapyramidal' in their course.

16.1.1. Extrapyramidal Motor Cortex

Cortical areas whose neurons yield descending extrapyramidal fibers are termed using various names such as **secondary, additional, supplementary**, or **extrapyramidal motor cortex**. These terms are not necessarily synonymous. Extrapyramidal motor areas often lie in close relationship to neocortical association areas. Under the heading of extrapyramidal motor areas is **premotor area 6** in the **frontal lobe, areas 3, 1**, and **2** along the postcentral gyrus of the parietal lobe, **areas 5 and 7** in the superior parietal lobule, and **area 22** and **parts of the middle and inferior gyri** of the **temporal lobe** (Fig. 16.1). Other extrapyramidal motor areas include the **amygdaloid body, insular lobe, subcallosal area**, motor areas of the **cingulate gyrus, anterior perforated substance**, and **hippocampal formation**.

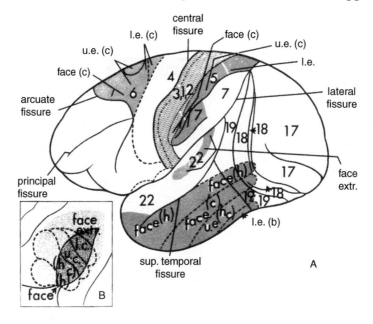

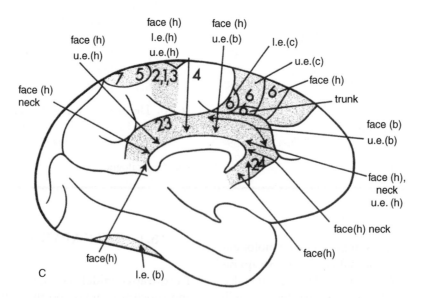

FIGURE 16.1. The extrapyramidal cortical areas on the lateral (**A**) and medial (**C**) surfaces and the insular region (**B**) of the monkey cerebral hemisphere (modified from Schneider, 1977). **Abbreviations:** (b), bilateral responses; (c), contralateral responses; (h), homolateral responses; l.e., lower extremity responses; u.e., upper extremity responses.

Extrapyramidal Cortical Stimulation

The pattern of movement that can be demonstrated following stimulation of extrapyramidal motor areas is similar to that found on stimulation of the primary motor area. However, extrapyramidal movements resulting from cortical stimulation are less precise and more generalized, usually involving a whole hand or all of one limb. These movements are likely to be ipsilateral, contralateral, or even bilateral rather than strictly contralateral as are movements resulting from stimulation of the primary motor area. To elicit movements from extrapyramidal motor areas requires a higher voltage and a lighter level of anesthesia.

Examples of Extrapyramidal Movements

Examples of extrapyramidal movements are movements visible during temporal lobe seizures, automatic or following movements of the eyes, the expression of emotion, and cortical automatic associated movements. Irritation of extrapyramidal motor areas in the middle and inferior gyri of the temporal lobe often underlies movements visible during **temporal lobe seizures**. The entire motor pattern is presumably manifest during the course of the seizure. **Following movements of the eyes** take place as one watches the flight of the ball in a game of tennis or ping-pong. These are involuntary, extrapyramidal movements of the eyes.

Movements from extrapyramidal motor areas underlie certain aspects of the **expression of emotion**. A patient who cannot voluntarily smile is able to do so at an amusing story as an expression of emotion. Another function of extrapyramidal motor areas (particularly in the temporal lobe) is that they provide for **cortical automatic associated movements** that often accompany and give character and color to precise, skilled voluntary movements. As the violinist makes fine finger movements attributable to the corticospinal system, they often softly tap their foot, have a pleasant expression on their face, and swing their body with the music. These accompanying movements are termed **cortical automatic associated movements**. The nature of such cortical automatic associated movements varies over the course of a lifetime providing individuality to our cortical movements. Such movements are as individualistic as the association areas of the cerebral cortex and are part of our personality. Hand movements when speaking, the habit of pulling on a necktie, gesturing, humming, or other mannerisms are examples of cortical automatic associated movements. Finally, extrapyramidal motor areas interplay with the basal nuclei and cerebellum in normal motor activity.

16.1.2. Basal Nuclei

Anatomically there are five subcortical masses of gray matter buried in the depths of the cerebral hemispheres. These five nuclei, the **caudate nucleus** (Cd), **putamen** (Pu), **globus pallidus** (GP), **claustrum** (Cl, Fig. 16.2) and the **amygdaloid body** (Amg, Fig. 16.2), belong to the telencephalon and were formerly termed the deep telencephalic nuclei. They are most clearly

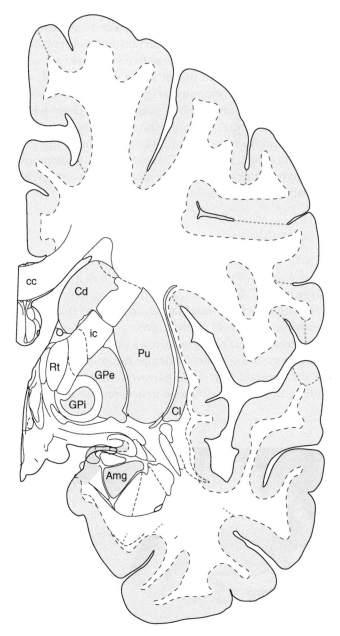

FIGURE 16.2. Subcortical masses of gray matter, the deep telencephalic nuclei, buried in the depths of the cerebral hemispheres in humans (modified from Mai, Assheuer and Paxinos, 2004). **Abbreviations:** cc, corpus callosum; Cd, caudate nucleus; Cl, claustrum; GPe, globus pallidus, external segment; GPi, globus pallidus, internal segment; ic, internal capsule; Pu, putamen; Rt, reticular nucleus.

visible on coronal sections through the cerebral hemi-spheres (Fig. 16.2). The *Terminologia Anatomica* (1998) uses the term **basal nuclei** to refer to three of these five nuclei (namely the caudate nucleus, putamen, and globus pallidus) but lists the claustrum and amygdaloid body as constituents of the **basal forebrain**. The amygdaloid body is a component of the limbic system. In the following account we will used the term **basal nuclei** to refer more broadly to the following five structures: the **caudate nucleus, putamen,** and **globus pallidus** along with the **substantia nigra** and **subthalamic nucleus** (parts of the mesencephalon and diencephalon, respectively) (Table 5.2). These five structures are constituents of the **basal nuclei** based on functional, not anatomical, or developmental grounds. Indeed the caudate, putamen and globus pallidus are deep telencephalic structures whereas the substantia nigra is a mesencephalic structure and the subthalamic nucleus is a diencephalic structure. The basal nuclei are an important part of the extrapyramidal motor system along with extrapyramidal areas in the cerebral cortex.

Caudate Nucleus

The first representative of the basal nuclei is the caudate nucleus. Each **caudate nucleus** has a relatively large head (Cd, Figs 16.3, 16.4) projecting from the lateral side into the **frontal (anterior) horn of the lateral ventricle** (Fig. 16.4) of each cerebral hemisphere. The large **head of the caudate** narrows into the **body of the caudate** near the **interventricular foramina**. The **body of the caudate** (Fig. 16.3) narrows considerably as it turns into the temporal lobe to become the **tail of the**

caudate that then curves along the roof of the ventricle and extends rostrally to fuse with the **amygdaloid body** (Fig. 16.3). This arrangement gives the caudate a C-shape that includes a large head, narrow body, a tail, and the amygdaloid body (Fig. 16.3). The caudate nucleus is one of several important C-shaped structures in the human brain (along with the cingulate gyrus and hippocampal gyrus, the corpus callosum, the fornix, the anterior horn, central part and inferior horn of the lateral ventricle, and the accompanying C-shaped choroid plexus. Accompanying the tail of the caudate throughout its course is a small, compact fiber bundle called the **stria terminalis** and a vein, termed the **vena terminalis** that makes the stria easy to identify. The human caudate nucleus has a total volume ranging from 4.6 to 5.3 cc. A characteristic feature of Huntington's disease is marked atrophy of the striatum, especially the caudate nucleus, due to the degeneration of the striatal projection neurons.

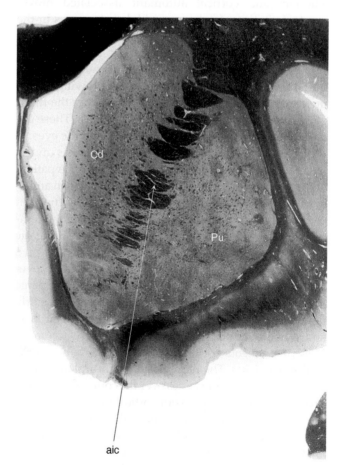

FIGURE 16.4. Coronal section through human cerebral hemisphere showing the head of the caudate nucleus (Cd) incompletely separated from the putamen (Pu) by the anterior limb of the internal capsule (aic) (modified from Mai, Assheuer and Paxinos, 2004).

FIGURE 16.3. The C-shaped human caudate nucleus and the amygdaloid body in the right cerebral hemisphere from the medial side (after Gluhbegovic and Williams, 1980).

Putamen

The ventral part of the head of the caudate is continuous with the corresponding part of the **putamen** (Pu, Fig. 16.4), the second representative of the basal nuclei. Functionally and structurally, the caudate nucleus and the putamen represent a common mass, incompletely separated in the human brain by a large band of fibers, the **anterior limb of the internal capsule** (Fig. 16.4). The putamen looks like the caudate both macro- and microscopically. The general term **striatum** refers collectively to the caudate and the putamen. A fiber bundle, the **lateral or external medullary lamina** (lml, Fig. 16.5), separates the putamen from the globus pallidus. Together the putamen and globus pallidus come under the heading of the **lentiform** [Latin, lens-shaped] **nucleus** (Table 5.2) because of their lens-shaped arrangement (Fig. 16.5). Viewing sections through

each cerebral hemisphere reveals the putamen beginning to disappear at caudal levels of the diencephalon.

Globus Pallidus

The **globus pallidus** [Latin, pale] – also called the pallidum – has myelinated fibers that give it a pale appearance and its name. On its lateral border, the globus pallidus is separable from the putamen by a band of fibers, the **lateral medullary lamina** (lml, Fig. 16.5). A **medial medullary lamina** (mml, Fig. 16.5) separates the globus pallidus into **external** and **internal** segments (GPe and GPi, Fig. 16.5). There may be three segments of the human globus pallidus. Viewed together the putamen and globus pallidus are lens-shaped and therefore called the **lentiform nucleus**. The caudate and lentiform nuclei (putamen and globus pallidus) collectively form the **corpus striatum** and all three are constituents of the basal nuclei. The internal segment of the globus pallidus is an important output nucleus of the basal nuclei.

Subthalamic Nucleus

The fourth representative of the basal nuclei is the subthalamic nucleus. The **subthalamic nucleus,** shaped like a biconvex lens, lies along the medial border of the internal capsule, inferior to the hypothalamic sulcus (Fig. 16.6) and lateral to the hypothalamus and **third ventricle** (Fig. 16.6). It lies in close relation to the rostral continuation of the substantia nigra and has reciprocal connections with the globus pallidus by way of a fiber

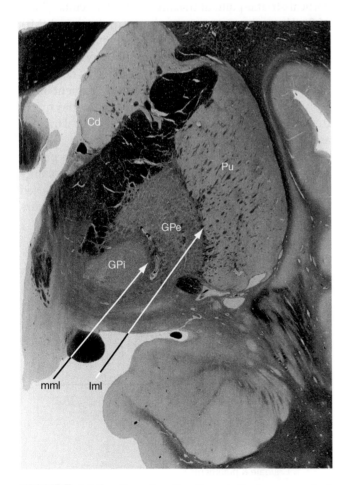

FIGURE 16.5. Coronal section through the human cerebral hemisphere at the level of the genu of the internal capsule showing the caudate nucleus (Cd), putamen (Pu), external (Gpe) and internal (Gpi) parts of the globus pallidus, and two intervening fiber bundles, the lateral (lml) and medial medullary lamina (modified from Mai, Assheuer and Paxinos, 2004).

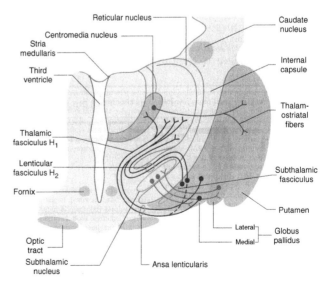

FIGURE 16.6. Efferent fibers from the basal nuclei including the thalamic fasciculus (H_1 bundle of Forel) and the ansa system (from Standring, 2005).

bundle, the **subthalamic fasciculus.** Structurally the **subthalamic nucleus, zona incerta,** and **nuclei of per-izonal fields** (of Forel), designated as **H, H1,** and **H2,** belong to the subthalamus.

Substantia Nigra

The last of the five nuclei belonging to the basal nuclei is the substantia nigra. In the unstained human brain, neurons in the **substantia nigra** are identified by their dark brown neuromelanin pigment, giving this cell mass a distinctive black color and its name. Neuromelanin is a waste product of catecholamine metabolism. The substantia nigra, extending through the midbrain and into the diencephalons, consists of four parts: **compact, reticular, lateral,** and **retrorubral.** Neurons of the human substantia nigra contain three different catecholamines of which dopamine (1000 ng/g) is the most prominent. Dopamine, synthesized in the pigmented cells of the compact part of the substantia nigra, makes its way along nigrostriate fibers to the striatum. Parkinson's disease results from a severe depletion of dopaminergic neurons in the substantia nigra which in turn leads to a severe depletion in striatal dopamine.

16.1.3. Afferents to the Basal Nuclei

Four groups of fibers reach the basal nuclei including the **corticostriates, thalamostriates, nigrostriates,** and **subthalamostriates** (Table 16.1). **Corticostriate fibers** from the neurons in somatosensory and prefrontal areas travel in the external and extreme capsule, before reaching the putamen, globus pallidus, and the caudate nucleus (Table 16.1). **Thalamostriate fibers** from large-celled neurons in the centromedian nucleus project to the caudate nucleus, from the small-celled centromedian nucleus to the putamen, and to the external

segment of the globus pallidus (Table 16.1). Thalamo-striate fibers are probably the source of considerable sensory input to the basal nuclei. **Nigrostriate fibers** from dopamine-containing cells of the **compact part** of the substantia nigra course in relation to the subthalamic nucleus and ascend before ending in the putamen and globus pallidus (Table 16.1). These nigrostriates provide a substantial dopaminergic input to the basal nuclei. In functional terms, the impact of the nigrostriatal inputs is likely to produce inhibition. **Subthalamostriates** pass from the subthalamic nucleus through the subthalamic fasciculus to the globus pallidus (Table 16.1).

16.1.4. Cortical-striatal-pallidal-thalamo-cortical Circuits

Underlying the functional aspects of the basal nuclei are cortical-striatal-pallidal-thalamo-cortical circuits (Table 16.2). These circuits connect primary motor area 4 (MI), premotor area 6 (PM), the supplementary motor area (SMA), and the prefrontal cortex (PFC) with the caudate and putamen via corticostriate fibers. These areas of the striatum connect with the internal segment of the globus pallidus (GPi) that then projects to the motor thalamus including the ventral anterior, ventral medial, and the ventral lateral nuclei (especially its anterior part). Completion of this circuit occurs via the projections of these nuclei of the motor thalamus over thalamo-cortical fibers with the same cortical areas we started with (primary motor area 4, premotor area 6, the supplementary motor area, and the prefrontal cortex.

16.1.5. Multisynaptic Descending Paths

These paths, arising from extrapyramidal cortical areas or from the basal nuclei, form a significant part of the extrapyramidal system.

TABLE 16.1.   Afferents to the basal nuclei

TABLE 16.2.   Cortical-striatal-pallidal-thalamo-cortical circuits

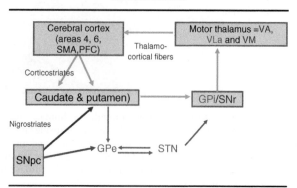

Cortical Origin

Multisynaptic descending motor paths arising in extrapyramidal cortical areas include **corticostriate**, **corticotegmental**, **corticorubral**, and **corticonigral** fibers. **Corticostriate fibers** from the primary motor cortex and from extrapyramidal motor areas pass in the external and internal capsule to the caudate nucleus, putamen, and globus pallidus. **Corticotegmental fibers** arise from extrapyramidal motor areas in front of area 19 and dorsal to the lateral sulcus, accompany the corticostriates in the external capsule, pass through or under the lentiform nucleus before they spread out in the midbrain and end there in accordance with a pattern of localization. Such fibers are likely to synapse in the subthalamus (direct or indirect corticotegmental fibers). Some of these fibers travel with the central tegmental tract to lower levels as Walberg's cortical component. **Corticorubral** and the **corticonigral fibers** arise from extrapyramidal motor areas rostral to area 19 and dorsal to the lateral sulcus accompanying corticostriate and corticospinal fibers through the internal capsule then to the red nucleus and substantia nigra of the midbrain.

Basal Nuclei

A prominent part of the extrapyramidal multisynaptic discharge paths are the **efferents of the basal nuclei**. Many short fibers interconnect the basal nuclei. The direction of conduction of impulses among the basal nuclei is from the caudate nucleus and putamen into the globus pallidus.

Efferent fibers from the basal nuclei include the **thalamic fasciculus** and the **ansa system**. The **ansa system** is a fan of fibers originating in the globus pallidus and projecting to the thalamus. This series of fibers is an important component of a complex system of cortical-striatal-pallidal-thalamo-cortical circuits that underlie the functioning of the basal nuclei. Three components of the ansa system include the **ansa lenticularis**, the **lenticular fasciculus** (H$_2$ bundle of Forel), and the **subthalamic fasciculus** (Fig. 16.6). Fibers of the **ansa lenticularis** arise from neurons in the lateral part of the internal segment of the globus pallidus (GPi), pass ventral to the internal capsule, and then enter the region of the subthalamus, particularly the H field of Forel. The **lenticular fasciculus**, which is dorsal to the fibers of the ansa, arises from the medial aspect of GPi, passes through the internal capsule and then enters a narrow region between the zona incerta and the subthalamic nucleus before it merges with the fibers of the ansa to enter the thalamic fasciculus. Some of these fibers may bypass the subthalamus and end in the brain stem. Thus, fibers of the **thalamic fasciculus** (H$_1$ bundle of Forel) are a continuation of the fibers of the ansa and lenticular fasciculus that run in a narrow zone between the thalamus and the zona incerta. In addition to these two sources of fibers from the globus pallidus, the thalamic fasciculus also receives fibers from the cerebellum and brain stem. Fibers of the thalamic fasciculus turn into the thalamus (Fig. 16.6) before ending in the **ventral anterior nucleus**. The ventral anterior nucleus sends axons to premotor area 6 – an extrapyramidal motor area – and to primary motor area 4. Fibers of the **subthalamic fasciculus** interconnect neurons in the globus pallidus with those in the subthalamic nucleus (Fig. 16.6). In its course, the subthalamic fasciculus crosses the internal capsule.

16.1.6. Common Discharge Paths

Fibers from neuronal cell bodies in the tegmentum of the midbrain and brain stem reticular formation descend and discharge their motor impulses to lower motoneurons at brain stem and spinal levels. Neurons giving rise to these **common discharge paths** receive input from extrapyramidal motor areas, the basal nuclei, and the cerebellum. Their function and organization follows the description of the cerebellum.

16.1.7. Somatotopic Organization of the Basal Nuclei

Microelectrode mapping during human stereotactic procedures has helped delineate somatotopic maps of the internal segment of the globus pallidus, the subthalamic nucleus, and the motor thalamus (VA, VLa, and VM). Perhaps in the basal nuclei of experimental primates and in humans there exists the representation of specific body parts in the context of a well-organized map of the body. Experimentally induced Parkinson's disease in nonhuman primates induces functionally distorted body maps. There may be some distortion of this somatotopic organization as the result of neurodegenerative diseases.

16.2. CEREBELLUM

Since Aristotle first mentioned it in the fourth century B.C., the function of the **cerebellum** has received considerable attention. Despite a great deal of investigation, many aspects of its function are still unclear. Rolando (1809) removed large parts of the cerebellum in a variety of animals and suggested that the cerebellum was the 'organ controlling locomotion'. Flourens (1824) concluded that the cerebellum coordinated motor activity. He noted that the results of injuries to the cerebellum were quite severe whereas superficial injuries were not so severe and that the nervous system demonstrated a great capacity to compensate for cerebellar injury. Much of our understanding of cerebellar histology is attributable to Cajal's pioneering studies. In the early 1900s, Sherrington emphasized the role of the cerebellum in maintaining the smoothness and effectiveness of movements.

16.2.1. External Features of the Cerebellum

The cerebellum has a wormlike **vermis** [Latin, worm] in the median plane that is vertical in orientation, and surrounded on either side by greatly expanded lateral parts called the **cerebellar hemispheres** (Figs 4.4, 16.7).

Cerebellar Zones

From medial to lateral, there are three **zones** of the cerebellum beginning with a **vermian zone** in the median plane, an adjacent **paravermal zone**, and most laterally, a **hemispheric zone**.

Cerebellar Folia and Fissures

Running transversely from ear to ear are the **cerebellar folia** that give a laminated appearance to the cerebellar

surface (Fig. 4.4). Cerebellar fissures separate the cerebellar folia from one another. These folia correspond to gyri of the cerebral cortex whereas the cerebellar fissures correspond to sulci of the cerebral cortex. Some of the cerebellar fissures are much deeper, helping to define the cerebellar lobes and lobules.

Cerebellar Lobes

Some of the deeper fissures are useful as landmarks in dividing the cerebellum into three **cerebellar lobes** including an **anterior, posterior**, and a **flocculonodular lobe** (Fig. 16.7). Each cerebellar lobe has a vermian part, paravermian part, and a hemispheric part and each is separated from the other lobes by a prominent fissure. The **primary fissure** (Fig. 16.7) separates the anterior lobe of the cerebellum from the posterior lobe whereas the **postnodular** and **posterolateral fissures** (Fig. 16.7) separate the flocculonodular lobe from the posterior lobe. This is actually a single continuous fissure that has different names for its medial and lateral parts. In general, the anterior lobe, based on its connections with the spinal cord, is termed the '**spinocerebellum**'. The posterior lobe with numerous connections to the cerebral cortex is termed the '**cerebrocerebellum**' whereas the flocculonodular lobe, with primarily vestibular connections, is termed the '**vestibulocerebellum**'.

Cerebellar Lobules

Each cerebellar lobe is divisible into smaller divisions called **lobules** (Fig. 16.8) that are easily identifiable on a median section of the cerebellum. One lobule of the posterior lobe nearest the postnodular and posterolateral

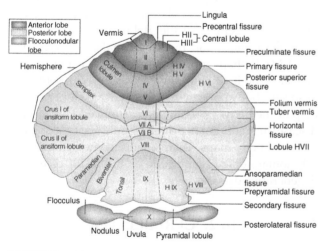

FIGURE 16.7.   External features of the cerebellum: a wormlike vermis in the median plane, and on either side greatly expanded lateral parts, the cerebellar hemispheres. The deeper fissures divide the cerebellum into three cerebellar lobes (from Manni and Petrosini, 2004).

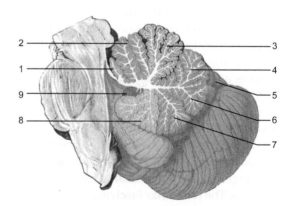

FIGURE 16.8.   Median section of the cerebellum illustrating the cerebellar lobules. Anterior lobe: 1. Lingula, 2. Central, 3. Culmen; Posterior lobe: 4. Simple, 5. Folium, 6. Tuber, 7. Pyramis, 8. Uvula; Flocculonodular lobe: 9. Nodule (from Standring, 2005).

fissure, the **tonsil of the cerebellum** (Fig. 16.9), is much larger than the other lobules in the posterior lobe. Under increased intracranial pressure, the cerebellar tonsil may protrude into the foramen magnum, a phenomenon termed **cerebellar tonsilar herniation** (Fig. 16.10). This causes compression of the medulla oblongata with cardiovascular and cardiorespiratory depression and other vagal disturbances.

Cerebellar Peduncles

The cerebellum presents three large bundles of fibers or **cerebellar peduncles** that interconnect the cerebellum

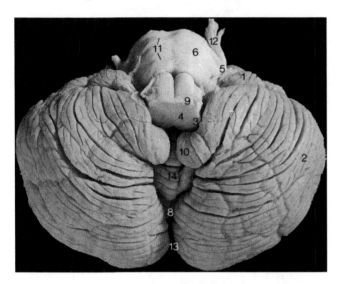

FIGURE 16.9. Inferior surface of the human cerebellum illustrating the cerebellar tonsils (10) that are hemispheric expansions of each uvula (from England and Wakely, 1991).

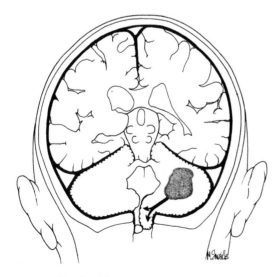

FIGURE 16.10. Cerebellar tonsilar herniation (from Nolte, 1988).

with other parts of the nervous system. The **superior cerebellar peduncle** (SCP) connects the cerebellum to the midbrain and thalamus (Fig. 16.11). The cerebellum receives fibers from the pons through the **middle cerebellar peduncle** (MCP) (Fig. 16.11). The **inferior cerebellar peduncle** (ICP) connects the cerebellum with the medulla oblongata and spinal cord (Fig. 16.11).

Gray and White Matter of the Cerebellum

The disposition of gray and white matter of the cerebellum resembles that of the cerebral hemispheres in that they both have an outer layer of cortex and a collection of neurons forming a subcortical gray matter. In the cerebellum, the outer layer of gray matter is the **cerebellar cortex** (Fig. 16.12) and the deeply lying collections of subcortical gray matter are termed the **cerebellar nuclei** (Fig. 16.13). Intervening between the cerebellar cortex and the deep cerebellar nuclei are masses of fiber bundles making up the **cerebellar white matter** (Fig. 16.13).

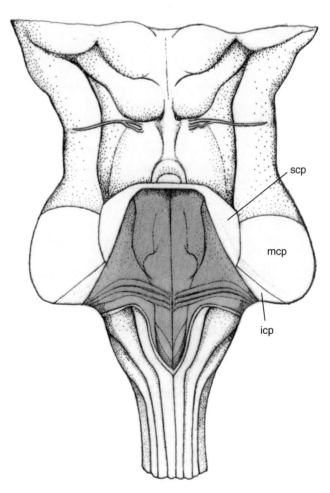

FIGURE 16.11. The three cerebellar peduncles: scp, superior cerebellar peduncle; mcp, middle cerebellar peduncle; icp, inferior cerebellar peduncle.

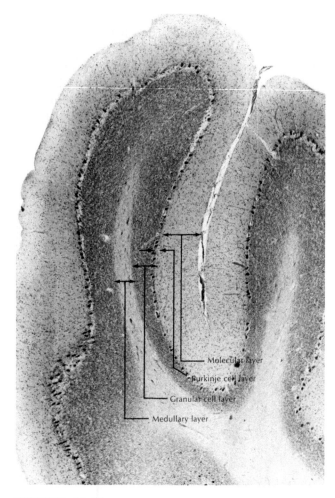

FIGURE 16.12. The three layers of the cerebellar cortex in humans (from DeArmond, Fusco and Dewey, 1989).

16.2.2. Cerebellar Cortex

Layers of the Cerebellar Cortex

The cerebellar cortex is surprisingly uniform with three layers: an outermost, rich synaptic region, the **molecular layer,** containing few neurons (Fig. 16.12); a deep layer of densely packed granule cells called the **granular layer** (Fig. 16.12); and an intervening **Purkinje layer** consisting of a single layer of neurons (Fig. 16.12). Purkinje cell axons project to the deep cerebellar nuclei whereas Purkinje cell dendrites enter the molecular layer.

Neurons in the Cerebellar Cortex

Four types of neurons characterize the cerebellar cortex. Piriform (pear) shaped **Purkinje cells** (Fig. 16.14) with dendrites that branch profusely in a plane perpendicular to the cerebellar folia are the characteristic cell of the cerebellar cortex. There are about 23 million Purkinje

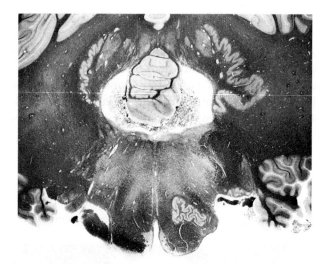

FIGURE 16.13. The deep cerebellar nuclei in humans: F, fastigial; G, globose; E, embol: form; and D, dentate (from DeArmond, Fusco and Dewey, 1989).

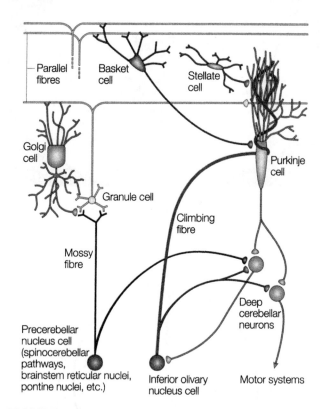

FIGURE 16.14. Neurons in the cerebellar cortex in humans (from Wang and Zoghbi, 2001).

cells in the human cerebellum. Some 160–180,000 dendritic spines occur per Purkinje cell. Each axon from the base of a Purkinje cell passes to the deep cerebellar nuclei (Fig. 16.14) and gives off recurrent collaterals that pass back to the Purkinje cell bodies.

Some 200 million cerebellar **stellate cells** (Fig. 16.14) occupy the superficial half of the molecular layer, where they are termed **outer stellate cells**. In the deeper part of the molecular layer, they are termed **basket cells** (Fig. 16.14). Basket cell dendrites lie in the same transverse plane as the Purkinje cell dendrites. Axons of the basket cells travel superficially to the Purkinje cell bodies in a plane perpendicular to the folia. At varying intervals, axons of the basket cells give off collaterals that form pericellular nests or baskets around the Purkinje cell bodies. One basket cell may contact as many as 20 Purkinje cells in one plane and 12 Purkinje cells in another plane. Therefore, one basket cell may influence as many as 240 Purkinje cells. With such an enormous overlap in this system, there is substantial opportunity for the basket cells to modulate or influence Purkinje cells.

An estimated 70 billion (and perhaps as many as 10^{11}) **granule cells** are densely packed in the deepest cerebellar layer, the **granular layer** (Figs 16.12, 16.14). Dendrites from the base of the granule cells break up into claw-like terminals. Axons of granule cells pass to the molecular layer and bifurcate at a T-shaped junction. These branches extend longitudinally, parallel to the long axis of the folia (Fig. 16.14). These fibers, like lines between telephone poles, form a system of parallel fibers that contact adjacent Purkinje cells. In its total length, a parallel fiber synapses with 300–460 Purkinje cells. Each Purkinje cell, because of its dendritic spread, may receive up to 200,000 synapses from different parallel fibers.

Cerebellar **Golgi neurons** (Fig. 16.14) are in the granular layer inferior to the Purkinje somata. Their large dendrites radiate into the molecular layer. Axons from the base of the Golgi neurons break up into a lavish arborization or nest of endings in the granular layer.

16.2.3. Deep Cerebellar Nuclei

The **deep cerebellar nuclei** are neuronal populations embedded in the cerebellar white matter. Since the deep cerebellar nuclei are in the roof of the fourth ventricle, they are termed the **roof** or **central nuclei**. From medial to lateral these are the **fastigial, globose, emboliform**, and **dentate nuclei** (Fig. 16.13). Collectively, the globose and emboliform nuclei are termed the **interposed nuclei**. Axons of the deep cerebellar nuclei form the main output or efferent system of the cerebellum.

16.2.4. Cerebellar White Matter

The white matter of the cerebellar cortex, surrounding the deep cerebellar nuclei and forming the central core of the cerebellar hemispheres, is thicker laterally than near the median plane. Median sections of the white matter of the cerebellum have a characteristic tree-like appearance called the **arbor vitae** (Fig. 16.15). Two types of fibers form the cerebellar white matter: **intrinsic** and **extrinsic fibers**. Several types of intrinsic fibers are in the cerebellum. Axons of Purkinje cells projecting on the deep cerebellar nuclei are an example of **projection fibers**. **Association fibers** extend from folia to folia, and **commissural fibers** extend from one cerebellar hemisphere to the other for interhemispheric relations. As the extrinsic fibers enter or leave the cerebellum, they form the **cerebellar peduncles**.

16.3. INPUT TO THE CEREBELLUM THROUGH THE PEDUNCLES

The coordination of movement by the cerebellum is dependent upon a wealth of input to the cerebellum from many sources.

16.3.1. Inferior Cerebellar Peduncle

Fiber bundles entering the inferior cerebellar peduncle include the following named tracts: dorsal spinocerebellar, dorsal external arcuate, ventral external arcuate, reticulocerebellar, olivocerebellar, and vestibulo-cerebellar. By way of these tracts, the cerebellum receives proprioceptive and tactile impulses from the neck, trunk, and limbs, cutaneous, corticospinal, and vestibular input as well as input from the spinal cord, brain stem reticular nuclei, cerebellum, basal nuclei, and cerebral cortex.

The **dorsal spinocerebellar tract** has its origin in the **thoracic nucleus** on the medial side of the base of the dorsal horn from C7 to L2 spinal levels. Processes of these neurons pass laterally, turn, and ascend in the ipsilateral lateral funiculus of the spinal cord superficial to the lateral corticospinal tract. These ascending dorsal spinocerebellar fibers enter the lower medulla oblongata and reach the level of the inferior cerebellar peduncle. After entering the inferior cerebellar peduncle, the dorsal spinocerebellar tract distributes primarily to the vermis of the anterior and posterior lobes. The dorsal spinocerebellar tract brings proprioceptive and tactile information from sacral, lumbar, thoracic, and lower cervical spinal levels.

The **dorsal external arcuate fibers** (cuneatocerebellar tract) originate in the **lateral cuneate nucleus** and then enter the inferior cerebellar peduncle. These fibers reach the vermis of the anterior lobe and perhaps the posterior lobe vermis (probably to the same region as

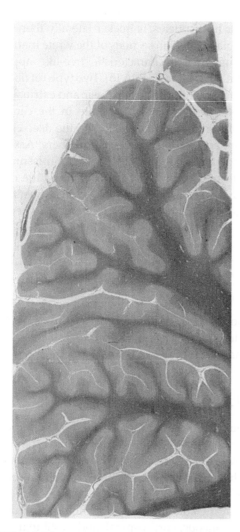

FIGURE 16.15. Median sections of the white matter of the cerebellum in humans have a characteristic tree-like appearance called the arbor vitae.

the dorsal spinocerebellar tract). Their function is to bring proprioceptive and tactile impulses from the neck and upper limb to the cerebellum.

The **ventral external arcuate tract** has its origin in the **arcuate nuclei** on the ventral aspect of the pyramids. Fibers from this nucleus travel along the surface of the medulla oblongata to reach the inferior cerebellar peduncle. Descending corticospinal fibers traveling in the medullary pyramids provide collaterals to the arcuate nuclei bringing corticospinal input to the cerebellar hemisphere of the posterior lobe.

Originating from neurons in the medial and lateral reticular nuclei of the brain stem reticular formation, the **reticulocerebellar tract** enters the inferior cerebellar peduncle, and perhaps the superior cerebellar peduncle, to distribute to the contralateral vermis, probably of the anterior lobe. Spinothalamic and other ascending sensory paths, including the spinovestibular

tract, provide collaterals to the reticular nuclei for relay of cutaneous impulses to the cerebellum.

The **olivocerebellar tract**, originating in the **inferior olivary nucleus**, collects at the hilus of the inferior olive, crosses the median plane, and enters the inferior cerebellar peduncle. The olivocerebellar tract distributes to the inferior and superior surface of the contralateral cerebellum and to medial and lateral parts of the cerebellar hemispheres. The inferior olivary nucleus receives afferents from a variety of sources including the spinal cord, brain stem reticular nuclei, cerebellum, basal nuclei, and cerebral cortex. This nucleus correlates impulses from these levels and then influences the Purkinje cells of the cerebellar cortex through a powerful and excitatory climbing fiber input that originates in the inferior olivary nucleus.

Vestibular nerve fibers enter the cerebellum on their way to the cerebellar cortex. Indirect vestibulocerebellar fibers synapse in the vestibular nuclei before reaching the cerebellar cortex. These **vestibulocerebellar fibers** pass to the cerebellum through the smaller inner part of the inferior cerebellar peduncle called the **juxtarestiform body** before ending in the flocculonodular lobe and the fastigial nuclei.

16.3.2. Middle Cerebellar Peduncle

The **pontocerebellar tract**, originating in the **pontine nuclei** (not the trigeminal pontine nuclei) and ending in the vermis, cerebellar cortex, and in all parts of the contralateral cerebellar hemisphere, forms the middle cerebellar peduncle (MCP). The MCP is the largest of the three cerebellar peduncles. Its size reflects the degree of development of cortical association areas. The pontocerebellar tract is part of a **corticopontocerebellar system** that puts the cerebellum under the influence of the association areas in the cerebral cortex. This path likely correlates sensations from muscle, tendon, and joints with auditory, vestibular, and visual impulses. Descending corticopontine tracts in this corticopontocerebellar system are in the outer and inner fifths of the cerebral crus in the base of the midbrain.

16.3.3. Superior Cerebellar Peduncle

Although primarily an efferent bundle, the superior cerebellar peduncle also has two incoming tracts. The **ventral spinocerebellar tract** arises from the **nucleus proprius** (spinal lamina III and IV) at all levels of the spinal cord, especially the enlargements, and ends in the vermis of the anterior lobe and in the ipsilateral nucleus fastigii. (Ventral spinocerebellar fibers from the feet area of the spinal cord enter the inferior cerebellar peduncle to reach the posterior vermis). The ventral spinocerebellar tract provides proprioceptive and tactile information from the limbs to the cerebellum. A second afferent bundle in the superior cerebellar peduncle is the **tectocerebellar tract** that arises from the tectum of the midbrain that corresponds to the superior and inferior colliculi, enters this peduncle, and ends in the vermis of the anterior and posterior lobes. Auditory and visual input reaches the cerebellum by means of this tract.

16.4. INPUT TO THE CEREBELLUM

To understand the cerebellum, one must appreciate the role played by the afferent fibers that enter the cerebellum as climbing or mossy fibers. One also needs to appreciate the relation of these incoming fibers to various neuronal elements in the cerebellar cortex, various neurons in the deep cerebellar nuclei, and the relations among various neuronal elements of the cerebellar cortex and deep nuclei.

16.4.1. Incoming Fibers to the Cerebellum

Climbing Fibers

All input entering the peduncles belongs to one of two general types of afferent fibers. These include the **climbing fibers** and the **mossy fibers** – both of which are excitatory. Climbing fibers, of inferior olivary origin, form the olivocerebellar tract and provide a precise point-to-point connection between the inferior olivary nucleus and the contralateral cerebellar hemisphere. Each climbing fiber ascends through the white matter of the cerebellum, spirals around a Purkinje cell, and ends by dividing repeatedly making many synapses with Purkinje cell apical dendrites. Climbing fibers provide collaterals to the deep cerebellar nuclei and exert a powerful, localized excitatory effect on Purkinje cell apical dendrites. The neurotransmitter **aspartate** mediates the excitatory action of the climbing fibers.

Mossy Fibers

In addition to climbing fibers, all other afferents entering the cerebellar peduncles enter as **mossy fibers**. Before reaching the cerebellar cortex, mossy fibers provide collaterals to the deep cerebellar nuclei. On entering the cerebellar cortex, they branch enormously and synapse with granule cells and on axonal terminals of Golgi neurons. One mossy fiber is likely to synapse with the dendritic claws of as many as 400 individual granule cells. The term **cerebellar glomerulus** refers to mossy fiber terminals plus granule cell dendritic claws. Mossy fibers are also excitatory to the cerebellar cortex and deep cerebellar nuclei. **Glutamate** appears to be the mossy fiber neurotransmitter.

Impulses entering the cerebellum, whether via mossy or climbing fibers, discharge and redischarge among neuronal elements in the cerebellar cortex. Output from Purkinje cells of the cerebellar cortex is strongly inhibitory onto the deep cerebellar nuclei by way of **corticobulbar projections**. There is an orderly, ipsilateral discharge from the zones of the cerebellar cortex to the deep cerebellar nuclei. The **vermal zone**, which receives **vestibular and ventral spinocerebellar input**, sends axons to the **nucleus fastigii** – the most medial of the cerebellar nuclei, Purkinje cells in the **paravermal zone** send axons to the **globose and emboliform nuclei**, and Purkinje cells in the **hemispheric zone** send axons to the dentate nuclei – the most lateral of the deep cerebellar nuclei.

16.5. CEREBELLAR OUTPUT

All output from the cerebellum originates in the deep cerebellar nuclei and is **excitatory**. Through these efferent fibers, the cerebellum is able to influence diverse areas of the nervous system involved in motor activity.

16.5.1. Output from the Fastigial Nuclei

The fastigial nuclei receive input from the vermian part of the cerebellar cortex and from vestibulocerebellar and ventral spinocerebellar fibers. Fibers leaving the **nucleus fastigii**, through the inferior cerebellar peduncle, influence brain stem reticular nuclei (**cerebelloreticular fibers**), trigeminal motor, oculomotor, abducent, and facial nuclei (**cerebellomotorius fibers**), ventral horn motoneurons supplying the neck musculature (**cerebellospinal fibers**), and the vestibular nuclei (**cerebellovestibular fibers**). These cerebellospinal fibers are termed the 'hook bundle' in that they arch up to the superior cerebellar peduncle and then hook posteriorly and inferiorly to leave the cerebellum through the smaller inner part of the inferior cerebellar peduncle called the **juxtarestiform body**.

16.5.2. Output from the Globose and Emboliform Nuclei

The globose and emboliform nuclei receive their input from the paravermal zone of the cerebellar cortex. Fibers from the **globose and emboliform nuclei** leave the cerebellum through the superior cerebellar peduncle to influence the inferior olive, nucleus ambiguus, and upper spinal cord. **Cerebello-olivary fibers** are part of the crossed descending component of the superior cerebellar peduncle that passes in the central tegmental tract. This fiber bundle provides proper tonus, coordination, and stabilization necessary for normal speech. Fibers in the crossed ascending component of the superior cerebellar peduncle influence the red nucleus (**cerebellorubral fibers**) and the tegmentum of the midbrain (**cerebellotegmental fibers**).

16.5.3. Output from the Dentate Nuclei

The dentate nuclei receive their input from the hemispheric zone of the cerebellar cortex. Fibers from neurons in the **dentate nuclei** leave the cerebellum in the superior cerebellar peduncle to influence the tegmentum of the midbrain (**dentatotegmental fibers**), red nucleus (**dentatorubral fibers**), and the posterior part of the ventral lateral nucleus (VL_p) of the thalamus (**dentatothalamic fibers**). Because the VL_p nucleus projects fibers to the primary motor cortex, these connections from the dentate nucleus permit the cerebellum to regulate and stabilize discharges of the corticospinal tract necessary for the proper tonus that underlies smooth, coordinated, fine, skilled, voluntary movements.

16.6. CEREBELLAR CIRCUITRY

In considering cerebellar circuitry, the following facts are noteworthy: (1) climbing fibers have an excitatory influence on Purkinje cells. One Purkinje cell fires many times per climbing fiber impulse. (2) Mossy fibers, also excitatory, likely excite as many as 400 granule cells. Each granule cell potentially influences 300–450 Purkinje cells. Therefore, the possibility exists for one mossy fiber to excite as many as 180,000 Purkinje cells. Each Purkinje cell may receive up to 200,000 synapses from different granule cells. (3) Parallel fibers of the granule cells excite basket cell dendrites. (4) Basket cells probably inhibit Purkinje somata. One basket cell may influence 250 Purkinje

cells. (5) Excitation of Golgi neurons occurs by parallel fibers of the granule cells and they, in turn, make inhibitory synapses with granule cells. This is an example of negative feedback. (6) Purkinje cell output is essentially inhibitory. (7) All neuronal elements intrinsic to the cerebellar cortex are inhibitory with the exception of the granule cells. Nowhere else in the nervous system is there such a dominance of inhibition as in the cerebellar cortex. (8) A vast divergence and convergence of incoming impulses to the cerebellar cortex takes place to such an extent that Cajal was lead to term this phenomenon 'avalanche conduction'. Despite our apparent understanding of cerebellar circuitry, it is particularly difficult to correlate this immense structural and functional data with signs and symptoms appearing after injury to the human cerebellum.

16.7. COMMON DISCHARGE PATHS

An essential component of the motor system are **common discharge paths** from neurons in the tegmentum of the midbrain and brain stem reticular formation whose fibers descend and influence lower motoneurons of the brain stem and spinal cord. Under the heading of common discharge paths are the **rubrospinal, associated tegmentospinal**, and **reticulospinal paths**.

Rubrospinal fibers originate in the red nucleus, decussate, and descend in a lateral position through the brain stem and into the spinal cord where they are ventral to the lateral corticospinal tract. Some rubrospinal fibers end on neurons in the brain stem reticular formation for relay to cranial nerve nuclei; most end at cervical levels. **Associated tegmentospinal fibers** originate in the tegmentum of the midbrain and descend on both sides of the brain stem ending in the lower spinal cord. These fibers supplement the rubrospinal fibers.

Neurons contributing to these common discharge paths receive input from diverse areas of the nervous system involved in motor activity including extrapyramidal motor areas, the basal nuclei, and cerebellum. Output from these neurons over the common discharge paths is the result of considerable integration and correlation of impulses arriving by way of corticotegmental fibers, the ansa system, cerebellorubral, and cerebellotegmental fibers. These common discharge paths are partly facilitatory and partly inhibitory.

16.8. CEREBELLAR FUNCTIONS

In addition to the enormous number of studies revealing the motor functions of the cerebellum, it appears that the cerebellum also has nonmotor functions.

16.8.1. Motor Functions of the Cerebellum

The cerebellum, receiving a wealth of exteroceptive and proprioceptive input from muscles, tendons, joints, and the vestibular system, uses this sensory input to assess and coordinate motor activity. Effective motor responses occur by the **production, maintenance**, and **regulation of muscle tone**. Normal muscle tone is the feeling of resistance or rigidity in the limbs by the examiner when passively moving each limb at several joints. Through the regulation of muscle tone, the cerebellum participates in the control of our posture and helps coordinate present and intended movements of the eyes, limbs, and trunk so that motor responses are smooth, correctly timed, and composed with a minimum of errors. The cerebellum permits alteration of intended acts under changing conditions so that corrections occur before the movement begins. This is especially true with regard to especially fine, coordinated, complex movements such as dancing, writing, and speaking. The cerebellum is not involved in the initiation of movements. Injury to the cerebellum degrades but does not abolish movement.

16.8.2. Nonmotor Functions of the Cerebellum

Studies in nonhuman primates and observations in humans have suggested a broader role for the cerebellum. First, the cerebellum participates in sensory integration, especially of visual and auditory input. Second, because a few cerebellar malformations have a correlation with behavioral abnormalities, some investigators consider the cerebellum to influence emotional and behavioral states. Third, stimulation or ablation of the cerebellum in experimental animals causes changes in blood pressure, heart rate, respiration, pupillary size, bladder activity, intrauterine pressure, and genital erection. Therefore, in these experimental animals the cerebellum probably influences autonomic functions. Fourth, because the vermis is a locus of sensorimotor integration and motor planning and increased vermis activation occurs in substance abusers during reward-related and other cognitive tasks, it is likely that parts of the cerebellar vermis may be involved in cocaine and other incentive-related behaviors. For all of these reasons, the nonmotor functions of the cerebellum are an ongoing focus of attention.

16.8.3. Studies Involving the Human Cerebellum

A beneficial effect in patients with intractable psychomotor seizures is observable after implantation of electrodes in the anterior lobe. Implanted cerebellar electrode systems are one method of treatment for patients with cerebral palsy and paraplegic spasticity. However, in a double blind study, no clinical efficacy resulted. Stereotaxic destructive procedures of the dentate nucleus are a means of surgical treatment for spasticity and dyskinesia, for cerebral palsy, including choreoathetosis and spastic quadriplegia, or in an effort to decrease hypertonicity in such patients.

16.8.4. Localization in the Cerebellum

Multiple, overlaid maps probably occur in the primate cerebellar nuclei with a medial to lateral pattern such that the trunk is represented in the fastigial nucleus, the proximal limbs in the globose and emboliform nuclei, and the distal limbs and head in the dentate nucleus. Neuroimaging studies in humans have confirmed the representation of the body across the cerebellar cortex characterized by the existence of two homunculi: one in the anterior lobe with an inverted order of body parts and one in the posterior lobe.

16.9. MANIFESTATIONS OF INJURIES TO THE MOTOR SYSTEM

Although the motor system is a functional entity, injuries in humans often involve specific parts of this system. In Chapter 15, we noted the effects of injury to the lower motoneurons and the upper motoneurons. In the remainder of this chapter, we will consider injuries to premotor area 6, the basal nuclei, and the cerebellum.

16.9.1. Injury to the Premotor Cortex

Premotor area 6 in the frontal lobe is anterior to primary motor area 4 and about six times larger than area 4 in humans. It includes both a dorsal premotor area (PMd) and a ventral premotor area (PMv) both of which have direct but weak influence on spinal cord lower motoneurons. Called the **aversive field** because of the turning movements that are obtainable on electrical stimulation of this non-primary motor area in humans, the premotor cortex is essentially a **motor association area**. The caudal part of the dorsal premotor area has connections with primary motor area 4 and may, through these connections, influence the generation of movements. The rostral part of PMd has connections with the prefrontal cortex.

Grasp Reflex

Manifestations of injury to the premotor cortex in humans include a forced grasping called a **grasp reflex**. Such patients grope about until an object touches their hand (tactile stimulus) then grasp it firmly but have difficulty relaxing their grasp. This takes place in the face of a normal proximal to distal sequence of muscle action.

Motor Apraxia

Another manifestation of injury to the premotor cortex is a **motor** (limb kinetic) **apraxia**. The term **apraxia**, meaning 'unable to do', is a disorder of purposeful, skilled movement caused by injury to the human cerebral cortex. A motor apraxia is clumsiness in performance of writing or drawing, threading a needle, or tying a shoe. In such cases, the pattern of skilled movements conceptualized by the brain is disturbed and the patient seems to have forgotten how to make a movement though they are able describe the movement in complete detail and the peripheral musculature is intact. The cause of this deficit is probably two fold – a reversed proximal to distal sequence in muscle activation by the cerebral cortex and a delay for preactivation of proximal arm movements.

Contralateral Weakness of Shoulder and Hip Musculature

A third effect of injury to the premotor cortex is weakness of all hip muscles and those in the shoulder concerned with abduction and elevation of the arm.

16.9.2. Injuries to the Basal Nuclei

The basal nuclei are a significant part of the extrapyramidal part of the motor system. These nuclei are responsible for the **automatic execution of learned motor plans**. They do not initiate movements (the pyramidal system does) nor do they correct movements in progress (the cerebellum does). Learning skilled movements involves the refining of relatively fast movements so that slow corrective action is unnecessary. A novice violinist gropes about to find the proper finger position on the strings for each note. With continued practice, the time necessary to find the proper finger position for each note diminishes. The means to skilled movements is in increasing the accuracy of the initial adjustment so that the later groping need be only within narrow limits. With additional practice, there will be no groping at all. Movements become smooth and coordinated as the basal nuclei automatically execute the fingering movements.

Many structural considerations are noteworthy with regard to injuries to the basal nuclei, including their deep location, irregular configuration, proximity to other structures (such as the internal capsule, thalamus, and the overlying white matter), and their position at a crossroads of fiber bundles of diverse origins. The signs and symptoms of injury to the basal nuclei go through a progressive series of stages over time usually ending in widespread damage to these nuclei. Variability is the rule without objective sensory, cognitive, or perceptual changes. Major abnormalities observed after injury to the basal nuclei include **disturbances of tone and posture**, **derangement of movements**, and the **loss of automatic movements**.

Disturbances of Tone and Posture

The manifestations of basal nuclei injury, which come under the heading of disturbances of tone and posture, include **hypertonia with rigidity**, **dystonia**, or **torsion spasms**.

Hypertonicity as Rigidity

Greatly increased tone, called **hypertonicity**, is manifest as a postural abnormality called **rigidity**. Rigidity is electrical activity at rest with constant and uniform resistance to passive movements and the inability to relax. Such increased tone is probably because of the inability of the uninjured extrapyramidal system to regulate the cerebellum. Without such regulation, the cerebellum has excessive and uncontrolled drive on lower motoneurons appearing as hypertonicity. Hypertonicity as rigidity is a principal feature of a number of disorders involving the basal nuclei including Parkinson's disease, the rigid form of Huntington's disease, and the rigid form of Wilson's disease.

Dystonia

Dystonia is a syndrome of sustained muscle contractions that cause abnormal postures or twisting and repetitive movements. Dystonia occurs in certain hereditary neurological syndromes. There is likely to be distortion of the limbs or trunk visible as a scoliosis or lordosis, distortion of the neck evident as a torticollis, or distortion of the face and mouth. Focal dystonias include writer's cramp, spasmodic dysphonia, blepharospasm, oromandibular dystonia, and hemifacial spasm. **Dystonia musculorums deformans** is an example of generalized dystonia with unusual, disabling, and painful postures of the entire body. It is of interest to note that the hand area of the somatosensory cortex of a flutist with dystonia differed from the hand representation found in a healthy flutist using noninvasive magnetic imaging. The nature of this sensory degradation and its role in dystonia is unknown. Experimental studies in an owl monkey model of focal dystonia (involving the hand) demonstrate cortical map abnormalities in these animals including a dramatic enlargement in the cortical receptive fields and a breakdown in cortical columnar architecture.

Torsion Spasm

Another example of the disturbances of tone and posture associated with injury to the basal nuclei is **torsion spasm** that is discernible as a proximal manifestation of dystonia involving the hip, shoulder, trunk, or neck. Such torsion spasms are usually generalized and unilateral.

Derangement of Movement

A second manifestation of injury to the basal nuclei is **derangement of movement** often evident as a decrease in or slowness of movements – a condition called **hypokinesia** (akinesia). More likely, however, with injury to the basal nuclei there will be a **hyperkinesia** or **abnormal involuntary movements**. The latter include tremors at rest, choreas, athetoid movements, myoclonus, and tics.

Tremors at Rest

Tremors at rest are rhythmic oscillations about a joint in excess of one beat per second. Such tremors usually occur in the waking state while the muscles are at rest and may increase with emotion and decrease with sleep and voluntary movements. An example is the pill-rolling tremor in Parkinson's disease in which the thumb rolls across the last two fingers together with wrist motion at the rate of 4–8 cycles per second. Surgical treatment for tremors includes stereotaxic lesions or chronic stimulation of VPL$_a$ (which corresponds to V.im according to Hassler's terminology). After this procedure, there is diminished influence of this part of the thalamus on the cerebral cortex. Both resting and postural tremors occur in Parkinson's disease. Tremors also occur in degenerative disorders such as Wilson's disease, spino-cerebellar heredodegeneration, dystonia musculorums deformans, Charcot–Marie–Tooth disease, and in the Déjerine–Sottas syndrome.

Choreas

Another example of derangement of movements that accompany injuries to the basal nuclei is **chorea** [Latin, a dance]. Choreas are unrelated, sudden, abrupt, and jerky movements that decrease in sleep. They are irregular, intermittent, spontaneous, and nonsequential movements. Chorea involving the face, head and especially the distal parts of the limbs is the characteristic feature in **Huntington's disease**, a progressive, hereditary movement disorder resulting from injury to the **caudate nucleus** with release of the globus pallidus from regulatory control. Drugs, metabolic and endocrine disorders and vascular accidents may lead to chorea. Syndenham's chorea may occur in children and adolescents as a complication of rheumatic fever.

Athetosis

A slightly different kind of abnormal involuntary movement is **athetosis** that involves irregular, slow, stereotyped, writhing, twisting, turning, and sinuous movements of the limbs and fingers. Athetoid movements are probably the result of injury to the putamen that then loses regulation over the globus pallidus. With considerable injury to the caudate nucleus and putamen, a mixture of chorea and athetosis, a condition termed **choreoathetosis**, may occur. In these cases, the injury causes interference with pallidal regulation over the ventral anterior nucleus of the thalamus and its regulation of the primary and premotor cortical areas. Primary athetosis may occur in status dysmyelinatus, Pelizaeus–Merzbacher disease, and in familial proximal choreoathetosis.

Myoclonus and Tics

Other examples of abnormal involuntary movements caused by injury to the basal nuclei include **myoclonus** and **tics**. The term **myoclonus** [Greek, muscle turmoil] refers to sudden involuntary contraction of a muscle or group of muscles that usually moves a joint at rest or at action. This sudden contraction is isolated and rhythmic. Examples include palatal myoclonus and diaphragmatic myoclonus. **Tics** include oral, vocal, respiratory, stuttering, or shrugging involuntary movements as well as blinking or sniffing, that may be generalized and involve the body such as in **Gilles de la Tourette syndrome**.

Loss of Automatic Movements

A third manifestation of injury to the basal nuclei is the **loss of automatic movements**. This example of hypokinesia is evident in the mask-like facial expression that accompanies Parkinson's disease in which there is damage to the pallidohypothalamic tract with a loss of regulation of emotional expression by the hypothalamus. Other examples of loss of automatic movements include infrequent blinking and the loss of arm swinging.

16.9.3. Injury to the Subthalamic Nucleus

Injury in or near the subthalamic nucleus of the subthalamus causes an involuntary movement disorder called **ballism** [Greek, to throw] that has a resemblance to a throwing motion. In ballism, the arm is usually involved. There is likely to be flinging, thrusting, or violent kicking often patterned, vigorous, and purposeless involving the proximal musculature of the limbs. If the ballism involves one side of the body, it is termed **hemiballismus**. In such cases, the injury often involves afferents to the subthalamus and efferent fibers from it, connections of the basal nuclei, and connections from the ventral lateral nuclear group of the thalamus. Injuries to the subthalamus permit rather than cause observed manifestations that occur contralateral to the injury. Primary cases of ballism include heredity bilateral ballism whereas secondary cases result from space-occupying lesions, trauma, vascular injuries, or infections.

16.9.4. Injury to the Cerebellum

The principal manifestations of cerebellar injury in humans include decreased muscle tone or **cerebellar hypotonia** with a resulting **cerebellar ataxia** as well as the presence of a **cerebellar tremor**, and **cerebellar nystagmus**.

Cerebellar Hypotonia

Cerebellar hypotonia appears as a weakness, fatigability, or an asthenia; the muscles involved seem flabby.

With cerebellar injuries, there is decreased resistance of the limb muscles to passive manipulation, perhaps because of depression of neuromuscular spindle primary afferents. Muscle responses to maintained stimuli also decrease, as do their responses to changes in length of muscles.

Cerebellar Ataxia

Because of the profound hypotonicity that results with cerebellar injury, a **cerebellar ataxia** [Greek, without order or arrangement] is usually evident. Ataxia caused by cerebellar injury refers to a loss of muscle coordination including **postural instability** and **disorders of movement**. Since ataxia is also found following dorsal funicular injury (sensory ataxia), it is clinically useful to clearly differentiate between these types of ataxia.

Cerebellar Ataxia is Manifested as Postural Instability

Postural instability is due to a loss of the muscle tone necessary for postural support. Patients so affected adopt a **widebased gait** to overcome this postural instability and may even stagger or fall to one side (the side of the injured cerebellum). A subtle test for cerebellar injury is to have the patient walk on their toes, walk heel to toe, backward, or in a circle.

Cerebellar Ataxia Appears as Disorders of Movement

Cerebellar ataxia includes **disorders of movement** especially complex, coordinated movements like tying a knot, speech, opening a jar, and writing, opening a lock, or dancing. There is likely to be **decomposition of movements**, **dysmetria**, and **adiadochokinesis**. **Decomposition of movements** refers to the condition in which muscles, acting on joints that normally move together, lose their synergy such that the intended movements are no longer coordinated or smooth but broken into their component parts. There is apt to be a **cerebellar ataxia of speech** in which the speech is slow, slurred, or jerky. Dysmetria [Greek, bad measure] refers to movements in which the velocity and timing is off such that there are errors of range, rate, and force. The finger to nose test is useful in testing for the presence of cerebellar injury. In this test, the patient rapidly moves their finger back and forth between their own nose and the finger of the examiner. Accuracy of such movements is lost with the upper limb under and overshooting its target. With **dysmetria**, patients cannot gauge distances, the speed, or power of a movement. Disorders of movement characteristic of cerebellar injury include a phenomenon called **adiadochokinesis** that means 'without successive movements'. If a patient with cerebellar injury slaps their thigh repeatedly, pronate and supinate their hands, or open and close their fists, such rapid alternating movements become very slow and no longer successive.

Cerebellar Tremor

Cerebellar tremor accompanying cerebellar injury is also termed an intention, action, or '**kinetic tremor**'. Such tremors include oscillatory movements observed at the same time as an active movement. Upon nearing the goal, the amplitude of the tremor increases.

Cerebellar Nystagmus

With cerebellar injury, there is likely to be a gaze-evoked or kinetic nystagmus. Such a cerebellar nystagmus is an involuntary, rhythmic, side-to-side movement of the eyes, with a clearly defined fast and slow phase. A cerebellar nystagmus is most marked when the patient is looking to the side of the injury.

16.9.5. Localization of Cerebellar Damage

Injury to the Posterior Vermis and Flocculonodular Lobe

Such injury may result from a cerebellar tumor such as a **medulloblastoma**. This tumor of childhood leads to a **postural instability** with the child adopting a **wide-base gait** to maintain their balance. If the tumor is in the median plane, the child often falls backward. The lower limbs are involved because both the dorsal and ventral spinocerebellar fibers, ending in the vermis of the posterior lobe, are likely to be injured. Since the spinocerebellar fibers synapse on the ipsilateral side, the cerebellar signs are also on the same side as the injury. Injuries to the posterior vermis and flocculonodular lobe often manifest a **cerebellar nystagmus**. If the injury or tumor is on one side, is irritative, or involves the nucleus fastigii or the cerebellovestibular connections from this nucleus, there is likely to be a nystagmus with the quick component to the side of the injury.

Injury to the Anterior Lobe of the Cerebellum

Clinical involvement of this cerebellar lobe is rare and difficult to detect until the injury enlarges and impinges on the superior cerebellar peduncle. After injury to the superior cerebellar peduncle, an **anesthesia of upward gaze** is likely to occur with decreased muscle tone to the ocular muscles. The flocculo-oculomotor path originates in the flocculus, travels through the superior cerebellar peduncle, and ends in the oculomotor nuclei to provide the appropriate amount of muscle tone needed for the ocular muscles. With decreased muscle tone, the eyes are simply too weak to stay up though they are not paralyzed.

Injury to the Posterior Lobe of the Cerebellum

Injuries to the hemispheric zone of the posterior lobe reveal a **hypotonicity** that is more severe than is found with injuries to other parts of the cerebellum. This hypotonicity leads to a **cerebellar ataxia** with postural instability – the patient falls to the side of the injury. A **rebound phenomenon** occurs such that the actively flexed arm of a patient will hit them in the face when released. There is **decomposition of movements, dysmetria** with past pointing, and **adiadochokinesis**. Cerebellar nystagmus is unlikely to occur with injuries to the hemispheric part of the posterior lobe.

Injury to the Cerebellar Nuclei

Unilateral injury to the nucleus fastigii of the cerebellum may result in a cerebellar nystagmus. Most of the symptoms occurring with fastigial injury also occur with injury to the flocculonodular lobe. Unilateral injury to the dentate nucleus of the cerebellum will cause a unilateral cerebellar tremor of the upper limb with a marked ataxia of the involved limb.

Injury to the Cerebellar Peduncles

Injuries to the **inferior cerebellar peduncle** result in a cerebellar ataxia with falling to the side of the injury, some hypotonicity, and perhaps the appearance of a cerebellar nystagmus. With involvement of the **middle cerebellar peduncle**, there is likely to be a cerebellar ataxia, hypotonicity of the limbs, and perhaps a hypotonicity of the cranial nerve nuclei. There may be a transient dysmetria and incoordination of the limbs. With involvement of the **superior cerebellar peduncle**, there is usually marked cerebellar ataxia and hypotonicity with falling to the side of the injury, marked cerebellar tremor, and the possible appearance of an asthenia of upward gaze if fibers of the flocculo-oculomotor path are involved.

16.10. DECORTICATE VERSUS DECEREBRATE RIGIDITY

Decorticate and decerebrate rigidity provides an excellent example of the rich interplay between the extrapyramidal cortical areas, basal nuclei, and cerebellum.

16.10.1. Decerebrate Rigidity

This condition results from injury to the rostral part of the transition between the midbrain and diencephalon. Such injuries usually involve considerable parts of the midbrain resulting in a **marked hypertonicity** manifested as a **rigidity** of the limbs **in extension** with **internal rotation of the limbs**. The cerebellum acts uninhibited as if cut off from regulation by the basal nuclei and extrapyramidal cortical areas. A common cause of decerebrate rigidity is injury at the junction of the midbrain and the diencephalon occurring as the result of a motorcycle accident.

16.10.2. Decorticate Rigidity

In decorticate rigidity there is destruction of much of the extrapyramidal motor cortex and lentiform nuclei (especially the putamen) perhaps including the corticostriate fibers. Extreme hypertonicity develops manifested as an extension of the lower limbs with flexion of the upper limbs. Self-inflicted gunshot wounds to the head often lead to decorticate rigidity. In decorticate rigidity some (but not all) of the cerebrum (some extrapyramidal motor cortex and part of the lentiform nuclei) is cut off from the brain stem.

16.11. EPILOGUE

In closing this discussion of the motor system, it is essential to appreciate that the regions involved in motor activity are a **functional unit** in that they **work together** to produce smooth, integrated complex movements. Discrepancies that appear in the clinical picture that vary from the classical description of injury to one part of the motor system merely serve to emphasize the complex interrelations that normally occur among the constituents of the motor system. The signs and symptoms of injury to the motor system in humans are more severe than those in experimental animals including nonhuman primates. In simplest terms, the motor system

functions as follows: ongoing reflex activity at the lower motoneuron level is under the influence of pyramidal and extrapyramidal impulses that converge onto these lower motoneurons. Several levels of regulation and control occur in the motor system: spinal cord, brain stem, thalamus, and basal nuclei with the highest level of regulation and control occurring at the level of the cerebral cortex. In essence, regions involved in motor activity are carefully integrated. As Sherrington noted such integration is **interaction for a purpose.**

FURTHER READING

Augood SJ, Waldvogel HJ, Munkle MC, Faull RL, Emson PC (1999) Localization of calcium-binding proteins and GABA transporter (GAT-1) messenger RNA in the human subthalamic nucleus. Neuroscience 88:521–534.

Blake DT, Byl NN, Merzenich MM (2002) Representation of the hand in the cerebral cortex. Behav Brain Res 135:179–184.

Byl NN, McKenzie A, Nagarajan SS (2000) Differences in somatosensory hand organization in a healthy flutist and a flutist with focal hand dystonia: a case report. J Hand Ther 13:302–309.

Chouinard PA, Paus T (2006) The primary motor and premotor areas of the human cerebral cortex. Neuroscientist 12:143–152.

de Lacalle S, Saper CB (1997) The cholinergic system in the primate brain: basal forebrain and pontine-tegmental cell groups. In: *Handbook of Chemical Neuroanatomy.* Björklund A, Hökfelt T, eds. Vol 13: The Primate Nervous System, Part I, pp 263–375. Elsevier, Amsterdam.

Dimitrova A, de Greiff A, Schoch B, Gerwig M, Frings M, Gizewski ER, Timmann D (2006) Activation of cerebellar nuclei comparing finger, foot and tongue movements as revealed by fMRI. Brain Res Bull 71:233–241.

Glickstein M (2006) Thinking about the cerebellum. Brain 129:288–290.

Graybiel AM (2000) The basal ganglia. Curr Biol 10:R509–511.

Graybiel AM (2004) Network-level neuroplasticity in cortico-basal ganglia pathways. Parkinsonism Relat Disord 10:293–296.

Graybiel AM (2005) The basal ganglia: learning new tricks and loving it. Curr Opin Neurobiol 15:638–644.

Grodd W, Hulsmann E, Ackermann H (2005) Functional MRI localizing in the cerebellum. Neurosurg Clin N Am 16:77–99.

Grodd W, Hulsmann E, Lotze M, Wildgruber D, Erb M (2001) Sensorimotor mapping of the human cerebellum: fMRI evidence of somatotopic organization. Hum Brain Mapp 13:55–73.

Herrero MT, Barcia C, Navarro JM (2002) Functional anatomy of thalamus and basal ganglia. Childs Nerv Syst 18:386–404.

Hoshi E, Tremblay L, Feger J, Carras PL, Strick PL (2005) The cerebellum communicates with the basal ganglia. Nat Neurosci 8:1491–1493.

Inoue T, Shimizu H, Yoshimoto T, Kabasawa H (2001) Spatial functional distribution in the corticospinal tract at the corona radiata: a three-dimensional anisotropy contrast study. Neurol Med Chir (Tokyo) 41:293–298.

Jankowska E, Edgley SA (2006) How can corticospinal tract neurons contribute to ipsilateral movements? A question with implications for recovery of motor functions. Neuroscientist 12:67–79.

Lewis DA, Sesack SR (1997) Dopamine systems in the primate brain. In: *Handbook of Chemical Neuroanatomy*. Björklund A, Hökfelt T, eds. Vol 13: The Primate Nervous System, Part I, pp 263–375. Elsevier, Amsterdam.

Marx JJ, Iannetti GD, Thomke F, Fitzek S, Urban PP, Stoeter P, Cruccu G, Dieterich M, Hopf HC (2005) Somatotopic organization of the corticospinal tract in the human brainstem: a MRI-based mapping analysis. Ann Neurol 57:824–831.

Morel A, Loup F, Magnin M, Jeanmonod D (2002) Neurochemical organization of the human basal ganglia: anatomofunctional territories defined by the distributions of calcium-binding proteins and SMI-32. J Comp Neurol 443:86–103.

Nitschke MF, Kleinschmidt A, Wessel K, Frahm J (1996) Somatotopic motor representation in the human anterior cerebellum. A high-resolution functional MRI study. Brain 119:1023–1029.

Romanelli P, Esposito V, Schaal DW, Heit G (2005) Somatotopy in the basal ganglia: experimental and clinical evidence for segregated sensorimotor channels. Brain Res Rev 48:112–128.

Schmahmann JD (2004) Disorders of the cerebellum: ataxia, dysmetria of thought, and the cerebellar cognitive affective syndrome. J Neuropsychiat Clin Neurosci 16:367–378.

Schmahmann JD, Caplan D (2006) Cognition, emotion and the cerebellum. Brain 129:290–292.

Schoch B, Dimitrova A, Gizewski ER, Timmann D (2006) Functional localization in the human cerebellum based on voxelwise statistical analysis: a study of 90 patients. NeuroImage 30:36–51.

Solov'ev SV (2006) The weight and linear dimensions of the human cerebellum. Neurosci Behav Physiol 36:479–481.

Utter AA, Basso MA (2007) The basal ganglia: An overview of circuits and function. Neurosci Biobehav Rev [Epub ahead of print, doi:10.1016/j.neubiorev.2006.11.003]

Voogd J (2003) The human cerebellum. J Chem Neuroanat 26:243–252.

Alcmaeon asserts that we smell by means of the nostrils when in respiration the air has been carried to the brain; that we distinguish flavors by the tongue, for since it is warm and soft it melts (substances) by its heat, and because of its yielding fineness it receives and passes on (the flavors).

Theophrastus (372/369–288/285 BC)

In humans, information about the central gustatory system is sparse at best, spread out, methodologically flawed, and inconsistent (Norgren, 1990). A decade later, this statement remains substantially true despite the advent of new imaging techniques.

Thomas C. Pritchard and Ralph Norgren (2004)

17

The Olfactory and Gustatory Systems

The sense of smell or **olfaction** and the sense of taste or **gustation** are difficult to separate. The term **flavor** refers to olfactory and gustatory sensations that seem to interact and occur simultaneously. The convergence of these two special senses probably involves the caudal part of the orbitofrontal cortex (OFC) and perhaps the adjoining anterior insula.

17.1. THE OLFACTORY SYSTEM

Olfaction or smell is a chemical sense. Olfactory impulses travel in fibers of the first cranial nerve, the **olfactory nerve** [I], functionally classified as a **special visceral afferent** (SVA) nerve. Although humans are capable of distinguishing a vast number of odors, there are seven primary odors. These include camphoric, musky, floral, minty, ethereal, pungent, and putrid.

In addition to these **odor qualities**, most humans are able to judge **odor strength** (weak, moderate, or strong) and are able to describe the **pleasurable aspects** or **hedonics** of odor (unpleasant, indifferent, or pleasant). At strong intensities, many odors usually considered pleasant are unpleasant. This is probably a result of over-stimulation or 'overdriving' of the olfactory system.

In comparison to other animals, humans are **microsomatic**, having limited ability to use olfactory clues. In general, women probably have better olfactory acuity. Variations in olfactory acuity occur during the menstrual cycle. Such acuity is highest at ovulation and lowest at menstruation. Fluctuations in olfactory sensitivity during the menstrual cycle are probably due, in part, to elevated blood estrogen levels which often lead to changes in the olfactory mucus layer that limits access of certain odorant molecules to the olfactory receptors.

17.1.1. Receptors

The primary function of the nasal cavities in most vertebrates is olfaction. In humans, however, the nasal cavities play a much more important role as a respiratory passage and as a system for conditioning the air. The **olfactory epithelium**, bounded by the superior nasal concha and the upper third of the nasal septum, is that part of the mucous membrane of the nasal cavities that has **olfactory receptor cells**. This is a pseudostratified columnar epithelium containing six or seven layers of cells and consisting of three cell types: **supporting cells**, **olfactory receptor cells**, and a layer of **basal cells**. The olfactory epithelium is thicker than the surrounding respiratory epithelium and has a distinctive dark yellow color because of the presence of olfactory pigment in the supporting cells and in Bowman's glands that lie beneath the basement membrane of the olfactory epithelium.

17.1.2. Primary Neurons

Receptor cells of the olfactory epithelium are also the primary neurons in the olfactory system. These cells are modified **bipolar neurons** (Fig. 17.1) with a dendritic process, projecting from the apical part of the bipolar neuron to the surface of the nasal cavity that ends in a ciliated, bulbous swelling called an **olfactory knob** (Fig. 17.1). Each olfactory cell may contain 1–20 cilia that lie at the nasal surface of the olfactory epithelium.

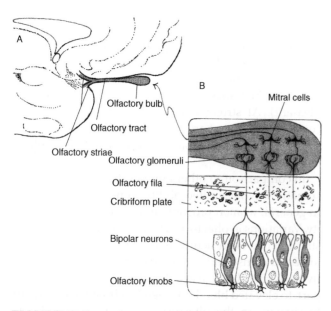

FIGURE 17.1. **A**, The human olfactory bulb, tract, and striae on the ventral surface of the cerebral hemisphere. **B**, Section of the olfactory epithelium, olfactory fila, and olfactory bulb.

Interaction with odorant molecules takes place at the olfactory receptor cell membrane. Molecules of a certain shape presumably fit into specific sites along the receptor membrane causing changes in membrane permeability. This causes current flow through the olfactory cell that leads to impulse conduction in the olfactory nerve.

17.1.3. Olfactory Fila and the Olfactory Nerve

An axonal process extends from the base of the first order bipolar neurons. Many of these nonmyelinated axons collect into macroscopically visible bundles called the **olfactory fila** (Fig. 17.1). These olfactory fila pass through the cribriform plate of the ethmoid bone where they are collectively termed the **olfactory nerve** [I]. Each odorous molecule causes a characteristic stimulus that is transformable into a patterned neural discharge in the fibers of the olfactory nerve. Information about the quality and intensity of an odor travels along the olfactory path to the brain. A one-to-one relationship appears to exist between the properties of the odoriferous stimulus and properties of the response to olfactory signals.

17.1.4. Olfactory Bulb – Secondary Olfactory Neurons

After passing upward and through the cribriform plate of the ethmoid bone, the olfactory fila enter at the tip of and along the ventral surface of the **olfactory bulb** (Fig. 17.1). The fila synapse with dendrites of **mitral cells** (Fig. 17.1) – the secondary neurons in this path located in the olfactory bulb at sites termed **olfactory glomeruli** (Fig. 17.1). The olfactory bulb has a laminated pattern in most mammals but these layers are unclear in humans. Axons of the mitral cells leave the olfactory bulb and pass to the base of the cerebral hemispheres (Fig. 17.1). The olfactory bulb is therefore a primary region of reception of olfactory impulses.

The human olfactory bulb, when quietly at rest, shows a constant background level of electrical activity. Organized rhythmic bursts occur when an odorant is available. Some of these bursts from the human olfactory bulb are extraordinarily distinctive. A most effective olfactory stimulus in humans is cigarette smoke. There is evidence for odor localization in humans, with different odorants yielding maximal responses in different areas in the bulb. The perception of an odor probably results from a combination of responses correlated from several areas in the bulb. The olfactory bulb likely plays a role in the preliminary discrimination of odors in humans.

17.1.5. Olfactory Stalk

As the axons of the mitral cells in each olfactory bulb pass back to the cerebral hemisphere, they form an **olfactory tract** (Figs 17.1, 17.2). As each olfactory tract joins the cerebral hemisphere, its fibers separate into two **olfactory striae** (Figs 17.1, 17.2) that bound a triangular area, the **olfactory trigone**. These two striae include a much smaller **medial stria** and a more prominent **lateral stria**. The area behind the olfactory trigone, called the **anterior perforated substance** (Fig. 17.2), is a site where numerous small blood vessels enter the brain. The anterior perforated substance corresponds essentially to the **olfactory tubercle** – to that part of the cerebral hemisphere of the frontal lobe, posterior and medial to the olfactory stalk, which has a trilaminar organization. This area of the cerebral cortex in the frontal lobe, which extends caudally to blend into the amygdaloid body, is the **prepiriform cortex**.

17.1.6. Medial Stria

Fibers of the olfactory stalk that enter the **medial stria** arise from the more medially located mitral cells. Since there are only a few scattered mitral cells in this location

in the human olfactory bulb, the human medial stria is poorly developed. Fibers of some mitral cells, perhaps as part of an intermediate olfactory stria of Crosby and Schnitzlein, enter the anterior commissure. Such fibers probably activate olfactory receptor cells in the contralateral olfactory bulb providing a means by which one olfactory bulb influences the contralateral bulb.

17.1.7. Lateral Stria

As the fibers of the **lateral olfactory stria** separate from the olfactory stalk, they synapse along the way in the lateral part of the **anterior olfactory nucleus**. Fibers of the **lateral olfactory stria** (Fig. 17.3) then proceed laterally and posteriorly to reach laterally lying **tertiary olfactory structures**. These include the lateral part of the **olfactory tubercle** and the **piriform cortex** (the anterior part of the temporal lobe medial to the rhinal sulcus). In humans, the olfactory tubercle is a trilaminar structure. Imaging studies of the human piriform cortex suggest it may be associated with odor memory. Other tertiary olfactory structures may include the **entorhinal cortex** (the anterior part of the parahippocampal gyrus that lies within the rhinal sulcus corresponding to Brodmann's area 28), and **amygdaloid body** (superficial nuclei including the anterior cortical amygdaloid nucleus, nucleus of the lateral olfactory tract and the periamygdaloid cortex) (Fig. 17.3). The

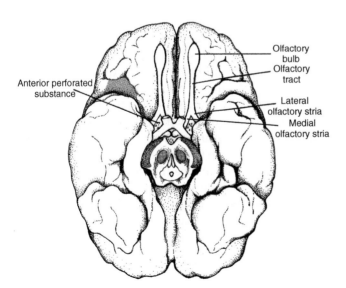

FIGURE 17.2. The ventral surface of the cerebral hemispheres showing the relations of the olfactory bulb, olfactory tract, and olfactory striae. Note the anterior perforated substance immediately behind the olfactory trigone. The shaded area on the lateral and posterior aspect of the orbital surface of the frontal lobe likely participates in olfactory discrimination in primates.

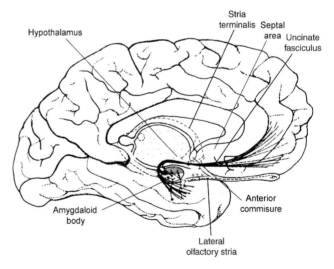

FIGURE 17.3. Distribution of the lateral olfactory stria to lateral olfactory areas. The discharge from these lateral olfactory areas to parts of the telencephalon and diencephalon is shown. The stria terminalis interconnects the superficial and deep amygdaloid nuclei with the subcallosal area and hypothalamus. Also shown is the uncinate fasciculus that connects lateral olfactory areas with the base of the frontal lobe.

entorhinal cortex receives fibers from the lateral stria in the monkey.

The **stria terminalis** (Fig. 17.3) is a fiber bundle interconnecting nuclei of the **amygdaloid body** (Fig. 17.3) with the **septal area** (Fig. 17.3) and the preoptic and anterior hypothalamic areas along the medial wall of each cerebral hemisphere. From the superficial (cortical and medial nuclei) and deep nuclei (basolateral and basomedial nuclei) of the amygdaloid body, the stria terminalis sends fibers along the tail of the caudate nucleus forward to the **anterior commissure** (Fig. 17.3). **Precommissural fibers** of the stria travel in front of the anterior commissure to reach the **septal area** (Fig. 17.3, the ventral part of the subcallosal area). Other **postcommissural fibers** from the stria travel behind the anterior commissure to enter the **hypothalamus** (Fig. 17.3). In addition to interconnections with the septal area and the hypothalamus through the stria terminalis, the amygdaloid body in nonhuman primates has many intrinsic fibers. These intrinsic fibers allow information to flow from the deep nuclei (lateral and basolateral) to the superficial nuclei (medial nucleus) or from the deep nuclei (basomedial) to the superficial nuclei (medial and cortical nuclei) and to the central nucleus.

Prominent electrical activity (rhythmic bursts in two broad frequency ranges) occurs in the human amygdaloid body in response to olfactory stimuli. Similar bursts occur in the human prepiriform cortex. This area consists of scattered clusters and bands of cells along the course of the lateral stria from the base of the olfactory stalk to the uncus and the underlying amygdaloid body. Patients who had undergone amygdalotomy in an attempt to eliminate or decrease intractable seizures showed judging errors of odor quality and an increase in 'different' judgments about the pleasurable quality of an odor (compared with distinct pleasant or distinctly unpleasant judgments). Such effects last no longer than one week even after bilateral amygdaloid damage. The amygdaloid body participates in determining the pleasant, unpleasant, or indifferent aspects of olfactory responses, that is, the **pleasurable aspects** of odor. The role of the amygdaloid body in this regard is of a discriminative or cortical nature. Electrophysiological studies in humans indicate that each odor is associated with a given group of frequency components. Odor discrimination in the human amygdaloid body is determined by the patterning of these components according to their amplitude rather than by which components are present. Similar classes of odors produce similar patterns of response.

The amygdaloid body and hypothalamus participate in certain aspects of feeding. In each instance, olfactory and gustatory inputs to these areas are involved in the behavioral responses to food. The hypothalamus appears to be involved in less discriminative eating such as eating in order to appease hunger and for the sake of survival. The primate amygdala participates in the control of behavioral responses to foods and thus appears to be involved in the more discriminative aspects of eating termed appetite. In this regard, one prefers a certain type of food based on its taste, smell, texture, temperature, and flavor as well as other qualities. Bilateral amygdalotomy in one case resulted in the patient being hyperphagic for some three and one-half months after the neurosurgical procedure.

17.1.8. Thalamic Neurons

One of the previously noted **tertiary olfactory structures**, the piriform cortex (the anterior part of the temporal lobe medial to the rhinal sulcus) has direct projections to the medial dorsal nucleus in primates. Therefore, olfactory input from this region reaches the thalamus. In nonhuman primates and in humans there is a thalamofrontal projection from the medial dorsal nucleus to considerable parts of the prefrontal cortex and to the orbital surface of the frontal lobe. The olfactory system, therefore, like other sensory systems, probably projects impulses to the cerebral cortex through a relay in the thalamus. The olfactory area on the orbital surface of the human frontal lobe is probably more extensive than that shown for the monkey in Fig. 17.2. Indeed, the orbital cortex is likely to occupy more area in the human brain than in the monkey brain. The base of the human frontal lobe sends impulses to the prefrontal cortex through the orbitofrontal fasciculus.

In addition to this olfactofrontal route through the thalamus, olfactory impulses trying to reach the frontal lobe may use other routes. Olfactory impulses travel from **tertiary olfactory structures**, such as the amygdaloid body, over association fibers of the **uncinate fasciculus** to the base of the frontal lobe (Fig. 17.3). A corticocortical fiber system is present in the rhesus monkey from the medial wall of the rhinal sulcus (the lateral boundary of the parahippocampal gyrus and the uncus) to the lateral and posterior orbital areas of the frontal lobe.

17.1.9. Cortical Neurons

Primary Olfactory Cortical Areas

Experimental studies in nonhuman primates and imaging studies in humans provide evidence for three

primary olfactory cortical areas including the **orbitofrontal cortex (OFC)**, the **inferior temporal gyrus**, and the **anterior insula**. In particular, the lateral and posterior aspects of the **orbitofrontal cortex** are likely involved in the fine discrimination of odors. Imaging studies in humans suggest that odor familiarity judgments activate the right orbitofrontal cortex whereas olfactory pleasurable judgments as well as unpleasant odors in particular activate the left orbitofrontal cortex. The **inferior temporal gyrus** has significant connections in primates with the anterior perforated substance. The human temporal lobe is involved in odor detection and in processing olfactory information. The right temporal lobe participates in olfactory memory functions (a nonverbal process). Patients who had undergone anterior temporal lobectomy had significant elevations in odor detection thresholds. The amygdaloid body, which receives olfactory input from the lateral stria, is interrelated by way of association fibers with the rostral part of the temporal lobe including the temporal pole, the insula, and the base of the frontal lobe. Thus, in addition to the orbitofrontal cortex and the inferior temporal gyrus, the **anterior insula**, particularly the agranular insula, is likely to play a role in olfaction. This part of the insula is at the caudal boundary of the orbitofrontal cortex.

17.1.10. Efferent Olfactory Connections

Efferents from the cerebral cortex, olfactory tubercle, prepiriform cortex, intralaminar thalamic nuclei, and perhaps the reticular nuclei of the midbrain, project to the olfactory bulbs. Functionally these efferents activate granule and periglomerular cells in the olfactory bulb. Such an inhibitory feedback loop helps contribute to the distinctiveness of olfactory responses, sharpening olfactory impulses, and allowing humans to differentiate thousands of different odors. As an odor lingers about, humans soon adapt to it and no longer notice it. This is probably due, in part, to efferent olfactory inhibition. These efferent olfactory connections also serve as a protective mechanism to prevent injury to olfactory receptors.

17.1.11. Injuries to the Olfactory System

Anosmia, Hyposmia, and Dysosmia

The most frequent reason for decreased olfactory acuity is the common cold. Any inflammation, infection (such as sinusitis), nasal obstruction, or swelling of the olfactory epithelium will decrease smell. Intracranial infections often affect smell. Of a serious and permanent nature are traumatic injuries, such as fractures of the cribriform plate of the ethmoid bone, that stretch or sever the olfactory nerves. In a report of 1167 patients with injuries to the head, some 7.5% had trauma to the olfactory system causing loss of smell, a condition called **anosmia**. Fifty percent of these patients recovered their sense of smell. A decreased sensitivity of smell is termed **hyposmia**. A distorted sense of smell is termed **dysosmia**. While most patients are not overly concerned with the gradual loss of smell, and anosmia is not life threatening, it does interfere with the enjoyment of life, and is likely to be professionally devastating to those employed as chefs, perfumers, tobacco blenders, or coffee and tea tasters.

The first intracranial tumor completely and successfully removed was an olfactory groove tumor (1885). **Olfactory groove meningiomas** are benign, slow growing tumors forming about 10% of all intracranial meningiomas. As they arise from the cranial dura along the cribriform plate of the ethmoid bone, they are in an ideal position to impinge on the olfactory fila, bulb, or stalk. In such cases, they often cause a reduction in our ability to smell or, if the tumor continues to grow and reaches a sufficient size, it is likely to cause a complete anosmia. In one study, the average size of 18 measured tumors was 6.1 cm in length by 4.5 cm in height. Without proper diagnosis and treatment, the olfactory meningioma will continue to grow along with the development of visual and mental symptoms in addition to anosmia.

Concomitant Loss of Smell and Taste

Many patients with anosmia because of trauma report a concomitant loss of taste. In such cases, no structural injury exists to explain the taste deficit. This is probably because other perceptions contribute to the sense of smell. Olfactory sensory stimuli, stimuli from taste receptors in the oral cavity, other sensory perceptions from the oral mucosa, and expectations aroused through visual stimuli contribute to the impression of smell. The tendency for patients to confuse the sensations of taste and smell is well known. Loss of smell often goes unreported by patients because they do not deem it significant or because they attribute their anosmia to other causes such as a sinus condition or exposure to air pollution. Although patients may be unaware or vaguely aware of their loss of smell, they all notice a flatness of taste.

Olfactory Auras or Hallucinations

Olfactory auras or hallucinations often accompany temporal lobe epilepsy. Such auras usually involve some disagreeable or unpleasant odors such as that of cigar smoke. These olfactory auras likely result from irritation of the **lateral olfactory areas** including the amygdaloid body. Conscious patients describe odor sensations on electrical stimulation of the uncus and amygdaloid body. Irritation of the ventral leaf of the uncinate fasciculus (Fig. 17.3) connecting the amygdaloid body with the frontal lobe is likely to yield olfactory auras that accompany seizures of temporal lobe origin. The **cingulum**, a prominent fiber bundle within the cingulate gyrus on the medial wall of each cerebral hemisphere connects with the amygdaloid body. Thus, irritation of the cingulum yields an aura of a sweet rose scent. Both an increase in strong responses and an increase in unpleasant responses occur in association with seizures. These effects are consistent with a general hyperexcitability in the amygdaloid body. When such hyperexcitability reaches its peak, it leads to the development of a seizure. In some patients, epileptic activity appears in the amygdaloid body on presentation of odoriferous stimuli. The use of the odor of jasmine or floral odors decreases epileptiform activity in some patients with temporal lobe seizures. EEG evoked responses to olfactory stimuli are recordable from the human amygdaloid body.

Specific Anosmia or Odor-blindness

Although some patients completely lose smell, many others may lose only one component of this sensation. This condition is termed **specific anosmia** or **odor-blindness**. In the olfactory system, perhaps as many as thirty odors may exist (in contrast to the three primary colors in the visual system). If a specific olfactory receptor protein represents each primary odor, then odor-blindness is explainable by the absence or defectiveness of a specific receptor protein.

Pheromones

Pheromones are chemical substances known to be present in the vaginal secretions of female rhesus monkeys. These substances act as sexual attractants and induce mating behavior in the male of the species. Although olfactory messages are particularly significant in lower animals for reproduction, in many primates olfactory communication by way of pheromones probably serves a variety of functions. The recent discovery of a second class of protein receptors in the mouse olfactory epithelium may hold the clue to further understanding of pheromones in humans. Genes encoding these receptors, called trace amine-associated receptors (TAARs), are present in humans. Perhaps these TAARs serve as olfactory receptors in humans. While these mouse data hint at the possibility of human pheromones, more research might move this hint to a more definitive answer.

17.2.   THE GUSTATORY SYSTEM

Taste or **gustation** is a chemical sense of the oral cavity functionally classified as a **special visceral afferent** (SVA) sensation. These SVA sensations travel in fibers of the facial [VII], glossopharyngeal [IX], and vagal [X] nerves. Gustatory stimuli consist of chemical molecules or ions in solution ingested into the mouth. The resulting stimuli play a role in reinforcing and eliciting human behavior.

17.2.1.   Receptors

Receptors for taste are the **taste buds** that appear in the oral cavity of the fetus at about three or four months of prenatal age – at a time when the fetus begins to swallow. The injection of saccharine into human amniotic fluid causes an increase in the rate of fetal swallowing. In the earliest hours of extrauterine life, newborns respond with facial expressions to food-related chemical stimuli. This observation and other studies of the development of taste preferences have led to the assumption that humans have a natural preference for sweet tastes.

Taste buds are modified epithelial structures in the oral cavity on the tongue and soft palate with a few on the lips, pharynx, and epiglottis particularly at birth. Most taste buds in adults are on the tongue and associated with specialized papillae of varying appearance (Fig. 17.4). These visible projections account for the tongue's characteristic rough and uneven surface. In their absence, the tongue is pale and smooth. The taste papillae exist in greater numbers at birth than at any other time in life. With advancing age, there is a gradual reduction in the number of papillae and the number of taste buds.

Most human papillae are specific to one class of taste stimuli. The tongue is divisible into different areas by taste sensitivities. Sensitivity is greatest at the tip of the tongue for sweet tastes, greatest at the sides of the tongue for salty and sour taste, and greatest at the back of the tongue for bitter taste. The soft palate is

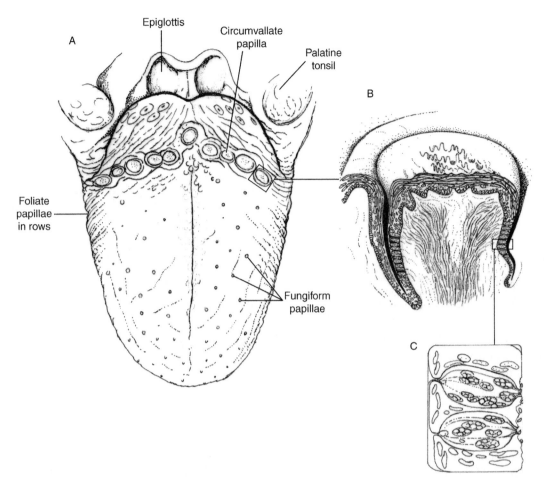

FIGURE 17.4. **A**, Dorsal surface and base of the human tongue including the epiglottis. The location of the specialized papillae is shown. **B**, Section through a circumvallate papilla illustrating the location and appearance of its taste buds. **C**, Drawing of an enlarged section of the side of two taste buds (after Henkin, 1976; Pfaffmann, 1978).

sensitive to sour and bitter. In contrast to the concept that individual papilla are sensitive to only one taste quality is the alternative view, based on experiments on single human fungiform papillae, that these papillae possess multiple sensitivity to compounds representative of the four primary taste qualities. Under this view, specificity, if it exists, is at the level of single taste buds or single receptors.

Qualities of Taste

There are four primary **qualities of taste** (salt, sweet, sour, bitter) each considered as a separate modality. These separate modalities have specific receptors and travel in specific fibers along the gustatory path. Tastes that are more complex stimulate a pattern of specific receptors. Another aspect of taste, in addition to these basic or primary tastes, is the pleasant or unpleasant quality that it arouses described as the **pleasurable** or **hedonic aspects of taste**.

Specialized Papillae with Taste Buds

Three different forms of specialized papillae containing taste buds are identifiable in humans. These include the **fungiform**, **foliate**, and **circumvallate papillae**. A fourth set of papillae termed the filiform papillae lack taste buds but have a dense distribution on the dorsal surface of the anterior two-thirds of the tongue. Taste buds on the anterior two-thirds of the tongue receive their innervation through the nervus intermedius part of the facial nerve [VII] (Fig. 17.5). The chorda tympani of the facial nerve joins the lingual nerve (a branch of the mandibular division [V_3] of the trigeminal nerve [V]) and travels with it to the tongue. Taste buds on the posterior third of the tongue receive their innervation through the glossopharyngeal nerve [IX] (Fig. 17.5). The borderline between these taste fields, termed the sulcus terminalis, runs about 2 cm in front of the circumvallate papillae. About 1280 taste buds are on the folds of the foliate papillae on the lateral border of the posterior

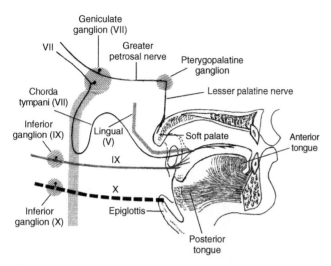

FIGURE 17.5. Sensory innervation of the taste buds on the soft palate, tongue (facial nerve-anterior tongue; glossopharyngeal nerve-posterior tongue), and the epiglottis (vagal nerve) (after Rollin, 1977).

tongue. These foliate papillae are sensitive to salty and sour taste. The foliate papillae at the base of the tongue and in front of the anterior palatine arch receive their innervation through fibers of the glossopharyngeal nerve. Taste buds on the soft palate and the tonsil receive their innervation by the facial nerve (Fig. 17.5) (greater petrosal nerve, pterygopalatine ganglion, lesser palatine nerves). Each mushroom-shaped fungiform papilla on the front of the tongue has three to five taste buds on its dorsal surface sensitive to sweet tastes. The fungiform papillae are scattered among these filiform papillae and are numerous at the anterior margin of the tongue. These papillae contain some taste buds into adult age. Finally, there are several to 12 large circumvallate papillae on the posterior tongue containing thousands of taste buds sensitive to bitter tastes.

Stimulus Removal of a Taste

Tongue movement enhances stimulus removal of a taste by forcing liquid in and out of the troughs that surround the circumvallate papillae. Serous glands near the circumvallate papillae at the base of the tongue (von Ebner's glands) produce a secretion to wash away the taste stimulus. Difficulty in stimulus removal leads to a lingering taste sensation commonly referred to as a 'bad taste'. Bitter aftertaste is most common in the back of the mouth.

Structure of Taste Buds

Taste buds are barrel-shaped structures about 50 μm in diameter. Each taste bud consists of 40 to 60 cells

arranged similar to the staves of a barrel (Fig. 17.4). Among these 40 to 60 cells are **taste cells**, **basal cells**, and **supporting cells**. Such cells have a lifespan of about 10 days and are in a constant state of flux – growing, developing, dying, and being replaced. The **taste bud** (Fig. 17.4), being open at the top, communicates with the tongue surface. This opening, called the **taste pore**, is a narrow channel. Each taste cell has numerous microvilli extending into the taste pore. Chemical stimuli react at the surface of the taste cell membrane. Such stimuli are absorbed to the microvilli surface causing conformational changes of the receptor membrane proteins. This causes depolarization and ultimately the development of a nerve impulse. Taste cells have many fine nerve endings that branch profusely near their base. A single taste fiber is likely to synapse with several taste cells in humans.

17.2.2. Primary Neurons

Primary neurons in the taste path are special visceral afferent neurons in several sensory ganglia that then provide fibers to the taste buds. Those in the geniculate ganglion of the facial nerve [VII] supply the anterior two-thirds of the tongue through the chorda tympani, those in the inferior ganglion of the glossopharyngeal nerve [IX] supply the posterior third of the tongue, and those in the inferior ganglion of the vagal nerve [X] supply taste buds on the epiglottis (Fig. 17.5). On stimulation of the chorda tympani, patients describe a metallic or sour taste on the anterior part of the tongue. Other investigators have reported the production of a sweet, sour, bitter, salty, or a metallic taste after chemical or mechanical stimulation of the human chorda tympani. The most common complaint after section of the chorda tympani is of a metallic taste on the tongue. Injury to both chorda results in a persistent metallic taste as well as a generally poorer taste and a dry mouth. Recordings from the 2000 fibers in the human chorda tympani are a useful means of studying the neural basis of taste sensation. Regeneration occurs in the human chorda tympani.

Processes of Primary Neurons

Peripheral processes of primary neurons in the gustatory path of humans supply the base of the taste cells. **Central processes** of these same primary neurons enter the central nervous system (at medullary and pontine levels) with their respective cranial nerve (Fig. 17.6) and ascend or descend to the medulla oblongata forming a

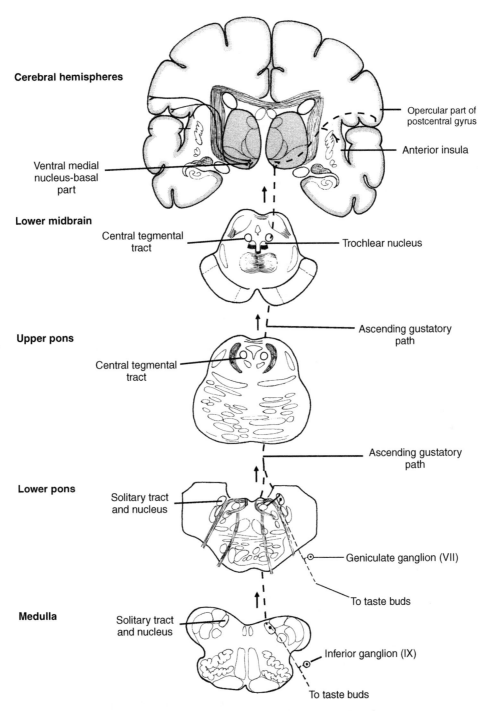

FIGURE 17.6. The origin, course and thalamic termination of the ascending gustatory path as it travels through the ipsilateral central tegmental tract to the primary gustatory cortex. This primary gustatory cortical area includes the base of the postcentral gyrus of the parietal lobe and the frontal operculum extending into the anterior insula.

fiber bundle referred to as the **solitary tract**. These central processes ultimately end in the nucleus of the solitary tract at or near the level of their entrance and at the levels of entrance of the facial, glossopharyngeal, and vagal nerves. All fibers carrying gustatory impulses, associated with these cranial nerves, participate in the formation of the **solitary tract** (Fig. 17.6). There is a slight decrease in the cross-sectional area of the human solitary tract with age in males; this suggests a decrease in taste function as well.

17.2.3. Secondary Neurons

Secondary neurons in the taste path are in the **solitary nucleus** (also called the **nucleus of the solitary tract)** that surrounds the **solitary tract**. In some cases the solitary tract surrounds the nucleus. The longest and most rostral subdivision of the nucleus of the solitary tract is the **interstitial nucleus of the solitary tract** (also called the **dorsal visceral gray** or **Nageotte's nucleus**). Most primary afferents (general visceral afferents [GVA] and special visceral afferents [SVA]) from the facial, glossopharyngeal, and vagal nerves (with some afferents from the trigeminal nerve) terminate in the nucleus of the solitary tract. Together the nucleus and solitary tract form a continuous complex whose rostral extent is at about the middle of the trigeminal pontine nucleus, in the middle third of the pons, and whose caudal extent is in the lower medulla. As fibers of each of these three cranial nerves enter the central nervous system, their visceral afferent components pass through or around the trigeminal spinal tract to enter and contribute to the solitary tract. The visceral afferent components of the facial nerve end in the rostral third of the nucleus of the solitary tract. Most of the glossopharyngeal visceral afferents end in the intermediate part of the nucleus of the solitary tract whereas the vagal visceral afferents ultimately end in the caudal third of the nucleus of the solitary tract. Thus central processes of primary gustatory neurons synapse with secondary gustatory neurons in the nucleus of the solitary tract. Central processes of these secondary gustatory neurons enter the ascending gustatory path (Fig. 17.6).

17.2.4. The Ascending Gustatory Path

The **ascending gustatory path** consists of fibers from neurons in the longest and most rostral subdivision of the nucleus of the solitary tract termed the **interstitial nucleus of the solitary tract** (also called the **dorsal visceral gray** or **Nageotte's nucleus** as noted above). In the monkey, an ipsilateral projection from the rostral or gustatory part of the nucleus of the solitary tract ascends as a component of the **central tegmental tract** before terminating in the parvocellular part of the ventral posterior medial nucleus (VPM$_{pc}$) of the dorsal thalamus. Clinicoanatomical studies confirm that the ascending gustatory path in humans (Fig. 17-6) is an ipsilateral path (21 patients in six separate studies) that travels in the pontine tegmentum (13 patients, two separate studies) and then enters the dorsomedial mesencephalic tegmentum (2 patients, one report). In one case report, however, the ascending gustatory path

was ipsilateral in the medulla but then crossed in the midbrain to reach the contralateral dorsal thalamus. It is perhaps these fibers that cross to the contralateral thalamus that lead to both contralateral and ipsilateral taste deficits in humans that have been reported following lesions to the thalamus or cortex. The ipsilateral ascending gustatory path is substantial in nonhuman primates and humans. There seems to be little support in the literature for an ascending gustatory path in humans that travels in the medial lemniscus.

17.2.5. Thalamic Neurons

Secondary ascending gustatory fibers are thought to end in the **basal ventral medial nucleus** (VM$_b$) of the human thalamus, an area corresponding to the parvocellular part of ventral posterior medial nucleus (VPM$_{pc}$) in monkeys. This thalamic taste area in monkeys has been well characterized by lesion and single cell recording studies. VPM$_{pc}$ receives tactile inputs from the tongue as well. Stimulation of the human VM$_b$ in patients who are awake during neurosurgical procedures elicits sensations of taste as well as general visceral sensations such as fullness of a hollow internal organ. Painful and nonpainful general somatic sensations can also be elicited from VM$_b$. Thus the human thalamic taste area mediates multiple sensations in addition to taste. Gustatory afferents also project to the posterior part of VPM adjacent to the centromedian nucleus (CM) in humans. The direct projection from nucleus of the solitary tract to the thalamus followed by thalamocortical projections to the cerebral cortex in nonhuman primates and humans is in contrast to the central gustatory system of nonprimates such as the rabbit, rat, cat, and hamster, which have intervening synapses along the way in the parabrachial nuclei of the pons. Also the ipsilateral projection to the thalamus is greater in primates, whereas both sides of the tongue in rats are equally represented in VM$_b$.

Thalamocortical Projections

Thalamocortical fibers leave the thalamic gustatory relay nucleus (the basal ventral medial nucleus or VM$_b$), pass through the internal capsule, and end in such cortical regions as the anterior insula, the opercular part of the frontal lobe, and the opercular part of the postcentral gyrus of the parietal lobe (Fig. 17-6). There appears to be some recognition of taste at the thalamic level in humans, particularly pleasant versus unpleasant aspects of taste.

17.2.6. Cortical Neurons

The **primary gustatory cortex** (PGC) includes parts of the **frontoparietal operculum** at the base of the central sulcus (the opercular part of the postcentral gyrus of the parietal lobe). The parietal opercular part of this PGC corresponds to **Brodmann's area 43** (Fig. 21.7). The primary gustatory cortex may extend to the opercular-insular junction and onto the floor of the anterior part of the **insular lobe**. This area is closely associated with the representation of the jaw, tongue, and intra-oral part of the sensory homunculus. Short association fibers coursing through the extreme capsule in primates link the anterior part of the insular lobe with the cortex at the base of the central sulcus, including the fronto-parietal operculum. The laterality of the human primary gustatory cortex (PGC) was studied by Onoda et al. (2005) in a series of patients whose chorda tympani had been severed unilaterally on the right side. The left chorda tympani was stimulated with NaCl using a device developed for measuring gustatory-evoked magnetic fields (GEMfs). The frequency and latency of PGC activation in each hemisphere was measured. Bilateral responses were observed more frequently than ipsilateral or contralateral responses suggesting to these authors that a unilateral taste signal diverges en route to the cerebral cortex before terminating in the PGC of both hemispheres.

Closely associated with the primary gustatory cortex is a **secondary gustatory cortex** in the anterior dorsal part of the insula. The latter area is involved in the subjective recognition of gustatory modalities. In nonhuman primates, gustatory impulses from the primary gustatory cortex project to the orbitofrontal cortex where they converge on neurons that also receive olfactory impulses. This convergence in the orbitofrontal cortex may provide the site for the cortical representation of flavor.

The discriminative aspects of taste require cortical participation. Along the primary gustatory cortex is a pattern of representation for sweet, sour, salt, and bitter (and perhaps umami). This area of the cerebral cortex is probably involved in the identification of specific substances, the comparison of one aspect of taste with another, and the appearance of gustatory hallucinations. Association fibers interconnect the primary gustatory areas and the adjacent gustatory association area with the amygdaloid body and hippocampal formation. Recent imaging studies in humans reveal that the pleasant and unpleasant aspects of taste activate the amygdaloid body. The basolateral amygdaloid nucleus receives some gustatory input and receives olfactory fibers, as does the anterior part of the insular lobe. Such olfactory-gustatory correlations occurring in the amygdaloid body come to expression as **appetite**–the more discriminative aspects of feeding. Therefore, injuries to the amygdaloid body or hippocampal formation often lead to perversions of appetite.

17.2.7. Injuries to the Gustatory System

Gustatory Auras

A **gustatory aura** involves the subjective experience of gustatory symptoms that precede a seizure or a migraine headache. Such gustatory auras may be illusions (a perversion or misperception of a taste – mistaking something for what it is not) or hallucinations (the subjective perception of gustatory sensation when no gustatory stimulus or sensation is present). A patient with an anterior insular tumor described a bitter taste in his mouth preceding his seizures. Another patient with vascular injury to the inferior part of the postcentral gyrus had seizures preceded by an aura of a sour-bitter taste. Migraine sufferers may experience gustatory symptoms that are present before their headaches start. Irritative injuries to the frontal opercular side of the rostral part of the insular lobe yield a **gustatory aura**.

Ageusia, Hypogeusia, and Parageusia

The loss of the ability to taste sweet, sour, bitter, salty, or umami is termed **ageusia** while a diminution or blunting of any of these tastes is termed **hypogeusia**. Perversions of taste or confusion about taste (sweet substances perceived as salty) is termed **parageusia**. The National Institute on Deafness and Other Communication Disorders (NIDCD) reports that while more than 200,000 people visit a physician for taste disorders every year many more taste disorders go unreported. While some people are born with disorders of taste, most develop them after an injury or illness. Any process that coats the surface of the tongue is likely to have an effect on taste. With age or with decreased salivary gland function, there is likely to be a diminution of taste. Excessive smoking, oral and peri-oral infections, dental procedures as well as oral appliances often affect taste, as do certain medications such as angiotensin-converting enzyme inhibitors that are associated with decreased sense of taste and a strongly metallic, bitter, or sweet taste. Patients receiving radiation therapy for cancers of the head and neck often develop disorders of taste.

Injury to the glossopharyngeal nerve results in the loss of taste on the posterior tongue and along the base of the tongue (including the foliate papillae). Surgical section of the chorda tympani causes unilateral ageusia on the anterior two-thirds of the tongue. Involvement of the facial nerve as in **Bell's palsy** may lead to gustatory disturbances. Pressure on the facial trunk with a growing cerebellopontine angle tumor often leads to taste deficits on the anterior tongue. Hypogeusia at the base of the tongue is often a component of glossopharyngeal neuralgia.

Familial dysautonomia is a rare, inherited disease with many neurological signs and symptoms. One of these is the inability to taste. Affected patients have an absence of fungiform and circumvallate papillae. This is the only consistent structural sign in this disease allowing a positive diagnosis of this condition (even in premature infants).

Flavor

As was noted in the discussion of injuries to the olfactory system, some patients with demonstrable olfactory deficits also describe a concomitant deficit of taste. Because these two senses often occur simultaneously and are hard to separate, a useful term for olfactory and gustatory sensations that interact is **flavor**. Event-related fMRI imaging reveals the brain representation of flavor is in the **lateral part of the anterior orbitofrontal cortex**. The **medial part of the orbitofrontal cortex** appears to be the cortical representation of the pleasantness of the olfactory and taste stimuli and their combinations. The simultaneous appearance of both olfactory and gustatory auras does occur but only rarely. One patient thought he had smelled and tasted peaches. On another occasion, the same patient reportedly described smelling and tasting roasted peanuts. McCabe and Rolls (2007) have made the interesting observation that umami is a rich and delicious flavor formed by convergence of taste and olfactory pathways in the human brain. Using functional brain imaging with fMRI these authors showed that the glutamate taste and savory odor combination produced much greater activation of the medial orbitofrontal cortex and pregenual cingulate cortex than the sum of the activations by the taste and olfactory components presented separately.

Miracle Fruit

A fascinating aspect of the subject of taste is the observation that gymnemic acid, a naturally occurring plant substance, blocks the sweet taste of sugar and saccharin in humans. This completely suppresses the human response to sweet substances without affecting responses to salty, acid, or bitter stimuli. Experiments in humans reveal that pre-rinsing with gymnemic acid makes stimuli that are primarily sweet tasting more difficult to identify. The red berries of a shrub native to West Africa, known as **miracle fruit**, also have an interesting property. After chewing these berries for some time, sour food and dilute mineral and organic acids have a sweet taste without affecting the bitter, salty, or sweet response. This observation is due to the presence of a taste-modifying protein, miraculin, in the fruit. Yamamoto et al. (2006) used taste-elicited magnetic fields to study the cortical representation of taste-modifying action of miracle fruit in humans. While the initial taste responses were localized to the area of the primary gustatory cortex in the fronto-parietal opercular/insular cortex, the sourness component of citric acid is greatly diminished at the level of subcortical relays. These authors propose the idea that the qualitative aspect of taste is processed in the primary taste area and the affective aspect is represented by a pattern of activation among different cortical areas.

FURTHER READING

Amaral DG, Insausti R, Cowan WM (1987) The entorhinal cortex of the monkey: I. Cytoarchitectonic organization. J Comp Neurol 264:326–355.

Beckstead RM, Morse JR, Norgren R (1980) The nucleus of the solitary tract in the monkey: projections to the thalamus and brain stem nuclei. J Comp Neurol 190:259–282.

Boyce JM, Shone GR (2006) Effects of ageing on smell and taste. Postgrad Med J 82:239–241.

Bromley SM (2000) Smell and taste disorders: a primary care approach. Am Fam Physician 61:427–436.

Brouwer JN, van der Wel H, Francke A, Henning GJ (1968) Miraculin, the sweetness-inducing protein from miracle fruit. Nature 220:373–374.

Costanzo RM, Miwa T (2006) Posttraumatic olfactory loss. Adv Otorhinolaryngol 63:99–107.

de Araujo IE, Rolls ET, Kringelbach ML, McGlone F, Phillips N (2003) Taste-olfactory convergence, and the representation of the pleasantness of flavour, in the human brain. Eur J Neurosci 18:2059–2068.

Doty RL, Bromley SM (2004) Effects of drugs on olfaction and taste. Otolaryngol Clin North Am 37:1229–1254.

Gent JF, Hettinger TP, Frank ME, Marks LE (1999) Taste confusions following gymnemic acid rinse. Chem Senses 24:393–403.

Hadley K, Orlandi RR, Fong KJ (2004) Basic anatomy and physiology of olfaction and taste. Otolaryngol Clin North Am 37:1115–1126.

Kringelbach ML, de Araujo IE, Rolls ET (2004) Taste-related activity in the human dorsolateral prefrontal cortex. NeuroImage 21:781–788.

Lenz FA, Gracely RH, Zirh TA, Leopold DA, Rowland LH, Dougherty PM (1997) Human thalamic nucleus mediating taste and multiple other sensations related to ingestive behavior. J Neurophysiol 77:3406–3409.

Liberles SD, Buck LB (2006) A second class of chemosensory receptors in the olfactory epithelium. Nature 442:645–650.

McCabe C, Rolls ET (2007) Umami: a delicious flavor formed by convergence of taste and olfactory pathways in the human brain. Eur J Neurosci 25:1855–1864.

Pearson H (2006) Mouse data hint at human pheromones. Nature 442:495.

O'Doherty J, Rolls ET, Francis S, Bowtell R, McGlone F (2001) Representation of pleasant and aversive taste in the human brain. J Neurophysiol 85:1315–1321.

Onoda K, Kobayakawa T, Ikeda M, Saito S, Kida A (2005) Laterality of human primary gustatory cortex studied by MEG. Chem Senses 30:657–666.

Raviv JR, Kern RC (2004) Chronic sinusitis and olfactory dysfunction. Otolaryngol Clin North Am 37:1143–1157.

Reiter ER, DiNardo LJ, Costanzo RM (2004) Effects of head injury on olfaction and taste. Otolaryngol Clin North Am 37:1167–1184.

Seiberling KA, Conley DB (2004) Aging and olfactory and taste function. Otolaryngol Clin North Am 37:1209–1228.

Shikama Y, Kato T, Nagaoka U, Hosoya T, Katagiri T, Yamaguchi K, Sasaki H (1996) Localization of the gustatory pathway in the human midbrain. Neurosci Lett 218:198–200.

Small DM (2006) Central gustatory processing in humans. Adv Otorhinolaryngol 63:191–220.

Small DM, Prescott J (2005) Odor/taste integration and the perception of flavor. Exp Brain Res 166:345–357.

Small DM, Zald DH, Jones-Gotman M, Zatorre RJ, Pardo JV, Frey S, Petrides M (1999) Human cortical gustatory areas: a review of functional neuroimaging data. NeuroReport 10:7–14.

Wilson DA, Kadohisa M, Fletcher ML (2006) Cortical contributions to olfaction: plasticity and perception. Semin Cell Dev Biol 17:462–470.

Wrobel BB, Leopold DA (2004) Clinical assessment of patients with smell and taste disorders. Otolaryngol Clin North Am 37:1127–1142.

Yamamoto C, Nagai H, Takahashi K, Nakagawa S, Yamaguchi M, Tonoike M, Yamamoto T (2006) Cortical representation of taste-modifying action of miracle fruit in humans. NeuroImage 33:1145–1151.

It is proposed that the hypothalamus, the anterior thalamic nuclei, the gyrus cinguli, the hippocampus, and their interconnections constitute a harmonious mechanism which may elaborate the functions of central emotion, as well as participate in emotional expression ... as a unit within the larger architectural mosaic of the brain.

James W. Papez, 1937

18

The Limbic System

18.1. HISTORICAL ASPECTS

Broca (1878) divided the human brain into two parts: an intelligent part, consisting of the cerebral cortex, and a brutal part that he referred to as '**le grand lobe limbique**'. This latter part forms a tennis racket-shaped ring or limbus [Latin, border] of gray matter whose handle corresponds to the olfactory bulb and stalk. The frame of this tennis racket corresponds to those nuclear groups that encircle medial parts of the cerebral hemisphere. This brutal part of the brain was termed the 'seat of those lower faculties which predominate in the beast'.

Papez (1937) published a paper entitled '**A Proposed Mechanism of Emotion**' suggesting that this wall of gray matter along the medial surface of each cerebral hemisphere was linked to the balance and control of emotion. He further suggested that this region constitutes 'a harmonious mechanism which may elaborate the functions of **central emotion** and participate in **emotional expression**'. Note the distinction between emotion and emotional expression, as we will conclude this chapter with comments related to the possible role of the limbic system in three aspects of human emotion. Papez also emphasized a **cyclic path** forming the functional basis for emotion. Since his idea was not well accepted, Papez never referred to the topic of emotion in his later writings.

In a discussion of 'motility, behavior, and the brain', Yakovlev (1948) postulated three neuronal coordinates of behavior: (1) an innermost **entopallium**, responsible for visceral function; (2) an intermediate **mesopallium**, responsible for emotional expression; and (3) an outermost **ectopallium** involved in effectuating or transforming emotions into acts.

MacLean (1949) described the Papez circuit as the '**visceral brain**'. He later proposed the term '**limbic system**' as the anatomical substrate of emotion and behavior. Since a great deal of emotional expression is of a visceral or autonomic nature, MacLean suggested this system dominates the sphere of visceral activity. He therefore linked the term 'limbic system' with medial hemispheric areas involved in the balance and control of emotion and emotional expression. Lastly, this author suggested that limbic epilepsy should provide a link between neurology, psychology, and psychiatry.

18.2. ANATOMY OF THE LIMBIC SYSTEM

The term limbic system includes a number of brain areas (Fig. 18.1) such as the **olfactory system**, the **septal area** (ventral part of the subcallosal area), and the **mamillary bodies** of the hypothalamus. Also part of the limbic system is the **anterior nuclear group** of the thalamus, the **hippocampal formation**, and the **amygdaloid body**. Other constituents of the limbic system are the **cingulate gyrus** with its associated fiber bundle the **cingulum**, and certain cortical areas termed the **limbic association areas** that influence structures of the limbic system. Others might include additional areas not named here. Still others regard certain of these limbic brain areas (such as the hippocampal formation and amygdaloid body) as independent functional systems rather than as conspicuous parts of the limbic system. The anatomic relationships among these structures are elegantly demonstrated using functional imaging methods in humans.

18.2.1. Olfactory System

Parts of the olfactory system included in the limbic system are the olfactory epithelium with olfactory cells and fila, each olfactory bulb, stalk, stria, and trigone. The area behind the olfactory trigone, penetrated by numerous small blood vessels and called the **anterior perforated substance** (Fig. 17.2), corresponds to the **tuberculum olfactorium**. The tuberculum includes that part of frontal cortex posterior and medial to the olfactory stalk that has a trilaminar organization. That

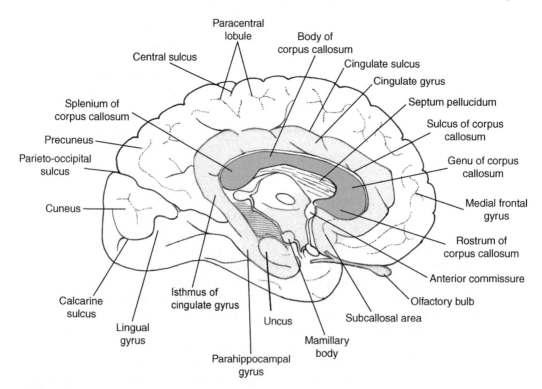

FIGURE 18.1. Limbic areas of the human brain (shaded) on the medial surface of the cerebral hemisphere.

part of the cerebral hemisphere of the frontal lobe, posterior and lateral to the olfactory stalk is the **prepiriform cortex**. This area of cerebral cortex extends caudally to blend into the amygdaloid body.

18.2.2. Septal Area

On the medial surface of each cerebral hemisphere, inferior to the rostrum of the corpus callosum, anterior to the lamina terminalis, and superior to the anterior perforated substance, is the **subcallosal area** (Fig. 18.1). The subcallosal area is divisible into dorsal and ventral parts. The dorsal part of the subcallosal area, termed the **subcallosal gyrus** is continuous with the cingulate gyrus whereas the ventral part of the subcallosal area includes the **septal area** that corresponds to **Brodmann's area 25**. The septal area includes the septal nuclei, the nucleus of the diagonal band of Broca, and the nucleus accumbens.

18.2.3. Mamillary Bodies of the Hypothalamus

The **mamillary bodies** (Fig. 18.1), which are rostral to the interpeduncular fossa, form a protuberance on the ventral surface of the hypothalamus. An extremely large fiber bundle of the limbic system, the **fornix**, ends in the mamillary bodies (Fig. 18.2). The mamillary bodies

of chronic alcoholics often show a loss of myelinated fibers with decreased levels of cerebrosides, cholesterol, and phospholipids.

18.2.4. Anterior Nuclei of the Thalamus

The **anterior nuclei**, at the rostral end of that structure, receive sensory impulses from the mamillary bodies of the hypothalamus by way of the **mamillothalamic tract** (Fig. 18.2). Impulses from the anterior nuclear group project to the cingulate gyrus over **thalamocingulate fibers** (Fig. 18.2). The anterior nuclear group is therefore a relay nucleus in the limbic system connecting two parts of the diencephalon – the thalamus and hypothalamus.

18.2.5. The Hippocampal Formation

The collateral sulcus separates the **parahippocampal gyrus** (Fig. 18.1), on the inferior surface of each cerebral hemisphere and the medial surface of each temporal lobe, from the occipitotemporal gyrus. Part of the parahippocampal gyrus projects into the **temporal horn of the lateral ventricle**. Several shallow grooves indent its anterior part giving it the appearance of the paw of an animal and giving rise to the term **pes hippocampi**.

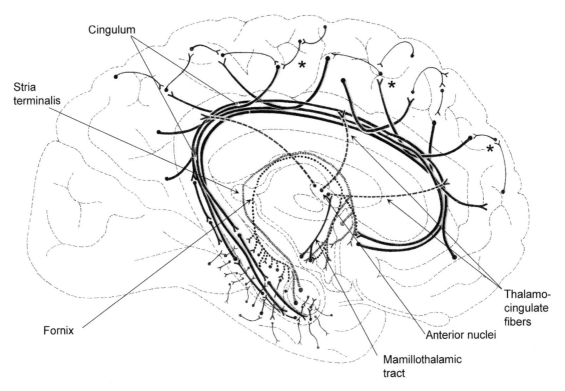

FIGURE 18.2. Some connections of the limbic areas of the human brain. Cortical areas (indicated by the asterisks) have a major influence on limbic structures.

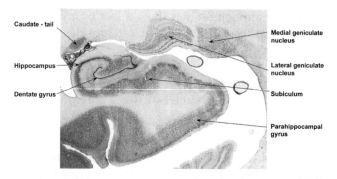

FIGURE 18.3. Coronal section through the parahippocampal gyrus (from DeArmond, Fusco, and Dewey 1989).

Coronal sections through the parahippocampal gyrus reveal four easily recognizable structures: the **dentate gyrus, hippocampus proper, subiculum**, and the **parahippocampal gyrus** (Figs 18.3, 18.4). These four structures are collectively termed the **hippocampal formation**. The hippocampus proper is more generally termed the **hippocampus**. It includes four cellular areas designated CA1, CA2, CA3, and CA4. The human CA1 is some 30 cells in thickness. The subiculum is actually a complex that includes the subiculum proper, presubiculum, prosubiculum, and parasubiculum. The parahippocampal gyrus includes an anterior part (entorhinal cortex and perirhinal cortex) and a posterior part.

The **entorhinal cortex**, as its name implies, lies within the rhinal sulcus and corresponds to Brodmann's area 28. The lateral boundary of the human entorhinal cortex is the collateral sulcus and its medial boundary is the uncus. While there are eight cytoarchitectonic subdivisions present in the monkey entorhinal cortex, the human entorhinal cortex may have from eight to twenty-seven such subdivisions. In humans, the entorhinal cortex is six-layered but represents an atypical or transitional type of cortex. The entorhinal cortex has interconnections with association areas of the cerebral cortex and is an early target in Alzheimer's disease. It would be interesting to determine what effect, if any, damage to these interconnections has on the early symptoms of Alzheimer's disease. In the monkey, the entorhinal cortex receives input from the olfactory bulb, cingulate cortex, insula, orbitofrontal cortex, superior temporal cortex, parahippocampal cortex, and the perirhinal cortex. This cortical input reaches the hippocampal formation by way of its interconnections with the entorhinal cortex. In this regard the entorhinal cortex is been termed the 'gateway to the hippocampus'. The entorhinal cortex is closely associated with the periamygdaloid and prepiriform cortices as well. Stimulation of the hippocampal formation in humans causes an alerting effect

on many cortical areas. Functionally the hippocampal formation is part of the limbic system playing an important role in some types of memory, particularly that described as declarative memory.

18.2.6. The Amygdaloid Body

On the medial aspect of the temporal lobe, at the anterior tip of the parahippocampal gyrus, and in the rostromedial wall of the temporal horn of the lateral ventricle is the **uncus**. Coronal sections through the uncus reveal the underlying **amygdaloid body** (Fig. 18.5) that eventually fuses with the tail of the caudate nucleus (Fig. 16.3). The amygdaloid body consists of a number of named nuclei (Table 18.1) broken down into several major nuclear groups: the **deep amygdaloid nuclei**, the **superficial amygdaloid nuclei**, and **other amygdaloid nuclei**. The terminology applied to the human amygdaloid body and used by different investigators can be challenging even to the experienced neuroscientist. The **deep amygdaloid nuclei** include the lateral, basolateral (basal) and basomedial (accessory basal) amygdaloid nuclei. The **superficial amygdaloid nuclei** include the anterior and posterior cortical nuclei, the medial nuclei, the nucleus of the lateral olfactory tract and the periamygdaloid cortex. The **other amygdaloid nuclei** include the anterior amygdaloid area, amygdalohippocampal area, the central nucleus, the cortico-amygdaloid transition area, and the interstitial (intercalated) amygdaloid nuclei. Figure 18.6 illustrates the amygdaloid nuclei as they appear on coronal sections through the brain.

In general, the nuclei of the amygdala have dense intrinsic and extrinsic nuclei. There are several functional systems with important connections involving the amygdala. These include (1) connections between the olfactory system (primary olfactory cortex or piriform cortex), amygdala and hypothalamus; (2) the autonomic nervous system with the central nucleus and the brain stem; and (3) the frontal and temporal cortices with the deep nuclei (lateral and basolateral amygdaloid nuclei). The deep nuclei have important and reciprocal connections with the nucleus basalis of Meynert. The intrinsic connections of the primate amygdala are such that information seems to flow in a lateral to medial direction, that is, from the deep nuclei to the superficial nuclei.

The principal efferent paths of the amygdala in nonhuman primates and humans are the **stria terminalis** and the **ventral amygdalofugal path**. The stria terminalis is an important fiber bundle projecting from the

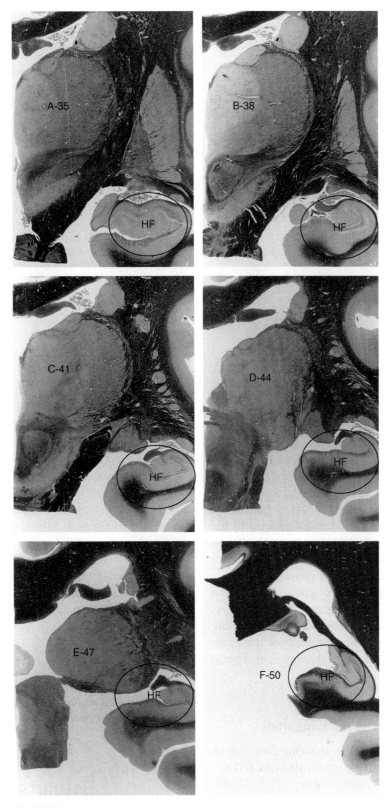

FIGURE 18.4. Diagram of a series of coronal sections through the human brain to demonstrate the nuclei of the hippocampal formation from rostral to caudal. (**A**, **B**, **C**, **D**, **E**, **F** are based on sections 35–50 respectively in Mai, Assheuer and Paxinos, 2004.)

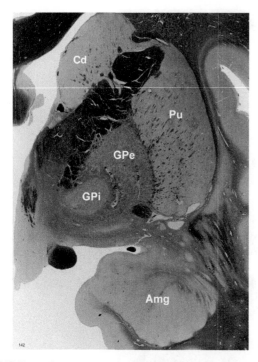

FIGURE 18.5. Human amygdaloid body (section 25 from Mai, Assheuer and Paxinos, 2004). **Abbreviations:** Amg, amygdaloid body; Cd, caudate nucleus; GPe, globus pallidus, external segment; GPi, globus pallidus, internal segment; Pu, putamen.

TABLE 18.1   Nuclei of the amygdaloid body in humans

Nuclear group	Abbreviation
I. Superficial nuclei (corticomedial complex)	
Cortical amygdaloid nucleus	Co
Anterior cortical nucleus	ACo
Posterior cortical nucleus	PCo
Medial amygdaloid nucleus	Me
Nucleus of the lateral olfactory tract	NLOT
Perimaygdaloid cortex	PAC
II. Deep nuclei (basolateral complex)	
Lateral amygdaloid nucleus	La
Basolateral (basal) amygdaloid nucleus	BL
Basomedial (accessory basal) amygdaloid nucleus	BM
Paralaminar nucleus	PL
III. Other nuclei and related areas	
Amygdalohippocampal area	AHi
Anterior amygdaloid area	AAA
Central amygdaloid nucleus	Ce
Interstitial (intercalated) amygdaloid nucleus	I
Amygdaloclaustral area	
Amygdalopiriform transition area	
Corticoamygdaloid transition area	

Modified from Crosby and Schnitzlein, 1982; Sims and Williams, 1990; Sovari et al., 1995; *Terminologia Anatomica*, 1998; and Mai, Assheuer and Paxinos, 2004.

superficial (cortical and medial nuclei) and deep nuclei (basolateral and basomedial nuclei) of the amygdaloid body to the hypothalamus and septal area. The **ventral amygdalofugal path** connects the medial amygdaloid nucleus with the hypothalamus. It leaves the dorsomedial aspect of the amygdala, travels inferior to the globus pallidus (adjacent to the ansa lenticularis and the nucleus basalis of Meynert) and enters the hypothalamus at the level of the anterior commissure.

These connections allow the amygdala to stand at a crossroads integrating exteroceptive, interoceptive, autonomic, and sensory information. This information, in turn, is able to influence diverse areas of the brain such as the cortex, thalamus, hypothalamus, and brain stem.

18.2.7. Cingulate Gyrus and Cingulum

The **cingulate gyrus**, ventral to the cingulate sulcus and dorsal to the trunk of the corpus callosum, follows the contour of the corpus callosum (Fig. 18.1). Caudal to the splenium of the corpus callosum, the cingulate gyrus narrows into the **cingulate isthmus**, which connects the cingulate gyrus with the parahippocampal gyrus of the temporal lobe (Fig. 18.1). The **cingulum** is a prominent association bundle in the cingulate gyrus

(Fig. 18.2). This bundle of long and short association fibers carries impulses to and from the parahippocampal gyrus. Thalamocingulate fibers from the anterior nuclei of the thalamus bring impulses to the cingulate gyrus. Through fibers of the cingulum, the cingulate gyrus connects to the subcallosal area (including the subcallosal gyrus and the septal area) and to limbic structures in the temporal lobe such as the hippocampal formation and amygdaloid body. The cingulum also connects the cingulate gyrus with cortical areas in the frontal, parietal, and occipital lobes.

18.2.8. Cortical Areas

Limbic association areas include parts of the anterior temporal, cingulate, insular, orbital frontal, and parahippocampal cortex.

18.3. CYCLIC PATHS OF THE LIMBIC SYSTEM

A cyclic path underlies the functioning and connects the principal components of the limbic system. This cyclic path is essentially a reinforcing system. Figure 18.2

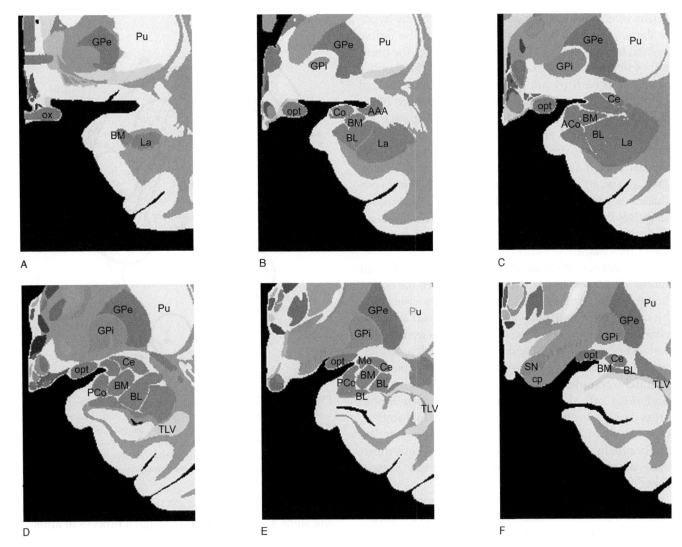

FIGURE 18.6. Series of coronal sections through the human brain to demonstrate the nuclei of the amygdaloid body from rostral to caudal. (**A, B, C, D, E, F** are based on sections 22–32 respectively in Mai, Assheuer and Paxinos, 2004.) **Abbreviations:** AAA, anterior amygdaloid area; ACo, anterior cortical amygdaloid nucleus; BL, basolateral amygdaloid nucleus; BM, basomedial amygdaloid nucleus; Ce, central amygdaloid nucleus; cp, cerebral peduncle; GPe, globus pallidus, external segment; GPi, globus pallidus, internal segment; La, lateral amygdaloid nucleus; Me, medial amygdaloid nucleus; opt, optic tract; ox, optic chiasma; PCo, posterior cortical amygdaloid nucleus; Pu, putamen; SN, substantia nigra; TLV, temporal horn of lateral ventricle. (This figure is reproduced in color in the color plate section.)

illustrates the links in the cyclic limbic path. Starting from the **mamillary body** of the hypothalamus, the **mamillothalamic tract** (bundle of Vicq d'Azyr) projects impulses to the **anterior nuclei** of the **thalamus**. From the anterior nuclear group, **thalamocingulate radiations** project to the **cingulate gyrus**. The cingulum, an association bundle in the cingulate gyrus, proceeds to the **parahippocampal gyrus**. Short association fibers interconnect this gyrus with the hippocampal formation and amygdaloid body.

The **fornix** (Fig. 18.2) completes this cyclic path as it arises in the hippocampal formation and passes posteriorly toward the medial wall of each cerebral hemisphere. Fibers of the fornix leave the caudal aspect of the hippocampus as a flat band of fibers that progressively thickens to form the crus of the fornix. Fibers of the crus then turn upward and forward to pass beneath the corpus callosum toward the medial wall of each hemisphere. In so doing they form the trunk of the fornix. The fibers of the trunk continue anteriorly and descend as the columns of the fornix with some of the fornix fibers passing in front of the anterior commissure, as precommissural fibers, to reach the **septal area** where they terminate. Other postcommissural fornix fibers end in the mamillary bodies of the hypo-thalamus. This prominent fiber bundle of the limbic system includes some 2.7 million fibers (twice the number of fibers in each optic nerve and twice the number of fibers in each pyramidal tract). Septohippocampal connections and hippocamposeptal connections occur through the fornix. Another link in the cyclic path is the **stria terminalis**, an efferent bundle of the amygdaloid body projecting from the superficial (cortical and medial nuclei) and deep nuclei (basolateral and baso-medial nuclei) of the amygdaloid body to the hypo-thalamus and septal area. Discharge and redischarge of impulses throughout this cyclic limbic circuit builds up and enhances these same impulses at various points along the circuit.

In humans, cortical areas have a major influence on limbic structures. In particular, this includes the cortex in the **frontal, parietal,** and **temporal lobes,** connected by way of the **cingulum** with the **subcallosal area,** the **cingulate gyrus,** and the **parahippocampal gyrus.** Such connections are afferent and efferent as well as excitatory and inhibitory. These interrelating systems are infinitely complex as is some of the behavior they underlie. Cortical interaction with and in these limbic areas is significant in humans. Cortical maturation, particularly of association areas, depends on genetics, environment, education, and experience. These influences are noteworthy when considering limbic system function.

There is a balance between the limbic system on the one hand and association areas of the cerebral cortex on the other hand, based on the connections between them. The outcome of the interaction between these areas results in the expression of emotion manifested as observable behavioral responses including autonomic and extrapyramidal responses. Although association areas of the cerebral cortex are intact, they may not function properly unless activated by the cyclic limbic paths.

18.4. SYNAPTIC ORGANIZATION OF HUMAN LIMBIC SYSTEM

A Case Study

Examination of the brain of a 64-year-old male with an eight-year history of progressive neurological deterioration including dementia, impaired cognition and affect, and autonomic dysfunction, demonstrates the synaptic organization of the limbic system in humans (Fig. 18.7). There was diffuse atherosclerosis of the cerebral blood

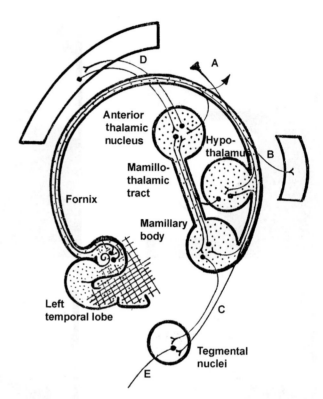

FIGURE 18.7. Anterograde transneuronal degeneration in the human limbic system (from Torch, Hirano and Solomon, 1977). **A,** commissure of the fornix; **B,** septal, preoptic, and parolfactory areas; **C,** brain stem tegmental nuclei; **D,** thalamocortical and corticothalamic fibers between the anterior nucleus and the cingulate gyrus; **E,** tegmentospinal tract and dorsal longitudinal fasciculus.

vessels that led to a cerebral infarction and the resulting signs and symptoms. Through a process of anterograde transneuronal degeneration, the cyclic limbic path was identifiable from the hippocampal formation to the anterior nuclei of the thalamus. The array of autonomic dysfunctions included polydipsia, polyuria, dehydration, tachycardia, hyperventilation, anorexia, severe cachexia, excessive gastric secretions, bowel and bladder incontinence, insomnia, lethargy, stupor, reversal of day/night sleep pattern, paranoia, agitation, mood changes, and decreased libido.

18.5. DESCENDING LIMBIC PATHS

The limbic system exerts its influence on lower levels of the nervous system through its descending efferent paths. These limbic efferents include the **habenulopeduncular tract, dorsal longitudinal fasciculus,** as well as the **mamillotegmental** and **tegmentospinal paths.**

The **habenulopeduncular tract** is a descending epithalamic discharge path to the tegmentum of the midbrain and spinal cord. Its discharge to spinal levels is through the **dorsal longitudinal fasciculus** (DLF). The dorsal longitudinal fasciculus extends from the hypothalamus interrelating olfactory and gustatory impulses involved with feeding, and then discharges such impulses to various preganglionic parasympathetic and motor nuclei of the brain stem. These connections form part of the efferent or output side of the limbic system.

The **dorsal longitudinal fasciculus** is a discrete bundle of fibers in the periaqueductal gray substance of the midbrain. As it descends through the brain stem, it ends in these nuclei: the accessory oculomotor nucleus, trigeminal motor nucleus, facial nucleus, superior and inferior salivatory nuclei, dorsal vagal nucleus, nucleus ambiguus, and hypoglossal nucleus. Through these connections, the dorsal longitudinal fasciculus discharges to nuclei related to eating (that influence salivary and digestive glands) and to nuclei that innervate non-striated muscle of the gastrointestinal tract (for peristalsis). This path also regulates neurons associated with overt emotional expression such as smiling or frowning (facial nucleus), tears and their production (lacrimal nucleus), as well as neurons underlying other accompanying movement and autonomic responses.

The **mamillotegmental** and **tegmentospinal** paths provide for discharge to preganglionic sympathetic neurons in the spinal cord. These are essentially **hypothalamotegmental paths** that travel in the ventrolateral part of the tegmentum of the midbrain discharging impulses from the hypothalamus to nuclei in the spinal cord that regulate body temperature.

18.6. FUNCTIONAL ASPECTS OF THE HUMAN LIMBIC SYSTEM

18.6.1. Emotion

The limbic system participates in aspects of two very complex brain functions: **emotion** and **memory**. The term **emotion** refers to our private world of feeling or mood. Humans are capable of experiencing and remembering a wide range of emotions including such emotions as happiness, fear, anger, sadness, and disgust. The limbic system modulates both internally and externally derived input in order to participate in the **elaboration of emotion** and in the memory of events with emotional significance. The efferent side of emotion and emotional memories is **emotional expression** that has somatic and autonomic components to it. The connections of the amygdala with the autonomic nervous

system are important with regard to these autonomic components of emotional expression. Emotion is what we feel and emotional expression is the demonstration of such feelings that are visible to others as our behavioral responses. This difference between our emotion and the expression of that emotion is likely due to the influences of overlying limbic association areas of the cerebral cortex that are able to monitor the expression of our emotions. It is possible for humans to control, monitor, and mask their emotions such that the expressions others see may not be an accurate reflection of an individual's true state of emotion. So while my students are all smiling at me during a lecture they may in fact be frowning on the inside because of a difficult test that I have just given. Thus in humans the expression of emotion may bear little relationship to the actual feelings or emotions themselves. Obviously, some individuals are very good at controlling or masking the expression of their emotions while others are much less so.

Injuries to various parts of the limbic system become apparent to others through our expression of emotion or behavior. Therefore, violent, aggressive, restless, assaultive, or destructive behaviors often follow limbic system injury. In addition, mood disturbances such as anxiety, fear, guilt, or altered affective tone is likely to be present. Such abnormal behavior usually has a distinctive autonomic character to it including oral tendencies, altered feeding and sexual behavior (either increased or decreased sexual behavior), disturbances of water balance, temperature regulation, altered sleep–wake cycle, respiratory and cardiovascular changes. Neuroendocrine disturbances often accompany limbic system disorders because of the interrelation between the limbic system and the hypothalamus. Those with limbic system disorders often lose the capacity to experience or express emotion.

In addition to an individual's **experience of emotion, memory of emotional events**, and the **expression of emotion**, there is another element, namely the **recognition of emotion**, in us and in others. All of these aspects of emotion likely involve different elements of the human limbic system linked together and functioning in concert. In this regard, we note that the human amygdala plays a vital role in recognizing facial expressions of fear and becomes active during facial expressions of contempt. The involvement of the hippocampal formation in the consolidation and recall of memories, and in this case, the recall of memories associated with negative consequences allows humans to make appropriate responses to aversive and stressful events. On the one hand the appropriate response may be nothing more than the conscious awareness that the situation, while unpleasant or uncomfortable, is not threatening or worthy of action. On the other hand, the appropriate response may consist of behaviors or expressions of emotion with both visceral and motor responses that move us away from threatening circumstances or people. Clearly human emotions are an extraordinarily complex topic as is the anatomical foundation upon which they are built and carried out. Papez may not have been too far off in his description of the limbic system and its component parts as 'a harmonious mechanism which may elaborate the functions of **central emotion** and participate in **emotional expression'**.

18.6.2. Memory

Another important function of the limbic system is that of cortical activation or reinforcement particularly as it relates to the learning of new information, the storage of learned information (**memory**), and the recall of that information. **Declarative memory** is that aspect of human memory involving the storage of facts and experiences that can be consciously discussed, or *declared*. Declarative memory may be **episodic** or **semantic**. **Episodic memory** relates to daily experiences, events or episodes that occur at a specific time and place. (I remember that I had cereal and fruit for breakfast this morning.) These events can be *recent* or *remote* (past). **Semantic memory**, also termed propositional memory, relates to facts and learned general knowledge of the world that is independent of time and place. (I remember that the United States has sent a man to the moon.) There are other interesting aspects of normal memory and of forgetting. For example, as time passes, our memory gradually fades or becomes less accessible, a phenomenon called **transience**. In Alzheimer's disease, impairment of *recent* memories is more likely than impairment of *remote* memories.

Memory is a complex function of neurons interlocked with one another at special regions of convergence and synapse. Disturbances of memory sometimes result from injury to brain areas where information processing occurs, where such information is stored, or in those circuits that underlie the recall of previously stored memories. Such disturbances may also be due to failure of the limbic system to activate these cortical areas. The discharge and redischarge through the hippocampal formation aids in the processing of information and memory consolidation. There is little doubt that the human hippocampus plays an important role in some types of memory, particularly declarative memory that can be consciously discussed or declared

(recalled). In this regard, the hippocampus is said to be part of the medial temporal lobe memory system which is downstream of the ventral ('what') pathway of the visual system (Table 8.2). The activation of cortical areas functioning in the recall of that information strengthens such recall. Confabulation is likely the result of lowered cortical activation with mental confusion combined with a memory deficit yielding poor judgment and the inability to differentiate reality from fiction. Such individuals often confabulate in order to cover up their memory deficit.

18.7. LIMBIC SYSTEM DISORDERS

Injuries to various parts of the limbic system often result in mental confusion and mental disability, disorientation, dementia, memory disturbances, and confabulation. Bilateral hippocampal injuries or injury to structures along the limbic path such as the mamillary bodies or the septal area (ventral part of the subcallosal area), are likely to cause poor activation of cortical areas which then lead to decreased mental activity with confusion.

Two characteristic disorders of the limbic system include **Korsakoff's syndrome** and the **Klüver–Bucy syndrome**. Mental manifestations accompanying Korsakoff's syndrome include amnesia, confabulation, and disorientation. In this syndrome, injuries are commonly in the hippocampi, septal area (there are likely to be hemorrhages into this area or aneurysms of the anterior communicating artery with bleeding and pressure on the area), or in the mamillary bodies (usually pressure on these structures from a tumor such as a craniopharyngioma). Since the hippocampal formation is involved in the recall of recent events, injuries causing Korsakoff's syndrome interfere with the reinforcing function of the limbic system leading to mental confusion and confabulation.

As early as 1888 Brown and Schäfer reported substantial behavioral changes in monkeys after bilateral temporal lobe removal. In 1937, Klüver and Bucy arrived at a group of symptoms that they considered typical in the bilateral temporal lobectomized monkey. These include: (1) **visual agnosia**; (2) **strong oral tendencies**; (3) **hypermetamorphosis**; (4) **profound changes in emotional behavior** or a **complete loss of emotional responses**; (5) **changes in sexual behavior**; and (6) **changes in feeding habits**. **Visual agnosia** refers to the monkey's inability to detect and recognize the character or meaning of an object based on visual criteria alone. The animals placed any objects in their mouths including a reflex hammer, tuning fork, and buttons thus manifesting **strong oral tendencies**. **Hypermetamorphosis** refers to the excessive tendency to take notice of and attend to every visual stimulus. The **changes in sexual behavior** include hypersexual behavior, whereas the changes in **feeding habits** include an increase in food consumption (vociferous appetite) and a lack of discrimination as to what the animal ate (talcum powder, banana peels, inappropriate objects). The first human case of Klüver–Bucy syndrome (1955) followed bilateral temporal lobectomy involving the uncus and the hippocampal formation. Most other cases in humans have involved bilateral injuries with atrophy or destruction of both temporal lobes including the uncus, amygdaloid bodies, and both hippocampi. The visual agnosia in monkeys is probably a psychic blindness in humans. In adult humans, childlike behavior, changes in sexual behavior, including hypersexual behavior, may occur. As an example, one Klüver–Bucy patient threw off her hospital gown, while continually writhing and gyrating mimicking sexual movements. This syndrome is an excellent example of a limbic system disorder in humans.

18.8. INJURIES TO LIMBIC CONSTITUENTS

The functional aspects of limbic constituents are often evident following injury to those structures in humans.

18.8.1. Septal Area

Anterior communicating artery aneurysms impinging on the septal area may result in an overall decrease in cortical functioning, particularly a poverty of associations. One is likely to see altered states of consciousness, changes in mood, personality, and poor judgment caused by decreased cortical functioning.

In humans, cortical control of the hypothalamus takes place through the subcallosal area. The **medial forebrain bundle** connects the septal area and the anterior perforated substance to the hypothalamus. This puts the hypothalamus on the efferent side of the connections for emotional expression. Irritative injuries to the septal area and medial forebrain bundle result in attacks of rage and other emotional changes. Permanent personality changes are also likely to occur. The septal area is involved to some extent in the regulation of water balance and body salts. Irritative injuries in and around the bed nucleus of the anterior commissure

in nonhuman primates often result in edema of the contralateral limb.

18.8.2. Hippocampal Formation

Cyclic discharge and redischarge through the limbic system, including structures of the hippocampal formation, strengthens recently made memories (consolidation) and favors recall of those memories. Since the hippocampal formation stimulates or activates cortical areas involved in the recall of events, it is not surprising that bilateral injuries to the hippocampal formation cause poor activation of the cerebral cortex leading to the inability to remember recent events, decreased mental activity, and mental confusion.

18.8.3. Amygdaloid Body

Injuries to the amygdaloid body yield perversions of appetite with a loss of discriminative eating. In humans, there appears to be an amygdalohypothalamic axis controlling reproductive functions under the regulation of the frontal lobe such that the amygdaloid body participates in the discriminative aspects of sexuality. Stereotaxic amygdalotomy may lead to increased sexual behavior postoperatively. Marked degeneration in the medial, cortical, and medial central amygdaloid nuclei occurs in those with senile dementia suggesting that the amygdaloid body may participate in the behavioral changes that occur in senile dementia. In another case, idiopathic localized amygdaloid degeneration was the cause of the patient's presenile dementia. In this case, changes in character and emotional disorders were predominant over memory disturbance and/or disorientation.

18.8.4. Seizures Involving the Limbic System

Limbic structures in the temporal lobe such as the amygdaloid body, hippocampal formation, and surrounding piriform cortex have a low seizure threshold when electrically stimulated. Therefore, these areas are often responsible for the origin or propagation of complex partial seizures.

18.9. PSYCHOSURGERY OF THE LIMBIC SYSTEM

Various points along the limbic system including the fornix, anterior commissure, thalamus, hypothalamus, cingulum, and amygdaloid body have served as targets for psychosurgical procedures in efforts to treat a variety of behavioral disorders. Over the years, the size and location of targets for psychosurgery have changed greatly. The procedure termed radical frontal leukotomy eventually became a more restricted frontal leukotomy including orbital undercutting then cingulectomy and finally, selective stereotaxic procedures. The latter have also undergone an evolution of sorts as well from inferomedial leukotomy, to subcaudate tractotomy, cingulotomy, amygdalotomy, thalamotomy, hypothala-lmotomy, and most recently, multitarget procedures (involving the amygdaloid body, cingulum, and substantia innominata).

18.9.1. Drug Resistant Epilepsy

A procedure carried out in some individuals with nonfocal general forms of epilepsy of nontemporal lobe origin that are resistant to drug treatment utilizes a bilateral fornico-amygdalotomy involving the fornix, anterior commissure, and amygdaloid body as well as a thalamotomy of the ventral anterior nucleus of the thalamus.

18.9.2. Violent, Aggressive, or Restless Behaviors

The **hypothalamus** was the stereotaxic target in many individuals with violent, aggressive, or restless behaviors with a marked calming effect in 95% of the cases. Bilateral **amygdalotomy** is often carried out in those with bouts of uncontrollable fear, violence, or both and in those with assaultive, destructive, or hyperoral behavior and pyromania. The **posteromedial hypothalamus** was the target in a series of 31 patients ranging from 3 to 15 years who displayed aggressive behavior toward themselves and others. All improved after the cryohypothalamotomy in that they remained calm. Neurosurgeons performing four-target limbic system surgery for violent, aggressive states may carry out bilateral destruction of the **cingulum** and **amygdaloid body**.

18.9.3. Schizophrenia

Multitarget limbic system surgery of the amygdaloid body, substantia innominata, and the cingulum may be useful in patients with schizophrenia. A demonstrable clinical improvement occurred in most of those

desperately ill persons who underwent bilateral destruction of the **amygdaloid body, substantia innominata**, and **cingulum**.

18.9.4. Intractable Pain

Those suffering with intractable pain and drug addiction have found relief following bilateral anterior cingulotomy. In some cases, the amygdaloid body is a target. Cingulotomy has a long history as a method of treatment for psychiatric illness and intractable pain with significant improvement reported in 70–80% of those treated in this manner.

18.9.5. Psychiatric Disorders and Abnormal Behavior

Chronic depression, obsessive-compulsive neuroses, and drug addiction are conditions for which bilateral **basofrontal tractotomy**, cingulotomy, or a combination of both procedures may be considered. These procedures 'favorably modify' such abnormal behaviors as neurosis, schizoaffective behavior, involutional depression, epile psychosis, manic depression, and psychopathic behavior. In the face of these behaviors, the cingulum bundle (bilaterally) is the target with 75% of those treated pronounced well, improved, or greatly improved. Patients with uncontrollable, violent explosions of temper, those easily excitable, or those with unsteadiness of mood have benefited from unilateral or bilateral **medial amygdalotomy**. Over 60% of such patients showed significant improvement with outbursts of temper less destructive than before.

FURTHER READING

Aggleton JP (1992) *The Amygdala: Neurobiological Aspects of Emotion, Memory, and Mental Dysfunction.* Wiley-Liss, New York.

Aggleton JP (1992) *The Amygdala: A Functional Analysis.* 2nd ed. Oxford University Press, New York.

Amaral DG, Cowan WM (1980) Subcortical afferents to the hippocampal formation in the monkey. J Comp Neurol 189:573–591.

Amaral DG, Insausti R, Cowan WM (1987) The entorhinal cortex of the monkey: I. Cytoarchitectonic organization. J Comp Neurol 264:326–355.

Amunts K, Kedo O, Kindler M, Pieperhoff P, Mohlberg H, Shah NJ, Habel U, Schneider F, Zilles K (2005) Cytoarchitectonic mapping of the human amygdala, hippocampal region and entorhinal cortex: intersubject variability and probability maps. Anat Embryol (Berl) 210:343–352.

Andersen P, Morris R, Amaral D, Bliss T, O'Keefe J (2007) *The Hippocampus Book*. Oxford University Press, New York.

Bilir E, Craven W, Hugg J, Gilliam F, Martin R, Faught E, Kuzniecky R (1998) Volumetric MRI of the limbic system: anatomic determinants. Neuroradiology 40:138–144.

Braak H, Braak E, Yilmazer D, Bohl J (1996) Functional anatomy of human hippocampal formation and related structures. J Child Neurol 11:265–275.

Broca P (1878) Anatomie comparée des circonvolutions cérébrales: le grande lobe limbique. Rev Anthropol 1:385–498.

Crosby EC, Schnitzlein HN (1982) *Comparative Correlative Neuroanatomy of the Vertebrate Telencephalon*. Macmillan, New York.

Dalgleish T (2004) The emotional brain. Nat Rev Neurosci 5:583–589.

De Olmos JS (2004) Amygdala. In: Paxinos G, Mai JK, eds. *The Human Nervous System*. 2nd ed. pp 739–868. Elsevier/Academic Press, Amsterdam.

Duvernoy HM (1998) *The Human Hippocampus*. 2nd ed. Springer, Berlin Heidelberg.

El-Falougy H, Benuska J (2006) History, anatomical nomenclature, comparative anatomy and functions of the hippocampal formation. Bratisl Lek Listy 107:103–106.

Gloor P (1997) *The Temporal Lobe and Limbic System*. Oxford University Press, New York.

Herman JP, Ostrander MM, Mueller NK, Figueiredo H (2005) Limbic system mechanisms of stress regulation: hypothalamo-pituitary-adrenocortical axis. Prog Neuropsychopharmacol Biol Psychiatry 29:1201–1213.

Insausti R, Amaral DG, Cowan WM (1987a) The entorhinal cortex of the monkey: II. Cortical afferents. J Comp Neurol 264:356–395.

Insausti R, Amaral DG, Cowan WM (1987b) The entorhinal cortex of the monkey: III. Subcortical afferents. J Comp Neurol 264:396–408.

Isaacson RL (1982) *The Limbic System*. 2nd ed. Plenum, New York.

Kalus P, Slotboom J, Gallinat J, Mahlberg R, Cattapan-Ludewig K, Wiest R, Nyffeler T, Buri C, Federspiel A, Kunz D, Schroth G, Kiefer C (2006) Examining the gateway to the limbic system with diffusion tensor imaging: the perforant pathway in dementia. NeuroImage 30:713–720.

Klinger J, Gloor P (1960) The connections of the amygdala and of the anterior temporal cortex in the human brain. J Comp Neurol 115:333–369.

Krimer LS, Hyde TM, Herman MM, Saunders RC (1997) The entorhinal cortex: an examination of cyto- and myeloarchitectonic organization in humans. Cereb Cortex 7:722–731.

Mai JK, Assheuer J, Paxinos G (2004) *Atlas of the Human Brain*. 2nd ed. Elsevier/Academic Press, Amsterdam.

Mayberg HS (1997) Limbic-cortical dysregulation: a proposed model of depression. J Neuropsychiatry Clin Neurosci 9:471–481.

MacLean PD (1952) Some psychiatric implications of physiological studies on frontotemporal portion of limbic system (visceral brain). Electroencephalogr. Clin. Neurophysiol 4:407–418.

Morgane PJ, Galler JR, Mokler DJ (2005) A review of systems and networks of the limbic forebrain/limbic midbrain. Prog Neurobiol 75:143–160.

Papez, JW (1937) A proposed mechanism of emotion. Arch Neurol and Psychiat 38:725–743.

Rosene DL, Van Hoesen GW (1987) The hippocampal formation of the primate brain: a review of some comparative aspects of cytoarchitecture and connections. In: *Cerebral Cortex*. Jones E, Peters A, eds. 6:345–456. Plenum Press, New York.

Sambataro F, Dimalta S, Di Giorgio A, Taurisano P, Blasi G, Scarabino T, Giannatempo G, Nardini M, Bertolino A (2006) Preferential responses in amygdala and insula during presentation of facial contempt and disgust. Eur J Neurosci 24:2355–2362.

Schnider A (2001) Spontaneous confabulation, reality monitoring, and the limbic system – a review. Brain Res Rev 36:150–160.

Sims KS, Williams RS (1990) The human amygdaloid complex: a cytologic and histochemical atlas using Nissl, myelin, acetylcholinesterase and nicotinamide adenine dinucleotide phosphate diaphorase staining. Neuroscience 36:449–472.

Sitoh YY, Tien RD (1997) The limbic system. An overview of the anatomy and its development. Neuroimaging Clin N Am 7:1–10.

Solodkin A, Van Hoesen GW (1996) Entorhinal cortex modules of the human brain. J Comp Neurol 365:610–617.

Sorvari H, Soininen H, Paljarvi L, Karkola K, Pitkanen A (1995) Distribution of parvalbumin-immunoreactive cells and fibers in the human amygdaloid complex. J Comp Neurol 360:185–212.

Suzuki WA, Amaral DG (1994) Topographic organization of the reciprocal connections between the monkey entorhinal cortex and the perirhinal and parahippocampal cortices. J Neurosci 14:1856–1877.

Sweatt JD (2003) *Mechanisms of Memory*. Academic Press, San Diego.

Terminologia Anatomica: *International Anatomical Terminology* (1998) Federative Committee on Anatomical Terminology. Georg Thieme, Stuttgart.

Torch WC, Hirano A, Solomon S (1977) Anterograde transneuronal degeneration in the limbic system: Clinical-anatomic correlation. Neurology 27:1157–1163.

Yakovlev PI (1948) Motility, behavior and the brain. J Nerv Ment Dis 107:313–335.

Here in this well-concealed spot, almost to be covered by a thumb-nail, lies the very mainspring of primitive existence – vegetative, emotional, reproductive – on which, with more or less success, man, chiefly, has come to super-impose a cortex of inhibitions. The symptoms arising from disturbances of this ancestral apparatus are beginning to stand out in their true significance.

Harvey Cushing, 1929

19

The Hypothalamus

The **hypothalamus** weighs about 4 g and occupies the ventral part of the diencephalon, inferior to the dorsal thalamus and the hypothalamic sulcus (Fig. 19.1). The hypothalamus of one hemisphere is separable from its counterpart in the other hemisphere by the third ventricle (Fig. 19.2). The rostral limit of the hypothalamus is the **lamina terminalis** (Figs 19.1, 19.3), a thin membrane forming the rostral wall of the third ventricle. The lamina terminalis interconnects the optic chiasma with the anterior commissure (Fig. 19.3). The hypothalamus extends anteriorly into the **preoptic area**, a telencephalic structure. The **preoptic area** is on either side of the third

ventricle, anterior to a line extending from the ipsilateral **interventricular foramen** to the anterior surface of the optic chiasma (Fig. 5.1). The caudal limit of the hypothalamus follows an imaginary line drawn from the posterior commissure to the caudal end of the mamillary body (Fig. 19.1). The lateral boundary of the hypothalamus is the subthalamus while its ventral border, forming the floor of the third ventricle, includes many external surface features such as the optic chiasma, infundibular stalk, tuber cinereum, and mamillary bodies (Fig. 19.3). These external features are useful landmarks for determining anterior-posterior levels of the hypothalamus.

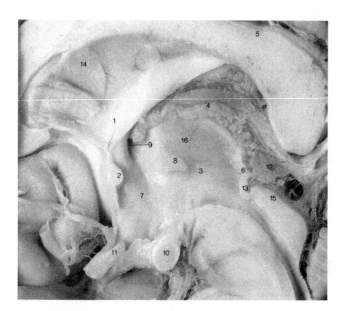

FIGURE 19.1. Medial surface of the human brain showing the position of the thalamus, hypothalamus, and epithalamus (from England and Wakely, 1991). **Key:** 1, anterior column of fornix; 2, anterior commissure; 3, cavity of third ventricle; 4, choroid plexus of third ventricle; 5, corpus callosum; 6, epithalamus; 7, hypothalamus; 8, interthalamic adhesion; 9, interventricular foramen; 10, mamillary body; 11, optic chiasma; 12, pineal gland; 13, posterior commissure; 14, septum pellucidum; 15, superior colliculus; 16, thalamus. (This Figure is reproduced in color in the color plate section).

19.1. HYPOTHALAMIC REGIONS

There are various ways of dividing the hypothalamus from anterior to posterior. All such schemes end up with a series of hypothalamic levels, regions, or areas each containing named nuclei. Based on the presence of **external features** visible along its ventral border

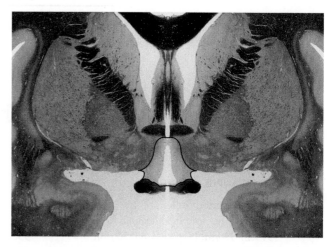

FIGURE 19.2. The human hypothalamus (outlined) in relation to the third ventricle (from Mai, Assheuer and Paxinos, 2004).

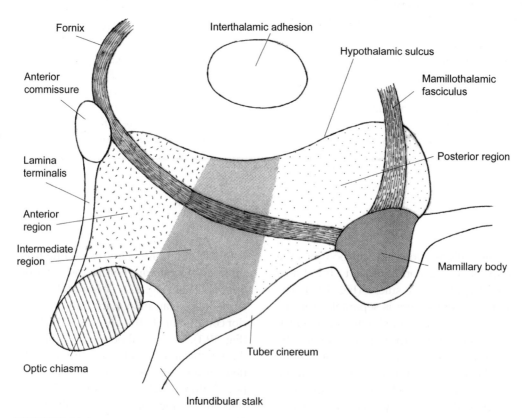

FIGURE 19.3. Medial surface of the human hypothalamus illustrating three regions of the hypothalamus and various related structures.

(optic chiasma, infundibular stalk, tuber cinereum, and mamillary bodies), the hypothalamus is externally divisible into anterior to posterior regions. Since the tuber cinereum merges anteriorly into the infundibular stalk of the pituitary gland, most authors combine the infundibular and tuberal regions into a single cone-shaped, tuberal region. Thus based on these external features, three hypothalamic regions are recognizable from anterior to posterior namely a **chiasmal**, **tuberal**, and **mamillary region**. The chiasmal region lies superior and anterior to the optic chiasma whereas the mamillary region is denominated by the mamillary bodies (Figs 19.1, 19.3). The tuberal region lies between the chiasmal region anteriorly and the mamillary region posteriorly.

Using internal features related to the hypothalamus, the *Terminologia Anatomica* (1998) defines five regions and one nuclear group (the posterior nucleus of the hypothalamus) that it does not assign to any regions. These regions include an **anterior hypothalamic region**, a **dorsal hypothalamic region**, an **intermediate hypothalamic region**, a **lateral hypothalamic region**, and a **posterior hypothalamic region**. In all, there are some thirty named nuclear groups in these five regions of the human hypothalamus (Table 19.1). Three of these regions, the **anterior** (chiasmal), **intermediate** (tuberal), and **posterior** (mamillary), are identifiable on Fig. 19.3.

19.2. HYPOTHALAMIC ZONES

In addition to these internal anterior-posterior regions, there are three internal longitudinal **zones** from medial to lateral identifiable in the human hypothalamus. These include a zone nearest the third ventricle called the **periventricular zone**, a **medial zone** adjoining the periventricular zone, and a **lateral zone** outside or lateral to the medial zone (Fig. 19.4). Most hypothalamic nuclei are in the medial hypothalamic zone. The lateral hypothalamic zone, extending throughout the length of the hypothalamus, consists of fiber bundles and some scattered neurons. Most of the former are fibers that pass through the lateral hypothalamic zone.

Any nuclear group or fiber path of the hypothalamus may be located according to its presence in a particular anterior-posterior region and in a medial-lateral zone. Using this method, the paraventricular nucleus would be located in the medial zone of the anterior (chiasmal) region. The **periventricular zone** of the hypothalamus has small-to medium-sized neurons interspersed between finely myelinated and nonmyelinated fibers. These fibers form the **diencephalic periventricular system** (DPS). This system of fibers connects caudal levels of the hypothalamus with the frontal cortex through the medial dorsal thalamic nucleus.

TABLE 19.1. Nuclei of the human hypothalamus

Terms in English	Abbreviation	Terms in English	Abbreviation
1. Anterior hypothalamic area or region	AHA	Arcuate nucleus (infundibular)	INF
Anterior hypothalamic nucleus	AHN	Periventricular nucleus	PVN
Anterior periventricular nucleus	Ape	Posterior periventricular nucleus	PPVN
Interstitial nuclei of anterior hypothalamus	INAH	Retrochiasmatic area (or region)	RCA
Lateral preoptic nucleus	LPN	Lateral tuberal nuclei	LTuN
Medial preoptic nucleus	MPN	Ventromedial nucleus	VMN
Median preoptic nucleus	MnPO		
Paraventricular nucleus	PVN		
Periventricular preoptic nucleus	Pe	**4. Lateral hypothalamic area**	LHA
Suprachiasmatic nucleus	SCN	Preoptic area	POA
Supra-optic nucleus	SON	Lateral tuberal nuclei	LTuN
Dorsolateral part	SON$_{de}$	Perifornical nucleus	PeF
Dorsomedial part	SON$_{dm}$	Tuberomamillary nucleus	TMN
Ventromedial part	SON$_{vm}$		
		5. Posterior hypothalamic area or region	PHA
2. Dorsal hypothalamic area or region	DHA	Dorsal premamillary nucleus	PMD
Dorsomedial nucleus	DMN	Lateral nucleus of the mamillary body	LMN
Endopeduncular nucleus	EPN	Medial nucleus of the mamillary body	MMN
Nucleus of the ansa lenticularis	NAL	Supramamillary nucleus	SMN
		Ventral premamillary nucleus	PMV
3. Intermediate hypothalamic area or region	IHA		
Dorsal nucleus	Do	**6. Posterior nucleus of hypothalamus**	PHN
Dorsomedial nucleus	DMN		

From *Terminologia Anatomica*, 1998.

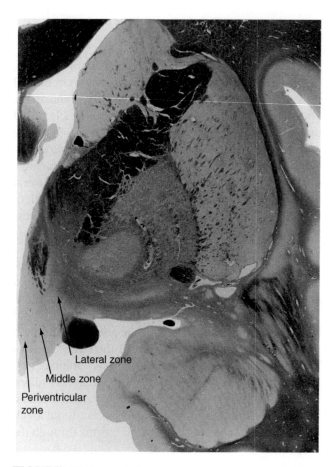

FIGURE 19.4. Medial to lateral zones of the hypothalamus; a narrow periventricular zone, a middle zone, and a lateral zone. Most of the nuclei are in the middle zone.

19.3. HYPOTHALAMIC NUCLEI

19.3.1. Anterior Hypothalamic Region

The anterior hypothalamic region contains nine named nuclei (Table 19.1). At the chiasmal level and over the lateral border of the optic tract are two large, magnocellular, neurosecretory nuclei – the **supra-optic nucleus** (SON) and the **paraventricular nucleus** (PVN) (Figs 19.5, 19.6). These nuclei, characterized by a conspicuous and dense network of capillaries, are responsible for the synthesis of oxytocin and vasopressin. Using chemical markers some five subnuclei are identifiable in the human paraventricular nucleus. Also at this level is an irregular mass of neurons called the **anterior hypothalamic nucleus** (Fig. 19.5). In experimental animals, the anterior hypothalamic nucleus is involved in the autonomic regulation (inhibitory) of arterial blood pressure and heart rate. The **suprachiasmatic nucleus (SCN)** (Fig. 19.6) is also present at this level. As the circadian pacemaker in the human brain

for sleep–wake states, the SCN exerts 'gentle' control over the sleep–wake cycle. This control is termed 'gentle' in the sense that humans are able to override the control mechanisms of this nucleus and nap during the day but also stay awake at night. GABA is the main neurotransmitter in the suprachiasmatic nucleus.

Almost 30 years ago, Gorski and his co-workers (1978) provided evidence for a morphological sex difference within the medial preoptic area of the rat brain. A few years later Swaab and Fliers (1985) identified a sexually dimorphic nucleus in the human brain that was 2.5 times larger in the male brain. The area in question likely corresponds to the interstitial nuclei of the anterior hypothalamus (INAH) under the present terminology (Table 19.1) which includes four cell groups (INAH 1–4), (Fig. 19.7) and the nucleus in question is likely to be INAH-1. Quantitative analysis of the volume of the four cell groups in the INAH of the brains of 22 humans by another group of investigators (Allen et al., 1989) found that two of the nuclei (INAH-1 and INAH-4, as well as the supra-optic nucleus) were not sexually dimorphic. They did observe gender-related differences in the other two nuclei in the INAH. These authors found INAH-3 to be 2.8 times larger in the male brain than in the female brain irrespective of age and INAH-2 to be twice as large in males. A third group of investigators (Byne et al., 2000) revisited this issue a few years ago (2000) and undertook a systematic examination of all four nuclei belonging to the interstitial nuclei of the anterior hypothalamus (Fig. 19.7) examining sexual variation in volume, neuronal number, and neuronal size. Only INAH-3 was found to occupy a significantly greater volume and contain more neurons in males than in females. This latter group has subsequently examined the INAH with regard to sex, sexual orientation, and HIV status and again found that only INAH-3 has more neurons and occupies a greater volume in presumed heterosexual males than females but tends to have a smaller volume in homosexual men. These studies suggest that, based on available data, sexual orientation cannot be reliably predicted on the basis of INAH-3 volume.

19.3.2. Dorsal Hypothalamic Region

The dorsal hypothalamic region contains three named nuclei (Table 19.1). These include the dorsomedial nucleus (Figs 19.5, 19.6), endopeduncular nucleus and the nucleus of the ansa lenticularis. The dorsomedial nucleus is also in the intermediate hypothalamic region.

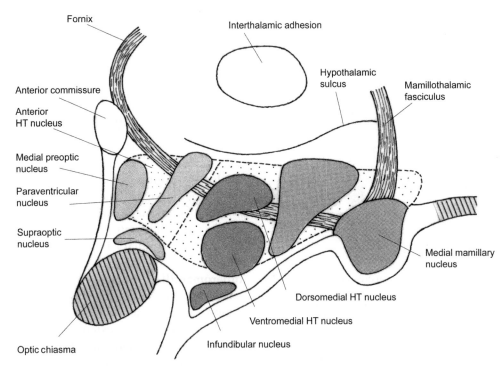

FIGURE 19.5. Medial surface of the human hypothalamus illustrating the major nuclei of the hypothalamus and their relation to one another.

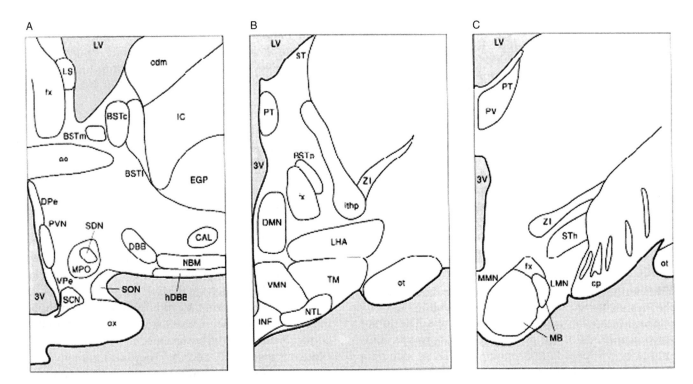

FIGURE 19.6. Coronal sections (**A**, **B**, **C**) through the human hypothalamus (from Swaab, 2003). **Abbreviations:** 3V, third ventricle; ac, anterior commissure; BST, bed nucleus of the stria terminalis, (c = centralis; m = medialis; l = lateralis; p = posterior); cp, cerebral peduncle; DPe, periventricular nucleus dorsal zone; fx, fornix; hDBB, horizontal limb of the diagonal band of Broca; INF, infundibular nucleus; MB, mamillary body i.e. MMN, medial mamillary nucleus + LMN, lateral mamillary nucleus; NBM, nucleus basalis of Meynert; OT, optic tract; OX, optic chiasma; PVN, paraventricular nucleus; SCN, suprachiasmatic nucleus; SDN, sexually dimorphic nucleus of the preoptic area; SON, supra-optic nucleus; VMN, ventromedial hypothalamic nucleus; VPe, periventricular nucleus ventral zone.

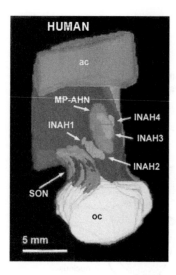

HUMAN

ac

MP-AHN

INAH1   INAH4

INAH3

INAH2

SON

oc

5 mm

FIGURE 19.7. Three-dimensional reconstruction of the medial preoptic-anterior hypothalamic continuum of the human hypothalamus. INAH3 occupies a larger volume and more neurons in presumed heterosexual men than women but there was no difference in the number of neurons in this nucleus based on sexual orientation. No sex difference in volume was detected for any other INAH (from Byne et al., 2001). **Abbreviations:** ac, anterior commissure; INAH, interstitial nucleus of the anterior hypothalamus; MP-AHN, medial preoptic-anterior hypothalamic nucleus; oc, optic chiasma; SON, supra-optic nucleus. (This Figure is reproduced in color in the color plate section).

19.3.3. Intermediate Hypothalamic Region

The intermediate hypothalamic region contains eight named nuclei (Table 19.1). This includes the **ventromedial nucleus** (Figs 19.5, 19.6), an ovoid mass rostral and dorsal to the mamillary bodies, and the **dorsomedial nucleus**. The dorsomedial nucleus is rostral and dorsal to the ventromedial nucleus. The human ventromedial nucleus contains a dense network of somatostatin cells and fibers in adults and infants. Opiod gene expression, substance P, and NADPH diaphorase characterize the ventromedial nucleus. Oxytocin binding sites are present in the ventromedial nucleus as well. A thin cell-sparse shell surrounds this nucleus with its two parts: a larger ventrolateral part and a smaller dorsomedial part. The ventromedial nucleus is involved in feeding behavior. Lesions of the ventromedial nucleus cause increased appetite obesity in humans. The dorsomedial nucleus of the human hypothalamus is composed of two populations of neurons. These include a larger but diffuse subdivision and a smaller but compact subdivision. The dorsomedial nucleus is involved in the regulation of homeostasis as it relates to feeding, reproduction, and sexual behavior. The human **arcuate** or **infundibular nucleus** (INF) (Figs 19.5, 19.6) contains neurons expressing estrogen receptors. It is likely

that these neurons are involved in reproduction and sexual behavior. In addition to these two functions, the arcuate nucleus likely functions in feeding behavior as well as growth and metabolism.

19.3.4. Lateral Hypothalamic Area

The lateral hypothalamic region in humans contains four nuclei (Table 19.1). One of these, the **perifornical nucleus** surrounds the fornix, a prominent bundle of fibers originating in the hippocampus and terminating, in part, in the mamillary bodies. The **lateral tuberal nuclei** are small, scattered nuclei that lie at the rostral end of the mamillary bodies. These small nuclei are involved in feeding behavior and metabolism. The **tuberomamillary nuclei** (TM) (Fig. 19.6) occupy the lateral hypothalamic area and posterior hypothalamic region. These magnocellular neurons surround the fornix and the caudal extent of the mamillary bodies. Neurons of the tuberomamillary nuclei are part of a histamine-containing neuronal system in the human brain. These histamine-containing neurons in the tuberomamillary nuclei are colocaized with GABA. The tuberomamillary nuclei show an age-related accumulation of lipofuscin pigment in humans. These nuclei participate in a wide variety of functions including arousal, sleep wakefulness, feeding, brain metabolism, and regulation of blood pressure. Reduction in the histamine content of neurons in the tuberomamillary nuclei occurs in Alzheimer's disease.

19.3.5. Posterior Hypothalamic Region

The posterior hypothalamic region in humans contains five named nuclei in humans (Table 19.1). The mamillary bodies are a conspicuous external feature of the human hypothalamus that distinguishes the location of the posterior hypothalamic region. Internally the mamillary bodies include a large **medial mamillary nucleus** (MMN) that constitutes its main mass, and a smaller **lateral mamillary nucleus** (LMN) (Figs 19.5, 19.6). The mamillary bodies form the caudal boundary of the hypothalamus. A capsule of fibers surrounds the mamillary bodies. Neurons in the human medial mamillary nucleus express substance-P mRNA. They also contain monamine oxidase. The **mamillotegmental** and its continuation as the **tegmentospinal** path provides for discharge to preganglionic sympathetic neurons in the spinal cord. The mamillotegmental path originates in the medial mamillary nucleus. The mamillary nuclei are an important link in the limbic system (mamillothalamic tract, anterior nucleus of the thalamus, cingulate gyrus and its underlying cingulum and the hippocampus) playing a role in memory and emotion.

19.3.6. Posterior Nucleus of the Hypothalamus

In addition to the mamillary bodies and other named nuclei (Table 19.1), the mamillary level also contains the **posterior nucleus of the hypothalamus** located dorsal to the mamillary bodies and extending dorsally to the hypothalamic sulcus and anteriorly to the tuberal region. The posterior nucleus of the hypothalamus is susceptible to age-related degenerative changes.

19.4. FIBER CONNECTIONS

A myriad of fiber connections exist with the hypothalamus. Only a few of these are described in the following paragraphs.

19.4.1. Medial Forebrain Bundle

The medial forebrain bundle has ascending and descending fibers connecting the hypothalamus with the septal area, anterior perforated substance, and medially lying olfactory areas. Olfactory impulses in the medial olfactory tract synapse in the septal region and the anterior perforated substance. These impulses travel over the medial forebrain bundle and project to the preoptic and anterior hypothalamic nuclei.

19.4.2. Stria Terminalis

The **stria terminalis** interconnects the preoptic and anterior hypothalamic nuclei with the amygdaloid body. From the superficial (cortical and medial nuclei) and deep nuclei (basolateral and basomedial nuclei) of the amygdaloid body, the stria terminalis sends its fibers to the hypothalamus and septal area.

19.4.3. Fornix

The **fornix** is a major fiber bundle on the medial aspect of each cerebral hemisphere connecting the hippocampal formation with the anterior hypothalamic area and the medial mamillary nucleus as well as the lateral hypothalamic region. Arising in the hippocampal formation, fibers of the fornix pass to the medial wall of each cerebral hemisphere. Some of the fornix fibers pass in front of the anterior commissure, as precommissural fibers, to reach the **septal area** where they terminate. Other postcommissural fornix fibers end in the mamillary bodies of the hypothalamus. This prominent fiber bundle includes some 2.7 million fibers (twice the number of fibers in each optic nerve and twice the number of fibers in each pyramidal tract). Septohippocampal connections and hippocamposeptal connections occur through the fornix as well.

19.4.4. Diencephalic Periventricular System (DPS)

The **diencephalic periventricular system** interconnects the hypothalamus with the thalamus. Visceral impulses from the hypothalamus that reach the medial dorsal nucleus of the thalamus correlate with somatic impulses from surrounding thalamic nuclei. This visceral and somatic information can then be associated with impulses from frontal and prefrontal cortical areas. Such information enters consciousness at the cortical level as a feeling or 'affective tone' which colors our responses and contributes to our personality and behavior.

19.4.5. Dorsal Longitudinal Fasciculus

One of two common discharge systems from the hypothalamus to the brain stem is the **dorsal longitudinal fasciculus** (DLF). This fiber bundle originates from much of the hypothalamic gray matter and as a discrete bundle runs through the periaqueductal gray substance of the brain stem. The DLF synapses with preganglionic parasympathetic nuclei and motor nuclei of the brain stem. In particular, the dorsal longitudinal fasciculus distributes to the accessory oculomotor nucleus, trigeminal motor nucleus, facial nucleus, superior and inferior salivatory nuclei, dorsal vagal nucleus, nucleus ambiguus, and the hypoglossal nucleus. Some ascending gustatory fibers are in the dorsal longitudinal fasciculus as well. Such impulses discharge to the mamillary body after relay in the ventral tegmental gray matter of the brain stem. Because of the relationship between the hypothalamus and these brain stem motor nuclei, irritation of the dorsal longitudinal fasciculus often leads to vomiting accompanied by nausea, poor appetite, weak spells, and pallor but not associated with headache. Such irritation often results from increased intracranial pressure.

19.4.6. Anterior and Posterior Hypothalamotegmental Tracts

From anterior and posterior hypothalamic areas arise fibers that pass through the lateral hypothalamic zone and make their way to the tegmentum of the midbrain. In particular, these fibers project to a region below the red nucleus (subrubral gray matter), to the reticular g-ray matter at inferior collicular levels, and that at pontine levels. From midbrain and pontine levels two paths descend through the brain stem, diverge from one another, and form two separate systems in the spinal cord. The first of these is the **ventral reticulospinal tract** – a crossed and uncrossed system of fibers descending through the brain stem to end bilaterally on preganglionic neurons of the intermediolateral nucleus at T1–L3 levels of the spinal cord. The ventral reticulospinal tract synapses with sympathetic neurons that innervate sweat glands of the upper and lower limbs.

The second path from midbrain and pontine levels is the **lateral reticulospinal tract** whose fibers descend through the lateral part of brain stem into the lateral funiculus of the spinal cord. This bundle of fibers ends on cells of the ipsilateral, **intermediolateral nucleus** at T1–T4 levels of the spinal cord. These sympathetic neurons pass to sympathetic neurons that innervate the sweat glands on the same side of the face.

19.4.7. Pallidohypothalamic Tract

The pallidohypothalamic tract is a small bundle of fibers that connect the hypothalamus with the globus pallidus.

19.4.8. Mamillothalamic Tract

The mamillothalamic tract is a prominent system of fibers connecting the mamillary body on the one hand with the anterior nuclei of the thalamus on the other. This system is a major link in the cyclic limbic path.

19.4.9. Hypothalamo-hypophyseal Tract

The **hypothalamo-hypophyseal tract** takes origin from the supra-optic and paraventricular nuclei of the hypothalamus. This tract passes over the optic chiasma and through the infundibular stalk before ending in the neurohypophysis (posterior lobe of the pituitary gland). Some 100,000 nonmyelinated fibers exist in the human hypothalamo-hypophyseal tract. This tract is functionally associated with the controlled release of vasopressin and oxytocin from the neurohypophysis.

19.4.10. Vascular Connections

This is a vascular connection between the hypothalamus and the hypophysis (pituitary gland) termed the **hypothalamo-hypophyseal portal system**. This system supplies blood to the adenohypophysis (anterior lobe of the pituitary gland). The internal carotid artery supplies vessels to the upper part of the pituitary stalk. Here a capillary bed is formed which drains into portal vessels. At the lower end of the stalk is another capillary bed that derives blood from the inferior hypophyseal arteries. Hence, a vascular connection exists between the hypothalamus and the anterior lobe of the pituitary. Releasing factors produced by the hypothalamus make their way into the anterior lobe of the pituitary by means of this vascular route. Such releasing factors control the output of hormones from the anterior lobe of the pituitary and thereby control the ultimate release of hormones from target organs. The hypothalamohypophyseal portal system is capable of regeneration if injured.

19.5. FUNCTIONS OF THE HYPOTHALAMUS

19.5.1. Water Balance – Water Intake and Loss

The hypothalamus participates in the regulation of water balance including water intake and loss. Injury to the hypothalamohypophyseal path and its connections to the neurohypophysis (posterior lobe of the pituitary gland) results in polydipsia and polyuria including the excretion of large amounts of dilute urine. This condition, called **diabetes insipidus** is due to the absence of production of ADH (vasopressin) from the supra-optic nucleus of the hypothalamus. Normally this hormone, produced in the hypothalamus and acting in the kidney, greatly increases the permeability of the renal tubules to water. The absence of ADH thus interferes with renal tubular reabsorption of water particularly in the distal tubules. Some patients with tumors of the hypothalamic region exhibit no thirst nor will they drink water unless strongly reminded to do so. Such an absence of thirst often necessitates prolonged intravenous fluid therapy.

Injuries to the septal area in and around the anterior commissure, or bilateral amygdaloid damage in nonhuman primates produces edema of the limb contralateral to the side of the injury. Such edema is due to a nonendocrine increase in capillary permeability that leads to the buildup of fluid in that limb. In humans, **preoptic pulmonary edema** may result from destructive injuries in the preoptic area. A final aspect of this water balance is the observation that arising from the hypothalamus and closely paralleling the dorsal longitudinal fasciculus is a path involved in bladder control.

19.5.2. Eating – Food Intake

The hypothalamus participates in basic aspects of eating for survival that involves the correlation of both olfactory and gustatory impulses. Olfactory impulses make their way to the hypothalamus by way of the medial forebrain bundle. Ascending gustatory impulses make their way into the hypothalamus over ascending fibers of the dorsal longitudinal fasciculus. Discharge from the hypothalamus is over the dorsal longitudinal fasciculus to preganglionic parasympathetic and motor nuclei of the brain stem that are associated with feeding. In this way, the hypothalamus is likely to influence salivary glands, glands of the gastrointestinal tract, and nonstriated muscle of that tract along with the muscles of mastication.

Children with injuries of the tuberal region often refuse to eat, becoming extremely emaciated. A recent review of hypothalamic tumors in children has shown an abnormality of body weight occurring in 76% of these children. More than 50% of these children with hypothalamic tumors showed failure to gain weight.

Anorexia nervosa is associated with an aversion to food, weight loss, and malnutrition that may lead to chronic starvation. This life-threatening condition is probably a combination of an emotional and an endocrine disturbance, perhaps a hypothalamic-pituitary dysfunction. Hypothalamic tumors in humans may yield anorexia.

19.5.3. Temperature Regulation

The **anterior hypothalamic area** probably regulates against heat in humans that occurs by sweating. The underlying physiological mechanism appears to be vasodilatation of blood vessels. One hazard of surgical procedures in and about the anterior hypothalamus is possible injury to this region causing increased internal body temperature. Irritative injuries often result in profuse sweating.

The **posterior hypothalamic area** is probably involved in regulating against cold. On exposure to cold, shivering and vasoconstriction result. Parkinson's patients have degenerative changes in the **posterolateral hypothalamus**. Such changes help explain the absence of shivering, impaired peripheral vasodilatation, and deficient sudomotor function in response to heat often accompanying Parkinson's disease.

19.5.4. Autonomic Regulation

The hypothalamus influences visceral activities through descending connections from the hypothalamus such as the **dorsal longitudinal fasciculus** (DLF) and the **anterior and posterior hypothalamotegmental paths**. The dorsal longitudinal fasciculus influences preganglionic parasympathetic neurons. Stimulation of this path causes decreased pupillary size, increased lacrimation, increased vasodilatation, increased salivation, decreased heart rate, and increased peristalsis. The anterior and posterior hypothalamotegmental paths influence the reticulospinal paths that then influence sympathetic neurons of the spinal cord. Hence, stimulation of these hypothalamotegmental paths result in increased sweating, increased pupillary size, increased heart rate, and decreased peristalsis.

19.5.5. Emotional Expression

Emotion is an extremely complex function of the human nervous system. The expression of that emotion takes place over the anatomical connections from various regions of the nervous system. The cerebral cortex, particularly the frontal lobe, and the limbic system both have inhibitory and regulatory functions over one another and over the hypothalamus and its output. Discharge and redischarge throughout the limbic system likely results in cortical activation and the arousal of emotion. Thalamocortical connections from the medial dorsal nucleus to the frontal lobe yield a subjective feeling or character to a given situation. This coupled with memory of past events influences our emotions. The expression of that emotion, however, is likely to be voluntary and controlled as manifested by our appearance, manner, words, actions, or facial expression. Under stress or under other extenuating circumstances, such emotional expression is likely to be involuntary and out of control. Extrapyramidal and autonomic responses, increased heart rate, pupillary changes, increased sweating, and respiration typify involuntary emotional expression.

A fascinating aspect of emotional expression is the observation of patients with fits of laughter caused by tumors of the floor of the third ventricle. The structural substrate for such fits of laughter is likely to be the hypothalamus. In a patient operated on under local anesthesia, a burst of laughter was provoked every time blood was sponged from the floor of the third ventricle.

19.5.6. Wakefulness and Sleep (Biological Rhythms)

Many factors are involved in the establishing levels of consciousness commonly referred to as wakefulness and sleep. Such factors include the enormous driving of the cerebral cortex over the ascending reticular activating system and the influence of cyclic limbic activation of cerebral cortical areas. There are complex interconnections between these areas of the nervous system, the hypothalamus, and the hippocampal formation. Our alertness and awareness varies during the day from a heightened state of acute awareness and emotional excitement to profound sleep. Our state of consciousness includes accompanying autonomic responses such as changes in respiration, heart rate, or body temperature. Injuries to the posterior hypothalamus including the region lateral and posterior to the mamillary bodies may result in pathological sleep. Such pathological sleep often becomes **narcolepsy**. Narcolepsy is a term used to describe attacks of irresistible sleep in the day. Injuries that involve a considerable part of the tegmentum of the midbrain result in a profound coma because of the interruption of ascending multisynaptic reticular paths as well as descending hypothalamotegmental and dorsal longitudinal fasciculus paths. One interesting aspect of the hypothalamus and sleep–wake disorders has been the recent discovery (1998) of the hypocretins (also called the orexins). These peptides, produced in the lateral hypothalamus, are deficient in the CSF of patients with sporadic narcolepsy-cataplexy and less commonly in familial narcolepsy. Moreover patients with narcolepsy have no or barely detectable hypocretin containing neurons in their hypothalamus. The exact role of these peptides in sleep–wake disorders is unclear.

19.5.7. Control of the Endocrine System

The hypothalamus controls the endocrine system through the release of hormones such as ADH and oxytocin, from the neurohypophysis (posterior lobe of the pituitary gland). Releasing factors produced in the hypothalamus and transported to the anterior lobe of the pituitary gland act there to yield hormone release. There are nine releasing factors associated with the hypothalamus. These releasing factors cause an increased synthesis or secretion of anterior pituitary hormones with the exception of certain substances that are inhibitory factors. Such inhibitory factors decrease synthesis or secretion of anterior pituitary hormones.

19.5.8. Reproduction

The hypothalamus participates in reproduction through its interrelationship with the hypophysis. Through this relationship, the hypothalamus influences the development of the gonads, ovulation, implantation, lactation, fetal growth and development, and parturition. As noted earlier, the human **arcuate** or **infundibular nucleus** (INF) contains neurons expressing estrogen receptors. It is likely that these neurons are involved in reproduction and sexual behavior.

FURTHER READING

Allen LS, Hines M, Shryne JE, Gorski RA (1989) Two sexually dimorphic cell groups in the human brain. J Neurosci 9:497–506.

Baumann CR, Bassetti CL (2005) Hypocretins (orexins) and sleep–wake disorders. Lancet Neurol 4:673–682.

Beersma DG, Gordijn MC (2006) Circadian control of the sleep–wake cycle. Physiol Behav 90:190–195.

Braak H, Braak E (1987) The hypothalamus of the human adult: chiasmatic region. Anat Embryol (Berl) 175:315–330.

Braak H, Braak E (1992) Anatomy of the human hypothalamus (chiasmatic and tuberal region). Prog Brain Res 93:3–14.

Byne W (1998) The medial preoptic and anterior hypothalamic regions of the rhesus monkey: cytoarchitectonic comparison with the human and evidence for sexual dimorphism. Brain Res 793:346–350.

Byne W, Lasco MS, Kemether E, Shinwari A, Edgar MA, Morgello S, Jones LB, Tobet S (2000) The interstitial nuclei of the human anterior hypothalamus: an investigation of sexual variation in volume and cell size, number and density. Brain Res 856:254–258.

Byne W, Tobet S, Mattiace LA, Lasco MS, Kemether E, Edgar MA, Morgello S, Buchsbaum MS, Jones LB (2001) The interstitial nuclei of the human anterior hypothalamus: an investigation of variation with sex, sexual orientation, and HIV status. Horm Behav 40:86–92.

Czeisler CA, Dijk DJ (2001) Human circadian physiology and sleep–wake regulation. In: *Handbook of Behavioral. Neurobiology: Circadian Clocks.* Takahashi S, Turek FW,

Moore RY, eds. 12:183–213. Kluwer Academic/Plenum Publishers, New York.

de Lecea L, Sutcliffe JG (2005) The hypocretins and sleep. FEBS J 272:5675–5688.

Gorski RA, Gordon JH, Shryne JE, Southam AM (1978) Evidence for a morphological sex difference within the medial preoptic area of the rat brain. Brain Res 148:333–346.

Guillemin R (2005) Hypothalamic hormones a.k.a. hypothalamic releasing factors. J Endocrinol 184:11–28.

Hestiantoro A, Swaab DF (2004) Changes in estrogen receptor-alpha and -beta in the infundibular nucleus of the human hypothalamus are related to the occurrence of Alzheimer's disease neuropathology. J Clin Endocrinol Metab 89:1912–1925.

Hofman MA (1997) Lifespan changes in the human hypothalamus. Exp Gerontol 32:559–575.

Hofman MA, Goudsmit E, Purba JS, Swaab DF (1990) Morphometric analysis of the supraoptic nucleus in the human brain. J Anat 172:259–270.

Ishunina TA, Kamphorst W, Swaab DF (2003) Changes in metabolic activity and estrogen receptors in the human medial mamillary nucleus: relation to sex, aging and Alzheimer's disease. Neurobiol Aging 24:817–828.

Ishunina TA, Unmehopa UA, van Heerikhuize JJ, Pool CW, Swaab DF (2001) Metabolic activity of the human ventromedial nucleus neurons in relation to sex and ageing. Brain Res 893:70–76.

Ishunina TA, van Heerikhuize JJ, Ravid R, Swaab DF (2003) Estrogen receptors and metabolic activity in the human tuberomamillary nucleus: changes in relation to sex, aging and Alzheimer's disease. Brain Res 988:84–96.

King BM (2006) The rise, fall, and resurrection of the ventromedial hypothalamus in the regulation of feeding behavior and body weight. Physiol Behav 87:221–244.

Kiss A, Mikkelsen JD (2005) Oxytocin – anatomy and functional assignments: a minireview. Endocr Regul 39:97–105.

Koutcherov Y, Mai JK, Ashwell KW, Paxinos G (2000) Organization of the human paraventricular hypothalamic nucleus. J Comp Neurol 423:299–318.

Koutcherov Y, Mai JK, Ashwell KW, Paxinos G (2002) Organization of human hypothalamus in fetal development. J Comp Neurol 446:301–324.

Koutcherov Y, Mai JK, Ashwell KW, Paxinos G (2004) Organisation of the human dorsomedial hypothalamic nucleus. NeuroReport 15:107–111.

Kruijver FP, Balesar R, Espila AM, Unmehopa UA, Swaab DF (2002) Estrogen receptor-alpha distribution in the human hypothalamus in relation to sex and endocrine status. J Comp Neurol 454:115–139.

LeVay S (1991) A difference in hypothalamic structure between heterosexual and homosexual men. Science 253:1034–1037.

Saper CB (2004) Hypothalamus. In: Paxinos G, Mai JK, eds. The Human Nervous System. 2nd ed. pp 513–550. Elsevier/Academic Press, Amsterdam.

Saper CB, Scammell TE, Lu J (2005) Hypothalamic regulation of sleep and circadian rhythms. Nature 437:1257–1263.

Swaab DF (1995) Ageing of the human hypothalamus. Horm Res 43:8–11.

Swaab DF (1995) Development of the human hypothalamus. Neurochem Res 20:509–519.

Swaab DF (1997) Neurobiology and neuropathology of the human hypothalamus. In: Handbook of Chemical Neuroanatomy. Björklund A, Hökfelt T, eds. Vol 13: The Primate Nervous System, Part I. pp 39–138. Elsevier, Amsterdam.

Swaab DF (1998) The human hypothalamo-neurohypophysial system in health and disease. Prog Brain Res 119:577–618.

Swaab DF (2003) The Human Hypothalamus: Basic and Clinical Aspects. Part I: Nuclei of the Human Hypothalamus. In: Aminoff MJ, Boller F, Swaab DF, Series eds. Handbook of Clinical Neurology. Vol 79. Elsevier, Amsterdam.

Swaab DF (2003) The Human Hypothalamus: Basic and Clinical Aspects. Part II: Neuropathology of the Human Hypothalamus and Adjacent Structures. In: Aminoff MJ, Boller F, Swaab DF, Series eds. Handbook of Clinical Neurology. Vol 80. Elsevier, Amsterdam.

Swaab DF (2006) The human hypothalamus in metabolic and episodic disorders. Prog Brain Res 153:3–45.

Swaab DF, Chung WC, Kruijver FP, Hofman MA, Ishunina TA (2002) Sexual differentiation of the human hypothalamus. Adv Exp Med Biol 511:75–100.

Swaab DF, Fliers E (1985) A sexually dimorphic nucleus in the human brain. Science 228:1112–1115.

Toni R, Malaguti A, Benfenati F, Martini L (2004) The human hypothalamus: a morpho-functional perspective. J Endocrinol Invest 27(Suppl 6):73–94.

All the neurons located outside the central nervous system that are concerned with the innervation of the viscera, except those which are afferent components of the cerebrospinal nerves, are included in the so-called autonomic nervous system. This system also includes the neurons located within the spinal cord and the brain stem through which the outlying efferent neurons are functionally connected with the central nervous system.

Albert Kuntz, 1953

20

The Autonomic Nervous System

20.1. HISTORICAL ASPECTS

Galen (ca 130–ca 200), who described nine of the twelve pairs of cranial nerves, originally described the autonomic nervous system. He traced the vagal nerve [X] into the thorax and abdomen and described its distribution to visceral organs. Galen advanced the theory of 'consent' or 'sympathy' between different parts of the body based on the concept that peripheral nerves were hollow, allowing animal spirits to flow through them from one body part to another, bringing about 'sympathy' between the various parts of the body. Bichat (1800) divided the life of the organism into both animal and vegetable parts. Therefore, there were vegetative and animal nerves which would now be termed visceral and somatic nerves. Reil (1807) introduced the term 'vegetative nervous system' and considered the sympathetic ganglia as independent nerve centers.

Gaskell (1886) envisioned visceral nerves as arising from homologous neuronal columns in the central nervous system. The development of the limbs and the nerves of the limbs broke these columns into three outflows. There is a cranial outflow, a thoracic outflow, and a sacral outflow. Later, Gaskell introduced the term 'involuntary nervous system'. Langley (1890) made extensive investigations of the autonomic nervous system and contributed many terms still in use today. He introduced such terms as preganglionic fibers, postganglionic fibers, and in 1893 introduced the terms 'autonomic system' dividing it into a tectal part, a bulbosacral part, and a sympathetic part. Later he combined the tectal and bulbosacral parts into one division and introduced the term 'parasympathetic'. He was later to publish the first book on the autonomic nervous system.

Dale (1933) proposed dividing the autonomic nervous system into adrenergic and cholinergic divisions based on the transmitter released by the terminals when the nerve fiber was stimulated. He observed that all preganglionic fibers are cholinergic. Albert Kuntz (1879–1957) was perhaps the most prolific American investigator in the field of the autonomic nervous system in the 20th century. He added greatly to our understanding of this system and was a leading authority on this system throughout his lifetime. Since his death, studies using enzyme histochemistry, fluorescent methods, immunocytochemistry, and electron microscopy have added greatly to our knowledge of the autonomic nervous system.

20.2. STRUCTURAL ASPECTS

The autonomic nervous system is the **general visceral efferent** part of the human nervous system including autonomic neurons and their associated ganglia that collectively innervate nonstriated muscle, cardiac muscle, and exocrine glands. Autonomic impulses are conducted away from (efferent) their cell bodies of origin and toward peripheral structures or effectors in the body viscera, including in the digestive, respiratory, urogenital, and endocrine systems as well as in the spleen, heart, and great vessels. Peripheral effectors receive such impulses and respond to them by contraction or secretion. Functional regulation of visceral structures involves the conduction of sensory impulses

from visceral structures back to the spinal cord and brain. Sensory neurons that receive impulses from the viscera including from nonstriated muscle, cardiac muscle, and exocrine glands are classified as visceral afferents. Hence, the autonomic nervous system is both a **visceral efferent** and a **visceral afferent** system.

Careful study of the (1) location of the cell bodies of origin of this system; (2) manner of distribution of its fibers; and (3) the termination of these fibers, will facilitate understanding of the autonomic nervous system.

20.2.1. Location of Autonomic Neurons of Origin

That part of the autonomic nervous system whose neuronal cell bodies are associated with cranial nerve nuclei (oculomotor, facial, glossopharyngeal, and vagal) and are at sacral spinal levels (S2 to S4) is the **craniosacral division** of the autonomic nervous system (Fig. 20.1). This division is also termed the **parasympathetic division** of the autonomic nervous system (Table 20.1).

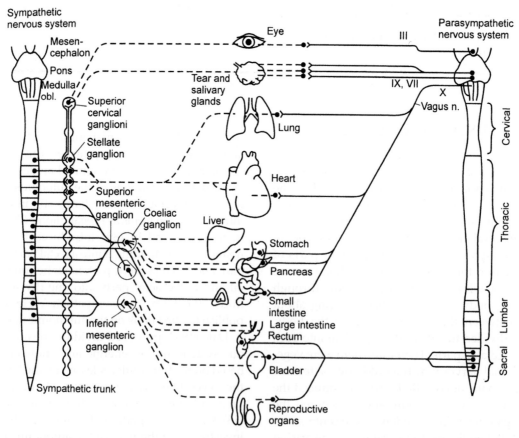

FIGURE 20.1. The sympathetic and parasympathetic components of the autonomic nervous system. Preganglionic fibers are solid lines whereas postganglionic fibers are dotted lines. The sympathetic outflow is from neurons at thoracolumbar levels of the spinal cord whereas the parasympathetic outflow is from cranial-sacral levels (from Jänig, 1997).

That part of the autonomic nervous system whose neurons of origin are at certain spinal levels, in particular thoracic levels T1 through T12 and lumbar levels L1 and L2, is the **thoracolumbar division** (Fig. 20.1). This division is also termed the **sympathetic division** of the autonomic nervous system (Table 20.1). Ganglia associated with the sympathetic division of the autonomic nervous system include the sympathetic trunk or chain of paravertebral ganglia and the prevertebral ganglia. The reason for this curious craniosacral and thoracolumbar arrangement of the neurons of origin of the autonomic nervous system is unclear. Fishes have a cranial parasympathetic system but no sacral parasympathetic system. Anurans, however, have also a sacral parasympathetic division.

20.2.2. Manner of Distribution of Autonomic Fibers

The **craniosacral (parasympathetic) division** distributes fibers through a two-neuron system (Fig. 20.1). From cell bodies in association with the oculomotor, facial, glossopharyngeal, and vagal nerves, axons pass through these respective cranial nerves (Table 20.2). Axons from sacral spinal levels (S2 to S4) pass out through the respective sacral spinal nerves (S2, S3, and S4). These axons pass almost to the organ innervated, but before reaching their site of termination, they synapse in course on parasympathetic neurons in special collections of neurons called **ganglia**. Axons of neurons that pass to the ganglia are termed 'preganglionic' in that these fibers come **before** the ganglia. All preganglionic fibers synapse in parasympathetic ganglia on their way to their eventual termination (Fig. 20.1). There are four of these parasympathetic ganglia in the head. These include the ciliary, pterygopalatine, submandibular, and otic ganglia (Table 20.2). Other parasympathetic ganglia are in the walls of organs. Therefore, the parasympathetic division of the autonomic nervous system has **long preganglionic fibers** that synapse in or near the organs innervated, and **short postganglionic fibers** that then must travel only a slight distance to reach their eventual termination (Fig. 20.1). Preganglionic and postganglionic fibers of the parasympathetic division are cholinergic. The discharge of such preganglionic and postganglionic fibers influences only a limited area.

The thoracolumbar or sympathetic division of the autonomic nervous system has its cell bodies of origin in the **intermediolateral nucleus** at spinal levels T1 through L1 or L2. The **intermediolateral nucleus** forms an extension of the gray matter lateral to the central canal called the **lateral horn** (Fig. 20.2). Some of these neurons retain a position near the central canal as an **intermediomedial nucleus**. The intermediolateral and the intermediomedial nuclei are the neurons of origin for preganglionic fibers of the sympathetic division

TABLE 20.1.   Comparison of the divisions of the autonomic nervous system

	Sympathetic	Parasympathetic
Somata of origin	**Thoracolumbar:** Lateral horn of all thoracic and first two lumbar spinal segments; intermediomedial and intermediolateral cell columns	**Craniosacral:** In the accessory oculomotor nucleus, inferior and superior salivatory nuclei, and in the dorsal vagal nucleus and the sacral parasympathetic gray in spinal segments S2 to S4
Preganglionic fibers	Short fibers with many collaterals	Long fibers with few collaterals
Preganglionic neurotransmitter	Acetylcholine	Acetylcholine
Postganglionic fibers	Long fibers	Short fibers with few branches
Location of postganglionic somata	Remote from organs they supply	Near or in effector they supply
Postganglionic neurotransmitter	Norepinephrine	Acetylcholine
Areas influenced	Widespread influence throughout the body	Only limited area of influence
Functions	Pupillary dilation. Increases heart rate, force of contraction and coronary flow. Decreased GI motility, contractility and tone. Sphincter contraction and inhibition of gastric secretion. Dilation of bronchi and bronchioles. Constricts pulmonary vessels and vessels to the liver, pancreas and spleen. Ejaculation. Vascular constriction, sweat gland secretion, and contraction of arrector pili muscles	Pupillary constriction. Increases glandular secretion. Decreases heart rate, constricts coronary vessels, decreases conduction velocity, increases peristalsis, increases nonstriated muscular tone. Relaxes sphincters, aids in urination and erection

TABLE 20.2   Cranial parasympathetic nuclei

Cranial nerve	Location of preganglionic neurons	Location of postganglionic neurons	Structures innervated	Function
Oculomotor	Accessory oculomotor nucleus (midbrain)	Ciliary and episcleral ganglia	Sphincter pupillae and ciliary muscles of the iris	Pupillary constriction accompanying light reflex and accommodation
Facial	Superior salivatory nucleus (lower pons)	Submandibular and pterygopalatine ganglia	Submandibular sublingual, and lacrimal glands. Oral cavity glands, nasal and palatine glands	Secretion of these glands
Glossopharyngeal	Inferior salivatory nucleus (medulla)	Otic ganglion	Parotid gland	Secretion of parotid gland
Vagus	Dorsal vagal nucleus (medulla)	Terminal ganglia in or near organs innervated	Heart, nonstriated muscle and glands of lungs, esophagus, stomach, small intestine, colon to splenic flexure, liver, kidneys, pancreas, gallbladder, bile ducts	Decrease heart rate, constrict coronary vessels, peristalsis, nonstriated muscle tone, relax sphincters, secretion of glands

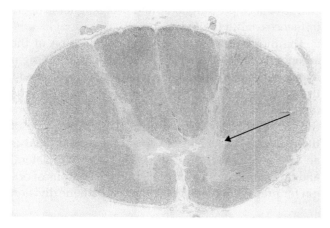

FIGURE 20.2.   Section of the upper thoracic level of the human spinal cord demonstrating the location of the intermediolateral cell column (arrow).

of the autonomic nervous system (Table 20.1). Preganglionic fibers arise from these primary neurons, travel outside the cord in the ventral roots and into the ventral rami of the respective spinal nerves. The ventral rami have two branches leading to a **sympathetic ganglion** and these ganglia connect in a vertical series forming the **sympathetic trunk**. After entering the ventral rami of the spinal nerves, the preganglionic fibers travel only a short distance, form **white rami communicantes**, and enter the sympathetic trunk. In the sympathetic trunk are neurons on which these short, preganglionic fibers synapse. Thus, fibers from a limited region of the spinal cord (in particular from T1 through L1 or L2 spinal levels) enter this distribution center (the sympathetic trunk) and are widely disbursed throughout the body. Fibers leaving the sympathetic trunk are postganglionic fibers in that they have already

synapsed on ganglionic neurons in the sympathetic chain. Postganglionic sympathetic fibers are long and are identifiable by the elaboration of adrenergic neurotransmitters at their terminals. Sympathetic fibers have a considerable influence throughout the body.

20.2.3.   Termination of Autonomic Fibers

Parasympathetic Fibers

Oculomotor Nerve

Parasympathetic fibers in the **oculomotor nerve** [III] originate from neurons at the level of the superior colliculus. The nucleus of their origin is the **accessory oculomotor** (Edinger–Westphal) **nucleus** (Table 20.2) medial to the oculomotor nucleus. Preganglionic fibers from these neuronal cell bodies travel in the oculomotor nerve before passing to secondary neurons in the **ciliary ganglion**. From the ciliary ganglion, postganglionic fibers pass to the sphincter pupillae and the ciliary muscle of the iris. These parasympathetic fibers are associated with pupillary constriction and accommodation of the lens (Table 20.2).

Facial Nerve

Parasympathetic fibers associated with the **facial nerve** [VII] have their cell bodies of origin in the lower third of the pons (Table 20.2). Primary parasympathetic neurons occur in a nuclear group known as the superior salivatory nucleus. This nucleus is on the **lateral side** of the intrapontine part of the facial nerve. Preganglionic parasympathetic fibers leave the superior salivatory nucleus and pass out of the brain stem with fibers of the facial nerve. Such preganglionic parasympathetic fibers pass in the chorda tympani and lingual nerve to reach secondary neurons in the submandibular ganglion. Postganglionic parasympathetic fibers leave the submandibular ganglion and supply the submandibular gland and the sublingual gland. Some preganglionic fibers associated with the superior salivatory nucleus pass to the pterygopalatine ganglion and ultimately innervate the lacrimal gland. These parasympathetic fibers associated with the facial nerve are involved in secretion of these glands.

Glossopharyngeal Nerve

The parasympathetic fibers related to the **glossopharyngeal nerve** [IX] occur in a nuclear group at the level of the entering fibers of the glossopharyngeal nerve (Table 20.2). This nucleus, the inferior salivatory nucleus, is at the rostral tip of the dorsal vagal nucleus. Preganglionic fibers leave this nucleus and travel in the glossopharyngeal nerve to reach the otic ganglion. Postganglionic fibers leave the secondary neuronal cell bodies in the otic ganglia to supply the parotid gland through the auriculotemporal branch of the mandibular nerve. When activated, these parasympathetic fibers produce a watery secretion of the parotid gland.

Vagal Nerve

Parasympathetic fibers in the **vagal nerve** [X] have their neurons of origin in the lower half of the medulla oblongata, dorsal and lateral to the hypoglossal nucleus. Primary parasympathetic preganglionic neurons are in the **dorsal vagal nucleus**. Preganglionic fibers leave this nucleus and travel in the vagal nerve and perhaps in the accessory nerve [XI] to reach secondary neurons in or near the organs innervated. This includes the heart, lungs, esophagus, and gastrointestinal tract to the descending colon. In these organs are intrinsic ganglia that serve as secondary neurons in this chain. Postganglionic fibers leave these intrinsic ganglia and pass to cardiac and nonstriated muscle and to glands associated with these visceral structures. Parasympathetic fibers associated with the vagus [X] function in decreasing heart rate and conduction velocity, promoting bronchial constriction, and increasing gastrointestinal motility (by promoting peristalsis and relaxing sphincters of the gastrointestinal tract).

Sacral Parasympathetic Nucleus

The **sacral parasympathetic nucleus** is at spinal levels S2 to S4. This nuclear group consists of primary parasympathetic neurons that supply preganglionic

fibers that form the **pelvic splanchnic nerves**. These preganglionic parasympathetic fibers supply secondary neurons in the organs they are to innervate. Ultimately, postganglionic fibers supply the rectum, urinary bladder, erectile tissue of the penis or clitoris, the testes or ovaries, uterus, uterine tubes, and parts of the large intestine from the splenic flexure to the anus. These parasympathetic fibers which have their cell bodies of origin at spinal levels, function in urinary detrusor contraction, trigone and sphincter relaxation, erection and vasodilation, glandular secretion of the gastrointestinal tract, and have an excitatory effect on colon motility. These fibers and neurons at sacral levels of the spinal cord form the **sacral part** of the **craniosacral** division of the autonomic nervous system.

Sympathetic Fibers

At thoracic and lumbar spinal cord levels are primary sympathetic neurons in the **intermediolateral and intermediomedial nuclei** (Table 20.1). These preganglionic sympathetic neurons form the **lateral horn** (not to be confused with the lateral subdivision of the ventral horn) from T1 to L1 or L2 spinal levels (Fig. 20.3). Preganglionic fibers leave these cell bodies of origin to pass to the paravertebral ganglia that form the sympathetic trunk and to the prevertebral ganglia by way of **white rami communicantes** (Fig. 20.3). Secondary sympathetic neurons are in ganglia of the sympathetic trunk and prevertebral ganglia. Postganglionic sympathetic fibers leave these ganglia and pass along blood vessels to supply sweat glands in skin, arrector pili muscles of hairs, many blood vessels, the heart, and lungs. These sympathetic fibers increase heart rate and conduction velocity as well as the force of ventricular contraction. They are also involved in blood vessel constriction, dilation of the pupils (Fig. 20.3) and bronchi, decreasing gastrointestinal motility, and gastrointestinal sphincter contraction, and in ejaculation.

20.3. COMPARISON OF THE SOMATIC EFFERENTS AND VISCERAL EFFERENTS

There are many distinctions between the **somatic efferent (voluntary) system** and the **general visceral efferent (autonomic or involuntary) system**. First, neurons of the somatic efferent system supply **voluntary, striated muscles** whereas neurons of the general visceral

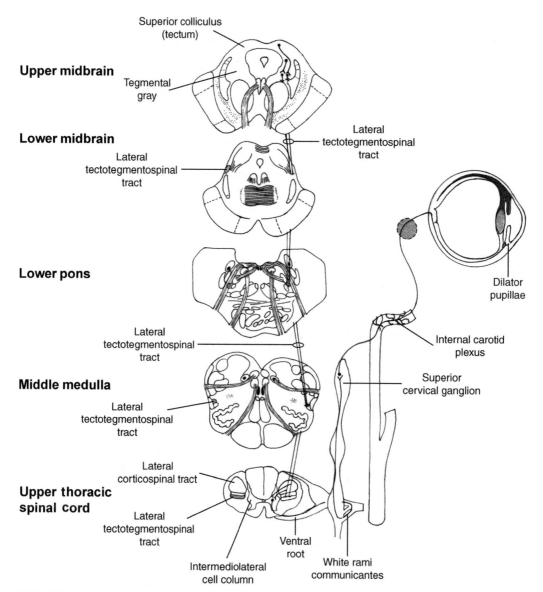

FIGURE 20.3. The origin, course, and termination of the lateral tectotegmentospinal tract. This path terminates on preganglionic neurons in the intermediolateral cell column at T1 and T2 cord levels. From these preganglionic neurons, fibers arise and exit the ventral roots from C8–T4 spinal cord levels to enter the sympathetic trunk through the white rami communicantes. These preganglionic fibers synapse in the superior cervical ganglion. Postganglionic fibers from this ganglion accompany the internal carotid artery as the internal carotid plexus. This plexus gives fibers that pass through the ciliary ganglion and short ciliary nerves to supply the dilator pupillae muscle (modified from DeJong, 1979).

efferent system supply **involuntary, nonstriated muscle, cardiac muscle**, and **exocrine glands**.

Second, somatic efferent neurons (both alpha and gamma motoneurons) occupy much of the ventral horn at all spinal levels and are in nuclear groups associated with the oculomotor, trochlear, abducent, and hypoglossal nerves. These nuclear groups are in line with one another. General visceral efferent neurons, however, are restricted to the lateral horn of the spinal

cord. In this location, they either form an **intermediolateral nucleus** in segments T1 to L1 or L2, or occur in the **sacral parasympathetic nucleus** in the lateral horn of segments S2 to S4. Other general visceral efferent nuclear groups are in association with certain cranial nerve nuclei including the **accessory oculomotor (Edinger–Westphal) nucleus** at rostral levels of the midbrain, the **superior salivatory nucleus** in the lower pons, the **inferior salivatory nucleus** and the **dorsal**

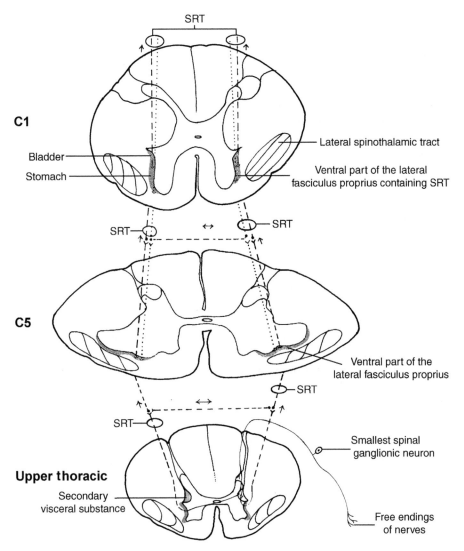

C1

Bladder

Stomach

Lateral spinothalamic tract

Ventral part of the lateral
fasciculus proprius containing SRT

C5

Ventral part of the
lateral fasciculus proprius

Upper thoracic

Secondary
visceral substance

Smallest spinal
ganglionic neuron

Free endings
of nerves

FIGURE 20.4. Origin and partial course of the human spinoreticulothalamic tract (SRT) from upper thoracic to C1 spinal levels. This bilateral, multisynaptic path carries chronic, visceral pain from spinal levels, ascends through the brain stem to the dorsal thalamus. Note the probable pattern of localization for fibers in the SRT carrying impulses from the bladder and stomach (adapted from Crosby, Humphrey, and Lauer, 1962).

vagal nucleus in the medulla oblongata. These four general visceral efferent nuclei contribute fibers to four or perhaps five cranial nerves – the oculomotor, facial, glossopharyngeal, and vagal nerves.

Third, the somatic efferent system is a **single-neuron system** from its cell bodies of origin in the spinal cord or brain stem to the termination of its fibers, with the cell bodies lying inside the central nervous system. The general visceral efferent system, however, is a **two-neuron system** with the first neuron in the system, the **preganglionic** neuron, **inside** the central nervous system and the **postganglionic**, or second neuron in the chain, **outside** the central nervous system. Fourth, the somatic efferent system is either a voluntary or a reflex system whereas the general visceral efferent system is exclusively a reflex system over which we do not normally exert voluntary control.

20.4. GENERAL VISCERAL AFFERENTS

The autonomic nervous system is not purely an efferent system in that it includes **general visceral afferents** as well. These visceral afferents are a prominent part of the sensory reflex arch that underlies much reflex activity of visceral structures. There must be some sensory input for the general visceral efferents to function in an appropriate manner. The general visceral afferents carry impulses to higher levels, particularly impulses associated with visceral pain such as ischemia, gut distention, pulling on mesenteries, excessive muscle contraction, and gastric hyperacidity. These are examples of poorly localized, chronic visceral pain.

The primary neurons of the visceral afferents are smaller neurons in all spinal ganglia (Fig. 20.4). Peripheral processes of these small spinal ganglionic neurons send thinly myelinated axons out into the periphery where they end as free nerve endings. The central processes of these spinal ganglionic neurons enter the spinal cord through the lateral division of the dorsal root. Here these fibers end in the **secondary visceral gray matter** from T1 to L1 or L2 as well as from S2 to S4 spinal levels (Fig. 20.4). From the secondary visceral gray matter, a secondary ascending visceral path termed the **spinoreticulothalamic tract** (Fig. 20.4) ascends on both sides of the spinal cord (a crossed and uncrossed path) and travels to the thalamus. This secondary ascending visceral path takes up a position in the **proprius bundles** at the gray matter–white matter junction. The proprius bundles surround the gray matter of the dorsal and ventral horns, both medially and laterally. A pattern is present here such that impulses from certain parts of the body travel in a certain part of the proprius bundles whereas impulses from other visceral structures are in a different position among the proprius bundles.

Apparently, there are also general visceral afferent impulses carried in the glossopharyngeal and vagal nerves that pass to the **nucleus of the solitary tract**. Visceral afferents distribute with or accompany the general visceral efferent fibers but never synapse in autonomic ganglion. Those fibers associated with reflex control of visceral activity probably accompany parasympathetic nerves whereas those fibers that convey visceral sensation presumably accompany sympathetic nerves. The exception appears to be the visceral pain fibers from certain pelvic viscera that accompany the pelvic parasympathetic nerves.

20.5. REGULATION OF THE AUTONOMIC NERVOUS SYSTEM

Both the hypothalamus and the cerebral cortex play prominent roles in the regulation of the autonomic nervous system. The hypothalamus is a significant correlation region that coordinates visceral activities responding to ever-changing conditions of external and internal environment. The hypothalamus exerts its regulation of the autonomic nervous system through two major output systems that leave the hypothalamus. These paths include the **dorsal longitudinal fasciculus** and the **hypothalamotegmental paths**. These paths descend to influence both parasympathetic and sympathetic neurons at brain stem and spinal levels.

Autonomic related areas of the cerebral cortex influence the autonomic nervous system. These include **area 25 of Brodmann**, the **posterior orbitofrontal cortex** (POF) and the **anterior insula**. That area inferior to the rostrum of the corpus callosum termed the subcallosal area is divisible into dorsal and ventral parts. The dorsal part of the subcallosal area, termed the **subcallosal gyrus**, is continuous with the cingulate gyrus whereas the ventral part of the subcallosal area includes the **septal area** and corresponds to **Brodmann's area 25**. These autonomic related cortices (ARC) influence the autonomic nervous system by way of the hypothalamus and through other connections to lower levels of the nervous system. Thus, there are corticohypothalamic fibers and there are corticothalamohypothalamic fibers, particularly those that pass to the medial dorsal nucleus (MD) of the thalamus. The amygdaloid body through amygdalohypothalamic fibers also exerts an influence on the hypothalamus. These paths are probably regulatory and provide necessary adjustments in autonomic function. Since both emotion and intellectual decisions are cortical in origin, it is apparent that these cortical influences exert their control through regulation of the autonomic nervous system.

20.6. DISORDERS OF THE AUTONOMIC NERVOUS SYSTEM

Numerous disorders threaten the autonomic nervous system. As one looks at the myriad of conditions that influence the autonomic nervous system, it is revealing to note that the autonomic nervous system is dispersed and widespread and that it participates in many functions. As an example, among disorders of the autonomic nervous system are disorders of cardiac control. Myocardial infarction and cardiac arrhythmias may have an autonomic component. Either a decrease (arterial hypertension) or increase (arterial hypotension) in arterial blood pressure beyond normal limits reflects an imbalance between the sympathetic and parasympathetic parts of the autonomic nervous system.

Some disorders of regional circulation lead to such conditions as migraine headaches or limb disorders. Abnormalities of body temperature regulation, including both hypo- and hyperthermia, are examples of autonomic dysfunctions. Increased or decreased sweating (called hyperhidrosis or anhidrosis), as well as changes in pupillary size or disturbances of micturition and sexual function come under the heading of autonomic disturbances. Gastrointestinal disorders, such as increased salivation, vomiting, ulcers or spastic colon, visceral pain such as phantom limb pain, causalgia or vascular pain are attributable to injury to the autonomic nervous system.

FURTHER READING

Appenzeller O (1999) The Autonomic Nervous System. Part I. Normal Functions. In: Vinken PJ, Bruyn GW, eds. *Handbook of Clinical Neurology.* Vol 74. Elsevier, Amsterdam.

Appenzeller O (1999) The Autonomic Nervous System. Part II. Dysfunctions. In: Vinken PJ, Bruyn GW, eds. *Handbook of Clinical Neurology.* Vol 75. Elsevier, Amsterdam.

Appenzeller O, Oribe, E (1997) *The Autonomic Nervous System: An Introduction to Basic and Clinical Concepts.* 5th ed. North Holland/Elsevier, Amsterdam.

Bielefeldt K, Christianson JA, Davis BM (2005) Basic and clinical aspects of visceral sensation: transmission in the CNS. Neurogastroenterol Motil 17:488–499.

Blessing WW (1997) *The Lower Brainstem and Bodily Homeostasis.* Oxford University Press, New York.

Chu CC, Tranel D, Damasio AR, Van Hoesen GW (1997) The autonomic-related cortex: pathology in Alzheimer's disease. Cereb Cortex 7:86–95.

Goldstein DS (2001) *The Autonomic Nervous System in Health and Disease.* Marcel Dekker, New York.

Goldstein DS, Robertson D, Esler M, Straus SE, Eisenhofer G (2002) Dysautonomias: clinical disorders of the autonomic nervous system. Ann Intern Med 137:753–763.

Jänig W (2006) *Integrative Action of the Autonomic Nervous System: Neurobiology of Homeostasis.* Cambridge University Press, Cambridge.

Jänig W, Habler HJ (2000) Specificity in the organization of the autonomic nervous system: a basis for precise neural regulation of homeostatic and protective body functions. Prog Brain Res 122:351–367.

Johnson RH, Spalding JMK (1974) Disorders of the autonomic nervous system. In: *Contemporary Neurology Series.* Vol 11. FA Davis, Philadelphia.

Kuntz A (1953) *The Autonomic Nervous System.* Lea & Febiger, Philadelphia.

Low PA (1993) Autonomic nervous system function. J Clin Neurophysiol 10:14–27.

Low PA (1997) *Clinical Autonomic Disorders: Evaluation and Management.* 2nd ed. Lippincott-Raven, Philadelphia.

Low PA (2003) Testing the autonomic nervous system. Semin Neurol 23:407–421.

Mathias C, Bannister R (2002) *Autonomic Failure: A Textbook of Clinical Disorders of the Autonomic Nervous System.* 4th ed. Oxford University Press, Oxford.

Pick J (1970) *The Autonomic Nervous System.* JB Lippincott, Philadelphia.

Robertson D, Low PA, Polinsky RJ (1996) *Primer on the Autonomic Nervous System.* Academic Press, San Diego.

Saper CB (2002) The central autonomic nervous system: conscious visceral perception and autonomic pattern generation. Annu Rev Neurosci 25:433–469.

Schondorf R (1993) New investigations of autonomic nervous system function. J Clin Neurophysiol 10:28–38.

Shields RW Jr (1993) Functional anatomy of the autonomic nervous system. J Clin Neurophysiol 10:2–13.

The most noble organ under heaven, the dwelling-house and seat of the soul, the habitation of wisdom, memory, judgment, reason, and in which man is most like unto God.

Robert Burton, 1652

21

General Features of the Cerebral Hemispheres

21.1. FACTS AND FIGURES

In the newborn, the two cerebral hemispheres weigh approximately 402 g in the male and 380 g in the female. By the end of the first year, the weight of the brain has doubled, and by the end of the sixth year, it has tripled to at least 90% of its adult weight at that time. This increase in weight is due mainly to an increase in the number of blood vessels, myelin layers, and supporting or glial elements. In adults, the cerebral hemispheres weigh about 1450 g in the male with a normal range of from 1200 to 1600 g. In the female, the cerebral hemispheres weigh about 1350 g, with a normal range between 1100 and 1500 g. The brain of an elephant weighs about 5000 g and that of the blue whale weighs about 6800 g.

The surface of the brain is folded such that two-thirds of its surface area lies buried and out of view in the depths of the cerebral fissures or sulci. When one considers the surface area of the brain per lobe, about 41% of the human brain consists of the frontal lobe, 17% consists of the occipital lobe, 21% consists of the temporal lobe, and the parietal and insular lobes make up about 21% of the surface area of the brain.

The **surface gray matter** of the cerebral hemispheres, or **cerebral cortex**, averages about 2.5 mm in thickness, with variation from one brain region to another. The **primary visual cortex** (area 17 of Brodmann) is about 1.5 mm in depth whereas the **primary motor cortex** (area 4 of Brodmann) is about 4.5 mm. There is variation in the thickness of the cerebral cortex, with it being thinner in the depths of the fissures than at the upper margins of the fissures. Some 10^{12} neurons may be present in both cerebral hemispheres (10^{12} = 1 trillion or 1 million × 1 million).

21.2. CORTICAL NEURONS

There are five neuronal types in the human cerebral cortex. These include the characteristic neuron of the cerebral cortex, the **pyramidal neuron** (Figs 21.1 to 21.4), which measures some 10 to 70 μm in diameter. The majority of neurons in the human cerebral cortex are pyramidal neurons. These excitatory neurons project to

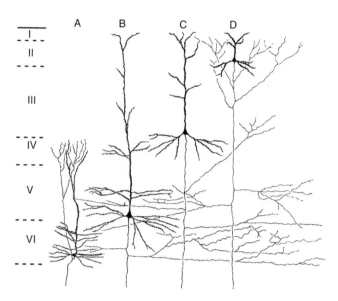

FIGURE 21.2. Size, shape and laminar distribution of pyramidal neurons in the human cerebral cortex (from Bryne and Roberts, 2004).

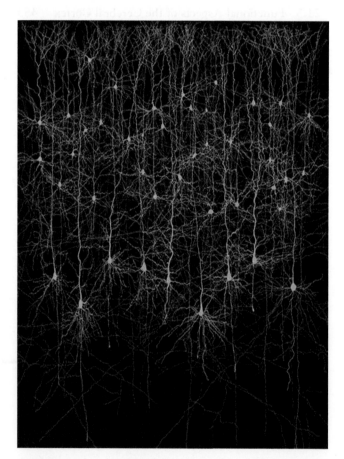

FIGURE 21.3. Pyramidal neurons in the cerebral cortex (from the Blue Brain Project, 2006).

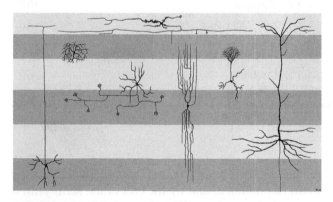

FIGURE 21.1. Five neuronal types in the human cerebral cortex (from Standring, 2005).

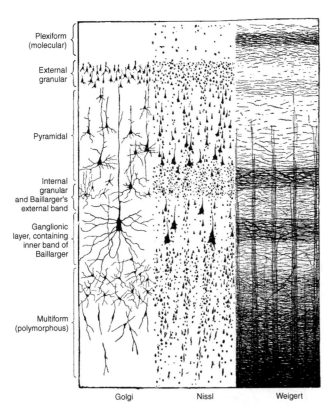

Plexiform (molecular)

External granular

Pyramidal

Internal granular and Baillarger's external band

Ganglionic layer, containing inner band of Baillarger

Multiform (polymorphous)

Golgi   Nissl   Weigert

FIGURE 21.4. Layers of the human cerebral cortex (from Standring, 2005).

other pyramidal neurons forming networks among themselves. A single pyramidal cell may have up to 200 synapses from other neurons on its cell body but upwards of 40,000 synapses on its axon and dendrites. The **stellate or granule neuron** is another neuronal type found in the cerebral cortex (Fig. 21.1). This neuronal type has a wide distribution and is probably correlative in function, interrelating different cortical layers. The **horizontal cells of Cajal** are in the superficial cortical layers (Fig. 21.1). These small neurons correlate adjoining areas within each cerebral hemisphere and have axons and dendrites that run parallel to the cortical surface. The **cells of Martinotti** are in various cortical layers (Fig. 21.1). Their axons pass to the cortical surface and they, too, are probably correlative in function. The fifth cortical neuronal type is the polymorph, multiform, or pleomorph neuron. These essentially modified pyramidal neurons are associative in function and occur in the innermost cortical layer.

21.3. CORTICAL LAYERS

Based on cell size and packing density (**cytoarchitectonics**) each cortical region is divisible into six cellular

layers (Figs 21.1, 21.2, 21.4). These cellular layers are useful in parceling the brain into areas (cortical parcellation). From superficial to deep, these cortical layers include the: (I) molecular layer, (II) external granular layer, (III) external pyramidal layer, (IV) internal granular layer, (V) internal pyramidal layer, and (VI) multiform layer (Fig. 21.4). Those cortical areas with six definite layers are termed **homotypic areas**. Those cortical areas in which six layers are obscure, such as in the primary motor cortex (where there appears to be only five layers) or where seven layers are identifiable, such as in the primary visual cortex, are termed **heterotypic areas**.

Based on the types of fibers present (radial, oblique, and horizontal fibers), their distribution, layering, and the amount of myelin present, collectively termed **myeloarchitectonics**, six major **cortical fiber layers** are identifiable in the human cerebral cortex. Related to layer II is the **stria of the external granular layer** (Fig. 21.4). This layer of fibers is associative in function with many association and commissural fibers. The **stria of the internal granular layer**, in layer IV, receives input from various specific thalamic nuclei (Fig. 21.4). Finally, the **stria of the internal pyramidal layer** is in the deep part of layer V (Fig. 21.4). The infragranular cortical layers have their output through the stria in layer V.

Using cellular patterns present on histological sections of the cerebral cortex, many cortical maps were devised which define various **areas** or **fields**. Campbell, in the early 1900s, defined some twenty regions of the human cerebral cortex. Brodmann in the early 1900s defined some 52 cortical regions (Figs 21.5, 21.6, 21.7, 21.8). The Vogts defined some 200 cortical areas based on cellular patterns. There have also been many maps devised which are based on the fiber arrangement in cortical layers. The Vogts also were leaders in this field, defining some 229 cortical areas.

No doubt, a distinctive pattern exists from one area to the next with regard to the distribution of fibers and cells. This pattern varies from one region to the next of the cerebral hemispheres with perhaps as many as 200 to 250 cortical areas or fields. Such cortical fields or areas are a morphological expression of functional differences between the various regions of the cerebral hemispheres. Contemporary methods of cortical parcellation have used a combination of anatomy, neurochemistry, and function.

There are different types of functional areas of the cerebral cortex including **receptive areas** and **projection areas**. The receptive areas relate to specific sensory modalities and include the primary somatosensory,

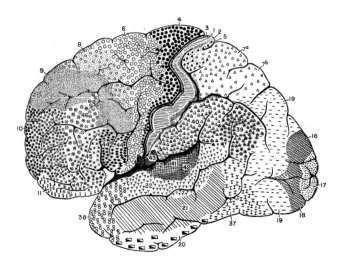

FIGURE 21.5. Cytoarchitectural map showing Brodmann's areas on the lateral surface of the human brain (modified after Brodmann, 1909).

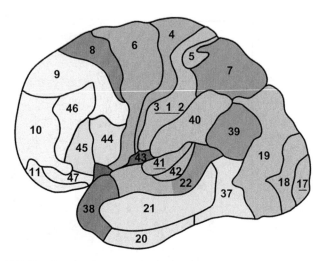

FIGURE 21.7. Modern rendition of the cytoarchitectural map showing Brodmann's areas on the lateral surface of the human brain (figure provided by Mark Dubin, University of Colorado, 2006).

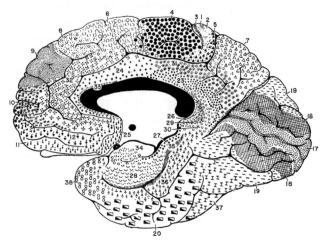

FIGURE 21.6. Cytoarchitectural map showing Brodmann's areas on the medial surface of the human brain (modified after Brodmann, 1909).

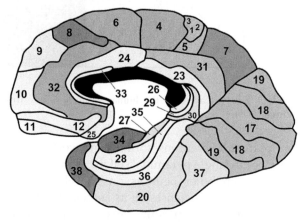

FIGURE 21.8. Modern rendition of the cytoarchitectural map showing Brodmann's areas on the medial surface of the human brain (figure provided by Mark Dubin, University of Colorado, 2006).

primary visual, primary gustatory, primary olfactory, primary vestibular, and primary auditory areas of the cerebral cortex. In addition to the primary and some well-defined secondary sensory receptive or motor projection areas, there are also **cortical association areas** related to correlating, interrelating, and interpreting information that reaches the cerebral cortex. The association areas receive impulses from various cortical regions. In the association areas are the ultimate functional capacities of the human brain. An association area may be under the influence of only one type of impulse from a single neighboring projection (a primary association area) or receptive area, or it may correlate two different types of sensory information (a secondary association area). There may also be

tertiary or **multimodal association areas** in the human brain that are regions of integration, association, and correlation. Indeed in terms of the associations that underlie the most complex of human capabilities there may well be **heteromodal and supramodal association areas** that provide the anatomical basis for our ability to read, write, speak, learn, think, reason, remember, and ultimately create new concepts and ideas.

21.4. CORTICAL COLUMNS (MICROARCHITECTURE)

The cell size and packing density of the cerebral cortex (**cytoarchitectonics**), permits the identification of cortical layers and cortical parcellation into regions and areas. In addition, there is also a columnar cortical organization (**microarchitecture**). Neurons in the somatosensory cortex are in vertical columns at right angles to the surface and extend through all cortical layers; they often share similar receptive fields and have specific functions. The width of these cortical columns has an upper limit in primates of the order of about 5 mm. Columns in SI are functionally interrelated by intrinsic corticocortical connections and by direct horizontal connections between columns. These intrinsic connections serve to connect parts of the cerebral cortex having different response properties yet lying in regions of similar regional representation.

21.5. FUNCTIONAL ASPECTS OF THE CEREBRAL CORTEX

In discussing functions of the cortical areas in humans, it is essential to understand that we are not dealing with properties inherent in the cerebral cortex itself, but are dealing with complete neuronal arcs or chains of neurons that project their impulses from specific areas of the nervous system to their terminations. The functional features of any cortical area are dependent on the connections to and from that area and the interrelations that exist between the various primary and secondary sensory receptive areas, the motor projection areas and the primary, secondary, and tertiary association areas.

21.6. CEREBRAL DOMINANCE, LATERALIZATION, AND ASYMMETRY

Cerebral dominance refers to the observation that one cerebral hemisphere of the brain is more involved in certain functions such as language, handedness, musical talents, visual-spatial abilities, attention and perhaps emotion. In some cases, there is a functional lateralization and in other cases, there is an accompanying structural asymmetry in the brain. The early studies of Broca and Wernicke demonstrated the lateralization of language function to the left hemisphere. Subsequent studies demonstrated a structural asymmetry in the temporal language region termed the planum temporale (PT) which corresponds to the posterior superior surface of the superior temporal gyrus. This asymmetry in a region corresponding to the core of **Wernicke's region**, an auditory association area,

was greater in right-handers than in left-handers. Both macroscopic and microscopic structural asymmetry exists in the primary motor area in humans with regard to handedness. In right-handers, MR morphometry reveals a deeper central sulcus on the left than on the right and vice versa in left-handers. Correspondingly, at the microscopic level the neuropil volume in Brodmann's area 4 is larger on the left than on the right. These anatomical asymmetries may reflect increased connectivity in area 4 along with an increase in the intrasulcal length of the posterior bank of the precentral gyrus in the left hemisphere of right-handers.

21.7. FRONTAL LOBE

21.7.1. Primary Motor Cortex

Superior to the lateral sulcus and in front of the central sulcus is the frontal lobe. In this region are many significant functional areas. This includes the **primary motor cortex** corresponding to **Brodmann's area 4**, which occupies the precentral gyrus on the lateral surface of the hemisphere, and that part of the paracentral lobule on the medial surface of the brain that lies anterior to the central sulcus (Figs 21.5 to 21.9) and extending inferiorly to the cingulate sulcus. On the lateral surface of the brain, the precentral sulcus bounds the

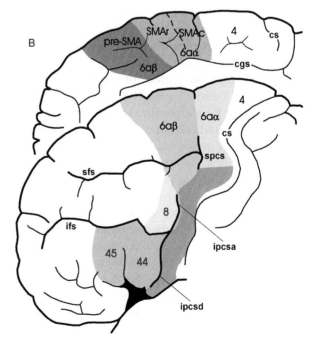

FIGURE 21.9. Several Brodmann's areas in the frontal lobe of the human brain: primary motor area 4, premotor area 6, Broca's motor speech region (areas 44 and 45) (from Paxinos and Mai, 2004).

precentral gyrus anteriorly and the central sulcus bounds it posteriorly. However, on the medial surface of the brain there is no reliable anatomical landmark for the anterior border of area 4. Studies in primates have demonstrated two spatially separate motor representations of the arm and hand in area 4. The long axes of these representations are oriented approximately parallel to the central sulcus. Stimulation in both areas causes the same movements and activates the same muscles at similar current strengths. This phenomenon is termed the **double representation hypothesis**. These two different areas or subareas of area 4 differ with regard to their somatosensory input. The caudal area relates to movements that use tactile feedback for their execution whereas the rostral area relates to movements that use proprioceptive feedback for their execution. These findings along with cytoarchitectural observations and quantitative distributions of transmitter-binding sites have led to the suggestion that area 4 in humans includes **area 4 anterior (4a)** and **area 4 posterior (4p)**. In addition to the motor representation across area 4, tactile and proprioceptive maps occur in 4a and 4p as well. Thus, the primary motor area in humans can be subdivided based on anatomy, neurochemistry, and function. The thumb, index and middle fingers have a representation in both 4a and 4p.

The Motor Homunculus

Electrical stimulation of the precentral gyrus and the anterior part of the paracentral lobule in conscious patients causes movements of the contralateral face, limbs, hands, or feet in that sequence thus revealing a cortical pattern. This pattern of representation from toes to tongue, oriented along the precentral gyrus and anterior part of the paracentral lobule, is usually illustrated as a distorted caricature, the **motor homunculus** (Fig. 21.10). The representation of the hands, digits, and face including the mouth and larynx is large. The hand area is on the middle third of the human precentral gyrus. The size of its representation reflects its extraordinary capabilities to sense temperatures, to participate in the active sense of touch in exploring the world around us, as well as carry out complex fine, skilled movement including communicating with our hands. The representation of the knee, leg, ankle, foot and bladder begin on the edge of each cerebral hemisphere and pass over onto the anterior paracentral lobule. The bladder region is just anterior to the region representing the ankle and the toes.

Injury to the primary motor cortex will produce a contralateral flaccid paralysis. If the bladder

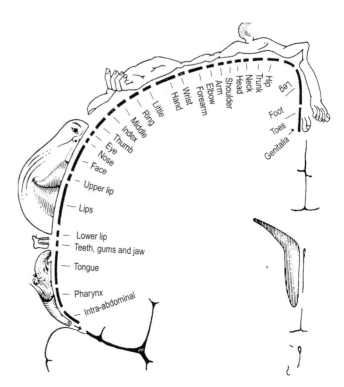

FIGURE 21.10. The motor homunculus across the precentral gyrus and paracentral lobule (after Penfield and Rasmussen, 1950).

representation on that part of area 4 that occupies the medial surface of each cerebral hemisphere is involved bilaterally, there will be an inability to start micturition and incontinence with dribbling because of overfilling of the bladder will result. The primary motor cortex discharges to lower motoneurons of the brain stem and spinal cord through components of the pyramidal system, both corticospinal and corticobulbar, for the initiation of voluntary motor activity.

21.7.2. Premotor Cortex

The **premotor cortex** corresponding to **Brodmann's area 6**, in the posterior part of the frontal lobe, is anterior to and parallels area 4, both on the medial and lateral surface of the cerebral hemisphere (Figs 21.5–21.9). The rostral border of area 6 on both medial and lateral surfaces of the human brain is unclear. On the medial surface of the brain area 6 extends inferiorly to the cingulate sulcus where its caudal border is Brodmann's area 24, part of the anterior cingulate cortex (Figs 21.6, 21.8). A motor pattern exists across the premotor cortex with responses that are contralateral, but not as discrete as the responses that follow stimulation of the primary motor cortex. In essence, the premotor cortex is an **extrapyramidal motor area** in that it supplements the activity of the primary motor cortex. The premotor cortex is also termed the frontal **aversive field** in that stimulation of this area causes rotation of the head, eyes, and trunk to the opposite side with complex synergistic movements of flexion and extension of the contralateral arm and leg. In addition to the motor representation across area 6, a tactile map occurs across this area as well.

Injury to the premotor cortex produces a **motor apraxia**. This purely cortical concept refers to the inability to carry out a sequence of motor activities in the presence of intact motor and sensory paths. Motor apraxia is evident as clumsiness in writing or drawing. A **grasp reflex** is likely to appear as well. Following damage to the primary motor cortex, the premotor cortex is able to substitute to some degree. This region also participates in **cortical automatic associated movements** that accompany fine skilled voluntary movements. A grasp reflex appearing in newborns and disappearing at two to four months is likely to be so marked as to suspend a child by his grasp. In older individuals, the primary motor cortex inhibits this reflex.

21.7.3. Supplementary Motor Area (SMA)

That part of the premotor area 6 that extends onto the medial hemisphere wall is termed the **supplementary motor area** (SMA). SMA is divisible into a rostral, **pre-SMA** and a caudal region (the **SMA proper**). Complex motor tasks activate the rostral SMA whereas simple motor tasks activate the caudal SMA proper. Pre-SMA is involved in the acquisition of new motor sequences whereas SMA proper participates in the acquisition of motor sequences that are essentially automatic. For example, the playing of a familiar piano piece by an experienced pianist (automatic movements) activates the SMA whereas the playing of an unfamiliar piece (perhaps more demanding and cognitive) activates laterally located area 6 along with pre-SMA on the medial hemisphere surface. SMA is somatotopically organized in humans with index finger responses anterior and dorsal to shoulder responses. Activation of SMA occurs during imagined movements whereas regional cerebral blood flow increases in this area during complex sequential movements.

21.7.4. Cingulate Motor Areas

In addition to the primary motor area 4, the premotor area 6, the SMA and their various subdivisions, the medial wall of the cerebral hemispheres contains motor areas buried in the cingulate sulcus. They do not extend onto the cingulate gyrus but include a rostral cingulate zone and a caudal cingulate zone. This cingulate region corresponds to Brodmann's areas 24 and 32 (Figs 21.6, 21.8).

21.7.5. Frontal Eye Fields

The **primary and secondary eye fields** occupy parts of **Brodmann's areas 6, 8** and **9** (Fig. 21.11). The **primary frontal eye field** is in the posterior part of the middle frontal gyrus. The **secondary frontal eye field** is in the dorsal part of the inferior frontal gyrus. A pattern for ocular movements exists across the frontal eye field such that stimulation here causes conjugate deviations of the eyes to the contralateral side, occasionally upward movement, convergence, divergence of the eyes and pupillary changes. Movements of the head and pupillary changes usually accompany movements of the eyes. Unilateral stimulation of the frontal eye fields in humans usually produces bilateral opening or closing of the lids.

The pattern on the secondary eye field is a mirror image of that on the primary eye fields. Processes of neurons in these eye fields project their fibers through the corticobulbar path to nuclei of the ocular muscles. The primary eye fields project fibers to the agonist muscles, whereas neurons in the secondary eye fields project fibers to the antagonistic muscles. Voluntary ocular movements obviously involve the cooperative action of eye fields in both cerebral hemispheres.

Two regions of the cerebral cortex in humans influence ocular movements. At the posterior end of the middle frontal gyrus and corresponding to Brodmann's areas 8 and 9 is a region responsible for voluntary eye

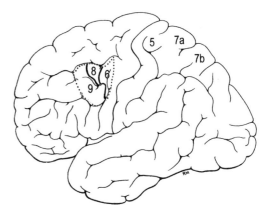

FIGURE 21.11. Frontal eye fields in the human brain corresponding to parts of Brodmann's areas 6, 8 and 9 (from Standring, 2005).

deviations called the **primary eye field**. There is also a **secondary** or **supplementary motor eye field** at the posterior end of the inferior frontal gyrus on its dorsal part. Stimulation of the human primary eye fields yields deviations of the eyes, including divergence, horizontal conjugate deviation to the contralateral side, and deviation up and to the contralateral side. The pattern on the secondary eye field is essentially a mirror image of that on the primary eye field. From the frontal eye fields fibers descend in the corticobulbar path, enter the genu of the internal capsule, cerebral crus at the base of the midbrain, and descend into the brain stem. Along the way, a few fibers turn off at oculomotor nuclear levels to supply the oculomotor nucleus as the remaining fibers descend and end in the abducent nucleus.

The secondary eye fields permit relaxation of the antagonistic muscles during voluntary eye deviations. Injuries to the frontal eye fields in humans are either irritative or destructive. Irritative injuries result in deviations of the eyes to the contralateral side, whereas destructive injuries result in deviation of the eyes to the same side. The effects of such injuries emphasize the close cooperation that exists between both cerebral hemispheres. Extrapyramidal cortical areas in the parietal and occipital lobes influence ocular movements, particularly following or automatic movements of the eyes.

Conjugate deviation of the eyes with pupillary dilatation of the abducting eye may appear with the onset of seizures. Such cases localize the epileptogenic focus to the contralateral frontal lobe. Two patients with bilateral ptosis and difficulty initiating lid elevation had vascular infarction of the right frontal lobe.

21.7.6. Motor Speech Region

Broca's region for **motor speech** occupies the triangular and opercular parts of the posterior end of the inferior frontal gyrus of the left cerebral hemisphere in 95% of humans. This region, corresponding to **Brodmann's areas 44 and 45** (Figs 21.7, 21.9), has connections with the primary motor cortex, especially with those parts of area 4 that supply muscles concerned with speech. **Area 44 or the pars opercularis** (which is dysgranular cortex) is separated from **area 45** or the **pars triangularis** (which is granular cortex) by the ascending or vertical ramus of the lateral sulcus whereas the anterior ramus of the lateral sulcus divides the pars triangularis from pars orbitalis (area 47) (Figs 5.9, 21.7).

Injury to the motor speech area causes an **expressive** or **motor aphasia** also described as **Broca's aphasia** or

in Broca's words as a loss of 'articulated speech'. The term **aphasia** refers to a weakening or loss of the ability to transmit ideas using language in any of its forms independent of disease of the vocal organs or their associated muscles. While there is no paralysis of the muscles related to speech, there is a loss of coordination of those muscles leading to a difficulty in expressing oneself. In **Broca's aphasia**, there is a low fluency associated with relatively well-preserved comprehension. The patient is likely to be unable to speak or continually repeats the last word said before the injury occurred. There may be difficulty with certain sounds or letters. The resulting deficit depends on the size, depth, severity, timing, and whether the cerebral hemisphere involved is the dominant cerebral hemisphere for language (the left cerebral hemisphere in approximately 95% of humans). Most patients with Broca's aphasia are able to sing, although they do not speak well. It is likely that the right cerebral hemisphere is dominant over the left for singing capacity. If injury leading to aphasia takes place at an early age, it is possible to build up speech patterns in the other cerebral hemisphere. This is an example of the plasticity of the nervous system and its ability to compensate when injured.

Imaging studies in humans with **persistent developmental stuttering** suggest a primary dysfunction in the left cerebral hemisphere with hyperactivation of the right cerebral hemisphere as a compensatory phenomenon. Underlying stuttering may be a timing problem reflecting an abnormality in complex coordination between brain areas (perhaps between prefrontal and motor speech areas) leading to a disruption in the fluency of verbal expression and the characteristic involuntary repetition or prolongation of sounds and syllables. Most cases of acquired stuttering following a stroke result from left cortical or bilateral cortical lesions.

21.7.7. Prefrontal Cortex

The '**prefrontal' cortex (PFC)** makes up about a third of the human neocortex and corresponds to three major regions on the various surfaces of the frontal lobe (Fig. 21.12). On the lateral surface of the frontal lobe, this involves the superior, middle and inferior frontal gyri anterior to the premotor cortex including **Brodmann's areas 8, 9, 10, 11, 44, 45, 46, and 47 (the lateral PFC)**. On the medial surface of the frontal lobe, this involves in part the anterior cingulate gyrus including **Brodmann's areas 8, 9, 10, 11, 12, 24, and 32 (the medial PFC)**.

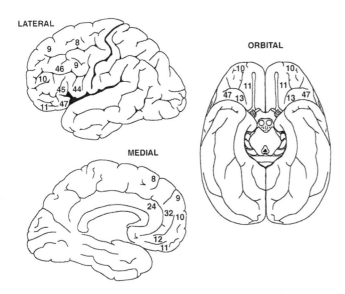

FIGURE 21.12. Human prefrontal areas numbered according to Brodmann (from Fuster, 2001).

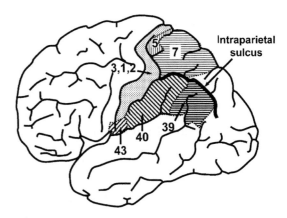

FIGURE 21.13. Brodmann's areas in the human parietal lobe (from Paxinos and Mai, 2004).

On the inferior or orbital surface of the frontal lobe, this includes **Brodmann's areas 10, 11, 13**, and **47 (the inferior or orbital PFC)** (Fig. 21.12). Collectively these three regions forming the prefrontal cortex constitute a complex association area that provides the structural basis for the most complex intellectual and moral functions of which humans are capable. These complex intellectual and moral functions include those attributes that are highly developed in educated, experienced, and cultured individuals. Such functions are modifiable by environment, memory, experiences, and the associations we have with others in our lifetime. It takes many years to develop and synthesize these complex associations. The more abstract aspects of human intellectual function, higher mental functions, creativity, and intangible activities of the nervous system are also properties of the prefrontal cortex.

Some understanding the function of these different prefrontal regions comes from studies of injuries to the human prefrontal cortex. For example, damage to the lateral prefrontal region results in a characteristic series of cognitive deficits in executive functions including the inability to formulate a plan of action (thinking) and then the inability to carry out that plan (behavior). Deficits in motility, attention, and emotion result from damage to the medial prefrontal region of the human brain.

The behavior evidenced after prefrontal injury depends on the state of development of the prefrontal cortex before the injury or damage to the cortex. By that, we mean that what results depends on what we had there and what we did with it. The resulting deficit

manifests a loss of something that was there. One cannot lose something you never had.

21.8. PARIETAL LOBE

The **parietal lobe** lies immediately behind the central sulcus and above the lateral sulcus on the lateral surface of the brain. Both the **postcentral gyrus** corresponding to Brodmann's areas 3, 1, and 2 (Fig. 21.13) and its accompanying sulcus parallel the central sulcus. The postcentral gyrus continues onto the medial surface of the brain as the posterior part of the paracentral lobule (Figs 21.6, 21.8) and it continues inferiorly to contribute to the frontoparietal operculum corresponding to Brodmann's area 43 (Fig. 21.13). An **intraparietal sulcus** (Fig. 5.6) begins in the middle part of the postcentral sulcus and runs posteriorly into the **parietal region** to separate it into **superior** and **inferior parietal lobules** (Fig. 5.6). The superior parietal lobule corresponds to Brodmann's areas 5 and 7 whereas the inferior parietal lobule includes Brodmann's areas 39 and 40. Part of the superior parietal lobule extends onto the medial surface of the brain (Figs 21.6, 21.8) down to the subparietal sulcus and the parieto-occipital sulcus. Parts of the inferior parietal lobule and the inferior frontal gyrus contribute to the upper lip of the lateral sulcus termed the **frontoparietal operculum** (Fig. 5.6). The lower part of the inferior parietal lobule includes the **supramarginal** and the **angular gyri** (Fig. 5.6). The former, corresponding to Brodmann's area 40 (Fig. 21.13) caps the posterior tip of the lateral sulcus (Fig. 5.6) and the latter, corresponding to Brodmann's area 39 caps the posterior tip of the superior temporal sulcus (Fig. 5.6). These two gyri are often indistinct, occurring instead as a region contiguous with their respective sulci.

21.8.1. Primary Somatosensory Cortex (SI)

A significant area in the **parietal lobe** is the **primary somatosensory cortex** or **SI** that corresponds to Brodmann's areas 3, 1, and 2 on the postcentral gyrus (Fig. 21.13) and the posterior part of the paracentral lobule of the parietal lobe. Area 1 is on the crown of the postcentral gyrus, area 2 intrasulcal in position forming the anterior wall of the postcentral sulcus, area 3 is divisible into areas 3a and 3b in nonhuman primates and humans, with area 3b being intrasulcal but forming the posterior wall of the central sulcus. Area 3a, also intrasulcal, forms the floor of the central sulcus, is a transition region between primary motor area 4 along the precentral gyrus and area 3b on the posterior bank of the central sulcus. Thus, there are four strip-like somatosensory areas in the human primary somatosensory area of the postcentral gyrus of the parietal lobe.

In addition to these four cytoarchitectonic subdivisions of SI, there is also a columnar cortical organization (microarchitecture) in the postcentral gyrus. Neurons in SI are in vertical columns at right angles to the surface and extend through all cortical layers; they often share similar receptive fields and have specific functions. Columns in SI are functionally interrelated by intrinsic corticocortical connections and by direct horizontal connections between columns. These intrinsic connections serve to connect parts of the cerebral cortex having different response properties yet lying in regions of similar regional representation.

The ability to recognize, localize and assess pain and thermal sensations requires cortical participation. Comparison and contrast of ongoing sensations and those previously experienced occurs in the frontal lobe thus providing a basis for immediate reactions and influencing intended behavior. Although sensations of thermal extremes and superficial pain enter consciousness at a thalamic level in humans, the primary somatosensory cortex is capable of discriminating and comparing qualities of pain and slight variations in temperature, accurately localizing these sensations with tactile and proprioceptive sensations.

Although impulses for visceral pain likely reach consciousness at a thalamic level, areas of the cerebral cortex are also involved in visceral sensation. The intralaminar nuclei, part of the 'diffuse thalamocortical projection system,' relay visceral impulses to the cerebral cortex. Their diffuseness presumably accounts for the poor localization of visceral pain. Electrical stimulation reveals that visceral sensory areas in the cerebral cortex in humans include the lower end of the **primary somatosensory cortex** and part of the **insular lobe**.

Contemporary imaging studies in humans have begun to clarify the role of cortical areas as we use our active sense of touch to explore the world around us. For example, in a study of the speed of moving stimuli across the skin of the hand (a rotating brush across the palmar surface of the fingers in contrast to rest), there was primarily activation in area 1 and secondarily in area 3b of the somatosensory cortex. In another study, the discrimination of curvatures or objects rich in curvatures led to an increase in regional cerebral blood flow outside of area 1 but perhaps in area 2 on the posterior bank of the postcentral sulcus. Positron emission tomography, cytoarchitectonic mapping, and a combination of sensory stimuli (passive or active touch, brush velocity, edge length, curvature, and roughness) reveal the probable hierarchical processing of tactile shape in sensory areas of the human brain. This involves areas 3b and 1 as the first steps in the process, area 2 being the next step, and the anterior part of the cortex lining the intraparietal sulcus (IPA) and the adjacent part of the supramarginal gyrus being the final step. The anterior part of the human intraparietal sulcus is involved in visually guided grasping. Lastly, in a fourth study, a pleasant and soft touch to the hand elicited by using velvet in comparison to a neutral stimulus produced activation in the orbitofrontal cortex. This suggests that the pleasant or rewarding aspects of touch ('that feels good') as well as the pleasant or rewarding aspects of taste ('that tastes good') or smell ('that smells good') all of which are represented in different parts of the orbitofrontal cortex, collectively contribute to our state of emotion.

The Sensory Homunculus

Electrical stimulation of the postcentral gyrus and the posterior paracentral lobule in conscious humans evokes abnormal sensations called **paresthesias**. These paresthesias include numbness and tingling on different areas of the contralateral body (sensations of pain or temperature are rare). In 163 patients, with 800 responses, coldness was reported 13 times, pain 11 times, and heat twice. If areas of the body receiving sensations are plotted along the postcentral gyrus and posterior paracentral lobule, a cortical pattern is evident. This pattern of representation, from toes to tongue, oriented along the postcentral gyrus and posterior paracentral lobule, is often illustrated as a distorted caricature, the **sensory homunculus** (Fig. 21.14). Cutaneous areas of greatest sensitivity have correspondingly larger cortical representation. Hence, the size of the parts of the homunculus is a reflection of the

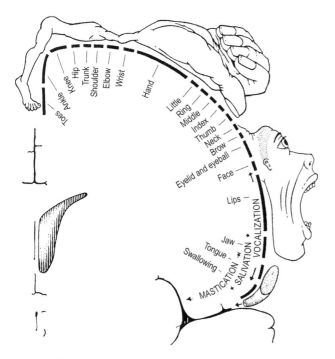

FIGURE 21.14. The sensory homunculus across the postcentral gyrus and paracentral lobule (after Penfield and Rasmussen, 1950).

cortical area devoted to individual parts of the body. The sensory homunculus has the same order as the motor homunculus in the frontal lobe. Movements also follow stimulation of the human postcentral gyrus, however, a slightly higher stimulus threshold is required and the corticospinal path must be intact. Such movements affirm the presence of corticospinal neuronal cell bodies that originate in the postcentral gyrus.

Certain aspects of the human sensory homunculus are of special interest. The toes, foot, ankle and leg have their representation in the posterior part of the paracentral lobule on the medial surface of the brain (Fig. 21.14). The representation of the thigh or hip occurs at the medial to superolateral surface junction followed in order by the trunk, neck, shoulder, arm, forearm, and hand (Fig. 21.14) across the superolateral surface of the brain. The genitalia and perineum likely have their representation on the medial surface of the human brain just ventral to the representation of the toes (though this representation is unclear). Recent recordings of cerebrocortical potentials evoked by stimulation of the dorsal nerve of the penis in humans suggest a representation of the penis with the hip and upper leg near the junction of the medial to superolateral surface. Additional studies should help to clarify the location of the cortical representation of the human genitalia and perineum. There is variation in the leg/body boundary in the

human brain. All five fingers have a separate medial to lateral representation along the wall of the central sulcus not on the lateral surface. The anterior surface of the body faces the central sulcus. A patient who had lost four fingers on his right hand underwent successful reimplantation of two fingers. Postimplantation cortical evoked fields were localizable in the primary somatosensory cortex demonstrating cortical reorganization of digit representation.

Experimental studies of the face area of SI in nonhuman primates demonstrate that oral soft tissues (gingiva, inner aspect of the lips) have their representation near the surface of the postcentral gyrus with the dental pulp along the floor of the central sulcus, coinciding with area 3a. Stimulation of dental pulp in humans causes painful sensations that are accurately localized suggesting involvement of the primary somatosensory cortex for localization of painful sensations.

The face area, occupying most of the lower half of the postcentral gyrus (Fig. 21.14), receives trigeminal impulses including those from the face and mouth. The original sensory homunculus depicted that part of the face area concerned with the upper face and head to be superior to that concerned with the lips and oral cavity. The back of the head has its representation caudally in the postcentral gyrus; the face faces the central sulcus. Sensory seizures, analyzed in many patients, indicate that a revision of the accepted homuncular pattern along the postcentral gyrus is in order. In particular, the thumb is adjacent to the lower lip and corner of the mouth (Fig. 7.6).

The Multiple Representation Hypothesis

In nonhuman primates, each cytoarchitectural component of the **primary somatosensory cortex** corresponding to **areas 3a, 3b, 1,** and **2 of Brodmann** is a separate map of the body, receiving inputs from different peripheral receptors (Fig. 21.15). Areas 3b and 1 represent cutaneous regions and are essentially mirror images, yet differ enough to suggest that each has a distinct role in cutaneous sensation. Area 2 appears to represent predominantly deep receptors, including those signaling the position of joints and other deep body sensations, whereas area 3a represents sensory input from muscles. Cortical representation along the postcentral gyrus reflects a composite of somatotopically organized areas instead of a continuous homunculus. Functional imaging studies of the human SI show that multiple digit representations occur in the four subdivisions of SI (3a, 3b, 1 and 2) resembling the **multiple representation**

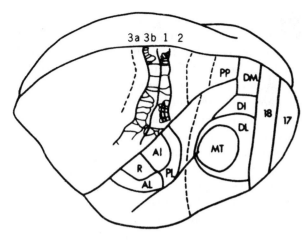

FIGURE 21.15. Multiple representations of the body within areas 3b and 1. Each architectonic field contains a representation (from Kaas et al., 1979).

hypothesis that is present in nonhuman primates (Fig. 21.15). The definition of these cytoarchitectural subdivisions during imaging studies in humans depends on approximations based on anatomical landmarks.

Thermal and Painful Auras

Loss of discriminative pain or thermal sensation may result from cortical injury. Thermal and painful sensations foreshadow the onset of an epileptic seizure, a phenomenon known as an **aura**. In 267 patients with focal seizures, only seven described painful sensations as a part of the seizure. A male patient had seizures marked by sensations of excruciating genital pain. These had been ongoing since the age of nine.

Sensory Jacksonian Seizures

An analysis of seizure patterns in 42 patients, involving subjective sensory experiences without objective signs, is a useful tool in shedding light on the function of the primary somatosensory cortex. Because they were alert and aware of their seizures, these patients were able to describe what and how they felt during the seizures. Such episodes or **sensory Jacksonian seizures**, involve abnormal, localizable, cutaneous sensations without apparent prior stimulation. The sensation spreads or progresses to adjacent cutaneous areas, in the body or along a limb, reflecting propagation of an epileptic discharge – an abnormal firing of neurons in the postcentral gyrus. This study supports the concept of a cortical sensory sequence with the thumb juxtaposed to the lower lip and corner of the mouth but not to the brow and corner of the eyelid as classically described (Fig. 7.6).

Injury to the Primary Somatosensory Cortex

Injuries to the primary somatosensory cortex result in the absence of proprioception and two point tactile sensation on the side contralateral to the injury. This often causes a loss of proprioception for the entire body or an individual limb, depending on the extent of the injury. If the injury is limited to the hand region of the primary somatosensory cortex of one cerebral hemisphere, the patient will not recognize the position of their fingers on the contralateral hand with their eyes closed. The patient without the aid of visual clues would be unable to identify objects such as a coin or key when placed in the hand. This deficit, referred to as **cortical astereognosis of the receptive type**, represents a loss of the ability to use tactile information in order to identify objects placed in the hand.

21.8.2. Secondary Somatosensory Cortex

Adjoining the primary somatosensory cortex or SI is a smaller region, the **secondary somatosensory cortex** or **SII**, on the superior bank of the lateral sulcus, continuing onto the parietal operculum of the human brain and corresponding to Brodmann's **area 43** (Fig. 21.13). More than 10 secondary somatosensory areas are identifiable in the nonhuman primate brain. SII in nonhuman primates receives fibers from the ventral posterior medial nucleus (VPM) and the caudal part of the ventral posterior lateral nucleus (VPL). Reciprocal corticothalamic fibers occur from SII in nonhuman primates to the ventral posterior nucleus (VP). Output from SII is to SI, to the primary motor cortex (corresponding to Brodmann's area 4), and to the premotor cortex (corresponding to Brodmann's area 6). SII neurons exhibit predominantly contralateral, moderate to well-defined receptive fields that are usually larger than the receptive fields of neurons in SI. Although a detailed somatotopic organization exists in SII, with the exception of parts of the digit representation, neighboring neurons in SII do not form a precise body map comparable to that in SII of nonhuman primates. In SII, there is a split in the representation for the face, thumb, and foot causing a dual representation of the same body part in the composite map of SII.

Injury to the Secondary Somatosensory Cortex

Unilateral or bilateral removal of monkey area SII causes a profound defect of tasks requiring tactile learning and retention. Even in the face of injury to the primary association cortex in the parietal lobe, patients are still able to recognize two points as two points and to identify the position of the parts of their body. However, the patient will be unable to name an object placed in the affected hand, based on somatosensory information. With this deficit, called **cortical astereognosis** patients lose the ability to use the somatosensory information.

21.8.3. Superior Parietal Lobule

That part of the parietal lobe including the primary (areas 3, 1, and 2) and secondary somatosensory cortex (area 43) is often termed the anterior parietal cortex (Fig. 21.13). The entire parietal cortex behind SI and SII is often termed the posterior parietal cortex (PPC). PPC in turn is divisible in the superior parietal lobule and the inferior parietal lobule. The superior parietal lobule lies superior to the intraparietal sulcus and includes Brodmann's areas 5 and 7 whereas the inferior parietal lobule lies inferior to the intraparietal sulcus and includes Brodmann's areas 39 and 40. Through long and short association fibers, this complex **sensory association region** receives input from visual, auditory, somatosensory, limbic, and vestibular cortical areas. Based on the serial processing of sensory information from primary to secondary sensory areas, the superior parietal lobule is likely to be a tertiary sensory association region in terms of its higher order processing. This region has connections with the pulvinar nuclei and the lateral posterior nucleus of the thalamus. Such connections permit the reinforcement, correlation, and integration of sensory information. The superior parietal lobule is involved in spatial orientation.

Injury to the Secondary Association Areas

Injuries to the parietal secondary association areas often lead to **rare visual illusions** that are modifications of the normal positions of objects. The sidewalk appears crooked, the chair or table upside down, the world upside down, or a patient is likely to describe their face as 'slipping out of position'. Such phenomena do not have any emotional component to them, as do such illusions that occur after injury to the temporal lobe.

Loss of Body Scheme

In addition to these rare visual illusions, **loss of body scheme** often results from damage to the parietal secondary association areas. Such patients do not recognize a body part as belonging to themselves. They have, in essence, an **agnosia for parts of the body**. This can occur for a digit, hand, or half of the face. If the injury is in the nondominant cerebral hemisphere, the patient may experience ignorance of the existence of paralyzed or hemiplegic limbs. Finally, a **phantom limb** is likely to occur with injury to parietal secondary association areas. In this condition, the cerebral cortex is probably receiving abnormal impulses from the stump of the amputated limb. The cerebral cortex builds up an illusion of the nonexistent limb. This is perhaps a proprioceptive not a visual illusion.

Sensory Apraxia

Injuries to the dominant parietal lobe often result in the phenomena of **sensory apraxia** in which those affected are able to follow simple commands, but cannot carry out a series of movements in which each step depends on the preceding movement. Asking a patient to go to the table, pick up a cup, pour water in it, drink the water, dry the cup, and return to the table is a formidable task. The inability to carry out such simple commands is probably the result of a proprioceptive loss in the sequence of motor sensory motor responses.

Sensory Extinction

With parietal lobe injury, the simultaneous application of the same sensory stimulus on symmetrical parts of the body causes a failure to perceive one stimulus on the side opposite the injured parietal lobe whereas the patient is able to perceive isolated single stimulation of each point on the two sides of the body. Thus, the patient is able to perceive the examiner touching the tips of both index fingers when the stimuli are applied individually, but not when applied simultaneously. This phenomenon, called **sensory extinction**, is a subtle symptom of parietal lobe injury due to a perceptual rivalry, such that the injured area loses out and the sensation coming from that side of the body is extinguished. Critchley (1953) has pointed out that within the brain no territory surpasses the parietal lobe in the rich variety of clinical phenomena that follow disease states.

21.8.4. Inferior Parietal Lobule: Language Areas

The inferior parietal lobule (Fig. 21.13) includes the **supramarginal gyrus** and the **angular gyrus** corresponding to **Brodmann's areas 40 and 39** respectively (Figs 21.7, 21.13). The intraparietal sulcus forms the dorsal border of the inferior parietal lobule. The supra-marginal gyrus caps the posterior tip of the lateral sulcus (Fig. 5.6) whereas the angular gyrus caps the posterior tip of the superior temporal sulcus (Fig. 5.6). These two gyri are often indistinct, occurring instead as a region contiguous with their respective sulci. The **angular gyrus**, corresponding to **Brodmann's area 39**, is functionally involved in language. Bilateral injury to this gyrus or a unilateral injury in the dominant cerebral hemisphere will lead to a **visual receptive aphasia**. Patients with such injuries cannot read nor are they able to recognize words as words. Therefore, they cannot write well, although they are able to talk and understand speech. They cannot name an object they are able to see, although they know how to use it, a condition termed **alexia without agraphia** or the inability to read although the ability to write remains.

The **supramarginal gyrus**, particularly in the left cerebral hemisphere (dominant for language in 95% of humans) and corresponding to **Brodmann's area 40**, plays a role in the **symbolism of language**. Injury to this region causes the loss of the ability to express oneself in language, a condition called **conductive aphasia**. Speech is broken but comprehension is good.

21.8.5. Primary Vestibular Cortex (2v)

The **primary vestibular cortex**, along the upper lip of the intraparietal sulcus (Figs 11.7, 21.7), has the cytoarchitectural appearance of **area 2**. This region functions in vestibular consciousness – the awareness of position or movement of our body caused by the effects of gravity and acceleration or deceleration. Other than **vertigo** (the abnormal sensation of self-motion or of the motion of external objects often described as 'dizziness'), there is no easily definable, discrete vestibular sensation. Instead, it appears that vestibular stimuli integrate with visual and somatosensory modalities. In the process of this integration, vestibular stimuli seem to lose their original, pure character. This overlap of vestibular and visual with other somatosensory modalities often allows each to compensate for a deficiency in the other. Injuries along the intraparietal sulcus lead to dizziness and a subjective feeling that the world is whirling about the patient. There is the illusion of rotation or of floating.

21.8.6. Mirror Representation of Others' Actions

One of the most interesting findings in recent years related to the human cerebral cortex is that some cortical motor areas that are active during the planning and execution of movements are also active when viewing the motor actions of others. This phenomenon is termed 'mirror representation' of the actions of others. The anterior part of the human intraparietal sulcus is involved in visually guided grasping as well as during the observation of others' actions. There is a complex network of cortical regions in the frontal, parietal, and temporal lobes forming this **mirror system** in the human brain.

21.8.7. Preoccipital Areas Involved in Following Ocular Movements

In the parietal lobe and in part of the occipital lobe, is an area involved in **following or automatic ocular movements**. In the parietal lobe, this area corresponds to **Brodmann's area 19** while in the occipital lobe it corresponds to **Brodmann's area 18** (Figs 21.7, 21.8). One is often not aware of these movements as the eyes fix on a moving object. As we watch a ping-pong or tennis match, our eyes move back and forth following the path of the ball; yet, after a while, we are unaware of the fact that our eyes are moving to follow the ball from one side of the table or court to the other. The eyes automatically follow the moving object until the movement ceases, or until we voluntarily turn them elsewhere. These ocular movements are extrapyramidal in type and obviously significant as emphasized by the presence of two different cortical fields (the frontal eye fields for voluntary ocular movement and preoccipital areas for following movements of the eyes) as well as two separate extrapyramidal paths for their production. Cortical stimulation of parietal area 19 and occipital area 18 yields deviation of the eyes upward, downward, or horizontal deviations. If the upper part of area 19, or the lower part of area 18, is stimulated, upward deviation of the eyes will occur. Presumably, there are connections from these two regions of the parietal and occipital lobes, respectively, to the tegmentum of the midbrain by way of the internal corticotectal fibers. In particular, these fibers pass to the rostromedial part of the superior colliculus. From the rostromedial superior colliculus, tecto-oculomotor fibers pass to the oculomotor and trochlear nuclear complexes of the midbrain. They end on neurons that supply muscles that raise the lids and lift the eyes in an upward direction. If the upper part of area 18, or the lower part of area 19, is stimulated, downward deviation of the eyes will occur. The anatomic basis for this is by way of **internal corticotectal fibers** that pass from the parietal and occipital lobes, respectively, to the caudolateral part of the superior colliculus. From this region, tecto-oculomotor fibers pass to the oculomotor nucleus, particularly to those neurons that supply muscles that lower the eyes. If the middle part of parietal area 18 and the middle part of occipital area 19 is stimulated, horizontal deviations of the eyes will occur. Underlying these deviations are the corticotegmental paths, which pass from the cerebral cortex to the tegmentum of the midbrain. From there, connections fibers project to the contralateral abducent nucleus and the abducent reticular gray matter to initiate a horizontal deviation of the eyes.

The integrity of areas 18 and 19 and the fiber paths from them can be tested by producing a **physiological nystagmus** with the aid of a rotating drum with black and white vertical stripes or the usage of a nystagmus tape that has vertical black and white stripes on it. The patient fixes his eyes on the black lines, follows them until they are no longer visible, and then fixes their eyes on the next black line. Following this process produces **nystagmus** as a normal physiological response. Nystagmus involves fine, rhythmic, rapid, back and forth movements of the eyes. Absence of such nystagmus is abnormal. Such nystagmus is an exaggerated repetitious sequence of following ocular movements, often referred to as **optokinetic nystagmus**, also called optomotor or railway nystagmus. The origin of the term railway nystagmus denotes these types of ocular movements often occur naturally in an individual watching the movement of telephone poles from a steadily moving train (or more likely today, an automobile).

The connections and functional aspects of ocular movements are exceedingly complex. We have provided only a glimpse of them in this discussion. In the future, we hope to better understand and appreciate the functional interrelations of the various areas of the brain involved in ocular movements. Bilateral destruction of the frontal eye fields makes it impossible for an individual to turn their eyes in any direction on command. The patient is still able to read by placing their finger under the words on a page and following their moving finger from word to word. In this regard, following ocular movements often substitute and compensate for the loss of frontal eye fields and the loss of voluntary ocular movements.

21.9. OCCIPITAL LOBE

21.9.1. Primary Visual Cortex (V1)

The **primary visual cortex** or **striate area** corresponding to **Brodmann's area 17** (Figs 12.9, 21.8) is functionally related to conscious vision. Anatomically, it involves the lips of the calcarine sulcus on the medial surface of the brain. The superior lip of the calcarine sulcus receives fibers projecting from the superior retinal quadrants whereas the inferior lip of the calcarine sulcus receives projections from the inferior retinal quadrants. The macular fibers project most posteriorly along the calcarine sulcus, followed by fibers from the paramacular areas, and finally, most anteriorly the peripheral retinal fibers project more anteriorly along the calcarine sulcus. Unilateral injury to the calcarine sulcus will lead to a contralateral homonymous hemianopia.

21.9.2. Secondary Visual Cortex

Adjoining the primary visual cortex is a **secondary visual cortex** corresponding to **Brodmann's area 18** (Figs 12.9, 21.8). This area, also termed the **parastriate area,** has connections with the primary visual cortex and with area 19 in the occipital lobe through short association fibers. Area 18 has connections with areas 18 and 19 in the contralateral cerebral hemisphere by way of commissural fibers. Long association fibers connect area 18 with the frontal lobe permitting this region to receive sensory, motor, and auditory impulses as well as those from the insular and temporal lobe. Area 18 interprets, associates, and facilitates the understanding of impulses that reach area 17. Information from area 18 in the dominant cerebral hemisphere correlates with that from area 18 in the nondominant cerebral hemisphere. Area 18 participates in color vision in humans. Area 18 and area 19 in the parietal lobe are involved with following or automatic eye movements. Stimulation of area 18 in humans yields unformed images, wheels, flashing lights, streaks, flickering lights, and colored spots. Injury to area 18 will result in the loss of following eye movements toward the contralateral side. Although the patient is still able see because the primary visual cortex is intact, they often present with a **visual agnosia** – the loss of power to recognize the import of visual stimuli in the absence of visual defects. The importance of the existence of long association fibers interrelating the frontal lobe with the occipital lobe is evident in a patient with an injury in the middle frontal gyrus who saw stars and flashes of light. This suggests that the occipital lobe was being stimulated when in fact the lesion irritated fibers in the frontal lobe that interrelate the frontal and the occipital lobes.

21.10. TEMPORAL LOBE

The human temporal lobe lies below the lateral sulcus. Its posterior boundary is an imaginary vertical line from the preoccipital notch to the parieto-occipital sulcus

(Fig. 5.9). The temporal lobe has three gyri: **superior, middle**, and **inferior temporal gyri**, separated by two sulci: **superior temporal** and **inferior temporal** (the latter was formerly termed the middle temporal sulcus) (Fig. 5.7).

21.10.1. Primary Auditory Cortex (AI)

Two transversely-oriented gyri, the anterior and posterior transverse temporal gyri are continuous with the superior temporal gyrus but folded and hidden from view when one observes the lateral surface of the cerebral hemisphere. The **anterior transverse temporal gyrus** (of Heschl), lying in the depth of the lateral sulcus and corresponding to **Brodmann's area 41** (Fig. 21.7), is the **primary auditory cortex** or **AI** (Figs 10.6, 21.16). The length and surface area of the left primary auditory cortex is greater than the right in newborns and adults. The circular insular sulcus limits area 41 medially while laterally area 41 does not reach the temporal operculum. Orderly projections from neurons of the medial geniculate body to the primary auditory cortex are termed **thalamotemporal projections** or **acoustic radiations** (Fig. 10.5) that reach the temporal lobe through the sublenticular part of the internal capsule. Neurons in each medial geniculate body project fibers to layers IIIb and IV of the auditory cortex. Area 41 is sensory or koniocortex with layers II–IV of densely arranged small neurons surrounded by fields with large to medium-sized IIIc pyramidal neurons. Sounds are heard in the primary auditory cortex and a pattern for tones of different pitch exists here in nonhuman primates and, presumably, also in humans.

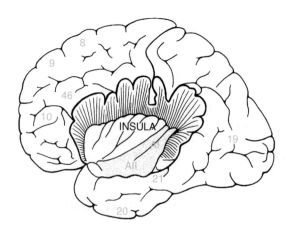

FIGURE 21.16. Following removal of the opercula of the frontal and parietal lobes, the insular lobe and the anterior and posterior transverse temporal gyri are revealed (from Standring, 2005).

21.10.2. Wernicke's Region

The posterior transverse temporal gyrus corresponds to **Brodmann's area 42** (Figs 21.5, 21.7) whereas the lateral most parts of both transverse temporal gyri that extend onto the opercular surface of the superior temporal gyrus correspond to **Brodmann's area 22**. Those parts of the superior temporal gyrus corresponding to **Brodmann's areas 42 and 22** constitute **Wernicke's region**, an **auditory association region**. In this region, sounds are appreciated, they become meaningful for understanding language, and are interpreted as words. Although areas 41 and 42 are in the lateral sulcus, maps of Brodmann's areas (Figs 21.5 and 21.7) often show these areas as though they were visible on the lateral surface even though they are not. The superior temporal gyrus that is visible on the lateral surface of the hemisphere consists primarily of area 22 and perhaps a small part of area 52 (Fig. 21.5). Area 52 is the highest number in the Brodmann system of mapping cortical areas.

Bilateral destruction of the primary auditory cortex, the bordering area 22, or both will result in **cortical deafness**. Cortical deafness is a total bilateral loss of hearing caused by bilateral cerebral injuries. Although such patients have bilateral cerebral injury, they often have a reasonably normal audiogram because tones of different pitch come to consciousness at a thalamic level. Destruction of only one primary auditory cortex will lead to subtle auditory deficits. Such patients have difficulty localizing sounds in the contralateral auditory field and perhaps a loss of discrimination of distorted, interrupted, or accelerated speech.

Destruction of the auditory association cortex (**Wernicke's region**) in the left cerebral hemisphere (dominant for language in 95% of humans) often yields a handicapping **auditory receptive aphasia**. Such patients have normal hearing and the normal production of words (fluent). Yet they are unable to interpret spoken sounds related to language or use words appropriately. Patients with injuries limited to the parietal operculum are often indifferent to loud and unpleasant noises.

21.10.3. Temporal Vestibular Cortex

On the superior temporal gyrus, anterior to the auditory areas, is a **temporal vestibular cortex** (Fig. 11.7). Irritative injuries of this region often cause the subjective feeling of vertigo (the abnormal sensation of self-motion or of external objects, often described as 'dizziness'). Such patients describe themselves as though they were whirling in space. The experimental application of cortical irritants to this region causes monkeys to somersault backwards repeatedly, a dozen or more times without stopping. Although the **primary vestibular cortex** is in the parietal lobe, the **temporal vestibular cortex** requires further clinical and experimental documentation with regard to its relationship with the parietal vestibular region (Fig. 11.7).

21.10.4. Midtemporal Areas Related to Memory

In the middle part of the inferior temporal gyrus, on the lateral and ventral surface, and in the region dorsal to the secondary auditory cortex, are visual and auditory association areas. These secondary association areas are richly interrelated with occipital, frontal, parietal, and insular cortices. Penfield and his associates suggested that these midtemporal areas participate in the recall of past events or so-called **remote memory**. This appears to be the case especially with events that have visual and auditory components visualized and heard at some time in the past, and those events accompanied by some emotional stress or intellectual conflict. These areas have connections with the frontal lobe and **limbic system** including the **amygdaloid body** and **hippocampal formation**. A destructive injury to these midtemporal regions often interferes with the recall of past events. Slight irritation of these midtemporal areas often causes the vague feeling of familiarity that falls under the term **déjà vu** [French, already seen]. These visual association areas are likely to form visual hallucinations consisting of events visualized in the past having an emotional connotation to them. These are usually formed visual hallucinations of events that have already occurred in the past. Irritation of auditory association areas in these midtemporal regions is likely to produce reminiscent speech or sounds, particularly a piece of music or a sound of someone's voice from our previous experience. The temporal lobe probably plays a vital role in processing olfactory and gustatory information. Irritations of midtemporal regions often cause olfactory and gustatory auras or hallucinations. The human temporal lobe is probably involved in odor detection and is involved in olfactory memory functions, particularly the right temporal lobe. Odor experiences often cause a retrieval of images, an unfolding of past memories, and the initiation of a complex feeling state. During his seizures, one patient experienced the sound of an airplane flying overhead, was able to look up and

see the airplane, and then saw the airplane crash ultimately being able to smell the burning rubber of the tires. This particular patient obviously displayed auditory, visual, and olfactory hallucinations.

21.10.5. Anomia

In the posterior temporal lobe, particularly parts of the fusiform and probably the lingual gyrus, corresponding to **Brodmann's areas 37 and 29** (Fig. 21.8), is a region connected by the inferior occipitofrontal fasciculus with the inferior and perhaps midtemporal regions and with other cortical areas along its course. This region connects with the inferolateral and part of the lateral area of the frontal lobe. Destruction of area 37 is reported to result in an **anomia** – difficulty remembering the names of familiar objects.

21.10.6. Prosopagnosia

In the posterior temporal and bordering occipital region are complex association areas. A bilateral infarct of the posterior cerebral artery territory often causes failure to recognize previously well-known people and common objects. A severe deficit in discriminating and identifying faces with the inability to recognize one's own pictures or mirror images is likely to occur with injury to both fusiform and lingual gyri. Brain imaging studies reveal the fusiform gyrus responds more readily to faces than to other objects and that the middle fusiform gyrus (MFG) and the inferior occipital gyrus (IOG) are activated by both detection and identification of faces. The visual association properties of these regions include the synthesis of specific faces, which requires the complex synthesis of visual input, memory, and other sensory phenomena. Such a failure to recognize faces is termed **prosopagnosia** or face blindness. Individuals with this condition can compensate quite well. They may learn to recognize people by their clothes, mannerisms, gait, hairstyle or voice. While they cannot recognize a person by their face, they are able to recognize facially expressed emotions (happy, sad, angry).

21.10.7. Psychomotor Seizures

During temporal lobe seizures, the extrapyramidal motor area in the middle and inferior temporal gyri function is stimulated causing movements of the face, arm, and leg. Hughlings Jackson first drew attention to the temporal lobe origin of epileptic dreamy states and used the term **psychomotor epilepsy** to designate seizure episodes with various degrees of impairment of consciousness. He noted that such seizure episodes include dreamy states, episodes of confusion, abrupt changes in mood, and automatic behavior with amnesia. He also noted an accompanying motor behavior, such as simple staring, incoherence of speech, repetitive licking or chewing, or other automatic movements. There is little doubt that the motor behavior accompanying temporal lobe or psychomotor epilepsy results from activation of the extrapyramidalmotor areas of the middle and inferior temporal gyri. Personality or behavioral changes often provide a clue in the diagnosis of temporal lobe epilepsy. The observation of facial asymmetry in patients with temporal lobe epilepsy is often a useful clinical sign enabling one to suspect the presence of a temporal focus contralateral side to the facial weakness. Weakness is not striking but does occur on the contralateral lower face in 73% of the patients with temporal lobe epilepsy.

21.11. INSULAR LOBE

The fifth lobe of the brain is the **insular lobe** (Fig. 21.16), hidden by the opercular parts of the frontal, parietal, and temporal lobes. The term operculum [Latin, lid or covering] refers to the fact that these parts of the respective lobes form a lid or cover the insular lobe. The insular lobe forms the floor of the lateral sulcus. On the surface of the insula, a **central sulcus of the insula** separates two groups of insular gyri. In front of the central sulcus are the **short gyri of the insula** and behind them is the **long gyrus of the insula**. The **circular sulcus of the insula** surrounds these sulci and gyri of the insula (Fig. 5.8).

The insular lobe in primates including humans has connections with the cerebral cortex of the frontal lobe (orbital cortex, frontal operculum, lateral premotor cortex, ventral granular cortex and medial area 6) and the parietal lobe (second somatosensory area and retroinsular area of the parietal lobe). There are also insular connections with the temporal lobe (temporal pole and superior temporal sulcus) and the insular lobe (an abundance of local intra-insular connections). There are insular projections to subdivisions of the cingulate gyrus and to the lateral, lateral basal, central, cortical and medial amygdaloid nuclei. There are also insular connections with nonamygdaloid areas such as the perirhinal cortex, entorhinal cortex, and the periamygdaloid cortex. The thalamic tase area projects fibers to the ipsilateral insular-opercular cortex.

Functionally the insula is a multifaceted sensory and motor area. It has visceral sensory and somatic sensory roles as well as visceral motor and somatic motor roles. The insula participates in the autonomic regulation of the gastrointestinal tract and the heart as well as functioning as a motor association area. This motor function includes a role in the recovery of motor function after stroke and in ocular movements. The insular lobe also serves as a vestibular area and as a language area including memory tasks related to language and auditory processing underlying speech. The insula is a limbic integration cortex with some involvement in Alzheimer's disease. Other insular functions include a possible role in the verbal component of memory and its role in selective visual attention. Adequate stimulation of the insular lobe demonstrates an extrapyramidal motor pattern. Experimentally produced seizures involving the insular lobe in nonhuman primates often yield circling or rotational movements that are similar to those observed in humans with lesions of the insula. The insular lobe is also an olfactory gustatory correlation region. Irritative injuries to the anterior insular lobe in humans often result in a gustatory aura preceding convulsions. Such gustatory auras characteristically include a bitter taste in the mouth. Irritation of the caudal end of the insular lobe often causes gastric or abdominal discomfort or epigastric pain preceding seizures. It may be that visceral structures from the mouth to the end of the gastrointestinal tract have a representation along the insula.

21.12. APHASIA

Certain functions are performed better or solely by one cerebral hemisphere. Language – a unique, human capacity – and handedness are two examples of unilateral hemispheric functions. Approximately 95% of humans have their left cerebral hemisphere dominant for language. Hemispheric language development may occur by about eight years of age. The right cerebral hemisphere retains the potential to develop language up to that time. Children who sustain cortical injuries often make a good recovery and use the remaining undamaged hemisphere to develop language providing the insult occurs before this early age. More than 50% of left-handers are language left. For this reason, many authors suggest the left cerebral hemisphere serves verbal activities, whereas the right cerebral hemisphere serves nonverbal activities. Therefore, cerebral dominance implies a predominantly unilateral function.

Aphasia is a disorder of previously intact language occurring secondary to unilateral brain injury independent of disease of the vocal organs. An injury of the left cerebral hemisphere will produce a severe aphasia. However, a similar injury, if in a symmetric part of the right cerebral hemisphere, may produce no language disturbance. About 40% of vascular injuries to the cerebral hemisphere result in language disturbances. Added to this are cases of trauma, tumors, degenerative disease, which will further increase the incidence of aphasia. Although the incidence of aphasia is frequent, its presence often escapes notice or is misinterpreted as confusion or psychoses. Individuals fluent in two or more languages usually show the same degree and type of impairment in all languages. Inadequate evidence exists to suggest that true aphasia follows nondominant cerebral hemisphere injury. Most aphasias are a remnant of cerebrovascular disease, especially involving the middle cerebral or internal carotid arteries.

21.12.1. Historic Aspects of Aphasia

The modern study of aphasia began with the observation by **Broca** that a patient of his who had lost the faculty of speech had incurred an injury involving the posterior part of the inferior frontal gyrus. Broca called attention to the fact that the left cerebral hemisphere was dominant for language, and the region injured in his patient is now termed **Broca's region for motor speech**. This region corresponds to Brodmann's areas 44 and 45 (Fig. 21.17). This particular patient had expressive or motor aphasia, a disorder of the expression of spoken language.

In the 1870s, Wernicke established the existence of aphasia in which there was a disorder of comprehension not of expression. Wernicke noted that these disorders were associated with injuries to the posterior part of the superior temporal gyrus corresponding to Brodmann's area 22 – an area often given the designation **Wernicke's region**. It appears that in his discussions of language areas of the brain, Wernicke was describing the inferior frontal gyrus, the parietal operculum, including the supramarginal and angular gyri, and the superior and middle temporal gyri – the entire area surrounding the lateral sulcus. This posterior part of the superior temporal gyrus (Wernicke's region) connects via a fiber bundle, the arcuate fasciculus, with Broca's area.

21.12.2. Broca's Aphasia

Broca's aphasia, also referred to as a motor aphasia, an expressive aphasia, a verbal aphasia, an efferent aphasia, or the loss of articulate speech, was the first variety described with a specific anatomic localization. The injury involves the posterior part of the inferior frontal gyrus (Fig. 21.18). Such patients have a nonfluent output; their words are sparse and poorly articulated with great effort. Their comprehension, however, is better than their speech and they readily comprehend written material. Such patients have the capacity to sing, because the right cerebral hemisphere is probably more responsible than the left for singing, but the degree of dominance is not as high for singing as it is for speech.

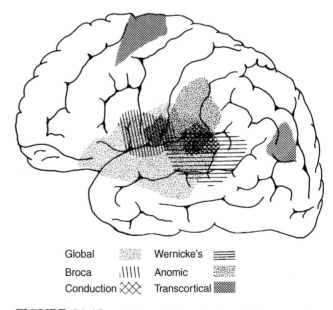

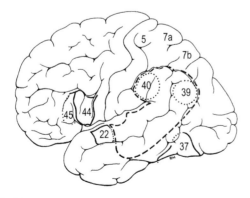

FIGURE 21.17. Broca's area and Wernicke's area on the lateral surface of the human brain (from Standring, 2005).

FIGURE 21.18. Radionuclide localization of infarcts producing different types of aphasia (from Kertesz et al., 1977).

21.12.3. Wernicke's Aphasia

Wernicke's aphasia is a type aphasia in which there is fluent speech output such that the expression of speech is satisfactory, as is articulation. The significant difficulty is in the comprehension of spoken language. Such patients understand nothing said to them and therefore they will repeat nothing. Reading is also disturbed, but the degree of disturbance parallels the disturbance of spoken comprehension. The theory is that most language initially develops as auditory language. Visual language develops next and is associated with the already learned auditory language. Therefore, disturbances of auditory language because of cerebral injury will disturb this visual-auditory interaction. Other terms for Wernicke's aphasia are **sensory aphasia, central aphasia, receptive aphasia**, and **auditory aphasia**. Injury in Wernicke's aphasia is usually limited to the posterior third of the superior temporal gyrus, in or near the secondary auditory cortex in the dominant cerebral hemisphere, corresponding to Brodmann's area 22 (Fig. 21.18).

21.12.4. Conductive Aphasia

Disconnection between Broca's and Wernicke's areas often causes **conductive aphasia**. The injury usually involves the parietal lobe, especially in the white matter, deep to the supramarginal gyrus. The injury usually involves the superior longitudinal fasciculus or other association bundles that join Broca's region for motor speech with Wernicke's area. Radioactive uptake studies indicate that conduction aphasia usually involves the upper lip of the lateral sulcus, including the parietal operculum, the supramarginal gyrus, and perhaps the precentral operculum, extending deeply to involve the insular lobe and the underlying white matter. In conductive aphasia, the fluent output that exists is less than that in Wernicke's aphasia. Speech is often broken into four or five word phases such as 'I don't know if I can,' or 'What did you say?' Comprehension is good, especially for normal conversation. However, such patients have great difficulty in repetition. They read silently for comprehension with ease but cannot read aloud.

21.12.5. Global Aphasia

Global aphasia, also termed **total aphasia**, results from large, severe injuries of the brain. Such patients cannot speak nor can they comprehend speech, read, write, repeat, or name. Recent radioactive uptake studies have indicated that injuries in global aphasia are anatomically limited to the inferior frontal, superior temporal, and the region of the lateral sulcus with extensive depth, involving the basal nuclei, large areas of insular cortex, and the subcortical white matter (Fig. 21.18).

21.13. ALEXIA

Alexia is a cortical disorder of reception in which there is the loss of the ability to comprehend written material manifested as an impairment of reading. Those afflicted with alexia have at one time had the ability to read, but no longer comprehend written material. Thus, alexia differs from dyslexia, a term used to describe a specific developmental reading disability in children and adults.

Hemialexia is the loss of reading ability in the nondominant visual field as the result of surgical section of the corpus callosum. Alexia, with agraphia, is acquired inability to read and write, probably localized to the left angular gyrus of the parietal lobe (the left cerebral hemisphere is dominant for language in 95% of humans), corresponding to **Brodmann's area 39** (Figs 21.7, 21.13, 21.17). Such patients can comprehend spoken words. Alexia without agraphia is a rare and uncommon disorder. Such patients write fluently, but cannot read what they have written. They cannot read words or letters, but can read numbers. The injury usually involves the dominant medial occipital lobe. The intact right primary visual cortex has probably lost its connection with the intact dominant visual language area, particularly the angular gyrus. Patients see words, but cannot make the associations necessary for their identification. Some patients are able to learn to trace letters with their fingers or use ocular movements to identify these letters and thus read, though slowly and inefficiently.

21.14. APRAXIA

Apraxia refers to the inability to carry out motor activities in the presence of intact motor and sensory systems, comprehension, attention, and cooperation. Many types of apraxia exist, one of which is a difficulty in skilled movement sequences. This is termed **ideomotor apraxia**. In response to verbal commands, such patients are unable to carry out a motor activity that they would normally perform with ease spontaneously. When asked to stick out their tongue, such patients cannot do so, but they often do so spontaneously to lick their lips. **Apraxia for dressing** refers to the inability to dress oneself, because of a difficulty in performing the required skilled movements. Apraxia for walking obviously makes walking very difficult.

Constructional apraxia is a special form of apraxia manifested by a difficulty in figure copy or constructing. Such patients have the inability to carry out tasks involving drawing or constructing and arranging blocks and objects. Some patients with constructional apraxia show difficulty in the use of tools, or a difficulty in tying knots or building block designs into patterns. Such three-dimensional constructional tasks are extremely difficult for the apraxic patient. Patients are unable to analyze special relationships and to execute simple constructional tasks under visual control. An excellent example of construction apraxia is a medical student with constructional apraxia who had a hard time finding his bearings in surgery. The patient said that he could not see how to manipulate the bones in order to restore their alignment.

21.15. GERSTMANN'S SYNDROME

Gerstmann's syndrome, first described by him in 1924, initially included only one component – **finger agnosia**. Since that initial description, other observers have identified three additional components of this syndrome. A **right-left disorientation** exists in which the patient cannot point to the right and left sides of their own body on command. They cannot do the reverse on a person sitting opposite them. They cannot move in the direction ordered to tell whether objects are to their right or left. A third component of Gerstmann's syndrome is **acalculia**, which is a disturbance of the ability to calculate. A final component of Gerstmann's syndrome is **agraphia**, or the difficulty with spelling and sentence construction. Agraphia is often so severe that the patients are unable to form letters or combine the letters into words. The injury yielding Gerstmann's syndrome is usually in the dominant parietal lobe. Some have suggested components of Gerstmann's syndrome are in children with learning disorders, but without dyslexia or hyperkinesia. Such children usually are able to read, but have difficulty with math, spelling, and handwriting.

21.16. AGNOSIA

A tremendous confusion exists regarding the concept of agnosia. Simply stated, this is a difficulty in recognition with a variety of disturbances of recognition occurring. **Color agnosia** refers to loss of the ability to recognize colors confined to half of the visual field. **Finger agnosia** refers to the inability to recognize parts of the hand including fingers themselves. Prosopagnosia refers to the loss of the ability to recognize familiar faces. Injuries yielding prosopagnosia are in the posterior temporal and occipital regions, especially involving the fusiform and lingula corresponding to **Brodmann's areas 37 and 19** (Fig. 21.7). A visual agnosia refers to the inability to use visual information to appreciate the character or meaning of an object. Such persons cannot name an object, cannot demonstrate its use, or cannot remember seeing it before. If they are allowed to use a different sense, such as touch, they are likely able to find the name and employ the object appropriately. Agnosia often results from focal injury to the occipital or the parieto-occipital region. They are likely to be disconnection syndromes in that they are disconnecting the somatosensory cortex from the motor speech area.

21.17. DYSLEXIA

Dyslexia, the most common neurobehavioral disorder affecting children, is a specific reading disability characterized by the inability to learn to comprehend the sounds related to the written language. Children with dyslexia may have difficulty in learning to read beyond that anticipated for their age and general intelligence. Dyslexia may be associated with cortical injury or abnormal cortical development of the temporo-parieto-occipital region, striate or extrastriate cortex. There may be a problem with a specific component of the language system such that the sounds of speech are inappropriately processed.

FURTHER READING

Amunts K, Schlaug G, Schleicher A, Steinmetz H, Dabringhaus A, Roland PE, Zilles K (1996) Asymmetry in the human motor cortex and handedness. NeuroImage 4:216–222.

Augustine JR (1985) The insular lobe in primates including humans. Neurol Res 7:2–10.

Augustine JR (1996) Circuitry and functional aspects of the insular lobe in primates including humans. Brain Res Rev 22:229–244.

Bodegård A, Geyer S, Grefkes C, Zilles K, Roland PE (2001) Hierarchical processing of tactile shape in the human brain. Neuron 31:317–328.

Bodegård A, Geyer S, Naito E, Zilles K, Roland PE (2000) Somatosensory areas in man activated by moving stimuli: cytoarchitectonic mapping and PET. NeuroReport 11:187–191.

Bodegård A, Ledberg A, Geyer S, Naito E, Zilles K, Roland PE (2000) Object shape differences reflected by somatosensory cortical activation. J Neurosci 20:(RC51)1–5.

Bradley WE, Farrell DF, Ojemann GA (1998) Human cerebrocortical potentials evoked by stimulation of the dorsal nerve of the penis. Somatosens Mot Res 15:118–127.

Braitenburg V, Schüz A (1998) *Cortex Statistics and Geometry of Neuronal Connectivity*. 2nd ed. Springer Verlag, Berlin.

Brandt T, Dieterich M (1999) The vestibular cortex. Its locations, functions, and disorders. Ann NY Acad Sci 871:293–312.

Burton H, Fabri M, Alloway K (1995) Cortical areas within the lateral sulcus connected to cutaneous representations in areas 3b and 1: a revised interpretation of the second somatosensory area in macaque monkeys. J Comp Neurol 355:539–562.

Disbrow E, Roberts T, Krubitzer L (2000) Somatotopic organization of cortical fields in the lateral sulcus of Homo sapiens: evidence for SII and PV. J Comp Neurol 418:1–21.

Eickhoff SB, Schleicher A, Zilles K, Amunts K (2006) The human parietal operculum. I. Cytoarchitectonic mapping of subdivisions. Cereb Cortex 16:254–267.

Eickhoff SB, Amunts K, Mohlberg H, Zilles K (2006) The human parietal operculum. II. Stereotaxic maps and correlation with functional imaging results. Cereb Cortex 16:268–279.

Fuster JM (2001) The prefrontal cortex – an update: time is of the essence. Neuron 30:319–333.

Gelnar PA, Krauss BR, Szeverenyi NM, Apkarian AV (1998) Fingertip representation in the human somatosensory cortex: an fMRI study. NeuroImage 7:261–283.

Geyer S, Ledberg A, Schleicher A, Kinomura S, Schormann T, Burgel U, Klingberg T, Larsson J, Zilles K, Roland PE (1996) Two different areas within the primary motor cortex of man. Nature 382:805–807.

Geyer S, Schleicher A, Zilles K (1999) Areas 3a, 3b, and 1 of human primary somatosensory cortex. 1. Microstructural organization and interindividual variability. NeuroImage 10:63–83.

Geyer S, Schormann T, Mohlberg H, Zilles K (2000) Areas 3a, 3b, and 1 of human primary somatosensory cortex. Part 2. Spatial normalization to standard anatomical space. NeuroImage 11:684–696.

Grodzinsky Y, Amunts K (2006) *Broca's Region*. Oxford University Press, New York.

Hadjikhani N, de Gelder B (2002) Neural basis of prosopagnosia: an fMRI study. Hum Brain Mapp 16:176–182.

Heilman KM, Valenstein E (1993) *Clinical Neuropsychology*. 3rd ed. Oxford University Press, New York.

Kaas JH (1999) The transformation of association cortex into sensory cortex. Brain Res Bull 50:425.

Kaas JH (2004) Evolution of somatosensory and motor cortex in primates. Anat Rec A Discov Mol Cell Evol Biol 281:1148–1156.

Krubitzer L, Clarey J, Tweedale R, Elston G, Calford MA (1995) Redefinition of somatosensory areas in the lateral sulcus of macaque monkeys. J Neurosci 15:3821–3839.

Lobel E, Kahane P, Leonards U, Grosbras M, Lehericy S, Le Bihan D, Berthoz A (2001) Localization of human frontal eye fields: anatomical and functional findings of functional magnetic resonance imaging and intracerebral electrical stimulation. J Neurosurg 95:804–815.

Mesulam MM (2004) The cholinergic innervation of the human cerebral cortex. Prog Brain Res 145:67–78.

Milea D, Lobel E, Lehéricy S, Duffau, Rivaud-Péchoux S, Berthoz A, Pierrot-Deseilligny C (2002) Intraoperative frontal eye field stimulation elicits ocular deviation and saccade suppression. NeuroReport 13:1359–1364.

Moore CI, Stern CE, Corkin S, Fischl B, Gray AC, Rosen BR, Dale AM (2000) Segregation of somatosensory activation in the human Rolandic cortex using fMRI. J Neurophysiol 84:558–569.

Mufson EJ, Sobreviela T, Kordower JH (1997) Chemical neuroanatomy of the primate insula cortex: relationship to cytoarchitectonics, connectivity, function and neurodegeneration. In: Björklund A, Hökfelt T, eds. *Handbook of Chemical Neuroanatomy*. Vol 13: The Primate Nervous System, Part I, pp 377–454. Elsevier, Amsterdam.

Overduin SA, Servos P (2004) Distributed digit somatotopy in primary somatosensory cortex. NeuroImage 23:462–472.

Parsons LM, Sergent J, Hodges DA, Fox PT (2005) The brain basis of piano performance. Neuropsychologia 43:199–215.

Pandya DN, Yeterian EH (1996) Comparison of prefrontal architecture and connections. Philos Trans R Soc Lond B Biol Sci 351:1423–1432.

Paus T (1996) Location and function of the human frontal eye-field: a selective review. Neuropsychologia 34:475–483.

Ramnani N, Owen AM (2004) Anterior prefrontal cortex: insights into function from anatomy and neuroimaging. Nature Reviews/Neuroscience 5:184–194.

Roland PE, Zilles K (1996) Functions and structures of the motor cortices in humans. Curr Opin Neurobiol 6:773–781.

Schiltz C, Sorger B, Caldara R, Ahmed F, Mayer E, Goebel R, Rossion B (2006) Impaired face discrimination in acquired prosopagnosia is associated with abnormal response to individual faces in the right middle fusiform gyrus. Cereb Cortex 16:574–586.

Schmitt FO, Worden FG, Adelman G, Dennis SG (1981) *The Organization of the Cerebral Cortex*. MIT Press, Cambridge.

Schüz A, Miller R (2002) *Cortical Areas: Unity and Diversity*. Taylor & Francis, London and New York.

Shaywitz SE (1998) Dyslexia. NEJM 338:307–312.

Shmuelof L, Zohary E (2006) A mirror representation of others' actions in the human anterior parietal cortex. J Neurosci 26:9736–9742.

Steinmetz H (1996) Structure, functional and cerebral asymmetry: in vivo morphometry of the planum temporale. Neurosci Biobehav Rev 20:587–591.

Vogt BA, Vogt LJ, Nimchinsky EA, Hof PR (1997) Primate cingulate cortex chemoarchitecture and its disruption in Alzheimer's disease. In: Björklund A, Hökfelt T, eds.

Handbook of Chemical Neuroanatomy. Vol 13: The Primate Nervous System, Part I, pp 455–528. Elsevier, Amsterdam.

Wiech K, Preissl H, Lutzenberger W, Kiefer RT, Topfner S, Haerle M, Schaller HE, Birbaumer N (2000) Cortical reorganization after digit-to-hand replantation. J Neurosurg 93:876–883.

Zilles (2004) Architecture of the human cerebral cortex. Regional and laminar organization. In: Paxinos G, Mai JK, eds. *The Human Nervous System*. 2nd ed. Chapter 27, pp 997–1055. Elsevier/Academic Press, Amsterdam.

Zilles K, Palomero-Gallagher N (2001) Cyto-, myelo-, and receptor architectonics of the human parietal cortex. NeuroImage 14:S8–20.

One thing about the cerebral circulation that has been agreed upon from the start is that it is unique in all essential aspects, beginning with the morphological.

Carl F. Schmidt, 1950

22

Blood Supply to the Central Nervous System

22.1. CEREBRAL CIRCULATION

Blood flow through the human brain is relatively constant despite changes in cardiac output and arterial blood pressure. Even though the brain makes up only about 2% of the body weight, about 15% of the resting cardiac output is destined for the brain. Each minute about 750–900 ml of blood reaches the brain. Inasmuch as **glucose** is the main fuel for energy metabolism in the brain and since the brain does not have significant carbohydrate reserves, a steady supply of glucose must be available to the brain. This amount of glucose available to the brain is about 434 μmol/min or about 10% of the circulating blood glucose at resting levels. The adult brain also needs about 49 ml/min of **oxygen** or about 20% of the circulating oxygen at resting levels. Almost all of this oxygen plays a role in the metabolism of this much-needed glucose. In addition to its requirement for oxygen and glucose, the brain also uses about 20 watts/min of **energy**. During exercise, when cardiac output is often as high as 15,000 ml/min, the brain continues to receive a stable amount of blood.

The term 'stroke' refers to all types of cerebrovascular disease, resulting from either 'leaks' or 'plugs' of the blood vessels to the brain. Some 700,000 Americans suffer a stroke each year with 162,672 stroke deaths in 2002. Stroke is the third leading cause of death and long-term disability in the United States (2002). Of the

2.5 million living stroke victims in the United States, a third are younger than 65 years of age and 10% require institutional care. The resulting emotional burden to family members is incalculable but the annual expense is about 1.2 billion dollars for medical care and over 3 billion dollars in lost earnings.

22.2. AORTIC ARCH, BRACHIOCEPHALIC TRUNK, AND SUBCLAVIAN VESSELS

Blood from the left ventricle of the heart reaches the brain by way of the **aortic arch** (Fig. 22.1) which gives rise to the **brachiocephalic trunk**, **left subclavian**, and **left common carotid arteries**. Minor variations in the branches of the arch of the aorta are frequent. The **aortic arch** (Fig. 22.1) is a likely site of atherosclerotic degeneration. Atheromatous plaques often narrow or occlude the opening of the vessels supplying the brain. This impairs cerebral circulation and produces neurological

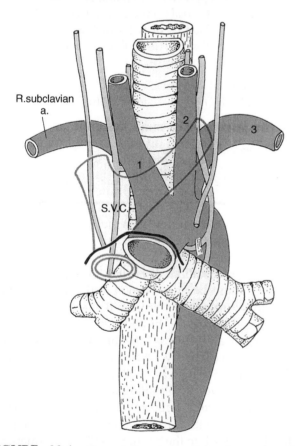

FIGURE 22.1. The arteries given off by the aortic arch (modified from O' Rahilly, 1986). **Key: 1**, brachiocephalic trunk; **2**, left common carotid; **3**, left subclavian; **S.V.C.**, superior vena cava; R. subclavian a., right subclavian artery.

deficits. The **brachiocephalic trunk** arises from behind the widest part of the sternum at the level of the third thoracic vertebra. As it ascends behind the right sterno-clavicular joint, it divides into a **right subclavian** and a **right common carotid artery** (Fig. 22.1). The trunk is at first in front of the trachea and then to the right of it. The right subclavian artery supplies the right upper limb. The **left subclavian artery** arises as a direct branch of the aortic arch behind the left common carotid artery and then ascends lateral to the trachea (Fig. 22.1) before it supplies the left upper limb. Each subclavian artery gives rise to a **vertebral artery** (Fig. 22.1). The origin of a vertebral vessel from the subclavian is another common site for the formation of atherosclerotic plaques that narrow or block the opening.

22.3. VERTEBRAL-BASILAR ARTERIAL SYSTEM

Each subclavian artery gives rise to a **vertebral artery** (Figs 22.1, 22.2). The left and right **vertebral arteries** ascend through the neck and enter the foramen magnum before reaching the cranial cavity. One vertebral artery is usually larger. In the neck, the vertebrals ascend in the osseous canals of the transverse processes of the cervical vertebrae. Trauma to the cervical spine producing transient dislocation of one vertebra on another (as might occur with a whiplash injury) often traumatizes the vertebrals in these canals. Chiropractic manipulation of the neck may cause vertebral trauma.

After entering the foramen magnum, the vertebrals ascend along the ventrolateral aspect of the medulla oblongata giving off numerous perforating vessels to that part of the brain stem. The left vertebral artery is frequently larger in caliber than the right; rarely one vertebral is absent. The blood that the vertebrals carry is vital to nuclei and fiber paths of the brain stem that participate in cardiorespiratory homeostasis and maintenance of consciousness among other functions. The vertebrals travel along the ventral surface of the medulla oblongata in a groove between the pyramid and olive before uniting at the ponto-medullary junction to form a single **basilar artery** (Fig. 22.2). The basilar artery, averaging some 32 mm in length, reaches the upper pons and opposite the interpeduncular perforated substance bifurcates into the **posterior cerebral arteries** (Fig. 22.2). The basilar artery in infants is straight but, with age, it lengthens and widens. If affected by atherosclerosis the basilar artery may become tortuous and

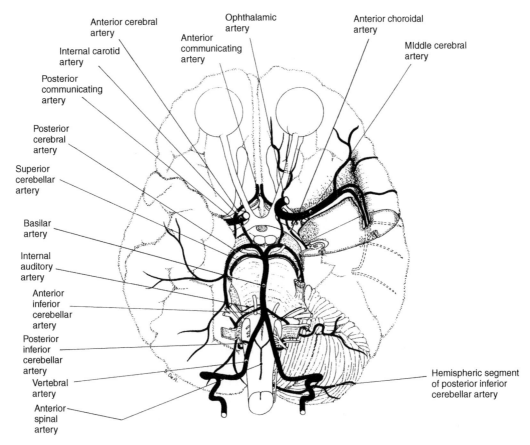

FIGURE 22.2. The inferior surface of the human brain illustrating the internal carotid, vertebral-basilar arteries, and the arterial circle (of Willis) (from DeArmond, Fusco, and Dewey, 1989).

lose its anatomical relationship with the basilar sulcus. This combination of the vertebral arteries and the basilar artery is the **vertebral-basilar system.** Some 15% of saccular aneurysms occur in the vertebral-basilar system – over 60% of these occur at the basilar bifurcation.

Branches of the intracranial part of the vertebral arteries and the basilar artery are of two types: **extrinsic or superficial arteries** and **intrinsic or penetrating arteries**. Only the superficial arteries are visible as they encircle the brain stem. The penetrating arteries are small in caliber and branch in clusters from the concealed surfaces of all the superficial arteries. These branches enter the pons, midbrain, or interpeduncular perforated substance. The upper centimeter of the basilar artery has an abundant number of these penetrating arteries.

22.3.1. Branches of the Vertebral Arteries

The pattern of branching of the superficial or extrinsic branches from the vertebral and basilar arteries is unpredictable. The named branches from the **vertebral arteries** are the **anterior** and **posterior spinal rami** and the **posterior inferior cerebellar arteries** (PICA)

(Figs 22.2, 22.3). Distinct arteries that fit the classical description of these vessels are uncommon.

Anterior Spinal Rami

Before entering the intracranial dura, each vertebral gives off a descending **anterior spinal ramus** that originates from the medial aspects of its parent vessel, curves medially, and unites at the medullary or upper cervical level (Fig. 22.2). These spinal rami are rarely of equal diameter and may not unite to form a single median trunk. Their primary brain stem distribution is by tiny penetrating branches to structures in the median plane of the medulla. These rami join on the surface of the cervical cord to form a single vessel which descends to about the fourth cervical segment before it is augmented by various spinal medullary vessels along the length of the cord.

Posterior Spinal Rami

The **posterior spinal rami** (Fig. 22.3) arise as distinct branches of the vertebral arteries, but may branch from

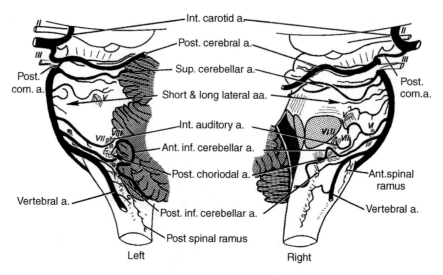

FIGURE 22.3.   The superficial arteries of the human brain stem (from Gillilan, 1964).
Abbreviations: a, artery; aa, arteries; Ant, anterior; com, communicating; III, oculomotor nerve;
inf, inferior; int, internal; Post, posterior; sup, superior.

the posterior inferior cerebellar artery. Though difficult to identify, the posterior rami supply the posterior medulla oblongata.

Posterior Inferior Cerebellar Artery (PICA)

The largest and most variable branch of each vertebral artery is a **posterior inferior cerebellar artery** (PICA) (Figs 22.2, 22.3) that supplies parts of the cerebellum and a wedge-shaped area in the lateral medulla.

22.4. BLOOD SUPPLY TO THE SPINAL CORD

22.4.1. Extramedullary Vessels

Before 1750 it was generally accepted that the arterial blood supply of the spinal cord was segmental, i.e. that along all 62 spinal roots, segmental arteries contributed to the vascularization of the cord. The segmental concept lingered on until the end of the World War II. Some authors continue to describe a segmental blood supply to the cord. Advances in imaging technology has led to our current understanding that the human cord receives a **nonsegmental blood supply** with only 9–12 arteries participating in the vascularization of the entire spinal cord.

Those segmentally arranged vessels that enter the intervertebral foramen and do not reach the spinal cord proper but end in the distal parts of the spinal roots or the spinal dura mater are termed **spinal radicular arteries**. Those nonsegmental vessels that enter the intervertebral foramen and contribute to the substance of the spinal cord by augmenting the anterior and posterior spinal arteries, as they lie on the surface of the cord and run parallel to the longitudinal extent of the cord, are known as **spinal medullary arteries**. These vessels course along the internal aspects of their respective roots but give no branches to these roots in course. As they enter the spinal dura mater, however, they often supply the distal parts of the spinal roots.

Based on the arrangement of vessels to the cord it is possible to divide the entire cord into three arterial zones: (1) **cervicothoracic**; (2) **midthoracic**; and (3) **thoracolumbar** (Fig. 22.4). The **cervicothoracic zone** consists of the entire cervical cord and the first two or three thoracic segments (Fig. 22.4) while the **midthoracic zone** includes the fourth through the eighth thoracic segments (Fig. 22.4) and the **thoracolumbar zone** corresponds to the last few thoracic levels and the lumbar enlargement (Fig. 22.4). The caudal end of the cord, the conus medullaris, receives by a rich anastomosis of vessels that collectively form the **anastomotic loop of the conus** (Figs 22.4, 22.7).

The spinal cord blood supply arises from a few parent vessels. These include a system of three vertical channels – a single **anterior spinal artery** and paired **posterior spinal arteries** that lie on the respective surfaces of the cord (Figs 22.2, 22.3, 22.4, 22.5). The **anterior spinal artery** lies in the spinal pia mater and the pia thickens over it. The anterior spinal artery proper supplies the upper four cervical segments as a single

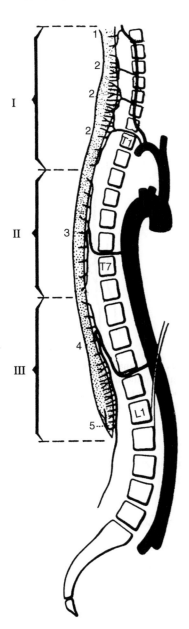

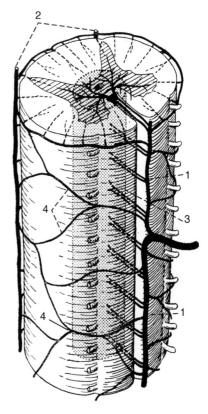

FIGURE 22.4. Arterial vascularization and zones of the human spinal cord (from Lazorthes et al., 1971). **Key:** I, cervicothoracic zone; II, midthoracic Zone; III, thoracolumbar zone; 1, anterior spinal artery; 2, artery of the cervical enlargement; 3, posterior spinal artery; 4, artery of the lumbar enlargement; 5, anastomotic loop of the conus medullaris.

FIGURE 22.5. The central and peripheral arterial systems of the human spinal cord (from Lazorthes, 1972). **Key:** 1, anterior spinal artery; 2, central arteries; 3, peripheral arteries; 4, posterior spinal arteries.

trunk formed by spinal rami from each vertebral (Figs 22.4, 22.6). From this point on it becomes an anastomotic channel as terminal branches of several nonsegmental spinal medullary arteries augment the anterior spinal artery. These nonsegmental spinal medullary arteries enter the vertebral canal in the intervertebral foramina.

The **spinal medullary arteries** vary in size and supply territories corresponding to several cord segments.

The anterior spinal artery is an **anastomotic channel** consisting of a series of spinal medullary vessels that augment the anterior spinal artery. This channel is a continuous structure originating from the medial aspects of the vertebral vessels and extending to the fourth cervical segment. From that point, it continues along the anterior surface of the cord as an anastomotic channel formed by terminal branches of successive nonsegmental spinal medullary arteries.

Fewer, but larger, nonsegmental spinal medullary arteries supply the anterior spinal artery. The **posterior spinal arteries**, on the other hand, receive more numerous but smaller nutritive contributions from various levels. Though certain regional tendencies exist, the origin of the nonsegmental spinal medullary arteries may be at any vertebral level. Though only a few of these vessels persist to feed the anterior spinal artery, a slightly greater number of vessels of smaller caliber supply both posterior spinal arteries. The nonsegmental spinal medullary arteries are more frequently on the left side than on the right in the thoracic and lumbar regions (because the aorta is on the left).

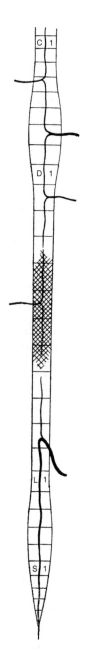

FIGURE 22.6. The anastomotic anterior spinal artery supplying the human spinal cord at different cord levels (C1, D1, L1, and S1). Hatched area represents the critical midthoracic zone (from Lazorthes, 1972).

A single nonsegmental spinal medullary artery may contribute to the anterior or to the posterior spinals or divide and contribute to both.

The anterior spinal artery proper along with its continuation as an anastomotic channel, distributes blood to the cord forming three well-defined vascular zones with minimal collateral exchange at the contiguous boundaries between these zones. These three zones are

FIGURE 22.7. The anastomotic loop of the conus medullaris formed by the union of the anterior spinal and two posterior spinal arteries in the shape of a triangle with an artery at each point of the triangle (from Lazorthes et al., 1971).

a cervicothoracic zone, midthoracic zone, and a thoracolumbar zone.

Cervicothoracic Zone

The anterior spinal artery proper supplies the upper four cervical segments as a single trunk formed by spinal rami from each vertebral (Figs 22.4, 22.6). These upper four segments have no other source of vessels. The lower four cervical segments and the first two thoracic segments (including the cervical enlargement) possess an independent blood supply. This includes two or three nonsegmental spinal medullary vessels from the intratransverse part of the vertebral arteries, and a single vessel arriving with C7 or C8 root. This lowest vessel, designated the **artery of the cervical enlargement** (Figs 22.4, 22.6, 22.7), arises from the deep cervical branch of the costocervical trunk.

Included among vessels providing collateral circulation to the cervicothoracic zone are the occipital artery, ascending cervical artery, and other branches of the costocervical trunk. The absence of spinal medullary vessels in the upper cervical cord is of considerable importance in cervical trauma. A **zone of poor collateral circulation** occurs near the fourth thoracic segment, the boundary between the cervicothoracic and midthoracic zones (Figs 22.4, 22.6).

Midthoracic Zone

This region, from the fourth through the eighth thoracic segments, receives its blood (Fig. 22.4) by way of a single vessel derived from the thoracic aorta. This vessel, named the **spinal branch of the posterior intercostal artery** (Figs 22.4, 22.6), enters the intervertebral foramen about the level of T7. Since the thoracic aorta is often the principal source of blood to the thoracic cord, disease of the aorta often secondarily causes injury to the spinal cord. Though at first glance it appears that there are few collateral channels in this region, selective aortic angiography has shown numerous communicating vessels in the thoracic cord. The functional significance of these collateral systems is impossible to assess. The posterior spinal artery in this region is smaller than at inferior levels and is often discontinuous.

Thoracolumbar Zone

This lower cord, including the last few thoracic segments and the lumbosacral enlargement, depends on a single vessel from the abdominal aorta, the **artery of the lumbar enlargement** (Figs 22.4, 22.6). This vessel enters the vertebral canal with one of the lowest four thoracic nerves (T9 to T12). The anterior spinal artery in this region displays a greater caliber. Though the lumbar enlargement receives its blood supply from only one vessel, this zone can receive blood from other vessels that under normal circumstances are not a source of blood to this zone.

The **posterior spinal arteries** become larger at the level of the twelfth thoracic vertebra although there may be only a single posterior spinal vessel in this region. The posterior spinal medullary arteries are frequently on roots that supply the caudal half of the cord. In the thoracic and lumbar regions, they are more often on the left than on the right. The **artery of the lumbar enlargement** always divides into an anterior and a posterior spinal medullary branch. The artery is on the left side in 80% of those examined. When it reaches the cord, it becomes the anterior spinal artery by branching into a small ascending and a large descending branch. When the artery of the lumbar enlargement enters the vertebral canal more rostrally (T7 or T8), a second spinal medullary vessel is usually found caudally.

Anastomotic Loop of the Conus

The caudal end of the spinal cord, the conus medullaris, receives its blood from a rich anastomosis of vessels. This **anastomotic loop of the conus** results from the union of the caudal end of the anterior spinal channel with the caudal ends of both posterior spinal arteries (Figs 22.4, 22.7). On each root of the cauda equina, inferior to the artery of the lumbar enlargement, one or more thin arteries converge on the anterior and posterior spinal arteries respectively.

22.4.2. Intramedullary Vessels

In addition to the arteries on the spinal cord, there is also a system of vessels in the cord that supply the spinal cord gray and white matter. This system is often termed the **intramedullary arterial system**. The two types of intramedullary arteries throughout the entire extent of the cord include a system of **central arteries** and a **peripheral arterial system** (Fig. 22.5).

Central Arteries to the Spinal Cord

As implied by their name, the **central arteries** supply the central part of the spinal gray matter and the adjacent parts of the white matter (Fig. 22.5). These central arteries are branches of the anterior spinal artery. From their origin, they penetrate the ventral median fissure and on reaching the ventral white commissure, they pass to one side of the cord. The adjacent central artery then continues to the other side of the cord. Arteries to the right side of the cord alternate with arteries to the left throughout the length of the anterior spinal artery. The number of these central arteries varies with the different regions of the spinal cord. They are larger in diameter and more numerous in the cervicothoracic and thoracolumbar zones. Each central artery supplies several neuronal groups and several arteries irrigate each neuronal group. Since the central arteries anastomose with one another, there exists an overlapping of the areas irrigated by the central arteries.

As the anterior and posterior spinal arteries extend along the length of the cord, they give off branches that ramify on the cord surface and interconnect with one another forming a pial arterial plexus (pial mesh or vasa corona) (Fig. 22.5). No distinction exists between the part of the pial mesh supplied by the anterior and that supplied by the posterior spinal artery. The proximal third of each emerging ventral root receives its blood supply from branches emanating from the pial arterial plexus. Branches of the posterior spinal arteries supply the proximal part of each dorsal root. The spinal radicular vessels supply the distal aspects of the roots. The region between these areas of differing blood supply (near the middle of each root) is an area of hypovascularity.

Peripheral Arteries to the Spinal Cord

The **peripheral arterial system** of the cord includes penetrating branches of this pial mesh that supply the bulk of the white matter and the greater part of the dorsal horns (Fig. 22.5). The terminal arterioles of the central and peripheral intramedullary vessels form an extensive capillary network. Because of the greater metabolic needs of the somata, the capillaries are more numerous in the gray matter than in the white matter. Hence, the gray matter in the freshly cut spinal cord is pink to reddish in color due to its vascularity.

22.4.3. Spinal Veins

Spinal veins run parallel to arteries in the substance of the cord, on its surface, and along the dorsal and ventral roots. An intrinsic group of veins drains the intramedullary part of the spinal cord into a system of extrinsic veins that lie outside the substance of the

cord. A single **anterior spinal vein** and one or two **posterior spinal veins** drain the intrinsic veins of the cord and the **coronal veins or plexus** (that correspond to the pial mesh) (Fig. 22.8). These anterior and posterior spinal veins drain into epidural veins and into intervertebral veins. The latter accompany the spinal nerves in the intervertebral foramina. They receive blood from the spinal cord and drain the internal and external vertebral venous plexuses. The intervertebral veins flow into the vertebral, posterior intercostal, lumbar, and lateral sacral veins.

The veins of the vertebral column include the internal and external (in or outside the vertebral canal) vertebral plexuses (of Batson). These vertebral plexuses are extradural but communicate with the veins that drain the neural tissue of the spinal cord on the one hand, and the systemic veins of the thorax, abdomen, and pelvis, and the dural sinuses of the skull, on the other hand.

22.5. BLOOD SUPPLY TO THE BRAIN STEM AND CEREBELLUM

22.5.1. Extrinsic or Superficial Branches

The **vertebral arteries** enter the foramen magnum, converge on the ventral surface of the medulla oblongata

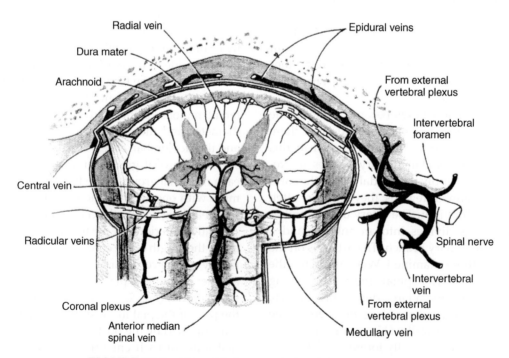

FIGURE 22.8. Veins of the human spinal cord (from Gillilan, 1970).

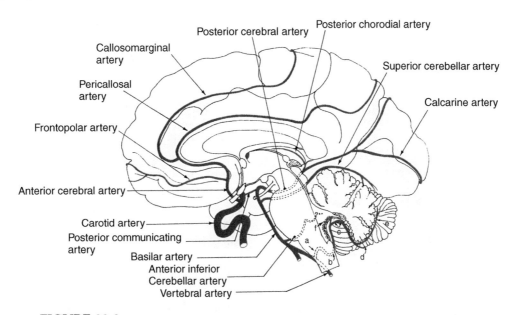

FIGURE 22.9. Arteries on the medial surface of the human brain (from DeArmond, Fusco, and Dewey, 1989).

in a groove between the pyramid and the inferior olive, and unite at the lower pons forming a single **basilar artery** (Figs 22.2, 22.9). In addition to the anterior and posterior spinal rami (described under the section on 'Blood Supply to the Spinal Cord'), the largest and most variable branch of each vertebral artery is a **posterior inferior cerebellar artery** (PICA) that supplies parts of the cerebellum and a wedge-shaped area in the lateral medulla oblongata (Figs 22.2, 22.3, 22.10, 22.11). The posterior inferior cerebellar arteries are asymmetrical in origin and caliber as they take a tortuous course over the brain stem, coursing transversely, then downward along the lateral aspect of the medulla oblongata making a caudal loop before ascending in a sulcus that separates the cerebellar tonsil from the medulla oblongata. The **posterior inferior cerebellar artery** then makes a large, cranially directed loop near the lateral apertures (a poorly defined gap between the cerebellum and medulla), bends inferiorly and medially where it gives rise to a medial and a lateral branch somewhere between the two loops (Fig. 22.3). The territory supplied by the medial PICA on the posterior-inferior aspects of the cerebellum includes a wedge or triangular area with a dorsal base and a ventral apex directed toward the fourth ventricle. This involves parts of the cerebellar tonsil and cerebellar vermis. Substantial communication exists across the median plane between the posterior inferior cerebellar arteries supplying the cerebellum. In the ascending part of the loop, as the vessel is parallel to the emerging glossopharyngeal and vagal roots, many penetrating vessels arise that enter

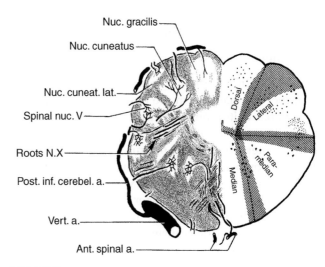

FIGURE 22.10. Arterial zones of the human brain stem at the level of the sensory decussation (from Gillilan, 1964). **Abbreviations:** a, artery; inf, inferior; lat, lateral; Nuc, nucleus; Post, posterior.

the lateral aspect of the medulla oblongata (Fig. 22.3) in company with the cranial nerves. Of all cerebellar arteries, infarction most frequently involves the posterior inferior cerebellar artery. The resulting signs and symptoms make up the **posterior inferior cerebellar artery syndrome** (lateral medullary or Wallenberg's syndrome) (Figs 22.10, 22.11). If the posterior inferior cerebellar artery does not supply the lateral medulla oblongata, obstruction of this vessel leads to a pure vertigo, with or without signs of cerebellar involvement.

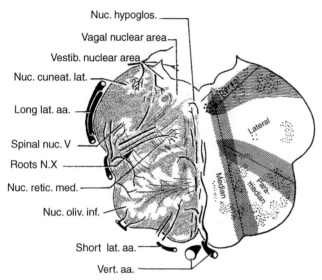

Nuc. hypoglos.
Vagal nuclear area
Vestib. nuclear area
Nuc. cuneat. lat.
Long lat. aa.
Spinal nuc. V
Roots N.X
Nuc. retic. med.
Nuc. oliv. inf.
Short lat. aa.
Vert. aa.

Lateral

Median

Para median

FIGURE 22.11. Arterial zones of the human brain stem at the level of the vagal and hypoglossal nuclei (from Gillilan, 1964). **Abbrevations:** a, artery aa, arteries; inf, inferior; lat, lateral; med, medial; nuc, nucleus; Post, posterior; retic, reticular; Vert, vertebral; Vestib, vestibular.

22.5.2. Branches of the Basilar Arteries

Labyrinthine Artery

The **labyrinthine artery** (internal auditory or internal acoustic) originates from the basilar artery (Figs 22.2, 22.3) or one of its branches (such as the anterior inferior cerebellar artery) but does not give off any branches until it enters the internal meatus with the facial [VII] and vestibulocochlear [VIII] nerves. Thereafter its branches supply the roots of the vestibulocochlear and facial nerves as well as the inner ear. Since it is an end artery, reduction of blood flow in it often yields disturbances of equilibrium causing nausea, vomiting, and vertigo. Occlusion of the branch to the cochlea will result in a sudden loss of hearing. These symptoms often provide important clues to disease of the vertebral-basilar system. The labyrinthine artery does not contribute blood to the brain stem.

Anterior Inferior Cerebellar Artery (AICA)

The **anterior inferior cerebellar artery**, a branch of the lower third of the basilar artery (or from the vertebral artery or a stem common to the basilar and the posterior inferior cerebellar artery) (Figs 22.2, 22.3, 22.9) passes laterally and posteriorly, inferior to the abducent and then the trigeminal nerve [V] before reaching the cerebellopontine angle. This vessel supplies the lateral parts of the medullary and pontine tegmentum, part of the middle cerebellar peduncle, flocculus, and

the anterior inferior part of the cerebellar hemisphere. The anterior inferior cerebellar artery, the posterior inferior cerebellar artery, or both supply the inferior cerebellar surface. The **artery to the choroid plexus** originates from either of these sources, from the vertebral, or from the basilar artery.

If the anterior inferior cerebellar artery is small, the posterior inferior cerebellar artery on the same side is large. Penetrating branches arise from the undersurface of the anterior inferior cerebellar artery, enter the pons with rootlets of the facial nerve, and distribute to the lateral part of the inferior pontine tegmentum. Because of its pontocerebellar territory of supply, infarction of the anterior inferior cerebellar artery often leads to rotational dizziness, vomiting, tinnitus, and dysarthria with ipsilateral dysmetria, facial palsy, trigeminal sensory involvement, Horner's syndrome (which includes miosis, mild ptosis and anhidrosis), and perhaps a lateral gaze palsy. Contralateral signs include pain and temperature sensory loss.

Superior Cerebellar Artery (SCA)

This is the last large branch of the basilar artery (Fig. 22.3) before its bifurcation into the **posterior cerebral arteries**. The **superior cerebellar artery** (Figs 22.2, 22.3, 22.9) has an anterior pontine segment (between the pons and clivus), ambient segment (lateral to the brain stem along the upper border of the pons), and a quadrigeminal segment (in the quadrigeminal cistern). The superior cerebellar artery divides into two main vessels: a **lateral or marginal branch** and a **medial branch** that course superior to the lateral. This vessel supplies perforating branches to the dorsolateral part of the upper brain stem, superior cerebellar peduncle, nuclei beneath the fourth ventricle, and some of the superior cerebellar surface. Cortical branches of all three cerebellar vessels (anterior, posterior, and superior) anastomose freely on the cerebellar surface. The oculomotor nerve [III] emerges between the superior cerebellar artery and the posterior cerebral artery (Figs 22.2, 22.3, 22.9) near their origin while the trochlear nerve [IV] passes between these vessels on the lateral aspect of the brain stem. Small, concealed rami pass from the undersurface of the superior cerebellar artery into the upper pons. One or more of its branches enter between the middle cerebellar peduncle and the superior cerebellar peduncle. Usually the inferior colliculus receives its arterial supply from this vessel through one or several small rami. The superior cerebellar artery may give branches that descend along the middle of the vermis and a larger branch that ends in the cerebellar dentate nucleus.

Posterior Cerebral Artery (PCA)

The terminal branches of the basilar artery are the **posterior cerebral arteries** (Figs 22.2, 22.3, 22.9). The posterior cerebrals encircle the brain stem at the level of the **tentorial notch**, passing superior to the oculomotor nerve and around the lateral aspect of the midbrain to reach the inferior and medial aspects of the temporal lobes before ending at the occipital poles. Small twigs in course supply rostral parts of the underlying **midbrain** (cerebral crus and superior colliculi). That segment of the posterior cerebral artery from the basilar bifurcation to the point at which the posterior communicating artery joins the posterior cerebral is the P_1 (midbrain) **segment**. Four relatively constant branches leave this segment of the posterior cerebral artery. That segment of the posterior cerebral artery after the posterior communicating artery and to the posterior aspect of the thalamus is the P_2 segment. This latter segment is divisible into anterior and posterior halves each about 25 mm long with the anterior half designated as P_2A and the posterior half as P_2B. P_2B begins at the posterior margin of the cerebral peduncle and follows along parallel and inferior to the optic tract to the posterior aspect of the thalamus. The P_3 **segment**, measuring some 20 mm in length, begins at the pulvinar and proceeds posteriorly to end in the anterior limit of the calcarine sulcus (Fig. 22.9). Because it supplies so many structures related to the visual system (extraocular nerves and nuclei, superior colliculi, lateral geniculate bodies, optic tracts, optic radiations, striate and extrastriate cortex) the posterior cerebral artery is termed the 'artery of the visual system'.

The posterior cerebral arteries approach one another posterior to the colliculi. The oculomotor nerve leaves the brain stem between the P_1 **segment** and the superior cerebellar artery making it vulnerable at this point to **aneurysms** (ballooning of the vessel wall) that may impinge on the nerve.

Brain Stem Branches of the Posterior Cerebral Artery

Branches of the P_1 segment include the thalamoperforating arteries, posterior choroidal artery (Fig. 22.9), quadrigeminal artery, and various rami to the base (cerebral crus) and tegmentum of the midbrain. One or more thalamoperforating branches enter the interpeduncular perforated substance. The posterior choroidal artery supplies parts of the thalamus, tela choroidea of the third ventricle, and the **choroid plexus** of the lateral ventricle. This vessel is most identifiable on the medial surface of the brain. The quadrigeminal artery supplies

the superior and inferior colliculi. Thalamogeniculate arteries, originating from the P_2 segment, reach the pulvinar, posterior thalamus, and lateral geniculate nucleus. The major part of the posterior cerebral artery distributes to the posteromedial cerebral cortex.

22.5.3. Intrinsic or Penetrating Branches

The intrinsic brain stem arterial patterns are constant and predictable. They are small in caliber and supply nuclei. The caliber of a penetrating artery is directly proportional to its length. The intrinsic vessels arise from all named and unnamed superficial arteries. After careful study of the course and distribution of these penetrating arteries at all levels of the medulla oblongata, pons, and midbrain it was apparent that these vessels fall into four major zones and that definable maps could be made of these zones.

Median Zone

The **median zone** (Figs 22.10, 22.11, 22.12, 22.13, 22.14, 22.15) receives its blood through intrinsic vessels from medial arteries on the ventral surface of the brain stem (Fig. 22.3). The most central arteries of this group tend to be the longest, supplying structures in the median plane such as the hypoglossal, abducent, trochlear, and oculomotor nuclei, and those roots that emerge near the median plane (but not the trochlear roots). The medullary pyramids, medial corticospinal fibers in the pons, medial bundles in the cerebral crus at the base of the midbrain, medial longitudinal fasciculus, medial lemniscus, and

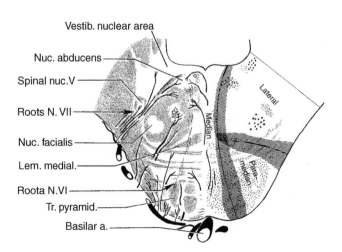

FIGURE 22.12. Arterial zones of the human brain stem at the level of the lower pons (from Gillilan, 1964). **Abbreviations:** a, artery; lem, lemniscus; nuc, nucleus; Tr, tract; Vestib, vestibular.

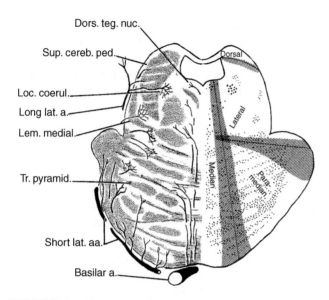

FIGURE 22.13. Arterial zones of the human brain stem at the level of the upper pons (from Gillilan, 1964). **Abbrevitions:** a, artery; aa, arteries; cereb, cerebellar; coerul, coruleus; Dors, dorsal; lat, lateral; lem, lemniscus; loc, locus; nuc, nucleus; ped, peduncle; pyramid, pyramidal; sup, superior; teg, tegmental; Tr, tract.

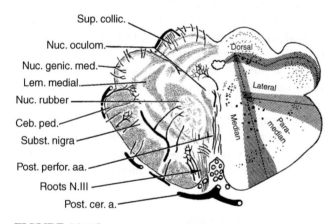

FIGURE 22.15. Arterial zones of the human brain stem at the level of the upper midbrain (from Gillilan, 1964). **Abbreviations:** aa, arteries; Ceb, cerebral; collic, colliculus; genic, geniculate; perforated; post, posterior; rber, red; Subst, substantia; sup, superior.

the medial part of the inferior olive are among the structures supplied by arteries of the median zone.

Paramedian Zone

Vessels to the **paramedian zone** (Figs 22.10 to 22.15) are branches of the vertebrals at medullary levels but they arise from the basilar and posterior cerebral arteries in the pons and midbrain. Included in the paramedian zone are such structures as most of the inferior olive, motor paths in the basilar pons, and sensory paths

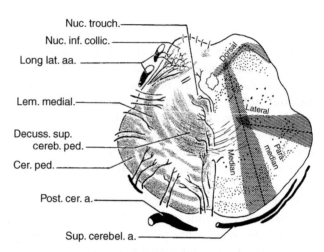

FIGURE 22.14. Arterial zones of the human brain stem at the level of the lower midbrain (from Gillilan, 1964). **Abbreviations:** a, artery; aa, arteries; cer, cerebral; cereb, cerebellar; collic, colliculus; decuss, discussation; inf, inferior; lat, lateral; lem, lemniscus; nuc, nucleus; ped, peduncle; sup, superior; troch, trochlear.

dorsolaterally. In the midbrain, paramedian arteries supply the outer two-thirds of the cerebral crus and most of the red nucleus.

Lateral Zone

The **lateral arterial zone** is throughout the brain stem (Figs 22.10 to 22.15). These arteries never extend to the median plane or to the floor of the fourth ventricle except in the upper pons. Structures in this zone are roots of the glossopharyngeal and vagal nerves, the dorsal vagal nucleus, nucleus and solitary tract, inferior and superior salivatory nuclei, nucleus ambiguus, trigeminal spinal tract and nucleus, spinothalamic tracts, vestibular nuclei, ventral trigeminothalamic tract, facial nucleus, and much of the inferior cerebellar peduncle. (The trigeminal motor and trigeminal pontine nuclei get their blood from a single artery, branches of which enter with the roots of this nerve.) Also in the lateral zone are the upper pontine tegmentum, the superior cerebellar peduncle, lateral lemniscus, and long ascending tracts. Most vascular injuries in the brain stem involve arteries of the lateral zone.

Dorsal Zone

Arteries of the **dorsal zone** make up the fourth and last group (Figs 22.10, 22.11, 22.3, 22.14, 22.15). They go to brain stem structures that form the roof of the ventricular system including the tectum, gracile nucleus, cuneate nucleus, lateral cuneate nucleus, and dorsal part of the inferior cerebellar peduncle. The dorsal vagal nucleus and nucleus of the solitary tract receive some arteries from the dorsal zone. No pial arteriolar plexus exists on the brain stem surface. In contrast to the variability of the patterns of the superficial brain stem arteries, the intrinsic arterial patterns are remarkably constant. Areas affected by the occlusion of a particular vessel correspond to the intrinsic arterial pattern.

22.5.4. Classical Brain Stem Syndromes

In describing classical brain stem syndromes, terminology based on the zone and segment of the brain stem involved is more descriptive and accurate than using the name of the suspected extrinsic artery. Though the **anterior spinal artery syndrome** involves the area of the lower medulla oblongata supplied by the median group of penetrating vessels, the actual extrinsic arteries involved are the anterior spinal rami or even the segment of the vertebral arteries from which the anterior spinal rami arise. In the following account, an effort is made to correlate clinically recognized syndromes with the involved anatomical structures and the resultant clinical signs and syndromes.

Median Medullary Syndrome

This syndrome, termed the anterior spinal artery syndrome, alternating hypoglossal hemiplegia, crossed hypoglossal paralysis or the syndrome of the pyramid and the hypoglossal nerve, involves structures in the median zone at medullary levels. The structures involved might include roots of the hypoglossal nerve [XII], fibers in the medullary pyramid, and perhaps the medial lemniscus (Fig. 22.10). If the long penetrating vessels are involved, the lesion will extend to the floor of the ventricle and likely involve the hypoglossal nucleus. Damage to the structures in this zone would lead to a contralateral hemiplegia, a loss of discriminative touch, pressure and proprioception on the contralateral side of the body (trunk and limbs), and an ipsilateral flaccid paralysis of the tongue.

Median Pontine Syndrome

This syndrome, termed Raymond's syndrome, the Millard–Gubler syndrome, or middle alternating hemiplegia, involves structures in the median zone at pontine levels. The structures involved include roots of the abducent nerve [VI] as well as the corticospinal and corticobulbar tracts, and perhaps the abducent nucleus (Fig. 22.12). Damage to the structures in this zone would lead to a contralateral hemiplegia and paralysis of the lower face and perhaps the tongue as well as an ipsilateral paralysis of the lateral rectus muscle.

Median Midbrain Syndrome

This syndrome, termed Weber's syndrome, superior alternating hemiplegia, the syndrome of the interpeduncular fossa, or the syndrome of the midbrain gray matter, involves structures in the median zone of the midbrain at the level of the superior colliculus. The structures involved include the roots of the oculomotor nerve [III] as they travel through the cerebral crus, the corticobulbar and perhaps the corticospinal tracts. The oculomotor nuclei may be involved bilaterally (Fig. 22.15). Damage to the structures in this zone would lead to a contralateral hemiplegia and paralysis of the lower half of the face as well as the tongue along with an ipsilateral paralysis of the oculomotor nerve.

Paramedian Midbrain Syndrome

This syndrome, termed Benedikt's syndrome, involves structures in the paramedian zone of the midbrain at the level of the superior colliculus. The structures involved include the roots of the oculomotor nerve [III] and the red nucleus (Fig.22.15). Damage to the structures in this zone would lead to an ipsilateral paralysis of the oculomotor nerve and a contralateral cerebellar tremor of the upper limb and a cerebellar ataxia.

Lateral Medullary Syndrome

This syndrome, termed the posterior inferior cerebellar artery syndrome or Wallenberg's syndrome, involves structures in the lateral zone at the level of the upper medulla. Structures involved include the intramedullary fibers and nuclei of the vagal nerve (X) and the glossopharyngeal nerve [IX], the trigeminal spinal tract and nucleus, the lateral tectotegmentospinal tract, the lateral spinothalamic tract, the ventral trigeminothalamic tract, the inferior cerebellar peduncle, and the inferior and medial vestibular nuclei (Fig. 22.11). Damage to this zone involves a whole host of structures. Signs and symptoms might include an ipsilateral laryngeal and pharyngeal paralysis, ipsilateral paralysis of the soft palate, ipsilateral loss of pain and temperature on the face, and an ipsilateral Horner's syndrome (which includes miosis, mild ptosis, and anhidrosis). There might also be a loss of pain and temperature on the contralateral limbs and trunk, contralateral loss of pain and diminution of temperature on the face, ipsilateral cerebellar signs and a nystagmus.

Inferolateral Pontine Syndrome

This syndrome, termed the anterior inferior cerebellar artery syndrome, involves structures in the inferior and lateral zone of the pons. The structures involved in damage to this zone include the roots and nucleus of the facial nerve [VII], the roots and nucleus of the cochlear nerve, the trigeminal spinal nucleus and tract, lateral spinothalamic tract, and the lateral tecto-tegmentospinal tract (Fig. 22.12). Damage to these structures would lead to an ipsilateral facial paralysis including loss of taste on the anterior two-thirds of the tongue, an ipsilateral loss of hearing, loss of pain and temperature on that side of the face and on the opposite side of the body and limbs, and an ipsilateral Horner's syndrome (which includes miosis, mild ptosis, and anhidrosis).

Superolateral Pontine Syndrome

This syndrome, termed the superior cerebellar artery syndrome, involves the superior cerebellar peduncle, the lateral spinothalamic tract, the ventral trigeminothalamic tract, and lateral tectotegmentospinal tract (Fig. 22.13). Damage to these structures would lead to an ipsilateral hypotonicity and cerebellar tremor of the arm and leg, contralateral loss of pain and temperature on the face and a contralateral loss of temperature on the body and limbs. There would also be an ipsilateral Horner's syndrome.

Dorsal Midbrain Syndrome

This syndrome, also termed Parinaud's syndrome or the tegmental syndrome of the midbrain, involves the superior colliculus (Fig. 22.15). The resulting deficit would be a paralysis of upward gaze. In late stages, there may be paralysis of downward gaze as well.

Basilar Artery Syndrome

Structures involved in this syndrome include nuclei or tracts in the pons, midbrain, and visual areas in the occipital cortex. The manifestations of this syndrome include vertigo, nausea, vomiting, headache, clouding or loss of consciousness, confusion, diplopia, ptosis, dorsal midbrain syndrome, nystagmus, mono or hemiparesis, facial palsy, dysarthria, dysphagia, and loss of some sensory modalities on one limb or side of the body. Various combinations of these signs and symptoms occur in transient episodes often causing a progressively severe neurological deficit.

Vertebral Artery Syndrome

Structures involved in this syndrome include nuclei and fibers in the medulla oblongata. This may be evident as a median medullary syndrome, a lateral medullary syndrome, both, or as spotty, focal manifestations involving all zones.

Vertebral-Basilar Artery Syndrome

Structures involved in this syndrome include nuclei and fibers in the medulla oblongata, pons, midbrain or the occipital cortex. The signs and symptoms are similar to those of a basilar artery syndrome, plus the vertebral artery syndrome.

22.6. COMMON CAROTID ARTERY

The **right common carotid artery** is a branch of the brachiocephalic trunk whereas the **left common carotid artery** is a direct branch of the aortic arch arising slightly to the left of the brachiocephalic trunk (Fig. 22.1). The right common carotid ascends initially in front of the trachea and then to its left where it enters the neck behind the left sternoclavicular joint. The **common carotid arteries**, with their adventitial sympathetic nerves, ascend to the upper border of the thyroid cartilage immediately inferior to the angle of the mandible, where each divides into **external and internal carotid arteries**.

22.6.1. External Carotid Artery

The **external carotid artery** gives off numerous branches in its ascent before it ends as the **maxillary** and **superficial temporal arteries**. A notable branch of the external carotid artery is the **middle meningeal artery** that arises from the maxillary artery and provides the bulk of the meningeal circulation. The middle meningeal artery enters the base of the skull through the foramen spinosum in the sphenoid bone and passes anteriorly to give branches to the cranial dura over the convexity of the brain. Fractures of the skull, particularly those near the **pterion,** where four bones of the skull meet, may tear the middle meningeal artery in that this point overlies the anterior branch of the middle meningeal artery. Branches of the external carotid artery are capable of providing an alternate route for blood to reach the brain should there be blockage of the internal carotid artery.

22.6.2. Internal Carotid Artery: Cervical, Petrous, and Cavernous Parts

Just distal to the carotid bifurcation, each internal carotid artery has a bulbous expansion, the **carotid sinus** innervated by the glossopharyngeal nerve [IX]. Next to the carotid sinus and similarly innervated is the **carotid body** – the largest collection of chemoreceptors in the body. In humans, the bodies are pale pink structures (small, flat, ovoid, weighing nearly 4 mg, and measuring about $2 \times 5 \times 6$ mm). Chemoreceptors respond to changes in composition of their extracellular fluid. Alterations in pH, oxygen tension, and to a lesser extent, carbon dioxide tension in arterial blood, will stimulate them. This increases the rate and depth of respiration. A prolonged decrease in oxygen to the carotid body, often present in pulmonary emphysema or at high altitudes, causes enlargement of the carotid bodies, or **carotid body hyperplasia**. Blood flow in a carotid body per unit of tissue exceeds that in any other body organ. The region near the carotid bifurcation and sinus is a common site for the **deposition of atherosclerotic plaques**, presumably because of the currents that often develop in the blood stream at orifices and angulations of arteries. By involving sympathetic fibers in the wall of the internal carotid artery, diseases such as atherosclerosis as well as trauma from angiography or ligation may result in a case of Horner's syndrome (which includes miosis, mild ptosis, and anhidrosis).

As the **internal carotid artery** ascends from its origin to its termination, it is divisible into four parts: **C_1 or cervical part, C_2 or petrous part, C_3 or cavernous part**, and the **C_4 or cerebral part**. After branching from the common carotid artery, the **cervical part of the internal carotid artery** ascends through the neck before entering the base of the skull through the external opening of the **carotid canal**. That part of the internal carotid that courses in the periosteal-lined carotid canal in the petrous temporal bone is the **petrous part of the internal carotid artery**. Before entering the cranial dura, the internal carotid artery changes direction and makes four loops in course (posterior, lateral, medial, and anterior loops). The petrous part of the internal carotid artery has two segments: a vertical or ascending segment separated at a bend or genu from a horizontal segment that then enters the cranium as it runs anteromedially beneath the trigeminal ganglion (which it supplies with small branches). As it crosses the foramen lacerum, it forms a **lateral loop** and ascends along the posterolateral aspect of the sella and medial aspect of the anterior clinoid to enter the cavernous sinus as the **cavernous part of the internal**

carotid, being the only artery in the body within a venous plexus. Rupture here often produces an **arteriovenous fistula**. A segment of the abducent nerve [VI] is also in the sinus adherent to the lateral aspect of the internal carotid. A medial loop of the internal carotid artery is often lateral to the posterior clinoid process as the vessel continues anteriorly. The **meningohypophyseal trunk**, the **artery of the inferior cavernous sinus**, and perhaps certain **capsular arteries** arise from the cavernous part of the internal carotid artery. Closely related on the lateral side of the internal carotid and between both dural leaves of the lateral wall of the cavernous sinus are the oculomotor and trochlear [IV] and the ophthalmic [V_1] and maxillary [V_2] branches of the trigeminal nerve [V]. A fourth loop, which is nearly horizontal, the **anterior loop**, is in the anterior part of the cavernous sinus.

22.7. BLOOD SUPPLY TO THE CEREBRAL HEMISPHERES

22.7.1. Internal Carotid Artery: Cerebral Part

After the **internal carotid** enters the cranial dura medial to the anterior clinoid it passes superior to the oculomotor nerve [III] but inferior to the optic [II] nerve. The internal carotid then becomes the **C_4 part** or **cerebral (intracranial or supraclinoid) part of the internal carotid artery** that ranges from 14 to 25 mm in length (average 19 mm). More than a third of all intracranial aneurysms occur on the cerebral part of the internal carotid. The cavernous and cerebral parts of the internal carotid are collectively termed the **carotid siphon** in that these two parts form an 'S'. The cerebral part of the internal carotid artery enters the subarachnoid space and terminates at its bifurcation into the **anterior** and **middle cerebral arteries**. Significant branches arise from this part of the internal carotid artery including the **ophthalmic, posterior communicating**, and **anterior choroidal arteries** (Fig. 22.2) along with small perforating branches such as the **superior hypophyseal arteries** these distribute to the infundibulum and body of the hypophysis (pituitary gland). Other perforating vessels from the cerebral part reach the optic chiasma, optic nerve, and perhaps the optic tract as well as the floor of the third ventricle, anterior perforated substance, and the uncus. Terminal branches of the cerebral part of the internal carotid artery are the **anterior** and **middle cerebral arteries** (Fig. 22.2). Branches of the external carotid artery are capable of providing an alternate route for blood to reach the brain should there be blockage of the internal carotid artery.

22.7.2. Branches of the Internal Carotid Artery

Ophthalmic Artery

The **ophthalmic artery** (Fig. 22.2) arises from the medial aspect of the cerebral part of the internal carotid artery, below the optic nerve, and at a point where the internal carotid leaves the cavernous sinus. This vessel then passes into the optic canal and orbit. There are three identifiable groups of ophthalmic vessels: (1) an **ocular group**; (2) an **orbital group**; and (3) an **extra-orbital group**. The **ocular group** includes the **central retinal** (a most important vessel of the optic nerve and of the eye whose branches are observable with the ophthalmoscope) and **ciliary arteries** whereas the **orbital group** includes the lacrimal and ocular muscle branches. The **extra-orbital group** consists of the anterior and posterior ethmoidal arteries, dorsal nasal, supra-orbital, and supratrochlear arteries.

Posterior Communicating Artery

Arising from the middle part of the cerebral segment and passing horizontally to join the posterior cerebral artery is the **posterior communicating artery** (PCA) (Fig. 22.2, 22.3, 22.9). The PCA passes posterior and medially above the oculomotor nerve [III] toward the interpeduncular fossa to join the posterior cerebral artery (the termination of the basilar artery). Embryologically the posterior communicating persists as the posterior cerebral artery but in the adult, the basilar artery connects with the posterior cerebral as the basilar artery terminates. The posterior communicating artery varies greatly in caliber from person to person. Often one is smaller than the other is. Sometimes both are thread-like.

Some 4–12 perforating branches arise from the **posterior communicating artery**. An **anterior group** supplies the hypothalamus, subthalamus, anterior third of the optic tract, and posterior limb of the internal capsule. A **posterior group** of perforating vessels enters the interpeduncular perforated space to supply the subthalamic nucleus. Occlusion of the latter vessel often leads to a contralateral hemiballismus. Since these perforating vessels do not anastomose, their occlusion leads to infarction.

When of sufficient size, these posterior communicating arteries act as channels that equilibrate pressure between the carotid and the vertebral-basilar system. Blood in these two systems does not normally mix. Reduction in pressure in one system results in large amounts of blood flow to the other system.

Anatomical relations between the posterior communicating artery and the oculomotor nerve are important (aneurysms).

Anterior Choroidal Artery

Arising from the cerebral part of the internal carotid some 2–4 mm distal to the posterior communicating artery and passing posteriorly with the optic tract along its inferior surface, around the cerebral peduncle and into the temporal horn, is the **anterior choroidal artery** (Fig. 22.2). The anterior choroidal artery gives numerous small branches to supply the base of the **midbrain** (middle third of cerebral crus, substantia nigra), **parts of the visual system** (optic tract, lateral geniculate body, optic radiations), **basal nuclei** (medial globus pallidus, tail of the caudate), **thalamus**, and the **internal capsule** (posterior limb and retrolenticular parts which include the auditory and optic radiations). The anterior choroidal also supplies the **medial aspect of the temporal lobe**, especially the uncus, piriform cortex, and part of the amygdaloid body, before joining the **choroid plexus** of the temporal horn and atrium.

Occlusion of the anterior choroidal artery often leads to a contralateral hemianopia, hemiplegia, and hemi-anesthesia although these signs are inconsistent depending on the presence of anastomoses between the anterior choroidal, posterior choroidal, and posterior communicating arteries.

Anterior Cerebral Artery (ACA)

The **anterior cerebral artery,** the smaller of the two terminal branches of the internal carotid artery (Fig. 22.9), passes medially and then superior to the optic chiasma or nerves before entering the longitudinal cerebral fissure. Immediately superior to the chiasma both anterior cerebral arteries join by way of the **anterior communicating artery** (ACoA) (Fig. 22.2). This communication between both carotid circulations averages 2.6 mm in length (range 0.3–7.0 mm) and is likely to be single, double, trebled or absent. The anterior communicating artery may have up to four branches that reach the dorsal surface of the optic chiasma, suprachiasmatic area, and anterior perforated substance, and perhaps the frontal lobe. These perforating vessels are likely to supply the fornix, corpus callosum, subcallosal area, and anterior part of the cingulum. Occlusion of these perforating vessels is likely to lead to hypothalamic sign symptoms and loss of memory for recent events ('recent memory').

That segment of the anterior cerebral artery from the carotid to the anterior communicating artery is the A_1 segment. Some two to 15 perforating vessels leave this segment of the anterior cerebral artery. Occlusion of the hypothalamic branches from A_1 may result in psychiatric symptoms without a motor deficit. The former often include feelings of anxiety and fear, disordered mentation, dizziness, and agitation.

The segment of the anterior cerebral artery after the anterior communication artery is the A_2 segment that has up to four perforating vessels including the recurrent branch of the anterior cerebral artery (medial striate artery or recurrent artery of Heubner). This recurrent branch doubles back on the anterior cerebral artery and accompanies it for a short distance before branching and entering the lateral sulcus or penetrating the anterior perforated substance (an area behind the olfactory trigone perforated by numerous blood vessels) at a point above the carotid bifurcation. The recurrent branch supplies the anterior parts of the caudate nucleus, putamen, globus pallidus, and internal capsule, as well as the uncinate fasciculus and olfactory trigone. Occlusion of the recurrent artery often causes a contralateral hemiparesis especially of the arm, face, palate, and tongue and perhaps aphasia if the injury is in the hemisphere dominant for language (the left cerebral hemisphere in 95% of humans). The other perforating branches of A_2 enter the gyrus rectus, inferior frontal area, anterior perforated substance, and the dorsal optic chiasma and suprachiasmatic area. An accessory middle cerebral artery, when present, originates from the anterior cerebral artery. Since it supplies similar deeply located structures, it may be a variant of the recurrent artery.

The anterior cerebrals run parallel to one another and successively forward, upward, and backward around the genu of the corpus callosum and onto its upper aspect to reach the back end or splenium of the corpus callosum. Along its course, the anterior cerebral gives off named branches such as the fronto-polar artery (Fig. 22.9) that supply the anterior part of the frontal lobe on its medial side.

The callosomarginal artery (Fig. 22.9) runs into the cingulate sulcus to the paracentral lobule. Some of these branches curve over the superior margin of the cerebral hemisphere onto its lateral surface. Obstruction of the callosomarginal branch often leads to infarction of the paracentral lobule causing paralysis of the contralateral leg with associated cortical sensory loss and perhaps urinary incontinence (if the cortical injury is bilateral to the bladder area on the medial surface of the brain). In such cases, there is sparing of the face, arm

and hand. The anterior cerebral artery is easily visible on arteriograms. Displacement from its normal position suggests pressure or traction on one cerebral hemisphere. This may be evidence of a tumor in the frontal lobe.

Middle Cerebral Artery (MCA)

The middle cerebral artery is the larger of two terminal branches of the internal carotid artery. In course, it has four segments: sphenoidal, insular, opercular, and cortical (Fig. 22.16). The M1 or sphenoidal (horizontal or cisternal) segment of the middle cerebral originates near the anterior aspect of the deep part of the lateral fissure, lateral to the optic chiasma, below the anterior perforated substance, and posterior to the point at which the olfactory tract divides into medial and lateral striae.

From its origin, the middle cerebral artery turns laterally into the depths of the medial aspect of the lateral sulcus, crosses over the limen insulae, and then enters the insular area where it divides into many branches that spread out on (and supply) the surface of the insular lobe. This is the M2 or insular (Sylvian) segment of the middle cerebral artery (Figs 22.16, 22.17). Collateral branches over 50 μm in diameter are a feature along the course of the insular segment of the middle cerebral artery that penetrates the surface of the insula. From the periphery of the insula, near its circular fissure, branches forming the M3 or opercular segment of the middle cerebral artery (Fig. 22.16) continue over the medial surface of the opercula, and emerge from the lateral sulcus by passing between the opposing lips of the frontoparietal and temporal opercula. These

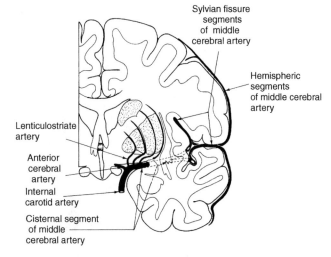

Sylvian fissure segments of middle cerebral artery

Hemispheric segments of middle cerebral artery

Lenticulostriate artery

Anterior cerebral artery

Internal carotid artery

Cisternal segment of middle cerebral artery

FIGURE 22.16. The course of the middle cerebral artery seen in a coronal section of the human brain (from DeArmond, Fusco, and Dewey, 1989).

branches then pass around the opercular parts of the frontal, parietal, and temporal lobes. At this point, branches forming the **M4 or cortical** (hemispheric) **segment of the middle cerebral artery** begin, become superficial, and extend over the lateral and inferior surfaces of the brain (Figs 22.16, 22.18).

Branches of the Middle Cerebral Artery

The middle cerebral artery proper, which averages some 15.8 ± 0.9 mm in length, divides into two or more *trunks* by bifurcation (superior and inferior trunks),

trifurcation (superior, middle, and inferior trunks), or division into multiple trunks. Some six to eight (range 6–11) stem arteries arise from the trunks and give rise to cortical branches that supply about a dozen cortical areas supplied by one or two (range 1–5) cortical branches. These 12 cortical areas include the orbitofrontal, prefrontal, precentral, central, anterior and posterior parietal, angular, anterior, middle and posterior temporal, temporopolar, and temporo-occipital vessels (Fig. 22.18). Each stem artery yields one to five cortical branches that supply these cortical areas. The middle cerebral artery supplies a larger territory than either the anterior or posterior cerebral. Indeed this vessel is twice the size of the anterior cerebral and carries about 80% of the blood received by each cerebral hemisphere.

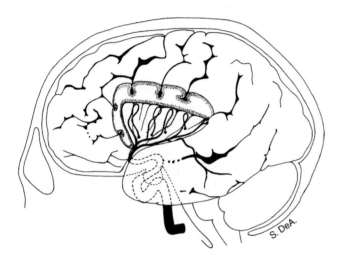

FIGURE 22.17. Branches of the middle cerebral artery on the surface of the insular lobe (from DeArmond, Fusco, and Dewey, 1989).

Perforating Branches of the Middle Cerebral Artery

This large and complex vessel has as many as 15 perforating vessels that arise from the middle cerebral artery proper or from one of its main trunks before entering the anterior perforated substance. These perforating vessels, often divided into medial and lateral groups, are collectively termed the **lenticulostriate arteries** (arteries of the corpus striatum or striatal branches) (Fig. 22.16). They supply the body and head of the caudate nucleus, putamen, and lateral segment of the globus pallidus, genu, and posterior limb of the internal capsule. A prominent perforating artery from the lateral group courses along the base of the lentiform

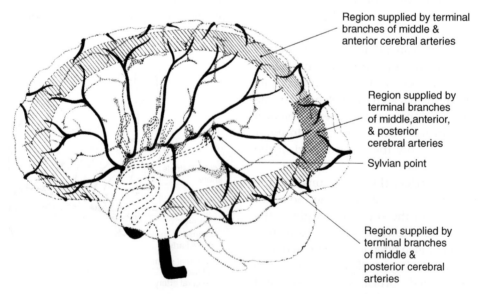

Region supplied by terminal branches of middle & anterior cerebral arteries

Region supplied by terminal branches of middle, anterior, & posterior cerebral arteries

Sylvian point

Region supplied by terminal branches of middle & posterior cerebral arteries

FIGURE 22.18. The cortical branches of the middle cerebral artery as they emerge from the lateral sulcus (from DeArmond, Fusco, and Dewey, 1989).

nucleus, crosses the upper part of the internal capsule, and then supplies the caudate nucleus. This long and unbranched vessel is often termed the 'artery of cerebral hemorrhage'.

Cortical Branches of the Middle Cerebral Artery (MCA)

In general, the middle cerebral artery and its branches supply parts of all five lobes of the brain. This includes most of the lateral surface of the frontal, parietal, and occipital lobes as well as all insular surfaces and opercular surfaces of the frontal, parietal, and temporal lobes. The MCA also supplies the superior and lateral aspect of the temporal lobe, lateral parts of the inferior surface of the temporal lobe and the orbital surface of the frontal lobe. The MCA supplies the temporal pole, uncus, and adjacent part of the parahippocampal gyrus. Branches of the middle cerebral artery do not reach the occipital or frontal poles or that part of the cerebral hemisphere near the longitudinal cerebral fissure.

Angular Artery

The most constant branch of the middle cerebral artery is the **angular artery (parieto-occipital artery)**. Considered the terminal branch of the middle cerebral artery, it divides on the cortical surface at the level of the angular gyrus into two branches to supply the inferior parietal lobule (supramarginal and angular gyri), part of the superior parietal lobule, the posterior part of the superior and middle temporal convolutions, and the superior occipital convolution.

Involvement of the MCA in Disease States

The middle cerebral artery and its branches are the most commonly occluded ('plugged') of any intracranial vessel by a thrombus or an embolus. The MCA is also one of the most common sites of **saccular aneurysms** usually on the distal part of its sphenoidal segment (Fig. 22.16). Most intracranial **arteriovenous malformations** receive part of their blood supply from the middle cerebral artery.

Because of the number of branches and the broad area supplied by the middle cerebral artery, the manifestations of occlusion of this vessel or one of its cortical branches vary widely depending on the affected cortical area. With involvement of the precentral gyrus, there may be a **contralateral hemiplegia** especially severe in the arm. With involvement of the premotor cortex there is likely to be a **grasp reflex, motor apraxia**

and **contralateral weakness of the shoulder and hip muscles**. With involvement of the motor speech area, there is likely to be a **motor aphasia**. With involvement of the prefrontal cortex, there are likely to be **personality changes** and if there is involvement of the visual radiations, there is likely to be a **homonymous hemianopia**. Involvement of the postcentral gyrus may lead to an absence **of proprioception and discriminative tactile sensation on the contralateral side** and **perhaps a loss of the ability to use tactile information in order to identify objects placed in the hand**. With involvement of the superior temporal gyrus there is likely to result an **auditory receptive aphasia**. Perhaps other signs such as **apraxias** (sensory, dressing, or constructional) or **Gerstmann's syndrome** (acalculia, agraphia, finger agnosia, and right-left disorientation) would follow damage to the parietal lobe.

22.7.3. Posterior Cerebral Artery and its Cerebral Supply

As noted earlier, the vertebrals unite at the pontomedullary junction to form a single **basilar artery**. At the upper pons, the **basilar artery** bifurcates, giving rise to the **posterior cerebral arteries** (Fig. 22.9). The subclavian, vertebral, and basilar arteries act as a unit to supply the posterior cerebral circulation. This is unique because nowhere else in the body do two major arteries (vertebrals) join to form a single vessel (the basilar) which then splits to form two major vessels (the posterior cerebrals). Thus, the terminal branches of the basilar artery are the **posterior cerebral arteries**. The posterior cerebrals encircle the brain stem at the level of the **tentorial notch**. They then pass superior to the oculomotor nerve [III] and laterally around the brain to reach the inferior and medial aspects of the temporal lobes before ending at the occipital poles. About one-third of the time one posterior cerebral artery (left more commonly) retains its embryological relationship as a major branch of the ipsilateral internal carotid artery. The narrow **posterior communicating artery** in these instances is between the posterior cerebral artery and the basilar artery, and the flow of blood from the basilar artery is into the larger posterior cerebral branch. Herniation of the temporal lobe often compresses the posterior cerebral artery with infarction of the occipital lobe. Such displacement is usually visible on arteriograms. The major part of the posterior cerebral artery supplies the posteromedial cerebral cortex.

Cerebral Branches of the Posterior Cerebral Artery

That segment of the posterior cerebral artery from the basilar bifurcation to the posterior communicating artery is the P_1 (midbrain) **segment** (Fig. 22.9). Some four perforating vessels leave this segment of the posterior cerebral artery. That segment of the posterior cerebral artery after the posterior communication artery is the P_2 segment (Fig. 22.9) whereas that segment of the posterior cerebral that runs in the calcarine fissure is the P_3 segment (Fig. 22.9). The posterior cerebral supplies the splenium of the corpus callosum, anterior and posterior parts of the temporal lobes, the parieto-occipital region, and posterior part of the cerebral hemispheres.

22.8. CEREBRAL ARTERIAL CIRCLE

At the base of the brain is a polygon formed by the paired anterior cerebral, internal carotid, posterior cerebrals, and posterior communicating arteries along with a single anterior communicating artery. This **cerebral arterial circle** (the circle of Willis) has a 'normal' configuration in only 50% of examined brains with variations primarily in the relative size of vessels (Fig. 22.2). This polygon provides an avenue for collateral circulation between vessels on both sides of the brain and between the carotid and vertebral-basilar circulation. The rate of flow is dependent upon pressure gradients and patency of the vessels. The efficiency of this system determines the degree of neurological deficit occurring with stenosis or occlusion of one or more arteries in the neck.

22.8.1. Types of Arteries Supplying the Brain

The pattern of arterial supply to the cerebral hemispheres is the same for the entire brain and is dependent on **three types of vessels**. (1) **Long circumferential branches** of parent arteries that course around the ventral and lateral aspects of the brain before reaching the dorsal surface. Here they anastomose with distal branches of other long circumferential vessels. They carry blood for long distances on the surfaces of the cerebral hemispheres or brain stem, while giving off numerous unnamed perforating branches that enter the substance of the brain. (2) **Short circumferential** or **lateral perforating branches** arise from a parent artery and travel for shorter distances before plunging into

the surface of the brain to supply underlying gray and white matter. (3) **Paramedian** or **medial perforating arteries** arise from parent vessels and enter the brain on either side of the median plane. The paramedian branches supply central nuclear groups close to the median plane. Short circumferential vessels supply a zone between the area nourished by the long circumferential and the area nourished by the paramedian arteries. Unlike the long circumferentials, the short circumferential and paramedian vessels have limited anastomosis.

22.9. EMBRYOLOGICAL CONSIDERATIONS

The development of the cerebral vasculature often determines the outcome of many cerebrovascular diseases in adult life. Unusual configurations of the **cerebral arterial circle** at the base of the brain, arteriovenous malformations, congenital saccular aneurysms, and string-like or absent vessels are abnormalities that result from defective development of the cerebral vessels.

In the eighth week of development, nearly all intracranial arteries that remain throughout life are present. There is also an increase in venous anastomoses and in the proliferation of capillaries in the neocortex. Variations in the cerebral arterial circle, including a string-like anterior or posterior cerebral artery, is likely to jeopardize the role of these vessels as potential collateral channels and lead to an increase in the incidence of brain infarction. Though the posterior cerebral arteries arise embryologically from the carotids, blood flow through the posterior cerebrals comes from the vertebrals in 90% of patients.

22.10. VASCULAR INJURIES

Many everyday activities (including childbirth, overhead work, turning the head while driving, and beauty parlor events) and sports activities (such as gymnastics, calisthenics, yoga, archery, driving, and swimming) as well as chiropractic manipulation of the cervical spine reportedly disrupt the cerebral circulation and are associated with vascular accidents of the brain or spinal cord. In vivo and autopsy evidence indicates that the vertebral arteries are compromised at the atlanto-axial articulation by head motion, particularly rotation (rotation and flexion, rotation and extension) but not tilt and extension of the head. During rotation of the head, the atlanto-axial joint on the side to which the head is turned remains fixed while on the opposite side, the atlas moves forward on the axis. Symptoms occur if the vertebral artery between the atlas and axis becomes narrow or stretches sufficiently to obstruct blood flow and if flow in the contralateral vertebral artery is in some way already compromised.

22.10.1. Brain Stem Vascular Injuries

Vascular disease here leads to an array of signs and symptoms. Chief among them are: dysarthria, dysphagia, motor and sensory deficits, disturbances of consciousness, drop attacks, attacks of vertigo, nausea, and vomiting, temporary loss of vision, ocular abnormalities (such as double vision).

22.10.2. Visualization of Brain Vessels

The vascular system of the brain is evident on radiographs with the introduction of radiopaque substances into the cerebral vessels. With this procedure, called **cerebral angiography** or **arteriography**, only the long circumferential conducting arteries are evident. Occasionally the largest of the short circumferential branches are visible but the unnamed, perforating vessels are not visible such that only 10% of intracranial arteries are visible using conventional angiography.

FURTHER READING

Cosson A, Tatu L, Vuillier F, Parratte B, Diop M, Monnier G (2003) Arterial vascularization of the human thalamus: extra-parenchymal arterial groups. Surg Radiol Anat 25:408–415.

Erdem A, Yasargil G, Roth P (1993) Microsurgical anatomy of the hippocampal arteries. J Neurosurg 79:256–265.

Fine AD, Cardoso A, Rhoton AL Jr (1999) Microsurgical anatomy of the extracranial–extradural origin of the posterior inferior cerebellar artery. J Neurosurg 91:645–652.

Gibo H, Carver CC, Rhoton AL Jr, Lenkey C, Mitchell RJ (1981) Microsurgical anatomy of the middle cerebral artery. J Neurosurg 54:151–169.

Gillilan LA (1958) The arterial blood supply of the human spinal cord. J Comp Neurol 110:75–103.

Gillilan LA (1964) The correlation of the blood supply to the human brain stem with clinical brain stem lesions. J Neuropathol Exp Neurol 23:78–108.

Gillilan LA (1970) Veins of the spinal cord. Anatomic details: suggested clinical applications. Neurology 20:860–868.

Gillilan LA (1972) Anatomy and embryology of the arterial system of the brain stem and cerebellum. In: Vinken PJ, Bruyn GW, eds. Handbook of Clinical Neurology. 11:24–44. Elsevier, Amsterdam.

Hassler O (1966) Blood supply to human spinal cord. A microangiographic study. Arch Neurol 15:302–307.

Lazorthes G, Gouaze A, Zadeh JO, Santini JJ, Lazorthes Y, Burdin P (1971) Arterial vascularization of the spinal cord. Recent studies of the anastomotic substitution pathways. J Neurosurg 35:253–262.

Lazorthes G (1972) Pathology, classification and clinical aspects of vascular diseases of the spinal cord. In: Vinken PJ, Bruyn GW, eds. Handbook of Clinical Neurology. 12:492–506. Elsevier, Amsterdam.

Oka K, Rhoton AL Jr, Barry M, Rodriguez R (1985) Microsurgical anatomy of the superficial veins of the cerebrum. Neurosurgery 17:711–748.

Ono M, Rhoton AL Jr, Peace D, Rodriguez RJ (1984) Microsurgical anatomy of the deep venous system of the brain. Neurosurgery 15:621–657.

Rhoton AL Jr (2000) The cerebellar arteries. Neurosurgery 47(Suppl 3):S29–68.

Rhoton AL Jr (2002) The cavernous sinus, the cavernous venous plexus, and the carotid collar. Neurosurgery 51 (Suppl 1):375–410.

Rhoton AL Jr (2002) The supratentorial arteries. Neurosurgery 51(Suppl 4):S53–120.

Rosner SS, Rhoton AL Jr, Ono M, Barry M (1984) Microsurgical anatomy of the anterior perforating arteries. J Neurosurg 61:468–485.

Schmahmann JD (2003) Vascular syndromes of the thalamus. Stroke 34:2264–2278.

Tanriover N, Rhoton AL Jr, Kawashima M, Ulm AJ, Yasuda A (2004) Microsurgical anatomy of the insula and the sylvian fissure. J Neurosurg 100:891–922.

Turnbull IM (1972) Blood supply of the spinal cord. In: Vinken PJ, Bruyn GW, eds. Handbook of Clinical Neurology. 12:478–491. Elsevier, Amsterdam.

Vinas FC, Lopez F, Dujovny M (1995) Microsurgical anatomy of the posterior choroidal arteries. Neurol Res 17:334–344.

It has long been known that there is an active formation of cerebrospinal fluid as evidenced by the rapidity with which the fluid reforms after it has been withdrawn either by lumbar or ventricular puncture. The endowment of the choroidal plexus with an elaborate blood-supply indicates that it is a structure with a special function. Since the work of Faivre (1854) and Luschka (1855) showing the secretory character of the cells, the chorioid plexuses have been regarded as glands from which at least part of the cerebrospinal fluid is formed.

Walter E. Dandy, 1913

23

The Meninges, Ventricular System and Cerebrospinal Fluid

23.1. THE CRANIAL MENINGES AND RELATED SPACES

23.1.1. Cranial Dura Mater

Three membranes or **meninges** surround and protect the cerebral hemispheres, brain stem, and spinal cord. A description of the spinal meninges and related spaces is available in Chapter 3, The Spinal Cord. The dense fibrous tissue of the **cranial dura mater** (Figs 23.1, 23.2, 23.3) supports the brain and acts as the lining layer of the cranial bones. The cranial dura is thick and includes an internal **meningeal layer** and an external **endosteal layer** that lies on the internal aspect of the skull (Fig. 23.3). These two layers are indistinguishable except where the dural venous sinuses intervene between them (Fig. 23.4). The meningeal and endosteal layers serve both meningeal and periosteal functions. The endosteal layer of the cranial dura firmly adheres to the inner aspect of the cranial bones, especially at the sutures and along the base of the skull. Thus the space external to the cranial dura, the epidural space, is a potential space. Following trauma to the skull, blood may enter this potential space and accumulate there forming an epidural hematoma. At foramina at the base of the skull, the meningeal layer provides sheaths for the cranial nerves. The meningeal layer of the cranial dura is continuous at the foramen magnum with the spinal dura mater. Thus, the spinal dura mater consists of only one layer – the continuation of the meningeal layer of the cranial dura. While there is a definable subdural space between the spinal dura mater and the spinal arachnoid mater, the subdural space in the cranial region is absent. Studies of the fine structure of the dura-arachnoid interface layer in humans reveal that the innermost part of the cranial dura and outermost part of the arachnoid are intimately fused.

23.1.2. Cranial Arachnoid

The **cranial arachnoid** (Figs 23.2, 23.3, 23.4) consists of an outer part, the arachnoid barrier layer and in inner

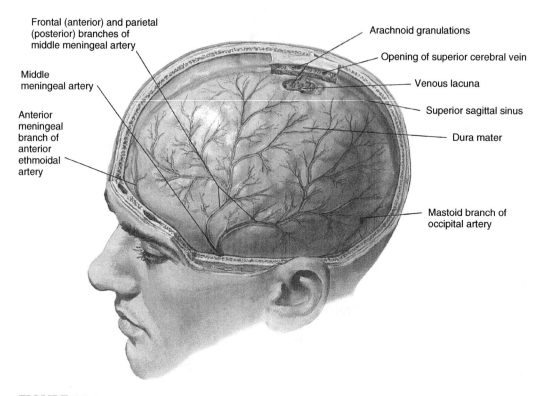

FIGURE 23.1. The cerebral dura mater covering the left cerebral hemisphere of the human brain (from Netter, 1989). (This figure is reproduced in color in the color plate section.)

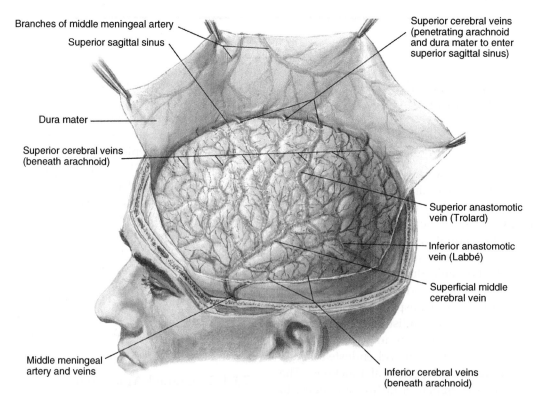

FIGURE 23.2. The cerebral dura mater of the left cerebral hemisphere cut and elevated to reveal the underlying cerebral arachnoid (from Netter, 1989). (This figure is reproduced in color in the color plate section.)

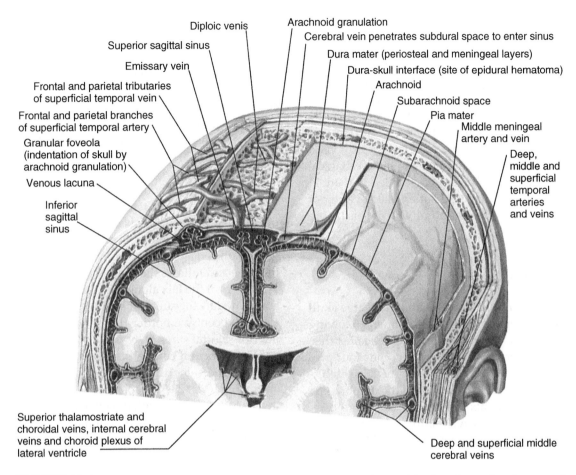

FIGURE 23.3. The layers of the human cranial meninges including two layers of the dura, the arachnoid mater and the cerebral pia mater (from Netter, 1989). (This figure is reproduced in color in the color plate section.)

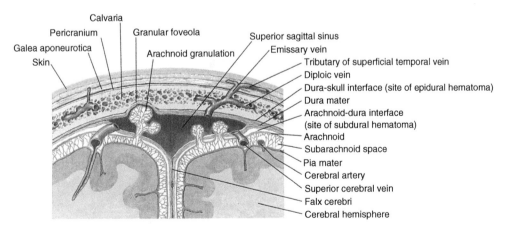

FIGURE 23.4. Coronal section through the superior sagittal sinus within the cerebral dura of the human brain (from Netter, 1989). (This figure is reproduced in color in the color plate section.)

part, the arachnoid trabeculae. The arachnoid barrier layer fuses with the inner or meningeal layer of the cranial dura. Structurally and functionally, this arachnoid barrier layer prevents cerebrospinal fluid in the subarachnoid space from reaching the dura. Delicate strands of the **arachnoid trabeculae** (extensions of fibroblasts), bridge the subarachnoid space and give it a spider web-like appearance. These trabeculae surround vessels in the subarachnoid space and attach to the pia on the surface of the brain. In the meshes of the arachnoid trabeculae (subarachnoid space), cerebrospinal fluid circulates. Although the arachnoid dips into the longitudinal cerebral fissure, it does not dip into the individual sulci of each cerebral hemisphere.

23.1.3. Cranial Pia Mater

The **cranial pia mater** (Figs 23.3, 23.4) is the most delicate layer of the cranial meninges. It closely follows every convolution on the brain surface and enters every cerebral sulcus. The cranial pia also lines the basal cisterns. The cranial arachnoid trabeculae join the pia to the arachnoid mater. Although very thin and apparently fragile, cranial pia nonetheless exhibits a high level of stiffness thus influencing the mechanical properties of the brain. The cerebral vessels that are in the subarachnoid space have a superficial covering of the cranial pia. Smaller vessels ramify on the pia before penetrating the brain.

23.1.4. Dural Projections

A freely moving brain encased in an immovable skull would result in serious injury to the brain following trauma to the head. To provide protection against this, the meningeal layer of the cranial dura sends four projections or folds into the cranial cavity: the **falx cerebri**, **tentorium cerebelli**, **falx cerebelli**, and the **diaphragma sellae** (Fig. 23.5). The falx cerebri and tentorium cerebelli divide the cranial cavity into two hemispheric compartments occupying the anterior and middle cranial fossa and a cerebellar compartment occupying the posterior cranial fossa. The frontal lobe occupies the anterior cranial fossa whereas the lateral parts of the middle cranial fossa contain the temporal lobes. The tentorium cerebelli forms the roof of the cerebellar compartment.

The longitudinal cerebral fissure separating the two cerebral hemispheres contains the crescent-shaped **falx cerebri** (Fig. 23.5). The falx is a single median structure that is narrow anteriorly where it attaches to the crista galli of the ethmoid bone but much broader posteriorly where it fuses with the tentorium cerebelli (Fig. 23.5) in the median plane. The falx cerebri has a superior border that is convex in shape and attached to the overlying skull and an inferior border that is free but concave in shape and follows the shape of the corpus callosum. The layers of the falx along the convex border divide to accommodate the superior sagittal sinus whereas those along the concave border divide to accommodate the inferior sagittal sinus.

The **tentorium cerebelli** supports the occipital lobe while at the same time forming a dural roof for the cerebellum (Fig. 23.5). It too is crescent shaped but lies in a horizontal plane with its convex border laterally or externally and its concave border medially or internally. The layers of the tentorium along the convex border

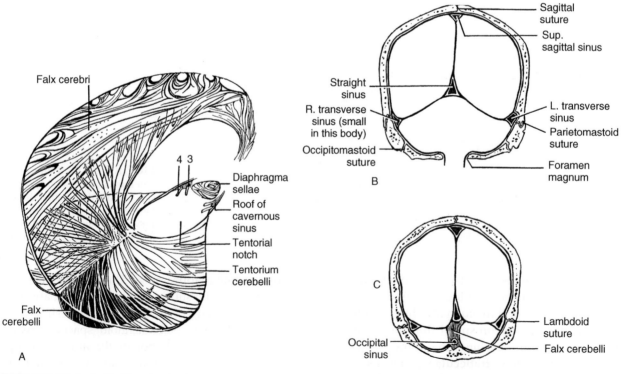

FIGURE 23.5. A, B, C. The dural folds and related sinuses of the human brain (from O'Rahilly, 1986).

divide to accommodate the transverse sinus. The concave borders do not meet in the median plane but bound an opening, the **tentorial notch** (Fig. 23.5). The midbrain occupies the tentorial notch. The tentorium cerebelli attaches to the posterior clinoid processes of the sphenoid bone. The compartment inferior to the tentorium contains not only the cerebellum but also the pons and medulla oblongata.

The **falx cerebelli** (Fig. 23.5) lies below the tentorium and attaches to the inferior aspect of the tentorium in the median plane and to the internal occipital crest. The posterior margin of the falx cerebelli contains the occipital sinus whereas the free anterior margin of the falx cerebelli projects between the two cerebellar hemispheres. The sella turcica lies on the upper surface of the body of the sphenoid bone. It normally contains the hypophysis.

The fourth dural projection, the **diaphragma sellae** (Fig. 23.5) is circular in shape forming a roof over the sella turcica and covering its contents. There is a centrally placed opening in the diaphragma sellae which transmits the infundibular stalk of the hypophysis.

23.1.5. Intracranial Herniations

Space-occupying intracranial mass lesions such as tumors or blood clots as well as disorders of cerebrospinal fluid (CSF) circulation may greatly increase intracranial pressure (ICP). The cranial cavity is divisible into compartments by the falx cerebri and tentorium cerebelli. In such cases, brain swelling occurs and part of the brain may herniate from one dural compartment to another. Three types of intracranial herniation are generally recognized: cingulate, uncal, and tonsillar. **Cingulate gyrus herniation** may occur as the result of pressure from a subdural hematoma. In this case one cingulate gyrus may slip under the falx cerebri and press on the opposite cingulate gyrus (Fig. 23.6). This can happen with no serious neurological complication. In the temporal lobe, the brain may herniate through the tentorial notch and compress the midbrain pushing it against the opposite tentorial edge. This may cause **herniation of the uncus** (Fig. 23.7) from one dural compartment to another by way of the tentorial notch. As the uncus herniates through the tentorial notch, it compresses the midbrain pushing it against the opposite tentorial edge. The oculomotor nerve [III] and posterior cerebral artery on the side of the expanding lesion are vulnerable to compression between the uncus and the free edge of the tentorium. Uncal herniation may produce alterations in consciousness including

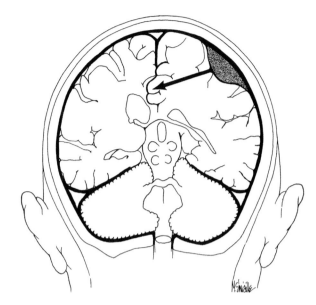

FIGURE 23.6. Because of increased intracranial pressure from a subdural hematoma, there is herniation of the cingulate gyrus under the falx cerebri with pressure on the contralateral cingulate gyrus (from Nolte, 1988).

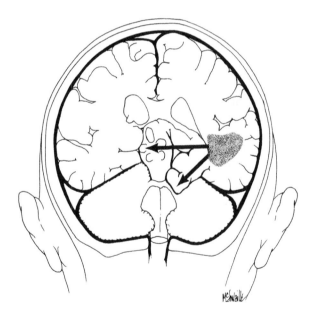

FIGURE 23.7. As a result of an expanding tumor in the one temporal lobe the most medial part of that temporal lobe, the uncus, has herniated (uncal herniation) through the tentorial notch and is pressing against the midbrain (from Nolte, 1988).

coma, often followed by death. Finally, pressure from a cerebellar lesion may cause one tonsil of the cerebellum to herniate through the foramen magnum (Fig. 23.8), a condition termed **cerebellar tonsillar herniation**. The result is compression of the medulla against the margin of the foramen. This may lead to cardiovascular and respiratory depression with rapid death.

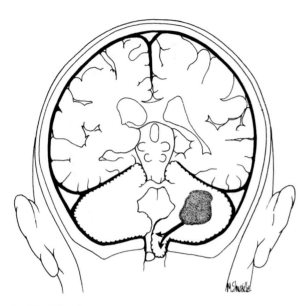

FIGURE 23.8. Because of an expanding cerebellar tumor, one cerebellar tonsil has herniated (cerebellar tonsilar herniation) through the foramen magnum compressing the medulla oblongata (from Nolte, 1988).

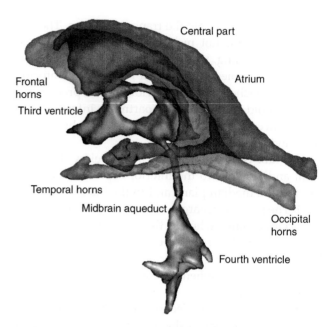

FIGURE 23.9. Three-dimensional rendering of an automatically segmented ventricular system. The left and right lateral ventricles are evident (from Martin Styner, University of North Carolina, 2006).

23.2. VENTRICULAR SYSTEM

23.2.1. Introduction

The **ventricles** are described in the order in which they are numbered from rostral to caudal. Each of the two **lateral ventricles** (ventricles 1 and 2) communicate with the **third ventricle** by an **interventricular foramen** on either side of the median plane. The **third ventricle** communicates with the **fourth ventricle** by way of the **aqueduct of the midbrain**. The fourth ventricle becomes continuous with the **central canal** of the medulla oblongata and spinal cord and opens by means of **apertures** into the subarachnoid space.

The neuroglial cells that line the ventricles of the brain and the **central canal** of the spinal cord are termed **ependymal cells**. The ependyma is nonciliated in adults. In the ventricles, vascular fringes of pia mater, known as the tela choroidea, invaginate their covering of modified ependyma and project into the ventricular cavities. The combination of vascular tela and cuboidal ependyma protruding into the ventricular cavities is termed the **choroid plexus**. The plexuses are invaginated into the cavities of both lateral as well as the third and fourth ventricles; they are concerned with the formation of cerebrospinal fluid.

The term 'blood–cerebrospinal fluid barrier' refers to the tissues that intervene between the blood and the cerebrospinal fluid including the capillary endothelium,

several homogeneous and fibrillary layers (identified by electron microscopy) and the ependyma of the **choroid plexus**. The chief elements in the barrier are tight junctions between the ependymal cells.

23.2.2. Lateral Ventricles

Each **lateral ventricle** is a cavity in the interior of a cerebral hemisphere and each communicates with the **third ventricle** by means of an **interventricular foramen**. Each lateral ventricle includes six parts. The first part is in front of the interventricular foramen and is the **frontal (anterior) horn** (Fig. 23.9). Behind this is the **central part** of the ventricle (Fig. 23.9) that includes the second, third, and fourth parts, respectively. The fourth part of the ventricle termed the atrium divides into a fifth part, or **occipital (posterior) horn**, and a sixth part, or **temporal (inferior) horn** (Fig. 23.9). The **frontal**, **occipital**, and **temporal horns** are in the frontal, occipital and temporal lobes of the cerebral hemispheres, respectively.

The **frontal horn**, **central part** and **temporal horn** are C-shaped, a characteristic of the fetal and the adult lateral ventricle. The **occipital horn** develops about halfway through fetal life as a backward extension and is variable in size in adults. Because of this unusual embryological development, the posterior horn may be separate from the other parts of the ventricular system.

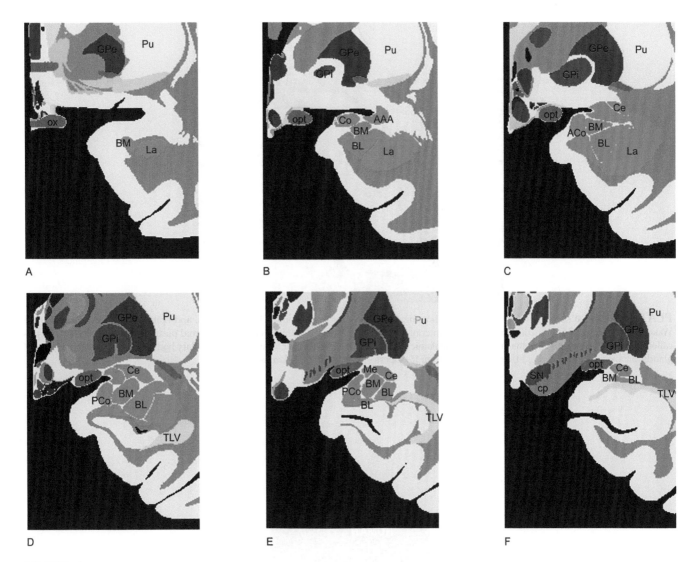

PLATE 1. Series of coronal sections through the human brain to demonstrate the nuclei of the amygdaloid body from rostral to caudal. (**A**, **B**, **C**, **D**, **E**, **F** are based on sections 22–32 respectively in Mai, Assheuer and Paxinos, 2004.) **Abbreviations:** AAA, anterior amygdaloid area; ACo, anterior cortical amygdaloid nucleus; BL, basolateral amygdaloid nucleus; BM, basomedial amygdaloid nucleus; Ce, central amygdaloid nucleus; cp, cerebral peduncle; GPe, globus pallidus, external segment; GPi, globus pallidus, internal segment; La, lateral amygdaloid nucleus; Me, medial amygdaloid nucleus; opt, optic tract; ox, optic chiasma; PCo, posterior cortical amygdaloid nucleus; Pu, putamen; SN, substantia nigra; TLV, temporal horn of lateral ventricle.

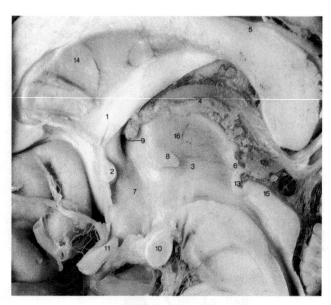

PLATE 2. Medial surface of the human brain showing the position of the thalamus, hypothalamus, and epithalamus (from England and Wakely, 1991). **Key:** 1, anterior column of fornix; 2, anterior commissure; 3, cavity of third ventricle; 4, choroid plexus of third ventricle; 5, corpus callosum; 6, epithalamus; 7, hypothalamus; 8, interthalamic adhesion; 9, interventricular foramen; 10, mamillary body; 11, optic chiasma; 12, pineal gland; 13, posterior commissure; 14, septum pellucidum; 15, superior colliculus; 16, thalamus.

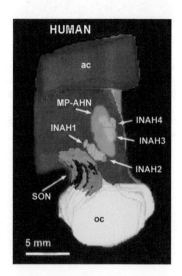

PLATE 3. Three-dimensional reconstruction of the medial preoptic-anterior hypothalamic continuum of the human hypothalamus. INAH3 occupies a larger volume and more neurons in presumed heterosexual men than women but there was no difference in the number of neurons in this nucleus based on sexual orientation. No sex difference in volume was detected for any other INAH (from Byne et al., 2001.) **Abbreviations:** ac, anterior commissure; INAH, interstitial nucleus of the anterior hypothalamus; MP-AHN, medial preoptic-anterior hypothalamic nucleus; oc, optic chiasma; SON, supra-optic nucleus.

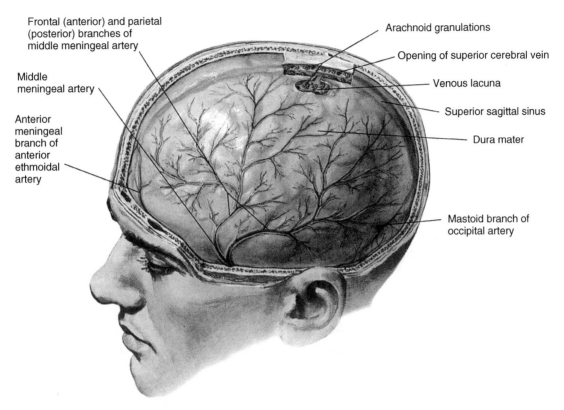

PLATE 4. The cerebral dura mater covering the left cerebral hemisphere of the human brain (from Netter, 1989).

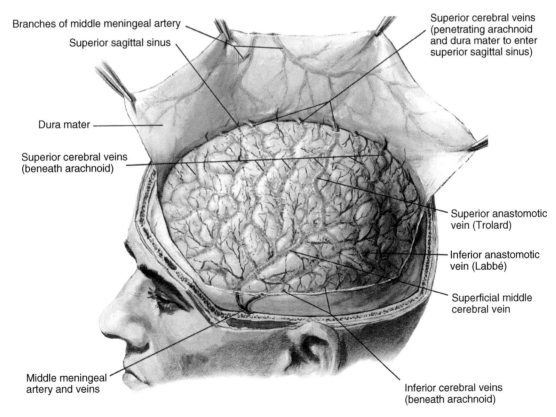

PLATE 5. The cerebral dura mater of the left cerebral hemisphere cut and elevated to reveal the underlying cerebral arachnoid (from Netter, 1989).

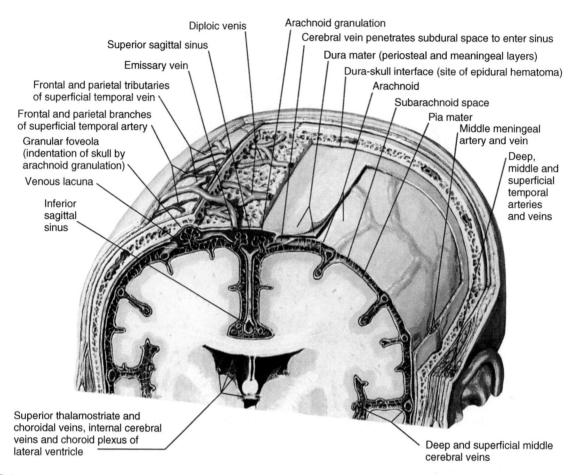

PLATE 6. The layers of the human cranial meninges including two layers of the dura, the arachnoid mater and the cerebral pia mater (from Netter, 1989).

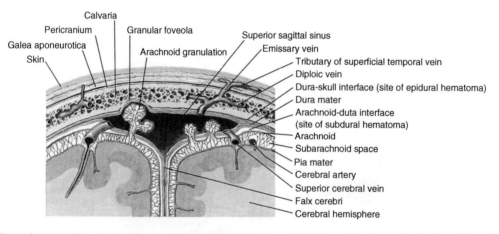

PLATE 7. Coronal section through the superior sagittal sinus within the cerebral dura of the human brain (from Netter, 1989).

Frontal (Anterior) Horn

The **frontal horn** of the lateral ventricle (Fig. 23.9) has as its inferior boundary the rostrum, its rostral boundary the genu, and its superior boundary the trunk, of the corpus callosum. Laterally the frontal horn of the lateral ventricle is limited by the bulging head of the caudate nucleus. A thin vertical partition, the **septum pellucidum** separates the frontal horn of one hemisphere from the frontal horn in the other hemisphere. Occasionally the septum has a space (cavity of the septum pellucidum) between its two layers.

Central Part

The **central part** of the lateral ventricle (Fig. 23.9) is inferior to the trunk of the corpus callosum. The central part is also on the medial aspect of the dorsal thalamus and body of the caudate nucleus. In the central part of the lateral ventricle is the choroid plexus.

Occipital (Posterior) Horn

The variable **occipital horn** of the lateral ventricle (Fig. 23.9) tapers into the occipital lobe of each cerebral hemisphere. The **occipital horns** are usually asymmetrical – the back part of one horn may appear as a separate entity.

Temporal (Inferior) Horn

The **temporal horn** of the lateral ventricle (Fig. 23.9) extends into the temporal lobe from the fourth part of the **central part** of the lateral ventricle. Fibers derived from the corpus callosum form its lateral boundary. Inferiorly is an elevation known as the **hippocampus**, partly covered by part of the **choroid plexus**. Superiorly, the **tail of the caudate nucleus** runs forward to end in the amygdaloid body.

Choroid Plexus of the Lateral Ventricle

The **choroid plexus** of each lateral ventricle is invaginated along a curved line known as the choroid fissure. The fissure extends from the **interventricular foramen** in front and arches over the posterior end of the dorsal thalamus, as far as the end of the **temporal horn**. The **choroid plexus** of the lateral ventricle is only in the **central part** of the lateral ventricle and the **temporal horn**. At the junction of the **central part** with the **temporal horn** it is best developed and is there known as the glomus choroideum. Calcified areas (corpora amylacea) are frequent in the glomus. The vessels of the plexus originate from the internal carotid (anterior choroidal artery) and from the posterior cerebral (posterior choroidal arteries). At the **interventricular foramina**, the **choroid plexuses** of both **lateral ventricles** become continuous and then join that of the **third ventricle**.

23.2.3. Third Ventricle

The **third ventricle** is a narrow cleft between both thalami (Fig. 23.9). Over a variable area, the thalami are frequently adherent to each other at the **interthalamic adhesion** (massa intermedia). The hypothalamus forms the floor of the third ventricle. In the anterior part of the third ventricle, the **optic chiasma** crosses its floor. The **lamina terminalis**, a delicate sheet that connects the optic chiasma with the corpus callosum (Fig. 5.1), forms the anterior wall of the third ventricle.

A thin layer of ependyma covered by pia mater forms the thin roof of the **third ventricle**. In the ventricles, vascular fringes of pia mater, known as the tela choroidea, invaginate their covering of modified ependyma and project into the ventricular cavities. From the inferior surface of this roof, the **choroid plexuses of the third ventricle** project downwards into the ventricular cavity. On either side of the median plane is a collection of choroid plexus invaginating the ependymal roof and protruding into the ventricular cavity. Pia mater and connective tissue occupy the space between the corpus callosum dorsally, and the dorsal thalamus, choroid plexus and pineal, ventrally. Through this area the internal cerebral veins pass.

Interventricular Foramina

The **third ventricle** communicates with the **lateral ventricles** by means of the **interventricular foramina** (Fig. 23.9). Each **interventricular foramen** is at the superior and anterior part of the third ventricle, at the rostral limit of the dorsal thalamus and at the site of the outgrowth of the cerebral hemisphere in the embryo. From it, a shallow groove, the **hypothalamic sulcus**, runs posteriorly toward the **aqueduct of the midbrain**. The sulcus marks the boundary between the dorsal thalamus (superior to the sulcus) and the subthalamus and hypothalamus (inferior to the sulcus).

Recesses of the Third Ventricle

The **third ventricle** presents several recesses: (1) the **optic recess** superior to the optic chiasma; (2) the

infundibular recess in the infundibulum of the neuro-hypophysis (posterior lobe of the pituitary gland); (3) a less well-defined recess rostral to the mamillary bodies; (4) the **pineal recess** in the stalk of the pineal body; (5) the **suprapineal recess**. Indentations occur in the outline of the third ventricle by the anterior and posterior commissures and by the optic chiasma.

Choroid Plexuses of the Third Ventricle

The **choroid plexus** of the third ventricle invaginates into the roof of the ventricle on either side of the median plane. At the interventricular foramina, they become continuous with those of the **lateral ventricles**. The posterior choroidal arteries supplying the lateral ventricles arise from the posterior cerebral artery.

23.2.4. Aqueduct of Midbrain

The **aqueduct of the midbrain** (of Sylvius) is a narrow channel in the midbrain about 1 cm in length. The aqueduct is widest in its central part and connects the third and fourth ventricles.

23.2.5. Fourth Ventricle

The **fourth ventricle** (Fig. 23.9) is a rhomboidal cavity in the posterior part of the pons and medulla. Superiorly it is narrow becoming continuous with the **aqueduct of the midbrain**. Inferiorly, it narrows into the **central canal** of the medulla oblongata, which, in turn, is continuous with the **central canal** of the spinal cord. In the widest part of the ventricle, on either side, is a **lateral recess** that permits cerebrospinal fluid to leave the ventricular system and enter the subarachnoid space. The superior and inferior cerebellar peduncles form the lateral boundaries of the ventricle.

Floor of the Fourth Ventricle

The anterior boundary or floor of the fourth ventricle is the pons superiorly and the medulla oblongata inferiorly. The nuclei of origin of the last eight cranial nerves directly or indirectly contribute to the floor of the fourth ventricle. A **dorsal median sulcus** divides the floor into right and left halves. A longitudinal groove (the sulcus limitans) subdivides each half of the floor of the fourth ventricle into medial (basal) and lateral (alar) parts. The **medial eminence**, an elevation on the floor of the fourth ventricle, extends the length of the pons and into the open medulla. In the caudal part of

the pons, the medial eminence presents a swelling called the **facial colliculus**. The **hypoglossal** and **vagal trigones** are caudal to the **facial colliculus** and form the caudal part of the medial eminence at medullary levels. The area lateral to the sulcus limitans overlies the vestibular nuclei associated with the vestibular nerve. The lowermost part of the floor of the fourth ventricle has the shape of the point of a pen (calamus scriptorius). Here are nuclei related to respiration, cardiovascular activity and the act of swallowing (deglutition).

Roof of the Fourth Ventricle

The extremely thin posterior boundary or roof of the fourth ventricle is under cover of the cerebellum. Sheets of white matter (the superior and inferior medullary vela), which are lined by ependyma and which stretch between both superior and inferior cerebellar peduncles form the roof of the fourth ventricle. The lower part of the roof presents a deficiency, the **median aperture** of the fourth ventricle. By means of this aperture, the ventricular cavity is in direct communication with the subarachnoid space. The **median aperture** lies inferior to the nodule of the cerebellum in the median plane. The extent of the median aperture varies considerably. The ends of the **lateral recesses** have similar openings, the **lateral apertures**. The lateral apertures are not well-defined openings but rather gaps between the cerebellum and medulla oblongata with choroid plexus protruding through them. By turning the fila of the glossopharyngeal and vagal nerves medially, one is able to see the choroid plexus projecting through the **lateral aperture** immediately inferior to the flocculus.

Clinical Correlation

Cerebrospinal fluid formed in the ventricles enters the subarachnoid space by means of the median and lateral apertures. With occlusion of the apertures, the ventricles become distended, a condition called **hydrocephaly** (the term **hydrocephalus** refers to a person afflicted with hydrocephaly).

Choroid Plexuses of the Fourth Ventricle

The **choroid plexus of the fourth ventricle** invaginates into the roof of the ventricle on either side of the median plane. A prolongation of each plexus protrudes through the corresponding **lateral aperture**. The vessels to the plexus arise from cerebellar branches of the vertebral and basilar arteries.

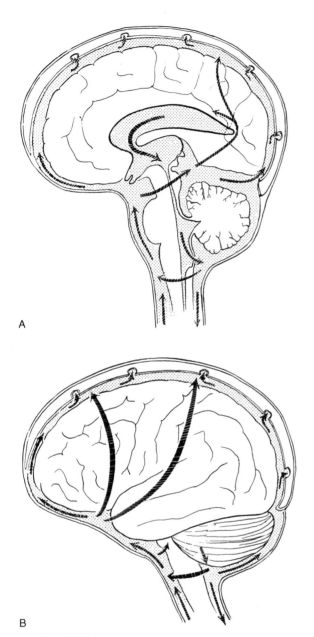

FIGURE 23.10. Major pathways of cerebrospinal fluid flow. **A**, Medial and **B**, lateral views (from Milhorat, 1972).

23.3. CEREBROSPINAL FLUID

A lumbar puncture is useful for obtaining specimens of cerebrospinal fluid (CSF) for testing. The total volume of the fluid is about 100 to 140 ml and its pressure is about 150 mm of saline (normal range: 70–180 mm) in the lateral recumbent position. The pressure is several times higher in the lumbar region when sitting, but is about atmospheric at the foramen magnum and is negative in the ventricles.

The **choroid plexuses** produce about 500 ml of cerebrospinal fluid each day. The total volume of the

ventricular system is about 100–140 ml. Complete turnover of this volume of cerebrospinal fluid occurs every 6–8 hours. Cerebrospinal fluid formed in the ventricles enters the subarachnoid space by means of the median and lateral apertures. Once in the subarachnoid space CSF circulates over the surface of the hemispheres to reach the sinuses of the cranial dura (Fig. 23.10). The arachnoid villi and arachnoid granulations are responsible for the bulk outflow of cerebrospinal fluid from the subarachnoid space into the venous sinuses of the cranial dura or into the spinal veins. **Cranial arachnoid villi** are fingerlike protrusions of arachnoid tissue into the dural venous sinuses. This close association of the arachnoid villi with the dural venous plexuses allows cerebrospinal fluid to leave the subarachnoid space through the arachnoid villi and enter the venous system. This CSF-blood interface consists of the arachnoid cell layer on one side and the endothelial cells of the dural venous sinuses on the other. Through this interface, the excess CSF produced each day is absorbed into the venous system.

Cerebrospinal fluid acts as a fluid buffer for the protection of the central nervous system. Cerebrospinal fluid also compensates for changes in blood volume in the cranium, allowing the cranial contents to remain at a constant volume. No one element of the cranial contents (brain, blood, or cerebrospinal fluid) can increase except at the expense of the others (Monroe–Kellie doctrine).

FURTHER READING

Apuzzo, MLJ (1998) *Surgery of the Third Ventricle*. Williams & Wilkins, Baltimore.

Barshes N, Demopoulos A, Engelhard HH (2005) Anatomy and physiology of the leptomeninges and CSF space. Cancer Treat Res 125:1–16.

Born CM, Meisenzahl EM, Frodl T, Pfluger T, Reiser M, Moller HJ, Leinsinger GL (2004) The septum pellucidum and its variants. An MRI study. Eur Arch Psychiatry Clin Neurosci 254:295–302.

Dunn LT (2002) Raised intracranial pressure. J Neurol Neurosurg Psychiatry 73(Suppl l):23–27.

Haines DE, Harkey HL, al-Mefty O (1993) The 'subdural' space: a new look at an outdated concept. Neurosurgery 32:111–120.

Kawashima M, Li X, Rhoton AL Jr, Ulm AJ, Oka H, Fujii K (2006) Surgical approaches to the atrium of the lateral ventricle: microsurgical anatomy. Surg Neurol 65:436–445.

Matsushima T, Rhoton AL Jr, Lenkey C (1982) Microsurgery of the fourth ventricle: Part 1. Microsurgical anatomy. Neurosurgery 11:631–667.

McLone DG (2004) The anatomy of the ventricular system. Neurosurg Clin N Am 15:33–38.

Rhoton AL Jr (2000) Cerebellum and fourth ventricle. Neurosurgery 47 (Suppl 3):S7–27.

Rhoton AL Jr (2000) Tentorial incisura. Neurosurgery 47 (Suppl 3):S131–153.

Rhoton AL Jr (2000) The foramen magnum. Neurosurgery 47 (Suppl 3):S155–193.

Rhoton AL Jr (2002) The cavernous sinus, the cavernous venous plexus, and the carotid collar. Neurosurgery 51 (Suppl 1):375–410.

Rhoton AL Jr (2002) The lateral and third ventricles. Neurosurgery 51 (Suppl 1):207–271.

Rhoton AL Jr (2002) The sellar region. Neurosurgery 51 (Suppl 1):335–374.

Switka A, Narkiewicz O, Dziewiatkowski J, Morys J (1999) The shape of the inferior horn of the lateral ventricle in relation to collateral and occipitotemporal sulci. Folia Morphol (Warsz) 58:69–80.

Timurkaynak E, Rhoton AL Jr, Barry M (1986) Microsurgical anatomy and operative approaches to the lateral ventricles. Neurosurgery 19:685–723.

Yamamoto I, Rhoton AL Jr, Peace DA (1981) Microsurgery of the third ventricle: Part I. Microsurgical anatomy. Neurosurgery 8:334–356.

References

Abadi RV, Sandikcioglu M (1975) Visual resolution in congenital pendular nystagmus. Am J Optom Physiol Opt 52:573–581.

Abbie A (1933) The blood supply of the lateral geniculate body with a note on the morphology of the choroidal arteries. J Anat 67:491–521.

Abbott NJ, Ronnback L, Hansson E (2006) Astrocyte–endothelial interactions at the blood–brain barrier. Nat Rev Neurosci 7:41–53.

Abe H, Rhoton AL Jr (2006) Microsurgical anatomy of the cochlear nuclei. Neurosurgery 58:728–739.

Abend WK (1977) Functional organization of the superior vestibular nucleus of the squirrel monkey. Brain Res 132:65–84.

Abend WK (1978) Response to constant angular accelerations of neurons in the monkey superior vestibular nucleus. Exp Brain Res 31:459–473.

Abramov I, et al. (1982) The retina of the newborn human infant. Science 217:265–267.

Accornero N, Berardelli A, Bini G, Cruccu G, and Manfredi M (1980) Corneal reflex elicited by electrical stimulation of the human cornea. Neurology 30:782–785.

Accornero N, et al. (1980) Corneal reflex elicited by electrical stimulation of the human cornea. Neurology 30:782–785.

Adams JC (1986) Neuronal morphology in the human cochlear nucleus. Arch Otolaryngol Head Neck Surg 112:1253–1261.

Adams JE (1976) Naloxone reversal of analgesia produced by brain stimulation in the human. Pain 2:161–166.

Aggleton JP (1992) *The Amygdala: A Functional Analysis.* 2nd ed. Oxford University Press, New York.

Aggleton JP (1992) *The Amygdala: Neurobiological Aspects of Emotion, Memory, and Mental Dysfunction.* Wiley-Liss, New York.

Aggleton JP, Desimone R, Mishkin M (1986) The origin, course, and termination of the hippocampothalamic projections in the macaque. J Comp Neurol 243:409–421.

Ahveninen J, Jaaskelainen IP, Raij T, Bonmassar G, Devore S, Hamalainen M, Levanen S, Lin FH, Sams M, Shinn-Cunningham BG, Witzel T, Belliveau JW (2006) Task-modulated 'what' and 'where' pathways in human auditory cortex. Proc Natl Acad Sci 103: 14608–14613.

Akil H, Richardson DE, Barchas JD (1979) Pain control by focal brain stimulation in man: relationship to enkephalins and endorphin. In: *Mechanisms of Pain and Analgesic Compounds.* Beers Jr RF, Bassett EG, eds, pp 239–247. Raven Press, New York.

Akil H, Watson SJ (1980) The role of endogenous opiates in pain control. In: *Pain and Society.* Kosterlitz HW, Terenius LY, eds. Dahlem Konferenzen. Life Sciences Research Report 17, pp 201–222. Verlag Chemie, Weinheim.

Alain C, Arnott SR, Hevenor S, Graham S, Grady CL (2001) 'What' and 'where' in the human auditory system. Proc Natl Acad Sci 98:12301–12306.

Albe-Fessard D (1965) Organization of somatic central projections. In: *Contributions to Sensory Physiology.* Neff WD, ed., 2:101–167. Academic Press, New York.

Albe-Fessard D (1973) Electrophysiological methods for the identification of thalamic nuclei. Z Neurol 205:15–28.

Albe-Fessard D, Arfel G, Derome P, Mondey M (1970) Electrophysiology of the human thalamus with special reference to trigeminal pain. In: *Trigeminal Neuralgia. Pathogenesis und Pathophysiology.* Hassler R, Walker AE, eds, pp 139–148. Thieme, Stuttgart.

Albe-Fessard D, Boivie J, Grant G, Levante A (1975) Labelling of cells in the medulla oblongata and the spinal cord of the monkey after injections of horseradish peroxidase in the thalamus. Neurosci Lett 1:75–80.

Albe-Fessard D, Levante A, Lamour Y (1974) Origin of spinothalamic tract in monkeys. Brain Res 65:503–509.

Albe-Fessard D, Levante A, Lamour Y (1974) Origin of spinothalamic and spinoreticular pathways in cats and monkeys. Adv Neurol 4:157–166.

Alexander MP, Warren RL (1988) Localization of callosal auditory pathways: A CT case study. Neurology 38:802–804.

Alkan A, Sigirci A, Ozveren MF, Kutlu R, Altinok T, Onal C, Sarac K (2004) The cisternal segment of the abducens nerve in man: three-dimensional MR imaging. Eur J Radiol 51:218–222.

Allen LS, Gorski RA (1991) Sexual dimorphism of the anterior commissure and massa intermedia of the human brain. J Comp Neurol 312:97–104.

Allen LS, Gorski RA (1992) Sexual orientation and the size of the anterior commissure in the human brain. Proc Natl Acad Sci USA 89:7199–7202.

Allen LS, Hines M, Shryne JE, Gorski RA (1989) Two sexually dimorphic cell groups in the human brain. J Neurosci 9:497–506.

Allen LS, Richey MF, Chai YM, Gorski RA (1991) Sex differences in the corpus callosum of the living human being. J Neurosci 11:933–942.

Allen NJ, Barres BA (2005) Signaling between glia and neurons: focus on synaptic plasticity. Curr Opin Neurobiol 15:542–548.

Alling C, Bostrom K (1980) Demyelination of the mamillary bodies in alcoholism. A combined morphological and biochemical study. Acta Neuropathol (Berl) 50:77–80.

Al-Sheikhli ARJ (1980) Pain in the ear – with special reference to referred pain. J Laryngol Otol 94:1433–1440.

Amano K, et al. (1978) Single neuron analysis of the human midbrain tegmentum. Rostral mesencephalic reticulotomy for pain relief. Appl Neurophysiol 41:66–78.

Amaral DG, Cowan WM (1980) Subcortical afferents to the hippocampal formation in the monkey. J Comp Neurol 189:573–591.

Amaral DG, Insausti R, Cowan WM (1987) The entorhinal cortex of the monkey: I. Cytoarchitectonic organization. J Comp Neurol 264:326–355.

Amaral DG, Sinnamon HP (1977) The locus coeruleus: neurobiology of a central noradrenergic nucleus. Prog Neurobiol 9:147–196.

Amarenco P, et al. (1990) Infarction in the territory of the medial branch of the posterior inferior cerebellar artery. J Neurol Neurosurg Psychiatry 53:731–735.

Amarenco P, Hauw J-J (1990) Cerebellar infarction in the territory of the anterior and inferior cerebellar artery. Brain 113:139–155.

Ambrosi G, Virgintino D, Benagiano V, Maiorano E, Bertossi M, Roncali L (1995) Glial cells and blood–brain barrier in the human cerebral cortex. Ital J Anat Embryol 100 (Suppl 1):177–184.

Aminoff MJ, Greenberg DA, Simon RP (2005) *Clinical Neurology.* 6th ed. Lange Medical Books/McGraw-Hill, New York.

Amoore JE (1963) Stereochemical theory of olfaction. Nature 198:271–272.

Amunts K, Jancke L, Mohlberg H, Steinmetz H, Zilles K (2000) Interhemispheric asymmetry of the human motor cortex related to handedness and gender. Neuropsychologia 38:304–312.

Amunts K, Kedo O, Kindler M, Pieperhoff P, Mohlberg H, Shah NJ, Habel U, Schneider F, Zilles K (2005) Cytoarchitectonic

mapping of the human amygdala, hippocampal region and entorhinal cortex: intersubject variability and probability maps. Anat Embryol (Berl) 210:343–352.

Amunts K, Schlaug G, Schleicher A, Steinmetz H, Dabringhaus A, Roland PE, Zilles K (1996) Asymmetry in the human motor cortex and handedness. NeuroImage 4:216–222.

Amunts K, Schleicher A, Zilles K (2002) Architectonic mapping of the human cerebral cortex. In: *Cortical Areas: Unity and Diversity.* Schüz A, Miller R, eds, pp 29–52. Taylor & Francis, London and New York.

Anderson DJ, Matthews B (1977) *Pain in the Trigeminal Region.* Elsevier, New York.

Anderson DR, Hoyt WF (1969) Ultrastructure of intraorbital portion of human and monkey optic nerve. Arch Ophthalmol 82:506–530.

Anderson HJ, et al. (1982) Early responses to neural injury: group report. In: *Repair and Regeneration of the Nervous System.* Nicholls JG, ed., pp 315–339. Springer-Verlag, New York.

Andersen P, Morris R, Amaral D, Bliss T, O'Keefe J (2007) *The Hippocampus Book.* Oxford University Press, New York.

Andersson SA, Finger S, Norrell U (1975) Cerebral units activated by tactile stimuli via a ventral spinal pathway in monkeys. Acta Physiol Scand 93:119–128.

Andres KH, During MV (1973) Morphology of cutaneous receptors. In: *Handbook of Sensory Physiology.* Iggo A, ed., 2:3–28. Springer-Verlag, New York.

Angeleri F, Ferro-Milano F, Parige S (1964) Electrical activity and reactivity of the rhinencephalic, pararhinencephalic and thalamic structures: prolonged implantation of electrodes in man. Electroencephalogr Clin Neurophysiol 16:100–129.

Angevine JB Jr (1973) Clinically relevant embryology of the vertebral column and spinal cord. Clin Neurosurg 20:95–113.

Anson BJ, Harper DG, Winch TR (1967) The vestibular system: anatomic considerations. Arch Otolaryngol 85:497–514.

Antel J (2005) Oligodendrocyte/myelin injury and repair as a function of the central nervous system environment. Clin Neurol Neurosurg 108:245–249.

Apkarian AV, Hodge CJ (1989) Primate spinothalamic pathways: I. A quantitative study of the cells of origin of the spinothalamic pathway. J Comp Neurol 288:447–473.

Apkarian AV, Hodge CJ (1989) Primate spinothalamic pathways: II. The cells of origin of the dorsolateral and ventral spinothalamic pathways. J Comp Neurol 288:474–492.

Apkarian AV, Hodge CJ (1989) Primate spinothalamic pathways: III. Thalamic terminations of the dorsolateral and ventral spinothalamic pathways. J Comp Neurol 288:493–511.

Appenzeller O (1999) The Autonomic Nervous System. Part I. Normal Functions. In: *Handbook of Clinical Neurology.* Vinken PJ, Bruyn GW, eds. Vol 74. Elsevier, Amsterdam.

Appenzeller O (1999) The Autonomic Nervous System. Part II. Dysfunctions. In: *Handbook of Clinical Neurology.* Vinken PJ, Bruyn GW, eds. Vol 75. Elsevier, Amsterdam.

Appenzeller O, Oribe, E (1997) *The Autonomic Nervous System: An Introduction to Basic and Clinical Concepts.* 5th ed. North-Holland/Elsevier, Amsterdam.

Applebaum AE, et al. (1979) Nuclei in which functionally identified spinothalamic tract neurons terminate. J Comp Neurol 188:575–586.

Apuzzo ML, Stieg PE, Starr P, Schwartz RB, Folkerth RD (1996) Case problems in neurological surgery: Surgery of the Soul's cistern. Neurosurgery 39:1022–1029.

Aquilonius S-M, Eckernäs S-Å, Gillberg P-G (1981) Topographical localization of choline acetyltransferase within the human spinal cord and a comparison with some other species. Brain Res 211:329–340.

Arey LB (1965) *Developmental Anatomy. A Textbook and Laboratory Manual of Embryology.* 7th ed. Saunders, Philadelphia.

Arey LB, Tremaine MJ, Monzingo FL (1935) The numerical and topographical relations of taste buds to human circumvallate papillae throughout the life span. Anat Rec 64:9–25.

Ariëns Kappers CU, Huber GC, Crosby EC (1936) *The Comparative Anatomy of the Nervous System of Vertebrates, Including Man.* Macmillan, New York.

Armstrong DM, Leroy S, Shields D, Terry RD (1985) Somatostatin-like immunoreactivity within neuritic plaques. Brain Res 338:71–79.

Arnold W (1987) Myelination of the human spiral ganglion. Acta Otolaryngol [Suppl] (Stockh) 436:76–84.

Atwell WJ (1930) A human embryo with seventeen pairs of somites. Contrib Embryol Carneg Instn 21:1–24.

Augood SJ, Waldvogel HJ, Munkle MC, Faull RL, Emson PC (1999) Localization of calcium-binding proteins and GABA transporter (GAT-1) messenger RNA in the human subthalamic nucleus. Neuroscience 88:521–534.

Augustine JR (1974) Certain experimentally demonstrated connections of the chief sensory trigeminal nucleus in the squirrel monkey, *Saimiri sciureus.* Ala J Med Sci 11:277–301.

Augustine JR (1985) The insular lobe in primates including humans. Neurol Res 7:2–10.

Augustine JR (1996) Circuitry and functional aspects of the insular lobe in primates including humans. Brain Res Rev 22:229–244.

Augustine JR, DesChamps EG, Ferguson JG Jr (1981) Functional organization of the oculomotor nucleus in the baboon. Am J Anat 161:393–403.

Augustine JR, White JF (1986) The accessory nerve nucleus in the baboon. Anat Rec 214:312–320.

Austin ES, Lindgren WD, Dietrich SL (1972) Spina bifida and myelomeningocele. Am Fam Physician 6:105–111.

Avery JK, Cox CF (1977) Role of nerves in teeth relative to pain and dentinogenesis. In: *Pain in the Trigeminal Region.* Anderson DJ, Matthews B, eds, pp 37–48. North-Holland/Elsevier, Amsterdam.

Aziz Q, Schnitzler A, Enck P (2000) Functional neuroimaging of visceral sensation. J Clin Neurophysiol 17:604–612.

Azulay A, Schwartz AS (1975) The role of the dorsal funiculus of the primate in tactile discrimination. Exp Neurol 46:315–331.

Babb TL, Wilson CL, Crandall PH (1982) Asymmetry and ventral course of the human geniculostriate pathway as determined by hippocampal visual evoked potentials and subsequent visual field defects after temporal lobectomy. Exp Brain Res 47:317–328.

Bahill AT, Adler D, Stark L (1975) Most naturally occurring human saccades have magnitudes of 15 degrees or less. Invest Ophthalmol 14:468–469.

Bakay L (1984) Olfactory meningiomas. The missed diagnosis. JAMA 251:53–55.

Bakay L, Cares HL (1972) Olfactory meningiomas. Report on a series of twenty-five cases. Acta Neurochir 26:1–12.

Balabam CD, Ito M, Watanabe E (1981) Demonstration of zonal projections from the cerebellar flocculus to vestibular nuclei in monkeys (*Macaca fuscata*). Neurosci Lett 27:101–105.

Balazsi AG, et al. (1984) The effect of age on the nerve fiber population of the human optic nerve. Am J Ophthalmol 97:760–766.

Balercia G, Kultas-Ilinsky K, Bentivoglio M, Ilinsky IA (1996) Neuronal and synaptic organization of the centromedian nucleus of the monkey thalamus: a quantitative ultrastructural study, with tract tracing and immunohistochemical observations. J Neurocytol 25:267–288.

Ballantyne J, Engström H (1969) Morphology of the vestibular ganglion cells. J Laryngol Otol 83:19–42.

Ballenger JJ (1977) *Diseases of the Nose, Throat and Ear.* 12th ed., pp 605–651 and 669–724. Lea & Febiger, Philadelphia.

Balliet R, Blood KMT, Bach-Y-Rita P (1985) Visual field rehabilitation in the cortically blind? J Neurol Neurosurg Psychiatry 48:1113–1124.

Baloh RW, Halmagyi GM (1996) *Disorders of the Vestibular System.* Oxford University Press, New York.

Baloh RW, Honrubia V (1979) *Clinical Neurophysiology of the Vestibular System.* FA Davis Co, Philadelphia.

Baloh RW, Honrubia V (2001) *Clinical Neurophysiology of the Vestibular System.* 3rd ed. Oxford University Press, New York.

Barba D, Alksne JF (1984) Success of microvascular decompression with and without prior surgical therapy for trigeminal neuralgia. J Neurosurg 60:104–107.

Barberá J, Barcia-Salorio JL, Broseta J (1979) Stereotaxic pontine spinothalamic tractotomy. Surg Neurol 11:111–114.

Bargmann W, Schadé JP (1963) The Rhinencephalon and Related Structures. Prog Brain Res Vol 3. Elsevier, Amsterdam.

Barker D (1966) The motor innervation of the mammalian muscle spindle. In: *Muscular Afferents and Motor Control.* First Nobel Symposium. Granit R, ed., pp 51–58. Wiley, New York.

Barron JD (1983) Axon reaction and central nervous system regeneration. In: *Nerve, Organ, and Tissue Regeneration: Research Perspectives.* Seil FJ, ed., pp 3–36. Academic Press, New York.

Barshes N, Demopoulos A, Engelhard HH (2005) Anatomy and physiology of the leptomeninges and CSF space. Cancer Treat Res 125:1–16.

Barson AJ (1970) The vertebral level of termination of the spinal cord during normal and abnormal development. J Anat 106:489–497.

Barson AJ, Sands J (1977) Regional and segmental characteristics of the human adult spinal cord. J Anat 123:797–803.

Bartelmez GW (1922) The origin of the otic and optic primordia in man. J Comp Neurol 34:201–232.

Bartelmez GW (1923) The subdivisions of the neural folds in man. J Comp Neurol 35:231–247.

Bartelmez GW (1954) The formation of neural crest from the primary optic vesicle in man. Contrib Embryol Carneg Instn 35:55–71.

Bartelmez GW, Dekaban AS (1962) The early development of the human brain. Contrib Embryol Carneg Instn 37:13–32.

Bartelmez GW, Evans HM (1926) Development of the human embryo during the period of somite formation including embryos with 2 to 16 pairs of somites. Contrib Embryol Carneg Instn 17:1–67.

Bartolomei J, Wong J, Awad IA, Dickman CA, Das K, Kalfas I, Rodts G (2000) Case problems conference: thoracic spinal cord hernia. Neurosurgery 46:1408–1415.

Basbaum AI (1980) The anatomy of pain and pain modulations. In: *Pain and Society*. Kosterlitz HW, Terenius LY, eds. Dahlem Konferenzen. Life Sciences Research Report 17, pp 93–122. Verlag Chemie, Weinheim.

Basbaum AI, Fields HL (1980) Pain control: a new role for the medullary reticular formation. IBRO Monograph Series 6:329–348.

Batjer HH, Stieg PE, Schwartz RB (1995) Case problems in neurological surgery: a case of acute vertigo with incidental aneurysms. Neurosurgery 36:827–833.

Batton RR III, Jayaraman A, Ruggiero D, Carpenter MB (1977) Fastigial efferent projections in the monkey: an autoradiographic study. J Comp Neurol 174:281–306.

Baumann CR, Bassetti CL (2005) Hypocretins (orexins) and sleep–wake disorders. Lancet Neurol 4:673–682.

Baumann N, Pham-Dinh D (2001) Biology of oligodendrocyte and myelin in the mammalian central nervous system. Physiol Rev 81:871–927.

Bazelon M, Fenichel GM, Randall J (1967) Studies on neuromelanin. I. A melanin system in the human adult brainstem. Neurology 17:512–519.

Bealer SL, Smith DV (1975) Multiple sensitivity to chemical stimuli in single human taste papillae. Physiol Behav 14:795–799.

Bealer SL, Smith DV (1975) Multiple sensitivity to chemical stimuli in single human taste papillae. Physiol Behav 14:795–799.

Bealer SL, Smith DV (1975) Multiple sensitivity to chemical stimuli in single human taste papillae. Physiol Behav 14:795–799.

Beck CHM (1976) Dual dorsal columns: a review. Can J Neurol Sci 3:1–7.

Becker LE, Armstrong DL, Chan F (1986) Dendritic atrophy in children with Down's syndrome. Ann Neurol 20:520–526.

Beckstead RM, Morse JR, Norgren R (1980) The nucleus of the solitary tract in the monkey: projections to the thalamus and brain stem nuclei. J Comp Neurol 190:259–282.

Beckstead RM, Norgren R (1979) An autoradiographic examination of the central distribution of the trigeminal, facial, glossopharyngeal, and vagal nerves in the monkey. J Comp Neurol 184:455–472.

Beersma DG, Gordijn MC (2006) Circadian control of the sleep–wake cycle. Physiol Behav 90:190–195.

Beitel RE, Dubner R (1976) Fatigue and adaptation in unmyelinated (C) polymodal nociceptors to mechanical and thermal stimuli applied to the monkey's face. Brain Res 112:402–406.

Bell JE (1979) Central nervous system defects in early human abortuses. Develop Med Child Neurol 21:321–332.

Bender MB, Weinstein EA (1943) Functional representation in the oculomotor and trochlear nuclei. Arch Neurol Psychiatry 49:98–106.

Benevento LA, Yoshida K (1981) The afferent and efferent organization of the lateral geniculo-prestriate pathways in the macaque monkey. J Comp Neurol 203:455–474.

Benjamin JW (1939) The nucleus of the oculomotor nerve with special reference to innervation of the pupil and fibers from the pretectal region. J Nerv Ment Dis 89:294–310.

Benjamin RM, Burton H (1968) Projection of taste nerve afferents to anterior opercular-insular cortex in squirrel monkey (*Saimiri sciureus*). Brain Res 7:221–231.

Benjamin RM, Jackson JC (1974) Unit discharges in the mediodorsal nucleus of the squirrel monkey evoked by electrical stimulation of the olfactory bulb. Brain Res 75:181–191.

Bennett GA, Hutchinson RC (1946) Experimental studies on the movements of the mammalian tongue. II. The protrusion mechanism of the tongue (dog). Anat Rec 94:57–83.

Berezhnaya LA (2006) Neuronal organization of the reticular nucleus of the thalamus in adult humans. Neurosci Behav Physiol 36:519–525.

Berger MS, Stieg PE, Danks RA, Schwartz RB, Folkerth RD (1997) Case problems in neurological surgery: Lesions in eloquent cortex. Neurosurgery 40:1059–1063.

Bergström B (1973) Morphology of the vestibular nerve. II. The number of myelinated vestibular nerve fibers in man at various ages. Acta Otolaryngol 76:173–179.

Bergström BL (1973) Morphology of the vestibular nerve. I. Anatomical studies of the vestibular nerve in man. Acta Otolaryngol 76:162–172.

Bergstrom L (1980) Pathology of congenital deafness. Present status and future priorities. Ann Otol Rhinol Laryngol 89:(Suppl 74):31–42.

Berkley KF (1980) Spatial relationships between the terminations of somatic sensory and motor pathways in the rostral brainstem of cats and monkeys. I. Ascending somatic sensory inputs to lateral diencephalon. J Comp Neurol 193:283–317.

Berlin CI (1980) Central deafness: fact or fiction? Birth Defects: Original Article Series 16:47–57.

Berry M (1982) Post-injury myelin-breakdown products inhibit axonal growth: a hypothesis to explain the failure of axonal regeneration in the mammalian central nervous system. Bibl Anat 23:1–11.

Besson J-MR (1980) Supraspinal modulation of the segmental transmission of pain. In: *Pain and Society*. Kosterlitz HW, Terenius LY, eds. Dahlem Konferenzen. Life Sciences Research Report 17, pp 161–182. Verlag Chemie, Weinheim.

Besson J-MR, Guilbaud G, Lombard MC (1974) Effects of bradykinin intra-arterial injection into the limbs upon bulbar and reticular unit activity. Adv Neurol 4:207–215.

Beuerman RW, Maurice DM, Tanelian DL (1977) Thermal stimulation of the cornea. In: *Pain in the Trigeminal Region*.

Anderson DJ, Matthews N, eds, pp 413–422. North-Holland/Elsevier, Amsterdam.

Beuerman RW, Tanelian DL (1979) Corneal pain evoked by thermal stimulation. Pain 7:1–14.

Biedenbach MA, Van Hassel HJ, Brown AC (1979) Tooth pulp-driven neurons in somatosensory cortex of primates: role in pain mechanisms including a review of the literature. Pain 7:31–50.

Bielefeldt K, Christianson JA, Davis BM (2005) Basic and clinical aspects of visceral sensation: transmission in the CNS. Neurogastroenterol Motil 17:488–499.

Bignami A, Dahl D (1982) Ten-nanometer filament proteins in neural development and aging. In: *The Aging Brain: Cellular and Molecular Mechanisms of Aging in the Nervous System*. Giacobini E, Filogamo G, Giacobini G, Vernadakis A, eds, pp 99–105. Raven Press, New York.

Binkofski F, Fink GR, Geyer S, Buccino G, Gruber O, Shah NJ, Taylor JG, Seitz RJ, Zilles K, Freund HJ (2002) Neural activity in human primary motor cortex areas 4a and 4p is modulated differentially by attention to action. J Neurophysiol 88:514–519.

Bijlani V, Keswani NH (1970) The salivatory nuclei in the brainstem of the monkey (Macaca mulatta) J Comp Neurol 139:375–384.

Bilir E, Craven W, Hugg J, Gilliam F, Martin R, Faught E, Kuzniecky R (1998) Volumetric MRI of the limbic system: anatomic determinants. Neuroradiology 40:138–144.

Birder LA, Perl ER (1994) Cutaneous sensory receptors. J Clin Neurophysiol 11:534–552.

Bisby MA (1982) Functions of retrograde axonal transport. Fed Proc 41:2307–2311.

Bischoff A (1979) Congenital insensitivity to pain with anhidrosis: a morphometric study of sural nerve and cutaneous receptors in the human prepuce. In: *Advances in Pain Research and Therapy*. Bonica JJ, Albe-Fessard, D eds, 3:53–65. Raven Press, New York.

Bizzi E (1975) Central control of eye and head movements in monkeys. In: *Basic Mechanisms of Ocular Motility and Their Clinical Implications*. Lennerstrand G, Bach-Y-Rita P, eds, pp 469–471. Pergamon Pr, Oxford and New York.

Björklund A, Stenevi U (1979) Regeneration of monoaminergic and cholinergic neurons in the mammalian central nervous system. Physiol Rev 59:62–100.

Black FO, Simmons FB, Wall C III (1980) Human vestibule-spinal responses to direct electrical eighth nerve stimulation. Acta Otolaryngol 90:86–92.

Black FO, Wall C III, O'Leary DP (1978) Computerized screening of the human vestibulospinal system. Ann Otol Rhinol Laryngol 87:853–860.

Blessing WW (1997) *The Lower Brainstem and Bodily Homeostasis*. Oxford University Press, New York.

Blessing WW, Gai WP (1997) Caudal pons and medulla oblongata. In: *Handbook of Chemical Neuroanatomy*. Björklund A, Hökfelt T, eds. Vol 13: The Primate Nervous System, Part I., pp 139–186. Elsevier, Amsterdam.

Blinkov SM, Glezer II (1968) *The Human Brain in Figures and Tables: A Quantitative Handbook*. Basic Books, New York.

Bloedel JR (1974) The substrate for integration in the central pain pathways. Clin Neurosurg 21:194–228.

Bloom F (1984) General Features of chemically identified neurons. In: *Handbook of Chemical Neuroanatomy*. Björklund A, Hökfelt T, eds. Vol 2: Classical Transmitters in the CNS, Part I., pp 1–22. Elsevier, New York.

Bloom FE, Morrison JH (1986) Neurotransmitters of the human brain. Human Neurobiol 5:145–146.

Bloom W, Fawcett D (1975) *A Textbook of Histology*. 10th ed. Saunders, Philadelphia.

Blum JJ, Reed MC (1988) The transport of organelles in axons. Math Biosci 90:233–245.

Blum M, Walker AE, Ruch TC (1943) Localization of taste in the thalamus of Macaca mulatta. Yale J Biol Med 16:175–191.

Blumenfeld, H (2001) *Neuroanatomy Through Clinical Cases*. Sinauer Associates, Sunderland, MA.

Bodegård A, Geyer S, Grefkes C, Zilles K, Roland PE (2001) Hierarchical processing of tactile shape in the human brain. Neuron 31:317–328.

Bodegård A, Geyer S, Naito E, Zilles K, Roland PE (2000) Somatosensory areas in man activated by moving stimuli: cytoarchitectonic mapping and PET. NeuroReport. 2000 Jan 17;11(1):187–191.

Bodegård A, Ledberg A, Geyer S, Naito E, Zilles K, Roland PE (2000) Object shape differences reflected by somatosensory cortical activation. J Neurosci 20:(RC51)1–5.

Bodian D (1967) Neurons, circuits, and neuroglia. In: *The Neurosciences; A Study Program*. Quarton GC, Melnechuk T, Schmitt FO, eds, pp 6–24. Rockefeller University Press, New York.

Boellaard JW, Schlote W (1986) Ultrastructural heterogeneity of neuronal lipofuscin in the normal human cerebral cortex. Acta Neuropathol (Berl) 71:285–294.

Bogen JE, Bogen GM (1969) The other side of the brain III: the corpus callosum and creativity. Bull LA Neurol Soc 34:191–216.

Boivie J (1978) Anatomical observations on the dorsal column nuclei, their thalamic projection and the cytoarchitecture of some somatosensory thalamic nuclei in the monkey. J Comp Neurol 178:17–48.

Boivie J (1979) An anatomical reinvestigation of the termination of the spinothalamic tract in the monkey. J Comp Neurol 186:343–370.

Boivie JJG, Perl ER (1975) Neural substrates of somatic sensation. In: *Neurophysiology. Physiology, Series One*. Hunt CC, ed., pp 303–411. Vol 3. MTP Int Rev Sci. Butterworths, London.

Bolande RP (1974) The neurocristopathies: a unifying concept of disease arising in neural crest maldevelopment. Hum Pathol 5:409–429.

Bolande RP (1981) Neurofibromatosis – the quintessential neurocristopathy: pathogenic concepts and relationships. Adv Neurol 29:67–75.

Boller F, Spinnler H (1994) The Frontal Lobes. In: *Handbook of Neurospsychology*. Boller F, Grafman J, series eds. Vol 9, Section 12. Elsevier, Amsterdam.

Bolton CF, Winkelmann RK, Dyck PJ (1966) A quantitative study of Meissner's corpuscles in man. Neurology 16:1–9.

Bondareff W, McLone DG (1973) The external glial limiting membrane in Macaca: ultrastructure of a laminated glioepithelium. Am J Anat 136:277–296.

Bone RA, Landrum JT, Tarsis SL (1985) Preliminary identification of the human macular pigment. Vision Res 25:1531–1535.

Bonica JJ (1975) The nature of pain in parturition. Clin Obstet Gynec 2:499–516.

Boothe RG, Dobson V, Teller DY (1985) Postnatal development of vision in human and nonhuman primates. Ann Rev Neurosci 8:495–545.

Born CM, Meisenzahl EM, Frodl T, Pfluger T, Reiser M, Moller HJ, Leinsinger GL (2004) The septum pellucidum and its variants. An MRI study. Eur Arch Psychiatry Clin Neurosci 254:295–302.

Börnstein WS (1940) Cortical representation of taste in man and monkey. I. Functional and anatomical relations of taste, olfaction, and somatic sensibility. Yale J Biol Med 12:719–736.

Börnstein WS (1940) Cortical representation of taste in man and monkey. II. The localization of the cortical taste area in man and a method of measuring impairment of taste in man. Yale J Biol Med 13:133–156.

Bortolami R, Veggetti A, Callegari E, Lucchi ML, Palmieri G (1977) Afferent fibers and sensory ganglion cells within the oculomotor nerve in some mammals and man. I. Anatomical investigations. Arch Ital Biol 115:355–385.

Boshes B, Padberg F (1953) Studies on the cervical spinal cord of man. Sensory pattern after interruption of the posterior columns. Neurology 3:90–101.

Bosley TM, et al. (1985) Ischemic lesions of the occipital cortex and optic radiations: Positron emission tomography. Neurology 35:470–484.

Bougousslavasky J, et al. (1987) Lingual and fusiform gyri in visual processing: a clinico pathologic study of superior altitudinal hemianopia. J Neurol Neurosurg Psychiatry 50:607–614.

Bowden DM, German DC, Poynter WD (1978) An autoradiographic, semistereotaxic mapping of major projections from locus coeruleus and adjacent nuclei in *Macaca mulatta*. Brain Res 145:257–276.

Bowmaker JK, Dartnall HJA (1980) Visual pigments of rods and cones in a human retina. J Physiol (Lond) 298:501–511.

Bowsher D (1957) Termination of the central pain pathway in man: the conscious appreciation of pain. Brain 80:606–622.

Bowsher D (1961) The termination of secondary somatosensory neurons within the thalamus of *Macaca mulatta*: an experimental degeneration study. J Comp Neurol 117:213–227.

Bowsher D (1974) Thalamic convergence and divergence of information generated by noxious stimulation. Adv Neurol 4:223–232.

Bowsher D (1975) Diencephalic projections from the midbrain reticular formation. Brain Res 95:211–220.

Bowsher D (1976) Role of the reticular formation in responses to noxious stimulation. Pain 2:361–378, 1976.

Bowsher D (2005) Pain. Pain 113:430.

Boya J, Calvo J, Prado A (1979) The origin of microglial cells. J Anat 129:177–186.

Boyce JM, Shone GR (2006) Effects of ageing on smell and taste. Postgrad Med J 82:239–241.

Boycott BB, Dowling JE (1969) Organization of the primate retina: light microscopy. With an appendix by Kolb H, Boycott BB, Dowling JE: A second type of midget bipolar cell in the primate retina. Philos Trans R Soc Lond (Biol) 255:109–184.

Braak H, Braak E (1987) The hypothalamus of the human adult: chiasmatic region. Anat Embryol (Berl) 175:315–330.

Braak H, Braak E (1992) Anatomy of the human hypothalamus (chiasmatic and tuberal region). Prog Brain Res 93:3–14.

Braak H, Braak E, Yilmazer D, Bohl J (1996) Functional anatomy of human hippocampal formation and related structures. J Child Neurol 11:265–275.

Bradley WH (1963) Central localization of gustatory perception: an experimental study. J Comp Neurol 121:417–423.

Bradley WE, Farrell DF, Ojemann GA (1998) Human cerebrocortical potentials evoked by stimulation of the dorsal nerve of the penis. Somatosens Mot Res 15:118–127.

Braitenburg V, Schüz A (1998) *Cortex Statistics and Geometry of Neuronal Connectivity*. 2nd ed. Springer-Verlag, Berlin.

Bramwell E (1927) A case of cortical deafness. Brain 50:579–580.

Brandt T, Daroff RB (1980) The multisensory physiological and pathological vertigo syndromes. Ann Neurol 7:195–203.

Brandt T, Dieterich M (1999) The vestibular cortex. Its locations, functions, and disorders. Ann NY Acad Sci 871:293–312.

Brandt T, Strupp M (2005) General vestibular testing. Clin Neurophysiol 116:406–426.

Brauer K, Leuba G, Garey LJ, Winkelmann E (1985) Morphology of axons in the human lateral geniculate nucleus: a Golgi study in prenatal and postnatal material. Brain Res 359:21–33.

Bray D, Bunge MB (1973) The growth cone in neurite extension. In: *Locomotion of Tissue Cells*. Ciba Foundation Symposium 14 (new series), pp 195–209. Elsevier/Excerpta Medica/North Holland, Amsterdam.

Bray D, Chapman (1985) Analysis of microspike movements on the neuronal growth cone. J Neurosci 12:3204–3213.

Bray D, Gilbert D (1981) Cytoskeletal elements in neurons Annu Rev Neurosci 4:505–523.

Bredberg G (1968) Cellular pattern and nerve supply of the human organ of Corti. Acta Otolaryngol (Suppl) 236:1–135.

Bredberg G (1984) Morphological aspects of the human cochlea in relation to cochlear implant. Adv Audiol 1:70–76.

Breinin GM (1971) The structure and function of extraocular muscles – an appraisal of the duality concept. Am J Ophthalmol 72:1–9.

Brewer JI (1938) A human embryo in the bilaminar blastodisc stage (The Edwards–Jones–Brewer ovum) Contrib Embryol Carneg Instn 27:85–93.

Bridgman CF (1970) Comparisons in structure of tendon organs in the rat, cat and man. J Comp Neurol 138:369–372.

Brimijoin WS (1982) Abnormalities of axonal transport. Are they a cause of peripheral nerve disease? Mayo Clin Proc 57:707–714.

Brizzee KR (1987) Neuron numbers and dendritic extent in normal aging and Alzheimer's disease. Neurobiol Aging 8:579–580.

Broca, P (1878) Anatomie comparée des circonvolutions cérébrales: le grande lobe limbique. Rev Anthropol 1:385–498.

Broca E (1958) Clinical aspects of cortical deafness. Laryngoscope 68:301–309.

Brodmann K (1908) Beiträge sur histologischen Lokalisation der grosshirnrinde. VI. Mitteilung: Die Cortexgliederung des Menschen. J Psychol Neurol (Lpz) 10:231–246.

Brouwer JN, van der Wel H, Francke A, Henning GJ (1968) Miraculin, the sweetness-inducing protein from miracle fruit. Nature 220:373–374.

Brown JW (1956) The development of the nucleus of the spinal tract of V in human fetuses of 14 to 21 weeks of menstrual age. J Comp Neurol 106:393–424.

Brown JW (1958) The development of subnucleus caudalis of the nucleus of the spinal tract of V. J Comp Neurol 110:105–134.

Brown JW (1962) Differentiation of the human subnucleus interpolaris and subnucleus rostralis of the nucleus of the spinal tract of the trigeminal nerve. J Comp Neurol 119:55–76.

Brown JW (1973) Differentiation of neurons in subnucleus caudalis of the nucleus of the spinal tract of V in human and rabbit embryos. Prog Brain Res 40:67–90.

Brown JW, Podosin R (1966) A syndrome of the neural crest. Arch Neurol 15:294–301.

Bruce MF, Sinclair DC (1980) The relationship between tactile thresholds and histology in the human finger. J Neurol Neurosurg Psychiat 43:235–242.

Bruce, A (1898) On the dorsal or so-called sensory nucleus of the glossopharyngeal nerve, and on the nuclei of origin of the trigeminal nerve. Brain 21:383–387.

Brugge JF (1975) Progress in neuroanatomy and neurophysiology of auditory cortex. In: *The Nervous System.* Tower DB, editor-in-chief. Eagles EL, volume ed. Vol 3: Human Communication and its Disorders, pp 97–111. Raven Press, New York.

Brugge JF, Volkov IO, Garell PC, Reale RA, Howard MA (2003) Functional connections between auditory cortex on Heschl's gyrus and on the lateral superior temporal gyrus in humans. J Neurophysiol 90:3750–3763.

Bryan RN, Coulter JD, Willis WD (1974) Cells of origin of the spinocervical tract in the monkey. Exp Neurol 42:574–586.

Bucher SF, Dieterich M, Wiesmann M, Weiss A, Zink R, Yousry TA, Brandt T (1998) Cerebral functional magnetic resonance imaging of vestibular, auditory, and nociceptive areas during galvanic stimulation. Ann Neurol 44:120–125.

Buchsbaum MS, Buchsbaum BR, Chokron S, Tang C, Wei TC, Byne W (2006) Thalamocortical circuits: fMRI assessment of the pulvinar and medial dorsal nucleus in normal volunteers. Neurosci Lett 404:282–287.

Bull TR (1965) Taste and chorda tympani. J Laryngol Otol 79:479–493.

Bullier J, Kennedy H (1983) Projection of the lateral geniculate nucleus onto cortical area V2 in the macaque monkey. Exp Brain Res 53:168–172.

Bunge RP (1968) Glial cells and the central myelin sheath. Physiol Rev 48:197–251.

Bunge RP (1970) Structure and function of neuroglia: some recent observations. In: *The Neurosciences; Second Study Program.* Schmitt FO, editor-in-chief, pp 782–797. Rockefeller University Press, New York.

Bunney WE Jr (Moderator) (1979) Basic and clinical studies of endorphins. Ann Intern Med 91:239–250.

Bunt AH, et al. (1975) Monkey retinal ganglion cells: morphometric analysis and tracing of axonal projections, with a consideration of the peroxidase technique. J Comp Neurol 164:265–286.

Bunt AH, Minckler DS (1977) Foveal sparing: new anatomical evidence for bilateral representation of the central retina. Arch Ophthalmol 95:1445–1447.

Bunt AH, Minckler DS, Johanson GW (1977) Demonstration of bilateral projection of the central retina of the monkey with horseradish peroxidase neuronography. J Comp Neurol 171:619–630.

Burchiel KJ, Wyler AR (1978) Ectopic action potential generation in peripheral trigeminal axons. Exp Neurol 62:269–281.

Burger LJ, Kalvin NH, Smith JL (1970) Acquired lesions of the fourth cranial nerve. Brain 93:567–574.

Burke D (1978) The fusimotor innervation of muscle spindle endings in man. TINS 1:89–92.

Burke D, Skuse NF, Stuart DG (1979) The regularity of muscle spindle discharge in man. J Physiol (Lond) 291:277–290.

Burkhalter A, Bernardo KL (1989) Organization of corticocortical connections in human visual cortex. Proc Natl Acad Sci 80:1071–1075.

Burnstock G (1976) Do some nerve cells release more than one transmitter? Neuroscience 1:239–248.

Burton H, Benjamin RM (1971) Central projections of the gustatory system. In: *Handbook of Sensory Physiology.* Beidler LM, ed. Vol IV: Chemical senses. Part 2, pp 148–164. Taste. Springer-Verlag, Berlin.

Burton H, Craig AD Jr (1979) Distribution of trigeminothalamic projection cells in cat and monkey. Brain Res 161:515–521.

Burton H, Fabri M, Alloway K (1995) Cortical areas within the lateral sulcus connected to cutaneous representations in areas 3b and 1: a revised interpretation of the second somatosensory area in macaque monkeys. J Comp Neurol 355:539–562.

Burton H, Loewy AD (1976) Descending projections from the marginal cell layer and other regions of the monkey spinal cord. Brain Res 116:485–491.

Burton H, Loewy AD (1977) Projections to the spinal cord from medullary somatosensory relay nuclei. J Comp Neurol 173:773–792.

Büttner U, Buettner UW (1978) Parietal cortex (2v) neuronal activity in the alert monkey during natural vestibular and optokinetic stimulation. Brain Res 153:392–397.

Büttner U, Büttner-Ennever JA, Henn V (1978) The vestibular thalamus: neurophysiological and anatomical studies in the monkey. In: *Vestibular Mechanisms in Health and Disease.* Hood JH, ed., pp 80–85. Academic Press, London.

Büttner U, Henn V (1976) Thalamic unit activity in the alert monkey during natural vestibular stimulation. Brain Res 103:127–132.

Büttner U, Lang W (1979) The vestibulocortical pathway: neurophysiological and anatomical studies in the monkey. Prog Brain Res 50:581–588.

Büttner U, Waespe W (1981) Vestibular nerve activity in the alert monkey during vestibular and optokinetic nystagmus. Exp Brain Res 41:310–315.

Büttner-Ennever JA, Henn V (1976) An autoradiographic study of the pathways from the pontine reticular formation involved in horizontal eye movements. Brain Res 108:155–164.

Byl NN, McKenzie A, Nagarajan SS (2000) Differences in somatosensory hand organization in a healthy flutist and a flutist with focal hand dystonia: a case report. J Hand Ther 13:302–309.

Byne W (1998) The medial preoptic and anterior hypothalamic regions of the rhesus monkey: cytoarchitectonic comparison with the human and evidence for sexual dimorphism. Brain Res 793:346–350.

Byne W, Lasco MS, Kemether E, Shinwari A, Edgar MA, Morgello S, Jones LB, Tobet S (2000) The interstitial nuclei of the human anterior hypothalamus: an investigation of sexual variation in volume and cell size, number and density. Brain Res 856:254–258.

Byne W, Tobet S, Mattiace LA, Lasco MS, Kemether E, Edgar MA, Morgello S, Buchsbaum MS, Jones LB (2001) The interstitial nuclei of the human anterior hypothalamus: an investigation of variation with sex, sexual orientation, and HIV status. Horm Behav 40:86–92.

Bystron I, Molnar Z, Otellin V, Blakemore C (2005) Tangential networks of precocious neurons and early axonal outgrowth in the embryonic human forebrain. J Neurosci 25:2781–2792.

Bystron I, Rakic P, Molnar Z, Blakemore C (2006) The first neurons of the human cerebral cortex. Nat Neurosci 9:880–886.

Cairns H (1952) Disturbances of consciousness with lesions of the brain-stem and diencephalon. Brain 75:109–146.

Calne DB (1979) Neurotransmitters, neuromodulators, and neurohormones. Neurology 29:1517–1521.

Calne DB, Pallis CA (1966) Vibratory sense: a critical review. Brain 89:723–746.

Calvin WH, Loeser JD, Howe JF (1977) A neurophysiological theory for the pain mechanism of tic douloureux. Pain 3:147–154.

Cammermeyer J (1970) The life history of the microglial cell: a light microscopic study. In: Neurosciences Research. Ehrenpreis S, Solnitzky OC, eds, 3:43–129. Academic Press, New York.

Campain R, Minckler J (1976) A note on the gross configurations of the human auditory cortex. Brain Lang 3:318–323.

Campbell WW (2005) DeJong's The Neurological Examinatio. 6th ed. Lippincott Williams & Wilkins, Philadelphia.

Caplan LR (1974) Ptosis. J Neurol Neurosurg Psychiatry 37:1–7.

Carbonetto S, Muller KJ (1982) Nerve fiber growth and the cellular response to axotomy. Cur Topics Devel Biol 17:33–76.

Cardello AV (1978) Chemical stimulation of single human fungiform taste papillae: sensitivity profiles and locus of stimulation. Sens Processes 2:173–190.

Carleton SC, Carpenter M (1984) Distribution of primary vestibular fibers in the brainstem and cerebellum of the monkey. Brain Res 294:281–298.

Carleton SC, Carpenter MB (1983) Afferent and efferent connections of the medial, inferior and lateral vestibular nuclei in the cat and monkey. Brain Res 278:29–51.

Carlstedt T, Cullheim S, Risling M (2004) Spinal Cord in Relation to the Peripheral Nervous System. In: The Human Nervous System. Paxinos G, Mai JK, eds. 2nd ed, Chapter 9, pp 250–263. Elsevier/Academic Press, Amsterdam.

Carpenter MB, et al. (1987) Vestibular and cochlear efferent neurons in the monkey identified by immunocytochemical methods. Brain Res 408:275–280.

Carpenter MB, Stein BM, Peter P (1972) Primary vestibulocerebellar fibers in the monkey: distribution of fibers arising from distinctive cell groups of the vestibular ganglia. Am J Anat 135:221–250.

Casey KL (1980) Reticular formation and pain: toward a unifying concept. Res Publ Assoc Res Nerv Ment Dis 58:93–105.

Casey KL, Keene JJ, Morrow T (1974) Bulboreticular and medial thalamic unit activity in relation to aversive behavior and pain. Adv Neurol 4:197–205.

Castiglioni AJ, Gallaway MC, Coulter JD (1978) Spinal projections from the midbrain in monkey. J Comp Neurol 178:329–346.

Cauna N (1956) Nerve supply and nerve endings in Meissner's corpuscles. Am J Anat 99:315–350.

Cauna N (1959) The mode of termination of the sensory nerves and its significance. J Comp Neurol 113:169–209.

Cauna N (1966) Fine structure of the receptor organs and its probable functional significance. In: Touch, Heat, and Pain. DeReuck AVS, Knight J, eds, pp 117–127. Ciba Foundation Symposium. Churchill, London.

Cauna N (1967) Light and electron microscopical structure of sensory end-organs in human skin. In: The Skin Senses. Kenshalo DR, ed. 1st Int Symp on Skin Senses, Fla State Univ. 1968, pp 15–37. Thomas, Springfield.

Cauna N (1973) The free penicillate nerve endings of the human hairy skin. J Anat 115:277–288.

Cauna N (1980) Fine morphological characteristics and microtopography of the free nerve endings of the human digital skin. Anat Rec 198:643–656.

Celesia GG (1976) Organization of auditory cortical areas in man. Brain 99:403–414.

Celesia GG (1979) Somatosensory evoked potentials recorded directly from human thalamus and Sm I cortical area. Arch Neurol 36:399–405.

Cervero F (1980) Deep and visceral pain. In: Pain and Society. Kosterlitz HW, Terenius LY, eds, pp 263–282. Life Sciences Research Report 17. Verlag Chemie, Weinheim.

Cervero F (1985) Visceral nociception: peripheral and central aspects of visceral nociceptive systems. Philos Trans R Soc Lond B Biol Sci 308:325–337.

Cervero F, Iggo A (1980) The substantia gelatinosa of the spinal cord: a critical review. Brain 103:717–772.

Cervero F, Jensen TS (2006) Pain. In: Handbook of Clinical Neurology. Aminoff MJ, Boller F, Swaab DF, series eds. Vol 81. Elsevier, Amsterdam.

Chambers WW, Liu C-N (1972) Anatomy of the spinal cord. In: *The Spinal Cord: Basic Aspects and Surgical Considera-tions.* Austin GM, editor. 2nd ed., pp 5–56. Thomas, Springfield.

Chatrian GE, Farrell DF, Canfield RC, Lettich E (1975) Congenital insensitivity to noxious stimuli. Arch Neurol 32:141–145.

Chatrian GE, Patersen MC, Lazarte JA (1960) Response to clicks from the human brain: some depth electrographic observations. Electroencephalogr Clin Neurophysiol 12:479–489.

Chaurasia BD (1977) Forebrain in human anencephaly. Anat Anz 142:471–478.

Chery-Croze S, Duclaux R (1980) Discrimination of painful stimuli in human beings: influence of stimulation area. J Neurophysiol 44:1–10.

Chi JG, Dooling EC, Gilles FH (1977) Gyral development of the human brain. Ann Neurol 1:86–93.

Chibuzo GA, Cummings JF (1980) Motor and sensory centers for the innervation of mandibular and sublingual salivary glands: a horseradish peroxidase study in the dog. Brain Res 189:301–313.

Choi BH, Lapham FW (1976) Interactions of neurons and astrocytes during growth and development of human fetal brain in vitro. Exp Mol Pathol 24:110–125.

Chouchkov Ch (1978) Cutaneous receptors. Adv Anat Embryol Cell Biol 54:1–61.

Chouchkov ChN (1973) The fine structure of small encapsulated receptors in human digital glabrous skin. J Anat 114:25–33.

Chouchkov ChN (1974) An electron microscopic study of the intraepidermal innervation of human glabrous skin. Acta Anat (Basel) 88:84–92.

Chouinard PA, Paus T (2006) The primary motor and premotor areas of the human cerebral cortex. Neuroscientist 12:143–152.

Chu CC, Tranel D, Damasio AR, Van Hoesen GW (1997) The autonomic-related cortex: pathology in Alzheimer's disease. Cereb Cortex 7:86–95.

Chung JM, et al. (1986) Classification of primate spinothalamic and somatosensory thalamic neurons based on cluster analysis. J Neurophysiol 56:308–327.

Chung JM, et al. (1986) Response characteristics of neurons in the ventral posterior lateral nucleus of the monkey thalamus. J Neurophysiol 56:370–390.

Clark FJ, Horch KW, Bach SM, Larson GF (1979) Contributions of cutaneous and joint receptors to static knee-position sense in man. J Neurophysiol 42:877–888.

Clemente CD (1964) Regeneration in the vertebrate central nervous system. Int Rev Neurobiol 6:257–301.

Clemente CD, Magoun HW (1959) The bulbar brain stem. In: *Introduction to Stereotaxis with an Atlas of the Human Brain.* Schaltenbrand G, Bailey P, eds, 1:70–77. Thieme, Stuttgart.

Cogan DG (1975) An introduction to eye movements. In: *The Nervous System.* Tower DB, editor-in-chief. Chase TN, volume ed. Vol 2: The Clinical Neurosciences, pp 481–486. Raven Press, New York.

Cogan DG, Kubik CS, Smith WL (1950) Unilateral internuclear ophthalmoplegia. Report of eight clinical cases with one postmortem study. Arch Ophthalmol 44:783–796.

Coggeshall RE (1979) Afferent fibers in the ventral root. Neurosurgery 4:443–448.

Coggeshall RE (1980) Law of separation of function of the spinal roots. Physiol Rev 60:716–755.

Coggeshall RE, Chung K, Chung JM, Langford LA (1981) Primary afferent axons in the tract of Lissauer in the monkey. J Comp Neurol 196:431–442.

Coggeshall RE, et al. (1975) Unmyelinated axons in human ventral roots, a possible explanation for the failure of dorsal rhizotomy to relieve pain. Brain 98:157–166.

Cohen AI (1967) Ultrastructural aspects of the human optic nerve. Invest Ophthalmol 6:294–308.

Cohen B, Goto K, Tokumasu K (1967) Return eye movements, an ocular compensatory reflex in the alert cat and monkey. Exp Neurol 17:172–185.

Cohen B, Komatsuzaki A (1972) Eye movements induced by stimulation of the pontine reticular formation: evidence for integration in oculomotor pathways. Exp Neurol 36:101–117.

Cohen B, Suzuki J-I (1963) Eye movements induced by ampullary nerve stimulation. Am J Physiol 204:347–351.

Cohen B, Uemura T (1975) Ocular changes in monkeys after lesions of the superior and medial vestibular nuclei and the vestibular nerve roots. In: *The Vestibular System.* Naunton RD, ed., pp 187–202. Academic Press, New York.

Coleman PD, Flood DG (1986) Dendritic proliferation in the aging brain as a compensatory repair mechanism. Prog Brain Res 70:227–237.

Coleman PD, Flood DG (1987) Neuron numbers and dendritic extent in normal aging and Alzheimer's disease. Neurobiol Aging 8:521–545.

Colletti V, Carner M, Miorelli V, Guida M, Colletti L, Fiorino F (2005) Auditory brainstem implant (ABI): new frontiers in adults and children. Otolaryngol Head Neck Surg 133:126–138.

Collins DF, Refshauge KM, Todd G, Gandevia SC (2005) Cutaneous receptors contribute to kinesthesia at the index finger, elbow, and knee. J Neurophysiol 94:1699–1706.

Collins P (2004) Development of the nervous system. In: *Gray's Anatomy.* Standring S, editor-in-chief. 39th ed., pp 241–274. Elsevier/Churchill Livingstone, London.

Collu R, Fraschini F (1972) The pineal gland – a neuroendocrine transducer. Adv Metab Disord 6:161–175.

Comfort A (1974) The likelihood of human pheromones. In: *Pheromones.* Birch M, ed. Frontiers of Biology 32:386–396.

Connolly M, Van Essen D (1984) The representation of the visual field in parvicellular and magnocellular layers of the lateral geniculate nucleus in the macaque monkey. J Comp Neurol 226:544–564.

Cook AW, Kawakami Y (1977) Commissural myelotomy. J Neurosurg 47:1–6.

Cooper MH, Beal JA (1978) The neurons and the synaptic endings in the primate basilar pontine gray. J Comp Neurol 180:17–42.

Cooper S, Daniel PM (1963) Muscle spindles in man; their morphology in the lumbricals and the deep muscles of the neck. Brain 86:563–586.

Cooper S, Daniel PM, Whitteridge D (1955) Muscle spindles and other sensory endings in the extrinsic eye muscles;

the physiology and anatomy of these receptors and of their connexions with the brain-stem. Brain 78:564–583.

Corkin S, Milner B, Rasmussen T (1970) Somatosensory thresholds contrasting effects of postcentral-gyrus and posterior parietal-lobe excisions. Arch Neurol 23:41–58.

Corner GW (1929) A well-preserved human embryo of 10 somites. Contrib Embryol Carneg Instn 20:81–101.

Cosson A, Tatu L, Vuillier F, Parratte B, Diop M, Monnier G (2003) Arterial vascularization of the human thalamus: extra-parenchymal arterial groups. Surg Radiol Anat 25:408–415.

Costanzo RM, Miwa T (2006) Posttraumatic olfactory loss. Adv Otorhinolaryngol 63:99–107.

Costunzo RM, Gardner EP (1981) Multiple-joint neurons in somatosensory cortex of awake monkeys. Brain Res 214:321–333.

Cotman CW, Nieto-Sampedro M (1984) Cell biology of synaptic plasticity. Science 225:1287–1293.

Courten C de, Garey LJ (1982) Morphology of the neurons in the human lateral geniculate nucleus and their normal development: a Golgi study. Exp Brain Res 47:159–171.

Covenas R, Martin F, Salinas P, Rivada E, Smith V, Aguilar LA, Diaz-Cabiale Z, Narvaez JA, Tramu G (2004) An immuno-cytochemical mapping of methionine-enkephalin-Arg(6)-Gly(7)-Leu(8) in the human brainstem. Neuroscience 128:843–859.

Cowan WM (1979) The development of the brain. Sci Am 241:113–133.

Craig AD (2002) How do you feel? Interoception: the sense of the physiological condition of the body. Nature Reviews Neuroscience 3:655–666.

Cras P, et al. (1990) A monoclonal antibody raised against Alzheimer cortex that specifically recognizes a subpopulation of microglial cells. J Histochem Cytochem 38:1201–1207.

Cras P, et al. (1990) Neuronal and microglial involvement in beta-amyloid protein deposition in Alzheimer's disease. Am J Pathol 137:241–246.

Craske B (1977) Perception of impossible limb positions induced by tendon vibration. Science 196:71–73.

Creutzfeldt OD (1972) Transfer functions of the retina. In: Recent Contributions to Neurophysiology. Cordeau J-P, Gloor P, eds. Electroencephalogr Clin Neurophysiol (Suppl) 31:159–169.

Critchley M (1953) Tactile thought, with special reference to the blind. Brain 76:19–35.

Critchley, M. (1953) The Parietal Lobes. Edward Arnold, London.

Crosby EC (1953) Relations of brain centers to normal and abnormal eye movements in the horizontal plane. J Comp Neurol 99:437–480.

Crosby EC, DeJonge BR (1963) Experimental and clinical studies of the central connections and central relations of the facial nerve. Ann Otol Rhinol Laryngol 72:735–755.

Crosby EC, Humphrey T (1941) Studies of the vertebrate telencephalon. II. The nuclear pattern of the anterior olfactory nucleus, tuberculum olfactorium and the amygdaloid complex in adult man. J Comp Neurol 74:309–352.

Crosby EC, Humphrey T, Lauer EW (1962) Correlative Anatomy of the Nervous System. Macmillan, New York.

Crosby EC, Lauer EW (1959) Anatomy of the midbrain. In: Introduction to Stereotaxis with an Atlas of the Human Brain. Schaltenbrand G, Bailey P, eds, 1:88–118. Thieme, Stuttgart.

Crosby EC, Schnitzlein HN (1982) Comparative Correlative Neuroanatomy of the Vertebrate Telencephalon. Macmillan, New York.

Crosby EC, Woodburne RT (1943) The nuclear pattern of the non-tectal portions of the midbrain and isthmus in primates. J Comp Neurol 78:441–482.

Crosby EC, Yoss RE (1954) The phylogenetic continuity of neural mechanisms as illustrated by the spinal tract of V and its nucleus. Res Publ Assoc Res Nerv Ment Dis 33:174–208.

Cross MJ, McCloskey DI (1973) Position sense following surgical removal of joints in man. Brain Res 55:443–445.

Crossman AR (2004) Basal ganglia. In: Gray's Anatomy. Standring S, editor-in-chief. 39th ed., pp 419–430. Elsevier/Churchill Livingstone, London.

Crossman AR (2004) Cerebral hemisphere. In: Gray's Anatomy. Standring S, editor-in-chief. 39th ed., pp 387–418. Elsevier/Churchill Livingstone, London.

Crossman AR (2004) Diencephalon. In: Gray's Anatomy. Standring S, editor-in-chief. 39th ed., pp 369–386. Elsevier/Churchill Livingstone, London.

Croze S, Duclaux R (1979) Burning and second pain: an alternative interpretation. In: Sensory Functions of the Skin of Humans. 2nd Int Symp on Skin Senses, pp 385–389. Fla State U. Plenum, New York.

Crutcher KA, Bingham WG Jr (1978) Descending monoaminergic pathways in the primate spinal cord. Am J Anat 163:159–164.

Cuello AC (1978) Endogenous opioid peptides in neurons of the human brain. Lancet 2:291–293.

Cuello AC, Polak JM, Pearse AGE (1976) Substance P: a naturally occurring transmitter in human spinal cord. Lancet 2:1054–1056.

Cumming WJK (1970) An anatomical review of the corpus callosum. Cortex 6:1–18.

Curcio CA, et al. (1987) Distribution of cones in human and monkey retina: individual variability and radial asymmetry. Science 236:579–582.

Curless RG (1977) Developmental patterns of peripheral nerve, myoneural junction and muscle: a review. Prog Neurobiol 9:197–209.

Curthoys IS, Markham CH, Curthoys EJ (1977) Semicircular duct and ampulla dimensions in cat, guinea pig and man. J Morphol 151:17–34.

Curtis MA, Kam M, Nannmark U, Anderson MF, Axell MZ, Wikkelso C, Holtås S, van Roon-Mom WM, Björk-Eriksson T, Nordborg C, Frisén J, Dragunow M, Faull RL, Eriksson PS (2007) Human neuroblasts migrate to the olfactory bulb via a lateral ventricular extension. Science 315: 1243–1249.

Cusick JF, Ackmann JJ, Larson SJ (1977) Mechanical and physiological effects of dentatotomy. J Neurosurg 46: 767–775.

Czeisler CA, Dijk DJ (2001) Human circadian physiology and sleep–wake regulation. In: Handbook of Behavioral. Neurobiology: Circadian Clocks. Takahashi S, Turek FW, Moore RY, eds, 12:183–213. Kluwer Academic/Plenum Publishers, New York.

Dahlstrom A, Fuxe K (1964) Evidence for the existence of monoamine-containing neurons in the central nervous system. I. Demonstration of monoamines in the cell bodies of brain stem neurons. Acta Physiol Scand (Suppl) 232:1–55.

Dalgleish T (2004) The emotional brain. Nat Rev Neurosci 5:583–589.

Dallos P (1981) Cochlear physiology. Annu Rev Psychol 32:153–190.

Darian-Smith I (1966) Neural mechanisms of facial sensation. Int Rev Neurobiol 9:301–395.

Darian-Smith I, Johnson KO (1977) Thermal sensibility and thermoreceptors. J Invest Dermatol 69:146–153.

Dartnall HJA, Bowmaker JK, Mollon JD (1983) Microspectrophotometry of human photoreceptors. In: *Colour Vision: Physiology and Psychophysics.* Mollon JD, Sharpe LT, eds, pp 69–80. Academic Press, London.

Das GD (1990) Neural transplantation: an historical perspective. Neurosci Biobehav Rev 14:389–401.

Dash MS, Deshpande SS (1976) Human skin nociceptors and their chemical response. In: *Advances in Pain Research and Therapy.* Bonica JJ, Albe-Fessard D, eds, 1:47–51. Raven Press, New York.

d'Avella D, Mingrino S (1979) Microsurgical anatomy of lumbosacral spinal roots. J Neurosurg 51:819–823 .

Davis JM, Zimmerman RA (1983) Injury of the carotid and vertebral arteries. Neuroradiology 25:55–69.

Davis KD, Kwan CL, Crawley AP, Mikulis DJ (1998) Functional MRI study of thalamic and cortical activations evoked by cutaneous heat, cold, and tactile stimuli. J Neurophysiol 80:1533–1546.

Davis L, Hart JT, Crain RC (1929) The pathway for visceral afferent impulses within the spinal cord. II. Experimental dilatation of the biliary ducts. Surg Gynecol Obstet 48:647–651.

Davis LE (1922) The pathway for visceral afferent impulses within the spinal cord. Am J Physiol 59:381–393.

Davison AN, Dobbing J (1966) Myelination as a vulnerable period in brain development. Br Med Bull 22:40–44.

Daw NW, Brunken WJ, Parkinson D (1989) The function of synaptic transmitters in the retina. Annu Rev Neurosci 12:205–225.

Dawson WW et al. (1989) Very large neurons of the inner retina of humans and other mammals. Retina 9:69–74.

Day BL, Fitzpatrick RC (2005) The vestibular system. Curr Biol 5:R583–R586.

de Araujo IE, Rolls ET, Kringelbach ML, McGlone F, Phillips N (2003) Taste–olfactory convergence, and the representation of the pleasantness of flavour, in the human brain. Eur J Neurosci 18:2059–2068.

de la Torre JC (1981) Spinal cord injury. Review of basic and applied research. Spine 6:315–335.

de Lacalle S, Saper CB (1997) The cholinergic system in the primate brain: basal forebrain and pontine-tegmental cell groups. In: *Handbook of Chemical Neuroanatomy.* Björklund A, Hökfelt T, eds. Vol 13: The Primate Nervous System, Part I., pp 263–375. Elsevier, Amsterdam.

de Lange EE, Vielvoye GJ, Voormolen JHC (1986) Arterial compression of the fifth cranial nerve causing trigeminal neuralgia: angiographic findings. Radiology 158:721–727.

de Lecea L, Sutcliffe JG (2005) The hypocretins and sleep. FEBS J 272:5675–5688.

De Monasterio FM (1978) Center and surround mechanisms of opponent-color X and Y ganglion cells of retina of macaques. J Neurophysiol 41:1418–1434.

De Monasterio FM (1978) Properties of concentrically organized X and Y ganglion cells of macaque retina. J Neurophysiol 41:1394–1417.

De Monasterio FM (1978) Properties of ganglion cells with atypical receptive-field organization in retina of macaques. J Neurophysiol 41:1435–1449.

De Monasterio FM, Gouras P (1975) Functional properties of ganglion cells of the rhesus monkey retina. J Physiol 251:167–195.

De Monasterio FM, Schein SJ, McCrane EP (1981) Staining of blue-sensitive cones of the macaque retina by a fluorescent dye. Science 213:1278–1281.

De Olmos JS (2004) Amygdala. In: *The Human Nervous System.* Paxinos G, Mai JK, eds. 2nd ed., pp 739–868. Elsevier/Academic Press, Amsterdam.

De Robertis EDP, Carrea R (1965) *Biology of Neuroglia.* Elsevier, New York.

De Valois RL, Abramov I, Jacobs GH (1966) Analysis of response patterns of LGN cells. J Opt Soc Am 56:966–977.

de Vellis J, Kukes G (1973) Regulation of glial cell functions by hormones and ions: a review. Tex Rep Biol Med 31:271–293.

de Waele C, Baudonniere PM, Lepecq JC, Tran Ba Huy P, Vidal PP (2001) Vestibular projections in the human cortex. Exp Brain Res 141:541–551.

DeArmond SJ, Fusco MM, Dewey MM (1989) *Structure of the Human Brain.* 3rd ed. Oxford University Press, New York.

Debbage PL (1986) The generation and regeneration of oligodendroglia: a short review. J Neurol Sci 72:319–336.

DeBettignies BH, Mahurin RK, Pirozzolo FJ (1990) Insight for impairment in independent living skills in Alzheimer's disease and multi-0infarct dementia. J Clin Exp Neuropsychol 12:355–363.

Deecke L, Schwarz DWF, Fredrickson JM (1973) The vestibular thalamus in the rhesus monkey. Adv Oto Rhinol Laryngol 19:210–219.

Deecke L, Schwarz DWF, Fredrickson JM (1974) Nucleus ventroposterior inferior (VPI) as the vestibular thalamic relay in the rhesus monkey. I. Field potential investigation. Exp Brain Res 20:88–100.

Deecke L, Schwarz DWF, Fredrickson JM (1977) Vestibular responses in the rhesus monkey ventroposterior thalamus. II. Vestibulo-proprioceptive convergence at thalamic neurons. Exp Brain Res 30:219–232.

DeJong RN (1979) *The Neurologic Examination: Incorporating the Fundamentals of Neuroanatomy and Neurophysiology.* 4th ed. Harper & Row, Hagerstown.

Dekaban A (1953) Human thalamus. An anatomical, developmental and pathological study. I. Division of the human adult thalamus into nuclei by use of the cyto-myelo-architectonic method. J Comp Neurol 99:639–684.

Dekaban AS (1963) Anencephaly in early human embryos. J Neuropathol Exp Neurol 22:533–548, 1963.

Dekaban AS, Sadowsky D (1978) Changes in brain weights during the span of human life: relation of brain weights to body heights and body weights. Ann Neurol 4:345–356.

Dekaban(1954) Human thalamus. An anatomical, developmental and pathological study. II. Development of the human thalamic nuclei. J Comp Neurol 100:63–97.

Del Bianco PL, Del Bene E, Sicuteri F (1974) Heart pain. Adv Neurol 4:375–381.

DeLong MR, Georgopoulos AP (1979) Physiology of the basal ganglia – a brief overview. Adv Neurol 23:137–153.

DeMyer W (1973) Anatomy and clinical neurology of the spinal cord. In: *Clinical Neurology.* Baker AB, Baker LH, eds, 3:1–24. Harper & Row, Hagerstown.

DeMyer W (1975) Congenital anomalies of the central nervous system. In: *The Nervous System.* Tower DB, editor-in-chief. Chase TN, volume ed. Vol 2: The Clinical Neurosciences, pp 347–357. Raven Press, New York.

Denny-Brown D, Yanagisawa N (1973) The function of the descending root of the fifth nerve. Brain 96:783–814.

Desmond ME (1982) Description of the occlusion of the spinal cord lumen in early human embryos. Anat Rec 204:89–93.

Desmond ME, O'Rahilly R (1981) The growth of the human brain during the embryonic period proper. 1. Linear axes. Anat Embryol (Berl) 162:137–151.

Desor JA, Maller O, Greene LS (1977) Preference for sweet in humans: infants, children, and adults. In: *Taste and Development: The Genesis of Sweet Preference.* Weiffenbach JM, ed., pp 161–172. DHEW Publ, Washington, DC.

DeYoe EA, Felleman DJ, Van Essen DC, McClendon E (1994) Multiple processing streams in occipitotemporal visual cortex. Nature 371:151–154.

Diamond IT (1974) Sound Localization – Symposium of the American Physiological Society. Fed Proc 33:1899–1932.

Dichgans J, Bizzi E, Morasso P, Tagliasco V (1973) Mechanisms underlying recovery of eye–head coordination following bilateral labyrinthectomy in monkeys. Exp Brain Res 18:548–562.

Dickmann Z, Chewe TH, Bonney WA, Noyes RW (1965) The human egg in the pronuclear stage. Anat Rec 152:293–302.

Dickson DW, Mattiace LA (1989) Astrocytes and microglia in human brain share an epitope recognized by a B-lymphocyte-specific monoclonal antibody (LN-1). Am J Pathol 135:135–147

Dieterich M (2004) Dizziness. Neurologist 10:154–164.

Dietert SE (1965) The demonstration of different types of muscle fibers in human extraocular muscle by electron microscopy and cholinesterase staining. Invest Ophthalmol 4:51–63.

Dietert SE: The demonstration of different types of muscle fibers in human extraocular muscle by electron microscopy and cholinesterase staining. Invest Ophthalmol 4:51–63, 1965.

Dimitrova A, de Greiff A, Schoch B, Gerwig M, Frings M, Gizewski ER, Timmann D (2006) Activation of cerebellar nuclei comparing finger, foot and tongue movements as revealed by fMRI. Brain Res Bull 71:233–241.

Disbrow E, Roberts T, Krubitzer L (2000) Somatotopic organization of cortical fields in the lateral sulcus of Homo sapiens: evidence for SII and PV. J Comp Neurol 418:1–21.

Djouhri L, Lawson SN (2004) Aβ-fiber nociceptive primary afferent neurons: a review of incidence and properties in relation to other afferent A-fiber neurons in mammals. Brain Res Brain Res Rev 46:131–145.

Dobbing J (175) Prenatal nutrition and neurological development. UCLA Forum Med Sci 18:401–420.

Dobbing J (1968) Vulnerable periods in developing brain. In: *Applied Neurochemistry.* Davison AN, Dobbing J, eds, pp 287–316. FA Davis Co, Philadelphia.

Dobbing J, Hopewell JW, Lynch A. (1971) Vulnerability of developing brain. VII. Permanent deficit of neurons in cerebral and cerebellar cortex following early mild undernutrition. Exp Neurol 32:439–447.

Dobbing J, Sands J (1971) Vulnerability of developing brain. IX. The effect of nutritional growth retardation on the timing of brain growth-spurt. Biol Neonate 19:363–378.

Dobbing J, Sands J (1979) Comparative aspects of the brain growth spurt. Early Hum Dev 3:79–83.

Dobelle WH, Stensaas SS, Miladejovsky MG, Smith JB (1973) A prosthesis for the deaf based on cortical stimulation. Ann Otol Rhinol Laryngol 82:445–463.

Doerr M, Krainick J-U, Thoden U (1978) Pain perception in man after long term spinal cord stimulation. J Neurol 217:261–270.

Domenech-Ratto G (1977) Development and peripheral innervation of the palatal muscles. Acta Anat (Basel) 97:4–14.

Donald DE, Shepherd JT (1979) Cardiac receptors: normal and disturbed function. Am J Cardiol 44:873–878.

Donaldson HH, Davis, DJ (1903) A description of charts showing the areas of cross sections of the human spinal cord at the level of each spinal nerve. J Comp Neurol 13:19–40.

Doran FSA (1967) The sites to which pain is referred from the common bile-duct in man and its implication for the theory of referred pain. Br J Surg 54:599–606.

Dorovini-Zis K, Dolman CL (1977) Gestational development of brain. Arch Path Lab Med 101:192–195.

Doty RL, Bromley SM (2004) Effects of drugs on olfaction and taste. Otolaryngol Clin North Am 37:1229–1254.

Douek E (1976) Olfaction In: *Scientific Foundations of Otolaryngology.* Hinchcliffe R, Harrison D, eds, pp 495–501. William Heinemann Medical Books, London.

Dowling JE (1970) Organization of vertebrate retinas: the Jonas M Friedenwald Memorial Lecture. Invest Ophthalmol 9:655–680.

Dowling JE (1987) *The Retina: An Approachable Part of the Brain,* pp 6, 14, 53, 70–71. Belknap Pr, Cambridge.

Dowling JE, Boycott BB (1966) Organization of the primate retina: electron microscopy. Proc R Soc Lond (Biol) 166:80–111.

Doyle LL, Lippes J, Winters HS, Margolis AJ (1966) Human ova in the fallopian tube. Am J Obstet Gynecol 95:115–117.

Dreyer DA, Loe PR, Metz CB, Whitsel BL (1975) Representation of head and face in postcentral gyrus of the macaque. J Neurophysiol 38:714–733.

Dreyer DA, Schneider RJ, Metz CB, Whitsel BL (1974) Differential contributions of spinal pathways to body representations in postcentral gyrus of *Macaca mulatta.* J Neurophysiol 37:119–145.

Droz B (1973) Renewal of synaptic proteins. Brain Res 62:383–394.

Droz B (1975) Synthetic machinery and axoplasmic transport: maintenance of neuronal connectivity. In: *The Nervous System.* Tower DB, editor-in-chief. Brady RO, volume ed. Vol 1: The Basic Neurosciences, pp 111–127. Raven Press, New York.

Duane DD (1977) A neurologic perspective of central auditory dysfunction. In: *Central Auditory Dysfunction.* Keith RW, ed., pp 1–42. Grune, New York.

Dublin WB (1974) Cytoarchiture of the cochlear nuclei. Arch Otolaryngol 100:355–359.

Dublin WB (1982) The cochlear nuclei revisited. Arch Otolaryngol head Neck Surg 90:744–760.

Dubner R, et al. (1984) Neural circuitry mediating nociception in the medullary and spinal dorsal horns. In: *Advances in Pain Research and Therapy.* Kruger L, Liebeskind JC, eds, 6:151–166. Raven Press, New York.

Dubner R, Gobel S, Price DD (1976) Peripheral and central trigeminal 'pain' pathways. In: *Advances in Pain Research and Therapy.* Bonica JJ, Albe-Fessard D, eds, 1:137–148. Raven Press, New York.

Dubner R, Hayes RL, Hoffman DS (1980) Neural and behavioral correlates of pain in the trigeminal system. Res Publ Assoc Res Nerv Ment Dis 58:63–72.

Dubner R, Sessle BJ, Storey AT (1978) *The Neural Basis of Oral and Facial Function.* Plenum, New York.

Dubner R, Sumino R, Starkman S: Responses of facial cutaneous thermosensitive and mechanosensitive afferent fibers in the monkey to noxious heat stimulation. Adv Neurol 4:61–71.

Dubner R, Sumino R, Wood WI (1975) A peripheral 'cold' fiber population responsive to innocuous and noxious thermal stimuli applied to monkey's face. J Neurophysiol 38:1373–1389.

Duckett S, Pearse AGE (1969) Histoenzymology of the developing human spinal cord. Anat Rec 163:59–66.

Duclaux R, Kenshalo DR Sr (1980) Response characteristics of cutaneous warm receptors in the monkey. J Neurophysiol 43:1–15.

Dunn JD, Matzke HA (1968) Efferent fiber connections of the marmoset (*Oedipomidas oedipus*) trigeminal nucleus caudalis. J Comp Neurol 133:429–438.

Dunn LT (2002) Raised intracranial pressure. J Neurol Neurosurg Psychiat 73(Supp l):i23–27.

Duvernoy HM (1998) *The Human Hippocampus.* 2nd ed. Springer, Berlin Heidelberg.

Dworken HJ, Biel FJ, Machella TE (1952) Supradiagphragmatic reference of pain from the colon. Gastroenterology 22:222–228.

Dykes RW (1978) The anatomy and physiology of the somatic sensory cortical regions. Prog Neurobiol 10:33–88.

Dykes RW (1975) Nociception. Brain Res 99:229–245.

Dzendolet E, Murphy C (1974) Electrical stimulation of human fungiform papillae. Chem Senses Flavor 1:9–15.

Earnest MP, Monroe PA, Yarnell PR (1977) Cortical deafness: demonstration of the pathologic anatomy by CT scan. Neurology 27:1172–1175.

Eccles JC (1964) *The Physiology of Synapses.* Academic Press, New York.

Eccles JC (1967) Functional organization of the spinal cord. Anesth 28:31–45.

Eccles JC (1986) Chemical transmission and Dale's principle. Prog Brain Res 68:3–13.

Eccles JC, Nicoll RA, Táboríková H, Willey TJ (1975) Medial reticular neurons projecting rostrally. J Neurophysiol 38:531–538.

Eccles JC, Schmidt RF, Willis WD (1963) Depolarization of central terminals of group 1b afferent fibers of muscle. J Neurophysiol 26:1–27.

Eccles JC, Schmidt RF, Willis WD (1963) The location and the mode of action of the presynaptic inhibitory pathways on to group I afferent fibers from muscle. J Neurophysiol 26:506–522.

Ecker A (1980) Prolonged relief of tic douloureux from partial root destruction is associated with localized analgesia. Ann Neurol 7:181–182.

Eckmiller R (1975) Differences in the activity of eye-position coded neurons in the alert monkey during fixation and tracking movements. In: *Basic Mechanisms of Ocular Motility and Their Clinical Implications.* Lennerstrand G, Bach-Y-Rita P, eds, pp 447–451. Pergamon Pr, Oxford and New York.

Edgerton BJ, House WF, Hitseleberger W (1982) Hearing by cochlear nucleus stimulation in humans. Ann Otol Rhinol Laryngol 91(2 Pt 3):117–124.

Edial (1970) The carotid body. Lancet 2:140–141.

Edial (1976) Any hope for the cut central axon? Lancet 1:1224–1225.

Efron R (1957) The conditioned inhibition of uncinate fits. Brain 80:251–262.

Ehrsson HH, Naito E, Geyer S, Amunts K, Zilles K, Forssberg H, Roland PE (2000) Simultaneous movements of upper and lower limbs are coordinated by motor representations that are shared by both limbs: a PET study. Eur J Neurosci 12:3385–3398.

Eickhoff SB, Amunts K, Mohlberg H, Zilles K (2006) The human parietal operculum. II. Stereotaxic maps and correlation with functional imaging results. Cereb Cortex 16:268–279.

Eickhoff SB, Grefkes C, Zilles K, Fink GR (2006) The somatotopic organization of cytoarchitectonic areas on the human parietal pperculum. Cereb Cortex [Epub ahead of print, doi:10.1093/cercor/bhl090]

Eickhoff SB, Lotze M, Wietek B, Amunts K, Enck P, Zilles K (2006) Segregation of visceral and somatosensory afferents: an fMRI and cytoarchitectonic mapping study. NeuroImage 31:1004–1014.

Eickhoff SB, Schleicher A, Zilles K, Amunts K (2006) The human parietal operculum. I. Cytoarchitectonic mapping of subdivisions. Cereb Cortex 16:254–267.

Eickhoff SB, Weiss PH, Amunts K, Fink GR, Zilles K (2006) Identifying human parieto-insular vestibular cortex using fMRI and cytoarchitectonic mapping. Hum Brain Mapp 27:611–621.

Eimas PD, Siqueland ER, Jusczyk P, Vigorito J (1971) Speech perception in infants. Science 171:303–306.

Eisenman JS, Masland WS (1977) The hypothalamus. In: *Scientific Approaches to Clinical Neurology.* Goldensohn ES, ed., 2:1886–1899. Lea & Febiger, Philadelphia.

Eldred E, Yellin H, DeSantis M, Smith CM (1977) Supplement to bibliography on muscle receptors: their morphology, pathology, physiology, and pharmacology. Exp Neurol 55:1–118.

Eldredge DH, Miller JD: Physiology of hearing. Annu Rev Physiol 33:281–310.

El-Falougy H, Benuska J (2006) History, anatomical nomenclature, comparative anatomy and functions of the hippocampal formation. Bratisl Lek Listy 107:103–106.

Elia JC (1968) *The Dizzy Patient*. CC Thomas, Springfield.

Eliasson S, Gisselsson L (1954) Electrical stimulation of the chorda tympani in human beings. Acta Otolaryngol (Suppl) 116:72–75.

Elliott HC (1943) Studies on the motor cells of the spinal cord. II. Distribution in the normal human fetal cord. Am J Anat 72:29–38.

Elliott HC (1945) Cross-sectional diameters and areas of the human spinal cord. Anat Rec 93:287–293.

Emmers R, Tasker RR (1975) *The Human Somesthetic Thalamus, With Maps for Physiological Target Localization During Stereotactic Neurosurgery*. Raven Press, New York.

Emmons WF, Rhoton AL Jr (1971) Subdivision of the trigeminal sensory root. Experimental study in the monkey. J Neurosurg 35:585–591.

Engelborghs S, Marien P, Martin JJ, De Deyn PP (1998) Functional anatomy, vascularisation and pathology of the human thalamus. Acta Neurol Belg 98:252–265.

Engh CA, Schofield BH (1972) A review of the central response to peripheral nerve injury and its significance in nerve regeneration. J Neurosurg 37:195–203.

England MA, Wakely J (1991) *Color Atlas of the Brain & Spinal Cord: An Introduction to normal neuroanatomy*. Mosby Year Book, St. Louis.

Engström B, Hillerdal M, Laurell G, Bagger-Sjöbäck D (1987) Selected pathological findings in the human cochlea. Acta Otolaryngol (Suppl) (Stockh) 436:110–116.

Engström H, Ades HW (1973) The ultrastructure of the organ of Corti. In: *The Ultrastructure of Sensory Organs*. Friedman I, ed., pp 85–151. North-Holland/Elsevier, Amsterdam.

Engström H, Bergström B, Rosenhall U (1974) Vestibular sensory epithelium. Arch Otolaryngol 100:411–418.

Engström H, Engström B (1978) Structure of the hairs on cochlear sensory cells. Hear Res 1:49–66.

Enoch JM, Laties AM (1971) An analysis of retinal receptor orientation. II. Predictions for psychophysical tests. Invest Ophthalmol 10:959–970.

Erdem A, Yasargil G, Roth P (1993) Microsurgical anatomy of the hippocampal arteries. J Neurosurg 79:256–265.

Erzurumlu RS, Killackey HP (1982) Critical and sensitive periods in neurobiology. Curr Top Dev Biol 17:207–240.

Essick CR (1912) The development of the nuclei pontis and the nucleus arcuatus in man. Am J Anat 13:25–54.

Evarts EV (1981) Functional studies of the motor cortex. In: *The Organization of the cerebral cortex. Proceedings of a Neurosciences Research Program Colloquium*. Schmitt FP, Wooden FG, Adelman G, Dennis SG, eds, pp 263–283. MIT Pr, Cambridge.

Evinger LC, Fuchs AF, Baker R (1977) Bilateral lesions of the medial longitudinal fasciculus in monkeys: effects on the horizontal and vertical components of voluntary and vestibular induced eye movements. Exp Brain Res 28:1–20, 1977.

Fabri M, Polonara G, Salvolini U, Manzoni T (2005) Bilateral cortical representation of the trunk midline in human first somatic sensory area. Hum Brain Mapp 25:287–296.

Falconer MA (1948) Relief of intractable pain of organic origin by frontal lobotomy. Res Publ Assoc Res Nerv Ment Dis 27:706–714.

Fallon JF, Simandl BK (1978) Evidence of a role for cell death in the disappearance of the embryonic human tail. Am J Anat 152:111–130.

Farber K, Kettenmann H (2005) Physiology of microglial cells. Brain Res Rev 48:133–143.

Fasold O, von Brevern M, Kuhberg M, Ploner CJ, Villringer A, Lempert T, Wenzel R (2002) Human vestibular cortex as identified with caloric stimulation in functional magnetic resonance imaging. NeuroImage 17:1384–1393.

Fedorow H, Tribl F, Halliday G, Gerlach M, Riederer P, Double KL (2005) Neuromelanin in human dopamine neurons: comparison with peripheral melanins and relevance to Parkinson's disease. Prog Neurobiol 75:109–124.

Feess-Higgins A, Larroche J-C (1987) *Development of the human foetal brain. An anatomical atlas*. Masson, Paris.

Feirabend HK, Choufoer H, Ploeger S, Holsheimer J, van Gool JD (2002) Morphometry of human superficial dorsal and dorsolateral column fibres: significance to spinal cord stimulation. Brain 125:1137–1149.

Felix D, Wiesendanger M (1970) Cortically induced inhibitions in the dorsal column nuclei of monkeys. Pflügers Arch 320:285–288.

Felten DL, Crutcher KA (1979) Neuronal–vascular relationships in the raphe nuclei, locus coeruleus, and substantia nigra in primates. Am J Anat 155:467–482.

Felten DL, Sladek JR Jr (1983) Monoamine distribution in primate brain V. Monoaminergic nuclei: anatomy, pathways and local organization. Brain Res Bull 10:171–284.

Ferguson IT (1978) Electrical study of jaw and orbicularis oculi reflexes after trigeminal nerve surgery. J Neurol Neurosurg Psychiat 41:819–823.

Fernández C, Goldberg JM (1976) Physiology of peripheral neurons innervating otolith organs of the squirrel monkey. I. Response to static tilts and to long-duration centrifugal force. J Neurophysiol 39:970–1008.

Ferraro A, Pacella BL, Barrera SE (1940) Effects of lesions of the medial vestibular nucleus: an anatomical and physiological study in macacus rhesus monkeys. J Comp Neurol 73:7–36.

Ferraro JA, Minckler J (1977) The brachium of the inferior colliculus. The human auditory pathways: a quantitative study. Brain Lang 4:156–164.

Ferraro JA, Minckler J (1977) The human lateral lemniscus and its nuclei. The human auditory pathways: a quantitative study. Brain Lang 4:277–294.

Ferrer I, Galofré E (1987) Dendritic spine anomalies in fetal alcohol syndrome. Neuropediatrics 18:161–163.

Fex J, Altschuler RA (1986) Neurotransmitter-related immunocytochemistry of the organ of Corti. Hear Res 22:249–263.

Fields HL (1981) An endorphin-mediated analgesia system: experimental and clinical observations. In: *Neurosecretion and Brain Peptides.* Martin JB, Reichlin S, Bick KL, eds, pp 199–212. Raven Press, New York.

Fields HL (1981) Pain II: new approaches to management. Ann Neurol 9:101–106.

Fillenz M, Widdicombe JG (1972) Receptors of the lungs and airways. In: *Handbook of Sensory Physiology.* Neil E, volume editor, III/1:81–112. Springer-Verlag, New York.

Finch C (1985) Comments on review by Gash et al. Applications of recombinant DNA techniques. Neurobiol Aging 6:156–157.

Fine AD, Cardoso A, Rhoton AL Jr (1999) Microsurgical anatomy of the extracranial–extradural origin of the posterior inferior cerebellar artery. J Neurosurg 91:645–652.

Firbas W (1978) The efferent innervation in the region of inner hair cells in the organ of Corti. Acta Otolaryngol 86:309–313.

Firszt JB, Ulmer JL, Gaggl W (2006) Differential representation of speech sounds in the human cerebral hemispheres. Anat Rec A Discov Mol Cell Evol Biol 288:345–357.

Fisch U, Wegmüller A (1974) Early diagnosis of acoustic neuromas. ORL 36:129–140.

Fisher CM (1965) Pure sensory stroke involving face, arm, and leg. Neurology 15:76–80.

Fisher CM (1978) Thalamic pure sensory stroke: A pathologic study. Neurology 28:1141–1144.

FitzGerald MJT (2004) Cerebellum. In: *Gray's Anatomy.* Standring S, editor-in-chief. 39th ed., pp 353–368. Elsevier/Churchill Livingstone, London.

Flores LP (2002) Occipital lobe morphological anatomy: anatomical and surgical aspects. Arq Neuropsiquiatr 60:566–571.

Foerster O (1931) The cerebral cortex in man. Lancet 2:309–312.

Foote SL (1997) The primate locus coeruleus: the chemical neuroanatomy of the nucleus, its efferent projections, and its target receptors. In: *Handbook of Chemical Neuroanatomy.* Björklund A, Hökfelt T, eds. Vol 13: The Primate Nervous System, Part I, pp 187–215. Elsevier, Amsterdam.

Foreman RD, et al. (1976) Effects of dorsal column stimulation on primate spinothalamic tract neurons. J Neurophysiol 39:534–546.

Foreman RD, et al. (1976) Inhibition of primate spinothalamic tract neurons by electrical stimulation of dorsal column or peripheral nerve. In: *Advances in Pain Research and Therapy.* Bonica JJ, Albe-Fessard D, eds, 1:405–410. Raven Press, New York.

Foreman RD, Weber RN (1980) Responses from neurons of the primate spinothalamic tract to electrical stimulation of afferents from the cardiopulmonary region and somatic structures. Brain Res 186:463–468.

Fortin M, Asselin MC, Gould PV, Parent A (1998) Calretinin-immunoreactive neurons in the human thalamus. Neuroscience 84:537–548.

Fox PT, Miezin FM, Allman JM, Van Essen DC, Raichle ME (1987) Retinotopic organization of human visual cortex mapped with positron-emission tomography. J Neurosci 7:913–922.

Fox PT, Raichle ME (1984) Stimulus rate dependence of regional cerebral blood flow in human striate cortex,

demonstrated by positron emission tomography. J Neurophysiol 51:1109–1120.

Frahm HD, Stephan H, Baron G (1984) Comparison of brain structure volumes in insectivora and primates. J Hirnforsch 25:537–557.

Franková H (1968) Comparison of the occurrence and variability of joint receptors in rhesus monkey and man. Folia Morphol (Praha) 16:83–92.

Frazier JE (1923–1924) The nomenclature of diseased states caused by certain vestigial structures in the neck. Br J Surg 11:131–136.

Frederick JM, et al. (1982) Dopaminergic neurons in the human retina. J Comp Neurol 210:65–79.

Frederick JM, Rayborn ME, Hollyfield JG (1984) Glycinergic neurons in the human retina. J Comp Neurol 227:159–172.

Fredrickson JM (1970) Vestibular nerve projection to association fields of the cerebral cortex in the monkey. In: *Vestibular Function on Earth and in Space.* Stahle J, ed., pp 289–299. Pergamon Pr, Oxford.

Fredrickson JM, Figge U, Scheid P, Kornhuber HH (1966) Vestibular nerve projection to the cerebral cortex of the rhesus monkey. Exp Brain Res 2:318–327.

Fredrickson JM, Kornhuber HH, Schwarz DWF (1974) Cortical projections of the vestibular nerve. In: *Vestibular System.* Kornhuber HH, ed. Part 1: Basic Mechanisms. Handbook Sensory Physiology VI/1:565–582. Springer-Verlag, Berlin-Heidelberg.

Freed WJ (1985) Repairing neuronal circuits with brain grafts: where can brain grafts be used as a therapy? Neurobiol Aging 6:153–156.

French BN (1982) The embryology of spinal dysraphism. Clin Neurosurg 30:295–340.

Friede RL (1976) *Developmental Neuropathology.* Springer-Verlag, Vienna and New York.

Friedman DP, Jones EG (1981) Thalamic input to areas 3a and 2 in monkeys. J Neurophysiol 45:59–85.

Frigon A, Rossignol S (2006) Functional plasticity following spinal cord lesions. Prog Brain Res 157:231–260.

Frisèn L (1979) Quadruple sectoranopia and sectorial optic atrophy: a syndrome of the distal anterior choroidal artery. J Neurol Neurosurg Psychiatry 42:590–594.

Frisèn L, Holmegaard L, Rosencrantz M (1978) Sectorial optic atrophy and homonymous, horizontal sectoranopia: a lateral choroidal artery syndrome? J Neurol Neurosurg Psychiatry 41:374–380.

Frot M, Garcia-Larrea L, Guenot M, Mauguiere F (2001) Responses of the supra-sylvian (SII) cortex in humans to painful and innocuous stimuli. A study using intracerebral recordings. Pain 94:65–73.

Frot M, Mauguiere F (2003) Dual representation of pain in the operculo-insular cortex in humans. Brain 126:438–450.

Frot M, Rambaud L, Guenot M, Mauguiere F (1999) Intracortical recordings of early pain-related CO_2-laser evoked potentials in the human second somatosensory (SII) area. Clin Neurophysiol 110:133–145.

Fuchs AF, Kimm J (1975) Unit activity in vestibular nucleus of the alert monkey during horizontal angular acceleration and eye movement. J Neurophysiol 38:1140–1161.

Fuchs AF, Luschei ES (1971) The activity of single trochlear nerve fibers during eye movements in the alert monkey. Exp Brain Res 13:78–89.

Fujioka M (1976/77) Some observations of the projection of VPMpc from the so-called 'frontal opercular pure taste area'. Appl Neurophysiol 39:165–170.

Furstenberg AC, Magielski JE (1955) A motor pattern in the nucleus ambiguus. Its clinical significance. Ann Otol Rhinol Laryngol 64:788–793.

Fuster JM (2001) The prefrontal cortex – an update: time is of the essence. Neuron 30:319–333.

Gacek RR (1961) The efferent cochlear bundle in man. Arch Otolaryngol 74:690–694.

Gacek RR (1972) Neuroanatomy of the auditory system. In: Foundations of Modern Auditory Theory. Tobias JV, ed., 2:239–262. Academic Press, New York.

Gacek RR (1978) Further observations on posterior ampullary nerve transection for positional vertigo. Ann Otol Rhinol Laryngol 87:300–305.

Gacek RR, Malmgren LT, Lyon MJ (1977) Localization of adductor and abductor motor nerve fibers to the larynx. Ann Otol Rhinol Laryngol 86:770–776.

Gage FH, Björklund A (1985) Problems specifically related to grafting to the aged brain. Neurobiol Aging 6:163–164.

Gainer H (1978) Intercellular transfer of proteins from glial cells to axons. TINS 1:93–96.

Gajdusek DC (1985) Interference with axonal transport of neurofilament as a mechanism of pathogenesis underlying Alzheimer's disease and many other degenerations of the CNS. In: Normal Aging, Alzheimer's Disease and Senile Dementia. Aspects on Etiology, Pathogenesis, Diagnosis and Treatment. Gottfries CG, ed., pp 51–67. Brussels, Editions de l'Universite de Bruxelles.

Galaburda A, Sanides F (1980) Cytoarchitectonic organization of the human auditory cortex. J Comp Neurol 190:597–610.

Galaburda AM, Lemay M, Kemper TL, Geschwind N (1978) Right–left asymmetries in the brain. Science 199:852–856.

Galambos R (1966) Glial cells. In: Neurosciences Research Symposium Summaries. Schmitt FO, Melnechuk T, eds, 1:375–436. MIT Pr, Cambridge.

Ganderia SC, McCloskey DI: Joint sense, muscle sense, and their combination as position sense, measured at the distal interphalangeal joint of the middle finger. J Physiol 260:387–407.

Gandolfi A, Horoupian DS, De Teresa RM (1981) Quantitative and cytometric analysis of the ventral cochlear nucleus in man. J Neurol Sci 50:443–455.

Garant PR, Feldman J, Cho MI, Cullen MR (1980) Ultrastructure of Merkel cells in the hard palate of the squirrel monkey (Saimiri sciureus). Am J Anat 157:155–167.

Garcha HS, Ettlinger G (1978) The effects of unilateral or bilateral removals of the second somatosensory cortex (area SII): a profound tactile disorder in monkeys. Cortex 14:319–326.

Garcia-Cabezas MA, Rico B, Sanchez-Gonzalez MA, Cavada C (2006) Distribution of the dopamine innervation in the macaque and human thalamus. NeuroImage 34:965–984.

Garcia-Rill E (1997) Disorders of the reticular activating system. Med Hypotheses 49:379–387.

Gardner E, Cuneo HM (1945) Lateral spinothalamic tract and associated tracts in man. Arch Neurol Psychiatry 53:423–430.

Gardner E, Gray DJ, O'Rahilly R (1975) Anatomy: A Regional Study of Human Structure. 4th ed. Saunders, Philadelphia.

Gardner EP, Constanzo RM (1980) Spatial integration of multiple-point stimuli in primary somatosensory cortical receptive fields of alert monkeys. J Neurophysiol 43:420–443.

Gardner EP, Costanzo RM (1980) Temporal integration of multiple-point stimuli in primary somatosensory cortical receptive fields of alert monkeys. J Neurophysiol 43:444–468.

Gardner EP, Costanzo RM (1981) Properties of kinesthetic neurons in somatosensory cortex of awake monkeys. Brain Res 214:301–319.

Gardner WJ (1968) Myelocele: rupture of the neural tube? Clin Neurosurg 15:57–79.

Gardner WJ (1973) The dysraphic states from syringomyelia to anencephaly. Excerpta Medica, Amsterdam.

Gardner, E, O'Rahilly R, Prolo D (1975) The Dandy–Walker and Arnold–Chiari malformations. Clinical, developmental, and teratological considerations. Arch Neurol 32:393–407.

Garfia A (1980) Glomus tissue in the vicinity of the human carotid sinus. J Anat 130:1–12.

Gartner S, Henkind P (1981) Aging and degeneration of the human macula. 1. Outer nuclear layer and photoreceptors. Br J Ophthalmol 65:23–28.

Gash DM, Collier TJ, Sladek JR (1985) Neural transplantation: a review of recent developments and potential applications to the aged brain. Neurobiol Aging 6:131–150.

Gasser HS, Erlanger J (1927) The rôle played by the sizes of the constituent fibers of a nerve trunk in determining the form of its action potential wave. Am J Physiol 80:522–547.

Gelnar PA, Krauss BR, Szeverenyi NM, Apkarian AV (1998) Fingertip representation in the human somatosensory cortex: an fMRI study. NeuroImage 7:261–283.

Gelfan S, Carter S (1967) Muscle sense in man. Exp Neurol 18:469–473.

Geller A, Gilles F, Shwachman H (1977) Degeneration of fasciculus gracilis in cystic fibrosis. Neurology 27:185–187.

Gellis M, Pool R (1977) Two-point discrimination distances in the normal hand and forearm. Application to various methods of fingertip reconstructions. Plast Reconstr Surg 59:57–63.

Gelnar PA, Krauss BR, Szeverenyi NM, Apkarian AV (1998) Fingertip representation in the human somatosensory cortex: an fMRI study. NeuroImage 7:261–283.

Geniec P, Morest DK (1971) The neuronal architecture of the human posterior colliculus: a study with the Golgi method. Acta Otolaryngol (Suppl) (Stockh) 295:5–32.

Gent JF, Hettinger TP, Frank ME, Marks LE (1999) Taste confusions following gymnemic acid rinse. Chem Senses 24:393–403.

George WC (1942) A presomite human embryo with chorda canal and prochordal plate. Contrib Embryol Carneg Instn 30:1–7.

Gerard RW (1923) Afferent impulses of the trigeminal nerve. The intramedullary course of the painful, thermal and tactile impulses. Arch Neurol Psychiatry 9:306–338.

Gerber MR, Connor JR (1989) Do oligodendrocytes mediate iron regulation in the human brain? Ann Neurol 26:95–98.

Gerebtzoff MA (1975) Localisation et organisation soma-totopique des colonnes sensitives, motrices et visceromotrices des nerfs craniens. Acta Otorhinolaryngol Belg 29:873–888.

Gerhart KD, Wilcox TK, Chung JM, Willis WD (1981) Inhibition of nociceptive and nonnociceptive responses of primate spinothalamic cells by stimulation in medial brain stem. J Neurophysiol 45:121–136.

German DC, Bowden DM (1975) Locus ceruleus in rhesus monkey (*Macaca mulatta*): a combined histochemical fluorescence, Nissl and silver study. J Comp Neurol 161:19–29.

Gershenbaum MR, Roisen FJ (1978) A scanning electron microscopic study of peripheral nerve degeneration and regeneration. Neuroscience 3:1241–1250.

Getz B, Sirnes T (1949) The localization within the dorsal motor vagal nucleus. An experimental investigation. J Comp Neurol 90:95–110.

Geyer S, Ledberg A, Schleicher A, Kinomura S, Schormann T, Burgel U, Klingberg T, Larsson J, Zilles K, Roland PE (1996) Two different areas within the primary motor cortex of man. Nature 382:805–807.

Geyer S, Schleicher A, Zilles K (1999) Areas 3a, 3b, and 1 of human primary somatosensory cortex. 1. Microstructural organization and interindividual variability. NeuroImage 10:63–83.

Geyer S, Schormann T, Mohlberg H, Zilles K (2000) Areas 3a, 3b, and 1 of human primary somatosensory cortex. Part 2. Spatial normalization to standard anatomical space. NeuroImage 11:684–696.

Ghabriel MN, Allt G (1981) Incisures of Schmidt–Lantermann. Prog Neurobiol 17:25–58.

Gibo H, et al. (1981) Microsurgical anatomy of the middle cerebral artery. J Neurosurg 54:151–169.

Gibo H, Carver CC, Rhoton AL Jr, Lenkey C, Mitchell RJ (1981) Microsurgical anatomy of the middle cerebral artery. J Neurosurg 54:151–169.

Gibo H, Lenkey C, Rhoton AL Jr (1981) Microsurgical anatomy of the supraclinoid portion of the internal carotid artery. J Neurosurg 55:560–574.

Giesler GL Jr, et al. (1981) Postsynaptic inhibition of primate spinothalamic neurons by stimulation of nucleus raphe magnus. Brain Res 204:184–188.

Giesler GL Jr, Liebeskind JC (1976) Inhibition of visceral pain by electrical stimulation of the periaqueductal gray matter. Pain 2:43–48.

Gildenberg PL (1976/1977) Percutaneous cervical cordotomy. Appl Neurophysiol 39:97–113.

Gildenberg PL, Murthy KSK (1980) Influence of dorsal column stimulation upon human thalamic somatosensory-evoked potential. Appl Neurophysiol 43:8–17.

Gilissen E, Iba-Zizen MT, Stievenart JL, Lopez A, Trad M, Cabanis EA, and Zilles K (1995) Is the length of the calcarine sulcus associated with the size of the human visual cortex? A morphometric study with magnetic resonance tomography. J Hirnforsch 36:451–459.

Gillilan LA (1958) The arterial blood supply of the human spinal cord. J Comp Neurol 110:75–103.

Gillilan LA (1964) The correlation of the blood supply to the human brain stem with clinical brain stem lesions. J Neuropathol Exp Neurol 23:78–108.

Gillilan LA (1970) Veins of the spinal cord. Anatomic details: suggested clinical applications. Neurology 20:860–868.

Gillilan LA (1972) Anatomy and embryology of the arterial system of the brain stem and cerebellum. In: *Handbook of Clinical Neurology.* Vinken PJ, Bruyn GW, eds, 11:24–44. Elsevier, Amsterdam.

Girgis M (1970) The rhinencephalon. Acta Anat (Basel) 76:157–199.

Giroud A (1977) Anencephaly. In: *Handbook of Clinical Neurology.* Vinken PJ, Bruyn GW, eds. Vol 30: Congenital Malformations of the Brain and Skull, Part I, pp 173–208. North-Holland Publishing Co, Amsterdam.

Giulian D, Vaca K, Noonan CA (1990) Secretion of neurotoxins by mononuclear phagocytes infected with HIV-1. Science 250:1593–1596.

Glees P (1953) The central pain tract. (Tractus spino-thalamicus.) Acta Neurovegetativa 7:160–174.

Glees P, Hasan M (1990) Ultrastructure of human cerebral macroglia and microglia: maturing and hydrocephalic frontal cortex. Neurosurg Rev 13:231–242.

Glickstein M (2006) Thinking about the cerebellum. Brain 129:288–290.

Glickstein M (1988) The discovery of the visual cortex. Sci Am 256:118–127.

Glickstein M, Whitteridge D (1987) Tatsuji Inouye and the mapping of the visual fields on the human cerebral cortex. TINS 10:350–353.

Gloor P (1997) *The Temporal Lobe and Limbic System.* Oxford University Press, New York.

Glueckert R, Pfaller K, Kinnefors A, Rask-Andersen H, Schrott-Fischer A (2005) Ultrastructure of the normal human organ of Corti. New anatomical findings in surgical specimens. Acta Otolaryngol 125:534–539.

Glueckert R, Pfaller K, Kinnefors A, Schrott-Fischer A, Rask-Andersen H (2005) High resolution scanning electron microscopy of the human organ of Corti. A study using freshly fixed surgical specimens. Hear Res 199:40–56.

Gluhbegovic N, Williams TH (1980) *The Human Brain: A Photographic Guide.* Harper & Row, Hagerstown.

Goldberg JM (1981) Peripheral vestibular receptors: functional aspects. Am J Otol 3:68–69.

Goldberg JM, Fernández C (1971) Physiology of peripheral neurons innervating semicircular canals of the squirrel monkey. I. Resting discharge and response to constant angular accelerations. J Neurophysiol 34:635–660.

Goldberg JM, Fernández C (1975) Vestibular mechanisms. Annu Rev Physiol 37:129–162.

Goldberg JM, Fernández C (1980) Efferent vestibular system in the squirrel monkey: anatomical location and influence on afferent activity. J Neurophysiol 43:986–1025.

Goldberg JM, Lindblom U (1979) Standardized method of determining vibratory perception thresholds for diagnosis and screening in neurological investigation. J Neurol Neurosurg Psychiat 42:793–803.

Goldblatt D (1975) Nutritional neurological disease. In: *The Nervous System.* Tower DB, editor-in-chief. Chase TN, volume ed. Vol 2: The Clinical Neurosciences, pp 387–393. Raven Press, New York.

Goldenberg PZ, Troiano RA, Kwon EE, Prineas JW (1990) Sera from MS patients and normal controls opsonize myelin. Neurosci Lett 109:353–356.

Goldman JE, Yen S-H (1986) Cytoskeletal protein abnormalities in neurodegenerative diseases. Ann Neurol 19:209–233.

Goldstein DS (2001) *The Autonomic Nervous System in Health and Disease.* Marcel Dekker, New York.

Goldstein DS, Robertson D, Esler M, Straus SE, Eisenhofer G (2002) Dysautonomias: clinical disorders of the autonomic nervous system. Ann Intern Med 137:753–763.

Goligher JC, Hughes ESR (1951) Sensibility of the rectum and colon. Its rôle in the mechanism of anal continence. Lancet, March 10, 543–548.

Goodale MA, Milner AD (1992) Separate visual pathways for perception and action. Trends Neurosci 15:20–25.

Goodale MA, Westwood DA (2004) An evolving view of duplex vision: separate but interacting cortical pathways for perception and action. Curr Opin Neurobiol 14:203–211.

Goodwin GM (1976) The sense of limb position and movement. Exerc Sport Sci Rev 4:87–124.

Goodwin GM, Luschei ES (1974) Effects of destroying spindle afferents from jaw muscles on mastication in monkeys. J Neurophysiol 37:967–981.

Goodwin GM, McCloskey DI, Matthews PBC (1972) The persistence of appreciable kinesthesia after paralysing joint afferents but preserving muscle afferents. Brain Res 37:326–329.

Goodwin PE, Johnson RM (1980) The loudness of tinnitus. Acta Otolaryngol (Stockh) 90:353–359.

Gorski RA, Gordon JH, Shryne JE, Southam AM (1978) Evidence for a morphological sex difference within the medial preoptic area of the rat brain. Brain Res 148:333–346.

Grafstein B, Forman DS (1980) Intracellular transport in neurons. Physiol Rev 60:1167–1283.

Grafton ST, Woods RP, Mazziotta JC (1993) Within-arm somatotopy in human motor areas determined by positron emission tomography imaging of cerebral blood flow. Exp Brain Res 95:172–176.

Graham J, Greenwood R, Lecky B (1980) Cortical deafness. A case report and review of the literature. J Neurol Sci 48:35–49.

Gramsch C, et al. (1979) Regional distribution of methionine-enkephalin- and beta-endorphin-like immunoreactivity in human brain and pituitary. Brain Res 171:261–170.

Granchrow D, Ornoy A (1979) Possible evidence for secondary degeneration of central nervous system in the pathogenesis of anencephaly and brain dysraphia. A study in young human fetuses. Virchow Arch (Pathol Anat) 384:285–294.

Grand W (1980) Microsurgical anatomy of the proximal middle cerebral artery and the internal carotid artery bifurcation. Neurosurgery 7:215–218.

Grant G (1966) Infarction localization in a case of Wallenberg's Syndrome. A neuroanatomical investigation with comments on structures responsible for nystagmus, impairment of taste and deglutition. J Hirnforsch 8:419–430.

Gray EG (1970) The fine structure of nerve. Comp Biochem Physiol 36:419–448.

Gray H (1985) *Anatomy of the Human Body.* 30th ed. Clemente CD, ed. Lea & Febiger, Philadelphia.

Graybiel A (1974) Measurement of otolith function in man. In: *Vestibular System.* Kornhuber HH, ed. Part 2: Psychophysics, Applied Aspects and General Interpretations, Handbook Sensory Physiol VI/2:233–266. Springer-Verlag, Berlin-Heidelberg.

Graybiel AM (1977) Organization of oculomotor pathways in the cat and rhesus monkey. In: *Control of Gaze in Brain Stem Neuro. Developments in Neuroscience.* Baker R, Berthoz A, eds, 1:79–88. North-Holland/Elsevier, Amsterdam.

Graybiel AM (2000) The basal ganglia. Curr Biol 10:R509–11.

Graybiel AM (2004) Network-level neuroplasticity in cortico-basal ganglia pathways. Parkinsonism Relat Disord 10:293–296.

Graybiel AM (2005) The basal ganglia: learning new tricks and loving it. Curr Opin Neurobiol 15:638–644.

Graybiel AM, Ragsdale CW Jr (1982) Pseudocholinesterase staining in the primary visual pathway of the macaque monkey. Nature 299:439–442.

Graziano A, Jones EG (2004) Widespread thalamic terminations of fibers arising in the superficial medullary dorsal horn of monkeys and their relation to calbindin immunoreactivity. J Neurosci 24:248–256.

Green AL, Wang S, Owen SL, Xie K, Liu X, Paterson DJ, Stein JF, Bain PG, Aziz TZ (2005) Deep brain stimulation can regulate arterial blood pressure in awake humans. NeuroReport 16:1741–1745.

Green GJ, Lessell S (1977) Acquired cerebral dyschromatopsia. Arch Ophthalmol 95:121–128.

Greenberg JH, et al. (1981) Metabolic mapping of functional activity in human subjects with the [18F]fluorodexyglucose technique. Science 212:678–680.

Gregg JM, Banerjee T, Ghia JN, Campbell R (1978) Radiofrequency thermoneurolysis of peripheral nerves for control of trigeminal neuralgia. Pain 5:231–243.

Griesen O, Neergaard EB (1975) Middle ear reflex activity in the startle reaction. Arch Otolaryngol 101:348–353.

Grisar TM (1986) Neuron–glia relationships in human and experimental epilepsy: a biochemical point of view. Adv Neurol 44:1045–1073.

Grodd W, Hulsmann E, Ackermann H (2005) Functional MRI localizing in the cerebellum. Neurosurg Clin N Am 16:77–99.

Grodd W, Hulsmann E, Lotze M, Wildgruber D, Erb M (2001) Sensorimotor mapping of the human cerebellum: fMRI evidence of somatotopic organization. Hum Brain Mapp 13:55–73.

Grodzinsky Y, Amunts K (2006) *Broca's Region.* Oxford University Press, New York.

Gross RE, Jones EG, Dostrovsky JO, Bergeron C, Lang AE, Lozano AM (2004) Histological analysis of the location of effective thalamic stimulation for tremor. Case report. J Neurosurg 100:547–552.

Gruetter R, et al. (1992) Direct measurement of brain glucose concentration in humans by 13C NMR spectroscopy. Proc Natl Acad Sci 89:1109–1112.

Gudmundsson K, Rhoton AL Jr, Rushton JG (1971) Detailed anatomy of the intracranial portion of the trigeminal nerve. J Neurosurg 35:592–600.

Guillemin R (2005) Hypothalamic hormones a.k.a. hypothalamic releasing factors. J Endocrinol 184:11–28.

Gussen R (1970) Pacinian corpuscles in the middle ear. J Laryngol Otol 84:71–76.

Guth L (1975) History of central nervous system regeneration research. Exp Neurol 48:3–15.

Guttmann L (1973) *Spinal Cord Injuries: Comprehensive Management and Research*. Blackwell, Oxford.

Ha H (1971) Cervicothalamic tract in the rhesus monkey. Exp Neurol 33:205–212.

Ha H, Wu RS, Contreras RA, Tan E-C (1978) Measurement of pain threshold by electrical stimulation of tooth pulp afferents in the monkey. Exp Neurol 61:260–269.

Haas A, Flammer J, Schneider U (1986) Influence of age on the visual fields of normal subjects. Am J Ophthalmol 101:199–203.

Hadjikhani N, de Gelder B (2002) Neural basis of prosopagnosia: an fMRI study. Hum Brain Mapp 16:176–182.

Hadley K, Orlandi RR, Fong KJ (2004) Basic anatomy and physiology of olfaction and taste. Otolaryngol Clin North Am 37:1115–1126.

Hadziselimovic H, Dilberovic F (1979) The hippocamp and the amygdaloid body in man. J Hirnforsch 20:263–268.

Hagbarth K-E (1979) Exteroceptive, proprioceptive, and sympathetic activity recorded with microelectrodes from human peripheral nerves. Mayo Clin Proc 54:353–365.

Haines DE, Harkey HL, al-Mefty O (1993) The 'subdural' space: a new look at an outdated concept. Neurosurgery 32:111–120.

Halgren E, Babb TL, Rausch R, Crandall PH (1977) Neurons in the human basolateral amygdala and hippocampal formation do not respond to odors. Neurosci Lett 4:331–335.

Hall AJ (1936) Some observations on the acts of closing and opening the eyes. Br J Ophthalmol 20:257–295.

Hall JG (1964) The cochlea and the cochlear nuclei in neonatal asphyxia. Acta Oto-Laryngol 59:(Suppl 194):1–93.

Hall SM (1978) The Schwann cell: a reappraisal of its role in the peripheral nervous system. Neuropathol Appl Neurobiol 4:165–176.

Halliday AM, Logue V (1972) Painful sensations evoked by electrical stimulation in the thalamus. In: *Neurophysiology Studied in Man*. Somjen GG, ed., pp 221–230. Excerpta Medica, Amsterdam.

Hallin RG, Torebjörk HE (1974) Activity in unmyelinated nerve fibers in man. Adv Neurol 4:19–27.

Halsey JH Jr, Allen N, Chamberlin HR (1971) The morphogenesis of hydranencephaly. J Neurol Sci 12:187–217.

Hamburger V (1975) Changing concepts in developmental neurobiology. Perspect Biol Med 18:162–178.

Hansen LA, Armstrong DM, Terry RD (1987) An immunohistochemical quantification of fibrous astrocytes in the aging human cerebral cortex. Neurobiol Aging 8:1–6.

Harbaugh RE (1985) Neural transplantation vs. central neurotransmitter augmentation in diseases of the nervous system. Neurobiol Aging 6:164–166.

Harding A, Halliday G, Caine D, Kril J (2000) Degeneration of anterior thalamic nuclei differentiates alcoholics with amnesia. Brain 123:141–154.

Hardman CD, McRitchie DA, Halliday GM, Cartwright HR, Morris JG (1996) Substantia nigra pars reticulata neurons in Parkinson's disease. Neurodegeneration 5:49–55.

Hardy J, et al. (1985) Transmitter deficits in Alzheimer's disease. Neurochem Int 7:545–563.

Hardy SGP, Leichnetz GR (1981) Cortical projections to the periaqueductal gray in the monkey: a retrograde and orthograde horseradish peroxidase study. Neurosci Lett 22:97–101.

Hardy TL, Bertrand G, Thompson CJ (1980) Organization and topography of sensory responses in the internal capsule and nucleus ventralis caudalis found during stereotactic surgery. Appl Neurophysiol 42:335–351.

Harkins SW, Chapman CR (1978) Cerebral evoked potentials to noxious dental stimulation: relationship to subjective pain report. Psychophysiol 15:248–252.

Harner SG, Laws ER Jr (1983) Clinical findings in patients with acoustic neurinoma. Mayo Clin Proc 58:721–728.

Haroutunian V, Davis KL (1985) Commentary on review by Gash et al. A clinical perspective on neural transplantation and potential applications to the aged brain. Neurobiol Aging 6:166–167.

Harrington DO (1981) *The Visual Fields: a Textbook and Atlas of Clinical Perimetry*. 5th ed. Mosby, St. Louis, pp 108–109.

Harris F, Jabbur SJ, Morse RW, Towe AL (1965) Influence of the cerebral cortex on the cuneate nucleus of the monkey. Nature 208:1215–1216.

Harris FS, Rhoton AL Jr (1976) Anatomy of the cavernous sinus. A microsurgical study. J Neurosurg 45:169–180.

Harris RE (1974) Viral teratogenesis: a review with experimental and clinical perspectives. Am J Obstet Gynecol 111:996–1008.

Hashimoto K (1972) Fine structure of Merkel cell in human oral mucosa. J Invest Dermatol 58:381–387.

Hassler O (1966) Blood supply to human spinal cord. A microangiographic study. Arch Neurol 15:302–307.

Hassler R (1959) Anatomy of the thalamus. In: *Introduction to Stereotaxis with an Atlas of the Human Brain*. Schaltenbrand G, Bailey P, eds, 1:230–290. Thieme, Stuttgart.

Hassler R (1970) Dichotomy of facial pain conduction in the diencephalon. In: *Trigeminal Neuralgia: Pathogenesis and Pathophysiology*, pp 123–138. Thieme, Stuttgart.

Hassler R (1982) Architectonic organization of the thalamic nuclei. In: *Stereotaxy of the Human Brain*. Schaltenbrand G, Walker AE, eds. 2nd ed., pp 140–180. Thieme, Stuttgart.

Haugen FP (1968) The autonomic nervous system and pain. Anesthesiology 29:785–792.

Haughton VM, Syvertsen A, Williams AL (1980) Soft-tissue anatomy within the spinal canal as seen on computed tomography. Radiology 134:649–655.

Hawkins JE Jr, Johnsson L-G (1976) Patterns of sensorineural degeneration in human ears exposed to noise. In: *Effects of Noise on Hearing – Critical Issues*. Henderson D, Hamernik RP, eds., pp 91–110. Raven Press, New York.

Hawrylyshyn PA, et al. (1978) Vestibulothalamic projections in man – a sixth primary sensory pathway. J Neurophysiol 41:394–401.

Hawrylyshyn PA, Rubin AM, Tasker RR, Organ LW, Fredrickson JM (1978) Vestibulothalamic projections in man – a sixth primary sensory pathway. J Neurophysiol 41:394–401.

Hayes GM, Woodroofe MN, Cuzner ML (1987) Microglia are the major cell type expressing MHC Class II in human white matter. J Neurol Sci 80:25–37.

Hayes NL, Rustioni A (1980) Spinothalamic and spinomedullary neurons in macaques: a single and double retrograde tracer study. Neuroscience 5:861–874.

Haymaker W (1969) *Bing's Local Diagnosis in Neurological Diseases*. 15th ed. Mosby, Saint Louis.

Haymaker W, Anderson E, Nauta WJH (1969) *The Hypothalamus*. Thomas, Springfield.

Hayward JN (1977) Functional and morphological aspects of hypothalamic neurons. Physiol Rev 57:574–658.

Hazell JWP (1980) Medical and audiological findings in subjective tinnitus. Clin Otolaryngol 5:75.

Heath CJ (1978) The somatic sensory neurons of pericentral cortex. Int Rev Physiol 17:193–237.

Hécaen H, Perenin MT, Jeannerod M (1984) The effects of cortical lesions in children: language and visual functions. In: *Early Brain Damage*. Almli CR, Finger S, eds. Vol 1: Research Orientations and Clinical Observations, pp 277–298. Academic Press, New York.

Heffner H, Masterton B (1975) Contribution of auditory cortex to sound localization in the monkey (Macaca mulatta). J Neurophysiol 38:1340–1358.

Heffner HE (1987) Ferrier and the study of auditory cortex. Arch Neurol 44:218–221.

Heffner HE, Heffner RS (1986) Hearing loss in Japanese macaques following bilateral auditory cortex lesions. J Neurophysiol 55:256–271.

Heidel KM, Benarroch EE, Gene R, Klein F, Meli F, Saadia D, Nogues MA (2002) Cardiovascular and respiratory consequences of bilateral involvement of the medullary intermediate reticular formation in syringobulbia. Clin Auton Res 12:450–456.

Heilman KM, Valenstein E (1993) *Clinical Neuropsychology*. 3rd ed. Oxford University Press, New York.

Helgason C, Caplan LR, Goodwin J, Hedges T III (1986) Anterior choroidal artery – territory infarction: report of cases and review. Arch Neurol 43:681–686.

Hellwig B (2002) Cyto- and myeloarchitectonics: their relationship and possible functional significance. In: *Cortical Areas: Unity and Diversity*. Schüz A, Miller R, eds, pp 15–28. Taylor & Francis, London and New York.

Hemmer R (1971) Meningoceles and myeloceles. Prog Neurol Surg 4:192–226.

Henkin RI (1976) Taste. In: *Scientific Foundations of Otolaryngology*. Hinchcliffe R, Harrison D, eds, pp 468–483. William Heinemann Medical Books Ltd, London.

Henkind P, Gottlieb MB (1973) Bilateral internal ophthalmoplegia in a patient with sarcoidosis. Br J Ophthalmol 57:792–796.

Henn FA (1976) Neurotransmission and glial cells: a functional relationship? J Neurosci Res 2:271–282.

Henn FA, Haljamäe, Hamberger A (1972) Glial cell function: active control of extracellular K+ concentration. Brain Res 43:437–443.

Henn V, Cohen B (1972) Eye muscle motor neurons with different functional characteristics. Brain Res 45:561–568.

Henn V, Cohen B (1976) Coding of information about rapid eye movements in the pontine reticular formation of alert monkeys. Brain Res 108:307–325.

Henry JA, Montuschi E (1978) Cardiac pain referred to site of previously experienced somatic pain. Br Med J 2:1605–1606.

Hensel H (1974) Thermoreceptors. Annu Rev Physiol 36:233–249.

Hensel H, Boman KKA (1960) Afferent impulses in cutaneous sensory nerves in human subjects. J Neurophysiol 23:564–578.

Hering H, Sheng M (2001) Dendritic spines: structure, dynamics and regulation. Nat Rev Neurosci 2:880–888.

Herman JP, Ostrander MM, Mueller NK, Figueiredo H (2005) Limbic system mechanisms of stress regulation: hypothalamo-pituitary-adrenocortical axis. Prog Neuropsychopharmacol Biol Psychiatry 29:1201–1213.

Herman LH, Fernando OU, Gurdjian ES (1966) The anterior choroidal artery: an anatomical study of its area of distribution. Anat Rec 154:95–101.

Herrero MT, Barcia C, Navarro JM (2002) Functional anatomy of thalamus and basal ganglia. Childs Nerv Syst 18:386–404.

Hertig AT, Rock J (1941) Two human ova of the pre-villous stage, having an ovulation age of about eleven and twelve days respectively. Contrib Embryol Carneg Instn 29:127–156.

Hertig AT, Rock J (1945) Two human ova of the pre-villous stage, having a development age of about seven and nine days respectively. Contrib Embryol Carneg Instn 31:65–84.

Hertig AT, Rock J (1949) Two human ova of the pre-villous state, having a developmental age of about eight and nine days respectively. Contrib Embryol Carneg Instn 33:169–186.

Hertig AT, Rock J, Adams EC (1956) A description of 34 human ova within the first 17 days of development. Am J Anat 98:435–493.

Hertig AT, Rock J, Adams EC, Mulligan WJ (1954) On the preimplantation stages of the human ovum: A description of four normal and four abnormal specimens ranging from the second to the fifth day of development. Contrib Embryol Carneg Instn 35:199–220.

Hertz L (1981) Functional interactions between astrocytes and neurons. Prog Clin Biol Res 59A:45–58.

Herzog AG, Kemper TL (1980) Amygdaloid changes in aging and dementia. Arch Neurol 37:625–629.

Hestiantoro A, Swaab DF (2004) Changes in estrogen receptor-alpha and -beta in the infundibular nucleus of the human hypothalamus are related to the occurrence of Alzheimer's disease neuropathology. J Clin Endocrinol Metab 89:1912–1925.

Heuser CH (1930) A human embryo with 14 pairs of somites. Contrib Embryol Carneg Instn 22:135–153.

Heuser CH (1932) A presomite human embryo with a definite chorda canal. Contrib Embryol Carneg Instn 23:251–257.

Heuser CH, Corner GW (1957) Developmental horizons in human embryos. Description of age group X-4 to 12 somites. Contrib Embryol Carneg Instn 36:85–99.

Heuser CH, Rock J., Hertig AT (1945) Two human embryos showing early stages of the definitive yolk sac. Contrib Embryol Carneg Instn 31:85–99.

Hickey TL (1977) Postnatal development of the human lateral geniculate nucleus: relationship to a critical period for the visual system. Science 198:836–838.

Hickey TL, Guillery RW (1979) Variability of laminar patterns in the human lateral geniculate nucleus. J Comp Neurol 183:221–246.

Hickey TL, Guillery RW (1981) A study of Golgi preparations from the human lateral geniculate nucleus. J Comp Neurol 200:545–577.

Hilaire G, Pasaro R (2003) Genesis and control of the respiratory rhythm in adult mammals. News Physiol Sci 18:23–28.

Hinchcliffe R, Harrison D (1976) *Scientific Foundations of Otolaryngology.* Year Bk Med, Chicago.

Hinojosa R, Seligsohn R, Lerner SA (1985) Ganglion cell counts in the cochleae of patients with normal audiograms. Acta Otolaryngol (Stockh) 99:8–13.

Hinton DR, Sadun AA, Blancks JC, Miller CA (1986) Optic-nerve degeneration in Alzheimer's disease. N Engl J Med 315:485–487.

Hirai T, Jones EG (1989) A new parcellation of the human thalamus on the basis of histochemical staining. Brain Res Rev 14:1–34.

Hirano A (1978) Neuronal and glial processes in neuropathology. J Neuropathol Exp Neurol 37:365–374.

Hirano A (1981) Structure of normal central myelinated fibers. In: *Demyelinating Disease: Basic and Clinical Electrophysiology.* Waxman SG, Ritchie JM, eds, pp 51–68. Raven Press, New York.

Hirshberg RM, Al-Chaer ED, Lawand NB, Westlund KN, Willis WD (1996) Is there a pathway in the posterior funiculus that signals visceral pain? Pain 67:291–305.

Hitchcock PG, Hickey TL (1980) Ocular dominance columns: evidence for their presence in humans. Brain Res 182:176–179.

Hockaday JM, Whitty CWM (1967) Patterns of referred pain in the normal subject. Brain 90:481–496.

Hofman MA (1997) Lifespan changes in the human hypothalamus. Exp Gerontol 32:559–575.

Hofman MA, Goudsmit E, Purba JS, Swaab DF (1990) Morphometric analysis of the supraoptic nucleus in the human brain. J Anat 172:259–270.

Hogg ID (1966) Observations of the development of the nucleus of Edinger–Westphal in man and the albino rat. J Comp Neurol 126:567–584.

Hökfelt T, Johansson O, Goldstein M (1984) Chemical anatomy of the brain. Science 225:1326–1334.

Hollyfield JG, Frederick JM, Rayborn ME (1983) Neurotransmitter properties of the newborn human retina. Invest Ophthalmol Vis Sci 24:893–897.

Holmgren I (1982) Synaptic organization of the dopaminergic neurons in the retina of the cynomolgus monkey. Invest Ophthalmol Vis Sci 22:8–24.

Honrubia V, Brazier MAB (1982) *Nystagmus and Vertigo: Clinical Approaches to the Patient with Dizziness.* UCLA Forum in Medical Sciences. Number 24. Academic Press, New York.

Hood JD (1962) Bone conduction: A review of the present position with especial reference to the contributions of Dr. Georg von Békésy. J Acoust Soc Am 34:1325–1332.

Hore J, Preston JB, Durkovic RG, Cheney PD (1976) Responses of cortical neurons (areas 3a and 4) to ramp stretch of hindlimb muscles in the baboon. J Neurophysiol 39:484–500.

Horn AKE, Büttner-Ennever JA, Suzuki Y, Henn V (1995) Histological identification of premotor neurons for horizontal saccades in monkey and man by parvalbumin immunostaining. J Comp Neurol 359:350–363.

Horne MK, Tracey DJ (1979) The afferents and projections of the ventroposterolateral thalamus in the monkey. Exp Brain Res 36:129–141.

Hornung JP (2003) The human raphe nuclei and the serotonergic system. J Chem Neuroanat 26:331–343.

Horton JC (1984) Cytochrome oxidase patches: a new cytoarchitectonic feature of monkey visual cortex. Philos Trans R Soc Lond B 304:199–253.

Horton JC, Hedley-Whyte ET (1984) Mapping of cytochrome oxidase patches and ocular dominance columns in human visual cortex. Philos Trans R Soc Lond B 304:255–272.

Horton JC, Hubel DH (1981) A regular patchy distribution of cytochrome oxidase staining in primary visual cortex of the macaque monkey. Nature 292:762–764.

Hoshi E, Tremblay L, Feger J, Carras PL, Strick PL (2005) The cerebellum communicates with the basal ganglia. Nat Neurosci 8:1491–1493.

Hoshino T, Kodama A (1979) Nerve supply to the inner sensory cells in a human cochlea. Arch Otorhinolaryngol 222:257–263.

Hosobuchi Y (1980) The majority of unmyelinated afferent axons in human ventral roots probably conduct pain. Pain 8:167–180.

Hosobuchi Y, Adams JE, Linchitz R (1977) Pain relief by electrical stimulation of the central gray matter in humans and its reversal by naloxone. Science 197:183–186.

Houk JC, Crago PE, Rymer WZ (1980) Functional properties of the Golgi tendon organs. Desmedt JE, ed. Basel, Karger, Prog Clin Neurophysiol 8:33–43.

House WF, Urban J (1973) Long term results of electrode implantation and electronic stimulation of the cochlea in man. Ann Otol Rhinol Laryngol 82:504–517.

Howard MA, Volkov IO, Mirsky R, Garell PC, Noh MD, Granner M, Damasio H, Steinschneider M, Reale RA, Hind JE, Brugge JF (2000) Auditory cortex on the human posterior superior temporal gyrus. J Comp Neurol 416:79–92.

Howe A, Neil E (1972) Arterial chemoreceptors. In: *Handbook of Sensory Physiology.* Neil E, volume ed., III/1:47–80. Springer-Verlag, New York.

Howell P, Marchbanks RJ, El-Yaniv N (1986) Middle ear muscle activity during vocalization in normal speakers and stutters. Acta Otolaryngol 102:396–402.

Hoyt WF (1969) Correlative functional anatomy of the optic chiasm. Clin Neurosurg 17:189–208.

Hu JW, Sessle BJ (1979) Trigeminal nociceptive and non-nociceptive neurones: brain stem intranuclear projections and modulation by orofacial, periaqueductal gray and nucleus raphe magnus stimuli. Brain Res 170:547–552.

Huang XF, Paxinos G (1995) Human intermediate reticular zone: a cyto- and chemoarchitectonic study. J Comp Neurol 360:571–588.

Hubbard JH (1972) The quality of nerve regeneration. Factors independent of the most skillful repair. Surg Clin North Am 52:1099–1108.

Hubel DH, Wiesel TN (1972) Laminar and columnar distribution of geniculo-cortical fibers in the macaque monkey. J Comp Neurol 146:421–450.

Hubel DH, Wiesel TN (1979) Brain mechanisms of vision. Sci Am 241:150–162.

Huber A (1962) Homonymous hemianopia after occipital lobectomy. Am J Ophthalmol 54:623–629.

Hudspeth AJ (2005) How the ear's works work: mechano-electrical transduction and amplification by hair cells. C R Biol 328:155–162.

Hufschmidt H-J, Spuler H (1962) Mono- and polysynaptic reflexes of the trigeminal muscles in human beings. J Neurol Neurosurg Psychiatry 25:332–335.

Hughes J, et al. (1975) Identification of two related pentapeptides from the brain with potent opiate agonist activity. Nature 258:577–579.

Hughes JR (1969) Neurophysiological mechanisms within the olfactory bulb. Int Congr Series No 206:12–22.

Hughes JR (1969) Neurophysiological mechanisms within the olfactory bulb. Excerpta Med Int Congr Series 206:12–22.

Hughes JR, Andy OJ (1979) The human amygdala. 1. Electrophysiological responses to odorants. Electroencephalogr Clin Neurophysiol 46:428–443.

Hughes JR, et al. (1972) Correlations between electrophysiological and subjective responses to odorants as recorded from the olfactory bulb, tract and amygdala of waking man. In: Neurophysiology Studied in Man. Somjen GG, ed., pp 260–279. Intern Congr Series No 253. Excerpta Medica, Amsterdam.

Huk AC, Dougherty RF, Heeger DJ (2002) Retinotopy and functional subdivision of human areas MT and MST. J Neurosci 22:7195–7205.

Hulshof JH (1986) The loudness of tinnitus. Acta Otolaryngol 102:40–43.

Humphrey T (1982) The central relations of the trigeminal nerve. In: Correlative Neurosurgery. Schneider RC, Kahn EA, Crosby EC, Taren JA, eds. 3rd ed., 2:1518–1532. Thomas, Springfield.

Humphrey T, Crosby EC (1938) The human olfactory bulb. Univ Hosp Bull Ann Arbor 4:61–62.

Hunt JR (1937) Geniculate neuralgia (neuralgia of the nervus facialis). A. Further contribution to the sensory system of the facial nerve and its neuralgic conditions. Arch Neurol Psychiat 37:253–285.

Hurlbert A (2003) Colour vision: primary visual cortex shows its influence. Curr Biol 13:R270–R272.

Hurst EM (1959) Some cortical association systems related to auditory functions. J Comp Neurol 112:103–119.

Hurvich LM (1981) Color Vision. Sinuaer Assocs, Sunderland.

Hutchins JB (1987) Review: acetylcholine as a neurotransmitter in the vertebrate retina. Exp Eye Res 45:1–38.

Hutchins JB, Hollyfield JG (1985) Acetylcholine receptors in the human retina. Invest Ophthalmol Vis Sci 26:1550–1557.

Hutchins JB, Hollyfield JG (1986) Human retinas synthesize and release acetylcholine. J Neurochem 47:81–87.

Huttenlocher PR, De Courten C (1987) The development of synapses in striate cortex of man. Hum Neurobiol 6:1–9.

Hyman SE (2005) Neurotransmitters. Curr Biol 15:R154–R158.

Hyvärinen J, Poranen A (1978) Movement-sensitive and direction and orientation-selective cutaneous receptive fields in the hand area of the post-central gyrus in monkeys. J Physiol 283:523–537.

Hyvärinen J, Poranen A (1978) Receptive field integration and submodality convergence in the hand area of the post-central gyrus of the alert monkey. J Physiol 283:539–556.

Hyvärinen J, Poranen A, Jokinen Y (1980) Influence of attentive behavior on neuronal responses to vibration in primary somatosensory cortex of the monkey. J Neurophysiol 43:870–882.

Iaria G, Petrides M (2007) Occipital sulci of the human brain: Variability and probability maps. J Comp Neurol 501:243–259.

Idé C, Munger BL (1980) The cytologic composition of primate laryngeal chemosensory corpuscles. Am J Anat 158:193–209.

Igarashi M, et al. (1978) Locomotor dysfunction after surgical lesions in the unilateral vestibular nuclei region in squirrel monkeys. Arch Otorhinolaryngol 221:89–95.

Iggo A (1966) Physiology of visceral afferent systems. Acta Neurovegetativa 28:121–134.

Iggo A (1972) Critical remarks on the gate control theory. In: Pain: Basic Principles – Pharmacology – Therapy. Janzen R, Keidel WD, Herz A, Steichele C, eds, pp 127–128. Churchill Livingstone, London.

Iggo A (1974) Activation of cutaneous nociceptors and their actions on dorsal horn neurons. Adv Neurol 4:1–9.

Iggo A (1976) Peripheral and spinal 'pain' mechanisms and their modulation. In: Advances in Pain Research and Therapy. Bonica JJ, Albe-Fessard D, eds, 1:381–394. Raven Press, New York.

Iggo A (1977) Cutaneous and subcutaneous sense organs. Br Med Bull 33:97–102.

Iggo A, Young DW (1975) Cutaneous thermoreceptors and thermal nociceptors. In: The Somatosensory System. Kornhuber HH, ed., pp 5–22. Thieme, Stuttgart.

Ignelzi RJ, Atkinson JH (1980) Pain and its modulation. Part 2. Efferent mechanisms. Neurosurgery 6:584–590.

Iivanainen M, Haltia M, Lydecken K (1977) Atelencephaly. Dev Med Child Neurol 19:663–668, 1977.

Imig TJ, et al. (1977) Organization of auditory cortex in the owl monkey (*Aotus trivirgatus*). J Comp Neurol 171:111–128.

Inbal R, et al. (1987) Collateral sprouting in skin and sensory recovery after nerve injury in man. Pain 28:141–154.

Ingalls NW (1920) A human embryo at the beginning of segmentation, with special reference to the vascular system. Contrib Embryol Carneg Instn 11:61–90.

Inglis JT, Kennedy PM, Wells C, Chua R (2002) The role of cutaneous receptors in the foot. Adv Exp Med Biol 508:111–117.

Inoue T, Shimizu H, Yoshimoto T, Kabasawa H (2001) Spatial functional distribution in the corticospinal tract at the corona radiata: a three-dimensional anisotropy contrast study. Neurol Med Chir (Tokyo) 41:293–298.

Insausti R, Amaral DG, Cowan WM (1987) The entorhinal cortex of the monkey: II. Cortical afferents. J Comp Neurol 264:356–395.

Insausti R, Amaral DG, Cowan WM (1987) The entorhinal cortex of the monkey: III. Subcortical afferents. J Comp Neurol 264:396–408.

Iqbal K, Grundke-Iqbal I, Merz PA, Wisniewski HM (1982) Age-associated neurofibrillary changes. In: *The Aging Brain: Cellular and Molecular Mechanisms of Aging in the Nervous System.* Giacobini E et al., eds, pp 247–257. Raven Press, New York.

Isaacson RL (1982) *The Limbic System.* 2nd ed. Plenum, New York.

Ishai A, Ungerleider LG, Martin A, Haxby JV (2000) The representation of objects in the human occipital and temporal cortex. J Cogn Neurosci 12 Suppl 2:35–51.

Ishii T, Murakami Y, Balogh K Jr (1967) Acetylcholinesterase activity in the efferent nerve fibers of the human inner ear. Ann Otol Rhinol Laryngol 76:69–82.

Ishunina TA, Kamphorst W, Swaab DF (2003) Changes in metabolic activity and estrogen receptors in the human medial mamillary nucleus: relation to sex, aging and Alzheimer's disease. Neurobiol Aging 24:817–828.

Ishunina TA, Unmehopa UA, van Heerikhuize JJ, Pool CW, Swaab DF (2001) Metabolic activity of the human ventromedial nucleus neurons in relation to sex and ageing. Brain Res 893:70–76.

Ishunina TA, van Heerikhuize JJ, Ravid R, Swaab DF (2003) Estrogen receptors and metabolic activity in the human tuberomamillary nucleus: changes in relation to sex, aging and Alzheimer's disease. Brain Res 988:84–96.

Issidorides MR, Mytilineou C, Whetsell WO Jr, Yahr MD (1978) Protein-rich cytoplasmic bodies of substantia nigra and locus ceruleus. A comparative study in Parkinsonian and normal brain. Arch Neurol 35:633–637.

Itaya SK, Van Hoesen GW, Benevento LA (1986) Direct retinal pathways to the limbic thalamus of the monkey. Exp Brain Res 61:607–613.

Itzev DE, Ovtscharoff WA, Marani E, Usunoff KG (2002) Neuromelanin-containing, catecholaminergic neurons in the human brain: ontogenetic aspects, development and aging. Biomed Rev 13:39–47.

Iurato S, Luciano L, Pannese E (1972) Efferent vestibular fibers in mammals: morphological and histochemical aspects. Prog Brain Res 37:429–443.

Iwamura Y, Tanaka M (1978) Postcentral neurons in hand region of area 2: their possible role in the form discrimination of tactile objects. Brain Res 150:662–666.

Iwamura Y, Tanaka M, Hikosake O (1980) Overlapping representation of fingers in the somatosensory cortex (area 2) of the conscious monkey. Brain Res 197:516–520.

Iwase A, Kitazawa Y, Ohno Y (1988) On age-related norms of the visual field. Jpn J Ophthalmol 32:429–437.

Iwatsubo T, Hasegawa M, Esaki Y, Ihara Y (1992) Lack of ubiquitin immunoreactivities at both ends of neuropil threads. Possible bidirectional growth of neuropil threads. Am J Pathol 140:277–282.

Jackson GD, Duncan JS (1996) *MRI Neuroanatomy: A New Angle on the Brain.* Churchill Livingstone, New York.

Jacob JM, McQuarrie IG (1991) Axotomy accelerates slow component b of axonal transport. J Neurobiol 22:570–582.

Jacobs MJ (1970) The development of the human motor trigeminal complex and accessory facial nucleus and their topographic relations with the facial and abducens nuclei. J Comp Neurol 138:161–194.

Jaffe GJ, Alvarado JA, Juster RP (1986) Age-related changes of the normal visual field. Arch Ophthalmol 104:1021–1025.

Jäger J, Henn V (1981) Habituation of the vestibule-ocular reflex (VOR) in the monkey during sinusoidal rotation in the dark. Exp Brain Res 41:108–114.

Jakobsen J, Brimijoin S, Sidenius P (1983) Axonal transport in neuropathy. Muscle Nerve 6:164–166.

James CCM, Lassman LP (1972) *Spinal Dysraphism: Spina Bifida Occulta.* Butterworth, London.

Jampel RS (1962) Extraocular muscle action from brain stimulation in the macaque. Invest Ophthalmol 1:565–578.

Jampel RS (1975) Ocular torsion and the function of the vertical extraocular muscles. Am J Ophthalmol 79:292–304.

Jänig W (2006) *Integrative Action of the Autonomic Nervous System: Neurobiology of Homeostasis.* Cambridge University Press, Cambridge.

Jänig W, Habler HJ (2000) Specificity in the organization of the autonomic nervous system: a basis for precise neural regulation of homeostatic and protective body functions. Prog Brain Res 122:351–367.

Jankowska E, Edgley SA (2006) How can corticospinal tract neurons contribute to ipsilateral movements? A question with implications for recovery of motor functions. Neuroscientist 12:67–79.

Jannetta PJ (1967) Gross (mesoscopic) description of the human trigeminal nerve and ganglion. J Neurosurg 26:109–111.

Järvilehto T (1977) Neural basis of cutaneous sensations analyzed by microelectrode measurements from human peripheral nerves – a review. Scand J Psychol 18:348–359.

Järvilehto T, Hamalainen H, Laurinen P (1976) Characteristics of single mechanoreceptive fibres innervating hairy skin of the human hand. Exp Brain Res 25:45–61.

Jay WM (1981) Visual field defects. Am Fam Physician 24:138–142.

Jefferson G (1935) Jacksonian epilepsy. A background and a post-script. Post Grad Med J 11:150–162.

Jefferson M (1952) Altered consciousness associated with brain-stem lesions. Brain 75:55–67.

Jeffery PL, Austin L (1973) Axoplasmic transport. Prog Neurobiol 2:207–255.

Jeppsson P-H (1969) Studies on the structure and innervation of taste buds. An experimental and clinical investigation. Acta Otolaryngol 259:1–95.

Jerge CR (1963) Organization and function of the trigeminal mesencephalic nucleus. J Neurophysiol 26:379–392.

Jerger J (1973) *Modern Developments in Audiology*. 2nd ed. Academic Press, New York.

Jerger J, Lovering L, Wertz M (1972) Auditory disorder following bilateral temporal lobe insult: report of a case. J Speech Hear Disord 37:523–535.

Jerger J, Weikers NJ, Sharbrough FW III, Jerger S (1969) Bilateral lesions of the temporal lobe. A case study. Acta Otolaryngol 258:1–51.

Joachim CL, Morris JH, Kosik KS, Selkoe DJ (1987) Tau antisera recognize neurofibrillary tangles in a range of neurodegenerative disorders. Ann Neurol 22:514–520.

Johansson RS (1978) Tactile sensibility in the human hand: receptive field characteristics of mechanoreceptive units in the glabrous skin area. J Physiol 281:101–123.

Johansson RS (1979) Tactile afferent units with small and well demarcated receptive fields in the glabrous skin area of the human hand. In: *Sensory Functions of the Skin of Humans*. Kenshalo DR, ed. 2nd Int Symp on Skin Senses, pp 129–145. Fla State U. Plenum, New York.

Johansson RS, Vallbo ÅB (1980) Spatial properties of the population of mechanoreceptive limits in the glabrous skin of the human hand. Brain Res 184:353–366.

Johnson BM, Miao M, Sadun AA (1987) Age-related decline of human optic nerve axon populations. Age 10:5–9.

Johnson KO (2001) The roles and functions of cutaneous mechanoreceptors. Curr Opin Neurobiol 11:455–461.

Johnson KO, Darian-Smith I, LaMotte C (1973) Peripheral neural determinants of temperature discrimination in man: a correlative study of responses to cooling skin. J Neurophysiol 36:347–370.

Johnson KO, Lamb GD (1981) Neural mechanisms of spatial tactile discrimination: neural patterns evolved by Braille-like dot patterns in the monkey. J Physiol 310:117–144.

Johnson KO, Phillips JR (1981) Tactile spatial resolution. I. Two-point discrimination, gap detection, grating resolution, and letter recognition. J Neurophysiol 46:1177–1192.

Johnson KO, Yoshioka T, Vega-Bermudez F (2000) Tactile functions of mechanoreceptive afferents innervating the hand. J Clin Neurophysiol 17:539–558.

Johnson KP (1974) Viral infections of the developing nervous system. Adv Neurol 6:53–67.

Johnson RH, Spalding JMK (1974) Disorders of the autonomic nervous system. In: *Contemporary Neurology Series*. Vol 11. FA Davis Co, Philadelphia.

Johnson RR, Wong JH, Awad IA, Kim JH, Bolognia JL, Sawaya R (1999) Case problems conference: Hemorrhagic lesion at the tentorial incisura in a patient with facial nevus. Neurosurgery 45(5):1216–1221.

Johnsson L-G, Hawkins JE Jr (1967) Otolithic membranes of the saccule and utricle in man. Science 157:1454–1456.

Johnston MC (1975) The neural crest in abnormalities of the face and brain. Birth Defects 11:1–18.

Jones DG (1983) Recent perspectives on the organization of central synapses. Anesth Analg 62:1100–1112.

Jones EG (1975) Lamination and differential distribution of thalamic afferents within the sensory–motor cortex of the squirrel monkey. J Com Neurol 160:167–204.

Jones EG (1997) A description of the human thalamus. In: *The Thalamus*. Steriade M, Jones EG, McCormick DA, eds, II:425–500. Chapter 9. Elsevier Science, Amsterdam.

Jones EG (1998) The thalamus of primates. In: *Handbook of Chemical Neuroanatomy*. Bloom FE, Björklund A, Hökfelt T, eds. Vol 14: The Primate Nervous System, Part II, pp 1–298. Elsevier, Amsterdam.

Jones EG (1998) Viewpoint: the core and matrix of thalamic organization. Neuroscience 85:331–345.

Jones EG (1998) A new view of specific and nonspecific thalamocortical connections. Adv Neurol 77:49–71.

Jones EG (2001) The thalamic matrix and thalamocortical synchrony. Trends Neurosci 24:595–601.

Jones EG (2002) A pain in the thalamus. J Pain 3:102–104.

Jones EG (2002) Thalamic organization and function after Cajal. Prog Brain Res 136:333–357.

Jones EG (2002) Thalamic circuitry and thalamocortical synchrony. Philos Trans R Soc Lond B Biol Sci 357:1659–1673.

Jones EG (2007) The human thalamus. In: Jones EG. *The Thalamus*. 2nd ed. Vol I, Chapter 18, pp 1396–1447. Cambridge University Press, Cambridge.

Jones EG, Burton H (1976) Areal differences in the laminar distribution of thalamic afferents in cortical fields of the insular, parietal and temporal regions of primates. J Comp Neurol 168:197–248.

Jones EG, Leavitt RY (1974) Retrograde axonal transport and the demonstration of non-specific projections to the cerebral cortex and striatum from thalamic intralaminar nuclei in the rat, cat and monkey. J Comp Neurol 154:349–77.

Jones EG, Peters A (1986) *Cerebral Cortex*. Vol 5: Sensory-Motor Areas and Aspects of Cortical Connectivity. Plenum Press, New York.

Jones EG, Peters A (1987) *Cerebral Cortex*. Vol 6: Further Aspects of Cortical Function, Including Hippocampus. Plenum Press, New York.

Jones EG, Porter R (1980) What is area 3a? Brain Res Rev 203:1–43.

Jones EG, Powell TPS (1969) Connexions of the somatic sensory cortex of the rhesus monkey. I. Ipsilateral cortical connexions. Brain 92:477–502.

Jones EG, Powell TPS (1969) Connexions of the somatic sensory cortex of the rhesus monkey. II. Contralateral cortical connexions. Brain 92:717–730.

Jones EG, Powell TPS (1970) Connexions of the somatic sensory cortex of the rhesus monkey. III. Thalamic connexions. Brain 93:37–56.

Jones EG, Wise SP, Coulter JD: Differential thalamic relationships of sensory–motor and parietal cortical fields in monkeys. J Comp Neurol 183:833–662.

Jones HO, Brewer JI (1941) A human embryo in the primitive-streak stage (Jones–Brewer ovum I). Contrib Embryol Carneg Instn 29:157–165.

Jozefowicz RF, Holloway, RG (1999) *Case Studies in Neuroscience*. FA Davis, Philadelphia.

Just T, Stave J, Pau HW, Guthoff R (2005) In vivo observation of papillae of the human tongue using confocal laser scanning microscopy. ORL J Otorhinolaryngol Relat Spec 67:207–212.

Kaas JH (1999) The transformation of association cortex into sensory cortex. Brain Res Bull 50:425.

Kaas JH (2004) Evolution of somatosensory and motor cortex in primates. Anat Rec A Discov Mol Cell Evol Biol 281:1148–1156.

Kaas JH (2004) Somatosensory System. In: *The Human Nervous System*. Paxinos G, Mai JK, eds. 2nd ed. Chapter 28, pp 1059–1092. Elsevier/Academic Press, Amsterdam.

Kaas JH, Hackett TA (1999) 'What' and 'where' processing in auditory cortex. Nat Neurosci 2:1045–1047.

Kaas JH, Huerta MF, Weber JT, Harting JK (1978) Patterns of retinal terminations and laminar organization of the lateral geniculate nucleus of primates. J Comp Neurol 182:517–554.

Kaas JH, Nelson RJ, Sur M, Merzenich MM (1981) Organization of somatosensory cortex in primates. In: *The Organization of the Cerebral Cortex. Proceedings of a Neurosciences Research Program Colloquium*. Schmitt FO, Worden FG, Adelman G, Dennis SG, eds, pp 237–261. MIT Pr, Cambridge.

Kaas JH, Nelson RJ, Sur M, Lin CS, Merzenich MM (1979) Multiple representations of the body within the primary somatosensory cortex of primates. Science 204:521–523.

Kahn EA (1947) The rôle of the dentate ligaments in spinal cord compression and the syndrome of lateral sclerosis. J Neurosurg 4:191–199.

Kahn EA, Taren JA, Newman MH (1982) Congenital anomalies of the brain, spinal cord, and their membranes. In: *Correlative Neurosurgery*. Schneider RC, Kahn EA, Crosby EC, Taren JA, eds. 3rd ed., 2:909–947. Thomas, Springfield.

Kalia M, Davies RO (1978) A neuroanatomical search for glossopharyngeal efferents to the carotid body using the retrograde transport of horseradish peroxidase. Brain Res 149:477–481.

Kalil K (1981) Projections of the cerebellar and dorsal column nuclei upon the thalamus of the rhesus monkey. J Comp Neurol 195:25–50.

Kalil K (1981) Projections of the cerebellar and dorsal column nuclei upon the thalamus of the rhesus monkey. J Comp Neurol 195:25–50.

Kallen B (1968) Early embryogenesis of the central nervous system with special reference to closure defects. Dev Med Child Neurol Suppl 16:44–53.

Kalter H (1968) *Teratology of the Central Nervous System*. Univ Chicago Press, Chicago.

Kalter H, Warkany J (1983) Congenital malformations: etiologic factors and their role in prevention. N Engl J Med 308:424–431, 491–497.

Kalus P, Slotboom J, Gallinat J, Mahlberg R, Cattapan-Ludewig K, Wiest R, Nyffeler T, Buri C, Federspiel A,

Kunz D, Schroth G, Kiefer C (2006) Examining the gateway to the limbic system with diffusion tensor imaging: the perforant pathway in dementia. NeuroImage 30:713–720.

Kanshepolsky J, Kelley JJ, Waggener JD (1973) A cortical auditory disorder. Clinical, audiologic and pathologic aspects. Neurology 23:699–705.

Karatas A, Caglar S, Savas A, Elhan A, Erdogan A (2005) Microsurgical anatomy of the dorsal cervical rootlets and dorsal root entry zones. Acta Neurochir (Wien) 147:195–199.

Katz J, Sommer A (1986) Asymmetry and variation in the normal hill of vision. Arch Ophthalmol 104:65–68.

Katzman R (1985) Commentary on review by Gash et al. Potential usefulness and current limitations of neural transplantation in the aged brain. Neurobiol Aging 6:152–153.

Kawabata I, Nomura Y (1981) The imprints of the human tectorial membrane: a SEM study. Acta Otolaryngol 91:29–35.

Kawashima M, Li X, Rhoton AL Jr, Ulm AJ, Oka H, Fujii K (2006) Surgical approaches to the atrium of the lateral ventricle: microsurgical anatomy. Surg Neurol 65:436–445.

Keele KD (1957) *Anatomies of Pain*. Thomas, Springfield.

Keeney AH (1951) *Chronology of Ophthalmic Development: An Outline Summary of the Anatomical and Functional Development of the Visual Mechanism Before and After Birth*. Thomas, Springfield.

Keidel WD, Neff WD (1974) Auditory System: Anatomy, Physiology (Ear). In: *Handbook of Sensory Physiology*. Vol V/1, 736 pp. Springer-Verlag, Berlin.

Keller EL (1974) Participation of medial pontine reticular formation in eye movement generation in monkey. J Neurophysiol 37:316–332.

Keller EL, Robinson DA (1972) Abducens unit behavior in the monkey during vergence movements. Vision Res 12:369–382.

Kemether EM, Buchsbaum MS, Byne W, Hazlett EA, Haznedar M, Brickman AM, Platholi J, Bloom R (2003) Magnetic resonance imaging of mediodorsal, pulvinar, and centromedian nuclei of the thalamus in patients with schizophrenia. Arch Gen Psychiatry 60:983–991.

Kennedy C (1985) Metabolic mapping of the primary visual pathway. Res Publ Assoc Res Nerv Ment Dis 63:61–72.

Kennedy H, Bullier J (1985) A double-labeling investigation of the afferent connectivity to cortical areas V1 and V2 of the macaque monkey. J Neurosci 5:2815–2830.

Kennedy WR (1970) Innervation of normal human muscle spindles. Neurology 20:463–475.

Kennedy WR (1970) What are muscle spindles? Med Times 98(5):159–170.

Kenshalo DR, Duclaux R (1977) Response characteristics of cutaneous cold receptors in the monkey. J Neurophysiol 40:319–332.

Kenshalo DR (1979) Sensory functions of the skin of humans. 2nd Int Symp on the Skin Senses, Fla State U. Plenum, New York.

Kenton B, et al. (1980) Peripheral fiber correlates to noxious thermal stimulation in humans. Neurosci Lett 17:301–306.

Kern EB (1972) Referred pain to the ear. Minn Med 55:896–898.

Kerr FW, Hendler N, Bowron P (1970) Viscerotopic organization of the vagus. J Comp Neurol 138:279–290.

Kerr FWL (1963) The divisional organization of afferent fibres of the trigeminal nerve. Brain 86:721–732.

Kerr FWL (1967) Correlated light and electron microscopic observations on the normal trigeminal ganglion and sensory root in man. J Neurosurg 26(Suppl):132–137.

Kerr FWL (1975) Neuroanatomical substrates of nociception in the spinal cord. Pain 1:325–356.

Kerr FWL (1975) The ventral spinothalamic tract and other ascending systems of the ventral funiculus of the spinal cord. J Comp Neurol 159:335–356.

Kerr FWL, Casey KL (1978) Pain. Neurosci Res Program Bull 16:1–207.

Kerr FWL, Fukushima T (1980) New observations on the nociceptive pathways in the central nervous system. Res Publ Assoc Res Nerv Ment Dis 58:47–61.

Kerr FWL, Hollowell OW (1964) Location of pupillomotor and accommodation fibres in the oculomotor nerve: experimental observations on paralytic mydriasis. J Neurol Neurosurg Psychiatry 27:473–481.

Kerr FWL, Kruger L, Schwassmann HO, Stern R (1968) Somatotopic organizations of mechanoreceptor units in the trigeminal nuclear complex of the macaque. J Comp Neurol 134:127–144.

Kerr FWL, Lysak WR (1964) Somatotopic organization of trigeminal-ganglion neurons. Arch Neurol 11:593–602.

Kerr FWL, Olafson RA (1961) Trigeminal and cervical volleys. Convergence on single units in the spinal gray at C-1 and C-2. Arch Neurol 5:171–178.

Kertesz A, Lesk D, McCabe P (1977) Isotope localization of infarcts in aphasia. Arch Neurol 34:590–601.

Keswani NH, Hollinshead WH (1955) The phrenic nucleus. III. Organization of the phrenic nucleus in the spinal cord of the cat and man. Staff Meetings of the Mayo Clinic 30:566–577.

Keverne EB (1978) Olfaction and taste – dual systems for sensory processing. TINS 1:32–34.

Khalfa S, Bougeard R, Morand N, Veuillet E, Isnard J, Guenot M, Ryvlin P, Fischer C, Collet L (2001) Evidence of peripheral auditory activity modulation by the auditory cortex in humans. Neuroscience 104:347–358.

Khurana RK, O'Donnell PP, Suter CM, Inayatullah M (1981) Bilateral deafness of vascular origin. Stroke 12:521–523.

Kidman AD, Hanwell MA, Cooper NA (1979) The role of the satellite cell in axonal metabolism. Int Congr Series 473:85–93.

Kido DK, Gomez DG, Pavese AM Jr, Potts DG (1976) Human spinal arachnoid villi and granulations. Neuroradiology 11:221–228.

Kiernan JA (1978) An explanation of axonal regeneration in peripheral nerves and its failure in the central nervous system. Med Hypotheses 4:15–26.

Kim JS (2003) Pure lateral medullary infarction: clinical–radiological correlation of 130 acute, consecutive patients. Brain 126:1864–1872.

Kim JS, Kim HG, Chung CS (1995) Medial medullary syndrome. Report of 18 new patients and a review of the literature. Stroke 26:1548–1552.

Kim JS, Kim J (2005) Pure midbrain infarction: clinical, radiologic, and pathophysiologic findings. Neurology 64:1227–1232.

Kim SH, Pohl PS, Luchies CW, Stylianou AP, Won Y (2003) Ipsilateral deficits of targeted movements after stroke. Arch Phys Med Rehabil 84:719–724.

Kimura RS (1975) The ultrastructure of the organ of Corti. Int Rev Cytol 42:173–222.

Kimura RS, Ota CY, Takahashi T (1979) Nerve fiber synapses on spiral ganglion cells in the human cochlea. Ann Otol Rhinol Laryngol 88:(Suppl 62):1–17.

Kimura RS, Schuknecht HF, Sando I (1964) Fine morphology of the sensory cells in the organ of Corti of man. Acta Otolaryngol 58:390–408.

King BM (2006) The rise, fall, and resurrection of the ventromedial hypothalamus in the regulation of feeding behavior and body weight. Physiol Behav 87:221–244.

King RB (1977) Anterior commissurotomy for intractable pain. J Neurosurg 47:7–11.

King WM, Lisberger SG, Fuchs AF (1976) Responses of fibers in medial longitudinal fasciculus (MLF) of alert monkeys during horizontal and vertical conjugate eye movements evoked by vestibular or visual stimuli. J Neurophysiol 39:1135–1149.

Kinney PC (1978) An experimental study of the central gustatory pathways in the monkey, *Macaca mulatta* and *Cercopithecus aethiops*. J Hirnforsch 19:21–43.

Kinomura S, Larsson J, Gulyas B, Roland PE (1996) Activation by attention of the human reticular formation and thalamic intralaminar nuclei. Science 271:512–515.

Kiss A, Mikkelsen JD (2005) Oxytocin – anatomy and functional assignments: a minireview. Endocr Regul 39:97–105.

Kiyosawa M, et al. (1986) Metabolic imaging in hemianopsia using positron emission tomography with 18F-deoxyfluoroglucose. Am J Ophthalmol 101:310–319.

Klein DC (1978) The pineal gland: a model of neuroendocrine regulation. In: *The Hypothalamus*. Reichlin S, Baldessarini RJ, Martin JB, eds, pp 303–327. Raven Press, New York.

Kline DG, LeBlanc HJ (1971) Survival following gunshot wound of the pons: neuroanatomic considerations. J Neurosurg 35:342–347.

Kline DG, Nulsen FE (1972) The neuroma in continuity. Its preoperative and operative management. Surg Clin North Am 52:1189–1209.

Kline LB, Kim JY, Ceballos R (1985) Radiation optic neuropathy. Ophthalmology 92:1118–1126.

Klinger J, Gloor P (1960) The connections of the amygdala and of the anterior temporal cortex in the human brain. J Comp Neurol 115:333–369.

Klinke R (1981) Neurotransmitters in the cochlea and the cochlear nucleus. Acta Otolaryngol 91:541–554.

Klinke R, Galley N (1974) Efferent innervation of the vestibular and auditory receptors. Physiol Rev 54:316–357.

Klüver H, Bucy PC (1937) Psychic blindness and other symptoms following bilateral temporal lobectomy in rhesus monkeys. Am J Physiol 119:352–353.

Kneisley LW, Biber MP, LaVail JH (1978) A study of the origin of brain stem projections to monkey spinal cord using the retrograde transport method. Exp Neurol 60:116–139.

Knibestöl M (1973) Stimulus–response functions of rapidly adapting mechanoreceptors in the human glabrous skin area. J Physiol 232:427–452.

Knibestöl M (1975) Stimulus–response functions of slowly adapting mechanoreceptors in the human glabrous skin area. J Physiol 245:63–80.

Knibestöl M, Vallbo ÅB (1970) Single unit analysis of mechanoreceptor activity from the human glabrous skin. Acta Physiol Scand 80:178–195.

Knorring L von, Almay BGL, Johansson F, Terenius L (1978) Pain perception and endorphin levels in cerebrospinal fluid. Pain 5:359–365.

Knosp E, Müller G, Perneczky A (1988) The paraclinoid carotid artery: anatomical aspects of a microneurosurgical approach. Neurosurgery 22:896–901.

Knudsen V, Kolze V (1972) Neurinoma of the Gasserian ganglion and the trigeminal root. Report of four cases. Acta Neurochir 26:159–164.

Knyihar E, Csillik B, Rakic P (1978) Transient synapses in the embryonic primate spinal cord. Science 202:1206–1209.

Koehler PJ, Endtz LJ, Te Velde J, Hekster REM (1986) Aware or non-aware: on the significance of awareness for the localization of the lesion responsible for homonymous hemianopia. J Neurol Sci 75:255–262.

Koehler RC, Gebremedhin D, Harder DR (2006) Role of astrocytes in cerebrovascular regulation. J Appl Physiol 100:307–317.

Koelle GB (1975) Microanatomy and pharmacology of cholinergic synapses. In: The Nervous System. Tower DB, editor-in-chief. Brady RO, volume ed. Vol 1: The Basic Neurosciences, pp 363–371. Raven Press, New York.

Konietzny F, Hensel H (1977) The dynamic response of warm units in human skin nerves. Pflügers Arch 370:111–114.

Konietzny F, Hensel H (1979) The neural basis of the sensory quality of warmth. In: Sensory Functions of the Skin of Humans. Kenshalo DR, ed. 2nd Int Symp on Skin Senses, pp 241–256. Fla State U. Plenum, New York.

Konigsmark BW (1973) Cellular organization of the cochlear nuclei in man. J Neuropath Exp Neurol 32:153–154.

Konigsmark BW, et al. (1973) Neuroanatomy of the auditory system. (Report on Workshop). Arch Otolaryngol 98:397–413.

Konigsmark BW, Murphy EA (1972) Volume of the ventral cochlear nucleus in man: its relationship to neuronal population and age. J Neuropathol Exp Neurol 31:304–316.

Koontz MA, Hendrickson AE, Ryan MK (1989) GABA-immunoreactive synaptic plexus in the nerve fiber layer of primate retina. Vis Neurosci 2:19–25.

Kosofsky BE, Molliver ME, Morrison JH, Foote SL (1984) The serotonin and norepinephrine innervation of primary visual cortex in the cynomolgus monkey (Macaca fascicularis). J Comp Neurol 230:168–178.

Koutcherov Y, Huang X-F, Halliday G, Paxinos G (2004) Organization of Human Brain Stem Nuclei. In: The Human Nervous System. Paxinos G, Mai JK, eds. 2nd ed., pp 267–320. Chapter 10. Elsevier/Academic Press, Amsterdam.

Koutcherov Y, Mai JK, Ashwell KW, Paxinos G (2000) Organization of the human paraventricular hypothalamic nucleus. J Comp Neurol 423:299–318.

Koutcherov Y, Mai JK, Ashwell KW, Paxinos G (2002) Organization of human hypothalamus in fetal development. J Comp Neurol 446:301–324.

Koutcherov Y, Mai JK, Ashwell KW, Paxinos G (2004) Organisation of the human dorsomedial hypothalamic nucleus. NeuroReport 15:107–111.

Krayenbuhl H, Yasargil MG (1975) Cranial chordomas. Prog Neurol Surg 6:380–434.

Kreisman NR, Zimmerman ID (1973) Representation of information about skin temperature in the discharge of single cortical neurons. Brain Res 55:343–353.

Kreutzberg GW (1982) Acute neural reaction to injury. In: Repair and Regeneration of the Nervous System. Nicholls JG, ed, pp 57–69. Springer-Verlag, New York.

Kreutzberg GW, Schubert P (1975) The cellular dynamics of intraneuronal transport. In: The Use of Axonal Transport for Studies of Neuronal Connectivity. Cowan WM, Cuénod M, eds, pp 83–112. Elsevier, New York.

Kriegler JS, Krishnan N, Singer M (1981) Trophic interactions of neurons and glia. In: Demyelinating Disease: Basic and Clinical Electrophysiology. Waxman SG, Ritchie JM, eds, pp 479–504. Raven Press, New York.

Krimer LS, Hyde TM, Herman MM, Saunders RC (1997) The entorhinal cortex: an examination of cyto- and myeloarchitectonic organization in humans. Cereb Cortex 7:722–731.

Kringelbach ML, de Araujo IE, Rolls ET (2004) Taste-related activity in the human dorsolateral prefrontal cortex. NeuroImage 21:781–788.

Kristensson K (1984) Retrograde signaling after nerve injury. Adv Neurochem 6:31–43.

Krmpotic'-Nemanic' J, Kostovic' I, Nemanic' D, Kelovic' Z (1979) The laminar organization of the prospective auditory cortex in the human fetus (11–13.5 weeks of gestation). Acta Otolaryngol 87:241–246.

Krnjevic K (1974) Chemical nature of synaptic transmission in vertebrates. Physiol Rev 54:418–540.

Krubitzer L, Clarey J, Tweedale R, Elston G, Calford MA (1995) Redefinition of somatosensory areas in the lateral sulcus of macaque monkeys. J Neurosci 15:3821–3839.

Kruijver FP, Balesar R, Espila AM, Unmehopa UA, Swaab DF (2002) Estrogen receptor-alpha distribution in the human hypothalamus in relation to sex and endocrine status. J Comp Neurol 454:115–139.

Kubota K, Negishi T, Masegi T (1975) Topological distribution of muscle spindles in the human tongue and its significance in proprioception. Bull Tokyo Med Dent Univ 22:235–242.

Kucera J. Dorovini-Zis K (1979) Types of human intrafusal muscle fibers. Muscle Nerve 2:437–451.

Kuhar MJ, Pert CB, Snyder S (1973) Regional distribution of opiate receptor binding in monkey and human brain. Nature 245:447–450.

Kulesza RJ Jr (2006) Cytoarchitecture of the human superior olivary complex: Medial and lateral superior olive. Hear Res. [Epub ahead of print, doi:10.1016/j.heares.2006. 12.006]

Kumar A (1982) Is spontaneous nystagmus a pathological sign? Laryngoscope 92:618–626.

Kumazawa T, Perl ER, Burgess PR, Whitehorn D (1975) Ascending projections from marginal zone (Lamina I) neurons of the spinal dorsal horn. J Comp Neurol 162:1–12.

Kumral E, Afsar N, Kirbas D, Balkir K, Ozdemirkiran T (2002) Spectrum of medial medullary infarction: clinical and magnetic resonance imaging findings. J Neurol 249:85–93.

Kunc Z (1964) Tractus spinalis nervi trigemini. Rozpr Cesk Akad Ved, Rada Mat Prir Ved 74:1–98.

Kunc Z, Fuchsová M, Novák M (1978) Excision of the spinal trigeminal tract. Electron microscopy. Acta Neurochir (Wien) 41:233–241.

Kunitomo K (1918) The development and reduction of the tail and of the caudal end of the spinal cord. Contrib Embryol Carneg Instn 8:161–198.

Kuntz A (1953) *The Autonomic Nervous System.* Lea & Febiger, Philadelphia.

Kupfer C, Chumbley L, Downer JDeC (1967) Quantitative histology of optic nerve, optic tract and lateral geniculate nucleus in man. J Anat 101:393–401.

Kurtzke JF (1982) The current neurologic burden of illness and injury in the United States. Neurology 32:1207–1214.

Kurtzke JF, Goldberg ID, Kurland LT (1973) The distribution of deaths from congenital malformations of the nervous system. Neurology 23:483–496.

Kuypers HGJM (1958) Corticobulbar connexions to the pons and lower brain-stem in man. An anatomical study. Brain 81:364–388.

Kuypers HGJM: (1960) Central cortical projections to motor and somato-sensory cell groups. An experimental study in the rhesus monkey. Brain 83:161–184.

Kuzemensky J (1976) Comparison of the cytoarchitectonic structure of the subthalamic nucleus in certain mammals. Folia Morphol (Praha) 24:129–140.

Kuzuhara S, Chou SM (1981) Preservation of the phrenic motoneurons in Werdnig–Hoffmann disease. Ann Neurol 9:506–510.

Kwon M, Lee JH, Kim JS (2005) Dysphagia in unilateral medullary infarction: lateral vs medial lesions. Neurology 65:714–718.

Lachman N, Acland RD, Rosse C (2002) Anatomical evidence for the absence of a morphologically distinct cranial root of the accessory nerve in man. Clin Anat 15:4–10.

Lack EE (1977) Carotid body hypertrophy in patients with cystic fibrosis and cyanotic congenital heart disease. Hum Pathol 8:39–51.

Lacour M, Roll JP, Appaix M (1976) Modifications and development of spinal reflexes in the alert baboon, (Papio) following an unilateral vestibular neurotomy. Brain Res 113:255–269.

Lacour M, Xerri C, Hugon M (1979) Compensation of postural reactions to fall in the vestibular neurectomized monkey. Role of the remaining labyrinthine afferences. Exp Brain Res 37:563–580.

Lacroix S, Havton LA, McKay H, Yang H, Brant A, Roberts J, Tuszynski MH (2004) Bilateral corticospinal projections arise from each motor cortex in the macaque monkey: a quantitative study. J Comp Neurol 473:147–161.

Laemle LK (1979) Neuronal populations of the human periaqueductal gray, nucleus lateralis. J Comp Neurol 186:93–108.

LaGuardia JJ, Cohrs RJ, Gilden DH (2000) Numbers of neurons and non-neuronal cells in human trigeminal ganglia. Neurol Res 22:565–566.

Laha RK, Jannetta PJ (1977) Glossopharyngeal neuralgia. J Neurosurg 47:316–320.

LaMotte C (1977) Distribution of the tract of Lissauer and the dorsal root fibers in the primate spinal cord. J Comp Neurol 172:529–562.

LaMotte RH (1979) Intensive and temporal determinants of thermal pain. In: *Sensory Functions of the Skin of Humans.* Kenshalo DR, ed. 2nd Int Symp on Skin Senses, pp 327–358. Fla State U. Plenum, New York.

Landis SC (1983) Neuronal growth cones. Annu Rev Physiol 45:567–580.

Lang J (1984) Clinical anatomy of the cerebellopontine angle and internal acoustic meatus. Adv Oto-Rhino-Laryngol 34:8–24.

Lang J, Kageyama I (1990) The ophthalmic artery and its branches, measurements and clinical importance. Surg Radiol Anat 12:83–90.

Lang W, Büttner-Ennever JA, Büttner U (1979) Vestibular projections to the monkey thalamus: an autoradiographic study. Brain Res 177:3–17.

Langer SZ (1978) Presynaptic receptors and neurotransmission. Med Biol 56:288–291.

Langford LA, Coggeshall RE (1981) unmyelinated axons in the posterior funiculi. Science 211:176–177.

Langman J. Welch GW (1966) Effect of vitamin A on development of the central nervous system. J Comp Neurol 128:1–16.

Lapresle J, Annabi A (1979) Olivopontocerebellar atrophy with velopharyngolaryngeal paralysis: a contribution to the somatotopy of the nucleus ambiguus. J Neuropathol Exp Neurol 38:401–406.

Larsell O, Jansen J (1972) *The Comparative Anatomy and Histology of the Cerebellum The Human Cerebellum, Cerebellar Connections, and Cerebellar Cortex.* U of Minn Pr, Minneapolis, pp 90–93, 135–137.

Larson SG, et al. (1974) Neurophysiological effects of dorsal column stimulation in man and monkey. J Neurosurg 41:217–223.

Lasansky A (1971) Nervous function at the cellular level: glia. Annu Rev Physiol 33:241–256.

Lasek RJ (1970) Protein transport in neurons. Int Rev Neurobiol 13:289–324.

Lasek RJ (1984) The structure of axoplasm. In: *Current Topics in Membranes and Transport.* Kleinzeller A, ed. PF Baker PF, guest ed. Vol 22: The Squid Axon, pp 39–53. Academic Press, New York.

Lasek RJ, McQuarrie IG, Wujek JR (1981) The central nervous system regeneration problem: neuron and environment. In: *Peripheral Nerve Regeneration: Experimental Basis and Clinical Implications.* Gorio A, et al., eds, pp 59–70. Raven Press, New York.

Laurence KM (1964) The natural history of spina bifida cystica: detailed analysis of 407 cases. Arch Dis Child 39:41–57.

Lauter JL, Herscovitch P, Formby C, Raichle ME (1985) Tonotopic organization in human auditory cortex revealed by positron emission tomography. Hear Res 20:199–205.

LaVail JH, LaVail MM (1972) Retrograde axonal transport in the central nervous system. Science 176:1416–1417.

Lawrence M, Johnsson L-G (1973) The role of the organ of Corti in auditory nerve stimulation. Ann Otol Rhinol Laryngol 82:464–472.

Lazorthes G (1972) Pathology, classification and clinical aspects of vascular diseases of the spinal cord. In: *Handbook of Clinical Neurology.* Vinken PJ, Bruyn GW, eds. Vol 12: Vascular Diseases of the Nervous System, 12:492–506. Elsevier, New York.

Lazorthes G, Gouaze A, Zadeh JO, Santini JJ, Lazorthes Y, Burdin P (1971) Arterial vascularization of the spinal cord. Recent studies of the anastomotic substitution pathways. J Neurosurg 35:253–262.

Le Gros Clark WE, Russell WR (1938) Cortical deafness without aphasia. Brain 61:375–383.

Lee SH, Kim DE, Song EC, Roh JK (2001) Sensory dermatomal representation in the medial lemniscus. Arch Neurol 58(4):649–651.

Leek BF (1972) Abdominal visceral receptors. In: *Handbook of Sensory Physiology.* Neil E, volume ed., III/1:113–160. Springer-Verlag, New York.

Lehericy S, Grand S, Pollak P, Poupon F, Le Bas JF, Limousin P, Jedynak P, Marsault C, Agid Y, Vidailhet M (2001) Clinical characteristics and topography of lesions in movement disorders due to thalamic lesions. Neurology 57:1055–1066.

Lehlová L, Vold ich L, Janisch R (1987) Correlative study of sensory cell density and cochlear length in humans. Hear Res 28:149–151.

Leigh RJ, Zee DS (2006) The neurology of eye movements. In: *Contemporary Neurology Series.* Vol 70. 4th ed. Oxford University Press, New York.

Leikola A (1976) The neural crest: migrating cells in embryonic development. Folia Morphol (Praha) 24:155–172.

Leino M (1984) 6–Methoxy-tetrahydro-carboline and melatonin in the human retina. Exp Eye Res 38:325–330.

Leisman G, Schwartz J (1977) Directional control of eye movement in reading: the return sweep. Int J Neurosci 8:17–21.

Leisman G, Schwartz J (1977) Ocular-motor function and information processing: implications for the reading process. Int J Neurosci 8:7–15.

Lele PP, Weddell G (1956) The relationship between neurohistology and corneal sensibility. Brain 79:119–154.

Lele PP, Weddell G (1959) Sensory nerves of the cornea and cutaneous sensibility. Exp Neurol 1:334–359.

Lemire RJ (1969) Variations in development of the caudal neural tube in human embryos (Horizons XIV–XXI). Teratology 2:361–369.

Lemire RJ (1982) Neural tube defects: clinical correlations. Clin Neurosurg 30:165–177.

Lemire RJ (1988) Neural tube defects. JAMA 259:558–562.

Lenarz T, Lim HH, Reuter G, Patrick JF, Lenarz M (2006) The auditory midbrain implant: a new auditory prosthesis for neural deafness-concept and device description. Otol Neurotol 27:838–843.

Lende RA, Popp AJ (1976) Sensory Jacksonian seizures. J Neurosurg 44:706–711.

Lennartsson B (1979) Muscle spindles in the human anterior digastric muscle. Acta Odontol Scand 37:329–333.

Lennie P (2003) Receptive fields. Curr Biol 13:R216–R219.

Leuba G, Garey LJ (1987) Evolution of neuronal numerical density in the developing and ageing human visual cortex. Hum Neurobiol 6:11–18.

LeVay S (1991) A difference in hypothalamic structure between heterosexual and homosexual men. Science 253:1034–1037.

LeVay S, Hubel DH, Wiesel TN (1975) The pattern of ocular dominance columns in macaque visual cortex revealed by a reduced silver stain. J Comp Neurol 159:559–576.

Leventhal AG, Ault SJ, Vitek DJ (1988) The nasotemporal division in primate retina: the neural basis of macular sparing and splitting. Science 240:66–67.

LeVere TE, LeVere ND (1985) Comments on review by D.M. Gash, T.J. Collier and J.R. Sladek. Transplants to the central nervous system as a therapy for brain pathology. Neurobiol Aging 6:151–152.

Lew JW, et al. (1977) Localization and characterization of phenylethanolamine N-methyl transferase in the brain of various mammalian species. Brain Res 119:199–210.

Lewis DA, Sesack SR (1997) Dopamine systems in the primate brain. In: *Handbook of Chemical Neuroanatomy.* Björklund A, Hökfelt T, eds. Vol 13: The Primate Nervous System, Part I, pp 263–375. Elsevier, Amsterdam.

Lewis GR, Pilcher R, Yemm R (1980) The effect of a stimulus strength on the jaw-jerk response in man. J Neurol Neurosurg Psychiat 43:699–704.

Lewis T (1942) *Pain.* Macmillan, New York.

Lewis WH, Hartmen, CG (1933) Early cleavage stages of the egg of the monkey (*Macacus rhesus*). Contrib Embryol Carneg Instn 24:189–201.

Liberles SD, Buck LB (2006) A second class of chemosensory receptors in the olfactory epithelium. Nature 442: 645–650.

Lichtman JW, et al. (1982) Factors involved in the restoration of specific neural connections: group report. In: *Repair and Regeneration of the Nervous System.* Nicholls JG, ed., pp 343–379. Springer-Verlag, New York.

Lidén G (1980) Impedance audiometry. Ann Otol Rhinol Laryngol 89:53–58.

Lieberman AR (1971) The axon reaction: a review of the principal features of perikaryal responses to axon injury. Int Rev Neurobiol 14:49–124.

Liedgren SRC, et al. (1976) Representation of vestibular afferents in somatosensory thalamic nuclei of the squirrel monkey (*Saimiri sciureus*). J Neurophysiol 39:601–612.

Liedgren SRC, Schwarz DWF (1976) Vestibular evoked potentials in thalamus and basal ganglia of the squirrel monkey (*Saimiri sciureus*). Acta Otolaryngol 81: 73–82.

Light AR, Perl E (1993) Peripheral sensory systems. In: *Peripheral Neuropathy.* Dyck PJ, Thomas PK, Griffin JW, Low PA, Poduslo JF, eds, pp 149–165. Saunders, Philadelphia.

Lim HH, Anderson DJ (2006) Auditory cortical responses to electrical stimulation of the inferior colliculus: implications for an auditory midbrain implant. J Neurophysiol 96:975–988.

Lin L-FH, et al. (1993) GDNF: a glial cell line-derived neurotrophic factor for midbrain dopaminergic neurons. Science 260:1130–1133.

Linberg KA, Fisher SK (1986) An ultrastructural study of interplexiform cell synapses in the human retina. J Comp Neurol 243:561–576.

Lindblom U, Meyerson BA (1976) Mechanoreceptive and nociceptive thresholds during dorsal column simulation in man. In: *Advances in Pain Research and Therapy*. Bonica JJ, Albe-Fessard D, eds, 1:469–474. Raven Press, New York.

Lindeman HH (1973) Anatomy of the otolith organs. Adv Otorhinolaryngol 20:405–433.

Lintl P, Braak H (1983) Loss of intracortical myelinated fibers: a distinctive age-related alteration in the human striate area. Acta Neuropathol (Berl) 61:178–182.

Liu CN, Chambers WW (1964) An experimental study of the cortico-spinal system in the monkey (Macaca mulatta). The spinal pathways and preterminal distribution of degenerating fibers following discrete lesions of the pre- and postcentral gyri and bulbar pyramid. J Comp Neurol 123:257–284.

Livingstone MS (1988) Art, illusion and the visual system. Sci Am 258:78–85.

Livingstone MS, Hubel D (1988) Segregation of form, color, movement, and depth: anatomy, physiology, and perception. Science 240:740–749.

Livingstone MS, Hubel DH (1984) Specificity of intrinsic connections in primate primary visual cortex. J Neurosci 4:2830–2835.

Livingstone MS, Hubel DH (1987) Connections between layer 4B of area 17 and thick cytochrome oxidase stripes of area 18 in the squirrel monkey. J Neurosci 11:3371–3377.

Lloyd DPC (1943) Reflex action in relation to pattern and peripheral source of afferent stimulation. J Neurophysiol 6:111–119.

Lobel E, Kahane P, Leonards U, Grosbras M, Lehericy S, Le Bihan D, Berthoz A (2001) Localization of human frontal eye fields: anatomical and functional findings of functional magnetic resonance imaging and intracerebral electrical stimulation. J Neurosurg 95:804–815.

Lockard BI, Kempe LG (1988) Position sense in the lateral funiculus? Neurol Res 10:81–86.

Lockard I (1948) Certain developmental relations and fiber connections of the triangular gyrus in primates. J Comp Neurol 89:349–386.

Locke S (1967) Thalamic connections to insular and opercular cortex of monkey. J Comp Neurol 129:219–240.

Locke S, Angevine JB Jr, Marin OSM (1962) Projection of the magnocellular medial geniculate nucleus in man. Brain 85:319–330.

Loeser JD (1972) Dorsal rhizotomy for the relief of chronic pain. J Neurosurg 36:745–750.

Lóken AC, Brodal A (1970) A somatotopical pattern in the human lateral vestibular nucleus. Arch Neurol 23:350–357.

Loo SK (1977) Fine structure of the olfactory epithelium in some primates. J Anat 123:135–145.

Lopes da Silva FH, Arnolds DEAT (1978) Physiology of the hippocampus and related structures. Annu Rev Physiol 40:185–216.

Lorke DE, Kwong WH, Chan WY, Yew DT (2003) Development of catecholaminergic neurons in the human medulla oblongata. Life Sci 73:1315–1331.

Low PA (1993) Autonomic nervous system function. J Clin Neurophysiol 10:14–27.

Low PA (1997) Clinical Autonomic Disorders: Evaluation and Management. 2nd ed. Lippincott-Raven, Philadelphia.

Low PA (2003) Testing the autonomic nervous system. Semin Neurol 23:407–421.

Lubinska L (1975) On axoplasmic flow. Int Rev Neurobiol 17:241–296.

Lund JS (1988) Anatomical organization of macaque monkey striate visual cortex. Annu Rev Neurosci 11:253–288.

Lundberg PO, Werner I (1972) Trigeminal sensory neuropathy in systemic lupus erythematosus. Acta Neurol Scand 48:330–340.

Luschei ES, Fuchs AF Activity of brain stem neurons during eye movements of alert monkeys. J Neurophysiol 35:445–461, 1972.

Luttenberg J (1965) Contribution to the fetal ontogenesis of the corpus callosum in man. II. Folia Morphol (Praha) 13:136–144.

Lynn GE, Gilroy J (1972) Neuro-audiological abnormalities in patients with temporal lobe tumors. J Neurol Sci 17:167–184.

Macchi G (1951) The ontogenetic development of the olfactory telencephalon in man. J Comp Neurol 95:245–305.

Macchi G, Jones EG (1997) Toward an agreement on terminology of nuclear and subnuclear divisions of the motor thalamus. J Neurosurg 86:670–685.

Mackel R, Kunesch E, Waldhor F, Struppler A (1983) Reinnervation of mechanoreceptors in the human glabrous skin following peripheral nerve repair. Brain Res 268:49–65.

Mackenzie I, Meighan S, Pollock EN (1933) On the projection of the retinal quadrants on the lateral geniculate bodies, and the relationship of the quadrants to the optic radiations. Trans Ophthalmol Soc UK 53:142–169.

Macko KA, et al. (1982) Mapping the primate visual system with [2–14C]deoxyglucose. Science 218:394–397, 1982.

MacLean PD (1949) Psychosomatic disease and the visceral brain; recent developments bearing on the Papez theory of emotion. Psychosom Med 11:338–353.

MacLean PD (1952) Some psychiatric implications of physiological studies on frontotemporal portion of limbic system (visceral brain). Electroencephalogr Clin Neurophysiol 4:407–418.

MacLean PD (1977) The triune brain in conflict. Psychother Psychosom 28:207–220.

Magnin M, Fuchs AF (1977) Discharge properties of neurons in the monkey thalamus tested with angular acceleration, eye movement and visual stimuli. Exp Brain Res 28:293–299.

Magoon EH, Robb RM (1981) Development of myelin in human optic nerve and tract. Arch Ophthalmol 99:655–659.

Magoun HW, Beaton LE (1942) The salivatory motor nuclei in the monkey. Am J Physiol 136:720–725.

Mahler HR, Gurd JW, Wang Y-J (1983) Molecular topography of the synapse. In: *The Nervous System.* Tower DB, editor-in-chief. Brady RO, volume ed. Vol 1: The Basic Neurosciences, pp 455–466. Raven Press, New York.

Mai JK, Assheuer J, Paxinos G (2004) *Atlas of the Human Brain.* 2nd ed. Elsevier/Academic Press, Amsterdam.

Mair RG, Bouffard JA, Engen T, Morton TH (1978) Olfactory sensitivity during the menstrual cycle. Sens Processes 2:90–98.

Maisonpierre PC, et al. (1990) Neurotrophin-3: a neurotrophic factor related to NGF and BDNF. Science 247: 1446–1451.

Malinovský L, Sommerová J (1973) Sensory nerve endings in the human labia minora pudendi and their variability. Folia Morphol (Praha) 21:351–353.

Manaker S, Winokur A, Rhodes CH, Rainbor TC (1985) Autoradiographic localization of thyrotropin-releasing hormone (TRH) receptors in human spinal cord. Neurology 35:328–332.

Mann DMA (1985) Commentary on review by DM Gash, TJ Collier and JR Sladek. The application of neural transplantation to degenerative diseases of the human nervous system. Neurobiol Aging 6:160–162.

Mann MD (1984) The growth of the brain and skull in children. Devel Brain Res 13:169–178.

Mannoji H, Yeger H, Becker LE (1986) A specific histochemical marker (lectin Ricinus communis agglutinin-1) for normal human microglia, and application to routine histopathology. Acta Neuropathol (Berl) 71:341–343.

Manolidis L, Baloyannis S (1987) The neuropeptides of the acoustic cortex. I. Substance P. Acta Otolaryngol (Stockh) 103:481–488.

Manolidis L, Baloyannis SJ (1985) Pathology of acoustic nerve Schwannoma in vivo and in vitro. Acta Otolaryngol (Stockh) 99:280–284.

Marani E, Schoen JH (2005) A reappraisal of the ascending systems in man, with emphasis on the medial lemniscus. Adv Anat Embryol Cell Biol 179:1–74.

Marani E, Usunoff KG (1998) The trigeminal motonucleus in man. Arch Physiol Biochem 106:346–354.

Marc RE, Liu W-LS (1985) (3H) Glycine-accumulating neurons of the human retina. J Comp Neurol 232:241–260.

Marg E (1988) Imaging visual function of the human brain. Am J Optom Physiol Opt 65:828–851.

Margolis RB, Krause SJ, Tait RC (1985) Lateralization of chronic pain. Pain 23:289–293.

Mariani AP (1982) Biplexiform cells: ganglion cells of the primate retina that contact photoreceptors. Science 216:1134–1136.

Marrone ACH (1985) Anatomo-surgical study of the human parieto-occipital artery (angular artery). Arch Anat Hist Embryol (Strasb) 68:79–92.

Marrone ACH, Severino AG (1988) Insular course of the branches of the middle cerebral artery. Folia Morphol 36:331–336.

Marsden CD, Rowland R (1965) The mammalian pons, olive and pyramid. J Comp Neurol 124:175–187.

Marshall J (1951) Sensory disturbances in cortical wounds with special reference to pain. J Neurol Neurosurg Psychiat 14:187–204.

Marshall J, Grindle J, Ansell PL, Borwein B (1979) Convolution in human rods: an ageing process. Br J Ophthalmol 63:181–187.

Martin KAC (1988) From enzymes to visual perception: a bridge too far? TINS 11:380–387.

Martin KAC (1988) From single cells to simple circuits in the cerebral cortex. Q J Exp Physiol 73:637–702.

Marx JJ, Iannetti GD, Thomke F, Fitzek S, Urban PP, Stoeter P, Cruccu G, Dieterich M, Hopf HC (2005) Somatotopic organization of the corticospinal tract in the human brainstem: a MRI-based mapping analysis. Ann Neurol 57:824–831.

Masland RH (2001) The fundamental plan of the retina. Nat Neurosci 4:877–886.

Masland RH (2004) Neuronal cell types. Curr Biol 14:R497–R500.

Massey SC (2006) Functional anatomy of the mammalian retina. In: *Retina.* Ryan SJ, editor-in-chief. 4th ed. Chapter 4, 1:43–82. Elsevier/Mosby, Philadelphia.

Matelli M, Rizzolatti G, Bettinardi V, Gilardi MC, Perani D, Rizzo G, Fazio F (1993) Activation of precentral and mesial motor areas during the execution of elementary proximal and distal arm movements: a PET study. NeuroReport 4:1295–1298.

Mathews GJ, Osterholm JL (1972) Painful traumatic neuromas. Surg Clin North Am 51:1313–1324.

Mathias CJ, Bannister R (1999) *Autonomic Failure. A Textbook of Clinical Disorders of the Autonomic Nervous System.* 3rd ed. Oxford University Press, Oxford.

Mathias C, Bannister R (2002) *Autonomic Failure: A Textbook of Clinical Disorders of the Autonomic Nervous System.* 4th ed. Oxford University Press, Oxford.

Matsushima T, Rhoton AL Jr, Lenkey C (1982) Microsurgery of the fourth ventricle: Part 1. Microsurgical anatomy. Neurosurgery 11:631–667.

Matthews PBC (1977) Muscle afferents and kinaesthesia. Br Med Bull 33:137–142.

Matthews PBC (1980) Developing views on the muscle spindle. In: *Spinal and Supraspinal Mechanisms of Voluntary Motor Control and Locomotion.* Desmedt JE, ed., 8:12–27. Karger, Basel.

Matthews PBC (1980) Developing views on the muscle spindle. Prog Clin Neurophysiol 8:12–27.

Mattiace LA, Davies P, Yen S-H, Dickson DW (1990) Microglia in cerebellar plaques in Alzheimer's disease. Acta Neuropathol 80:493–498.

Mayberg HS (1997) Limbic-cortical dysregulation: a proposed model of depression. J Neuropsychiatry Clin Neurosci 9:471–481.

Mayer DJ, Price DD (1976) Central nervous system mechanisms of analgesia. Pain 2:379–404.

Mayo Clinic and Mayo Foundation (1976) *Clinical Examinations in Neurology.* 4th ed. Saunders, Philadelphia.

Mazziotta JC, Phelps ME, Carson RE, Kuhl DE (1982) Tomographic mapping of human cerebral metabolism: Auditory stimulation. Neurology 32:921–937.

McCabe BF (1965) The quick component of nystagmus. A presentation of a theory of its origin and mechanism involving the dynamic rhythmic inhibition of the slow component, based upon a comprehensive review of prior work and additional experimental evidence. Laryngoscope 75:1619–1646.

McCloskey DI (1978) Kinesthetic sensibility. Physiol Rev 58:763–820.

McCormick DA (1992) Neurotransmitter actions in the thalamus and cerebral cortex and their role in neuromodulation of thalamocortical activity. Prog Neurobiol 39:337–388.

McCotter RE (1915) A note on the course and distribution of the nervus terminalis in man. Anat Rec 9:243–246.

McCotter RE (1916) Regarding the length and extent of the human medulla spinalis. Anat Rec 10:559–564.

McCrary JA III (1977) Light reflex anatomy and the afferent pupil defect. Trans Am Acad Ophthalmol Otolaryngol 83:820–826.

McCutcheon NB, Saunders J (1972) Human taste papilla stimulation: stability of quality judgments over time. Science 175:214–216.

McDougal DB Jr (1984) Transport of transmitter-related enzymes: changes after injury. Adv Neurochem 6:105–118.

McElveen JT Jr, et al. (1985) Electrical stimulation of cochlear nucleus in man. Am J Otol (Suppl Issue) Nov:88–91.

McFadden D (Chairman, Working Group 89) (1982) *Tinnitus: Facts, Theories, and Treatments.* National Academy Pr, Washington.

McIntyre AD, Robinson RG (1959) Pathway for the jaw-jerk in man. Brain 82:468–474.

McKenzie J (1962) The development of the sternomastoid and trapezius muscles. Contrib Embryol Carneg Instn 37:121–129.

McLardy T (1950) Thalamic projection to frontal cortex in man. J Neurol Neurosurg Psychiatry 13:198–202.

McLaurin JA, Yong VW (1995) Oligodendrocytes and myelin. Neurol Clin 13:23–49.

McLone DG (2004) The anatomy of the ventricular system. Neurosurg Clin N Am 15:33–38.

McLone DG, Stieg PE, Scott RM, Barnett F, Barnes PD, Folkerth R (1998) Case problems in neurological surgery: Cerebellar epilepsy. Neurosurgery 42:1106–1111.

McMartin D (1983) Effects of age on axoplasmic transport in peripheral nerves. In: *Brain Aging: Neuropathology and Neuropharmacology.* Cervos-Navarro J, Sarkander H-I, eds, pp 351–361. Raven Press, New York.

McMasters RE, Weiss AH, Carpenter MB (1966) Vestibular projections to the nuclei of the extraocular muscles. Degeneration resulting from discrete partial lesions of the vestibular nuclei in the monkey. Am J Anat 118: 163–194.

McQuarrie IG (1988) Cytoskeleton of the regenerating axon. In: *Current Issues in Neural Regeneration Research.* Reier PJ, Bunge RP, Seil FJ, eds, pp 23–32. AR Liss, New York.

Meadows JC (1974) Disturbed perception of colours associated with localized cerebral lesions. Brain 97:615–632.

Mefford I, Oke A, Adams RN, Jonsson G (1977) Epinephrine localization in human brain stem. Neurosci Lett 5:141–145.

Mehler WR (1962) The anatomy of the so-called 'pain tract' in man: an analysis of the course and distribution of the ascending fibers of the fasciculus anterolateralis. In: *Basic Research in Paraplegia.* French JD, Porter RW, eds, pp 26–55. Thomas, Springfield.

Mehler WR (1966) The posterior thalamic region in man. Confin Neurol 27:18–29.

Mehler WR (1969) Some neurological species differences – a posteriori. Ann NY Acad Sci 167:424–468.

Mehler WR (1974) Central pain and the spinothalamic tract. Adv Neurol 4:127–146.

Mehler WR, Feferman ME, Nauta WJH (1960) Ascending axon degeneration following anterolateral cordotomy. An experimental study in the monkey. Brain 83:718–750.

Melzack R (1980) Psychologic aspects of pain. Res Publ Assoc Res Nerv Ment Dis 58:143–154.

Melzack R, Wall PD (1965) Pain mechanisms: a new theory. Science 150:971–979.

Mense S, Schmidt RF (1977) Muscle pain: which receptors are responsible for the transmission of noxious stimuli? In: *Physiological Aspects of Clinical Neurology.* Rose FC, ed., pp 265–278. Blackwell, Oxford.

Merton PA (1964) Human position sense and sense of effort. Symp Soc Exp Biol 18:387–400.

Merzenich MM, Brugge JF (1973) Representation of the cochlear partition on the superior temporal plane of the macaque monkey. Brain Res 50:275–296.

Messlinger K (1996) Functional morphology of nociceptive and other fine sensory endings (free nerve endings) in different tissues. Prog Brain Res 113:273–298.

Mesulam M-M (1979) Tracing neural connections of human brain with selective silver impregnation. Observations on geniculocalcarine, spinothalamic, and entorhinal pathways. Arch Neurol 36:814–818.

Mesulam MM (1995) Cholinergic pathways and the ascending reticular activating system of the human brain. Ann NY Acad Sci 757:169–179.

Mesulam MM (2004) The cholinergic innervation of the human cerebral cortex. Prog Brain Res 145:67–78.

Mesulam MM, Geula C, Bothwell MA, Hersh LB (1989) Human reticular formation: cholinergic neurons of the pedunculopontine and laterodorsal tegmental nuclei and some cytochemical comparisons to forebrain cholinergic neurons. J Comp Neurol 283:611–633.

Mesulam M-M, Mufson EJ, Levey AI, Wainer BH (1984) Atlas of cholinergic neurons in the forebrain and upper brainstem of the macaque based on monoclonal choline acetyltransferase immunohistochemistry and acetylcholinesterase histochemistry. Neuroscience 12:669–686.

Mesulam M-M, Pandya DN (1973) The projections of the medial geniculate complex within the Sylvian fissure of the rhesus monkey. Brain Res 60:315–333.

Meyer JE, Oot RF, Lindfors KK (1986) CT appearance of clival chordomas. J Comput Assist Tomogr 10:34–38.

Meyer M, Allison AC (1949) An experimental investigation of the connexions of the olfactory tracts in the monkey. J Neurol Neurosurg Psychiatry 12:274–286.

Meyerson BA, Boëthius J, Carlsson AM (1978) Percutaneous central gray stimulation for cancer pain. Appl Neurophysiol 41:57–65.

Meyerson BA, Linderoth B (2000) Mechanisms of spinal cord stimulation in neuropathic pain. Neurol Res 22:285–292.

Meyerson BA, Linderoth B (2006) Mode of action of spinal cord stimulation in neuropathic pain. J Pain Symptom Manage 31(Suppl 4):S6–12.

Miceli G (1982) The processing of speech sounds in a patient with cortical auditory disorder. Neuropsychologia 20:5–20.

Michel J-P, et al. (1986) Substance P-immunoreactive astrocytes related to deep white matter and striatal blood vessels in human brain. Brain Res 377:383–387.

Milea D, Lobel E, Lehéricy S, Duffau, Rivaud-Péchoux S, Berthoz A, Pierrot-Deseilligny C (2002) Intraoperative frontal eye field stimulation elicits ocular deviation and saccade suppression. NeuroReport 13:1359–1364.

Miles TS (1979) Features peculiar to the trigeminal innervation. Can J Neurol Sci 6:95–103.

Milhorat TH (1972) Hydrocephalus and the Cerebrospinal Fluid. Williams & Wilkins, Baltimore.

Miller BL, Cummings JL (1999) The Human Frontal Lobes: Functions and Disorders. The Guilford Press, New York.

Miller C, Haugh M, Kahn J, Anderton B (1986) The cytoskeleton and neurofibrillary tangles in Alzheimer's disease. TINS 9:76–81.

Miller JE (1971) Recent histologic and electron microscopic findings in extraocular muscle. Trans Am Acad Ophthalmol Otolaryngol 75:1175–1185.

Miller JE (1975) Aging changes in extraocular muscle. In: Basic Mechanisms of Ocular Motility and Their Clinical Implications. Lennerstrand G, Bach-Y-Rita P, eds, pp 47–61. Pergamon Pr, Oxford and New York.

Miller MR, Ralston HJ III, Kasahara M (1958) The pattern of cutaneous innervation of the human hand. Am J Anat 102:183–217.

Miller SH, Rusenas I (1976) Changes in primate Pacinian corpuscles following volar pad excision and skin grafting. Plast Reconstr Surg 57:627–636.

Mills CK (1891) The localization of the auditory centre. Brain 14:465–472.

Minckler J, et al. (1977) The human auditory pathways: a quantitative study. Brain Lang 4:152–155.

Minneapolis Pain Seminar (1975) Electrical stimulation of the human nervous system for the control of pain. Surg Neurol 4:61–204.

Mishkin M (1979) Analogous neural models for tactual and visual learning. Neuropsychologia 17:139–151.

Mitchell GAG, Warwick R (1955) The dorsal vagal nucleus. Acta Anat (Basel) 25:371–395.

Miyamoto T, Fukushima K, Takada T, De Waele C, Vidal PP (2005) Saccular projections in the human cerebral cortex. Ann NY Acad Sci 1039:124–131.

Modesti LM, Waszak M (1975) Firing pattern of cells in human thalamus during dorsal column stimulation. Appl Neurophysiol 38:251–258.

Mohn G, Van Hof-van Duin J (1986) Development of the binocular and monocular visual fields of human infants during the first year of life. Clin Vision Sci 1:51–64.

Mokrasch LC, Bear RS, Schmitt FO (1972) Myelin. In: Neurosciences Research Symposium Summaries. Schmitt FO, Adelman G, Melnechuk T, Worden FG, eds, 6:439–598. MIT Pr, Cambridge.

Molenaar I, Kuypers HGJM (1978) Cells of origin of propriospinal fibers and of fibers ascending to supraspinal

levels. A HRP study in cat and rhesus monkey. Brain Res 152:429–450.

Molina-Negro P, Bertrand RA (1982) Vestibular Neurotology. Symposium on Vestibular Neurotology, Montreal, September 9–12, 1980. Adv Otorhinolaryngol 28:1–148.

Møller AR (1972) The middle ear. In: Foundations of Modern Auditory Theory. Tobias JV, ed., 2:133–194. Academic Press, New York.

Møller AR (1978) Neurophysiological basis of discrimination of speech sounds. Audiology 17:1–9.

Moller AR (2006) History of cochlear implants and auditory brainstem implants. Adv Otorhinolaryngol 64:1–10.

Moller AR (2006) Physiological basis for cochlear and auditory brainstem implants. Adv Otorhinolaryngol 64:206–223.

Møller M (1978) Presence of a pineal nerve (nervus pinealis) in the human fetus; a light and electron microscopical study of the innervation of the pineal gland. Brain Res 154:1–12.

Monagle, RD, Brody H (1974) The effects of age upon the main nucleus of the inferior olive in the human. J Comp Neurol 155:61–66.

Montagna W (1977) Morphology of cutaneous sensory receptors. J Invest Dermatol 69:4–7.

Moore CI, Stern CE, Corkin S, Fischl B, Gray AC, Rosen BR, Dale AM (2000) Segregation of somatosensory activation in the human rolandic cortex using fMRI. J Neurophysiol 84:558–569.

Moore JK, Osen KK The cochlear nuclei in man. Am J Anat 154:393–418.

Morel A, Loup F, Magnin M, Jeanmonod D (2002) Neurochemical organization of the human basal ganglia: anatomofunctional territories defined by the distributions of calcium-binding proteins and SMI-32. J Comp Neurol 443:86–103.

Morel A, Magnin M, Jeanmonod D (1997) Multiarchitectonic and stereotactic atlas of the human thalamus. J Comp Neurol. 1997 Nov 3;387(4):588–630.

Morgane PJ, Galler JR, Mokler DJ (2005) A review of systems and networks of the limbic forebrain/limbic midbrain. Prog Neurobiol 75:143–160.

Mori S, Leblond CP (1969) Identification of microglia in light and electron microscopy. J Comp Neurol 135:57–80.

Morillo A, Cooper I (1955) Occlusion of the anterior choroidal artery. Am J Ophthalmol 40:796–801.

Morin F (1955) A new spinal pathway for cutaneous impulses. Am J Physiol 183:245–252.

Morin F, Schwartz HG, O'Leary JL (1952) Experimental study of the spinothalamic and related tracts. Acta Psychiatr Neurol Scand 26:371–396.

Moruzzi G, Magoun, HW (1949) Brain stem reticular formation and activation of the EEG. Electroenceph Clin Neurophysiol 1:455–473.

Moses HL, Ganote CE, Beaver DL, Schuffman SS (1966) Light and electron microscopic studies of pigment in human and rhesus monkey substantia nigra and locus coeruleus Anat Rec 155:167–184.

Mosinger JL, Yazulla S, Studholme KM (1986) GABA-like immunoreactivity in the vertebrate retina: a species comparison Exp Eye Res 42:631–644.

Moskowitz N (1969) Comparative aspects of some features of the central auditory system of primates. Ann NY Acad Sci 167:357–369.

Moskowitz N, Liu J-C (1972) Cortical projections of the spiral ganglion of the squirrel monkey. J Comp Neurol 144:335–344.

Moss MB, Rosene DL (1985) Neural transplantation: a panacea? Neurobiol Aging 6:168–169.

Mountcastle VB (2005) *The Sensory Hand: Neural Mechanisms of Somatic Sensation.* Harvard University Press, Cambridge.

Mountcastle VB, Henneman E (1952) The representation of tactile sensibility in the thalamus of the monkey. J Comp Neurol 97:409–431.

Mountcastle VB, Powell TPS (1959) Central nervous mechanisms subserving position sense and kinesthesia. Bull Hopkins Hosp 105:173–200.

Mufson EJ, Sobreviela T, Kordower JH (1997) Chemical neuroanatomy of the primate insula cortex: relationship to cytoarchitectonics, connectivity, function and neurodegeneration. In: *Handbook of Chemical Neuroanatomy.* Björklund A, Hökfelt T, eds. Vol 13: The Primate Nervous System, Part I, pp 377–454. Elsevier, Amsterdam.

Müller F, O'Rahilly R (2004) Embryonic development of the central nervous system. In: *The Human Nervous System.* Paxinos G, Mai JK, eds. 2nd ed., pp 22–48. Elsevier/Academic Press, Amsterdam.

Müller F, O'Rahilly R (1980) The early development of the nervous system in staged insectivore and primate embryos. J Comp Neurol 193:741–751.

Müller F, O'Rahilly R (1983) The first appearance of the major divisions of the human brain at stage 9. Anat Embryol (Berl) 168:419–432.

Müller F, O'Rahilly R (1984) Cerebral dysraphia (future anencephaly) in a human twin embryo at stage 13. Teratology 30:167–177.

Müller F, O'Rahilly R (1985) The first appearance of the neural tube and optic primordium in the human embryo at stage 10. Anat Embryol 172:157–169.

Müller F, O'Rahilly R (1986) The development of the human brain and the closure of the rostral neuropore at stage 11. Anat Embryol 175:205–222.

Müller F, O'Rahilly R (1987) The development of the human brain, the closure of the caudal neuropore, and the beginning of secondary neurulation at stage 12. Anat Embryol 176:413–430.

Müller F, O'Rahilly R (1988) The development of the human brain from a closed neural tube at stage 13. Anat Embryol 177:203–224.

Müller F, O'Rahilly R (1991) Development of anencephaly and its variants. Am J Anat 190:193–218.

Müller F, O'Rahilly R (1997) The timing and sequence of appearance of neuromeres and their derivatives in staged human embryos. Acta Anat (Basel) 158: 83–99.

Müller F, O'Rahilly R (2003) Segmentation in staged human embryos: the occipitocervical region revisited. J Anat 203:297–315.

Müller F, O'Rahilly R (2003) The prechordal plate, the rostral end of the notochord and nearby median features in staged human embryos. Cells Tissues Organs 173:1–20.

Müller F, O'Rahilly R (2004) Olfactory structures in staged human embryos. Cells Tissues Organs 178:93–116.

Müller F, O'Rahilly R (2004) The primitive streak, the caudal eminence and related structures in staged human embryos. Cells Tissues Organs 177:2–20.

Müller F, O'Rahilly R (2006) The amygdaloid complex and the medial and lateral ventricular eminences in staged human embryos. J Anat 208:547–564.

Müller-Preuss P, Ploog D (1981) Inhibition of auditory cortical neurons during phonation. Brain Res 215:61–76.

Mumford JM, Bowsher D (1976) Pain and protopathic sensibility. A review with particular reference to the teeth. Pain 2:223–243.

Munger BL (1971) Patterns of organization of peripheral sensory receptors. In: *Handbook of Sensory Physiology.* Loewenstein WR, ed., I:523–556. Springer-Verlag, New York.

Munger BL (1977) Neural-epithelial interactions in sensory receptors. J Invest Dermatol 69:27–40.

Munkle MC, Waldvogel HJ, Faull RL (1999) Calcium-binding protein immunoreactivity delineates the intralaminar nuclei of the thalamus in the human brain. Neuroscience 90:485–491.

Munkle MC, Waldvogel HJ, Faull RL (2000) The distribution of calbindin, calretinin and parvalbumin immunoreactivity in the human thalamus. J Chem Neuroanat 19:155–173.

Murabe Y, Sano Y (1982) Morphological studies on neuroglia VI. Postnatal development of microglia cells. Cell Tiss Res 225:469–485.

Murphy C, Cain WS, Bartoshuk LM (1977) Mutual action of taste and olfaction. Sens Processes 1:204–211.

Myrianthopoulos NC (1979) Our load of central nervous system malformations. Birth Defects: Original Article Series 15:1–18.

Nadol JB Jr (1981) Reciprocal synapses at the base of outer hair cells in the organ of Corti of man. Ann Otol Rhinol Laryngol 90:12–17.

Nageotte J (1906) The pars intermedia or nervus intermedius of Wrisberg, and the bulbo-pontine gustatory nucleus in man. Rev Neurol Psychiatr 4:473–488.

Nakano KK (1973) Anencephaly: a review. Dev Med Child Neurol 15:383–400.

Nashold B, Somjen G, Friedman H (1972) Paresthesias and EEG potentials evoked by stimulation of the dorsal funiculi in man. Exp Neurol 36:273–287.

Nashold BS Jr, Friedman H (1977) Pain and sensation: some recent observations in man. In: *Scientific Approaches to Clinical Neurology.* Goldensohn ES, Appel SH, eds, pp 1980–1988. Lea & Febiger, Philadelphia.

Nashold BS Jr, Friedman H (1972) Dorsal column stimulation for control of pain. Preliminary report on 30 patients. J Neurosurg 36:590–597.

Nashold BS Jr, Ostdahl RH (1979) Dorsal root entry zone lesions for pain relief. J Neurosurg 51:59–69.

Nashold BS Jr, Wilson WP, Boone E (1979) Depth recordings and stimulation of the human brain: a twenty year experience. In: *Functional Neurosurgery.* Rasmussen T, Marino R, eds, pp 181–195. Raven Press, New York.

Nashold BS Jr, Wilson WP, Slaughter DG (1969) Sensations evoked by stimulation in the midbrain of man. J Neurosurg 30:14–24.

Nathan H, Goldhammer Y (1973) The rootlets of the trochlear nerve. Anatomical observations in human brains. Acta Anat (Basel) 84:590–596.

Nathan PW (1976) The gate-control theory of pain: a critical review. Brain 99:123–158.

Nathan PW (1981) Gastric sensation: report of a case. Pain 10:259–262.

Nathan PW, Rudge P (1974) Testing the gate-control theory of pain in man. J Neurol Neurosurg Psychiatry 37:1366–1372.

Nathan PW, Smith MC (1959) Fasciculi proprii of the spinal cord in man: review of present knowledge. Brain 82:610–668.

Nathan PW, Smith MC (1982) The rubrospinal and central tegmental tracts in man. Brain 105:223–269.

Nathan PW, Smith MC, Deacon P (1990) The corticospinal tracts in man. Course and location of fibres at different segmental levels. Brain 113:303–324.

Nathan PW, Smith M, Deacon P (2001) The crossing of the spinothalamic tract. Brain 124:793–803.

Natkin E, Harrington GW, Mandel MA (1975) Anginal pain referred to the teeth. Report of a case. Oral Surg 40:678–680.

Nauta HJ, Hewitt E, Westlund KN, Willis WD Jr (1997) Surgical interruption of a midline dorsal column visceral pain pathway. Case report and review of the literature. J Neurosurg 86:538–542.

Nauta HJ, Soukup VM, Fabian RH, Lin JT, Grady JJ, Williams CG, Campbell GA, Westlund KN, Willis WD Jr (2000) Punctate midline myelotomy for the relief of visceral cancer pain. J Neurosurg 92:125–130.

Nelson RJ, Kaas JH (1981) Connections of the ventroposterior nucleus of the thalamus with the body surface representations in cortical areas 3b and 1 of the cynomolgus macaque, (Macaca fascicularis). J Comp Neurol 199:29–64.

Nelson RJ, Sur M, Felleman DJ, Kaas JH (1980) Representations of the body surface in postcentral parietal cortex of *Macaca fascicularis.* J Comp Neurol 192:611–643.

Neves G, Lagnado L (1999) The retina. Curr Biol 9: R674–R677.

Newburger PE, Sallan SE (1981) Chronic pain: principles of management. J Pediatr 98:180–189.

Newman EA (2003) New roles for astrocytes: regulation of synaptic transmission. Trends Neurosci 26:536–542.

Newman N, Gay AJ, Heilbrun MP (1971) Disjugate ocular bobbing: its relation to midbrain, pontine, and medullary function in a surviving patient. Neurology 21:633–637.

Newman PK, Terenty TR, Foster JB (1981) Some observations on the pathogenesis of syringomyelia. J Neurol Neurosurg Psychiatry 44:964–969.

Newman RP, Kinkel WR, Jacobs L (1984) Altitudinal hemianopia caused by occipital infarctions. Clinical and computerized tomographic correlations. Arch Neurol 41:413–418.

Newman SA, Miller NR (1983) Optic tract syndrome: neuronophthalmologic considerations. Arch Ophthalmol 10:1241–1250.

Niebroj-Dobosz I, Fidzianska A, Rafalowska J, Sawicka E (1980) Correlative biochemical and morphological studies of myelination in human ontogenesis. I. Myelination of the spinal cord. Acta Neuropathol (Berl) 49:145–152.

Nielsen DW, Slepecky N (1986) Stereocilia. In: *Neurobiology of Hearing: The Cochlea.* Altschuler RA, Hoffman DW, Bobbin RP, eds, pp 23–46. Raven Press, New York.

Niimi K, Kuwahara E (1973) The dorsal thalamus of the cat and comparison with monkey and man. J Hirnforsch 14:303–325.

Nishio J, Matsuya T, Machida J, Miyazaki T (1976) The motor nerve supply of the velopharyngeal muscles. Cleft Palate J 13:20–30.

Nitschke MF, Kleinschmidt A, Wessel K, Frahm J (1996) Somatotopic motor representation in the human anterior cerebellum. A high-resolution functional MRI study. Brain 119:1023–1029.

Noback CR, Moss ML (1956) Differential growth of the human brain. J Comp Neurol 105:539–551.

Nobin A, Björklund A (1973) Topography of the monoamine neuron systems in the human brain as revealed in fetuses. Acta Physiol Scand (Suppl) 388:1–40.

Nomina Anatomica. 5th ed. (1983) Williams & Wilkins, Baltimore.

Nomura Y, Kirikae I (1967) Innervation of the human cochlea. Ann Otol Rhinol Laryngol 76:57–68.

Nord SG, Ross GS (1973) Responses of trigeminal units in the monkey bulbar lateral reticular formation to noxious and non-noxious stimulation of the face: experimental and theoretical considerations. Brain Res 58:385–399, 1973.

Norrsell U (1980) Behavioral studies of the somatosensory system. Physiol Rev 60:327–354.

North RB, Fischell TA, Long DM (1977/78) Chronic dorsal column stimulation via percutaneously inserted epidural electrodes. Preliminary results in 31 patients. Appl Neurophysiol 40:184–191.

Norton AC, Kruger L (1973) *The Dorsal Column System of the Spinal Cord. Its Anatomy, Physiology, Phylogeny and Sensory Function.* 5th ed. Brain Info Service/Brain Res Institute, Los Angeles.

Nozue M, Ryu H, Uemura K (1987) Causes of vertigo, tinnitus, and sensorineural hearing loss, especially in relation to neurovascular compression. In: *The Vestibular System: Neurophysiologic and Clinical Research.* Graham MD, Kemink JL, eds, pp 205–209. Raven Press, New York.

Nuttall AL (1986) Physiology of hair cells. In: *Neurobiology of Hearing: The Cochlea.* Altschuler RA, Hoffman DW, Bobbin RP, eds, pp 47–75. Raven Press, New York.

O'Rahilly R, Müller F (2001) *Human Embryology & Teratology.* 3rd ed. Wiley-Liss, New York.

O'Rahilly R, Müller F (2001) Prenatal development of the brain. In: *Ultrasonography of the Prenatal & Neonatal Brain.* Timor-Tritsch I, Monteagudo A, Cohen H, eds. McGraw-Hill, New York.

O'Rahilly R, Müller F (2006) *The Embryonic Human Brain: An Atlas of Developmental Stage.* 3rd ed. Wiley-Liss, New York.

Oberheim NA, Wang X, Goldman S, Nedergaard M (2006) Astrocytic complexity distinguishes the human brain. Trends Neurosci 29:547–553.

Ochs S (1984) Axoplasmic transport in relation to nerve fiber regeneration. Adv Neurochem 6:1–12.

Ochs S, Erdman J, Jersild RA Jr, McAdoo V (1978) Routing of transported materials in the dorsal root and nerve fiber branches of the dorsal root ganglion. J Neurobiol 9:465–481.

Ochs S, Iqbal Z (1982) The role of calcium in axoplasmic transport in nerve. In: *Calcium and Cell Function*, 3:325–355. Academic Press, New York.

Ochs S, Worth RM (1978) Axoplasmic transport in normal and pathological systems. In: *Physiology and Pathobiology of Axons.* Waxman SG, ed., pp 251–264. Raven Press, New York.

Ödkvist LM, Liedgren SRC, Aschan G (1977) Cerebral cortex and vestibular nerve. Adv Otorhinolaryngol 22:125–135.

Ödkvist LM, Schwarz DWF, Fredrickson JM, Hassler R (1974) Projection of the vestibular nerve to the area 3a arm field in the squirrel monkey (*Saimiri sciureus*). Exp Brain Res 21:97–105.

Ödkvist LM, Schwarz DWF, Rubin AM, Fredrickson JM (1973) A comparative study of vestibulocortical projection. Int J Equilib Res 3:17–19.

O'Doherty J, Rolls ET, Francis S, Bowtell R, McGlone F (2001) Representation of pleasant and aversive taste in the human brain. J Neurophysiol 85:1315–1321.

Ogden TE (1984) Nerve fiber layer of the primate retina: morphometric analysis. Invest Ophthalmol Vis Sci 25:19–29.

Ohtsubo K, et al. (1990) Three-dimensional structure of Alzheimer's neurofibrillary tangles of the aged human brain revealed by the quick-freeze, deep-etch and replica method. Acta Neuropathol 79:480–485, 1990.

Oka K, Rhoton AL Jr, Barry M, Rodriguez R (1985) Microsurgical anatomy of the superficial veins of the cerebrum. Neurosurgery 17:711–748.

Okabe S, Hirokawa N (1991) Actin dynamics in growth cones. J Neurosci 11:1918–1929.

Okado N, Kakimi S, Kojima T (1979) Synaptogenesis in the cervical cord of the human embryo: sequence of synapse formation in a spinal reflex pathway. J Comp Neurol 184:491–518.

Olszewski J (1950) On the anatomical and functional organization of the spinal trigeminal nucleus. J Comp Neurol 92:401–413.

Olszewski J, Baxter D (1982) *Cytoarchitecture of the Human Brain Stem.* 2nd ed. Karger, Basel.

Ongerboer de Visser BW (1980) The corneal reflex: electrophysiological and anatomical data in man. Prog Neurobiol 15:71–83.

Ono M, Kubic S, Abernathey CD (1990) *Atlas of the Cerebral Sulci.* Thieme, New York.

Ono M, Rhoton AL Jr, Peace D, Rodriguez RJ (1984) Microsurgical anatomy of the deep venous system of the brain. Neurosurgery 15:621–57.

Onoda K, Kobayakawa T, Ikeda M, Saito S, Kida A (2005) Laterality of human primary gustatory cortex studied by MEG. Chem Senses 30:657–666.

Onofrio BM, Campa HK (1972) Evaluation of rhizotomy. Review of 12 years experience. J Neurosurg 36:751–755.

Onuf B (1902) On the arrangement and function of the cell groups of the sacral region of the spinal cord in man. Arch Neurol Psychopathol 3:387–412.

O'Rahilly R (1963) The early development of the otic vesicle in staged human embryos. J Embryol Exp Morphol 11:740–755.

O'Rahilly R (1965) The optic, vestibulocochlear, and terminal-vomeronasal neural crest in staged human embryos. In: *Second Symposium on Eye Structure.* Rohen JW, ed., pp 557–564. Schattauer Verlag, Stuttgart.

O'Rahilly R (1966) The early development of the eye in staged human embryos. Contrib Embryol Carneg Instn 38:1–42.

O'Rahilly R (1968) The development of the epiphysis cerebri and the subcommissural complex in staged human embryos. Anat Rec.160:488–489.

O'Rahilly R (1970) The manifestation of the axes of the human embryo. Z Anat Entwicklungsgesch 132:50–57.

O'Rahilly R (1973) *Developmental Stages in Human Embryos. Including a Survey of the Carnegie Collection. Part A: Embryos of the First Three Weeks (Stages 1–9).* Publication 631. Carnegie Inst, Washington, DC.

O'Rahilly R (1983) The timing and sequence of events in the development of the human eye and ear during the embryonic period proper. Anat Embryol 168:87–99.

O'Rahilly R (1997) Making planes plain. Clin Anat 10:128–129.

O'Rahilly R, Gardner E (1971) The timing end sequence of events in the development of the human nervous system during the embryonic period proper. Z Anat Entwicklungsgesch 134:1–12.

O'Rahilly R, Gardner E (1979) The initial development of the human brain. Acta Anat (Basel) 104:123–133.

O'Rahilly R, Müller F (1981) The first appearance of the human nervous system at stage 8. Anat Embryol (Berl) 163:1–13.

O'Rahilly R, Müller F (1994) Neurulation in the normal human embryo. Ciba Found Symp 181:70–82.

O'Rahilly R, Müller F (1999) Minireview: summary of the initial development of the human nervous system. Teratology 60(1):39–41.

O'Rahilly R, Müller F (2000) Prenatal ages and stages – measures and errors. Teratology 61:382–384.

O'Rahilly R, Müller F (2002) The two sites of fusion of the neural folds and the two neuropores in the human embryo. Teratology 65:162–170.

O'Rahilly R, Müller F (2003) Somites, spinal ganglia, and centra. Enumeration and interrelationships in staged human embryos, and implications for neural tube defects. Cells Tissues Organs 173:75–92.

O'Rahilly R, Müller F, Hutchins GM, Moore GW (1984) Computer ranking of the sequence of appearance of 100 features of the brain and related structures in staged human embryos during the first 5 weeks of development. Am J Anat 171:243–257.

O'Reilly GV, et al. (1983) Pseudotumor cerebri: computed tomography of resolving papilledema. J Comput Assist Tomogr 7:364–366.

Orkand RK (1982) Signalling between neuronal and glial cells. In: *Neuronal–glial Cell Interrelationships.* Sears TA, ed., pp 147–158. Springer-Verlag, New York.

Orona CJ (1990) Temporality and identity loss due to Alzheimer's disease. Soc Sci Med 30:1247–1256.

Osol G, Schwartz B (1984) Melatonin in the human retina. Exp Eye Res 38:213–215.

Osterberg G (1935) Topography of the layer of rods and cones in the human retina. Acta Ophthalmol (Suppl) (Copenh) 6:11–102.

Ostrowsky K, Magnin M, Ryvlin P, Isnard J, Guenot M, Mauguiere F (2002) Representation of pain and somatic sensation in the human insula: a study of responses to direct electrical cortical stimulation. Cereb Cortex 12:376–385.

Ota CY, Kimura RS (1980) Ultrastructural study of the human spiral ganglion. Acta Otolaryngol 89:53–62.

Overduin AS, Servos P (2004) Distributed digit somatotopy in primary somatosensory cortex. NeuroImage 23:462–472.

Owens E, Telleen CC (1981) Speech perception with hearing aids and cochlear implants. Arch Otolaryngol 107:160–163.

Özdamar Ö, Kraus N, Curry F (1982) Auditory brain stem and middle latency responses in a patient with cortical deafness. Electroencephalogr Clin Neurophysiol 53:224–230.

Paggio GF, Mountcastle VB (1963) The functional properties of ventrobasal thalamic neurons studied in unanesthetized monkeys. J Neurophysiol 26:775–806.

Pagin CA, Maspes PE (1976) Microneurosurgical treatment of trigeminal neuralgia by selective juxtapontine rhizotomy of the portio major sparing the intermediate fibers. In: *Advances in Pain Research and Therapy.* Bonica JJ, Albe-Fessard D, eds, 1:849–853. Raven Press, New York.

Paintal AS (1972) Cardiovascular receptors. In: *Handbook of Sensory Physiology.* Neil E, volume ed., III/1:1–45. Springer-Verlag, New York.

Palay SL (1956) Synapses in the central nervous system. J Biophys Biochem Cytol 2:193–202.

Palay SL (1967) Principles of cellular organization in the nervous system. In: *The Neurosciences; A Study Program.* Quarton GC, Melnechuk T, Schmitt FO, eds, pp 24–31. Rockefeller University Press, New York.

Palay SL, Chan-Palay V (1976) A guide to the synaptic analysis of the neuropil. Cold Spring Harbor Symp Quant Biol 40:1–16.

Palay SL, Sotelo C, Peters A. Orkand PM (1968) The axon hillock and the initial segment. J Cell Biol 38:193–201.

Pamulova L, Linder B, Rask-Andersen H (2006) Innervation of the apical turn of the human cochlea: a light microscopic and transmission electron microscopic investigation. Otol Neurotol 27:270–275.

Pandya DN, Hallett M, Mukherjee SK (1969) Intra- and interhemispheric connections of the neocortical auditory system in the rhesus monkey. Brain Res 14:49–65.

Pandya DN, Sanides F (1973) Architectonic parcellation of the temporal operculum in rhesus monkey and its projection pattern. Z Anat Entwickl-Gesch 139:127–161.

Pandya DN, Yeterian EH (1996) Comparison of prefrontal architecture and connections. Philos Trans R Soc Lond B Biol Sci 351:1423–1432.

Paparella MM, Oda M, Hiraide F, Brady D (1972) Pathology of sensorineural hearing loss in otitis media. Ann Otol Rhinol Laryngol 81:632–647.

Papez, JW (1937) A proposed mechanism of emotion. Arch Neurol and Psychiat 38:725–743.

Pappagallo M (2005) *The Neurological Basis Of Pain.* McGraw-Hill, New York.

Pappas GD (1975) Ultrastructural basis of synaptic transmission. In: *The Nervous System.* Tower DB, editor-in-chief. Brady RO, volume ed. Vol 1: The Basic Neurosciences, pp 19–30. Raven Press, New York.

Pappas GD, Purpura DP (1972) *Structure and Function of Synapses.* Raven Press, New York.

Paran N, Mattern CF, Henkin RI (1975) The ultrastructure of the taste bud of the human fungiform papilla. Cell Tissue Res 161:1–10.

Parent M, Parent A (2005) Single-axon tracing and three-dimensional reconstruction of centre median-parafascicular thalamic neurons in primates. J Comp Neurol 481:127–144.

Parker GJM, Luzzi S, Alexander DC, Wheeler-Kingshott CAM, Ciccarelli O, Lambon Ralph MA (2004) Lateralization of ventral and dorsal auditory-language pathways in the human brain. NeuroImage 24:656–666.

Parkins CW (1985) The bionic ear: principles and current status of cochlear prostheses. Neurosurgery 16:853–865.

Parsons LM, Sergent J, Hodges DA, Fox PT (2005) The brain basis of piano performance. Neuropsychologia 43:199–215.

Paton D, Hyman BN, Justice J Jr (1977) *Introduction to Ophthalmoscopy.* A Scope Publication. Upjohn, Kalamazo.

Patton HD, Ruch TC (1946) The relation of the front of the pre- and post-central gyrus to taste in the monkey and chimpanzee. Fed Proc 5:79.

Paul RL, Merzenich M, Goodman H (1972) Representation of slowly and rapidly adapting cutaneous mechanoreceptors of the hand in Brodmann's areas 3 and 1 of Macaca mulatta. Brain Res 36:229–249.

Paullus WS, Pait TG, Rhoton AL Jr (1977) Microsurgical exposure of the petrous portion of the carotid artery. J Neurosurg 47:713–726.

Paus T (1996) Location and function of the human frontal eye-field: a selective review. Neuropsychologia 34:475–483.

Paus T (2000) Functional anatomy of arousal and attention systems in the human brain. Prog Brain Res 126:65–77.

Paxinos G, Huang X-F (2004) *Atlas of the Human Brainstem.* Academic Press, San Diego.

Payne HA, et al. (1978) Recovery from primary pontine hemorrhage. Ann Neurol 4:557–558.

Pearce JM (2006) Glossopharyngeal neuralgia. Eur Neurol 55:49–52.

Pearce JMS (2003) The nucleus of Theodor Meynert (1833–1892). J Neurol Neurosurg Psychiatry 74:1358.

Pearce JMS (2006) Brodmann's cortical maps. J Neurol Neurosurg Psychiatry 76:259.

Pearlman AL, Birch J, Meadows JC (1979) Cerebral color blindness: an acquired defect in hue discrimination. Ann Neurol 5:253–261.

Pearson AA (1938) The spinal accessory nerve in human embryos. J Comp Neurol 68:243–266, 1938.

Pearson AA (1939) The hypoglossal nerve in human embryos. J Comp Neurol 71:21–39, 1939.

Pearson AA (1941) The development of the olfactory nerve in man. J Comp Neurol 75:199–21.

Pearson AA (1943) The trochlear nerve in human fetuses. J Comp Neurol 78:29–43, 1943.

Pearson AA (1944) The oculomotor nucleus in the human fetus. J Comp Neurol 80:47–63.

Pearson AA (1946) The development of the motor nuclei of the facial nerve in man. J Comp Neurol 85:461–476.

Pearson AA (1947) The roots of the facial nerve in human embryos and fetuses. J Comp Neurol 87:139–159, 1947.

Pearson AA (1949) Further observations on the mesencephalic root of the trigeminal nerve. J Comp Neurol 91:147–194.

Pearson AA (1949) The development and connections of the mesencephalic root of the trigeminal nerve in man. J Comp Neurol 90:1–46.

Pearson AA (1952) Role of gelatinous substance of spinal cord in conduction of pain. AMA Arch Neurol Psychiatr 68:515–529.

Pearson AA, Sauter RW (1971) Observations on the caudal end of the spinal cord. Am J Anat 131:463–470.

Pellerin L (2005) How astrocytes feed hungry neurons. Mol Neurobiol 32:59–72.

Pelletier VA, Poulos DA, Lende RA (1974) Functional localization in the trigeminal root. J Neurosurg 40:504–513.

Penfield W (1947) Some observations on the cerebral cortex of man. Proc Soc Lond (Biol) 134:329–347.

Penfield W (1957) Vestibular sensation and the cerebral cortex. Ann Otol Rhinol Laryngol 66:691–698.

Penfield W (1958) Functional localization in temporal and deep sylvian areas. Res Publ Assoc Res Nerv Ment Dis 36:210–226.

Penfield W (1958) The Excitable Cortex in Conscious Man. Thomas, Springfield.

Penfield W, Boldrey E (1937) Somatic motor and sensory representation in the cerebral cortex of man as studied by electrical stimulation. Brain 60:389–443.

Penfield W, Jasper H (1954) Epilepsy and the Functional Anatomy of the Human Brain, pp 26, 115, 127, 129, 168–169, 371–372, 406, 439, 448, 450, 527, 829. Little, Brown, Boston.

Penfield W, Perot P (1963) The brain's record of auditory and visual experience. A final summary and discussion. Brain 86:595–696.

Penfield W, Rasmussen T (1950) The Cerebral Cortex of Man. A Clinical Study of Localization of Function. Macmillan, New York.

Peng N, Wei CC, Oparil S, Wyss JM (2000) The organum vasculosum of the lamina terminalis regulates noradrenaline release in the anterior hypothalamic nucleus. Neuroscience 99:149–156.

Perl ER (1980) Afferent basis of nociception and pain: evidence from the characteristics of sensory receptors and their projections to the spinal dorsal horn. Res Publ Assoc Res Nerv Ment Dis 58:19–45.

Perl ER (1984) Pain and nociception. In: Handbook of physiology. The nervous system, Vol 3, Darian-Smith I, ed., pp 915–975. Bethesda, MD: American Physiological Society.

Perl ER (1992) Function of dorsal root ganglion neurons: an overview. In: Sensory neurons: diversity, development and plasticity, Scott S, ed., pp 3–23. New York: Oxford. Wien, Austria: Springer.

Perlmutter D, Rhoton AL Jr (1976) Microsurgical anatomy of the anterior cerebral-anterior communicating-recurrent artery complex. J Neurosurg 45:259–272.

Perrot X, Ryvlin P, Isnard J, Guenot M, Catenoix H, Fischer C, Mauguiere F, Collet L (2006) Evidence for corticofugal modulation of peripheral auditory activity in humans. Cereb Cortex 16:941–948.

Perry G, et al. (1986) Electron microscopic localization of Alzheimer neurofibrillary tangle components recognized by an antiserum to paired helical filaments. J Neuropath Exp Neurol 45:161–168.

Perry G, et al. (1991) Neuropil threads of Alzheimer's disease show a marked alteration of the normal cytoskeleton. J Neurosci 11:1748–1755.

Perry G, Rizzuto N, Autilio-Gambetti L, Gambetti P (1985) Paired helical filaments from Alzheimer disease patients contain cytoskeletal components. Proc Natl Acad Sci USA 82:3916–3920.

Perry VH, Oehler R, Cowey A (1984) Retinal ganglion cells that project to the dorsal lateral geniculate nucleus in the macaque monkey. Neuroscience 12:1101–1123.

Pert A, Yaksh T (1974) Sites of morphine induced analgesia in the primate brain: relation to pain pathways. Brain Res 80:135–140.

Peters A, Palay SL, Webster HdeF (1976) The Fine Structure of the Nervous System: The Neurons and Supporting Cells. Saunders, Philadelphia.

Petit TL, LeBoutillier JC, Alfano DP, Becker LE (1984) Synaptic development in the human fetus: a morphometric analysis of normal and Down's Syndrome neocortex. Exp Neurol 83:13–23.

Pettegrew RK, Windle WF (1976) Factors in recovery from spinal cord injury. Exp Neurol 53:815–829.

Peters A, Jones EG (1984) Cerebral Cortex. Vol 1: Cellular Components of the Cerebral Cortex. Plenum Press, New York.

Peters A, Jones EG (1984) Cerebral Cortex. Vol 2: Functional Properties of Cortical Cells. Plenum Press, New York.

Peters A, Jones EG (1984) Cerebral Cortex. Vol 3: Visual Cortex. Plenum Press, New York.

Peters A, Jones EG (1985) Cerebral Cortex. Vol 4: Association and Auditory Cortices. Plenum Press, New York.

Peters A, Jones EG (1988) Cerebral Cortex. Vol 7: Development and Maturation of Cerebral Cortex. Plenum Press, New York.

Pfaffmann C (1978) Neurophysiological mechanisms of taste. Am J Clin Nutr 31:1058–1067.

Pfaffmann C, Frank M. Norgren R (1979) Neural mechanisms and behavioral aspects of taste. Annu Rev Psychol 30:283–325.

Phelps ME, Kuhl DE, Mazziotta JC (1981) Metabolic mapping of the brain's response to visual stimulation: studies in humans. Science 211:1445–1448.

Philibert B, Veuillet E, Collet L (1998) Functional asymmetries of crossed and uncrossed medial olivocochlear efferent pathways in humans. Neurosci Lett 253:99–102.

Phillips CG, Powell TPS, Wiesendanger M (1971) Projection from low-threshold muscle afferents of hand and forearm to area 3a of baboon's cortex. J Physiol 217:419–446.

Phillips DD, Hibbs RG, Ellison JP, Shapiro H (1972) An electron microscopic study of central and peripheral nodes of Ranvier. J Anat 111:229–238.

Phillips JR, Johnson KO (1981) Tactile spatial resolution. II. Neural representation of bars, edges, and gratings in monkey primary afferents. J Neurophysiol 46:1192–1203.

Phillips JR, Johnson KO (1981) Tactile spatial resolution. III. A continuum mechanics model of skin predicting mechanoreceptor responses to bars, edges, and gratings. J Neurophysiol 46:1204–1225.

Phillis JW (1980) Substance P in the central nervous system. In: *The Role of Peptides in Neuronal Function*. Barker JL, Smith Jr TG, eds, pp 615–652. Marcel Dekker Inc, New York and Basel.

Picard N, Strick PL (1996) Motor areas of the medial wall: a review of their location and functional activation. Cereb Cortex 6:342–353.

Picard N, Strick PL (2001) Imaging the premotor areas. Curr Opin Neurobiol 11:663–672.

Pick J (1970) *The Autonomic Nervous System*. Lippincott, Philadelphia.

Picker S, Goldring S (1982) Electrophysiological properties of human glia. TINS 5:73–76.

Pierson RJ, Carpenter MB (1974) Anatomical analysis of pupillary reflex pathways in the rhesus monkey. J Comp Neurol 158:121–144.

Pineda A (1978) Complications of dorsal column stimulation. J Neurosurg 48:64–68.

Pitts DG (1982) The effects of aging on selected visual functions: dark adaptation, visual acuity, stereopsis, and brightness contrast. In: *Aging and Human Visual Function*. Sekuler R, Kline D, Dismukes K, eds, pp 131–159. AR Liss, New York.

Plum F, Posner JB (1982) The pathologic physiology of signs and symptoms of coma. In: *The Diagnosis of Stupor and Coma*. Plum F, Posner JB, eds, pp 1–86. FA Davis Co, Philadelphia.

Podoshin L, Fradis M (1975) Hearing loss after head injury. Arch Otolaryngol 101:15–18.

Pollack JG, Hickey TL (1979) The distribution of retino-collicular axon terminals in rhesus monkey. J Comp Neurol 185:587–602.

Pollak A, Felix H, Schrott A (1987) Methodological aspects of quantitative study of spiral ganglion cells. Acta Otolaryngol (Suppl) (Stockh) 436:37–42.

Pollen DA (1973) Focal epilepsy and the neuroglial impairment hypothesis. In: *Epilepsy: its Phenomena in Man*. Brazier MAB, ed., pp 29–35. Academic Press, New York.

Pollen DA, Trachtenberg MC (1970) Neuroglia: gliosis and focal epilepsy. Science 167:1252–1253.

Pompeiano O (1973) Reticular formation. In: *Handbook of Sensory Physiology*. Iggo A, ed., II:381–488. Springer-Verlag, New York.

Porter R, Lemon R (1993) *Corticospinal Function and Voluntary Movement*. Oxford University Press, New York.

Potter H, Nauta WJH (1979) A note on the problem of olfactory associations of the orbitofrontal cortex in the monkey. Neuroscience 4:361–367.

Powell TPS (1977) The somatic sensory cortex. Br Med Bull 33:129–135.

Powell TPS, Mountcastle VB (1959) Some aspects of the functional organization of the cortex of the postcentral gyrus of the monkey: a correlation of finding obtained in a single unit analysis with cytoarchitecture. Bull Hopkins Hosp 105:133–162.

Powell TPS, Mountcastle VB (1959) The cytoarchitecture of the postcentral gyrus of the monkey *Macaca mulatta*. Bull Hopkins Hosp 105:108–132.

Pribram KH, Chow KL, Semmes J (1953) Limit and organization of the cortical projection from the medial thalamic nucleus in monkey. J Comp Neurol 98:433–448.

Price DD, Dubner R (1977) Neurons that subserve the sensory-discriminative aspects of pain. Pain 3:307–338.

Price DD, Dubner R, Hu JW (1976) Trigeminothalamic neurons in nucleus caudalis responsive to tactile, thermal, and nociceptive stimulation of monkey's face. J Neurophysiol 39:936–953.

Price DD, Hayashi H, Dubner R, Ruda MA (1979) Functional relationships between neurons of marginal and substantia gelatinosa layers of primate dorsal horn. J Neurophysiol 42:1590–1608.

Price DD, Hayes RL, Ruda M, Dubner R (1978) Spatial and temporal transformations of input to spinothalamic tract neurons and their relation to somatic sensations. J Neurophysiol 41:933–947.

Price DD, Mayer DJ (1974) Physiological laminar organization of the dorsal horn of *M. mulatta*. Brain Res 79:321–325.

Prosiegel M, Holing R, Heintze M, Wagner-Sonntag E, Wiseman K (2005) The localization of central pattern generators for swallowing in humans – a clinical-anatomical study on patients with unilateral paresis of the vagal nerve, Avellis' syndrome, Wallenberg's syndrome, posterior fossa tumours and cerebellar hemorrhage. Acta Neurochir Suppl 93:85–88.

Pruzansky S (1975) Anomalies of face and brain. Birth Defects: Original Article Series 11:183–204.

Purpura DP (1974) Dendritic spine 'dysgenesis' and mental retardation. Science 186:1126–1128.

Rademacher J (2002) Topographical variability of cytoarchitectonic areas. In: *Cortical Areas: Unity and Diversity*. Schüz A, Miller R, eds, pp 53–78. Taylor & Francis, London and New York.

Rademacher J, Burgel U, Geyer S, Schormann T, Schleicher A, Freund HJ, Zilles K (2001) Variability and asymmetry in the human precentral motor system. A cytoarchitectonic and myeloarchitectonic brain mapping study. Brain 124:2232–2258.

Radius RL (1983) Pressure-induced fast axonal transport abnormalities and the anatomy at the lamina cribrosa in primate eyes. Invest Ophthalmol Vis Sci 24:343–346.

Radius RL, Anderson DR (1979) The course of axons through the retina and optic nerve head. Arch Ophthalmol 97:1154–1158.

Radius RL, Anderson DR (1979) The histology of retinal nerve fiber layer bundles and bundle defects. Arch Ophthalmol 97:948–950.

Radpour S, Gacek RR (1980) Facial nerve nucleus in the cat. Further study. Laryngoscope 90:685–692.

Rae ASL (1954) The form and structure of the human claustrum. J Comp Neurol 100:15–39.

Raeva SN (2006) The role of the parafascicular complex (CM-Pf) of the human thalamus in the neuronal mechanisms of selective attention. Neurosci Behav Physiol 36:287–295.

Rafel E, et al. (1980) Congenital insensitivity to pain with anhidrosis. Muscle Nerve 3:216–220.

Raine CS (1984) Morphology of myelin and myelination. In: *Myelin.* Morell P, ed. 2nd ed., pp 1–50. Plenum, New York.

Raisman G, Matthews MR (1972) Degeneration and regeneration of synapses. In: *The Structure and Function of Nervous Tissue.* Bourne GH, ed., 4:61–104. Academic Press, New York.

Rakic P (1975) Cell migration and neuronal ectopias in the brain. Birth Defects: Original Article Series 11:95–129.

Rakic P (1982) The role of neuronal–glial cell interaction during brain development. In: *Neuronal–glial Cell Interrelationships.* Sears TA, ed., pp 25–38. Springer-Verlag, New York.

Rakic P (1983) Geniculo-cortical connections in primates: normal and experimentally altered development. Prog Brain Res 58:393–404.

Rakic P (1985) DNA synthesis and cell division in the adult primate brain. Ann NY Acad Sci 457:193–211.

Rakic P, Goldman-Rakic PS, Gallager D (1988) Quantitative autoradiography of major neurotransmitter receptors in the monkey striate and extrastriate cortex. J Neurosci 8:3670–3690.

Rakic P, Yakovlev PI (1968) Development of the corpus callosum and cavum septi in man. J Comp Neurol 132:45–72.

Ralston DD, Ralston HJ 3rd (1985) The terminations of corticospinal tract axons in the macaque monkey. J Comp Neurol 242:325–337.

Ralston HJ III (1974) On the neuronal organization of the spinal cord. In: *Essays on the Nervous System.* Bellair R, Gray EG, eds, pp 179–190. Clarendon Pr, Oxford.

Ralston HJ III (2005) Pain and the primate thalamus. Prog Brain Res 149:1–10.

Ralston HJ III, Miller MR, Kasahara M (1960) Nerve endings in human fasciae, tendons, ligaments, periosteum, and joint synovial membrane. Anat Rec 136:137–147.

Ralston HJ III, Ralston DD (1979) The distribution of dorsal root axons in laminae I, II and III of the macaque spinal cord: a quantitative electron microscope study. J Comp Neurol 184:643–684.

Ramnani N, Owen AM (2004) Anterior prefrontal cortex: insights into function from anatomy and neuroimaging. Nature Reviews/Neuroscience 5:184–194.

Randolph M, Semmes J (1974) Behavioral consequences of selective subtotal ablations in the postcentral gyrus of *Macaca mulatta.* Brain Res 70:55–70.

Rapaport DH, Stone J (1984) The area centralis of the retina in the cat and other mammals: focal point for function and development of the visual system. Neuroscience 11:289–301.

Rasmussen AT, Peyton WT (1941) The location of the lateral spinothalamic tract in the brain stem of man. Surgery 10:699–710.

Rasmussen AT, Peyton WT (1948) The course and termination of the medial lemniscus in man. J Comp Neurol 88:411–424.

Rasmussen GL (1967) Efferent connections of the cochlear nucleus. In: *Sensorineural Hearing Processes and Disorders.* Graham AB, ed., pp 61–75. Little, Brown, Boston.

Rasmussen GL, Windle WF (1960) *Neural Mechanisms of the Auditory and Vestibular Systems.* Thomas, Springfield.

Rausch R, Serafetinides EA (1975) Human temporal lobe and olfaction. In: *Olfaction and Taste V.* Denton DA, Coghlan JP, eds, pp 321–324. Academic Press, New York.

Rausch R, Serafetinides EA, Crandall PH (1977) Olfactory memory in patients with anterior temporal lobectomy. Cortex 13:445–452.

Raviv JR, Kern RC (2004) Chronic sinusitis and olfactory dysfunction. Otolaryngol Clin North Am 37: 1143–1157.

Reed CL, Klatzky RL, Halgren E (2005) What vs. where in touch: an fMRI study. NeuroImage 25:718–726.

Reger SN (1978) Selected hearing impairment associated with pinealoma. Ann Otol Rhinol Laryngol 87:834–836.

Reh TA, Moshiri A (2006) The development of the retina. In: *Retina.* Ryan SJ, editor-in-chief. 4th ed. Chapter 1, 1:2–21. Elsevier/Mosby, Philadelphia.

Reichlin S, Baldessarini RJ, Martin JB (1978) *The Hypothalamus.* Res Publ Assoc Res Nerv Ment Dis. Vol 56. Raven Press, New York.

Reid JD (1960) Effects of flexion–extension movements of the head and spine upon the spinal cord and nerve roots. J Neurol Neurosurg Psychiatry 23:214–221.

Reiter ER, DiNardo LJ, Costanzo RM (2004) Effects of head injury on olfaction and taste. Otolaryngol Clin North Am 37:1167–1184.

Reiter RJ (1978) Evidence for an endocrine function of the human pineal gland. J Neural Transm Suppl 13:247–249.

Réthelyi M, Trevino DL, Perl ER (1979) Distribution of primary afferent fibers within the sacrococcygeal dorsal horn: an autoradiographic study. J Comp Neurol 185:603–622.

Rexed B (1964) Some aspects of the cytoarchitectonics and synaptology of the spinal cord. Prog Brain Res 11:58–92.

Rhodes RH (1978) Development of the human optic disc: light microscopy. Am J Anat 153:601–616.

Rhodes RH (1979) A light microscopic study of the developing human neural retina. Am J Anat 154:195–210.

Rhoton AL Jr (1978) Microsurgical neurovascular decompression for trigeminal neuralgia and hemifacial spasm. J Fla Med Assoc 65:425–428.

Rhoton AL Jr (2000) Cerebellum and fourth ventricle. Neurosurgery 47 (Suppl 3):S7–27.

Rhoton AL Jr (2000) The cerebellar arteries. Neurosurgery 47 (Suppl 3):S29–68.

Rhoton AL Jr (2000) Tentorial incisura. Neurosurgery 47 (Suppl 3):S131–153.

Rhoton AL Jr (2000) The foramen magnum. Neurosurgery 47 (Suppl 3):S155–193.

Rhoton AL Jr (2002) The cavernous sinus, the cavernous venous plexus, and the carotid collar. Neurosurgery 51 (Suppl 1):375–410.

Rhoton AL Jr (2002) The cerebrum. Neurosurgery 51 (Suppl 1):1–51.

Rhoton AL Jr (2002) The lateral and third ventricles. Neurosurgery 51(Suppl 4):S207–271.

Rhoton AL Jr (2002) The sellar region. Neurosurgery 51 (Suppl 1):335–374.

Rhoton AL Jr (2002) The supratentorial arteries. Neurosurgery 51 (Suppl 4):S53–120.

Rhoton AL Jr, Fujii K, Fradd B (1979) Microsurgical anatomy of the anterior choroidal artery. Surg Neurol 12:171–187.

Rhoton AL Jr, O'Leary JL, Ferguson JP (1966) The trigeminal, facial, vagal, and glossopharyngeal nerves in the monkey. Afferent connections. Arch Neurol 14: 530–540.

Richardson DE (1974) Thalamotomy for control of chronic pain. Acta Neurochir (Suppl) 21:77–88.

Richardson DE (1976) Brain stimulation for pain control. IEEE Trans Biomed Eng 23:304–306.

Richardson DE, Akil H (1977) Long term results of periventricular gray self-stimulation. Neurosurgery 1:199–202.

Richardson DE, Akil H (1977) Pain reduction by electrical brain stimulation in man. Part 1: Acute administration in periaqueductal and periventricular sites. J Neurosurg 47:178–183.

Richardson DE, Akil H (1977) Pain reduction by electrical brain stimulation in man. Part 2: Chronic self-administration in the periventricular gray matter. J Neurosurg 47:184–194.

Richter DW (1996) Neural regulation of respiration: rhythmogenesis and afferent control. In: *Comprehensive Human Physiology*. Greger R, Windhorst U, eds, pp 2079–2095. Springer-Verlag, Berlin.

Richter E (1980) Quantitative study of human Scarpa's ganglion and vestibular sensory epithelia. Acta Otolaryngol 90:199–208.

Rieck RW, Ansari MS, Whetsell WO Jr, Deutch AY, Kessler RM (2004) Distribution of dopamine D2-like receptors in the human thalamus: autoradiographic and PET studies. Neuropsychopharmacology 29:362–372.

Riga D, Riga S, Halalau F, Schneider F (2006) Brain lipopigment accumulation in normal and pathological aging. Ann NYAcad Sci 1067:158–163.

Riley HA (1943) *An Atlas of the Basal Ganglia, Brain Stem and Spinal Cord: Based on Myelin-stained Material*. Williams & Wilkins, Baltimore.

Risse GL, et al. (1978) The anterior commissure in man: functional variation in a multisensory system. Neuropsychologia 16:23–31.

Ritchie J (1973) Pain from distension of the pelvic colon by inflating a balloon in the irritable colon syndrome. Gut 14:125–132.

Roberts M, Hanaway J (1970) *Atlas of the Human Brain in Section*. Lea & Febiger, Philadelphia.

Roberts TDM (1976) Vestibular physiology In: *Scientific Foundations of Otolaryngology*. Hinchcliffe R, Harrison D, eds, pp 371–382. Year Bk Med, Chicago.

Roberts TS, Akert K (1963) 1. Insular and opercular cortex and its thalamic projection in macaca mulatta. Schweiz Arch Neurol Neurochir Psychiatr 92:1–43.

Robertson D, Low PA, Polinsky RJ, (1996) *Primer on the Autonomic Nervous System*. Academic Pr, San Diego.

Robertson JD (1985) The synapse: morphological and chemical correlates of function. In: *Neurosciences Research Symposium Summaries*. Schmitt FO, Melnechuk T, eds, 1:463–541. MIT Pr, Cambridge.

Robinson CJ, Burton H (1980) Somatotopographic organization in the second somatosensory area of *M. fascicularis*. J Comp Neurol 192:43–67.

Robinson DA, O'Meara DM, Scott AB, Collins CC (1969). Mechanical components of human eye movements. J Appl Physiol 26:548–553.

Robinson RJ, Tizard JPM (1966) The central nervous system in the new-born. Br Med Bull 22:49–55.

Rodieck RW, Binmoeller KF, Dineen J (1985) Parasol and midget ganglion cells of the human retina. J Comp Neurol 233:115–132.

Roessmann U (1982) The embryology and neuropathology of congenital malformations. Clin Neurosurg 30:157–164.

Rogers J, Zornetzer SF, Simon ML (1985) Therapeutic applications of neural transplant technology. Neurobiol Aging 6:169–172.

Roland PE (1975) Do muscular receptors in man evoke sensations of tension and kinaesthesia? Brain Res 99:162–165.

Roland PE, Larsen B, Lassen NA, Skinhoj E (1980) Supplementary motor area and other cortical areas in organization of voluntary movements in man. J Neurophysiol 43:118–136.

Roland PE, Nielsen VK (1980) Vibratory thresholds in the hands. Comparison of patients with normal subjects. Arch Neurol 37:775–779.

Roland PE, Zilles K (1996) Functions and structures of the motor cortices in humans. Curr Opin Neurobiol 6:773–781.

Rollin H (1977) Course of the peripheral gustatory nerves. Ann Otol Rhinol Laryngol 86:1–8.

Romanelli P, Esposito V, Schaal DW, Heit G (2005) Somatotopy in the basal ganglia: experimental and clinical evidence for segregated sensorimotor channels. Brain Res Rev 48:112–128.

Romanes GJ (1941) Cell columns in the spinal cord of a human foetus of fourteen weeks. J Anat 75:145–153.

Romanes GJ (1964) The motor pools of the spinal cord. Prog Brain Res 11:93–119.

Romaniuk K, Nixon KC (1977) Protracted gingival pain associated with free nerve endings – report of a case. J Periodontol 48:303–305.

Romanski LM, Tian B, Fritz J, Mishkin M, Goldman-Rakic PS, Rauschecker JP (1999) Dual streams of auditory afferents target multiple domains in the primate prefrontal cortex. Nat Neurosci 2:1131–1136.

Romijn HJ (1978) The pineal, a tranquillizing organ? Life Sci 23:2257–2274.

Ron S, Robinson DA (1973) Eye movements evoked by cerebellar stimulation in the alert monkey. J Neurophysiol 36:1004–1022.

Rorstad OP, Senterman MK, Hoyte KM, Martin JB (1980) Immunoreactive and biologically active somatostatin-like material in the human retina. Brain Res 199:488–492.

Rosenberg G (1958) Effect of age on peripheral vibratory perception. J Am Geriat Soc 6:471–481.

Rosenberg NL, Koller R (1981) Computerized tomography and pure sensory stroke. Neurology 31:217–220.

Rosenhall U (1972) Vestibular macular mapping in man. Ann Otol Rhinol Laryngol 81:339–351.

Rosenhall U (1973) Degenerative patterns in the aging human vestibular neuron-epithelia. Acta Otolaryngol 76:208–220.

Rosenhall U, Rubin W (1975) Degenerative changes in the human vestibular sensory epithelia. Acta Otolaryngol 79:67–80.

Rosner SS, Rhoton AL Jr, Ono M, Barry M (1984) Microsurgical anatomy of the anterior perforating arteries. J Neurosurg 61:468–485.

Ross MD, Johnsson L-G, Peacor D, Allard LF (1976) Observations on normal and degenerating human otoconia. Ann Otol Rhinol Laryngol 85:310–326.

Rossi GF, Zanchetti A (1957) The brain stem reticular formation. Anatomy and physiology. Arch Ital Biol 95:199–435.

Routal RV, Pal GP (2000) Location of the spinal nucleus of the accessory nerve in the human spinal cord. J Anat 196:263–268.

Roy S, Zhang B, Lee VM, Trojanowski JQ (2005) Axonal transport defects: a common theme in neurodegenerative diseases. Acta Neuropathol (Berl) 109:5–13.

Rubenstein R, et al. (1986) Paired helical filaments associated with Alzheimer disease are readily soluble structures. Brain Res 372:80–88.

Rubino PA, Rhoton AL Jr, Tong X, Oliveira E (2005) Three-dimensional relationships of the optic radiation. Neurosurgery 57:219–227.

Ruch TC, Patton H (1946) The relation of the deep opercular cortex to taste. Fed Proc 5:89–90.

Ruge D, Wiltse LL (1977) *Spinal Disorders: Diagnosis and Treatment*. Lea & Febiger, Philadelphia.

Rushton JG, Stevens JC, Miller RH (1981) Glossopharyngeal (vagoglossopharyngeal) neuralgia. Arch Neurol 38:201–205.

Russell GV (1955) The nucleus locus coeruleus (dorsolateralis tegmenti). Tex Rep Biol Med 13:939–988.

Russell WR, Whitty CWM (1953) Studies in traumatic epilepsy. 2. Focal motor and somatic sensory fits: a study of 85 cases. J Neurol Neurosurg Psychiat 16:73–97.

Rustioni A (1976) Spinal neurons project to the dorsal column nuclei of rhesus monkeys. Science 196:656–658.

Rustioni A, Hayes NL, O'Neill S (1979) Dorsal column nuclei and ascending spinal afferents in macaques. Brain 102:95–125.

Rymer WZ, D'Almeida A (1980) Joint position sense: the effects of muscle contraction. Brain 103:1–22.

Sacco RL, Freddo L, Bello JA, Odel JG, Onesti ST, Mohr JP (1993) Wallenberg's lateral medullary syndrome. Clinical-magnetic resonance imaging correlations. Arch Neurol 50:609–614.

Sadikot AF, Parent A, Francois C (1992a) Efferent connections of the centromedian and parafascicular thalamic nuclei in the squirrel monkey: a PHA-L study of subcortical projections. J Comp Neurol 315:137–159.

Sadikot AF, Parent A, Smith Y, Bolam JP (1992b) Efferent connections of the centromedian and parafascicular thalamic nuclei in the squirrel monkey: a light and electron microscopic study of the thalamostriatal projection in relation to striatal heterogeneity. J Comp Neurol 320:228–242.

Sadjadpour K, Brodal A (1968) The vestibular nuclei in man: a morphological study in the light of experimental findings in the cat. J Hirnforsch 10:299–323.

Sadun AA, Lessell S (1985) Brightness-sense and optic nerve disease. Arch Ophthalmol 102:39–43.

Sadun AA, Smith LEH, Kenyon KR (1983) Paraphenylenediamine: a new method for tracing human visual pathways. J Neuropathol Exp Neurol 42:200–206.

Saeki N, Rhoton AL Jr (1977) Microsurgical anatomy of the upper basilar artery and the posterior circle of Willis. J Neurosurg 46:563–578.

Sagar SM, Marshall PE (1988) Somatostatin-like immunoreactive material in associational ganglion cells of human retina. Neuroscience 27:507–516.

Salt AN, Konishi T (1986) The cochlear fluids: perilymph and endolymph. In: *Neurobiology of Hearing: The Cochlea*. Altschuler RA, Hoffman DW, Bobbin RP, eds, pp 109–122. Raven Press, New York.

Sambataro F, Dimalta S, Di Giorgio A, Taurisano P, Blasi G, Scarabino T, Giannatempo G, Nardini M, Bertolino A (2006) Preferential responses in amygdala and insula during presentation of facial contempt and disgust. Eur J Neurosci 24:2355–2362.

Sanchez-Gonzalez MA, Garcia-Cabezas MA, Rico B, Cavada C. The primate thalamus is a key target for brain dopamine. J Neurosci 25:6076–6083.

Sandahl B, Ulmsten U, Andersson K-E (1980) Local application of ketocaine for treatment of referred pain in primary dysmenorrhea. Acta Obstet Gynecol Scand 59:259–290.

Sandell JH (1986) NADPH diaphorase histochemistry in the macaque striate cortex. J Comp Neurol 251:388–397.

Sando I (1965) The anatomical interrelationships of the cochlear nerve fibers. Acta Otolaryngol 59:417–436.

Saper CB (2002) The central autonomic nervous system: conscious visceral perception and autonomic pattern generation. Annu Rev Neurosci 25:433–469.

Saper CB (2004) Hypothalamus. In: *The Human Nervous System*. Paxinos G, Mai JK, eds. 2nd ed., pp 513–550. Elsevier/Academic Press, Amsterdam.

Saper CB, Scammell TE, Lu J (2005) Hypothalamic regulation of sleep and circadian rhythms. Nature 437:1257–1263.

Sarnat HB (2005) Ontogeny of the reticular formation: its possible relation to the myoclonic epilepsies. Adv Neurol 95:15–22.

Sauer B (1983) Lamina boundaries of the human striate area compared with automatically-obtained grey level index profiles. J Hirnforsch 24:79–87.

Sauer B (1983) Quantitative analysis of the laminae of the striate area in man. An application of automatic image analysis. J Hirnforsch 24:89–97.

Sauer FC (1935) The cellular structure of the neural tube. J Comp Neurol 63:13–23.

Saunders RL, Sachs E Jr (1970) Relation of the accessory rootlets of the trigeminal nerve to its motor root. A microsurgical autopsy study. J Neurosurg 33:317–324.

Saunders RL, Weider D (1985) Tympanic membrane sensation. Brain 108:387–404.

Schachern PA, Paparella MM, Duvall AJ III, Choo YB (1984) The human round window membrane: An electron microscopic study. Arch Otolaryngol 110:15–21.

Schachter M (1981) Enkephalins and endorphins. Br J Hosp Med 25:128–136.

Scharenberg K, Liss L (1965) The histologic structure of the human pineal body. Prog Brain Res 10:193–217.

Scheibel AB (1979) The hippocampus: organizational patterns in health and senescence. Mech Ageing Dev 9:89–102.

Scheibel ME, Scheibel AB (1968) On the nature of dendritic spines – report of a workshop. Commun Behav Biol Part A 1:231–265.

Schelper RL, Adryan EK Jr (1986) Monocytes become macrophages; they do not become microglia: A light and electron microscopic autoradiographic study using 125-Iododeoxyuridine. J Neuropathol Exp Neurol 45:1–19.

Schiller PH, Malpeli JG (1977) Properties and tectal projections of monkey retinal ganglion cells. J Neurophysiol 40:428–445.

Schiltz C, Sorger B, Caldara R, Ahmed F, Mayer E, Goebel R, Rossion B (2006) Impaired face discrimination in acquired prosopagnosia is associated with abnormal response to individual faces in the right middle fusiform gyrus. Cereb Cortex 16:574–586.

Schimrigk K, Rüttinger H (1980) The touch corpuscles of the plantar surface of the big toe. Histological and histometrical investigations with respect to age. Eur Neurol 19:49–60.

Schlaepfer WW (1987) Neurofilaments: structure, metabolism and implications in disease. J Neuropathol Exp Neurol 46:117–129.

Schmahmann JD (2004) Disorders of the cerebellum: ataxia, dysmetria of thought, and the cerebellar cognitive affective syndrome. J Neuropsychiat Clin Neurosci 16:367–378.

Schmahmann JD (2003) Vascular syndromes of the thalamus. Stroke 34:2264–2278.

Schmahmann JD, Caplan D (2006) Cognition, emotion and the cerebellum. Brain 129:290–292.

Schmidt RF (1972) The gate control theory of pain: an unlikely hypothesis. In: *Pain: Basic Principles – Pharmacology – Therapy.* Janzen R, Keidel WD, Herz A, Steichele C, eds, pp 124–127. Churchill Livingstone, London.

Schmidt SY, Peisch RD (1986) Melanin concentration in normal human retinal pigment epithelium: regional variation and age-related reduction. Invest Ophthalmol Vis Sci 27:1063–1067.

Schmitt FO, Worden FG, Adelman G, Dennis SG (1981) *The Organization of the Cerebral Cortex.* MIT Press, Cambridge.

Schnapf JL, Baylor DA (1987) How photoreceptor cells respond to light. Sci Am 256:40–47.

Schneider KA, Richter MC, Kastner S (2004) Retinotopic organization and functional subdivisions of the human lateral geniculate nucleus: a high-resolution functional magnetic resonance imaging study. J Neurosci 24:8975–8985.

Schneider RC, Crosby EC (1971) Surgery of the cranial nerves. In: *Practice of Surgery.* Mark V, ed., pp 1–103. Harper & Row, Hagerstown.

Schneider RC, Crosby EC (1980) Motion sickness: Part I – A theory. Aviat Space Environ Med 51:61–64.

Schneider RC, Crosby EC (1980) Motion sickness: Part II – A clinical study based on surgery of cerebral hemisphere lesions. Aviat Space Environ Med 51:65–73.

Schneider RC, Crosby EC (1980) Motion sickness: Part III – A clinical study based on surgery of posterior fossa tumors. Aviat Space Environ Med 51:74–85.

Schneider RC, Crosby EC, Russo RH, Gosch HH (1973) Traumatic spinal cord syndromes and their management. Clin Neurosurg 20:424–492.

Schneider RC, Kahn EA, Crosby EC, Taren J (1982) *Correlative Neurosurge.* 3rd ed. Thomas, Springfield.

Schneider RJ, Kulics AT, Ducker TB (1977) Proprioceptive pathways of the spinal cord. J Neurol Neurosurg Psychiatry 40:417–433.

Schneider RJ, Paul RL (1976) Body representation in the anterolateral funiculus as reconstructed by percutaneous cordotomies for benign intractable pain. In: *Advances in Pain Research and Therapy.* Bonica JJ, Albe-Fessard D, eds, 1:267–270. Raven Press, New York.

Schnider A (2001) Spontaneous confabulation, reality monitoring, and the limbic system – a review. Brain Res Rev 36:150–160.

Schnitzler A, Ploner M (2000) Neurophysiology and functional neuroanatomy of pain perception. J Clin Neurophysiol 17:592–603.

Schoch B, Dimitrova A, Gizewski ER, Timmann D (2006) Functional localization in the human cerebellum based on voxelwise statistical analysis: a study of 90 patients. NeuroImage 30:36–51.

Schoenen J (1991) Clinical anatomy of the spinal cord. Neurol Clin 9:503–532.

Schoenen J, Faull RLM (2004) Spinal cord: cyto- and chemoarchitecture. In: *The Human Nervous System.* Paxinos G, Mai JK, eds. 2nd ed. Chapter 7, pp 190–232. Elsevier/Academic Press, Amsterdam.

Schoenen J, Grant G (2004) Spinal cord: connections. In: *The Human Nervous System* Paxinos G, Mai JK, eds. 2nd ed. Chapter 8, pp 233–249. Elsevier/Academic Press, Amsterdam.

Scholten JM (1967) Considerations on oligodendroglia. Part 1. An attempt at synthesis. Psychiatr Neurol Neurochir 70:435–451.

Scholtz CL, Swettenham K, Brown A, Mann DMA (1981) A histoquantitative study of the striate cortex and lateral geniculate body in normal, blind and demented subjects. Neuropathol Appl Neurobiol 7:103–114.

Schondorf R (1993) New investigations of autonomic nervous system function. J Clin Neurophysiol 10:28–38.

Schossberger P (1974) Vasculature of the spinal cord: a review. Part I. Anatomy and Physiology. Bull Los Angeles Neurol Soc 39:71–85.

Schossberger P (1974) Vasculature of the spinal cord: a review. Part II. Clinical Considerations. Bull Los Angeles Neurol Soc 39:86–97.

Schrøder HD (1981) Onuf's nucleus X: a morphological study of a human spinal nucleus. Anat Embryol 162:443–453.

Schrott-Fischer A, Kammen-Jolly K, Scholtz AW, Gluckert R, Eybalin M (2002) Patterns of GABA-like immunoreactivity in efferent fibers of the human cochlea. Hear Res 174:75–85.

Schuknecht HF, Kitamura K (1981) Vestibular neuritis. Second Louis H. Clerf Lecture. Ann Otol Rhinol Laryngol (Suppl) 90 (Suppl 78):1–19.

Schut L (1996) Case problems in neurological surgery: Management of a pediatric hypothalamic mass. Neurosurgery 38:806–811.

Schüz A (2002) Introduction: Homogeneity and heterogeneity of cortical structure: a theme and its variations. In: *Cortical Areas: Unity and Diversity.* Schüz A, Miller R, eds, pp 1–11. Taylor & Francis, London and New York.

Schüz A, Miller R (2002) *Cortical Areas: Unity and Diversity.* Taylor & Francis, London and New York.

Schvarcz JR (1978) Spinal cord stereotactic techniques re trigeminal nucleotomy and extralemniscal myelotomy. Appl Neurophysiol 41:99–112.

Schwartz HG, Roulhac GE, Lam RL, O'Leary JL (1951) Organization of the fasciculus solitarius in man. J Comp Neurol 94:221–237.

Schwarz DWF, Deecke L, Fredrickson JM (1973) Cortical projection of group I muscle afferents to areas 2, 3a, and the vestibular field in the rhesus monkey. Exp Brain Res 17:516–526.

Schwarz DWF, Fredrickson JM (1971) Rhesus monkey vestibular cortex: a bimodal primary projection field. Science 172:280–281.

Schwarz DWF, Fredrickson JM (1974) The clinical significance of vestibular projection to the parietal lobe: a review. Can J Otolaryngol 3:381–392.

Schwarz DWF, Fredrickson JM, Deecke L (1973) Structure and connections of the rhesus vestibular cortex. Adv Otorhinolaryngol 19:206–209.

Schwyn RC, Fox CA (1974) The primate substantia nigra: a Golgi and electron microscopic study. J Hirnforsch 15:95–126.

Scott GI (1957) *Traquair's Clinical Perimetry.* 7th ed., p 73. Henry Kimpton, London.

Seiberling KA, Conley DB (2004) Aging and olfactory and taste function. Otolaryngol Clin North Am 37:1209–1228.

Seltzer B, Pandya DN (1980) Converging visual and somatic sensory cortical input to the intraparietal sulcus of the rhesus monkey. Brain Res 192:339–351.

Sensenig EC (1951) The early development of the meninges of the spinal cord in human embryos. Contrib Embryol Carneg Instn 34:147–157, 1951.

Sensenig EC: (1957) The development of the occipital and cervical segments and their associated structures in human embryos. Contrib Embryol Carneg Instn 36:141–151.

Sergent J (1993) Mapping the musician brain. Hum Brain Mapp 1:20–38.

Sessle BJ (1978) Oral-facial pain: old puzzles, new postulates. Int Dent J 28:28–42.

Sessle BJ, Dubner R, Greenwood LF, Lucier GE (1976) Descending influences of periaqueductal gray matter and somatosensory cerebral cortex on neurones in trigeminal brain stem nuclei. Can J Physiol Pharmacol 54:66–69.

Sessle BJ, Dubner R, Hu JW, Lucier GE (1977) Modulation of trigeminothalamic relay and non-relay neurones by noxious, tactile and periaqueductal gray stimuli: implications in perceptual and reflex aspects of nociception. In: *Pain in the Trigeminal Region.* Anderson DJ, Matthews B., eds, pp 285–294. Elsevier, Amsterdam.

Shacklett DE, et al. (1984) Congruous and incongruous sectorial visual field defects with lesions of the lateral geniculate nucleus. Am J Ophthalmol 98:283–290.

Shanta TR, Evans JA (1972) The relationship of epidural anesthesia to neural membranes and arachnoid villi. Anesthesiology 37:543–55.

Shapley R, Perry VH (1986) Cat and monkey retinal ganglion cells and their visual functional roles. TINS 9:229–235.

Sharpe JA, Hoyt WF, Rosenberg MA (1975) Convergence-evoked nystagmus. Congenital and acquired forms. Arch Neurol 32:191–194.

Shaw C, Cynader M (1986) Laminar distribution of receptors in monkey (*Macaca fascicularis*) geniculostriate system. J Comp Neurol 248:301–312.

Shaw KM (1977) The pineal gland: a review of the biochemistry, physiology and pharmacological potential of melatonin and other pineal substances. Adv Drug Res 11:75–96.

Shaywitz SE (1998) Dyslexia. NEJM 338:307–312.

Shealy CN, Mortimer JT, Hugfors NR (1970) Dorsal column electroanalgesia. J Neurosurg 32:560–564.

Shearer DE, Dustman RE, Emmerson RY (1987) Hydrocephalus: electrophysiological correlates. Am J EEG Technol 27:199–212.

Shelanski ML, Liem RKH (1979) Neurofilaments. J Neurochem 33:5–13.

Shende MC, King RB (1967) Excitability changes of trigeminal primary afferent preterminals in brain-stem nuclear complex of squirrel monkey (*Saimiri sciureus*). J Neurophysiol 30:949–963.

Shenkin HA, Lewey FH (1944) Taste aura preceding convulsions in a lesion of the parietal operculum. J Nerv Ment Dis 10:352–354.

Shepherd GM (1972) The neuron doctrine: a revision of functional concepts. Yale J Biol Med 45:584–599.

Shepherd GM (1978) Microcircuits in the nervous system. Sci Am 238:92–103.

Shepherd GM, Getchell TV, Kauer JS (1975) Analysis of structure and function in the olfactory pathway. In: *The Basic Neurosciences.* Brady RO, ed., 1:207–220. Raven Press, New York.

Sherman DL, Brophy PJ (2005) Mechanisms of axon ensheathment and myelin growth. Nat Rev Neurosci 6:683–90.

Sherman SM, Guillery RW (2001) *Exploring the Thalamus.* Academic Press, San Diego.

Sherrington CS (1906) *The Integrative Action of the Nervous System.* Yale University Press, Newhaven.

Shettles LB (1955) Further observations on living human oocytes and ova. Am J Obstet Gynecol 69:365–371.

Shettles LB (1960) *Ovum Humanum.* Hafner, New York.

Shibasaki H, Yamashita Y, Motomura S (1978) Suppression of congenital nystagmus. J Neurol Neurosurg Psychiat 41:1078–1083.

Shields RW Jr (1993) Functional anatomy of the autonomic nervous system. J Clin Neurophysiol 10:2–13.

Shikama Y, Kato T, Nagaoka U, Hosoya T, Katagiri T, Yamaguchi K, Sasaki H (1996) Localization of the gustatory pathway in the human midbrain. Neurosci Lett 218:198–200.

Shinohara H, Tanaka O (1988) Development of the notochord in human embryos: ultrastructural, histochemical, and immunohistochemical studies. Anat Rec 220:171–178.

Shipp S (2003) The functional logic of cortico-pulvinar connections. Philos Trans R Soc Lond B Biol Sci 358:1605–1624.

Shmuelof L, Zohary E (2006) A mirror representation of others' actions in the human anterior parietal cortex. J Neurosci 26:9736–9742.

Shraberg D, Weisberg L (1978) The Kluver–Bucy syndrome in man. J Nerv Ment Dis 166:130–134.

Sibley RK, et al. (1980) Neuroendocrine (Merkel cell) carcinoma of the skin. A histologic and ultrastructural study of two cases. Am J Surg Pathol 4:211–221.

Sicuteri F, Franchi G, Michelacci S (1974) Biochemical mechanism of ischemic pain. Adv Neurol 4:39–44.

Siegfried J (1977) 500 percutaneous thermocoagulations of Gasserian ganglion for trigeminal pain. Surg Neurol 8:126–131.

Simmons FB (1966) Electrical stimulation of the auditory nerve in man. Arch Otolaryngol 84:2–54.

Simmons FB, Mongeon CJ, Lewis WR, Huntington DA (1964) Electrical stimulation of acoustical nerve and inferior colliculus: results in man. Arch Otolaryngol 79:559–567.

Sims KS, Williams RS (1990) The human amygdaloid complex: a cytologic and histochemical atlas using Nissl, myelin, acetylcholinesterase and nicotinamide adenine dinucleotide phosphate diaphorase staining. Neuroscience 36:449–472.

Sindou M, Quoex C, Baleydier C (1974) Fiber organization at the posterior spinal cord-rootlet junction in man. J Comp Neurol 153:15–26.

Singer T, Seymour B, O'Doherty J, Kaube H, Dolan RJ, Frith CD (2004) Empathy for pain involves the affective but not sensory components of pain. Science 303:1157–1162.

Singhabhandhu B, Gray SW, Bryant MF, Skandalakis JE (1973) Carotid body tumors. Am Surg 39:501–508.

Sinha UK, Terr LI, Galey FR, Linthicum FH Jr (1987) Computer-aided three-dimensional reconstruction of the cochlear nerve root. Arch Otolaryngol Head Neck Surg 113:651–655.

Sireteanu R, Kellerer R, Boergen K-P (1984) The development of peripheral visual acuity in human infants: a preliminary study. Hum Neurobiol 3:81–85.

Sitoh YY, Tien RD (1997) The limbic system. An overview of the anatomy and its development. Neuroimaging Clin N Am 7:1–10.

Sittig O (1925) A clinical study of sensory Jacksonian fits. Brain 48:233–254.

Sjöstrand FS (1961) Electron microscopy of the retina. In: The Structure of the Eye. Smelser GK, ed., pp 1–28. Academic Press, New York.

Sjöstrand FS (1963) The structure and formation of the myelin sheath. In: Mechanisms of Demyelination. Rose AS, Pearson CM, eds, pp 1–43. McGraw-Hill, New York.

Skavenski AA, Robinson DA (1973) Role of abducens neurons in vestibuloocular reflex. J Neurophysiol 36:724–738.

Small DM (2006) Central gustatory processing in humans. Adv Otorhinolaryngol 63:191–220.

Small DM, Prescott J (2005) Odor/taste integration and the perception of flavor. Exp Brain Res 166:345–57.

Small DM, Zald DH, Jones-Gotman M, Zatorre RJ, Pardo JV, Frey S, Petrides M (1999) Human cortical gustatory areas: a review of functional neuroimaging data. NeuroReport 10:7–14.

Smialowski A (1972) Preoptic area in macaque brain. Acta Biol Cracovensia 15:9–13.

Smith BH (1978) Changing concepts of neuroglial function. Neurosurgery 2:175–180.

Smith BH, Kornblith PL (1982) Axoplasmic transport and neurological surgery. Neurosurgery 10:268–276.

Smith DE (1977) The effect of deafferentation on the development of brain and spinal nuclei. Prog Neurobiol 8:349–367.

Smith DG, Mumford JM (1980) Petrous angle and trigeminal neuralgia. Pain 8:269–277.

Smith DW (1970) Recognizable Patterns of Human Malformation, Genetic, Embryologic and Clinical Aspects. Saunders, Philadelphia.

Smith KR Jr (1970) The ultrastructure of the human Haarscheibe and Merkel cell. J Invest Dermatol 54:150–159.

Smith MC (1976) Retrograde cell changes in human spinal cord after anterolateral cordotomies. Location and identification after different periods of survival. In: Advances in Pain Research and Therapy. Bonica JJ, Albe-Fessard D, eds, 1:91–98. Raven Press, New York.

Smith RL (1975) Axonal projections and connections of the principal sensory trigeminal nucleus in the monkey. J Comp Neurol 163:347–376.

Smith SJ (1988) Neuronal cytomechanics: the actin-based motility of growth cones. Science 242:708–715.

Smolen AJ, Truex RC (1977) The dorsal motor nucleus of the vagus nerve of the cat: localization of preganglionic neurons by quantitative histological methods. Anat Rec 189:555–566.

Snider RS (1959) The cerebellar nuclei. In: Introduction to Stereotaxis with an Atlas of the Human Brain. Schaltenbrand G, Bailey P, eds,1:78–87. Thieme, Stuttgart.

Snow RB, Fraser RAR (1987) Cerebellopontine angle tumor causing contralateral trigeminal neuralgia: a case report. Neurosurgery 21:84–86.

Snyder R (1977) The organization of the dorsal root entry zone in cats and monkeys. J Comp Neurol 174:47–70.

Sobue G, Brown MJ, Kim SU, Pleasure D (1984) Axolemma is a mitogen for human Schwann cells. Ann Neurol 15:449–452.

Solodkin A, Van Hoesen GW (1996) Entorhinal cortex modules of the human brain. J Comp Neurol 365:610–627.

Solov'ev SV (2006) The weight and linear dimensions of the human cerebellum. Neurosci Behav Physiol 36:479–481.

Somjen GG (1975) Electrophysiology of neuroglia. Annu Rev Physiol 37:163–190.

Sorvari H, Soininen H, Paljarvi L, Karkola K, Pitkanen A (1995) Distribution of parvalbumin-immunoreactive cells and fibers in the human amygdaloid complex. J Comp Neurol 360:185–212.

Spector RH, Glaser JS, David NJ, Vining DQ (1981) Occipital lobe infarctions: perimetry and computed tomography. Neurology 31:1098–1106.

Sperry RW (1964) The great cerebral commissure. Sci Am 210(1):42–52.

Spillane JD (1975) *An Atlas of Clinical Neurology.* 2nd ed. Oxford, London.

Spinal Cord Injury. Hope Through Research. (1981) NIH Publication No. 81–160. February.

Spoendlin H (1979) Sensory neural organization of the cochlea. J Laryngol Otol 93:853–877.

Standring S (2004) Editor-in-chief. *Gray's Anatomy.* 39th ed. Section 2, Neuroanatomy: Chapter 18: Spinal Cord, pp 307–326. Elsevier/Churchill Livingstone, London.

Standring S (2004) Editor-in-chief. *Gray's Anatomy.* 39th ed. Section 2, Neuroanatomy: Chapter 19. Brain Stem, pp 327–352. Elsevier/Churchill Livingstone, London.

Standring S (2004) Editor-in-chief. *Gray's Anatomy.* 39th ed. Section 2, Neuroanatomy: Chapter 21, Diencephalon, pp 369–386. Elsevier/Churchill Livingstone, London.

Standring S (2004) Editor-in-chief. *Gray's Anatomy.* 39th ed. Section 2, Neuroanatomy: Chapter 22. Cerebral Hemisphere, pp 387–418. Elsevier/Churchill Livingstone, London.

Starr MA (1888) Ophthalmoplegia externa partialis. J Nerv Ment Dis 15:301–316.

Steel KP (1983) Review. The tectorial membrane of mammals. Hear Res 9:327–359.

Steel KP (1986) Tectorial membrane. In: *Neurobiology of Hearing: The Cochlea.* Altschuler RA, Hoffman DW, Bobbin RP, eds, pp 139–148. Raven Press, New York.

Stein BM, Carpenter MB (1967) Central projections of portions of the vestibular ganglia innervating specific parts of the labyrinth in the rhesus monkey. Am J Anat 120:281–318.

Stein DG (1985) Fetal brain tissue transplant techniques: a cautionary note. Neurobiol Aging 6:157–160.

Stein SC, Schut L (1979) Hydrocephalus in myelomeningocele. Child's Brain 5:413–419.

Stein Z, Susser M (1987) Early nutrition, fetal growth, and mental function: observations in our species. In: *Current Topics in Nutrition and Disease.* Rassin DK, Haber B, Drujan BD, eds. Vol 16: Basic and Clinical Aspects of Nutrition and Brain Development, pp 323–338. AR Liss, New York.

Steiner JE (1973) The gustofacial response: observation on normal and anencephalic newborn infants. In: *Fourth Symposium on Oral Sensation and Perception.* Bosma JF, ed. DHEW Publ No (NIH) 73-546. NIH, Bethesda.

Steiner JE (1977) Facial expressions of the neonate infant indicating the hedonics of food-related chemical stimuli. In: *Taste and Development: The Genesis of Sweet Preference.* Weiffenbach JM, ed. DHEW Publ No (NIH) 77-1068. NIH, Bethesda.

Steinman RM, Haddad GM, Skavenski AA, Wyman D (1973) Miniature eye movement. Science 181:810–819.

Steinmetz H (1996) Structure, functional and cerebral asymmetry: in vivo morphometry of the planum temporale. Neurosci Biobehav Rev 20:587–591.

Stensaas SS, Eddington DK, Dobelle WH (1974) The topography and variability of the primary visual cortex in man. J Neurosurg 40:747–755.

Stephan H, Frahm H, Baron G (1981) New and revised data on volumes of brain structures in insectivores and primates. Folia Primatol 35:1–29.

Steriade M (1996) Arousal: revisiting the reticular activating system. Science 27:225–226.

Steriade M, Glenn LL (1982) Neocortical and caudate projections of intralaminar thalamic neurons and their synaptic excitation from midbrain reticular core. Neurophysiol 48:352–371.

Steriade M, Jones EG, McCormick DA (1997) *The Thalamus.* 2 Vols. Elsevier Science, Amsterdam.

Sterkers JM, Perre J, Viala P, Foncin J-F (1987) The origin of acoustic neuromas. Acta Otolaryngol (Stockh) 103:427–431.

Sterling P (1973) Referred cutaneous sensation. Exp Neurol 41:451–456.

Sternbach RA (1978) *The Psychology of Pain.* Raven Press, New York.

Stevens B (2003) Glia: much more than the neuron's sidekick. Curr Biol 13:R469–R472.

Stewart RM, Rosenberg RN (1979) Physiology of glia: glial-neuronal interactions. Int Rev Neurobiol 21:275–309.

Stilwell DL Jr (1957) The innervation of deep structures of the hand. Am J Anat 101:75–99.

St-John WM, Paton JF (2004) Role of pontile mechanisms in the neurogenesis of eupnea. Respir Physiol Neurobiol 143:321–332.

Stojanovic MP, Abdi S (2002) Spinal cord stimulation. Pain Physician 5:156–166.

Stone J, Johnston E (1981) The topography of primate retina: a study of the human, bushbaby, and new- and old-world monkeys. J Comp Neurol 196:205–223.

Straatsma BR, Foos RY, Heckenlively JR, Taylor GN (1981) Myelinated retinal nerve fibers. Am J Ophthalmol 91:25–38.

Streeter GL (1919) Factors involved in the formation of the filum terminale. Am J Anat 25:1–11.

Streeter GL (1920) Weight, sitting height, head size, foot length, and menstrual age of the human embryo. Contrib Embryol Carneg Instn 11:143–170.

Streeter GL (1927) Archetypes and symbolism. Science 65:405–412.

Streeter GL (1942) Developmental horizons in human embryos. Description of age group XI, 13–20 somites, and age group XII, 21 to 29 somites. Contrib Embryol Carneg Instn 30:211–245.

Streeter GL (1945) Developmental horizons in human embryos. Description of age group XIII, embryos about 4 or 5 millimeters long, and age group XIV, period of indentation of the lens vesicle. Contrib Embryol Carneg Instn 31:27–63.

Streeter GL (1948) Developmental horizons in human embryos. Descriptions of age groups XV, XVI, XVII, and XVIII, being the third issue of a survey of the Carnegie Collection. Contrib Embryol Carneg Instn 32:133–203.

Streeter GL (1951) Developmental horizons in human embryos. Description of age groups XIX, XX, XXI, XXII, and XXIII, being the fifth issue of a survey of the Carnegie Collection. (Prepared for Publication by CH Heuser and GW Corner). Contrib Embryol Carneg Instn 34:165–196.

Streit WJ and Kreutzberg GW (1987) Lectin binding by resting and reactive microglia. J Neurocytol 16:249–260.

Strominger NL, Hurwitz JL (1976) Anatomical aspects of the superior olivary complex. J Comp Neurol 170:485–498.

Strominger NL, Strominger AI (1971) Ascending brain stem projections of the anteroventral cochlear nucleus in the rhesus monkey. J Comp Neurol 143:217–242.

Sturrock RR (1983) Problems of glial identification and quantification in the aging central nervous system. In: *Brain Aging: Neuropathology and Neuropharmacology.* Cervos-Navarro J, Surkander H-I, eds, pp 179–209. Raven Press, New York.

Sumino R, Dubner R, Starkman S (1973) Responses of small myelinated 'warm' fibers to noxious heat stimuli applied to the monkey's face. Brain Res 62:260–263.

Summary of a Subcommittee Report Commissioned by the National Advisory Council of the National Institute of Neurological and Communicative Disorders and Stroke, 1975 (1976) The current status of research on growth and regeneration in the central nervous system. Surg Neurol 5:157–160.

Sunderland S (1969) Anatomical features of nerve trunks in relation to nerve injury and nerve repair. Clin Neurosurg 17:38–62.

Sunderland S (1979) The painful nerve lesion: a prologue. Adv Pain Res Therap 3:35–37.

Sung JH (1981) Tangled masses of regenerated central nerve fibers (non-myelinated central neuromas) in the central nervous system. J Neuropathol Exp Neurol 40:645–657.

Sur M (1980) Receptive fields of neurons in areas 3b and 1 of somatosensory cortex in monkeys. Brain Res 198:465–471.

Sur M, Wall JT, Kaas JH (1981) Modular segregation of functional cell classes within the postcentral somatosensory cortex of monkeys. Science 212:1059–1061.

Susac JO, Hoyt WF (1977) Inferior branch palsy of the oculomotor nerve. Ann Neurol 2:336–339.

Susac JO, Hoyt WF, Daroff RB, Lawrence W (1970) Clinical spectrum of ocular bobbing. J Neurol Neurosurg Psychiatry 33:771–775.

Suzuki J-I, Cohen B (1964) Head, eye, body and limb movements from semicircular canal nerves. Exp Neurol 10:393–405.

Suzuki J-I, Cohen B, Bender MB (1964) Compensatory eye movements induced by vertical semicircular canal stimulation. Exp Neurol 9:137–160.

Suzuki WA, Amaral DG (1994) Topographic organization of the reciprocal connections between the monkey entorhinal cortex and the perirhinal and parahippocampal cortices. J Neurosci 14:1856–1877.

Swaab DF (1995) Ageing of the human hypothalamus. Horm Res 43:8–11.

Swaab DF (1995) Development of the human hypothalamus. Neurochem Res 20:509–519.

Swaab DF (1997) Neurobiology and neuropathology of the human hypothalamus. In: *Handbook of Chemical Neuroanatomy.* Björklund A, Hökfelt T, eds. Vol 13: The Primate Nervous System, Part I, pp 39–138. Elsevier, Amsterdam.

Swaab DF (1998) The human hypothalamo-neurohypophysial system in health and disease. Prog Brain Res 119:577–618.

Swaab DF (2003) The Human Hypothalamus: Basic and Clinical Aspects. Part I: Nuclei of the Human Hypothalamus. In: *Handbook of Clinical Neurology.* Aminoff MJ, Boller F, Swaab DF, series eds. Swaab DF, volume author. Vol 79. Elsevier, Amsterdam.

Swaab DF (2003) The Human Hypothalamus: Basic and Clinical Aspects. Part II: Neuropathology of the Human Hypothalamus and Adjacent Structures. In: *Handbook of Clinical Neurology.* Aminoff MJ, Boller F, Swaab DF, series eds. Vol 80. Elsevier, Amsterdam.

Swaab DF (2006) The human hypothalamus in metabolic and episodic disorders. Prog Brain Res 153:3–45.

Swaab DF, Chung WC, Kruijver FP, Hofman MA, Ishunina TA (2002) Sexual differentiation of the human hypothalamus. Adv Exp Med Biol 511:75–100.

Swaab DF, Fliers E (1985) A sexually dimorphic nucleus in the human brain. Science 228:1112–1115.

Swaab DF, Schadé JP (1974) *Integrative Hypothalamic Activity.* Prog Brain Res, Vol 41. Elsevier, New York.

Swanson AG, Buchan GC, Alvord EC Jr (1963) Absence of Lissauer's tract and small dorsal root axons in familial, congenital, universal insensitivity to pain. Trans Am Neurol Assoc 88:99–103.

Swarbrick ET, et al. (1980) Site of pain from the irritable bowel. Lancet, August 30, 443–446.

Swash M, Fox KP (1972) Muscle spindle innervation in man. J Anat 112:61–80.

Sweatt JD (2003) *Mechanisms of Memory.* Academic Press, San Diego.

Sweet WH (1975) Pain: mechanisms and treatment. In: *The Nervous System.* Tower DB, editor-in-chief. Chase TN, volume ed. Vol 2: The Clinical Neurosciences, pp 487–500. Raven Press, New York.

Sweet WH (1976) Treatment of facial pain by percutaneous differential thermal trigeminal rhizotomy. Prog Neurol Surg 7:153–179.

Sweet WH, Wepsic JG (1974) Controlled thermocoagulation of trigeminal ganglion and rootlets for differential destruction of pain fibers. Part 1: Trigeminal neuralgia. J Neurosurg 39:143–156.

Swinyard CA, Sansaricq C (1969) The birth defect and the pediatric problems it presents. Med Clin North Am 53:488–492.

Switka A, Narkiewicz O, Dziewiatkowski J, Morys J (1999) The shape of the inferior horn of the lateral ventricle in relation to collateral and occipitotemporal sulci. Folia Morphol (Warsz) 58:69–80.

Takahashi M, Wilson, G, Hanafee W (1968) The anterior inferior cerebellar artery: its radiographic anatomy and significance in the diagnosis of extra-axial tumors of the posterior fossa. Radiology 90:281–287.

Takeuchi Y, Sano Y (1984) Serotonin nerve fibers in the primary visual cortex of the monkey. Quantitative and immunoelectromicroscopical analysis. Anat Embryol 169:1–8.

Tanabe T, et al. (1975) An olfactory projection area in orbitofrontal cortex of the monkey. J Neurophysiol 38:1269–1283.

Tanabe T, Iino M, Ooshima Y, Takagi SF (1974) An olfactory area in the prefrontal lobe. Brain Res 80:127–130.

Tanabe T, Iino M, Takagi SF (1975) Discrimination of odors in olfactory bulb, pyriform-amygdaloid areas, and orbitofrontal cortex of the monkey. J Neurophysiol 38:1284–1296.

Tanaka K, Sakai N, Terayama Y (1979) Organ of Corti in the human fetus: scanning and transmission electronmicroscope studies. Ann Otol Rhinol Laryngol 88:749–758.

Tanriover N, Rhoton AL Jr, Kawashima M, Ulm AJ, Yasuda A (2004) Microsurgical anatomy of the insula and the sylvian fissure. J Neurosurg 100:891–922.

Tapp E (1979) The histology and pathology of the human pineal gland. Prog Brain Res 52:481–500.

Taren JA (1964) The positions of the cutaneous components of the facial glossopharyngeal and vagal nerves in the spinal tract of V. J Comp Neurol 122:389–397.

Taren JA, Davis R, Crosby EC (1969) Target physiologic corroboration in stereotaxic cervical cordotomy. J Neurosurg 30:569–584.

Taren JA, Kahn EA (1982) Trigeminal neuralgia. In: Correlative Neurosurgery. Schneider RC, Kahn EA, Crosby EC, Taren JA, eds. 3rd ed., 2:1532–1549. Thomas, Springfield.

Tarkkanen A, et al. (1983) Substance P immunoreactivity in normal human retina and in retinoblastoma. Ophthalmic Res 15:300–306.

Tarlov IM (1940) Sensory and motor roots of the glossopharyngeal nerve and the vagus-spinal accessory complex. Arch Neurol Psychiatr 44:1018–1021.

Tasker RR (1976/77) Cutaneous cervical cordotomy. Appl Neurophysiol 39:114–121.

Tasker RR, Hawrylyshyn P, Rowe IH, Organ LW (1977) Computerized graphic display of results of subcortical stimulation during stereotactic surgery. Acta Neurochir (Suppl) 24:85–98.

Tasker RR, Organ LW (1972) Mapping of the somatosensory and auditory pathways in the upper midbrain and thalamus of man. Int Congr Series No 253:169–187.

Tasker RR, Organ LW (1973) Stimulation-mapping of the upper human auditory pathway. J Neurosurg 38:320–325.

Tasker RR, Organ LW, Hawrylyshyn P (1976/77) Sensory organization of the human thalamus. Appl Neurophysiol 39:139–153.

Tasker RR, Organ LW, Hawrylyshyn PA (1982) The Thalamus and Midbrain of Man: A Physiological Atlas Using Electrical Stimulation, pp 200–215. Thomas, Springfield.

Tasker RR, Organ LW, Rowe IH, Hawrylyshyn P (1976) Human spinothalamic tract – stimulation mapping in the spinal cord and brainstem. In: Advances in Pain Research and Therapy. Bonica JJ, Albe-Fessard D, eds, 1:251–257. Raven Press, New York.

Tatarek NE (2005) Variation in the human cervical neural canal. Spine J 5:623–631.

Taylor MM, Lederman SJ, Gibson RH (1973) Tactual perception of texture. In: Handbook of Perception. Biology of Perceptual Systems. Carterette EC, Friedman MP, eds, III:251–172. Academic Press, New York.

Taylor P, Brown JH (1989) Acetylcholine. In: Basic Neurochemistry: Molecular, Cellular, and Medical Aspects. Siegel GJ, Agranoff BW, Albers RW, Molinoff PB, eds. 4th ed., pp 203–231. Raven Press, New York.

Terenius L, Wahlström A (1979) Endorphins and clinical pain, an overview. Adv Exp Med Biol 116:261–277.

Terminologia Anatomica: International Anatomical Terminology (1998) Federative Committee on Anatomical Terminology. Thieme, Stuttgart.

Terr LI, Edgerton BJ (1985) Physical effects of the choroid plexus on the cochlear nuclei in man. Acta Otolaryngol (Stockh) 100:210–217.

Terr LI, Edgerton BJ (1985) Surface topography of the cochlear nuclei in humans: Two-and three-dimensional analysis. Hear Res 17:51–59.

Terr LI, Edgerton BJ (1985) Three-dimensional reconstruction of the cochlear nuclear complex in humans. Arch Otolaryngol Head Neck Surg 111:495–501.

Terr LI, House WF (1988) Neurons of the inferior medullary velum in the cerebellopontine angle. Ann Otol Rhinol Laryngol 97:52–54.

Tew JM Jr, Keller JT, Williams DS (1978) Application of stereotactic principles to treatment of trigeminal neuralgia. Appl Neurophysiol 41:146–156.

Tew JM Jr, Keller JT, Williams DS (1979) Functional surgery of the trigeminal nerve: treatment of trigeminal neuralgia. In: Functional Neurosurgery. Rasmussen T, Marino R, eds, pp 129–141. Raven Press, New York.

Thawley SE (1978) Disorder of taste and smell. So Med J 71:267–270.

Thoden U, Doerr M, Dieckmann G, Krainick J-U (1979) Medial thalamic permanent electrodes for pain control in man: an electrophysiological and clinical study. Electroencephalogr Clin Neurophysiol 47:582–591.

Thoenen H, Edgar D (1985) Neurotrophic factors. Science 229:238–242.

Thomas CK, et al. (1987) Patterns of reinnervation and motor unit recruitment in human muscles after complete ulnar and median nerve section and resuture. J Neurol Neurosurg Psychiatry 50:259–268.

Thompson HS, Montague P, Cox TA, Corbett JJ (1982) The relationship between visual acuity, pupillary defect, and visual field loss. Am J Ophthalmol 93:681–688.

Tigges J, Tigges M (1979) Ocular dominance columns in the striate cortex of chimpanzee. Brain Res 166:386–390.

Timurkaynak E, Rhoton AL Jr, Barry M (1986) Microsurgical anatomy and operative approaches to the lateral ventricles. Neurosurgery 19:685–723.

Tiwari RK, King RB (1974) Fiber projections from trigeminal nucleus caudalis in primate (squirrel monkey and baboon). J Comp Neurol 158:191–206.

Tokita T, Taguchi T, Matuoka T (1972) A study on labyrinthine ataxia with special reference to proprioceptive reflexes. Acta Otolaryngol 74:104–112.

Tokumasu K, Goto K, Cohen B (1969) Eye movements from vestibular nuclei stimulation in monkeys. Ann Otol Rhinol Laryngol 78:1105–1119.

Tomasch J (1954) Size, distribution, and number of fibres in the human corpus callosum. Anat Rec 119:119–135.

Tomasch J, Ebnessajjade D (1961) The human nucleus ambiguus: a quantitative study. Anat Rec 141:247–252.

Tomasch J, Malpass AJ (1958) The human motor trigeminal nucleus: a quantitative study. Anat Rec 130:91–102.

Toncray JE, Krieg WJS (1946) The nuclei of the human thalamus: a comparative approach. J Comp Neurol 85:421–459.

Toni I, Shah NJ, Fink GR, Thoenissen D, Passingham RE, Zilles K (2002) Multiple movement representations in the human brain: an event-related fMRI study. J Cogn Neurosci 14:769–784.

Toni R, Malaguti A, Benfenati F, Martini L (2004) The human hypothalamus: a morpho-functional perspective. J Endocrinol Invest 27(Suppl 6):73–94.

Tonndorf J (1977) Cochlear prostheses: a state-of-the-art review. Ann Otol Rhinol Laryngol (Suppl) 86:(Suppl 44):1–20.

Tonndorf J (1981) Stereociliary dysfunction, a cause of sensory hearing loss, recruitment, poor speech discrimination and tinnitus. Acta Otolaryngol 91:469–479.

Tootell RBH, Hamilton SL, Silverman MS, Switkes E (1988) Functional anatomy of macaque striate cortex. I. Ocular dominance, binocular interactions, and baseline conditions. J Neurosci 8:1500–1530.

Tootell RBH, Silverman MS, De Valois RL, Jacobs GH (1983) Functional organization of the second cortical visual area in primates. Science 220:737–739.

Torch WC, Hirano A, Solomon S (1977) Anterograde transneuronal degeneration in the limbic system: Clinical-anatomic correlation. Neurology 27:1157–1163.

Torebjörk HE (1979) Activity in C nociceptors and sensation. In: Sensory Functions of the Skin of Humans. Kenshalo DR, ed. 2nd Int Symp on Skin Senses, pp 313–321. Fla State U. Plenum, New York.

Torebjörk HE, Hallin RG (1974) Identification of afferent C units in intact human skin nerves. Brain Res 67:387–403.

Torebjörk HE, Hallin RG (1978) Recordings of impulses in unmyelinated nerve fibres in man; afferent C fibre activity. Acta Anesthesiol Scand (Suppl) 70:124–129.

Tornqvist K, Ehinger B (1988) Peptide immunoreactive neurons in the human retina. Invest Ophthalmol Vis Sci 29:680–686.

Torok N, Kumar A (1978) An experimental evidence of etiology in postural vertigo. ORL 40:32–42.

Torrealba F, Carrasco MA (2004) A review on electron microscopy and neurotransmitter systems. Brain Res Rev 47:5–17.

Torvik A (1976) Central chromatolysis and the axon reaction: a reappraisal. Neuropathol Appl Neurobiol 2:423–432.

Tracey D (1978) Joint receptors – changing ideas. TINS 1:63–65.

Tracey I, Iannetti GD (2006) Brainstem functional imaging in humans. Suppl Clin Neurophysiol 58:52–67.

Tramo MJ, et al. (1985) Vertebral artery injury and cerebellar stroke while swimming: case report. Stroke 16:1039–1042.

Tran TD, Lam K, Hoshiyama M, Kakigi R (2001) A new method for measuring the conduction velocities of A beta-, A delta- and C-fibers following electric and CO_2 laser stimulation in humans. Neurosci Lett 301:187–190.

Tran-Dinh H (1986) The accessory middle cerebral artery – a variant of the recurrent artery of Heubner (A. centralis longa)? Acta Anat 126:167–171.

Treede RD, Kenshalo DR, Gracely RH, Jones AK (1999) The cortical representation of pain. Pain 79:105–111.

Trevino DL, Carstens E (1975) Confirmation of the location of spinothalamic neurons in the cat and monkey by the retrograde transport of horseradish peroxidase. Brain Res 98:177–182.

Trobe JD, Acosta PC, Krischer JP, Trick GL (1981) Confrontation visual field techniques in the detection of anterior visual pathway lesions. Ann Neurol 10:28–34.

Truex RC (1973) Functional neuroanatomy of the spinal cord: clinical implications. Clin Neurosurg 20:29–55.

Truex RC, Taylor MJ, Smythe MQ, Gudenberg PL (1965) The lateral cervical nucleus of cat, dog and man. J Comp Neurol 139:93–104.

Tsukita S, Ishikawa H (1979) Morphological evidence for the involvement of the smooth endoplasmic reticulum in axonal transport. Brain Res 174:315–318.

Tubbs RS, Oakes WJ (1998) Relationships of the cisternal segment of the trochlear nerve. J Neurosurg 89:1015–1019.

Turnbull IM (1972) Blood supply of the spinal cord. In: Handbook of Clinical Neurology. Vinken PJ, Bruyn GW, eds, 12:478–491. Elsevier, Amsterdam.

Turnbull IM (1973) Blood supply of the spinal cord: normal and pathological considerations. Clinical Neurosurg 20:56–84.

Turner BH, Gupta KC, Mishkin M (1978) The locus and cytoarchitecture of the projection areas of the olfactory bulb in Macaca mulatta. J Comp Neurol 177:381–396.

Tyler WJ, Murthy VN (2004) Synaptic vesicles. Curr Biol 14:R294–R297.

Uchizono K, Nomura Y, Kosaka K (1967) Localization of acetylcholinesterase activity in the efferent fiber of the human cochlea. J Physiol Soc Japan 29:241–242.

Ueda S, Sano Y (1986) Distributional pattern of serotonin-immunoreactive nerve fibers in the lateral geniculate nucleus of the rat, cat and monkey (Macaca fuscata). Cell Tissue Res 243:249–253.

Umansky F, et al. (1984) Microsurgical anatomy of the proximal segments of the middle cerebral artery. J Neurosurg 61:458–467.

Umansky F, et al. (1985) The perforating branches of the middle cerebral artery. J Neurosurg 62:261–268.

Umansky F, et al. (1988) Anomalies and variations of the middle cerebral artery: a microanatomical study. Neurosurgery 22:1023–1027.

Ungerleider LG, Haxby JV (1994) 'What' and 'where' in the human brain. Curr Opin Neurobiol 4:157–165.

Urban BJ, Nashold BS Jr (1978) Percutaneous epidural stimulation of the spinal cord for relief of pain. Long-term results. J Neurosurg 48:323–328.

Urban PP, Wicht S, Vucorevic G, Fitzek S, Marx J, Thomke F, Mika-Gruttner A, Fitzek C, Stoeter P, Hopf HC (2001) The course of corticofacial projections in the human brainstem. Brain 124:1866–1886.

Usunoff KG, Itzev DE, Lolov SR, Wree A (2003) Pedunculopontine tegmental nucleus. Part 1: Cytoarchitecture, transmitters, development and connections. Biomed Rev 14:95–120.

Usunoff KG, Itzev DE, Ovtscharoff WA, Marani E (2002) Neuromelanin in the human brain: a review and atlas of pigmented cells in the substantia nigra. Arch Physiol Biochem 110:257–369.

Usunoff KG, Marani E, Schoen JH (1997) The trigeminal system in man. Adv Anat Embryol Cell Biol 136:1–126.

Usunoff KG, Popratiloff A, Schmitt O, Wree A (2006) Functional neuroanatomy of pain. Adv Anat Embryol Cell Biol 184:1–115.

Utter AA, Basso MA (2007) The basal ganglia: An overview of circuits and function. Neurosci Biobehav Rev [Epub ahead of print, doi:10.1016/j.neubiorev.2006.11.003]

Vale W, Rivier C, Brown M (1977) Regulatory peptides of the hypothalamus. Annu Rev Physiol 39:473–527.

Vallbo ÅB (1971) Single unit recording from human peripheral nerves: muscle receptor discharge in resting muscles and during voluntary contractions. In: *Neurophysiology Studied in Man.* Somjen GG, ed. Int Congr Series No 253:283–297. Excerpta Medica, Amsterdam.

Vallbo ÅB, Hagbarth K-E, Torebjörk HE, Wallin BG Somatosensory, proprioceptive, and sympathetic activity in human peripheral nerves. Physiol Rev 59:919–957.

Vallbo ÅB, Johansson RS (1984) Properties of cutaneous mechanoreceptors in the human hand related to touch sensation. Hum Neurobiol 3:3–14.

Vallbo ÅB, Olausson H, Wessberg J (1999) Unmyelinated afferents constitute a second system coding tactile stimuli of the human hairy skin. J Neurophysiol 81:2753–2763.

Van Buren JM, Borke RC (1972) *Variations and Connections of the Human Thalamus.* Springer, Berlin.

Van Buren JM, Borke RC (1972) *Variations and Connections of the Human Thalamus. 1. The Nuclei and Cerebral Connections of the Human Thalamus.* Springer-Verlag, New York.

Van der Kloot W (1988) Acetylcholine quanta are released from vesicles by exocytosis (and why some think not). Neuroscience 24:1–7.

Van der Kooy D (1987) The reticular core of the brain-stem and its descending pathways: anatomy and function. In: *Epilepsy and the Reticular Formation: the Role of the Reticular Core in Convulsive Seizures.* Fromm GH, Faingold C, Browning RA, Burnham WM, eds. Vol 27: Neurology and Neurobiology, pp 9–23. Alan R. Liss, New York.

Van der Werf YD, Witter MP, Uylings HB, Jolles J (2000) Neuropsychology of infarctions in the thalamus: a review. Neuropsychologia 38:613–627.

Van der Werf YD, Witter MP, Groenewegen HJ (2002) The intralaminar and midline nuclei of the thalamus. Anatomical and functional evidence for participation in processes of arousal and awareness. Brain Res Rev 39:107–140.

Van Hees J, Gybels JM (1972) Pain related to single afferent C fibers from human skin. Brain Res 48:397–400.

Van Hoesen GW, Mesulam M-M, Haaxma R (1976) Temporal cortical projection to the olfactory tubercle in the rhesus monkey. Brain Res 109:375–381.

Van Hoesen GW, Pandya DN (1975) Some connections of the entorhinal (area 28) and perirhinal (area 35) cortices of the rhesus monkey. I. Temporal lobe afferents. Brain Res 95:1–24.

Van Loveren HR, et al. (1991) The Dolenc technique for cavernous sinus exploration (cadaveric prosection). J Neurosurg 74:837–844.

Vandenabeele F, Creemers J, Lambrichts I (1996) Ultrastructure of the human spinal arachnoid mater and dura mater. J Anat 189:417–430.

Varon SS, Somjen GG (1979) Neuron–glia interactions. Neurosci Res Program Bull 17:9–239.

Vassilopoulos D, Emery AEH (1977) Quantitative histochemistry of the spinal motor neurone nucleus during human fetal development. J Neurol Sci 32:275–281.

Vaughn JE, Peters A (1968) A third neuroglial cell type: an electron microscopic study. J Comp Neurol 133:269–288.

Vautin RG, Dow BM (1985) Color cell groups in the foveal striate cortex of the behaving macaque. J Neurophysiol 54:273–292.

Vega JA, Haro JJ, Del Valle ME (1996) Immunohistochemistry of human cutaneous Meissner and Pacinian corpuscles. Microsc Res Tech 34:351–361.

Veraa RP, Grafstein B (1981) Cellular mechanisms for recovery from nervous system injury: a conference report. Exp Neurol 71:6–75.

Verrillo RT (1980) Age related changes in the sensitivity to vibration. J Gerontol 35:185–193.

Vierck CJ Jr (1973) Alterations of spatio-tactile discrimination after lesions of primate spinal cord. Brain Res 58:69–79.

Vierck CJ Jr (1974) Tactile movement detection and discrimination following dorsal column lesions in monkeys. Exp Brain Res 20:331–346.

Vierck CJ Jr (1977) Absolute and differential sensitivities to touch stimuli after spinal cord lesions in monkeys. Brain Res 134:529–539.

Vierck CJ Jr, Luck MM (1979) Loss and recovery of reactivity to noxious stimuli in monkeys with primary spinothalamic cordotomies, followed by secondary and tertiary lesions of other cord sectors. Brain 102:233–248.

Vijayashankar N, Brody H (1977) A study of aging in the human abducens nucleus. J Comp Neurol 173:433–438.

Vijayashankar N, Brody H (1977) Aging in the human brain stem. A study of the nucleus of the trochlear nerve. Acta Anat (Basel) 99:169–172.

Vijayashankar N, Brody H (1979) A quantitative study of the pigmented neurons in the nuclei locus coeruleus and subcoeruleus in man as related to aging. J Neuropathol Exp Neurol 38:490–497.

Villiger E, Ludwig E, Rasmussen AT (1951) *Atlas of Cross Section Anatomy of the Brain; Guide to the Study of the Morphology and Fiber Tracts of the Human Brain.* McGraw-Hill, New York.

Vinas FC, Lopez F, Dujovny M (1995) Microsurgical anatomy of the posterior choroidal arteries. Neurol Res 17:334–344.

Vogt BA (2005) Pain and emotion interactions in subregions of the cingulate gyrus. Nat Rev Neurosci 6:533–544.

Vogt BA, Pandya DN, Rosene DL (1987) Cingulate cortex of the rhesus monkey: I. Cytoarchitecture and thalamic afferents. J Comp Neurol 262:256–270.

Vogt BA, Vogt LJ, Nimchinsky EA, Hof PR (1997) Primate cingulate cortex chemoarchitecture and its disruption in Alzheimer's disease. In: *Handbook of Chemical Neuroanatomy.* Björklund A, Hökfelt T, eds. Vol 13: The Primate Nervous System, Part I, pp 455–528. Elsevier, Amsterdam.

Volterra A, Meldolesi J (2005) Astrocytes, from brain glue to communication elements: the revolution continues. Nat Rev Neurosci 6:626–640.

Von Bonin G, Green JR (1949) Connections between orbital cortex and diencephalon in the macaque. J Comp Neurol 90:243–254.

Von Noorden GK, Crawford MLJ, Levacy RA (1983) The lateral geniculate nucleus in human anisometropic amblyopia. Invest Ophthalmol Vis Sci 24:788–790.

Voogd J (2003) The human cerebellum. J Chem Neuroanat 26:243–252.

Voris HC (1957) Variations in the spinothalamic tract in man. J Neurosurg 14:55–60.

Waite PME, Ashwell KWS (2004) Trigeminal sensory system. In: The Human Nervous System. Paxinos G, Mai JK, eds. 2nd ed. Chapter 29, pp 1093–1124. Elsevier/Academic Press, Amsterdam.

Walberg F (1952) The lateral reticular nucleus of the medulla oblongata in mammals. A comparative-anatomical study. J Comp Neurol 96:283–343.

Walker AE (1940) The medial thalamic nucleus. A comparative anatomical, physiological and clinical study of the nucleus medialis dorsalis thalami. J Comp Neurol 73:87–115.

Walker AE (1940) The spinothalamic tract in man. AMA Arch Neurol Psychiatr 43:284–298.

Wall PD (1978) The gate control theory of pain mechanisms. A re-examination and re-statement. Brain 101:1–18.

Wallach H (1948) Brightness constancy and the nature of achromatic colors. J Exp Psychol 38:310–324.

Walloch R, DeWeese D, Brummett R, Vernon J (1973) Electrical stimulation of the inner ear. Ann Otol Rhinol Laryngol 82:473–485.

Walsh EG (1957) An investigation of sound localization in patients with neurological abnormalities. Brain 80:222–250.

Walsh FB (1957) Clinical Neuro-Ophthalmology. 2nd ed., pp 1059–1062. Williams & Wilkins, Baltimore.

Waltregny A, Trillet F, Geurts A (1977) Auditory evoked potentials recorded from chronic implanted gyrus of Heschl in man. Acta Neurochir (Suppl) 24:163–173.

Walz W, Hertz L (1983) Functional interactions between neurons and astrocytes. II. Potassium homeostasis at the cellular level. Prog Neurobiol 20:133–183.

Wanet-Defalque M-C, et al. (1988) High metabolic activity in the visual cortex of early blind subjects. Brain Res 446:368–373.

Wang MB (1971) An assessment of the neural taste code. In: Research in Physiology a liber memorialis in honor of Prof Chandler McCuskey Brooks. Kao FF, Koizumi K, Vassalle M, eds, pp 483–488. Aulo Gaggi Publ, Bologna.

Wannamaker BB, et al. (1973) Isolation and ultrastructure of human synaptic complexes. J Neurobiol 4:543–555.

Warkany J (1971) Congenital Malformations; Notes and Comments. Year Book Medical Publishers, Chicago.

Warr WB (1980) Efferent components of the auditory system. Ann Otol Rhinol Laryngol (Suppl) 89:(Suppl 74):114–120.

Wartenberg R (1939) A 'numeral test' in transverse lesions of the spinal cord. Am J Med Sci 198:393–396.

Warwick R (1953) Observations upon certain reputed accessory nuclei of the oculomotor complex. J Anat 87:46–53.

Warwick R (1953) Representation of the extra-ocular muscles in the oculomotor nuclei of the monkey. J Comp Neurol 98:449–504.

Warwick R (1954) The ocular parasympathetic nerve supply and its mesencephalic sources. J Anat 88:71–93.

Warwick R (1955) The so-called nucleus of convergence. Brain 78:92–114.

Warwick R (1964) Oculomotor organization. In: The Oculomotor System. Bender MB, ed., pp 173–204. Hoeber, New York.

Watkins BA, et al. (1990) Specific tropism of HIV-1 for microglial cells in primary human brain cultures. Science 249:549–553.

Watson WE (1974) Physiology of neuroglia. Physiol Rev 54:245–271.

Watts JW, Freeman W (1948) Frontal lobotomy in the treatment of unbearable pain. Res Publ Assoc Res Nerv Ment Dis 27:715–722.

Waxman S (1978) Physiology and Pathobiology of Axons. Raven Press, New York.

Waxman SG (1978) Prerequisites for conduction in demyelinated fibers. Neurology 29:27–33.

Weed LH (1914) A reconstruction of the nuclear masses in the lower portion of the human brain-stem. Publication 292. Carnegie Inst, Washington, DC.

Weidman TA, Sohal GS (1977) Cell and fiber composition of the trochlear nerve. Brain Res 125:340–344.

Weinberg E (1928) The mesencephalic root of the fifth nerve. A comparative anatomical study. J Comp Neurol 46:249–405.

Weinstein EA, Bender MB (1940) Dissociation of deep sensibility at different levels of the central nervous system. Arch Neurol Psychiat 43:488–497.

Weisbert JA, Rustioni A (1977) Cortical cells projecting to the dorsal column nuclei of rhesus monkeys. Exp Brain Res 28:521–528.

Weiss N, Lawson HC, Greenspan JD, Ohara S, Lenz FA (2005) Studies of the human ascending pain pathways. Thalamus Related Syst 3:71–86.

Weiss P, Hiscoe HB (1948) Experiments on the mechanism of nerve growth. J Exp Zool 107:315–395.

Wen GY, Wisniewski HM (1984) Substructure of neurofilaments. Acta Neuropathol (Berl) 64:339–343.

Wen GY, Wisniewski HM (1987) High resolution analysis of paired helical filaments in Alzheimer's disease. J Electron Microsc Tech 5:347–355.

Wendell-Smith CP, Blunt MJ, Baldwin F (1966) The ultrastructural characterization of macroglial cell types. J Comp Neurol 127:219–240.

Wenthold RJ (1980) Neurochemistry of the auditory system. Ann Otol Rhinol Laryngol (Suppl) 89:(Suppl 74):121–131.

Werner G (1977) Cutaneous stimulus registration and information processing in the somesthetic cortex. J Invest Dermatol 69:172–180.

Werner JS, Donnelly SK, Kliegl R (1987) Aging and human macular pigment density. Appended with translations from the work of Max Schultze and Ewald Hering. Vision Res 27:257–268.

Wersäll J (1972) Morphology of the vestibular receptors in mammals. Prog Brain Res 37:3–17.

Wersäll J, Bagger-Sjöbäck D (1974) Morphology of the vestibular sense organ. In: *Vestibular System. Part 1: Basic Mechanisms.* Kornhuber HH, ed., Handbook Sensory Physiology VI/1:123–170. Springer-Verlag, Berlin-Heidelberg.

Weston JA (1982) Neural crest cell development. In: *Embryonic Development, Part B: Cellular Aspects.* Burger MM, Weber R, eds, pp 359–379. AR Liss, New York.

Wetterberg L (1978) Melatonin in humans. Physiological and clinical studies. J Neural Transm Suppl 13:289–310.

White JC (1952) Conduction of visceral pain. New Engl J Med 246:686–691.

White JC (1954) Conduction of pain in man. Observations on its afferent pathways within the spinal cord and visceral nerves. AMA Arch Neurol Psychiatr 71:1–23.

White LE, Andrews TJ, Hulette C, Richards A, Groelle M, Paydarfar J, Purves D (1997a) Structure of the human sensorimotor system. I: Morphology and Cytoarchitecture of the central sulcus. Cereb Cortex 7:18–30.

White LE, Andrews TJ, Hulette C, Richards A, Groelle M, Paydarfar J, Purves D (1997b) Structure of the human sensorimotor system. II: Lateral symmetry. Cereb Cortex 7:31–47.

Whitfield IC (1967) *The Auditory Pathway.* Williams & Wilkins, Baltimore.

Whitsel BL, Dreyer DA, Roppolo JR (1971) Determinants of body representation in postcentral gyrus of macaques. J Neurophysiol 34:1018–1034.

Wiech K, Preissl H, Lutzenberger W, Kiefer RT, Topfner S, Haerle M, Schaller HE, Birbaumer N (2000) Cortical reorganization after digit-to-hand replantation. J Neurosurg 93:876–883.

Wiechmann AF, Hollyfield JG (1987) Localization of hydroxyindole-O-methyltransferase-like immunoreactivity in photoreceptors and cone bipolar cells in the human retina: a light and electron microscope study. J Comp Neurol 258:253–266.

Wiesel TN, Hubel DH (1966) Spatial and chromatic interactions in the lateral geniculate body of the rhesus monkey. J Neurophysiol 29:1115–1156.

Wigley C (2004) Nervous system. In: *Gray's Anatomy.* Standring S, editor-in-chief. 39th ed., pp 43–68. Elsevier/Churchill Livingstone, London.

Wilkins RH (1995) Case problems in neurological surgery: A problem of cervical pain. Neurosurgery 36:158–165.

Willer JC, Boureau F, Albe-Fessard D (1978) Role of large diameter cutaneous afferents in transmission of nociceptive messages: electrophysiological study in man. Brain Res 152:358–364.

Williams PL, Hall SM (1971) Chronic Wallerian degeneration – an in vivo and ultrastructural study. J Anat 109:487–503.

Williams PL, Warwick R (1975) *Functional Neuroanatomy of Man: Being the Neurology Section from Gray's Anatomy.* 35th British ed. Saunders, Philadelphia.

Willis WD (1976) Spinothalamic system: physiological aspects. In: *Advances in Pain Research and Therapy.* Bonica JJ, Albe-Fessard D, eds, 1:215–223. Raven Press, New York.

Willis WD (1979) Supraspinal control of ascending pathways. Prog Brain Res 50:163–174.

Willis WD (1980) Neurophysiology of nociception and pain in the spinal cord. Res Publ Assoc Res Nerv Ment Dis 58:77–92.

Willis WD Jr (1979) Studies of the spinothalamic tract. Tex Rep Biol Med 38:1–45.

Willis WD, Al-Chaer ED, Quast MJ, Westlund KN (1999) A visceral pain pathway in the dorsal column of the spinal cord. Proc Natl Acad Sci USA. 96:7675–7679.

Willis WD, Kenshalo DR Jr, Leonard RB (1979) The cells of origin of the primate spinothalamic tract. J Comp Neurol 188:543–574.

Willis WD, Westlund KN (1997) Neuroanatomy of the pain system and of the pathways that modulate pain. J Clin Neurophysiol 14:2–31.

Willis WD Jr, Westlund KN (2004) Pain system. In: *The Human Nervous System.* Paxinos G, Mai JK, eds. 2nd ed. Chapter 30, pp 1125–1170. Elsevier/Academic Press, Amsterdam.

Willis WD Jr, Zhang X, Honda CN, Giesler GJ Jr (2002) A critical review of the role of the proposed VMpo nucleus in pain. J Pain 3:79–94.

Wilson DA, Kadohisa M, Fletcher ML (2006) Cortical contributions to olfaction: plasticity and perception. Semin Cell Dev Biol 17:462–470.

Wilson JG (1911) The nerves and nerve endings in the membrana tympani of man. Am J Anat 11:101–112.

Wilson JG, Fraser FC (1977) *Handbook of Teratology.* Vol 1. General Principles and Etiology. Plenum, New York.

Windle WF (1970) Development of neural elements in human embryos of four to seven weeks gestation. Exp Neurol 28:[Suppl 5] 44–83.

Windle WF, Fitzgerald JE (1937) Development of the spinal reflex mechanism in human embryos. J Comp Neurol 67:493–509.

Windle WF, Fitzgerald JE (1942) Development of the human mesencephalic trigeminal root and related neurons. J Comp Neurol 77:597–608.

Winer JA (1984) The human medial geniculate body. Hear Res 15:225–247.

Winick M (1970) Nutrition and nerve cell growth. Fed Proc 29:1510–1515.

Winick M (1975) Nutritional disorders during brain development. In: *The Nervous System.* Tower DB, editor-in-chief. Chase TN, volume ed. Vol 2: The Clinical Neurosciences, pp 381–386. Raven Press, New York.

Winick M, Rosso P (1969) The effect of severe early malnutrition on cellular growth of human brain. Pediatr Res 3:181–184.

Winkelmann RK (1977) The Merkel cell system and a comparison between it and the neurosecretory or APUD cell system. J Invest Dermatol 69:41–46.

Winkler C (1981) *Anatomie du Systéme Nerveux. I. Opera omnia.* Vol VI. Bohn, Harlem.

Wirth A, Cavallacci G, Genovesi-Ebert F (1984) The advantages of an inverted retina: a physiological approach to a teleological question. Dev Ophthalmol 9:20–28.

Wise SP, Tanji J (1981) Neuronal responses in sensorimotor cortex to ramp displacements and maintained positions imposed on hindlimb of the unanesthetized monkey. J Neurophysiol 45:482–500.

Wisniewski HM, et al. (1985) Morphology and biochemistry of Alzheimer's disease. In: *Senile Dementia of the Alzheimer Type. Proceedings of the Fifth Tarbox Symposium.* Hutton JT, Kenny AD, eds, pp 263–274. The Norman Rockwell Conference on Alzheimer's Disease. AR Liss, New York.

Wisniewski HM, Merz PA, Iqbal K (1984) Ultrastructure of paired helical filaments of Alzheimer's neurofibrillary tangle. J Neuropathol Exp Neurol 43:643–656.

Wisniewski KE, Laure-Kamionowska M, Connell F, Wen GY (1986) Neuronal density and synaptogenesis in the postnatal stage of brain maturation in Down syndrome. In: *The Neurobiology of Down Syndrome.* Epstein CJ, ed., pp 29–44. Raven Press, New York.

Wong-Riley MTT (1989) Cytochrome oxidase: an endogenous metabolic marker for neuronal activity. TINS 12:94–101.

Wood P, Bunge RP (1984) The biology of the oligodendrocyte. Adv Neurochem 5:1–46.

Woodburne RT (1939) Certain phylogenetic anatomical relations of localizing significance for the mammalian central nervous system. J Comp Neurol 71:215–257.

Woolsey CN (1964) Cortical localization as defined by evoked potential and electrical stimulation studies. In: *Cerebral Localization and Organization.* Schaltenbrand G, Woolsey CN, eds, pp 17–26. U of Wis Pr, Madison.

Woolsey CN (1972) Tonotopic organization of the auditory cortex. In: *Physiology of the Auditory System, A Workshop.* Sachs MB, ed., pp 271–282. National Educational Consultants Inc, Baltimore.

Woolsey CN, Erickson TC, Gilson WE (1979) Localization in somatic sensory and motor areas of human cerebral cortex as determined by direct recording of evoked potentials and electrical stimulation. J Neurosurg 51:476–506.

Wolpaw JR (2006) The education and re-education of the spinal cord. Prog Brain Res 157:261–280.

Wright A (1984) Dimensions of the cochlear stereocilia in man and the guinea pig. Hear Res 13:89–98.

Wright A, et al. (1987) Hair cell distributions in the normal human cochlea. A report of a European working group. Acta Otolaryngol (Suppl) (Stockh) 436:15–24.

Wright A, et al. (1987) Hair cell distributions in the normal human cochlea. Acta Otolaryngol (Suppl) (Stockh) 444:1–48.

Wright RH (1978) Specific anosmia: a clue to the olfactory code or to something much more important? Chem Senses Flavor 3:235–239.

Wright RH, Hughes JR, Hendrix DE (1967) Olfactory coding. Nature 216:404–406.

Wrobel BB, Leopold DA (2004) Clinical assessment of patients with smell and taste disorders. Otolaryngol Clin North Am 37:1127–1142.

Wurtman RJ, Moskowitz MA (1977) The pineal organ (first of two parts). N Engl J Med 296:1329–1333.

Wurtman RJ, Moskowitz MA (1977) The pineal organ (second of two parts). N Engl J Med 296:1383–1386.

Yahr MD (1976) *The Basal Ganglia.* Res Publ Assoc Res Nerv Ment Dis, Vol 55. Raven Press, New York.

Yakovlev, PI (1948) Motility, behavior and the brain. J Nerv Ment Dis 107: 313–335.

Yamada E (1966) Some observations on the fine structure of the human retina. Fukuoka Acta Med 57:163–183.

Yamada E (1969) Some structural features of the fovea centralis in the human retina. Arch Ophthalmol 82:151–159.

Yamada T, McGeer PL (1990) Oligodendroglial microtubular masses: an abnormality observed in some human neurodegenerative diseases. Neurosci Lett 120:163–166.

Yamamoto I, Rhoton AL Jr, Peace DA (1981) Microsurgery of the third ventricle: Part I. Microsurgical anatomy. Neurosurgery 8:334–356.

Yamamoto T, Iwasaki Y, Konno H (1984) Experimental sensory ganglionectomy by way of suicide axoplasmic transport. J Neurosurg 60:108–114.

Yarosh CA, Hoffman DS, Strick PL (2004) Deficits in movements of the wrist ipsilateral to a stroke in hemiparetic subjects. J Neurophysiol 92:3276–3285.

Yazulla S (1986) GABAergic mechanisms in the retina. Prog Retina Res 5:1–52.

Yelnik J, Percheron G (1979) Subthalamic neurons in primates: a quantitative and comparative analysis. Neuroscience 4:1717–1743.

Yeterian EH, Pandya DN (1989) Thalamic connections of the cortex of the superior temporal sulcus in the rhesus monkey. J Comp Neurol 282:80–97.

Ylikoski J (1982) Morphologic features of the normal and 'pathologic' vestibular nerve of man. Am J Otol 3:270–273.

Ylikoski J, Collan Y, Palva T (1979) Vestibular sensory epithelium in Meniere's disease. Arch Otolaryngol 105:486–491.

Ylikoski J, Palva T, Collan Y (1978) Eighth nerve in acoustic neuromas: special reference to superior vestibular nerve function and histopathology. Arch Otolaryngol 104:532–537.

Yokoh Y (1969) The early development of the nervous system in man Acta Anat (Basel) 71:492–518.

Yokoh Y (1975) Early development of the cerebral vesicles in man. Acta Anat (Basel) 91:455–461.

Yokota T, Hashimoto S (1976) Periaqueductal gray and tooth pulp afferent interaction on units in caudal medulla oblongata. Brain Res 117:508–512.

York GK, Gabor AJ, Dreyfus PM (1979) Paroxysmal genital pain: An unusual manifestation of epilepsy. Neurology 29:516–519.

Yoshida K, Benevento LA (1981) The projection from the dorsal lateral geniculate nucleus of the thalamus to extrastriate visual association cortex in the macaque monkey. Neurosci Lett 22:103–108.

Yoss RE (1953) Studies of the spinal cord. Part 3. Pathways for deep pain within the spinal cord and brain. Neurology 3:163–175.

Young PA, Young PH (1997) *Basic Clinical Neuroanatomy.* Lippincott Williams & Wilkins, Philadelphia.

Young RF (1978) Unmyelinated fibers in the trigeminal motor root. Possible relationship to the results of trigeminal rhizotomy. J Neurosurg 49:538–543.

Young RF, King RB (1973) Fiber spectrum of the trigeminal sensory root of the baboon determined by electron microscopy. J Neurosurg 38:65–72.

Young RF, Kruger L (1981) Axonal transport studies of the trigeminal nerve roots of the cat: with special reference to afferent contributions in the portio minor. J Neurosurg 54:208–212.

Young RF, Stevens R (1979) Unmyelinated axons in the trigeminal motor root of human and cat. J Comp Neurol 183:205–214.

Young, RF (1978) Evaluation of dorsal column stimulation in the treatment of chronic pain. Neurosurgery 3:373–379.

Younge BR, Sutula F (1977) Analysis of trochlear nerve palsies: diagnosis, etiology, and treatment. Mayo Clin Proc 52:11–18.

Yu J, Chambers WW, Liu CN (1978) Cerebral influence on postural effects of cerebellar vermal zonal lesions or eighth nerve section in monkeys. Acta Neurobiol Exp 38:87–95.

Yukie M, Iwai E (1981) Direct projection from the dorsal lateral geniculate nucleus to the prestriate cortex in macaque monkeys. J Comp Neurol 201:81–97.

Yukie M, Iwai E (1985) Laminar origin of direct projection from cortex area V1 to V4 in the rhesus monkey. Brain Res 346:383–386.

Zahn JR (1978) Incidence and characteristics of voluntary nystagmus. J Neurol Neurosurg Psychiatry 41:617–623.

Zander E, Weddell G (1951) Observations on the innervation of the cornea. J Anat 85:68–99

Zarbin MA, Wamsley JK, Palacios JM, Kuhar MJ (1986) Autoradiographic localization of high affinity GABA, benzodiazepine, dopaminergic, adrenergic and muscarinic cholinergic receptors in the rat, monkey and human retina. Brain Res 374:75–92, 1986.

Zec N, Filiano JJ, Kinney HC (1997) Anatomic relationships of the human arcuate nucleus of the medulla: a DiI-labeling study. J Neuropathol Exp Neurol 56:509–522.

Zec N, Kinney HC (2001) Anatomic relationships of the human nucleus paragigantocellularis lateralis: a DiI labeling study. Auton Neurosci 89:110–124.

Zec N, Kinney HC (2003) Anatomic relationships of the human nucleus of the solitary tract in the medulla oblongata: a DiI labeling study. Auton Neurosci 105:131–144.

Zeki S, Watson JD, Lueck CJ, Friston KJ, Kennard C, Frackowiak RS (1991) A direct demonstration of functional specialization in human visual cortex. J Neurosci 11:641–649.

Zeki SM (1973) Colour coding in rhesus monkey prestriate cortex. Brain Res 53:422–427.

Zeki SM (1978) The cortical projections of foveal striate cortex in the rhesus monkey. J Physiol 277:227–244.

Zeki SM (1978) The third visual complex of rhesus monkey prestriate cortex. J Physiol 277:245–272.

Zeki SM (1978) Uniformity and diversity of structure and function in rhesus monkey prestriate visual cortex. J Physiol 277:273–290.

Zihl J, Cramon D von (1985) Visual field recovery from scotoma in patients with postgeniculate damage. A review of 55 cases. Brain 108:335–365.

Zilles (2004) Architecture of the human cerebral cortex. Regional and laminar organization. In: *The Human Nervous System*. Paxinos G, Mai JK, eds. 2nd ed. Chapter 27, pp 997–1055. Elsevier/Academic Press, Amsterdam.

Zilles K, Werners R, Büsching U, Schleicher A (1986) Ontogenesis of the laminar structure in areas 17 and 18 of the human visual cortex. A quantitative study. Anat Embryol 174:339–353.

Zimmermann M (1976) Neurophysiology of nociception. Int Rev Physiol 10:179–221.

Zlatos J, Cierny G (1975) Statistical model map of the spinal cord and its use. Appl Neurophysiol 38:225–239.

Zwislocki JJ (1981) Sound analysis in the ear: a history of discoveries. Am Sci 69:184–192.

Index

Printed and bound by CPI Group (UK) Ltd, Croydon, CR0 4YY

03/10/2024

01040316-0019